# Recent Advances in Mineral Processing Plant Design

Edited by
Deepak Malhotra, Patrick R. Taylor,
Erik Spiller, and Marc LeVier

Published by
SME
Society for
Mining, Metallurgy
& Exploration

Your most
precious resource.

Society for Mining, Metallurgy, and Exploration, Inc. (SME)
12999 East Adam Aircraft Circle
Englewood, Colorado 80112
(303) 948-4200 / (800) 763-3132
www.smenet.org

SME advances the worldwide mining and minerals community through information exchange and professional development. With members in more than 70 countries, SME is the world's largest association of mining and minerals professionals.



ISBN 13: 978-0-87335-316-8
Ebook: 978-0-87335-318-2

**Library of Congress Cataloging-in-Publication Data**

Recent advances in mineral processing plant design / edited by Deepak Malhotra ... [et al.].
p. cm.
Includes bibliographical references and index.
ISBN 978-0-87335-316-8
1. Ore-dressing plants--Design and construction. I. Malhotra, Deepak.
TN504.R43 2009
622'.79—dc22

2009030031

**On the cover:** Photo courtesy of Taseko Mines Limited showing the SAG mill and discharge screen at the Gibraltar Mine (central British Columbia, Canada) prior to commissioning. The SAG mill is a 34-foot diameter, 10 MW mill that was designed and procured by Farnell-Thompson.

# Contents

## Extractive Metallurgy



## Comminution (Crushing and Grinding)



## Financial and Feasibility Studies



## Particle Separation



## Other Separation Processes

## Process Development

# Session Chairs

### Plenary Session

Deepak Malhotra
Resource Development Inc.
Wheat Ridge, CO

Erik Spiller
Colorado School of Mines
Golden, CO

### Research and Characterization

John Herbst
Metso Minerals
Kona, HI

J. Mark Richardson
Contract Support Services
Red Bluff, CA

### Engineering Commissioning, Start Up and Case Studies

Mike Blois
CH2MHill
Englewood, CO

Matt Bender
CH2MHill
Englewood, CO

### Sampling, Bench and Pilot Plant Test Work

Will Goodal
MiniAssist Pty Ltd
Australia

Dave Meadows
FLSmidth Minerals
Bethlehem, PA

### Flotation

Jan Miller
University of Utah
Salt Lake City, UT

Frank Traczyk
FLSmidth Minerals
Park City, UT

### Extractive Metallurgy

Corby Anderson
Montana Tech of University of Montana
Butte, MT

Hank Rawlins
Montana Tech of University of Montana
Butte, MT

### Comminution (Crushing and Grinding)

Rob McIvor
Metcom Technologies
Ishpeming, MI

Ken Major
KWM Consulting
Maple Ridge, BC, Canada

### Financial and Feasibility Studies

Ryan Bennett
Resource Capital Funding
Denver, CO

Nick Michael
SRK Consulting
Littleton, CO

### Particle Separation

Erik Spiller
Colorado School of Mines
Golden, CO

Mark Erickson
Newmont Mining Corp.
Englewood, CO

### Other Separation Processes

Rob Dunne
Newmont Mining Corp.
Centennial, CO

Ed Bentzen
Lyntek Inc.
Lakewood, CO

### Process Development

Mark Jorgensen
Rio Tinto
Salt Lake City, UT

Ed Dowling
Anatolia Minerals
Englewood, CO

# Preface

Engineers are constantly striving to build better and more reliable plants. In order to do so, they need to keep abreast of the new technologies and ongoing equipment improvements that can have significant impact on the plant design. One of the functions of the professional organization is to provide a forum for the engineers in the various disciplines to get together on a periodic basis to review the recent development in the various plant design areas. This symposium is the platform to update the recent development in plant design since the last symposium, which was held in Canada in 2002. Professor Andy Mular has provided a historical prospective of how this idea of plant design symposium was born in the forward to this book. The organizing committee extends its sincere appreciation to the plenary speakers who were the organizers for the plant design symposium in 2002; session chairs who were instrumental in finding authors and reviewing the papers; and the authors and presenters without whom this symposium would not have been possible. The efforts and support of the two organizations, SME and TMS, are greatly appreciated. It would not be possible to organize an event of this magnitude without the efforts of SME's Dave Kanagy, Tara Davis, Carol Cudworth, Jane Olivier, AnnMarie Estrada, Heather Bonic, and others.

# INTRODUCTORY REMARKS

Andrew L. Mular

P. Eng (Ret) and Professor Emeritus, Mineral Process Engineering
Norman B Keevil Institute of Mining Engineering, University of BC, Vancouver, BC, V6T 1Z4, Canada

## INTRODUCTION

In 1963 the Chairman of Metallurgical Engineering at Queen's University of Kingston, Ontario, Canada encouraged me to apply for a position in his department. This led to an appointment as an Assistant Professor of Metallurgical Engineering with a mandate to teach undergraduate and graduate courses in Mineral Process Engineering. Graduating metallurgists take positions available to them in the mining industry and mill "metallurgists" were in strong demand. Discussions with industry-oriented colleagues at the Department of Energy, Mines and Resources (Mines Branch) and with senior operators and plant design personnel at conferences such as the Conference of Metallurgists provided details of jobs available Canada-wide. This helped to fortify course content. At the undergraduate level, courses on principles of unit operations and processes were clearly necessary. A final year course on plant design would be a useful adjunct. Unfortunately, mineral processing plant design textbooks were not available. A graduate level course that dealt with Experimental Design, Plant Data Acquisition and Adjustment, Mathematical Modeling, Process Optimization and Computer Control had some relevance to plant design problems. However, a practical plant design textbook was lacking.

A plant design course should incorporate methodology to estimate major equipment costs and corresponding capital and operating costs. Hence, as an interim measure and with financial support from the Mines Branch, data were obtained from manufacturers and design firms. Acceptable costing techniques (for example, see American Association of Cost Engineers Handbook) were developed via statistical methods. A notebook entitled Mineral Processing Equipment Costs and Preliminary Capital Cost Estimations was written in 1970–71 and published by the CIM in 1972. It was updated three times over the years. The last version incorporated Mineral Project Evaluation Techniques. It is entitled CAPCOSTS and was released in early 1999 by the Canadian Mineral Processors of CIM. Unfortunately, this modest contribution did not resolve the issue, namely, that we did not have a textbook on plant design.

## SME-AIME PLANT DESIGN TEXTS

In 1972, I joined the Mining Department at the University of British Columbia and frequently attended SME meetings. I spoke with university professors, senior operators, researchers and plant design specialists seeking their ideas about the best way to generate a practical plant design text for 4th year university courses. University professors would hesitate to teach courses on the subject unless students were provided with relevant reading material. One possibility was to seek a 3-year sabbatical leave with a mine-mill design firm and draft a preliminary text. Thankfully, most agreed that the symposia route would be best. Colleagues felt such compilations would be of value to engineers, designers, manufacturers, construction firms, consultants and industry in general and that the SME-AIME would likely support it. After joining appropriate SME committees the idea to have SME-AIME sponsor plant design symposia and publish "practical textbooks" was discussed in earnest. In general, committee members were 100% in support!

## PLANT DESIGN SYMPOSIA HELD PRIOR TO 2009

A plant design symposium held prior to 2009, always a team effort, began with a structure (sections and corresponding "ideal" topics) which would lead to a "practical textbook" for use in Plant Design courses. Profs and students should benefit. Benefits will undoubtedly accrue to operators, consultants, design firms, and industry because of structure improvements made by the organizing committee, an editorial committee, section co-editors and/or session co-chairmen and authors. A call for papers preceded a symposium date by about 1 to 2.5 years with the latter being preferred. For the first effort a structure was drafted by a Chairman and Co-Chairman. The structure was revised and finalized by the editorial committee. An application to hold the symposium was submitted to the AIME Board of Directors meeting in New York in the mid-1970s and approved. The Call for Papers was mailed out by SME. The resulting symposium entitled *Mineral Processing Plant Design* was held in 1978 and was a resounding success (Marianne Snedeker helped a lot). Symposium volumes sold out and in 1980 a revised version was published.

Because of the importance of comminution circuits the second symposium, namely *Design and Installation of Comminution Circuits*, was initiated in 1980 in a manner somewhat similar to the first except that members of the editorial committee were asked to serve as Session Co-Editors. They were requested to seek evaluations of papers for their respective sessions. Deadlines were met and in 1982 the symposium was held in Oahu, Honolulu. The attendance was superb and the symposium was a great success!

A third symposium entitled *Design and Installation of Concentration and Dewatering Circuits* was initiated in 1984. In this case, an editorial board evaluated papers with a board member chosen to chair the board and act as an umpire if necessary. The symposium was held in 1986, and was most successful.

In late 1997 a colleague, who was to serve as an AIME trustee from SME, mentioned that the plant design series that was started in 1978 had not been completed. In early 1998, Mr. Gary Howell, Executive Director of SME, suggested that I chair another symposium dealing with Plant Design. Mr. Howell forwarded an application form which I filled out and submitted to SME-AIME to hold an international symposium entitled *Mineral Processing Plant Design, Practice, and*

*Control.* Strong support for the symposium was offered by the Canadian Mineral Processors of CIM. A structure based on previous volumes was generated and expanded to include sections and topics that had not been previously incorporated. This was circulated via E-mail to a large General Committee (senior operators, design firms, construction engineers, consultants, cost engineers, mining engineers) selected from the USA, Canada and several South American countries to get recommendations on content. The structure was revised accordingly, potential authors were listed and an Organizing Committee was selected. The Organizing Committee reviewed the revised structure sections and possible author list. Semi-Final revisions were prepared and 42 Section Co-Editors for 21 sections were contacted and sent the revised copy of structure with recommended authors. In 1999 a Call for Papers had been issued in Mining Engineering and various other journals. Section Co-Editors made final choices subject to agreement with the Organizing Committee. Potential authors were contacted by co-editors and encouraged to follow a stringent timetable to ensure that proceedings would be published in time for distribution at the symposium. The symposium was held in Vancouver, BC, Canada in 2002. By way of symposium statistics, there were 21 sections and 42 Section Co-Editors (2 per section) who also served as Session Chairmen. The number of papers totaled 138 and 262 authors were involved (an average of about 2 authors per paper). The symposium was judged a tremendous success. SME reviewers of the two Volumes (2500 pages) praised the content. They stressed that "the proceedings are a new standard text for university-level instruction and a valuable guidebook for operators considering new construction, plant renovations and/or expansions. It's a practical, quick reference for engineers, consultants, suppliers, manufacturers or anyone involved in the design or operation of a minerals processing plant." Papers, as judged by volume and section co-editors, were relevant and meritorious. A minor drawback is the weight of volumes!

## MINERAL PROCESSING PLANT DESIGN 2009

The symposium was a total surprise prior to scanning of the sections and titles of papers. After scanning, the symposium was judged to be a most welcome surprise! **First**, academic and international involvement was noted along with strong participation by North American design specialists. Many of the symposium contributors participated in at least one of the previous symposia and several contributed to all of them. Papers in a given section appear to fit reasonably well. A number of sections in basic university-level plant design courses should benefit from updates. **Second**, reading the papers should be exciting; especially those which are likely to be useful for updating portions of a design course at the university level.

The organizing committee and session chairmen deserve kudos for the strong efforts that brought the Mineral Processing Plant Design 2009 symposium to a successful conclusion. Sections/Topics are well chosen for corresponding content. Authors deserve the highest praise as was the case for previous symposia. It's a pleasure to say that I know personally or have met a number of the authors who contributed updates.

My years from 1963 to 2009 (as an Assistant to Associate to Full to Emeritus Professor of Mineral Process Engineering, including 5 years as a department head prior to appointment as an Emeritus, were devoted mainly towards the education of Mineral Process Engineers. The need for an undergraduate final-year plant design course was recognized in 1963. Practical textbooks on the subject were needed. SME-AIME provided the opportunity to be heavily involved with the development of four plant design "texts" by many authors that willingly geared their subject matter towards the teaching function. All four texts are still in use.

One concern has been that colleagues would not recognize the need for updates in the future. This is no longer of concern—there is now a strong nucleus of technical personnel who will take up the banner at appropriate intervals! Another concern is that some basic structure is still missing in practical textbooks on plant design. However, many of you currently involved in the design and construction of complete mining to processing to final-product facilities could scan the sections of all *FIVE* current texts, find the missing content and think of future plant design symposia! Don't forget to update the Do's and Don'ts of Installation (see *Design and Installation of Comminution Circuits*).

Thanks for the opportunity to add Introductory Remarks for the conference and *Recent Advances in Mineral Processing Plant Design.*

# Plenary Session

# PLANT DESIGN: WHAT NOT TO DO

Leonard (Len) Harris

Resource Development Inc., 11475 West I-70 Frontage Road North, Wheat Ridge, Colorado, USA

## ABSTRACT

It has been the author's experience since he began his career as an indentured cadet metallurgist at Mount Morgan Ltd. in Australia in 1943 that many mistakes have been made in the past when designing mineral processing plants. Unfortunately, many of the same mistakes continue today by engineering firms and in some cases, by the operators themselves.

A list of items that should not be done prior to and at the design stage, is presented and discussed in some detail for consideration by those concerned with this side of the business.

## INTRODUCTION

Over the past 66 years that the author has been involved in the tasks involved with plant design and operation of mineral dressing plants, many mistakes have been made. Costly mistakes caused by the design and erection of plants without a confirmed ore supply; faulty, inadequate or insufficient mineralogical and metallurgical test work; misinterpretation of test work results; jumping to conclusions; accepting opinions and assumptions; test work with non representative samples; installation of faulty equipment ; collapse of installed equipment; collapse of entire buildings; over or under estimation of processing units and power requirements; severe maintenance problems by overcrowding equipment installations; electrical fires; non duplication of essential equipment; failure to use ground contours; lack of equipment safety guards; faulty stockpile and bin discharge systems; lack of safety and operator training; lack of adequate security measures and inefficient pollution control systems. These mistakes have resulted in unnecessary capital expenses, loss of production, and worker accidents.

The same mistakes seem to be committed time after time and there is a very evident lack of internal control both by the engineers and operators to prevent these mistakes from being repeated from project to project.

## DON'T DESIGN AND BUILD THE MILL UNTIL YOU HAVE A MINE

This is the number 1 "don't." It sounds very obvious doesn't it? But the author is aware of several cases where it occurred and several other cases that were "near misses."

The reasons why it happens is underestimated ore reserves, lower ore grades than originally estimated, mining problems or higher capital and/or operating costs than originally visualized. The cases of "near misses" are usually associated with lower grade ore than originally estimated and the problem was overcome by higher sales prices coming into effect after the plant was built.

A good example of this problem was encountered several years ago when the writer was employed with Cerro Corporation in New York City. The company owned a fluorspar mine near Paducah, Kentucky. A very elegant and expensive mill employing a heavy media section followed by flotation to produce a high grade fluorspar concentrate was actually built and afterwards it was discovered that the water inflow into the mine could not be controlled making mining an impossible task. Remnants of the mill can be seen in a corn field just off the highway near Paducah.

## DON'T SKIMP ON TEST WORK

Adequate and reliable metallurgical test work is vital to obtain the results needed to develop the flow sheet and design criteria for any given plant. Such test work will most certainly involve bench scale test work and mineralogical examinations, and if deemed necessary, pilot plant operation in some cases. Tests with finely ground material are required in all cases to obtain the maximum extraction or recovery figures which can be expected from any given ore. Tests with coarsely ground or crushed material are also often needed when heap leach operation is under consideration. In all cases, interpretation of the results requires close examination and in some cases, some "gut feeling." Also of importance is to not jump to conclusions, employ assumptions and opinions and when uncertain as to the results, to repeat the tests and/or to do more tests. The author has seen many cases where test work results have been reported where the assay of the head sample cannot be reconciled with a head assay calculated from the assay of the product (s) and the tailings produced from the test work. These results should never have been published until the problem is resolved by re assaying the products or repeating the tests.

Crushing and grinding test work is of vital importance to calculate the size of the equipment and the power needs to produce the desired product. Such tests are required to determine abrasion index, crushing work index and ball mill work index. If the ore is homogeneous, it requires relatively simple test work. If however, the ore is hetrogeneous in nature consisting of soft and hard material, it is much more difficult to get meaningful results from small scale test work.

Don't be content when a wide range of extractions are obtained from testing samples from a particular ore body. A good example is that experienced with test work on Newmont's Gold Quarry project in Nevada. Hundreds of bottle roll tests were conducted with finely ground ore samples (80% - 200 mesh) Gold extractions varied between 65% and 95% and averaged about 90%. Most people were prepared to use the average figure for the feasibility study on this project except for the then Chairman of the Board Plato Malozemoff who wanted to know why some of the tests showed such low extractions. He noted that there "are no secrets in nature and it is merely a matter of unlocking them." He would not release the project for development until the secret was unlocked. This was done after an extensive mineralogical study by the late Donald Hausen, chief mineralogist for Newmont, who at that time was able to identify microscopic particles of gold completely encapsulated in chert. Finer grinding of the ore was required to release these encapsulated particles and this need was incorporated in the plant design.

Another example of unlocking secrets is that experienced by the author in the 1960s at the Cerro de Pasco mine in Peru. Samples of copper ore were tested in the company laboratory with good success producing a good grade concentrate of +20% Cu and recoveries exceeding 90%. The results were accepted without question and the plant was installed and started up. From day one, the concentrate grade did not exceed 10% Cu. Everything under the sun was tried unsuccessfully to increase the grade. Finally the problem was passed to the outside laboratory of American Cyanamid in Lima and the author was sent there to monitor the test work. Similar results were obtained as in the in-house laboratory producing a concentrate grade of 20%. Finally, a sample of the concentrate produced in the plant was examined under the microscope by the chief of the Cyanamid laboratory Pete Cadwell (now deceased). It was not until the magnification was increased to X900 that the secret was revealed. The concentrate contained a high percentage of framboidal (colloidal) pyrite coated with a very thin rim of copper. The ore contained soluble copper which did not have sufficient time in the short laboratory grinding mills to coat the pyrite. In the plant, the grinding time was much longer allowing for the precipitation of copper on the pyrite which floated as though it was a copper mineral. This study was covered (among other cases) by Pete Cadwell in the Cyanamid brochure "Microscopy for the Mill Man."

Don't discard the use of "gut feeling" when viewing the results of test work. A good example of this "art of gut feeling" was experienced with grinding tests on Newmont's Gold Quarry ore. A 700 ton sample of ore was sent offsite to test autogenous and semi-autogenous grinding tests. The tests were monitored by a Newmont engineer and showed that either autogenous or semi-autogenous grinding could be employed. The author who was VP of metallurgical operations at that time, caucused with Pete Crescenzo who was VP of engineering at that time and instead of blindly accepting the results, and by using "gut feeling," it was decided to build the foundations of the mill strong enough to carry the load of a semi-autogenous mill operating with a 10% ball load and start the mill up in the autogenous mode. Within a week it was obvious that the mill could not treat the required tonnage and a ball load was added with instant success.

Another example of not blindly believing the results of test work is that experienced with tests on Maggie Creek gold ore from Newmont's mine in Nevada. There was absolutely no evidence from column leach tests carried out in the laboratory that there was any need to agglomerate the ore before heap leaching. Commercial operation was commenced and gold extraction was very low. Trenches dug into the heap did not reveal any dry spots or any other evidence of channeling but low extractions continued. Finally, the operators "took the bull by the horns" and added cement to the lime silo. Immediate improvement in extraction resulted parallelizing the figures obtained in the laboratory column leach tests.

Don't jump to conclusions. There is a tendency in the industry to jump to conclusions when it comes to the treatment of sulfidic gold ores. The belief often assumed is that because the sulfide content of the ore is high, the material is refractory. A typical case is that of Sherrgold's gold deposit in Quebec, Canada (a JV with Newmont) some years ago. This ore was highly sulfidic but was not refractory and was successfully milled for several years with recoveries exceeding 90% for gold. Another case is that of Alamos Gold's Mulatos gold heap leaching project in Mexico where the ore is mainly sulfidic and had been considered by the previous mine owners as refractory. When Alamos took over the project, test work revealed that the ore was not refractory but it was size dependent i.e., the finer it was crushed, the greater was the extraction. Thus the crushing plant at Mulatos was designed to crush the ore to –3/8 inch.

Similarly, the ore treated at the Newmont JV heap leaching plant at Zarafshan in Uzbekistan required crushing to 1/8 inch in four stages to get acceptable gold recoveries a fact which was not considered originally (based on what the local engineers told the writer). Subsequent test work conducted in Newmont's metallurgical laboratory unconditionally showed the need to finely crush this ore.

Another example of jumping to conclusions was experienced several years ago by the writer at Tsumeb Corporation's mine in Namibia where very complex lead/copper ore was treated by flotation. It was assumed that longer flotation times would result in higher recovery of the minerals left in the tailings and an extensive laboratory campaign was commenced at an outside facility. The tests were a bust as all the minerals that would float had already been floated in the plant. Gravity and magnetic separation test work did, however, give good results and shaking tables were installed in the plant with great success.

Be careful with accepting false information, opinions and assumptions. Just stick to the facts. Good examples are those noted above and at the Newmont JV in Yanacocha in Peru where the ore is dump leached without crushing. Again, test work was extremely important in this case with bench scale testing. There was little or no difference in the extractions obtained from finely ground ore and ore crushed to 1/8 inch, ¼ inch, ½ inch, 1 inch, 2 inches, 3 inches and 4 inches which indicated that it might be possible to treat this ore by dump leaching (i.e., using uncrushed run-of-mine ore). Discussions with the JV partners revealed that in the opinion of one of them was that this was not possible and the ore should be crushed before leaching. As there was no hurry to develop this project, (because of the Shining Path terrorist situation in Peru at that time) the writer suggested that a pilot plant operation be conducted with run-of mine ore. A pilot heap was constructed and placed under leach. 85% of the gold in the ore pile was extracted in 5 days. These are the facts and run-of-mine leaching was introduced and continues today.

The writer also warns on accepting assumptions as fact. In his early training at the Mount Morgan Technical College, one of his teachers recommended not to use assumptions whenever possible and if they are used, they should be clearly identified as such. Good advice which has never been forgotten and religiously adhered to by the writer throughout his career.

A good example is that of a French company with a gold deposit in the Sudan on which the writer consulted. Because of its high grade of 9gpt Au, milling was assumed the way to treat this ore. The writer asked if heap leaching had been considered and was informed that it had not been considered as it was assumed that heap leaching was confined to the treatment of lower grade ore in the order of 1gpt Au or less. Laboratory column test work was conducted which revealed that the ore could be efficiently treated at crushing to 4 inch and obtain the same extraction as with finely ground ore. Obviously, the capital costs are much less for a heap leaching plant than for a milling plant[1]. There are other examples of using assumptions known to and personally experienced by the writer involving the installation of mills treating gold and uranium ores. After the fact, testing revealed that it would have been much more economical to have installed heap leaching operations.

A basic consideration in all cases of test work is the "results obtained are only as good as the sample tested." If the sample is not representative of the ore body, then the results are useless. A good case known to the writer is that of a copper/gold mine in Argentina. Test work had shown that a recovery of 70% of the gold could be obtained by gravity methods and a gravity plant was built to recover this amount. When the plant was built and put into operation, less than 5% of the gold in the plant feed reported as a gravity concentrate. Obviously, the sample tested was not representative of the ore body.

Finally, the writer notes that if metallurgical test work is to be conducted in an outside laboratory, one should not leave it entirely to the outside investigators. The best way is to place an in-house investigator in their laboratory to monitor the work. At the least, an in-house investigator should visit the outside laboratory frequently while the work is ongoing.

## DON'T FORGET YOUR ECONOMICS 101 CLASSES

Generally speaking if a base metal flotation plant is to be installed, the plan would be to produce the highest grade concentrate at the highest recovery—right? No not necessarily. If the company owns a smelter nearby (as was the case with Cerro de Pasco Corp. in Peru), it may be better from an economic point of view, to produce lower grade concentrates at higher recovery. If on the other hand, the concentrates are to be sold to distant smelters, it would probably be better to produce high grade concentrates even at lower recovery to offset the cost of transportation.

An interesting case of obtaining the best economics could be found at the previously Newmont owned San Manuel copper/molybdenum concentrator in Arizona. Laboratory test work revealed that higher molybdenum recovery could be produced in the concentrator at the expense of some loss in copper recovery. Depending on the sales prices of the two metals, it was more economical to produce more molybdenum when its sales price was high and copper sales price was low.

Another interesting case was that experienced at Cerro de Pasco Corp.'s Mahr Tunel copper/lead concentrator in Peru. Separate lead and copper concentrates were produced at this plant for many years but with difficulty with respect to separation of the galena and chalcopyrite minerals resulting in low recoveries of both minerals. The writer and the then general superintendent of concentrators of Cerro de Pasco Corp, Jun Kitagawa discussed the idea of producing a bulk copper/lead concentrate for shipping to the nearby La Oroya smelter where the separation of lead and copper could be made pyrometallurgically in the lead smelter.. This resulted in higher copper and lead recoveries in the concentrator and overall better economics.

## DON'T LEAVE CERTAIN IMPORTANT PEOPLE OFF THE DESIGN TEAM

In many cases, the plant design team consists of people of various professional disciplines employed by an outside engineering firm. Sometimes they are well experienced and sometimes they are not. For example, there are very few people who are experienced today in the "out dated or old fashioned" Merrill Crowe process for gold recovery from cyanide leach solutions and in particular, there appears to be an acute shortage of people who can design grinding circuits and to calculate the horsepower needed to power grinding mills. This can result in over estimation or under estimation of the tonnage which can be treated through the grinding circuit. An example encountered by the writer was a SAG mill followed by two balls mills installation to treat a medium hard ore at a Newmont gold deposit in Nevada. The in-house engineering group had calculated certain horse power requirements for the three mills which was deemed to be insufficient to do the job by an outside engineering firm. A consultant was called in who noted that the motors for the three mills were over sized in his opinion. There are many very competent grinding consultants throughout the world. The question is how to get the services of the "best available consultant." Obviously in this case, the consultant was not the "best" but he was "available." It was decided to go with the in-house estimates which turned out to be the correct decision.

A very important don't as far as plant design is concerned, is to omit to place the plant superintendent and the maintenance on the design team. These are the people who will operate and maintain the plant and it is vital to have their input on all matters pertaining to its design.

## WHEN PURCHASING USED EQUIPMENT, DON'T SEND A BOY TO DO A MAN'S JOB

Very often, used equipment can be purchased instead of purchasing new equipment. One should however, be aware of the pitfalls Sometimes it can cost more to rehabilitate the equipment that to buy new equipment. Sometimes there are faults such as cracks which are not visible to the inexperienced eye of a "boy." It is imperative that a "man" consisting of an experienced mechanical engineer be sent to inspect the equipment before purchasing.

**DON'T JAM THE EQUIPMENT INTO THE SMALLEST POSSIBLE SPACE**

There is a tendency to use the smallest possible footprint when placing the equipment in a new or expanded plant. The reason of course, it to reduce capital costs involved with site grubbing, concrete flooring, foundations, etc. Such practice however, often makes it very difficult for maintenance when the use of mobile cranes and fork lifts are needed. This is where the assistance of the maintenance superintendent is needed on the design team.

The writer knows of a case in Peru in a copper concentrator where pumps were jammed into corners making the changing of impellors and pump cases a very difficult job.

When the writer was a shift boss in a lead/zinc/silver mill in Australia, it was a very difficult job to get to the thickener underflow on/off valves. One had to crawl along a tunnel to get to the valves. At his very next job as a shift boss at a uranium plant in South Australia, the thickeners were placed on posts which allowed a person to walk upright under the thickener to close or open the valves.

In almost every case where the writer has been involved, there was a time when the mine owners wanted to expand the plant to increase throughput. If the equipment is jammed very close in the original plant, it can be very difficult to install additional equipment such as ball mills, flotation machines, filters, etc.

**DON'T INSTALL FAULTY ELECTRICAL SWITCHGEAR OR LOCATED IT IN TIGHT CORNERS OR POTENTIAL WET AREAS**

This should be very obvious to all concerned but the writer has seen many cases where such things have happened and in some cases, fires have resulted with disastrous results.

A good practice carried out by the Anglo group in South Africa, is to place the electrical cables in vertical trays rather than in horizontal trays as installed in most plants throughout the world. This prevents the build up of dirt and dust and rocks on top of the cables which can cause cable covering degradation and possible electrical shorts.

**DON'T PLACE THE EQUIPMENT ON FAULTY FOUNDATIONS OR ON WEAK GROUND**

Again, this should be very obvious but it has happened and continues to happen. One can ask, "why is that and why should it happen? The obvious answer is carelessness. Good engineering practices have not been followed in the case of structures and adequate ground geotechnical studies have not been followed in the case of equipment placement.

The writer knows of several cases where crusher installations have collapsed and of the case of the entire mill collapsing at the Granduc mine in B.C. Canada a few years ago.

**DON'T INSTALL UNWORKABLE CHUTES BELOW ORE STOCKPILES AND FINE ORE BINS**

The writer has seen many instances of unworkable ore chutes and in particular at Gold Quarry in Nevada and in Texada Mines and Mount Isa in Australia.

At Gold Quarry at the primary crushed ore stockpile the ore froze in the winter time in the chute area forcing the use of front end loaders to move the ore from the stockpile to the conveyor belt feeding the secondary crushers.

The same thing happened at Texada Mines at Lake McLeod except that in this case, it was not a case of freezing (ambient temperatures varied between 110-120 degrees F. for most of the year) The problem was simply bad design which caused "rat holing" in the stock pile which resulted in the use of bulldozers to move the raw salt to the discharge chutes in order to feed the wash plant.

At the same operation, a tunnel was installed under the stockpile at Cape Cuvier where salt was loaded to ships at the rate of 1,500 tonnes per hour. The chutes located in the tunnel were of different design than those at the wash plant and they operated very satisfactorily.

At Mount Isa the writer blew the side out of a feed bin when using dynamite to free a "rat hole." This happened on his very last shift at this plant and has not been forgotten by the plant superintendent at that time.

**DON'T BE STINGY ON INSTALLING DUPLICATE VITAL EQUIPMENT**

The writer first experienced the use of duplicate equipment installations at the Mount Isa concentrator .A spare bank of flotation cells was installed in both the lead and zinc flotation sections and above every bank of cells were spare impellors hanging on small cranes ready to be dropped into the cells when the operating mechanism failed. Likewise, every pump was duplicated and the sumps were divided and fed by flexible rubber pipes so as to allow quick pump changes without loss of production.

Such installations cost more of course, but in the writer's opinion, are well worth the added capital expense to allow for maximum plant throughput and ease of operation.

**DON'T INSTALL DEFECTIVE EQUIPMENT**

This would seem to be an obvious statement but it has been the writer's experience that it has been a relatively common event in the past and which continues today.

Equipment often comes out of the factories which has faulty welding, untightened bolts, bolts without nuts, cracks, dents, cuts, misshaped, rusted, seized bearings, etc. It is up to the Purchasing Agent and his assistants to carefully inspect all equipment before receiving it from the manufacturers. If it is a big plant with lots of equipment, periodic visits by a company employee to the manufacturer's shop should be made to inspect the fabrication and make sure it is being done according to the engineering plans and drawings. The writer knows of a very recent case where a brand new diesel power generator was installed at a Mexican gold mine and failed while still under warranty. A large agglomeration drum was recently installed at the same operation and a riding ring cracked soon after start up.

**DON'T HESITATE TO USE THE NATURAL CONTOURS OF THE LAND**

In former times, it was common to carefully inspect the site of a new plant so as to capitalize on downhill situations to minimize pumping requirements. It was especially common to allow for gravity feeding of tailings dams.

**DON'T FAIL TO INSTALL ADEQUATE SAFETY GUARDS ON MACHINERY AND MOTORS AND PREPARE SAFETY MANUALS ALONG WITH THE PLANT DESIGN**

Of particular interest is to place guards on drive mechanisms where many a life and limb have been lost in the past. Also of importance is to have a good lock out system to prevent start up of machinery while people are working on it. Over the years the writer has been employed in the industry, he knows of many deaths and serious injuries caused by accidental start up of equipment which was not properly locked out.

Also of importance is to place guards along conveyor belts. It was common in the past for operators to ride on moving conveyor belts and sometimes with disastrous results. Guards should be placed to prevent people from carrying out this dangerous practice.

As a general rule the company should start preparing safety manuals along with the designing of the plant. Don't wait on this matter and consult with the plant design team and make such training plans to put in place as soon as staffing commences.

Also of importance is to form safety committees. Members of the committees should come from management, supervision and operations down to the lowest person on the payroll. To quote Plato Malezemoff, "nobody has a monopoly on brains."

Also of importance is to install a first aid post and clinic on site staffed by a medical doctor and shift nurses and an ambulance permanently on site. In several places overseas such an installation and its staffing is dictated by law. In any event, the plant design team should plan for the design and installation of such facilities.

**DON'T WAIT TO THE LAST MINUTE TO PREPARE OPERATOR TRAINING MANUALS**

Similarly, get the operator training manuals out as early as possible so the new hires can undergo training immediately when they come on board.

Also to be considered is to send some of the key operators (and perhaps the superintendents) to other plants to learn how things are done elsewhere in similar type plants.

Keep in mind that with plants located either domestically or overseas, it is wise to hire locals as much as possible as operators and supervisors and to teach ex pats the local culture. (See "Cultural Issues Are Important in Implementation of Mining Ventures," L.Harris and D. Malhotra- Mining Engineering September 2008).

**DON'T HESITATE TO DESIGN THE PLANT TO BE A PLEASANT PLACE TO WORK IN**

Many people of the writer's generation will recall the "Spartan" facilities in place in mineral dressing plants in the past such as the lack of air conditioning and heating of the offices and control rooms and the plants themselves.

There are places throughout the world where such conditions still exist but if one wants to keep good people on the payroll, such facilities must be first class. Again, this is a point which should be considered carefully by the plant design team.

For anyone who has been a shift boss in a mill, he or she knows the pain of having to go up and down steps in the plant. In the Anglo plants in South Africa, a walkway on one level circles the plant making it very easy for the shift boss to come out of his office and walk around the plant to see what is happening. This is a small but important point from an operator's point of view.

Also of importance in larger plants consisting of several stories, is the need to install elevators to handle personnel and move spare parts, reagents and equipment to the various levels.

**DON'T OVER AUTOMATE THE OPERATIONS OF THE PLANT**

Automation of the operations of mineral dressing plants has been a very good thing with respect to reducing the number of operators needed, maximizing throughput and recoveries and minimizing operating costs. The principle idea of automation is of course, to supply the operator with the tools to do his job better. This is done by installing on-line analyzers, particle size monitors, etc. Even crushing plants can be automated which was first successfully introduced in the Brenda copper mill in B.C., Canada many years ago.

One needs to be careful however, that the plant is not over automated without overriding capabilities which actually hinders the operator from doing his job. The writer was first exposed to over automation when he visited the Mount Morgan gold tailings re treatment plant in Australia in the 1960s. This plant was highly automated but the failing was that the operator could not override the automated systems when the circuit was in problems which turned out to be a very negative feature.

On the other hand, a simple on line analyzer was installed at Newmont's Similkameen copper concentrator in B.C., Canada which simply indicated and recorded the information on a continuous basis in the shift boss's office which allowed him to made quick changes to the circuit as required.

**DON'T NEGLECT THE ENVIRONMENT**

Keep in mind that strict rules and regulations are now in place all parts the world to protect the environment. In general, these rules are based on certain guidelines set down several years ago by the International Finance Corporation (IFC) and the World Bank and cover limits on several elements contained in solid, liquid and gaseous discharges from mineral dressing plants. Of particular interest are the heavy metals such as mercury, arsenic, lead, zinc, copper, iron, etc. and soluble compounds such as sulfates, chlorides, fluorides, etc. All lenders worldwide now insist on the borrowers signing off on the World Bank guidelines. Keep in mind that the mining industry in general and milling plants in particular, have a very negative perception among the public which leads them to believe that mining in general is dirty, dangerous and deceptive. It cannot be denied that pollution of lakes and rivers and the air at many plants existed in the past which has resulted in millions of dollars to clean up the mess and have also resulted in many lawsuits. Gordon Parker, former chairman of Newmont aptly described the situation as SOOP—sins of our past. Like preventive maintenance, the modern idea is to prevent pollution prior to it occurring or in other words, pay for it now or pay many times more in the future to clean it up[1].

Of particular interest to the mineral dressing plant designers is how to dispose of tailings and how to control dust and gaseous emissions.

With respect to tailings handling, one has several choices for disposal including the use of conventional tailings dams or ponds (so called "bath tubs"), sub aerial deposition, deposition to the sea, underground mine fill, paste production before deposition to underground mine fill or tailings ponds and filtration before disposal.

The use of conventional tailing ponds or dams has resulted in the past in rupturing of the walls caused by the pressure of the high hydraulic head of solution in the pond. Heavy rain can and has, aggravated this problem resulting in pond breakaways causing death and injury to innocent bystanders. Also of concern is the possibility of leakage of solutions to ground water supplies which indeed has been reported at several locations known to the writer.

In the mid 1980s Don East of Knight Piesold and Company brought to the industry the concept of "sub-aerial tailings deposition" which deposited the tailings in layers in the tailings ponds and employed under drainage to remove water. Thus the problems caused by the high hydraulic head are overcome. Many tailings ponds now employ this concept with great success.

Disposal of tailings directly to the sea has been practiced at several places worldwide in the past including two operations in Peru which are familiar to the writer. Pollution of the sea did away with such practices. In more recent years, disposal to deep- off shore areas in the sea through pipelines is practiced at Newmont's Batu Hijau and Minahausa in Indonesia and formerly by BHP on Vancouver Island in Canada. Negative perception by the public towards such practice is very high even though technical studies show the method causes no harm.

Underground mine fill using tailings is practiced at many mines worldwide. The normal procedure is to deslime the tailings, mix it with Portland cement and transport the mix underground to fill voids together with some coarse rock. A unique procedure was developed several years ago at O'okiep Copper Company's Carolusberg mine in South Africa where unslimed tailings was mixed with finely ground copper reverberatory slag and gypsum which completely eliminated the use of Portland cement with savings of millions of dollars. Certain modifications were made in the slag in the smelting furnace before using it for this purpose[2].

Several years ago a former Chairman of Gold Fields of South Africa, Robin Plumbridge, warned on backfilling with gold plant tailings if the price of gold was to "rise to $1,000 per ounce in the future." Several operations decided that it was not such a good idea to use such tailings in the backfill.

The process for producing a paste from tailings was originally aimed at producing a high density low moisture material for mine back fill to eliminate the need to drain the fill when placed in the underground stopes. It was a huge success for this purpose and many mill managers saw the benefit of using it to store in tailings ponds thus increasing the capacity of the pond and maximizing water reuse in the mill. The writer is presently engaged as a consultant dealing with paste deposition in two small gold plants in Peru.

Filtration of tailings is also being practiced at several modern plants especially in mountainous areas where it is difficult to build the more conventional tailings ponds. This is more expensive both from capital and operating costs point of view than any other tailings deposition method but in some cases, it is the safest way.

Safe disposition of tailings is a very important design factor and can be a very challenging decision for the design team to make.

As a young boy the writer saw many workers in Mount Morgan slowly dying of silicosis. Crusher operators and underground miners were the most affected. Respirators were not known at that time and dust collection at crushers and ore passes underground were non existent. Even today in the developing countries, little attention is paid to installing dust collectors. The sight of people dying from silicosis is a frightening sight and the writer does not want to see it anymore. The design team should place much attention to this matter and ensure the installation of adequate wet or dry dust collection systems.

The best way to operate mineral dressing plants is to have no water discharge so as to prevent any type of contamination of the surrounding areas. This is not possible however, in high rainfall areas and many plants are saddled with very expensive water treatment plants such as the Minera Yanacocha's gold ore treatment plants in Peru. At this site, excess water is produced in large quantities especially during the wet season. Cyanide destruction is carried out by using chlorine gas but there are of course, other cyanide destruction methods practiced throughout the world such as the very well established Inco process using sulfur dioxide and copper sulfate. Caro's acid and hydrogen peroxide are also widely used for this purpose.

At many mine sites acid rock drainage (ARD) is produced as runoff from waste dumps containing sulfide minerals mainly pyrite. Such ARD completely polluted the Dee River at Mount Morgan and killed all the fish where the writer fished as a boy. Although this problem is really a mine problem, very often it falls under the responsibilities of the mill manager. Much attention has been focused on this matter in recent years and SME has a complete session devoted to it at its annual meetings entitled "Global Acid Rock Drainage (GARD)." The technology to treat ARD has been completely developed and many plants have been installed throughout the world. The most common technology used is neutralization with lime, compaction of the sludge for disposal in ponds or in some cases, on top of the waste piles.

As far as gaseous emissions are concerned, the major problem is to remove sulfur dioxide in he gases resulting from the pretreatment of refractory gold ores by roasting. At Jerritt Canyon and at Gold Quarry in Nevada, Freeport and Newmont developed roasters which use roasting air enriched with oxygen thus producing a very low volume of off-take gas which could be scrubbed with caustic soda or converted to sulfuric acid. At present, there is an acute shortage of sulfuric acid throughout the world and it behooves plant operators to carefully look at the possibility of using roasters when treating refractory gold ores and producing such acid from roaster off gases. A unique process was developed at the copper smelter by Palabora Mining Company in South Africa where the concentrator tailings high in dolomite is used to scrub the off-gases to remove sulfur dioxide.

In some case, the roaster off-take gases contain other contaminants such as mercury and arsenic but fortunately, processes has been developed to remove these contaminates from the gases before removing the sulfur dioxide.

In the plant itself there are cases of mercury being emitted to atmosphere from drying ovens handing Merrill Crowe precipitates and in the electro winning section which require strict ventilation and scrubbing installations.

Gold plants normally have an assay office attached to the facilities and many who have worked in these offices have suffered from lead intoxification from lead fumes emitting from the cupellation of the lead button. Even today the writer has observed inefficient hooding for gas collection and scrubbing in many assay offices throughout the world especially in developing countries. His advice is to cover this matter carefully at the design stage.

## DON'T FORGET TO INSTALL A FOOL PROOF SECURITY SYSTEM

All mineral dressing plants producing base metals concentrates or precious metals or diamonds or gem minerals or industrial minerals or washed coal located domestically or overseas, require some level of security to protect the employees from harm and the stealing of product and spare parts or reagents.

It is obvious that plants producing high value products like gold, diamond and silver require the highest level of security especially when located in areas where political terrorists, eco terrorists, gangsters, bandits and anti mining groups operate. The writer has had fist hand experience with this matter when he was general manager of Minera Yanacocha in Peru when this country was in the grips of two dangerous political terrorist groups namely, the Shining Path and the Tupac Amaru organization.

Another hot political terrorist area was Namibia where Tsumeb Corporation Limited operated several plants for many years.

Eco terrorism is another form of terrorism showing its ugly face in recent times and is a playing havoc at several mine sites particularly in Latin America[3].

One can never discard the possibility of stealing in gold plants and the writer has seen many instances of this. Last year a gold plant in Mexico was attacked and the gold bullion stolen twice in two months. The year before a gold plant in Peru was attacked in the wee hours and several guards were killed and a large quantity of bullion was stolen. In the frozen north of Canada, a brazen attack was made in mid winter by a person mounted on a snow mobile. The writer has also been witness to several events where the gold was stolen by the plant security guards.

If the plant being designed for operating anywhere in the world, the design team will need to come up with levels of security needed to handle each individual case. There will be a need in all cases to consult with experts in this field.

## DON'T BE AFRAID TO CHECK OUT WHAT OTHERS ARE DOING

Part of the design team's responsibilities is to plan to use the latest and best available technology in the design of the plant they are working on. The team therefore, should check with other operators and equipment suppliers to get the latest and best for the new plant. To do this, the team needs to visit other similar plants and discuss the matter with everyone and his brother. The One-Mine system recently introduced by SME is a digital library that contains thousands of technical papers and is a good source of possible information for the design team.

## DON'T REPEAT THE MISTAKES OF THE PAST

At first blush, this would appear to be a difficult task but it is not really so difficult if approached correctly. The correct way is the way Newmont handed this matter several years ago when the company set up a small committee composed of Ken Brunk, Larry Buter and the writer to carefully review all plans and other information associated with new installations or expansions to see that former mistakes had not been repeated. The system worked very well but for some unknown reason, was discontinued after about a year of operation.

The writer wishes to thank SME for giving him the chance to express his views on plant design and the important issues that should not be ignored.

## REFERENCES

Harris, L. and Phillip, T.P. Heap Leaching Versus Milling of Gold Ores. Technical Paper.

Harris, L. 1991. Mining and the Well Informed Citizen. Henry Krumb Lecture.

Atkinson, R.J, Hannaford, A.L., Harris, L., and Phillip, T.P. August 1989. Using Smelter Slag in Mine Backfill. *Mining Mag.*

Harris, L. 2007. *Why Is There So Much Opposition To Mining In Latin America? Is It a Conspiracy Or Naivety Or a Combination of Both?* Presented at the 2007 Diplomats Ball Seminar in Denver, Colorado, USA, June 2007 and at the New York City, New York, USA SME Local Section 2008.

# A FEW GENERAL RULES FOR PLANT DESIGN

Doug Halbe

Doug Halbe Consultant, P.C., 1773 Michigan Avenue, Salt Lake City, Utah, USA

## ABSTRACT

Metallurgists and engineers involved in the design of a new plant have a responsibility for more than just the direct technical and engineering aspects of the project. This paper discusses some of the more general things that need to be considered on your next design project. What problems and concerns has the author seen over the years—generally of a non-technical nature—that can go so wrong that they overwhelm the clever modeling and design and control systems on which the designers' and owners' metallurgists have spent so much time?

## INTRODUCTION

For a number of years I have spent more and more time as a sort of senior advisor on new projects, or in working with junior companies starting in the early stages of metallurgical studies of their exploration discoveries.

What I would like to do today is to try to pass on to you some of the things relating to plant design that I have noted particularly over these years—not specifics, but more general things: things that you need to keep in mind, or consider occasionally during your next design project, and the one after that, and after that. What problems and concerns have I seen—primarily of a non-technical nature—that can go so wrong that they overwhelm the clever modeling and design and control systems on which the designers and owner's metallurgists have spent so much time?

As one of the early reviewers of this paper (Allard 2009) commented: This should be of particular value to "...younger engineers that probably have not had the chance to look up from their computers to observe the consequences of [their] design decisions."

These are presented as some rules dealing with process development and project design. This is an area in which I have worked for many years; and this is the primary interest of most of you. Thus these rules are written from and for our point of view. However, it's important to note that there are a lot of other players involved in the development and commissioning of a new project—and many of these players have significantly different interests and points of view than we do. Who are these other players? Corporate executives, stockholders, bankers, venture capitalists, vendors, and plant operators.

As we go through some rules for our work, we need to remember that other players may see things somewhat differently than we do; and we need to make sure that we don't lose sight of this important concern. Remember in the third grade, your report card had a place for "Plays well with others" (and you got a "Needs Improvement")? This is relevant in the mineral processing plant design business, too.

Our interaction with these other players is an important consideration in process plant design, so I'm going to throw in occasionally some comments to help you keep this in mind.

## FIRST RULE

This first rule I more than occasionally find seriously neglected. It's one that is absolutely basic—and one that everybody knows about and agrees is critical. It just somehow doesn't always seem to get the attention that it deserves—and projects fail because of this.

**The most critical single item in process design is understanding the feed material the plant will be treating: What is the mine going to be sending to the mill, and how do each of these feed types react metallurgically?**

I'm not sure why this happens. We all know, beyond any doubt, the absolute critical nature of carrying out metallurgical testwork and then designing the plant based on samples or composites that accurately represent the various parts of the entire deposit. Yet, this particular problem is fairly often the cause of major start-up problems and even project failure. Terry McNulty (McNulty 1998), in his original paper on the subject noted as one of the common problems of poor start-ups and plant failures "Pilot-scale testing was incomplete or may have been conducted on non-representative samples."

Right from the time that an exploration project begins to look as if it might warrant development, a project metallurgist needs to start working with the project geologists: looking at core; going over geologic cross-sections; discussing with the geologists how you think different portions of the deposit might give different metallurgical results and, from them, how this correlates with their interpretation of the various geologic zones.

Every sample or composite selected for metallurgical testwork should be the product of discussions between the project metallurgist and the geologic staff: overall composites chosen to represent (insofar as possible) the entire deposit, those for each of the major geologic/metallurgical zones, and those that represent smaller zones with potentially differing metallurgical characteristics. A good example is Twin Creek's pressure oxidation development testwork program (Simmons 1996). On-going work at Bingham Canyon (Rose et al 2009) is a more general application of this concept in an existing operation.

Geologists don't generally know much about metallurgical processing. That's not surprising. Metallurgists would be pretty terrible at logging core. When I was Manager of Metallurgy for Homestake Mining Company, the company geologists would get together once or twice a year in meetings for discussion of on-going exploration projects, exploration plans, budgets and similar. The Vice-President of Exploration, in what I quickly determined was a very wise move, asked me to talk at each of these meetings about metallurgy: how various processing methods related to mineralogy, how plants are designed, what kind of samples and composites are needed, and what sort of information metallurgists need from geologists to do this work.

These discussions (and they moved quickly from formal talks to great discussions) made, I strongly believe, a significant improvement in subsequent geologist-metallurgist interaction. The geologists certainly improved their understanding of what metallurgists need, and our small metallurgical group had a much better working understanding of the problems and constraints of the exploration group as new projects developed.

Nothing new here—a lot of you already do this. If you don't, you should.

Later in this talk I will mention a project I worked on and note that continuous testwork was carried out on surrogate samples. The more observant of you will immediately note that I clearly don't follow my own best advice.

While that has, on more than one occasion, been true, we all know that for some projects it is economically impossible to obtain large enough representative samples or composites to provide pilot or continuous testing. This is a factor we often simply have to deal with. In the case discussed below, for the design of a roasting project in Australia, we used a procedure that many of you have used in these circumstances. We picked the surrogate with care, based on comparable mineralogy, and then correlated the results of static roasting tests on the surrogate sample with those of the actual deposit composites, and made the assumption that this correlation would generally hold in the continuous tests on the surrogate. We also built in a safely factor. In this case, we were working with a well-established process with a good history of lab and continuous testing through to a working process plant, and this helped validate the use of this procedure. The vendor's historical data also provided guidance on factors to consider in scale-up correlations. As discussed further below, the plant reached our design specifications.

A word of caution, here, though: the less well established the process the less valid this procedure becomes.

One more thought on this topic: while writing this paper I went through the abstracts for the Denver SME meeting held earlier this year and counted the papers that dealt somehow with any aspect of plant feed characterization. There were eight of them, all sort of generally helpful to the subject we are discussing, but none that really dealt with the crux of the problem: when metallurgical testwork starts, what samples and composites do you need in order to design a circuit that will effectively handle the various types of feed material you will receive?

There were, by my count, a total of 123 papers from minerals and metallurgical processing and from geology as related to ore characterization. Eight out of 123 that "sort of" deal with the problem of sample selection seems to me to say that we may be missing something here.

There are two encouraging notes here. First, I'm encouraged by the titles of a number of the papers in this symposium in the Research/Characterization and Sampling sessions that seem to deal with this subject, and am looking forward to hearing them.

Second, those people who are invaluable to us in process design, the vendors of equipment and services, seem to have noted this problem. Not long ago, for a client, I ran across a service offered by SGS in what they refer to as Geometallurgy (SGS 2009). Although I haven't actually worked with this program, I have looked it over, and reviewed a proposal of theirs. This is a fairly structured approach, and I believe that such a structured approach is useful if not essential. This is one example I've seen; there well may be other companies offering such a service. If I were a senior executive in geology or metallurgy I would certainly want to check this area of potential assistance carefully.

How does this rule appear to some of the players we mentioned earlier, for example the project owners, typically project management? From our point of view, the need for good sampling and metallurgical testing is self-evident, and absolutely critical to a successful project. From the point of view of management, though, these costs come at the worst possible time. Budgets are tight; probably project funding is not yet in place. The metallurgists want more tests, on more samples, to explain the last tests that were designed to explain the previous tests. Correct? Your job is not just to whine for more time and money, but to develop and present a persuasive case to management as to why these expenditures are required. We'll talk more about how best to do this a bit later.

## SECOND RULE

A number of you have heard me talk about this before. It's often overlooked until the financial people put together the cash flow projections after receipt of the pre-feasibility study—which is, in my opinion, a little late to begin to think about it.

**Planning for the estimation of the start-up parameters for a new project should begin during process development and testwork—and management should be kept aware of this estimate. It should not be left as the last, brief step, before completion of the cash flow study.**

There is a section at this symposium on Financial and Feasibility Studies. One of the key items associated with a feasibility study is a cash flow projection. You and others will spend a very large amount of time in estimating the project's capital and operating costs. Once these are entered into the cash flow projection, you will find that someone will have to make an estimate of how long it will take the project to come up to full design capacity and to projected recovery.

There have been, over the years, a number of papers dealing with this problem (e.g., Davis 1985, Glypton 2002, Nice 2002, Mackey and Nessett 2003, Twigge-Molecey 2003, McNulty 2004)—some of them containing a number of concrete examples showing these two parameters as a function of time—for good, and for bad-to-horrible startups. Terry McNulty and I, both individually and together, have looked at the problem of estimating the time required for start-up (McNulty 1998, Halbe 2003). We've developed some potentially useful, sort of semi-quantitative guidelines, but they are at best somewhat subjective, and they could, with some effort, probably be used to get whatever answer you (or your boss, or your financial people or corporate management) require.

I understand completely that your initial, best engineering estimate of time required to reach full tonnage and recovery will (1) be what you hope is conservative, (2) probably be unacceptably long for corporate purposes, and (3) thus be significantly reduced. And I certainly understand the corporate reasons for that.

What is the answer here? How do you make a reasonable, engineering-based estimate of start-up time? I do not have a quick and easy answer. At this point I do have one preliminary suggestion. You will run, in your feasibility study and cash flow analysis, sensitivities: capex, operating costs, recovery, head grade, metal prices, and perhaps a few others. I don't recall ever having seen start-up time sensitivities run. Might be worth including these—from most optimistic to some worst-case, horrible examples from those papers noted earlier.

Terry tells me that he has reviewed feasibility studies for new processes for several clients, and attempted to provide a more quantitative estimate of months to full tonnage and design recovery by a dispassionate third party. Here are a couple of examples, direct quotations from Terry:

> "A hydrometallurgical project was extensively piloted, the mineral deposit was well-understood, and everything seemed to have been covered, but three of the unit operations required equipment not previously used in that operation. Also, operating conditions were severe and process conditions had to be tightly controlled. I worked out a set of probability curves that led me to conclude that about 30 months would pass after commissioning before the plant was producing at the annual target metal production rate.
>
> A recent project was based on milling ore which had been milled in the late-1970s and the processing flowsheet was totally conventional. On the assumption of sound equipment selection and careful engineering/design/construction, coupled with a comprehensive training program with capable and motivated personnel, I predicted consistent production at design rate with acceptable quality in the 10th month after commissioning. However, production for the first year would only be 80% of design."
>
> —(McNulty 2009)

The possibility of an independent analysis of project start-up time is an area, in my opinion, that perhaps should be carefully considered by companies developing projects, most particularly those looking at new technology. The examples above show that a quantitative analysis is possible. While Terry has carried out several of these analyses, none of the projects studied have been constructed yet, so, unfortunately, we have no comparison of projected with actual results. I very much like the concept, though, and recommend you consider it.

Terry tells me that there should be enough information to carry out such an analysis toward the end of the pre-feasibility study, and very possibly well before. If such an analysis were carried out at this point, and the results indicated a 30-month ramp-up, engineers and management might well wish to hold a long, sincere meeting, and see what could be done to improve the ramp up. The steps (time and dollars) suggested might provide better economics than a 30-month start-up.

## THIRD RULE—HYDROMETALLURGY

This next rule is directly from Terry McNulty, who has been involved in a large number of hydrometallurgical projects over the course of his career.

**If the process chemistry is novel, be sure you completely understand it. Even if the process chemistry is not novel, be sure you understand *all* of the reactions that will take place.**

This rule sounds almost silly. However, with the help of a few friends I was able to assemble a short list of significant projects within recent memory that failed because of neglect of this rule. It's not necessary to mention specific projects here, but those readers with extensive experience in the business will be able to think of several. If you are involved in development of a hydrometallurgical process, don't know of any such examples, and don't even really believe that this could happen, talk to someone who has been around a bit longer.

## FOURTH RULE

As evidenced by the SME annual meeting earlier this year, as well as at this meeting, there exists a plethora of fascinating and potentially tremendously useful new developments in equipment and in measurement and control for operating plants: all kinds of wonderful new stuff to design into your new plant.

What about new stuff? Here I'm speaking of new control systems, new measurement systems, new (or one size larger than the largest one ever built) pieces of equipment, new unit operations or even processes. I'm almost hesitant to put up this next rule, particularly in the light of all of these new developments we have been hearing about. However, I'm an engineer, and as an engineer I have a professional obligation to give you my best professional opinion. This next rule needs to be used with caution, but it does need to be at least considered.

In process development the opportunity to use something new frequently arises: a new unit operation, a new type of measurement system, the biggest piece of this type of equipment ever built, maybe even a new process.

Here's the rule that I have given clients in the past when this situation comes up:

**If the use of some new, or leading edge, process, or new type of equipment, or anything else new is absolutely essential to the economic viability of the plant being designed, go ahead. If not, don't.**

My experience, over lots of start-ups, is that there are plenty of things that can go wrong during start-up with a circuit using plain old ordinary equipment and processes and measurement/control systems that you've used many times before. You'll have your hands full getting the plant up to design tonnage and recovery without having to deal with the problems that are inevitably associated with new or leading edge stuff.

Does this mean that I am opposed to "new stuff"? Absolutely not. However, the place, if you possibly can, to install and test out new stuff is in an existing plant, not a brand new one. If you work for a large company with a number of operations, try it in one of your existing plants. If not, maybe you can arrange to spend some time in the plant of another company that has installed the "new stuff" you are interested in. Fortunately, we are an industry where this sort of professional exchange of information is generally freely done. Obviously, your company has to be prepared to reciprocate—both in supplying information back once you start up your new stuff, and in honoring reasonable requests to share this sort of information with others (even competitors).

If these options are not appropriate or possible, extensive testing or piloting will unquestionably be necessary.

New stuff is not bad stuff—quite the opposite. But plan on spending a big bunch of time and effort getting this sort of equipment or process operating up to design.

A quick example to put this into context, perhaps. Ultra-fine grinding. Is this a technology that is tremendously useful in some instances—in fact critical to some processes? Yes, and we are delighted to have this as another tool in mineral processing and hydrometallurgy. Are we still on a pretty steep portion of the learning curve in application and control? I'd say so.

## FIFTH RULE

The next rule is based very much on my experience in design and start-up of plants.

**The things that you spend the most time and effort on and the potential problems that you plan for -- never happen.**

There are two aspects of this general rule. First, in most process design work for new plants there are one or more areas that are critical, or complicated, or maybe even (because you decided to ignore what I just told you a moment ago) new stuff. This might be the flotation reagent scheme for a highly variable complex polymetallic deposit, or maybe SAG milling of a heterogeneous deposit, or a relatively new hydrometallurgical process. This is where you will, no doubt, spend the vast amount of your time and money: ore characterization (I hope), testwork, piloting, and circuit design.

The rest of the circuit will be pretty much routine—"been there, done that."

I recall the design and start-up of the gold circuit at Windarra, in Western Australia. This involved flotation of an auriferous pyrite-arsenopyrite concentrate, followed by roasting, and then leaching and CIP (Quadrio and Kellett 1984). (A bit of an aside here: The referenced paper tells not only what was designed and built, but also what was wrong and had to be fixed. We could, I believe, beneficially use more papers following this format.)

Obviously the vast majority of our effort was on the gold roasting circuit: development of a surrogate concentrate because we could not get enough sample for continuous roasting tests; then extensive roasting testwork both batch and continuous, visits and detailed discussions with vendors, and visits to a number of gold roasting operations.

Actually the roaster started up generally pretty well. The authors of the referenced paper (the mill superintendent and the Dorr-Oliver project engineer, respectively) note and discuss the start-up problems experienced. There were some roaster problems that we worked through, with the assistance of the vendor. The other problems likely could have been avoided, though, if we had given them sufficient thought or some additional testwork or study: problems with zinc precipitation from the calcine leach solution due to high levels of copper and zinc, extremely rapid build-up of gypsum in the quench liquor pipeline and on thickener launder walls, pumping problems of thickener underflow due to unexpected (and un-tested for) high viscosity.

These non-roaster problems turned out to have relatively easy fixes for us—but this isn't always so. Think about, for example, the poor material transfer points you've had to live with because there wasn't enough space to modify them to work properly.

It's a good idea, early on, to resolve to spend significant time on the whole circuit—not just the new, fun part.

The second part covered by this rule comes after the design is finalized, the project approved, and construction is well underway. Near the end of construction you will have assembled your staff and operators and will be going through training. I've always found that this is a great time, after all of your experienced staff and operators are familiar with the circuit, to spend a chunk of time with them poring over the flowsheet and plans, and saying "What can go wrong—especially during start-up?"

With some experienced mill people this can be highly useful—and even fun.

Everyone who has designed a plant has had to face the comments of those who are condemned to operate it: "Boy, what a dumb design. If I had been involved in the design...." Sound familiar? The time to deal with this—beneficially—is before start-up. "Okay, gang. The design is fixed, and we are not—repeat not—going to have any change orders. This is what we are going to start up with, and we are all going to have to live that. The engineering company says that it will work, but you guys (and ladies) don't think so. If you're right, and it doesn't work, how can we deal with it?"

I'm sure that a number of you do this in your projects. I've found that this always picks up some areas of potential concern that have been overlooked—and some great ways of dealing with these problems: maybe a design for some sort of a quickly-rigged minor equipment change, maybe a change in operational procedures, conceivably even a minor piping change by the mod squad right after start-up.

Almost as important, this really gives the crews a buy-in and sense of ownership of the plant.

Summarizing the above, if you locate and anticipate potential problems and plan how to deal with them, these almost assuredly won't be the difficult start-up problems that you have to fight your way through.

## SIXTH RULE

This rule I learned late in my career, rather than earlier—much to my regret.

First an actual example, then the rule. Recently I was working as a consultant for a highly competent corporate executive, with lots of successful projects on his record. In an early meeting, he told the project manager that he needed the results of the preliminary study by December 15 The project manager, with whom I had never worked before, said "Terry, we cannot satisfactorily complete the study in that amount of time. Terry's reply was "I've promised the board, so I have to have it by then." Dennis responded: "I can have for you by that time sufficient information for you to make a preliminary general presentation to the board, but it won't be complete, and it won't be final. You may have the final report by January 30." My thought, sitting quietly in the meeting: "I think I'll enjoy working with this project manager." Terry, who is a very sensible and experienced executive, considered this comment for a moment, (thinking probably of what he was going to tell the board), and acquiesced.

I'm sure that all of you have gone through this sort of thing, and I'm sure you are all thinking: "Sure, but if the report really has to be completed by that time, you have to agree to do it, somehow.

Thus, our next rule:

**You can have it fast.**
**You can have it cheap.**
**You can have it correct.**

**Pick any two.**

If you haven't run across this rule before, think about it a bit. You can have the report or the study fast and cheap, but it likely won't be entirely correct or accurate. You can have it cheap and correct, but it won't be fast. You can have it fast and correct, but it won't be cheap.

This fits in very well with our comment on the first rule—about ore types and management's reaction to your expressed need for more samples and more testing. If you are asked, as you no doubt have been in the past and will no doubt be in the future, to do something in an unrealistic time frame, or with an insufficient budget, or with insufficient testing, you need to make absolutely sure that the entire project team, including the project VP, clearly understands the implications of this rule.

## LAST RULE

Many of you who have retained me as a consultant in the past will remember this one. I learned it many, many years ago, and I've always referred to this as the Bechtel Rules. If it did originate with Bechtel, my thanks, or perhaps apologies, to that organization.

**THE BECHTEL RULES**

**Rule 1: The client is always right.**

**Rule 2: If the client is wrong, refer to Rule 1.**

Whether you work for an engineering company and your client is a mining company; or you are a company engineer and your client is your boss, the VP of Engineering, this rule holds. Some decisions on the project may not make much sense to you, but they probably do from the owner's or corporate view point. If you think a decision is wrong, for heaven's sake say so clearly and back up your opinion with supporting evidence. If you are overruled, accept it and keep on track.

You may know more about the project, but the client knows external constraints that you don't. Remember, if you are tempted to forget this rule: The client has expertise in, among other areas, financing, development of a project and keeping the shareholders happy. These areas of expertise are very useful, since they are all requisites to keeping the rest of the project team employed.

## COMMENT

Let's finish up with a final look at project development work as we would like it to be carried out, by what are really the four key sets of players. Perhaps in your next project you might want to sit down with these players (maybe during the kick-off meeting—or maybe even before) and talk about the ideas below. This, again, is direct from Geoff Allard (Allard 2009), and it seems to me like a pretty good summary of how the game can best be played. He calls it:

**Why can't we all just get along?**

**Owners:** You don't have the time, experience or desire to engineer or operate the plant. If you do, you don't want the cut in pay to do it. Take the time to explain what you want and need and expect from the engineers. If you don't know, ask. When the engineer says you are wrong, listen to the explanation and ask questions. Maybe you are wrong, maybe not; then make a decision.

Tell the operators they are getting the plant the company can afford and still have sufficient funds to pay their salaries. *Ask* them to make it work for you.

**Operators:** You are there because you can make things work. Insist on being involved from the very beginning with everything: accessibility, safety, maintenance. Talk to the engineers; put their number on your speed dial. If something goes wrong tell them about it; they may have a simple solution or can identify an operational problem.

If you don't tell the engineers what does not work, they will continue to design plants the same way.

**Engineering company:** Every client wants an operating, fully functional plant at the end of the job. Don't bid or design anything else. If there are items missing from the bid request, identify them and bid them as an option. Help the client get the plant he needs.

**Engineers (This is us.):** Don't kid anyone; we do this because we enjoy it. Be thankful someone wants to pay us to do it. If the owner is wrong, tell him. He/she is a big kid and can handle the criticism. If he/she isn't, write up the meeting notes and do it his/her way. If you can't handle that, get a real estate license.

Talk to operators every chance you can get, learn from them. They know a lot more about what works than you. The difference between a good plant and a great plant are the operators. Make sure they have input.

Good luck on your next plant design and start up, and the one after that, and after.

## ACKNOWEDGEMENTS

Major credit is due to Terry McNulty and Geoff Allard, who read the original draft of this talk and responded with detailed comments and suggestions for significant additions. They then continued to review, comment and offer constructive suggestions through several further drafts. Thanks.

## REFERENCES

Allard, Geoff. 2009. Personal communication. (www.allard-enginering.com).

Davis, David. 1985. New projects: beware of false economies. *Harvard Business Review*, March-April, pp. 95-101.

Gypton, Chris. 2002. How Have We Done?—Feasibility Performance since 1980. *E/MJ*, January, pp. 40-46.

Halbe, Doug. 2003. Business Aspects and Future Technical Outlook for Hydrometallurgy. *Hydrometallurgy 2003*. The Minerals, Metals, & Materials Society, pp. 1091-1104.

Mackey, Phillip J. and Nesset, Jan E. 2003. The Impact of Commissioning and Start-up Performance on a Mining/Metallurgical Project. *Proceedings 35th Annual Meeting of the Canadian Mineral Processors*, pp. 331-347.

McNulty, Terry. 1998. Innovative Technology: Its Development and Commercialization. *Managing Innovation in the Minerals Industry*. Edited by M.C. Kuhn. Society of Mining Engineers, pp. 1-14.

McNulty, Terry. 2004. Minimization of Delays in Plant Startups. *Plant Operators' Forum*. Society for Mining, Metallurgy and Exploration, pp. 113-120.

McNulty, Terry. 2009. Personal communication.

Nice, R.W. 2002. 'Sensible' Ramp-up Times for PAL and Other Projects. *ALTA Nickel/Cobalt Conference (Australia)*.

Quadrio, A.J. and Kellett, K. R. 1984. Recovery of Gold from Lancefield Arsenical Sulphide Ore. *Gold Mining Metallurgy and Geology*. Regional Conference Australasian Institute of Mining and Metallurgy.

Rose, J. et al. 2009. Quantitative Mineralogical Study of Ore Domains at Bingham Canyon, Utah, USA. SME Preprint 09-108.

SGS 2009. Geometallurgy Overview. www.met.sgs.com/geometallurgical_mapping.

Simmons, Gary. 1996. Pressure Oxidation Process Development for Treating Carbonaceous Ores at Twin Creeks. *Randol Gold Forum, 1996*, pp. 199-208.

Twigge-Molecey, C. 2003. Knowledge, technology, profit. *Hydrometallurgy 2003*, The Minerals, Metals, & Materials Society, pp. 1-17.

# GRINDING CIRCUITS: WHAT TO LOOK FOR IN DESIGN

Derek Barratt

DJB Consultants, Inc., 1400-355 Burrard Street, Vancouver, BC, Canada

## ABSTRACT

This paper is a sequel to the three papers that were co-authored with Jerry Hanks (1) and Mark Sherman (2) in the 2002 Symposium on Mineral Processing, Design, Practice and Control published by the SME. The objective of this paper is to describe recent developments which affect the design process for comminution circuits without going into all the details that formed the substance of those earlier papers. Nevertheless, the 2002 papers are as relevant today as when they were published. In order to place everything in perspective, this paper will address:

- The impact of new technologies on "conventional" circuit design;
- Failures and solutions with respect to vendor supply;
- Available methods of comminution testing and how best to compare test results before defining design criteria;
- A strategy for testing samples within mineable blocks, "geo-comminution" and its use in defining design criteria for input to a mine block model and a mine production plan;
- Principles, limitations and solutions with respect to sizing grinding equipment;
- Properly designed comparisons for different flowsheets: CAPEX, OPEX, ROI;
- Optimization of grinding circuit operations with respect to "Mine-to-Mill," mine plan, processing objectives, and equipment modifications;
- Conclusion and recommendation.

## THE IMPACT OF NEW TECHNOLOGIES

At the last "SAG" conference (SAG' 2006), the demise of large SAG mills was being predicted in favour of the use of High Pressure Grinding Rolls (HPGR) to prepare ball mill feed. It was premised that delivery times (up to three years) for large SAG mills and their motors was too long and that Crushing/HPGR/Ball mill circuits could be designed and commissioned in less time. Lack of foundry space and machining facilities was cited as the principal cause of extended delivery periods.

The record shows that one new Crushing/HPGR/Ball mill circuit has been commissioned since 2006 (it was already in the works), that HPGR has been added to an existing Crushing/Ball mill circuit to increase feed rate, and that between 2007 and early 2008, at least 30 (undocumented) 40 ft diameter SAG mills were either on order or were being considered in project studies. Because of the financial "downturn" in 2008 and the associated decrease in commodity prices, the commercial viability of many of these projects has been compromised, many have been cancelled or are on hold, a few are going ahead, albeit it at a reduced pace (e.g., Los Bronces Expansion and the Esperanza project in Chile), and a few are still being studied (e.g., Olympic Dam in Australia and Pebble in Alaska). One significant project which is going ahead is the CITIC Iron Ore project in Australia with five 40 ft diameter SAG mills, each with a 28 MW gearless motor. At least one foundry in Europe has added "pot" space, but there is still concern about availability of machining facilities for large diameter flanges.

One of the side effects associated with the impact of HPGR has been the psychological limit imposed on practical mill diameter; 40 ft in the case of SAG mills and 26 ft in the case of ball mills. The argument in favour of HPGR stresses a reduction in the required power for the associated ball mills (hence smaller units, e.g., 24 ft diameter) and the flexibility in plant design of adding units if required. The argument in favour of SAG milling stresses its ability to cope with variations in ore competency and feed size analyses and its immunity from wide variations in feed moisture content.

## FAILURES AND SOLUTIONS

Whereas historical failures have been associated primarily with fabricated heads and shells of SAG mills, due to failure of welds and imperfections in castings for heads and trunnions, there have been few, if any, that have not been caught by testing and inspection since the advent in the mid 1980's of more stringent specifications and QA/QC/ inspection methods, coupled with newer designs (DEM techniques) and improved foundry and welding techniques, at least in western Europe and the Americas. Some western manufacturers are making use of facilities in the former communist countries of the Russian Federation and eastern Europe. The jury is still out with respect to China.

It has not gone unnoticed that, since 1997, there have been significant failures and large insurance claims associated with large gearless drives (a.k.a. ring motors) which have resulted in extended shut downs, value of lost production, and expensive remediation: the first, at Cadia Hill, on the first 40 ft diameter SAG mill; the second at Collahuasi, also on a 40 ft diameter mill; the third at Antamina on a 38 ft diameter SAG mill, and the fourth at Escondida (Laguna Seca), also on a 38 ft diameter mill. The causes (where known) can be summarized as follows:

**Figure 1. SAG Mill Ring Motor, Freeport Indonesia 38 ft dia mill**

| | |
|---|---|
| Cadia Hill (1998)<br>19.5 MW | Lack of structural rigidity in the stator frame with mechanical resonance causing intense vibration within the frequency range of motor operation.<br>Remedy: addition of structural steel and frame to the stator casing and re-examination of design. A similar, but less intense problem was experienced with the two 36 ft diameter, 12.86 MW motors at Minera Pelambres in Chile with a similar remedy. |
| Collahuasi (2005)<br>21 MW | Stress concentrations in stator lamination key bar positioning plates which caused failure resulting in contact with the rotor (i.e., closing of the air gap).<br>Remedy: re-design and installation of new plates but, more importantly, installation of a new stator at a later date. |
| Antamina (2007/8)<br>20.1 MW | Partial discharge damage to stator coils as well as loosening of stator laminations.<br>Remedy: operation at a reduced speed (ostensibly, to limit applied voltage) and installation of a new stator; re-design of the clamping bolts for stator laminations and re-examination of the application of coil insulation. |

**Why Did These Failures Happen?**

Human error/omission in the design of components, QA/QC/inspection of components, co-ordination of design with other components (e.g., motor with mill and foundations), commercial/cost/competitive considerations post the advent of two vendors instead of one; new generation of designers, supervisors, inspectors, inexperience, absence of a third party review of the mill supply and installation, etc., are all possible and interacting factors.

In summary at present: risk management in the design and manufacture of mills has improved so that there have been no structural failures in the last 15 years but, during the same period, changes in motor design and manufacturing have caused preventable failures. These have precipitated a return to proven practices for design, QA/QC, and inspection.

## AVAILABLE METHODS OF COMMINUTION TESTING

Since 2002, only one widely used new method of testing samples for comminution has come on-stream. It is the "SMC" test which is a less costly derivative, and a modification of, the JK drop weight test. This test produces results for:

- DWI, an intermediate energy value per unit volume;
- A × b, or the appearance function as used in the JK SimMet simulation protocol; estimated by an empirical relationship to DWI;

The SMC test does not produce a value for abrasion, $t_a$, but at least one JK practitioner uses an estimate of $t_a$ which is based on a broad correlation of $t_a$ with A × b. The SMC test can be used in simulations to predict mill feed rate in the same manner as with the JK drop weight test based on a default or applied breakage rate model. Estimates of specific power consumption can be calculated for a broad range of samples and for a particular selection of mill sizes; i.e., the simulation does not calculate kWh/t unless a particular mill size (for SAG mill and ball mill) is used to estimate the applied power and mill feed rate. The JK SimMet protocol relies on breakage rates to establish the capacity limit and the appearance function to establish material balance of a particular circuit design. It is not power-based.

In contrast, the SPI protocol is power-based, but it can also establish a material balance for a particular circuit design. The SPI test reports a SAG power index, expressed in minutes which relates directly to kWh/t, and which is used, in combination with the Bond Work Index for ball milling (either from the full test procedure or an abbreviated one), in simulations to predict mill feed rate, specific power consumption for SAG milling and secondary ball milling, and transfer sizes based on the input of particular mill sizes to estimate power draws.

The "Millpower2000" simulation protocol, which has been developed by DJB Consultants, Inc., is power-based using Bond Work Indices for:

- Low energy impact crushing, $Wi_C$;
- Rod milling to a defined mesh, $Wi_{RM}$;
- Ball milling to a defined mesh, $Wi_{BM}$.

Millpower2000, unlike JK SimMet (based on either the JK drop weight test or the SMC test) and CEET (based on the SPI test), can be used to:

- Predict specific power consumptions for SAG milling and secondary ball milling based on the selected mill feed rate and consequent power demand for each before it calculates the required mill sizes, selection of transfer size and estimation of operating conditions; e.g., ball charge volume, total mill charge volume, mill speed, and pulp density, or to:
- Predict mill feed rate based on pre-selection of mill sizes and operating conditions, as well as available power which is based on calculation of total specific power consumption and iteration or calculation of transfer size to balance the circuit between SAG milling and secondary ball milling;
- Predict the probability of the inclusion of pebble crushing in a circuit, but it cannot:
- Create a material balance for a particular circuit design.

These three testing and simulation protocols are commonly used in the design of grinding circuits, but very often consultants are called upon to compare results of predicted mill feed rate which may not be directly comparable because of differences in the selection of samples that are presented to the tests, or differences which may be apparent from actual test protocols; e.g., particle size or shape which is presented and/or geotechnical parameters; e.g., fracture frequency, or differences in machine characteristics.

## THE IMPORTANCE OF SAMPLING

DJB Consultants, Inc. has constructed a sampling protocol which can be used to ensure that results from the SMC, JK DWT, SPI, and Millpower2000 test protocols are comparable on the basis of testing the same sample interval from drill core for each. This has been done in the absence of any requirement from NI 43-101, JORC or any other code. Physical examination of this whole core by the comminution consultant is considered to be imperative. Testing of contiguous intervals in a particular drill core is preferred in order to describe variation by mine bench. DJB Consultants, Inc. has also constructed a data base of test results for $Wi_C$, $Wi_{RM}$, $Wi_{BM}$, A × b, $t_a$, and SPI, supplemented by unconfined compressive strength (UCS) and point load index (PLI). This database covers:

| | DDH Core Types & Diameter |
|---|---|
| • 979 samples of $Wi_C$; | whole PQ (85 mm), whole HQ (64 mm) |
| • 1413 samples of $Wi_{RM}$; | stage-crushed PQ, HQ or NQ |
| • 2531 samples of $Wi_{BM}$; | stage-crushed PQ, HQ or NQ |
| • 1255 samples of Ai; | stage-crushed PQ, HQ or NQ |
| • 1784 samples of A × b; | whole PQ or HQ for JK DWT, ¼ NQ (48 mm) for SMC |
| • 331 samples of $t_a$ (260 inferred); | whole PQ or HQ for JK DWT, ¼ NQ (48 mm) for SMC |
| • 852 samples of SPI; | HQ or NQ |
| • 122 samples of UCS; | PQ or HQ |
| • 110 samples of PLI; | determined while logging core |
| • >4000 samples of RQD, R-value, fracture frequency, and/or core recovery. | determined while logging core |

This database has been used to compare test results on a "parameter vs. parameter" basis in order to demonstrate any anomalies vs. expected relationships on the basis of:

- Lithology and Alteration
- Primary vs. Secondary mineralization
- Hypergene vs. Supergene zones
- Geotechnical parameters; e.g., UCS, PLI, or fracture frequency vs. $Wi_C$ or A × b.

DJB Consultants, Inc. has also found that differences in the application of machine characteristics in simulations can lead to differences in the prediction of mill feed rate, even though individual test results have been obtained on the same sample intervals (see section on mill sizing).

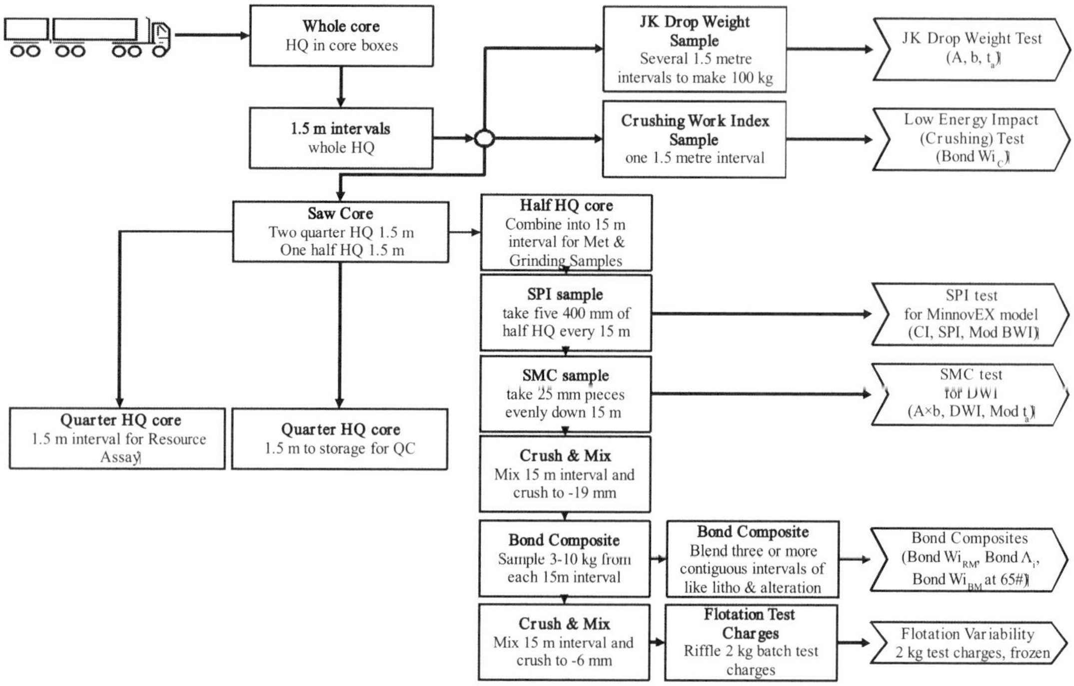


**Figure 2. Sample preparation protocol**

In summary, the preferred use of Millpower2000 in conjunction with JK SimMet is considered by DJB Consultants, Inc. to offer the best estimates to the industry, in the absence of pilot plant testwork, of specific power consumption, material balance, applied power, operating conditions, and mill sizing provided the relevant comminution test results are obtained on a comparable basis. In this manner, design criteria for a grinding circuit can be established with confidence.

## A STRATEGY FOR TESTING SAMPLES WITHIN MINEABLE BLOCKS, "GEO-COMMINUTION"

Coupled with the suggested sampling protocol in the previous section, the question arises with reference to "how many" samples should be tested? There is no guidance from NI 43-101, JORC, etc. Factors which influence the decision are:

- Level of confidence required; Pre-feasibility vs. Definitive feasibility;
- The size of the mineable resources; mining method;
- Processing rate;
- Geology within the "payback" pit or developed underground reserve with respect to different types of lithology and alteration and significant types thereof;
- Available drill core, either resource and/or dedicated for comminution tests; number of holes;
- Half-core or whole core; NQ, HQ, or PQ;
- Cost of drilling; access to site;
- Cost of testing; shipping of samples from site;
- Schedules; available time for drilling and testing vs. project study and/or engineering schedule.

DJB Consultants, Inc.'s preferred strategy is to exploit the benefits offered by studying variability within a mineable resource and establishing that variability on a "time" basis; i.e., with respect to the mine production schedule.

By testing a defined number of samples, variability can be displayed on a cumulative frequency basis for:

- Specific power consumption for SAG milling, $E_{SAG}$, kWh/t;
- Specific power consumption for ball milling, $E_{BM}$, kWh/t;
- Specific power consumption for pebble crushing, $E_{Peb}$, kWh/t;
- Specific power consumption for mills and pebble crushing, $E_{Total}$, kWh/t;
- Mill feed rate, tph;
- Mill throughput, tpd;
- Transfer size, $T_{80}$, microns;
- SAG mill speed;
- Ball mill speed;
- Product size, $P_{80}$, microns;
- All of the above on the basis of each significant lithology or combination of lithology and alteration;

- Geotechnical parameters in the ore zones: RQD, UCS, PLI, R-value, and fracture frequency as guides to the selection of primary crushers, top size, and the proportion of fines (minus 32 or minus 38 mm) in the SAG mill feed.

On this basis, testing of composites should be avoided because it would mask variability. The frequency graphs can be used to establish a "design" specific power consumption, $E_{Total}$, and a "design" mill feed rate, based on the applied and available mill motor power. The next stage in "geo-comminution" is to establish processing rates and their variability within the mine block model. This is done by assigning values of $E_{Total}$ to each block, or group of associated blocks, based on the location of the $E_{Total}$ values (i.e., sampled intervals) relative to the location of the mineable blocks and using geostatistical methods, e.g., nearest neighbour or inverse square distance, for this assignment. The mine planning engineer can then use these three-dimensional $E_{Total}$ values to compare the resultant processing rates with mine bench capability and use this comparative process to iterate and optimize mine design on the basis of applied kWh/year.

Returning to the question of "how many" samples should be taken for comminution tests in relation to cumulative frequency and mine size, DJB Consultants, Inc. suggests that:

- A sample should be geostatistically relevant to one month of production during the "payback" period for a large tonnage open pit porphyry copper project in which mill feed rate can typically vary by ± 25 % as shovels move from bench to bench; i.e., 70 samples for five years' production.
- A sample should be geostatistically relevant to one month of production for a small to medium size underground gold project for which variability might be reduced due to consistency of hanging and foot wall lithologies, i.e., 60 samples.

## PRINCIPLES, LIMITATIONS, AND SOLUTIONS: SIZING GRINDING EQUIPMENT

As has been seen in the foregoing, a guiding principle in sizing mills is to install enough power in the primary mill to meet production targets during the payback period and beyond based on output from the mine production plan and enough power in the secondary mill for it to operate at maximum percent of critical speed and ball volume to cope with variations in transfer size and minimize variation in product size by varying mill speed.

Once the installed power and operating conditions for the primary mill have been set, it remains to select the proper mill diameter in relation to the effective grinding length (EGL). The general equation for power demand in Millpower2000 is:

$$\text{kW (Net)} = P_N \times \rho_c \times D^{2.5} \times L$$

Where

PN = power number calculated from mill power measurements and which takes into account mill speed and all aspects of the charge motion within the mill cylinder and both conical ends: impact breakage, attrition, abrasion, friction and rotation within the mill charge, losses due to heat and sound emanating from within the mill, wind loss, shape of the mill charge, the imprecise position of the centre of gravity of the mill charge, size analysis of the mill charge, and no-load power.

$\rho_c$ = mill charge density, t/m$^3$

D = effective mill diameter (inside shell liners), m

L = effective grinding length (EGL) expressed as the length along the mill cylinder at the mill shell in between the feed end liner plate and the inboard side of the discharge grate plate, m

and:

$$\rho_c = \left[\frac{\%\ \text{v/v balls}}{\%\ \text{v/v total}} \times \rho_{ball} + \frac{\%\ \text{v/v ore}}{\%\ \text{v/v total}} \times \rho_{ore}\right] \times \left(1 - \frac{\%\ \text{voids}}{100}\right) + \rho_{pulp} \times \frac{\%\ \text{voids}}{100}$$

Where:

% v/v = volumetric percentage of the total mill volume

$\rho$ = specific gravity

Selection of the optimum mill diameter is important for the following reasons:

- The exponent "2.5" obviously means that mill diameter is a more significant contributor to power demand for high aspect primary mills than EGL;
- Power demand is also a function of the highest ball charge volume which is being contemplated for power-efficient operation since it affects the values of "ρ";
- Material transport through the primary mill is a function of the exponent "2.0" on mill diameter and grate open area is selected as a percentage of the total area at the discharge end, e.g., from 8 % to 10 % in sections as close to the mill periphery as possible, avoiding intermediate grate sections and minimizing the risk of the pulp flowback;
- Grate open area, coupled with slot width and pulp lifter design, govern the rate at which pebbles and pulp are discharged; these rates should be the subject of empirical studies in plants for mills of different diameters. DJB Consultants, Inc. has been able to participate in at least one such study which has provided input to JK SimMet simulations and material for scale-up in order to predict mill feed rates for comparison with Millpower2000 predictions.

The following tables illustrate the input parameters, output parameters, and limitations of the Millpower and JK SimMet protocols with respect to design considerations and prediction of mill feed rate:

**Table 1. Input parameters and limitations of Millpower2000 and JK SimMet**

| Parameters | Input Parameters Required to Predict Mill Feed Rate | |
|---|---|---|
| | Millpower2000 | JK SimMet |
| Test Data | $Wi_C$, $Wi_{RM}$, $Wi_{BM}$, S.G. | A × b, S.G. |
| Feed Size Analysis | $F_{80}$, microns, only | Yes, full analysis |
| Product Size Analysis | $P_{80}$, microns, only | Yes, full analysis |
| **Primary Mill** | | |
| Mill Diameter | Yes | Yes |
| Mill EGL | Yes | Yes |
| Type of Drive | Yes | No |
| No. of Mills | Yes | Yes |
| Operating Availability, % | Yes | Yes |
| Anticipated Ball Charge, % v/v | Yes | Yes |
| Total Mill Charge, % v/v | Yes | No |
| Mill Speed, % C.S. | Yes | Yes |
| Mill Pulp Density, % Solids w/w | Yes | Yes |
| Discharge Functions | No | Yes |
| Grate Open Area | No | Yes |
| Grate Slot Width | No | Yes |
| Target Pebble Discharge Rate | | |
| - $m^2/(t/h)$ | Yes | N/A |
| - $(m^3/h)/m^2$ | N/A | Yes |
| **Secondary Ball Mill** | | |
| Test Data | $Wi_{RM}$, $Wi_{BM}$ | $Wi_{BM}$ |
| Mill Diameter | Yes | Yes |
| Mill EGL | Yes | Yes |
| Type of Drive | Yes | No |
| No. of Mills | Yes | Yes |
| Anticipated Ball Charge, % v/v | Yes | Yes |
| Mill Speed, % C.S. | Yes | Yes |
| Cyclone Functions (Phantom) | Included | Recommended |

**Table 2. Output parameters and limitations of Millpower2000 and JK SimMet**

| Parameters | Output Parameters | |
|---|---|---|
| | Millpower2000 | JK SimMet |
| Specific Power Consumption, kWh/t | | |
| - $E_{Total}$ ($E_{SSBM}$ + 10%) | Yes | No |
| - $E_{SAG}$ | Yes | Yes |
| - $E_{BM}$ | Yes | Yes |
| - $E_{Peb}$ | Yes | Yes |
| Power Demand | | |
| - Primary Mill | Yes | Yes |
| - Ball Mill | Yes | Yes |
| - Transfer Size, $T_{80}$, microns | Yes | Yes |
| Operating work Indices | | |
| - Overall | Yes | No |
| - Primary Mill | Yes | Yes |
| - Ball Mill | Yes | Yes |
| Mass Balance | No | Yes |
| Stream Size Analyses | No | Yes |
| Breakage Rates | No | Yes |
| Pebble Discharge Rates | No | Yes |
| Total Mill Charge, % v/v | Input | Yes |

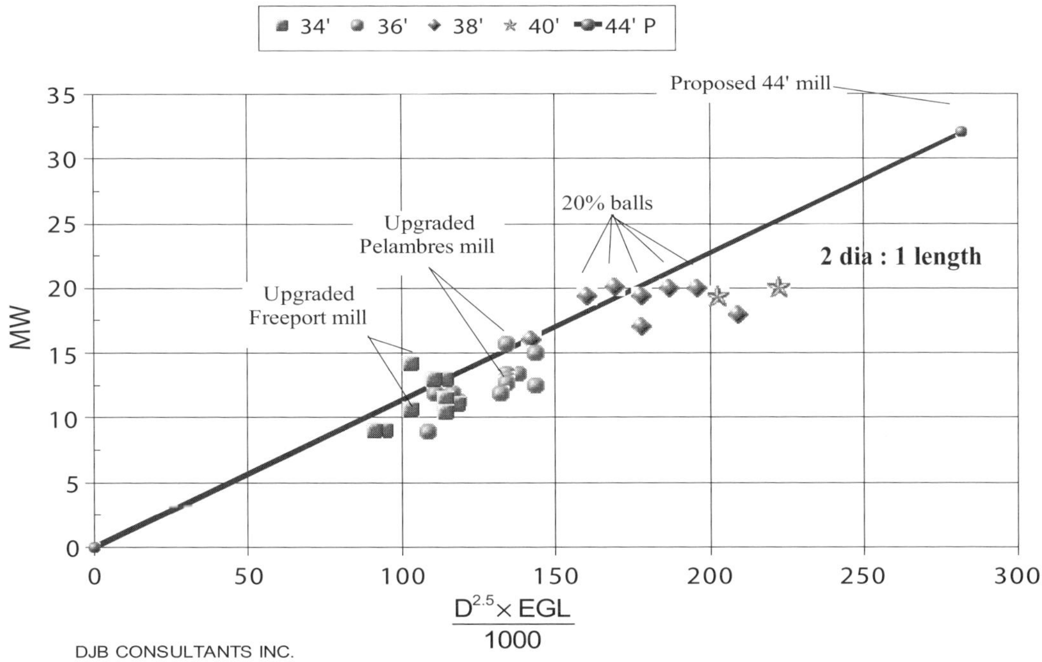


**Figure 3. Large mill power**

Table nos. 1 and 2, above, demonstrate that Millpower2000 and JK SimMet provide a great deal of complementary results when used together. DJB Consultants, Inc. frequently makes use of experienced JK SimMet consultants to provide a more complete design picture than could be generated by any single design tool.

Limitations in the use of Millpower2000 are focused on its inability to predict a circuit mass balance and stream size analyses, including pebbles, and to simulate the effects of different feed size analyses and pre-crushing of primary crusher discharge. On the other hand, Millpower2000 can predict on a timely basis mill sizes, operating conditions, and power demands for fixed mill feed rates and product sizes, $P_{80}$ (for example: 700 simulations require 3 seconds on a typical laptop). Millpower2000 also allows for the input of the drive efficiency according to drive type.

Use of Millpower2000 assumes that the primary mill is power-efficient with respect to the design of shell liners and lifters, discharge grates, and pulp dischargers. It also assumes that there is no restriction with respect to the discharge of pulp and pebbles. This raises the question of applying empirical limits for the discharge of pebbles and pulp in relation to predicted mill feed rates so that a practical upper limit of mill feed rate can be set (which should be automatic in the use of JK SimMet). Millpower2000 also assumes a "standard" feed size distribution based on Fred Bond's $\sqrt{2}$ series. Pre-crushing and scalped feeds require manual treatment.

Use of JK SimMet also assumes that the primary mill is power efficient and that there is no restriction with respect to the discharge of pebbles and pulp unless empirical limits for the discharge of pebbles and pulp are applied. For simulations of mill feed rate between Millpower2000 and JK SimMet to be comparable, such empirical limits should be identical.

Limitations in the use of JK SimMet are associated with a perceived bias in predictions of lower mill feed rates for harder ore types compared to those predicted by Millpower2000. Obviously, such comparison has to be based on comparable samples for testwork and operating parameters in order to minimize differences before researching other potential causes which may be associated with sample size and inherent imperfections, e.g., $Wi_C$ vs. A × b (SMC), vs. A × b (JK DWT), and geotechnical parameters, e.g., fracture frequency and the competence of fracture filling in a particular size class. In this context, the particle size which is presented to the SMC test is usually quarter NQ core: i.e., 24 mm radius × 38 mm arc, compared to 75 mm x 51 mm or larger for $Wi_C$ and across a range of size classes from 12.7 mm to 65 mm or coarser for A × b (JK DWT); the premise being that increased ore competency is biased towards finer size classes, e.g., JK tech's commonly reported plot of $E_{CS}$ (kWh/t) vs. size class on full drop weight tests. Occasionally, physical examination of competent ores has indicated the reverse.

Another possibly more important limitation in the use of JK SimMet is associated with the depth of plant surveys and updates in respect of harder ore types e.g., A × b < 35 and larger mill diameters (>34 ft or 10.36 m) under optimized operating conditions (e.g., higher ball charge volumes, > 12 % v/v, up to 20 % v/v, higher mill speeds, up to 80 % C.S., and optimized pebble crushing circuits). Currently, simulations are conducted with either a default breakage rate model obtained from the existing plant survey database or the breakage rate model that was obtained from the 40 ft diameter SAG mill at Cadia during 1998. Clearly, there is a need to survey and model 36 ft, 38 ft, and 40 ft diameter mills that are operating on harder porphyry and volcanogenic copper ores, and Archaen nickel or gold ores.

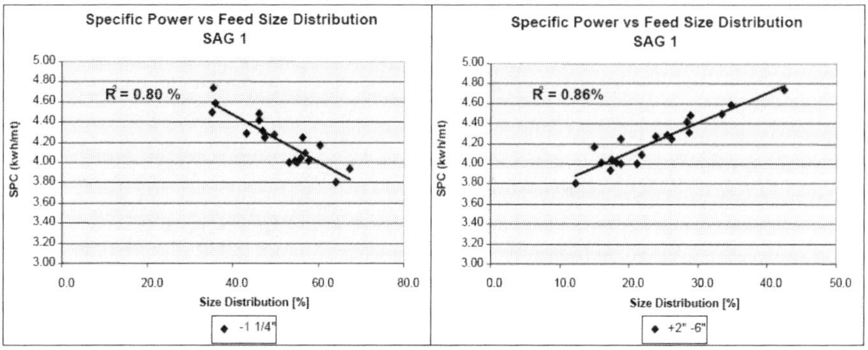


**Figure 4. Effect of feed size on SAG specific power consumption**

A principal limitation of JK SimMet is its inability to calculate mill size in a simulation; mill sizes have to be applied from an external source and, in the case of a fixed mill feed rate, iterated. Power demand for a particular mill size is estimated in a separate program on the basis of a 5 % loss in drive efficiency regardless of drive type.

In summary, Millpower2000 is a versatile tool for estimating mill feed rates, mill sizes, and operating conditions for mills in the conceptual, pre-feasibility, feasibility, and design stages of a project. It can be supplemented, with a co-ordinated approach, with JK SimMet to create material balances and stream size analyses for feasibility studies and plant design.

Selected mill diameter is a function of the required power demand, operating conditions, and grate open area. A plot of operating mills (Power demand vs. Mill diameter) shows a trend towards higher ball charge volumes against a design line of a minimum ratio of mill diameter : mill EGL = 2 : 1 for high aspect mills.

## OPTIMIZATION OF GRINDING CIRCUITS

Since 2002, considerable emphasis has been placed by the industry on optimization of mill feed rate. The expertise of the mill metallurgist has been extended to the mine geologist, the blast-hole and mine planning engineers, the electrical engineer, and the mill maintenance engineer, all under the umbrella of the "Mine-to-Mill" fit-for-purpose operating strategy. Such co-operation has not been seen in a generation or two, but it has obvious benefits for SAG mill operation:

- Mining operations are charged with delivering an optimum top size to the primary crusher which minimizes delays caused by oversized rocks and use of a rock breaker;
- Blast-hole patterns, hole design, time delays, and powder factors are structured to maximize energy intensity and fragmentation and minimize the necessity for secondary blasting;
- Ore delivery to the primary crusher is scheduled to choke feed the crusher consistent with total duties of the truck/shovel/loader fleet;
- Pre-crushing of extreme hard ores, e.g., a proportion of primary sulphide, creates fines for blending in to the coarse ore stockpile feed as minus 32 mm or minus 38 mm fines; such a pre-crushing strategy is not to be confused with pre-crushing as a flowsheet option for which a particular size fraction is crushed to fines;
- The coarse ore stockpile is designed with at least three, preferably four, variable speed feeders so that the SAG mill operator has flexibility in constructing mill feed with respect to coarse, mid-size, and fine size fractions for optimum effect with the aid of calibrated images (e.g., Split or Wipfrag cameras);
- Rated power of the SAG mill motor and the design speed of the motor are compatible with the desired power demand as dictated by the maximum expected ball charge/total mill charge volumes and the operating speed range of the SAG mill, e.g., harder ores require higher mill speeds in the range 75 % C.S. to 80 % C.S., whereas softer ores require lower mill speeds in the range 60 % C.S. to 70 % C.S. (see "Tent diagram");
- Ball diameter and ball charge volume are selected to optimize ore breakage rates, e.g., number of balls in the charge vs. mass of a ball vs. rock strengths of the major ore lithologies;
- Shell liner and lifter design and simulations in terms of the number of rows and spacing of lifters, the trend toward using uni-directional Hi-Hi instead of bi-directional or Hi-Lo, relief angle vs. lifter height and top width, can be used to optimize ball trajectories, media/liner wear rates, power draw, specific power consumption (kWh/t or power-efficiency), bearing pressure, and mill feed rates;
- The trend toward uni-directional operation with curved pulp lifters, either the committed "Dominion" or the retrofitted "hockey stick" designs, contributes to efficient evacuation of pulp and pebbles with minimal wear due to the use of ceramic surfaces in critical areas, e.g., rapid change of direction, and leads to a steeper hydraulic gradient, more efficient grinding, and an efficient design of discharge grates;
- The trend toward SABC-B operation, either as full SABC-B or partial SABC-A/B, takes into account the available ball mill power, the classifying mesh (either trommel or screen) on the SAG mill, and the effect on mill feed rates and final product $P_{80}$, microns; breakage rates of coarser size fractions for SABC-B are most likely higher than for SABC-A; trommel design by DEM should ensure its structural integrity;

**Figure 5. Curved grate lifters**

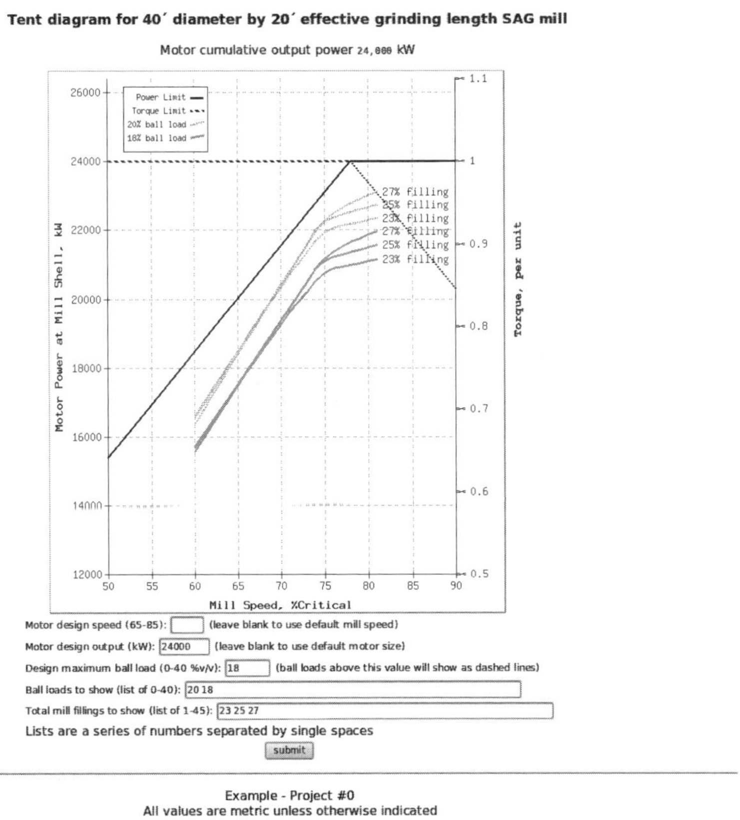


**Figure 6. Tent Diagram for a 40' Dia SAG Mill**

- The importance of a pebble crushing plant which is de-coupled from the grinding circuit ensures optimal operating availability of the plant;
- Material transport and pulp velocity through the secondary ball mills can be optimized through proper selection of cyclones, cyclone feed pulp density and pressure, and flow measurement sensors so that circulating loads are controlled;
- Shell liner and lifter design and simulations, in terms of the number of rows and lifter shape, can be used to optimize ball trajectories, media/liner wear rates, power draw, specific power consumption, and grindability in the ball mills.

## CONCLUSION

In summary, a holistic approach to plant design and operation for grinding circuits is essential for plant designers, equipment suppliers, and operators alike because it encompasses the efficiency of the whole front-end, from mine to entry into the recovery process, and governs the success of a project in terms of its cash flow and return on investment; it is the old story of "tonnes, tonnes, and more tonnes—."

## RECOMMENDATION

Design engineers, equipment suppliers, plant operators, mine operators, geologists, maintenance engineers, electrical engineers, and metallurgical engineers: talk to each other for success and choose/use consultants wisely.

## ACKNOWEDGEMENTS

The author would like to acknowledge the following persons who endured prolonged discussion and debate during the preparation of this paper: Messrs. Mac Brodie, Alex Doll, and Mark Richardson. Also, thanks are due to Anglo American, Xstrata, Teck Cominco, Antofagasta Minerals, Codelco-Teniente, and Antamina for access to information from their grinding operations and projects.

## REFERENCES

Hanks, J., and Barratt, D.J. 2002. Sampling a Mineral Deposit for Metallurgical Testing and the Design of Comminution and Mineral Separation Processes. Mineral Processing Plant Design, Practice, and Control, eds. A.L.Mular, D.N.Halbe, and D.J.Barratt, Proceedings, Vol.1.SME.

Barratt, D.J. and Sherman, M. 2002. Factors which Influence the Selection of Comminution Circuits. Mineral Processing Plant Design, Practice, and Control, eds. A.L.Mular, D.N.Halbe, and D.J.Barratt, Proceedings, Vol.1.SME.

Barratt, D.J. and Sherman, M. 2002. Selection and Sizing of Autogenous and Semi-Autogenous Mills. Mineral Processing Plant Design, Practice, and Control, eds. A.L.Mular, D.N.Halbe, and D.J.Barratt, Proceedings, Vol.1.SME.

Barratt, D.J. and Doll, A.G. 2008. Testwork Programs that Deliver Multiple Data Sets of Comminution Parameters for Use in Mine Planning and Project Engineering. Procemin2008 V International Mineral Processing Seminar, eds. Kuyvenhoven, R., Gomez, C. and Casali, A. Proceedings. Santiago, Chile.

# THE ROLE OF INNOVATION IN MINERAL PROCESSING AND METALLURGICAL PLANT DESIGN

John O. Marsden

John O. Marsden LLC, dba Metallurgium™, 10645 N. Tatun Bld., Suite 200-550, Phoenix, AZ 85028, USA

## ABSTRACT

Innovation is not optional in mineral processing and metallurgical plant design; it is an essential component of the plant design process. The only way to ensure that prior mistakes are avoided is to incorporate and embrace innovation within plant design activities. But innovation is also required to allow the mineral industry to create value from increasingly low grade and more complex ore deposits in the future. There are many ways that innovation occurs during plant design, and these are explored in this paper. The innovation process, and the balance between risk and rewards, must be managed. An outline of how innovation can be managed and driven effectively is provided.

## INTRODUCTION

Innovation is concisely and simply defined as "the introduction of a new thing; the alteration of something established" [1]. Mineral processing and metallurgical plant design would not be where it is today, and will not get the mining industry where it needs to be tomorrow, without aggressive innovation applied to all areas of practice. Also, innovation is critical for individual operators and mining companies to remain competitive over time in a highly competitive and challenging industry [2]. Innovation must be driven by innovators, and innovators by their nature must be positive, optimistic, enthusiastic and persistent. Conservatism is the enemy of innovation, but prudence is the friend of effective implementation. Innovation introduces risk. Risk must be understood and, where possible, quantified prior to making investment decisions, and then the risk must be managed during implementation. How do we strike a balance between innovation and conservatism; prudence and reckless optimism?

### Why Innovate?

When mineral processors and metallurgists think of innovation, they generally start talking about a recently implemented process flowsheet—how it was (or will be) either wildly successful or a dismal failure. New hydrometallurgical nickel laterite flowsheets often come up in this context. This is innovation on the largest, and most challenging, scale in mineral processing and metallurgy. But innovation on a smaller scale is just as important. In fact, innovation is fundamental for (almost) every mineral processing and metallurgical plant design. No two orebodies are exactly alike, and no single plant fits two orebodies perfectly. Sometimes, very occasionally, an existing plant design might be good enough to fit a different deposit and that's fine, but improvements are always possible. In such cases, let's count our blessings, adopt with pride, and move on. No major innovation required. Everywhere else, we must innovate, and that means taking the next best and closest plant designs we can find, then altering and adapting them to fit the new circumstances (i.e., our new orebody).

In fact, there is an even greater imperative to innovate than simply to adapt or alter an existing plant design to fit a new application. All plant designs are flawed, and if innovation is not incorporated aggressively and pro-actively at all stages of the design process, then we are doomed to repeat the same flaws. I have heard it said that it is better to repeat old flaws that we have learned to live with than to create new ones that we haven't. If this is true, then we will not be able to evolve mineral processing plants to deal with increasingly lower grade and more complex orebodies and resources that we will be required to process in the future.

The focus of this paper is on the role of innovation in traditional areas of mineral processing plant design—this means comminution, physical and surface chemical methods of separation, solid-liquid separation, and related processes. However, where relevant, hydrometallurgical examples have been used to help illustrate specific points since many of the same principles and guidelines apply.

## THE BUSINESS CASE FOR INNOVATION

In all cases, the drivers for innovation are to improve the overall economics of a project by; (i) decreasing capital (possibly at the expense of an increase in operating cost), (ii) decreasing operating cost (possibly at the expense of an increase in capital cost), (iii) improving metal recovery, (iv) improving product quality, (v) improving some other attribute of the process (e.g., safety, health or environmental considerations), or (vi) combinations of these. By its nature, innovation introduces risk. Any change or alteration in the way something is done has an element of the unknown which adds risk. In most cases, the risk can be quantified, or at least partially-quantified, for example;

- risk of increased downtime (lower availability) of equipment and processes,
- risk of delay in start-up and ramp up of production
- risk of throughput rate less than design
- risk of lower metal recovery
- risk of higher installed capital cost or operating cost
- risk of technical failure of equipment or a process step (quantifiable in terms of downtime, start-up delay, loss of production, replacement cost, etc)
- safety, health, and/or environmental risks

The innovation process must be managed, and this is a trade-off between the benefits of innovation and the identified risks associated with the innovation. Figure 1 shows a schematic illustration of the risk-reward profile of innovation in plant design. The y-axis shows the estimated benefit (value creation potential) and the x-axis the relative risk. The easy decisions fall into the upper left and lower right quadrants. The upper left quadrant considers opportunities with high rewards and low risk. These are no-brainer ("Yes") decisions, but they are generally rare. The lower right quadrant considers opportunities that have low rewards and high risk.

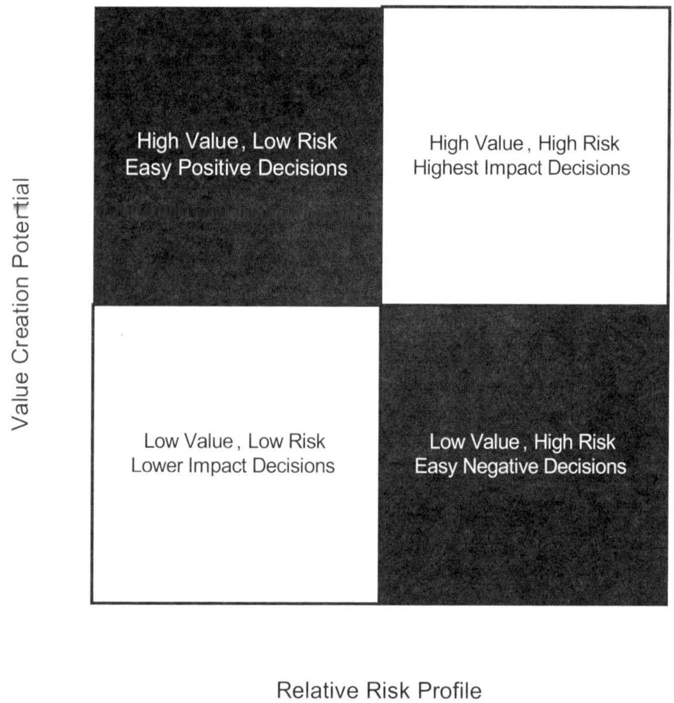


**Figure 1. Schematic representation of relative risk and value creation potential for innovation**

While options to manage the risk may be evaluated, generally these are easy "No Go" decisions. The opportunities represented in the lower left quadrant are low risk, with low reward. A process to allow this type of innovation to occur routinely should be developed early in the project development. This type of innovation is most effectively driven by client-side staff dedicated full time to the project and working in conjunction with EPC/EPCM contractor staff. This innovation is collaborative, builds on experience gained at other operations and projects and is ongoing throughout all phases of the project execution.

The opportunities in the upper right quadrant are more complex, and should be the main focus of attention at an early stage in any project. These represent the high value, but higher risk opportunities. Sometimes these are referred to as "strategic" opportunities because it may be difficult to fully and accurately quantify the benefits and the risks. In some cases, they truly are strategic as they meet a specific need within the company or project, such as a new flowsheet requirement for a new class of ore deposit or an industry shaping step change technology. Significant effort and resources are required to evaluate and, when appropriate, to justify these opportunities. To be successful, they must be client-driven, and must have ownership at a senior level within the client (mining company) organization. Operating staff must be involved at all stages of the development, justification and implementation of these opportunities.

Another way to look at this is to consider two broad types of innovation; incremental improvements (or alterations to established processes and/or equipment) and step change developments (new processes or methodologies). Incremental improvements generally fall into the category of low reward and low risk (lower left quadrant in Figure 1). Step change developments fall into the upper right quadrant—high reward and higher risk.

A key assumption used in the estimation of the financial benefit of a particular innovation is the discount rate. This seems to be a constant matter of debate, but one thing is clear; the cost of capital must be considered in these investment decisions. While there is no clear consensus or standard within the industry for what discount rate to apply, a few general rules can be provided, as follows;

1. The discount rate should be at least the cost of capital for the company (typically 6-9%, pre-financial crisis).
2. The discount rate should be applied to the incremental economics for the portion of the process plant directly affected by the innovation.

3. The discount rate may be linked to the level of risk and scale of innovation to be applied; low risk, small scale innovation investment decisions should use lower discount rates, whereas higher risk, larger scale innovations may require higher discount rates to be used.
4. The discount rate doesn't have to be the same as the discount rate used for the overall project; in fact it shouldn't be the same as the investment risk is different.

## TYPES OF INNOVATION

We have already talked about innovation on a grand scale (i.e., process selection) and small scale (at all stages of detailed plant design). Before we go any further, we need to define what innovation means in the context of mineral processing plant design. Innovation can occur in different ways at various stages of plant design, as follows:

- Process development
- Process selection
- Flowsheet design
- Equipment selection and design
- Detailed engineering design
- Process commissioning and optimization

The first three categories in this list (process development, process selection and flowsheet design) involve the broader, large scale aspects of plant design—factors that affect the basic physical and chemical processes to be used for metal/mineral extraction, the conditions under which they will be applied, and the order in which they will be combined. It is within these steps that macro-scale innovation takes place, for example; roasting versus pressure oxidation for gold-bearing sulfide concentrates, milling and flotation versus crushing and heap leaching for chalcocite copper ores, and flotation versus gravity concentration for cassiterite recovery.

The latter three categories (equipment selection and design, detailed engineering design, and process optimization) encompass more detailed aspects of plant design which must take into account the process design criteria, including; throughput (average and instantaneous), availability, scale up factors, factors of safety, and catch-up/turn-up capacity factors (e.g., design certain systems in the plant for a multiple of 1.15–1.25 times the expected average throughput or capacity), and unit operation efficiencies. However, many other factors must also be considered, including; layout, maintainability, operating cost, surge capacity requirements, safety, health and hygiene, social, environmental and sustainable development issues. Innovation is often just as important in these aspects of plant design.

Table 1 shows more detailed breakdowns within each of the categories listed above. For each category, the opportunity for innovation has been ranked (high, medium or low), and a relative indication of the potential value created by the innovation has been provided. Finally, the relative risk associated with implementing innovations within the particular area has been ranked. While such a ranking is imperfect and is to some extent subjective, this gives a good indication of the major areas in which innovation can occur. There are several conclusions that can be reached from careful study of Table 1. Firstly, the major opportunities for innovation occur early in plant design, i.e., process development and evaluation. At this stage, the risk is low and, provided an effective process to manage risk is identified and implemented early, the risk can be managed through implementation. Key guidelines for the early stage process development and evaluation include the following;

- perform extensive and effective benchmarking on similar deposits and projects worldwide
- wherever possible, replicate or duplicate what has been done before, if it fits the new orebody—***adopt and adapt with pride***,
- utilize key experts (internal and external) to review, evaluate, critique and rank options (including innovative aspects and opportunities),
- quantify the benefits and risks associated with each option identified,
- make as many decisions on options and alternatives (including innovation) as you can early in the development of the project,
- do not carry an option forward to the process selection step unless you have completed a risk assessment and you know the risk can be managed and how it will be managed.

The second conclusion is that, without a doubt, the process selection step carries the highest risk. This is the point at which tough choices must be made between process options. This generally occurs prior to commencing a feasibility study for the project, although in some cases two or more options may be carried through a final feasibility study before making the choice. This latter approach has a number of advantages; capital and operating costs can be more accurately defined for the various options (generally to within ±15%), more detailed risk assessments can be conducted, and other factors and uncertainties can be better defined (e.g., permitting, safety aspects etc.). However, the cost for the study will be significantly higher and the study may take longer to complete (not good for "fast-tracking" projects). More importantly, perhaps, is that the information may be adequate at an earlier stage in the project (i.e., at the scoping or preliminary feasibility study stage) to make the decision with an acceptable level of confidence.

*Example: At Freeport-McMoRan's Cerro Verde copper operation near Arequipa, Peru, a new concentrator was commissioned in late 2007 to treat a large primary sulfide resource. High pressure grinding rolls (HPGR's) were selected as an alternative to SAG milling. Prior to making the investment decision, both flowsheet options were carried through to the completion of the full feasibility study for the project, with capital and operating costs for both options completed to approximately ±15% level of accuracy. At the time this was being considered, the use of HPGR's in this application would be the first use of HPGR's to treat copper ore at a*

**Table 1. Assessment of innovation opportunities within process plant design**

| Process Plant Design Activity | *Opportunity for Innovation (Hi/Med/Low)* | *Opportunity for Value Creation (Hi/Med/Low)* | *Innovation Risk (Hi/Med/Low)* |
|---|---|---|---|
| ***Process development*** | | | |
| • Benchmarking similar deposits | H | H | L |
| • Utilization of external resources (breadth and depth of experience) | H | H | L |
| • Review and evaluation of options | H | H | L |
| • Trade off studies (type and quantity) | H | H | L |
| • Material characterization (extent and type of methods applied) | H | H | L |
| • Ore type classification | H | H | L |
| • Metallurgical testing and pilot plant testing | H | H | L |
| • Process scale up | H | L | H |
| ***Process selection*** | | | |
| • Comminution (crushing, grinding) | H | H | H |
| • Physical separation (gravity, magnetic, electrostatic) | H | H | H |
| • Chemical alteration (oxidation, reduction) | H | H | H |
| • Chemical separation (leaching, precipitation, adsorption, ion exchange) | H | H | H |
| • Surface chemical separation (flotation, amalgamation) | H | H | H |
| ***Flowsheet design*** | | | |
| • Comminution (blast fragmentation, crushing, grinding) | H | H | M |
| • Size separation (screening, cycloning) | M | M | L |
| • Materials handling (conveying, pumping, piping) | M | H | L |
| • Solid-liquid separation (thickening, filtering, drying) | M | M | H |
| • Combination of processes (order, sequence) | M | M | M |
| • Surge capacity and storage between unit operations | M | M | H |
| • Alternatives considered | H | M | H |
| • Tailings management; recirculation, storage and disposal | M | M | M |
| ***Equipment selection and design*** | | | |
| • Review and evaluation of options | H | M | L |
| • Qualifications of supplier(s) (large global vs. small local, niche applications) | L | L | M |
| • Similar application(s) of equipment | L | L | L |
| • Increased size/scale of application | M | M | M |
| • Modification of existing for specific use | M | M | M |
| • New equipment, new application | H | H | H |
| ***Detailed engineering design*** | | | |
| • Process design criteria | M | M | H |
| • Plant layout | H | M | L |
| • Civil | L | L | L |
| • Structural | L | L | L |
| • Mechanical | M | M | M |
| • Piping | L | L | M |
| • Electrical | L | L | M |
| • Instrumentation and control | L | L | M |
| • Materials of construction | L | L | H |
| • Design for safety, health and hygiene, environmental responsibility | L | L | M |
| ***Process optimization*** | | | |
| • Making what has been installed work (even if it doesn't at first) | L | L | L |
| • Improving the performance of installed equipment and processes | L | L | L |
| • Replacing under-designed components of the plant | M | L | M |
| • Adjusting process chemistry | L | M | L |

*large scale. Also, this followed a test of HPGR technology at Sierrita in 1995 that was a technical success, but perceived by some as a commercial failure; the unit was withdrawn from service following a mechanical failure due to prevailing economic conditions and the marginal economics of the specific application in which it was applied. In order to manage the risk associated with the proposed HPGR application at Cerro Verde, a detailed evaluation and risk assessment was conducted, visits were made to many HPGR installations in the iron ore and diamonds industries to learn about the application of the technology, and detailed pilot plant testing was conducted at two HPGR manufacturers facilities in Germany. During the feasibility study, 2.2 meter diameter by 1.6 m long Krupp-Polysius units were indicated by the process design engineering and analysis. However, the size of the units were scaled up to 2.4 meter diameter to provide a significant factor of safety [2]. The Cerro Verde circuit ramped up to design throughput and is currently operating above design throughput rate. The HPGR units are operating effectively and, while tire liner wear life is somewhat below expectations (4,000 hrs versus 6,000 hrs), improvements continue to be made which are expected to improve tire liner wear life over time up to the design expectations.*

The third conclusion is that there are significant opportunities for innovation within flowsheet design and equipment selection/design, but for many of the activities the risk associated with innovation is high. For example, there are always options to reduce capital and operating costs by introducing innovation in solid-liquid separation, surge capacity allowances, and elimination of process steps, but the risk associated with such changes is high.

*Example: At the Cerro Verde copper concentrator mentioned in the example above, the plant was originally designed and laid out to include a stockpile between the secondary crushers and the HPGR tertiary crushing units. This stockpile was eliminated during value-engineering and as a result of dynamic simulation of the crushing and grinding circuit, which indicated that the small benefit in operating time (and plant availability) provided by the stockpile did not outweigh the additional capital and operating cost. Storage bins with significantly lower volumetric capacity were considered to be adequate surge capacity. The expected reduction in plant availability was assessed and deemed to be acceptable, and the risk of additional decrease in availability was reviewed. In practice, this has proven to be an acceptable risk-benefit trade off, with plant availability achieved close to expectations.*

*Another example: In 1986, during the original design of the Gold Fields' Chimney Creek operation near Winnemucca, Nevada, the design team came up with a really innovative and clever double-thickener flowsheet concept. This incorporated the best aspects of counter-current decantation, carbon-in-pulp (CIP) and carbon-in-solution (CIS) for most effective recovery of gold and silver by cyanidation. This innovative circuit incorporated some aspects of other plant designs (e.g., Pinson, Nevada) but took these to the next level in the new plant. The result was the production of very clean loaded carbon for stripping, the minimization of gold (and silver) tenors in the CIP circuit, and reduced loss of cyanide to the tailings facility. The various flowsheet options were modeled using a simple Excel spreadsheet approach. The risk associated with the new flowsheet was low, but the benefits were significant. The Chimney Creek plant had an extremely fast start-up and achieved design production rates within weeks [4].*

The fourth conclusion is that there are many opportunities for innovation during equipment selection/design and detailed engineering design. When taken individually, the value created is generally low, but collectively such innovations can create tremendous value. Also, the value of these innovations typically pays off in terms of lower operating costs and improved plant availability and maintainability down the road. With a few exceptions, the risk associated with innovation at this stage is generally low. The exceptions are the process design criteria and materials of construction where there is little margin for error. These require application of the right expertise and review to ensure that the risk is well-managed.

*Examples. At Freeport-McMoRan's Candelaria copper concentrator near Copiapo in Chile, the plant was expanded from 30,400 mtpd to 60,800 mtpd in 1997, following successful commissioning of the plant in 1994-1995. An important consideration for the major expansion of the concentrator facilities was that as much as possible should be duplicated to minimize re-engineering and to complete the expansion as quickly as possible. However, a number of opportunities for innovation and improvement were identified early, as follows:*

- *the use of ten (10) 120 cubic meter Wemco cells for rougher flotation cells instead of the fourteen (14) 85 cubic meter Wemco cells used in the original plant (At the time the decision was made there were none of these cells in commercial operation and this would be the largest commercial applications of these bigger cells),*

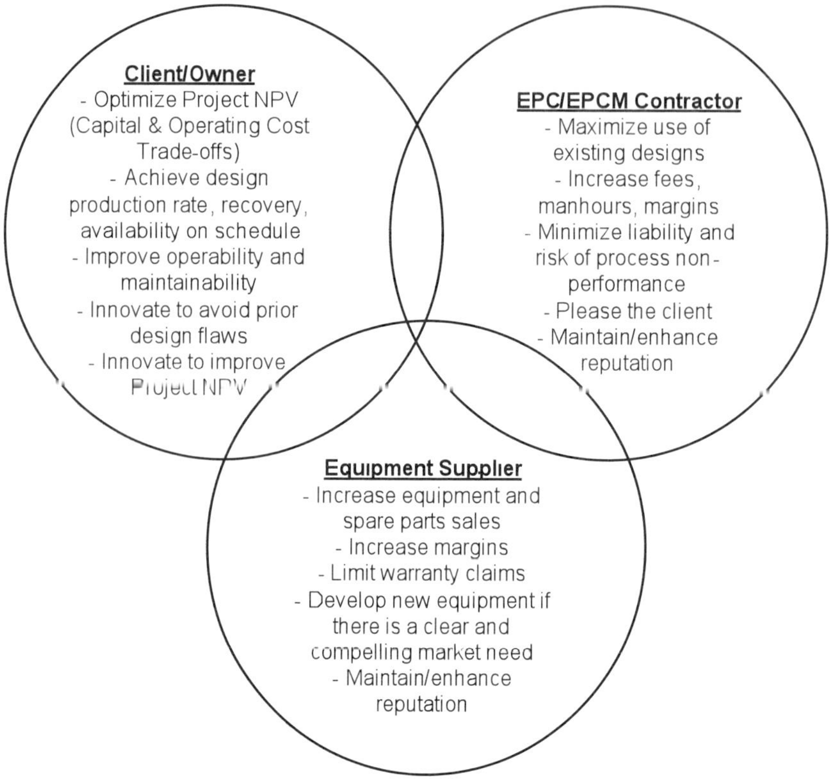


**Figure 3. Stakeholders in process plant design and their major objectives**

- *the use of a Metso Vertimill VTM 800 (800 hp) for rougher concentrate regrinding in place of conventional ball milling (1,500 hp),*
- *the use of three trains of tailings pumps to handle underflow from two tailings thickeners by multiplexing the pump intake piping (requiring only one new train of pumps), and*
- *the use of pressurized tailings distribution manifolds rather than drop boxes to manage tailings flow.*

*These changes, along with other innovations resulted in a total net present value increase of more than $15 million for the expansion project. All of the innovations listed above were executed successfully, without delays to the project or the production ramp up, and the changes met their original goals. The project was completed 8 months ahead of schedule and $30 million under budget. Candelaria could have duplicated everything, made life simple for themselves (plant operators and maintenance staff), and avoided the worry and risk of implementing several new components of the plant, but would have spent more capital and foregone the additional value creation [5, 6].*

Innovation during equipment selection/design and detailed engineering design is a complex process involving the client representatives, the EPC or EPCM contractor, and equipment suppliers. The goals of each of these stakeholders are summarized in Figure 3. These goals are not always aligned. It is the responsibility of the client (project owner) to make sure that a clear set of over-arching goals for the project is established early and applied to all decision-making for the project, including the use of innovative processes and equipment.

*Example: A good example of equipment innovation was the installation of four Metso VTM 1250 Vertimills at Chino in New Mexico in 1995. Up until that time, vertical stirred mills had seen limited application in the copper industry—they were an unproven commodity. At the same time vertical stirred mills offered the opportunity for more efficient grinding at fine sizes (below about 500 microns) and lower wear steel consumption. The project at Chino was based upon decreasing the flotation feed size to increase recovery by about 6.5% and to improve final concentrate grade by 1%. A commercial trial was conducted at Chino using a 450 hp Tower mill unit, on loan from another mining company. Based on the results of these trials, plant operators and company engineers worked closely with Metso staff to make a number of key changes to the VTM mill design, including; (i) reducing the height-to-diameter ratio (making the shaft more stable and robust, and improving the operability of the mill), (ii) reducing overhead requirements for easier shaft removal and mill maintenance, (iii) changes to the door design to allow rapid dumping of the mill charge for maintenance and incorporating the use of hydraulics for safer door handling by the operators, and (iv) changes to the slurry recycle system and entry point into the mill. This is a great illustration of the benefits of excellent communication and coordination between the plant operator/metallurgist and equipment suppliers to ensure effective innovation in equipment selection and design [7].*

Finally, process optimization provides opportunities to fix things that weren't quite right in the final design and to get the most out of the installed equipment. These activities may require innovation to address, but the value creation is low (no upside) and the risk is generally low—the plant and equipment is already installed and generally requires "tweaking."

*Example: During the design and development of the double-vee clarifier technology at the East Driefontein gold mine in South Africa in 1983-1984, there were a number of issues that had to be resolved during the commissioning and optimization. These included the need to adjust spacing between the inner cone and outer cones of the clarifier, the need to add slurry recirculation capability to build up a bed of flocculated slurry clusters, and the development of sludge probe to measure the depth to the top of the flocculated slurry bed [8]. These changes were enablers for the equipment to be successfully commissioned and the technology would not have been a success without these innovations.*

The point here is not necessarily to explain the technical details of this development, but to provide an example of the context for innovation ***after*** the development and implementation of an innovative design. This required commitment and dedicated effort from the development and commissioning team, including plant operators, engineers, metallurgists and maintenance staff. No innovation can be successful without this commitment and dedication.

## MANAGING THE INNOVATION PROCESS

There are two ways to manage innovation; selectively and randomly. The selective process is planned and well-thought out. Brainstorming sessions are used early on to identify options and opportunities for innovation. The key areas for innovation (deviation from conventional technology, processes, equipment) are identified early, options and alternatives are listed, ranked and prioritized, and key stakeholders have been involved from an early stage. In some cases formal stage-gate processes are used with filters at important decision points in the project. The risks and benefits for each innovation decision are quantified. A formal review process is conducted prior to making a final "go, no go" decision.

Random innovation is un-planned and un-managed, although there may be recognition among senior client staff that innovation will occur during the course of a project. One or two key technical resources involved in the project come up with innovative options. These are developed on an ad-hoc basis, with little review by others and no formal approval process. The options considered and the evaluation is generally limited to the opinions, perspective and experience of one or two people. The rewards are quantified, but the risks are less well-understood and poorly quantified. As a result, the innovations incorporated into plant design may not be focused on the biggest value-adding opportunities, and the risk-benefit analysis may be flawed. This approach is not all bad—it's probably better than no innovation at all and, if driven by the right person or group, it can be quite effective. Every innovation needs a champion; a person who will carry the idea through detailed design into fully-effective commercial implementation, and if the champion is good then the results probably will be also.

In today's business environment, a perfect selective innovation process is not always practical or possible to implement. However, the innovation process must be managed. In order to accomplish this, the following guidelines provide some of the more important aspects to incorporate effective innovation in plant design:

- Involve key client-side staff early in the process development phase (before flowsheet design, detailed engineering and equipment selection), including the Plant or Process Manager, Operations Superintendent, Chief Metallurgist, Maintenance Superintendent
- Bring in outside expertise to assist with plant design, including brainstorming sessions, flowsheet review, risk-rewards review for innovative aspects of design
- Establish criteria for the evaluation of innovations early and communicate this to the team (financial and non-financial)
- Conduct as extensive benchmarking as you can to evaluate new technology applications or changes to existing technology before you incorporate it
- Make sure you know what you are asking the EPC/EPCM contractor to do for you (and write it down)—if you don't know, they won't either
- Every innovation should be justified in writing with expected benefits and costs quantified compared with the "base case"
- Develop a list of what will be standard and what could be innovative in the plant; plot them on the chart in Figure 1
- Don't sweat the small stuff (lower left quadrant in Figure 1)—let it happen, encourage it, but make sure competent people are doing it
- Put in place a review team and a process to review the plant design, including innovative aspects

For all significant innovation initiatives, prepare a ranking that shows the relative risk, the potential consequences of a failure or sub-design performance, the scale of the innovation, and the net present value for the innovation. The relative risk, scale of innovation and value creation rankings are illustrated in Figure 4, with examples of each. One way to look at the innovation initiatives is to rank them using the product of the relative risk and scale of innovation, and then compare the ranked opportunities with the value created by each initiative. This allows the operator to quickly make informed decisions on opportunities that are clearly favorable or unfavorable. It also allows decisions to be made based on the resources available to adequately manage the risk associated with all of the opportunities that are carried forward into implementation.

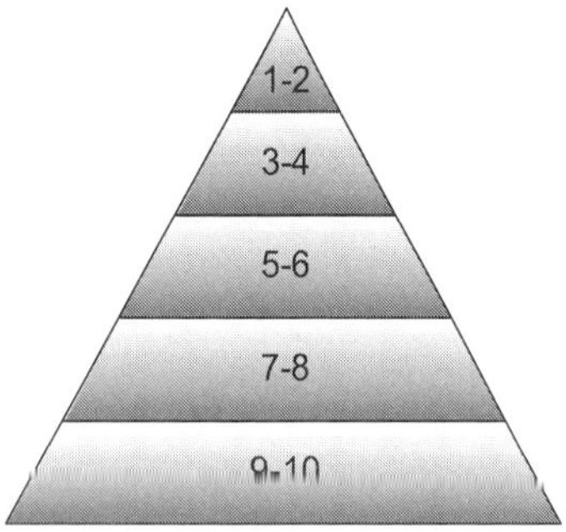


Relative Risk Factor

**Example of Relative Risk Factors**
(Note: Varies by project and/or company)

1. Fully-proven technology; operating >2 yrs; used by multiple plants
2. Applied commercially in one (or more) plant; comprehensive ore characterization and ore type testing for proposed application
3. Proven at demonstration scale in final configuration, all major ore types tested
4. Proven at continuous pilot plant scale; extensive ore characterization for proposed application
5. Proven at continuous pilot plant scale; some ore characterization for proposed application (incomplete)
6. Design based batch test work; some ore characterization for proposed application (incomplete)
7. Design based on batch test work; pre-feasibility level engineering
8. Design based on batch test work; scoping assessment
9. Limited metallurgical work, very limited ore characterization; scoping level assessment
10. Concept/idea

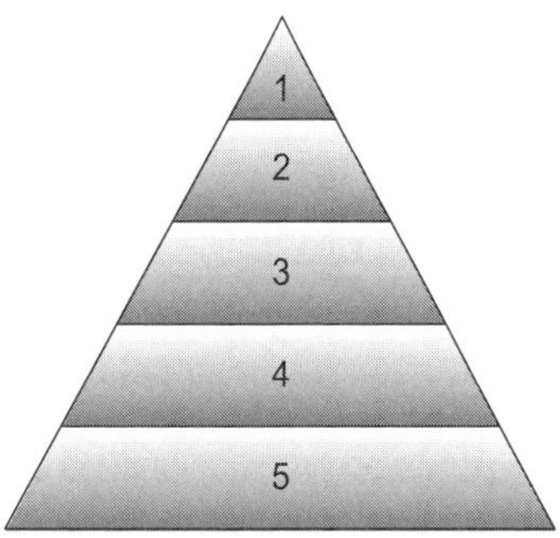


Scale of Innovation

**Examples of Scale of Innovation**
(Note: varies by project and/or company)

1. Change that affects one piece of equipment (increase in size or functionality) or change to ancilliary process/equipment; no significant impact on overall process
2. Significant changes to equipment and or unit operation used in an existing flowsheet(s) application
3. New equipment and/or new unit operations in flowsheet
4. Major change to existing flowsheet or design; Affects a significant portion of the plant; Significant increase in size of major plant components
5. Entirely new flowhseet; Affects the entire plant/process; Jumps outside the normal boundaries of size increase or scale-up

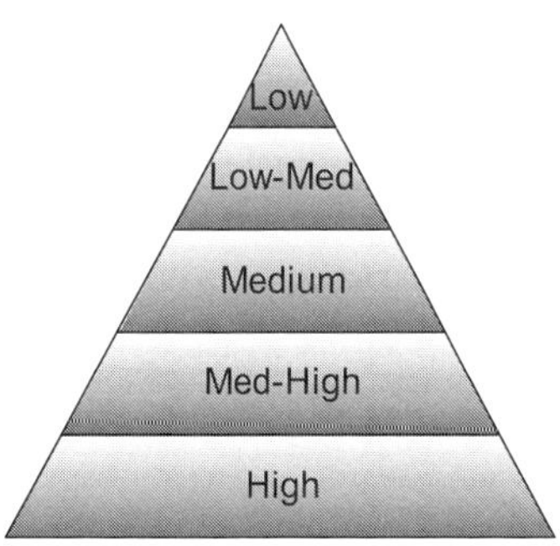


Relative Value Creation

**Example of Value Creation Ranges**
(Note: varies by project and/or& company)

NPV @10% discount rate)

| | |
|---|---|
| Low | <$100k |
| Low-Medium | $100k - $1 MM |
| Medium | $1-5 MM |
| Medium-High | $5-20 MM |
| High | >$20 MM |

**Figure 4. Example of ranking of relative risk, scale of innovation and value creation for innovation**

In some cases, the net present value of innovation opportunities is risk-adjusted, meaning that the net present value is multiplied by a probability of success. While this approach may be useful in certain circumstances, the drawback is that the probability of success is hard to quantify and ends up being highly subjective.

## CONCLUSIONS

Innovation is an essential component of mineral processing and metallurgical plant design. However, with innovation comes risk, and the risk must be managed to avoid plant start-up difficulties and operating performance below expectations. The process of incorporating innovation into plant design can, and should, be managed. This doesn't need to be a highly bureaucratic or burdensome activity, but a well-planned and well-thought out strategy that lays out how innovation will be incorporated into plant design is essential. The key aspects of this strategy are; 1) ensure that the criteria for making decisions between plant design options (including innovation) are established early in the project and communicated to all involved, 2) make sure that full advantage is taken of existing plant design information and expertise (don't re-invent the wheel; adopt and adapt with pride), 3) make sure that the right expertise is utilized to review and evaluate innovative aspects of plant design, 4) make decisions about process options and alternatives as early in the plant design as possible, 5) make sure that the risks are understood and quantified to the maximum extent possible, and 6) risk-adjusted value judgements can be misleading—don't let the accountants talk you out of something that you know is the right thing to do.

## ACKNOWLEDGMENTS

The author would like to thank Jim Vanderbeek for his comments and innovative contributions, and Freeport-McMoRan Copper & Gold Inc. for permission to use several of their projects as case examples in this paper.

## REFERENCES

[1] The New Shorter Oxford English Dictionary.

[2] Marsden, J.O. (2004) Technology Development and Competitive Advantage: Sustainable or Short Term? Proc. Plant Operators' Forum—Improving and Optimizing Operations: Things That Actually Work, Eds. E.C. Dowling & J.O. Marsden. SME, Littleton, CO. pp 127-137.

[3] Vanderbeek, J.L., Linde, T.B., Brack, W.S. & Marsden, J.O. (2006) HPGR implementation at Cerro Verde. Proc. International Autogenous and Semi-Autogenous Grinding Technology 2006. Vol. IV. Eds. M. J. Allan, K. Major, B.C. Flintoff, B. Klein, & A.L. Mular. Dept. Min. & Min. Proc. Eng., University of British Columbia, Vancouver, Canada. pp. 45-61.

[4] Mansanti, J.G., Arnold, J.R., Gourdie, J.H. and Marsden, J.O. (1989) The Double Thickener Circuit at Gold Fields' Chimney Creek Mine. Min. & Met. Proc. J., November 1989. pp 179-185.

[5] Marsden, J. O. & Ogonowski, D. L. (1999) The Candelaria concentrator expansion project. SME Annual Meeting Preprint No. 99-139. Society for Mining, Metallurgy & Exploration, Inc., Littleton, Colorado.

[6] Marsden, J .O. (1998) Operating ideas that have saved me some money—copper milling at Candelaria. Proc. SME Plant Operators Forum, Orlando, Florida. Ed. Halbe, D. N., Society for Mining, Metallurgy & Exploration, Inc. pp 110-124.

[7] Vanderbeek, J. L. (1996) Grinding Circuit Evolution at Chino Mines Company. Proc. International Autogenous and Semi-Autogenous Grinding Technology 1996. Vol. 1. Eds. A. L. Mular, D. J. Barratt, & D. A. Knight. Min. & Min. Proc. Eng. Dept., University of British Columbia, Vancouver, Canada. pp. 25-40.

[8] Marsden, J.O. & Van Der Westhuizen, J.H.L. (1985) Double-vee Clarification of Pregnant Gold-Bearing Solution at the East Driefontein Gold Mine. Mine Metallurgical Managers' Assoc. of South Africa. Volume 86/2. South African Inst. Min. & Met., Johannesburg, South Africa. pp 29-49.

# Research and Characterization

# A CASE STUDY IN THE FLOTATION CHARACTERIZATION OF AN ORE BODY AND ITS USE IN CIRCUIT DESIGN AND RISK ASSESSMENT

Dave Bulled and James Connolly

SGS-Mineral Services, 1140 Sheppard Avenue West #6, Toronto, Ontario, M3K 2A2, Canada

## ABSTRACT

This paper describes a procedure for the characterization of the flotation response of an ore body involving all the blocks in a proposed mine plan, and its practical use in the design to feasibility level of a flotation circuit, including recommended equipment sizes. Flotation kinetic data was produced from standard laboratory rougher flotation tests at specified reagent conditions on 83 samples of drill core, and a small number of rougher-cleaner tests on core composites to determine the change in kinetics resulting from regrind.

The samples were selected from across the ore body to determine the variability by grade, ore type and location. The flotation recovery and rate data was distributed using recovery/feed grade relationships and geostatistics to about 58,000 blocks in the mine plan representing a deposit of over 875 million metric tons. Simulation of the selected plant design produced a forecast of results to be made for every block and awareness of the proposed mining block schedule allowed the trend in the results to be forecast over the mine life.

The statistical method for the distribution of data assigned a measure to the uncertainty in the estimation of block kinetics resulting from having a given number of samples, so that the forecast results could also include a measure of uncertainty and quantify the technical risk in the flotation project at feasibility level.

## INTRODUCTION

The traditional approach to flotation plant design involves the extensive testing of a single large composite sample or a small number of composite samples that are reputed to represent the ore body. Laboratory tests are conducted to establish a process with potential flowsheets and reagent suites. The process is then confirmed by pilot plant work on a large bulk sample or a few samples. The size of equipment required to achieve a specified throughput, recovery and product grade is then calculated from one of a variety of scale-up models. Since the test work has been extensive, it is accepted as being reasonably precise in predicting results; however during the operation of the resulting plant the design is sometimes found to be inadequate, e.g. Mount Todd, N.T. Australia (Mathews, 1999). It is then suspected that the flaw in the design process lies in the samples not being sufficiently representative of the ore body, since using only a single or small number of composite samples does not recognize the variability of the flotation response of the ore, nor does it allow for the lack of ore body representation used in the design. It had not been possible to estimate the risk or error in the expected results.

This paper outlines an alternative procedure that tests many drill core samples from a molybdenite ore body by a standard technique that allows the determination of flotation kinetic parameters for each sample. The samples can be widely spaced across the resource, cover the expected major lithologies and have a wide range of molybdenite feed grade. These parameters are then distributed to all the blocks in a mine plan using a combination of statistical techniques. The average kinetic values of blocks to be treated are used with the FLEET software program to scale-up to a plant design with an optimum flowsheet and equipment selection. The variability of expected plant results are then examined by simulating the plant design with the parameters of each of the blocks. Blocks are assembled into expected production years and simulations used to forecast year-by-year flotation circuit results in terms of expected average grade and recovery. The statistical variability of the kinetics for each year is used with the FLEET program to determine the uncertainty in the expected average results of the selected plant design.

## SAMPLE SELECTION

In this study 83 samples were selected for flotation test work that covered the variation in lithology, alteration and molybdenite content and were spatially distributed across the ore body. The samples represented three lithologies and three alterations as shown in Table 1.

**Table 1: Breakdown of samples by lithology and alteration types**

| Lithology | | Potassic | Phyllic | Argillic | Total # samples |
|---|---|---|---|---|---|
| Code | Description | | | | |
| 20 | **Quartz porphyry** | 49 | 3 | 1 | 53 |
| 40 | **Rhyolite** | 4 | 0 | 3 | 7 |
| 50 | **Sedimentary** | 18 | 0 | 5 | 23 |
| | **Total # samples** | 71 | 3 | 9 | **83** |

The samples were made up from 50ft lengths of core and selected from 50 different drill holes. They were on average about 400 ft apart horizontally and 100 ft apart vertically, with two coincident samples. They were concentrated at the top south of the deposit in the planned early years of mining as follows:

- 70% from Years 1 to 5
- 20% from Years 6 to 15
- 10% from Years 16 onwards.

There were none in the lower quarter of the deposit as shown in Figure 1.

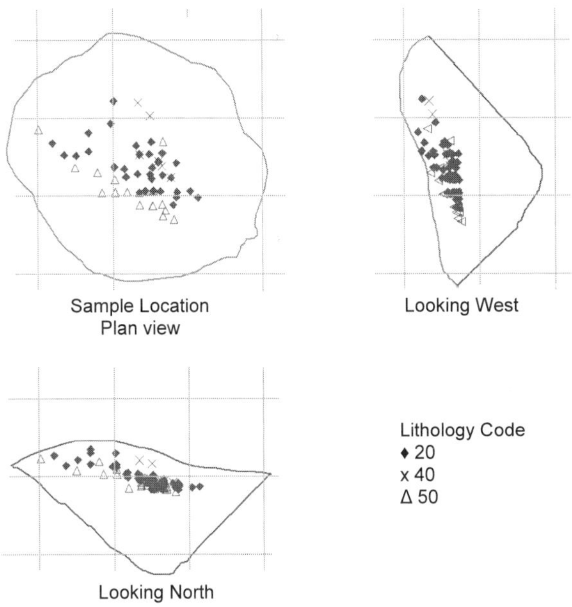


**Figure 1: Schematic view of sample points relative to pit design**

## LABORATORY FLOTATION TEST WORK

All 83 samples were subjected to a standard rougher laboratory flotation test known as the MinnovEX Flotation Test, MFT. The objective of the MFT is to measure the pulp kinetics applicable for each drill core sample at the reagent suite tested. The MFT uses size-by-size analysis; the protocol for the MFT is well described in the literature (MinnovEX, 2004). In this case the flotation conditions as determined from prior process development work were applied to each sample at a target grind of 80% passing 150 microns with lime to pH 9 with sodium silicate as dispersant, diesel as collector and a frother, for a time that was adequate for recovery of all attached particles. Pulp kinetic parameters for each test were derived for each mineral species from the sizing and chemical analyses of the feed, individual concentrates, combined concentrate and tailings samples (Dobby and Savassi, 2005). These parameters are a description of each mineral species present in terms of:

- the ultimate recovery achievable by true flotation, $R_{max}$ (that is, flotation by attachment only, after the contribution from hydraulic entrainment has been deducted)
- the average rate constant, $k_{avg}$ and a measure of the rate constant distribution, alpha.

$R_{max}$ and $k_{avg}$ are also expressed in terms of any change with respect to a small change in grind, i.e. $R_{max\_slope}$ and $k_{avg\_slope}$ so that any change in $P_{80}$ can be investigated as described in the MFT protocol.

Rougher-Cleaner tests were conducted on composite samples at a regrind of 80% passing 20 microns with further sodium silicate, diesel and frother additions at pH approximately 9 to 10. Regrind prior to cleaner flotation changes the pulp kinetics; hence the kinetics exhibited by the rougher concentrate is modified (mathematically) prior to feeding the first cleaner. The results of the rougher-cleaner tests can be used to estimate the modification so that a complete set of kinetic parameters is available for each mineral that can be applied in the simulation of full-scale plant performance. A range of samples should be tested to investigate the variability in the modifiers; typically the higher the NSG $R_{max}$, the lower the multiplier modifier for this parameter, since there is normally a greater effect of regrinding on liberation (Bulled, 2007).

## ANALYSIS OF TEST RESULTS

The kinetic parameter data extracted from the MFTs were analyzed to determine relationships that could assist in estimating kinetic parameters for each block.

The $R_{max}$ for molybdenite showed a distinct relationship with head grade, across all lithologies, which was even more marked if samples with low $R_{max}$, below 70%, were eliminated, as shown in Figure 1. It was obvious that the low $R_{max}$ samples (mostly from quartz porphyry and all with potassic alteration) were all located in an upper north-west section of the ore body away from the remainder. The flotation response of this sector was different and was considered to be due to oxidised ore. These samples and the mine blocks in this region were therefore treated separately for distribution of the data. The rest of the samples were considered to be sulphide ore.

Excluding the oxidised ore, the remainder (sulphide ore) had a relationship for $R_{max}$ to head grade (%Mo) shown on Figure 2, with no discernable difference by lithology:

$$R_{max} = 100 - 1.767 * (\%Mo)^{-0.54} \qquad (1)$$

The standard error in estimating from this function was an absolute value of 4%.

No other useful relationships could be found but the lithologies had different average values for the FLEET parameters "alpha" and "$R_{max\_slope}$".

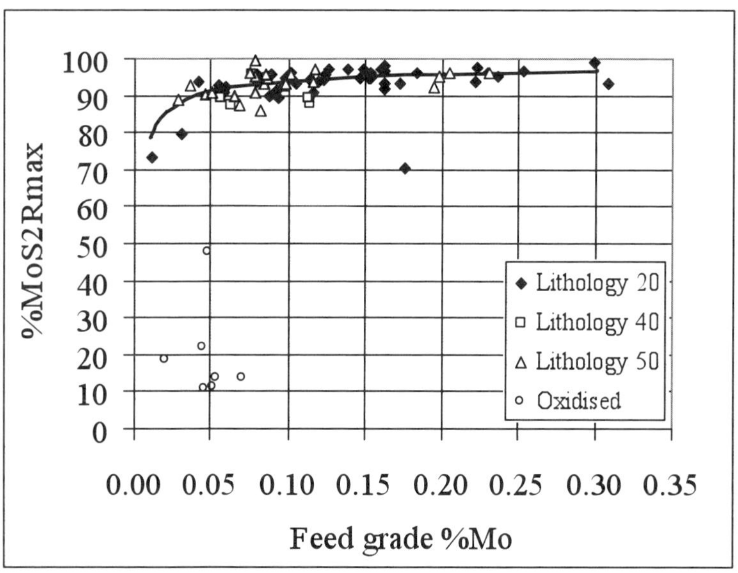


**Figure 2: MoS2 Rmax vs. Mo head grade**

A relationship for the chalcopyrite $R_{max}$ to head grade was found to be as equation (2), with no effect of lithology, as shown in Figure 3:

$$R_{max} = 100 * (1 - \exp(-76 * \%Cu) \qquad (2)$$

The correlation was rather poor and the standard error in estimating from this function was an absolute value of 19%. No other useful relationships could be found but the lithologies had different average values for the FLEET parameters $k_{avg}$, and alpha.

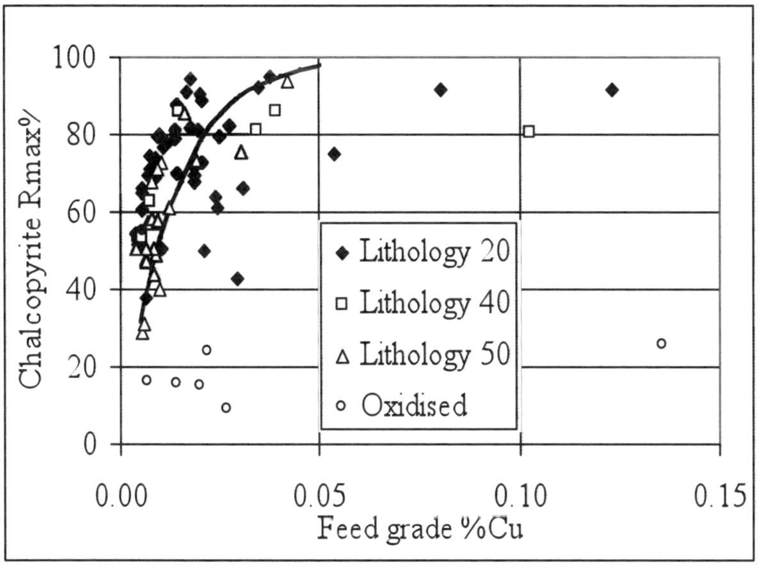


**Figure 3: Chalcopyrite Rmax vs. Cu head grade**

No relationships were found for sphalerite or pyrite but the lithologies had different average values for each parameter.

A relationship for the Non-sulphide gangue, NSG $R_{max}$ to %Mo head grade was found to be as in equation (3) with no effect of lithology as shown in Figure 4:

$R_{max} = 5.8 * \%Mo + 2.2$, with an absolute standard error of 0.35% (3)

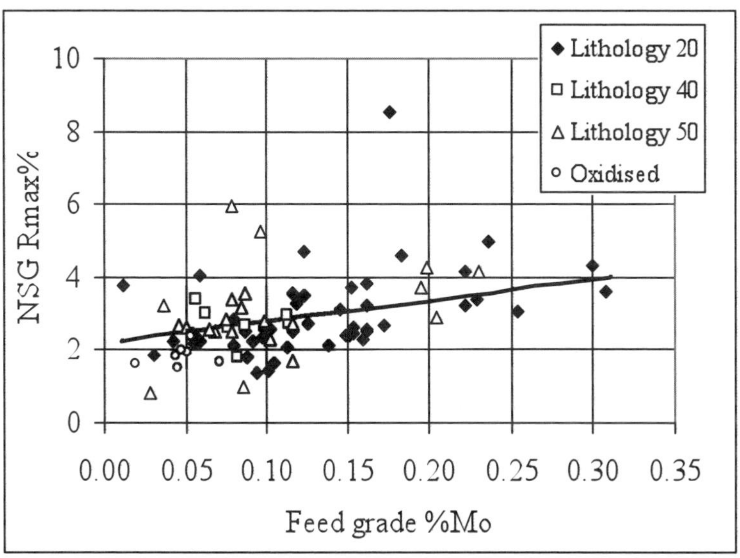


**Figure 4: NSG Rmax vs. Mo head grade**

No other useful relationships could be found but the lithologies had different average values for the other FLEET parameters.

The kinetics modifiers established from rougher-cleaner tests for flotation after regrind reflected a distinct reduction in pyrite and NSG $R_{max}$ and large slow-down in the $k_{avg}$ for all minerals except molybdenite. A single set of modifiers, as shown in Table 2, were used as an approximation for all feed samples in simulation of plant results using the FLEET modeling software as described in the literature (Bulled, 2007).

**Table 2: List of modifiers used for cleaning after regrind**

| Multiplier for | MoS2 | Chalcopyrite | Sphalerite | Pyrite | NSG |
|---|---|---|---|---|---|
| Rmax | 0.995 | 0.8 | 0.8 | 0.4 | 0.15 |
| kavg | 0.8 | 0.25 | 0.42 | 0.18 | 0.14 |

## DISTRIBUTION OF KINETICS DATA ON THE MINE PLAN

The ore body was characterized for flotation performance by the distribution of the measured kinetics from the 83 samples across the blocks identified on a mine plan. The mine plan comprised 57,863 blocks of 75 x 75 x 50 ft each (about 875 million tons) above cut-off grade to be mined over about 32 years. The mine plan as supplied included the following information for each block:

- assay data for %Mo, %Cu, %Zn, and %S and a standard error estimate for the %Mo resulting from previous geostatistical distribution of assay data to each block
- lithology of each block as allocated from the geological plan
- expected year of mine production.

The objective of data distribution is to assign every block with the best estimate of each flotation parameter and an estimation of the uncertainty in this value (quoted as a statistical "standard error"). The flotation characteristics of the whole ore body (or particular sections such as early production years) can then be considered for plant design and forecasting of plant results, rather than just including individual or composite samples, as in the traditional approach.

Since the most significant parameters in forecasting plant flotation results in this case are the molybdenite maximum attainable recovery, $R_{max}$ and average rate constant, $k_{avg}$, efforts were concentrated on the best distribution of these parameters across the mine plan. The $R_{max}$ value for chalcopyrite and NSG were assigned to blocks using equations (2) and (3) and lithology average values were used for all other parameters and minerals species.

Since the data analysis could determine no dependence of the flotation parameters, $MoS_2$ $R_{max}$ and $k_{avg}$, on lithology type, it was not considered in the distribution of this data. However, the $R_{max}$ values from the sample data did indicate distinctly separate regions of sulphide and oxidized ore. It was considered that the ore body could best be described as having areas that were purely sulphide within the region of sulphide samples and purely oxidized within that separate area of samples but that an intermediate situation was assumed to occur where the block has a certain proportion oxide, declining as one moves from pure oxide to pure sulphide.

Consequently it was decided to distribute the data from these two types of ore sample separately to the blocks. Parameters of blocks were then determined from separate oxide and sulphide components that were then combined to make a final estimate.

As shown by equation 1, the value of $R_{max}$ in the sulphide component of a block could be estimated from a relationship with the head grade. However, this is merely one estimate which has a standard error of 4%. An alternative estimate for the $R_{max}$ of the sulphide component that block can be made by assuming that there is some other factor that we have not been able to determine from the information available, e.g. some function of mineralogy such as liberation that is determined by lithology or alteration, and that this factor has a regional basis. On this assumption, we can make a further estimate for $R_{max}$ of a block from the values of neighboring samples, as is traditionally done for assay values by geostatistical techniques. The two estimates for $R_{max}$ can then be combined to give a better estimate with more confidence. Since no relationship of $R_{max}$ to head grade could be determined for the oxidized ore samples, the geostatistical method is all that we have available for estimating the $R_{max}$ for the oxidized component of any block.

Estimation of block $R_{max}$ values was made using the geostatistical distance weighting method known as Kriging (Preece, 2006) i.e. weighting available sample data according to their distance from the block in such a way that samples close to the block have more influence on the block estimate than other samples further away. Kriging requires the construction of geostatistical variograms for $R_{max}$ by plotting the semi-variance of differences in the value of pairs of samples of equal distance apart against that distance.

The establishment of a model for the variogram allows for the estimation of the $R_{max}$ and the standard error of each estimate to be made for each block. The range of influence is determined by the shortest distance apart for pairs having attained the maximum variance, known as the "sill". In this study the range for $R_{max}$ values was 800ft. The value where the curve cuts the y-axis is referred to as the nugget effect and is a measure of the inherent errors in sampling and measurement of individual data points, 10% in this case (estimated from the results of the coincident samples).

Since the data analysis indicated no significant $R_{max}$ differences could be determined between the 3 lithologies, a single variogram was developed for all samples so that a block of any one type was interpolated using samples from any other type. The variogram for $R_{max}$ of the sulphide samples is plotted in Figure 5.

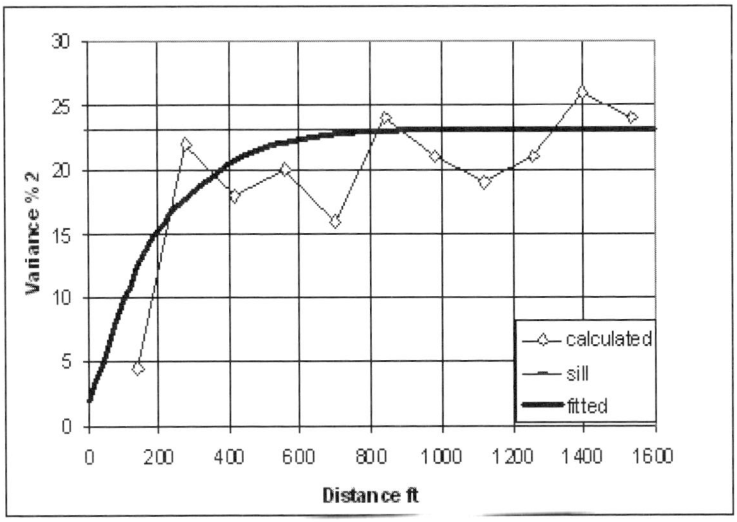


**Figure 5: Average Variogram for sulphide MoS2 $R_{max}$**

The estimated $R_{max}$ values and the uncertainty (standard error) of each estimate were determined from the fitted variogram model.

These two estimates of the $R_{max}$ for the sulphide component (i.e. from the head grade relationship and the geostatistics) were then combined by weighting in inverse proportion to the error variance of each estimate. On average the weighting of the head grade estimate was 49% and the "kriged" estimate was 51%.

There were insufficient samples from oxidized ore to draw a variogram, so the sulphide fitted model was used for estimations with adjustment for overall variability. The oxide component of each block was estimated from oxidized ore samples only. The sulphide and oxide components were then combined in proportion to their weighting in the block to give a final estimate of $R_{max}$ and the standard error of that estimate.

The $k_{avg}$ values cannot be distributed across the blocks directly by any statistical method since $k_{avg}$ has rate constant units of $min^{-1}$. Consequently nominal production values, kgNOM, and a "service variable" were calculated for each sample and these values were distributed using there own variograms and kriging, where:

$$\text{kgNOM} = \text{head grade \%MoS2} * R_{max} * k_{avg} \quad (4)$$

$$\text{"service variable"} = \text{head grade \%MoS2} * R_{max} \quad (5)$$

The $k_{avg}$ for the block was then determined by:

$$k_{avg} = \text{kg NOM} / \text{"service variable"} \quad (6)$$

The use of this technique rather than dividing kgNOM by the estimated block values of head grade and $R_{max}$ ensures that the estimated $k_{avg}$ for each block falls within the range of measured values and there are no abnormal outliers. This procedure was used for the sulphide and oxide components separately and then they were combined. The average $k_{avg}$ for the ore body was found to be 2.24 $min^{-1}$.

Variability in MoS2 $R_{max}$ across the blocks was very high as shown in Figure 6 mirroring that expected from the range of values reported in the MFT results (and illustrated in Figure 2); all molybdenum is presented as equivalent MoS2 values from here on in this paper in the absence of any assays for oxidized molybdenum.

The standard errors in the estimates had an average absolute value of 3.7% (ranging from 1.5 to 10.6%) for $R_{max}$ and an average relative value of 39% (ranging from 17 to 76%) for $k_{avg}$; the large amount of uncertainty is because many of the blocks were far from any samples. The standard errors to be applied to all other parameters were those from either equations (2) and (3) or those determined from the population of test data for each lithology.

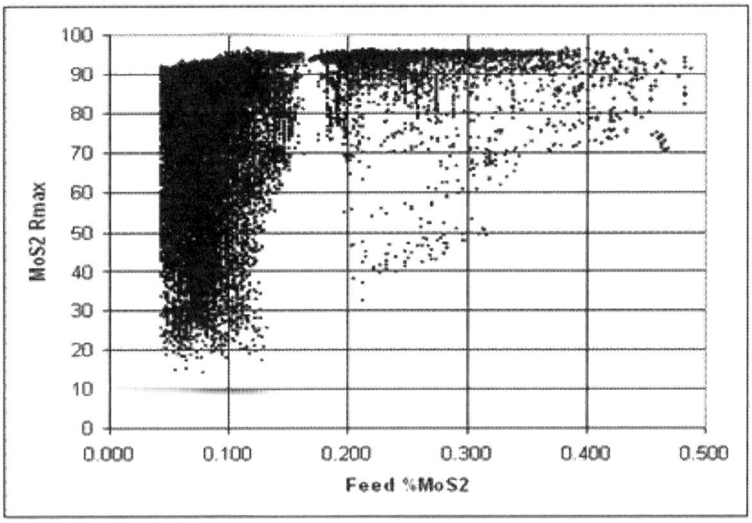


**Figure 6: Estimated $R_{max}$ vs. head grade for all the blocks**

Since blocks have been identified by year on the mining plan, the geostatistical analysis also allowed the determination of annual average flotation kinetics values and their precision by the same Kriging technique, making it possible to predict the throughput in each production year. It is also possible to extend the analysis to calculate how many more samples need to be tested from within the volume of the blocks mined in a production year in order to improve the precision to any desired level.

The average head grade, $R_{max}$, $k_{avg}$ and the proportion of oxidized ore mined was calculated for each year of a 32-year mine life for a treatment rate of 55,000 metric tpd, assuming that the lower grade blocks were stockpiled to treat only the target tonnage each year. Parameters were plotted against production year and shown in Figures 7 and 8; the very low recovery expected from year 13 to 17 corresponds to the spike in the mining of oxidized ore. It may be advantageous to reschedule the mining of blocks to ensure a smooth concentrate production program. Standard error (uncertainty) in the forecast was found to increase with time into the future, since most of the samples were situated in the area of the ore that was scheduled to be mined first. There was also a spike in the uncertainty when oxidized ore is to be mined, since that is an area with a low sampling density.

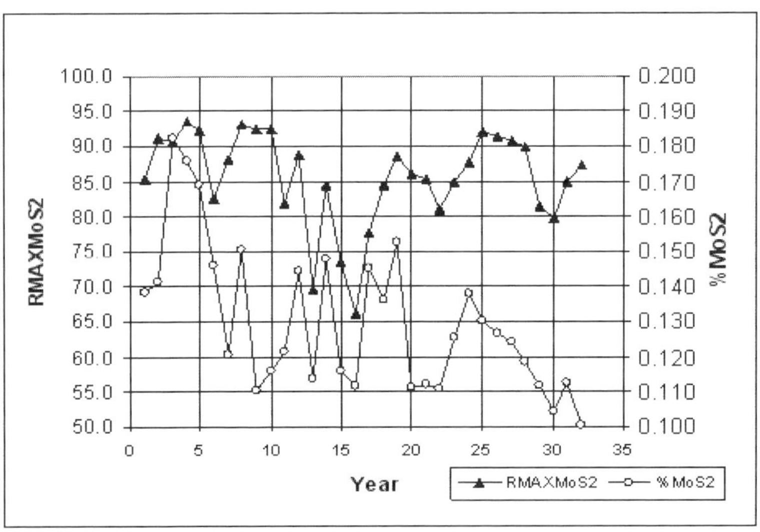


**Figure 7: Change of $R_{max}$ and head grade with time during the life of mine**

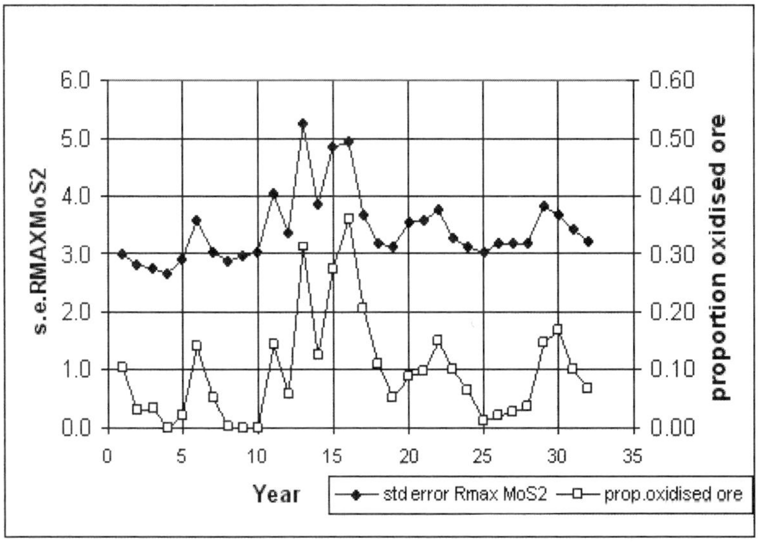


**Figure 8: Increase in uncertainty in $R_{max}$ (standard error of the individual blocks) and change in proportion of oxidised ore with year of life of mine**

## FLOTATION PLANT DESIGN

A brief description of FLEET simulation software is a pre-requisite to understanding the design approach used in this case study. A more extensive description is available in the literature (Dobby et al, 2002, 2005). FLEET was used in both flowsheet design and equipment size selection. A FLEET system comprises three components:

- In the FLEET Flowsheet component, the configuration of the circuit and the equipment specifications are defined in terms of number of cells and dimensions. Various flowsheets are then simulated to determine an optimum (Bulled, 2007).
- The FLEET Input parameters define how the circuit is being operated in terms of stage froth recoveries across individual units (pull-rates), stream % solids (water recoveries across units) and the cleaner circuit modifier factors.
- The FLEET Dataset defines the feed material entering a flotation circuit in terms of feed rate (tph), feed grind ($P_{80}$), head grade and flotation kinetic parameters at that grind. After each flotation stage the residual kinetic parameters are calculated and used in the subsequent stage with modification to account for regrinding or depressant addition as required. Values are combined whenever streams are joined, e.g. in recycle flows. For a design or production forecasting study, the dataset can comprise tens of thousands of lines, where each line represents a block of ore (of say 4 to 5 hours of production). This allows the effect of variations in feed rate, feed grind and feed ore characteristics (head grade, flotation kinetics) to be determined.

FLEET simulations were used to investigate the performance of a full-scale flotation plant treating ore as represented by the average kinetics of approximately 14,000 blocks to be treated in the first 10 years of planned operation, with the reagent conditions as tested in the laboratory. The proposed circuit to treat 55,000 metric tpd of ore is presented as a FLEET diagram in Figure 9 comprising:

- Two lines of eight rougher cells of 250m$^3$ each
- a regrind of the rougher concentrate to $P_{80}$ = 20 microns
- One line of four 1$^{st}$ cleaner cells of 30m$^3$ each
- One line of eight cleaner-scavenger cells of 30m$^3$ each
- One line of four 2$^{nd}$ cleaner cells of 30m$^3$ each
- Three further stages of columns with diameters deceasing in size (3.2m, 2.8m and 2.4m) at each stage.

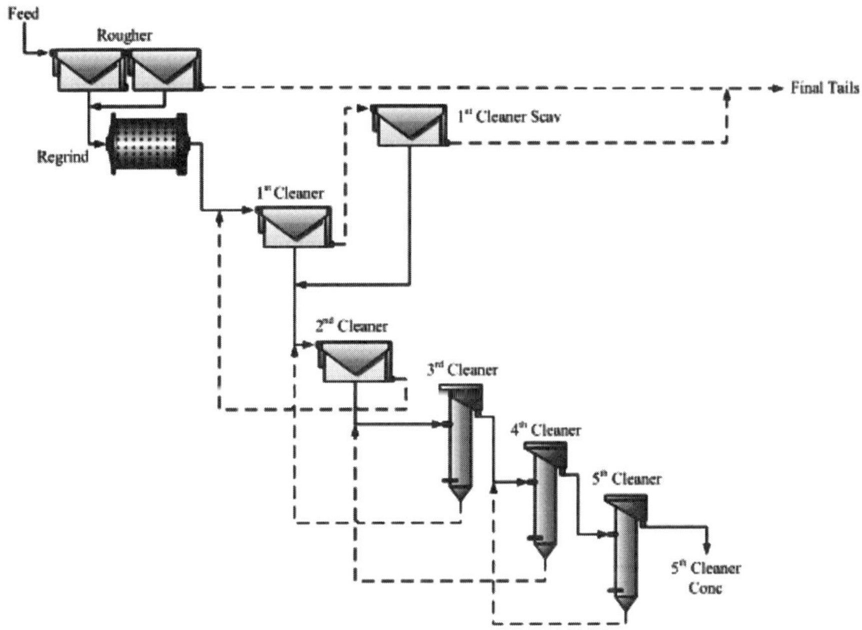


**Figure 9: FLEET circuit**

**FLEET SIMULATION RESULTS FOR FULL SCALE PLANT DESIGNS**

The FLEET model for the circuit was used to simulate the expected results for each of the blocks to be treated in the first 10 years of operation in terms of final concentrate grade and recovery when operating the plant under standard conditions. Variability of the simulated molybdenite recovery across the blocks is very high as would be expected from the range in $R_{max}$ values (illustrated in Figure 6) (Dobby et al, 2004). Typically the recovery of molybdenite to final concentrate is about 4% less than $R_{max}$.

Simulated MoS2 grade of the final concentrate varies with recovery and molybdenite head grade. Most of the contamination in the final concentrate is expected to be chalcopyrite, as shown in Figure 10, occurring when the ratio of chalcopyrite to molybdenite in the block is high or when the MoS2 $R_{max}$ is low (due to oxidized ore).

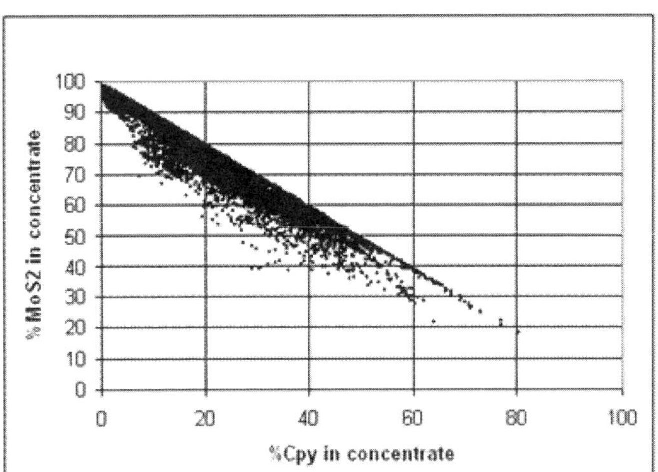


**Figure 10: Simulated %MoS2 in concentrate vs. % Chalcopyrite in concentrate**

Since the plant cannot be expected to operate under these standard conditions at all times, simulations were conducted with varying stage froth recovery scenarios on the annual average dataset of kinetic parameters for each or the first 10 years of operation (as summarized in Figure 11) in order to illustrate the available grade/recovery relationship for each year.

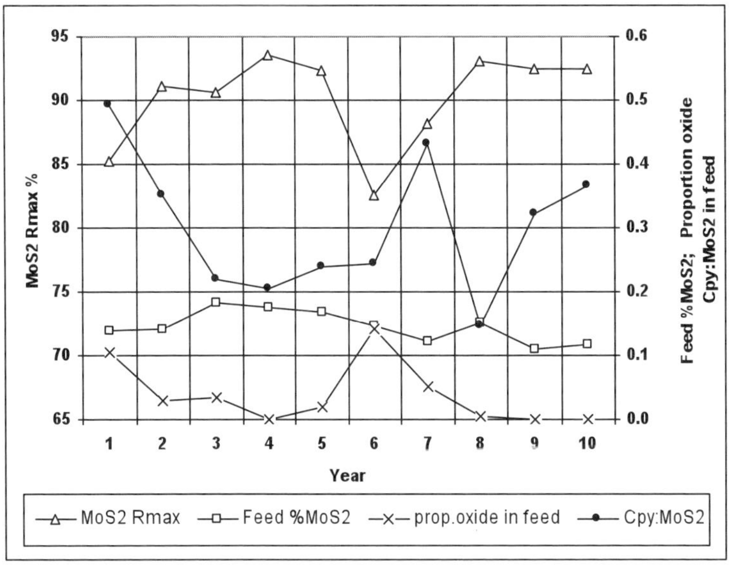


**Figure 11: Average kinetics parameters for Years 1 to 10**

The results are presented as curves for each year in Figure 12 showing the recovery that must be sacrificed to achieve higher grade. Not only will recovery vary from year to year (as expected by the changes in MoS2 $R_{max}$) but MoS2 concentrate grade also changes. MoS2 recovery is limited by the value of $R_{max}$; no increase in froth recovery or residence time will take the recovery in Year 6 up to the other years without complete sacrifice of concentrate grade since the $R_{max}$ for that year is only 82.5%. The low concentrate grade, particularly in Years 1 and 7 is a result of increased contamination by chalcopyrite when the copper content of the feed is high relative to MoS2. Since all of these simulations were generated from flotation test work that included no chalcopyrite depression, it is evident that a chalcopyrite removal system such as Cytec 7260 or Flomin 910 addition must be investigated and further simulations conducted based on the results of such testing. It appears that such an approach will be required in years with high copper feed grades to produce an acceptable plant product. A ferric chloride leach has since been incorporated into the project design to guarantee molybdenum concentrate specifications for copper, zinc and lead will be met.

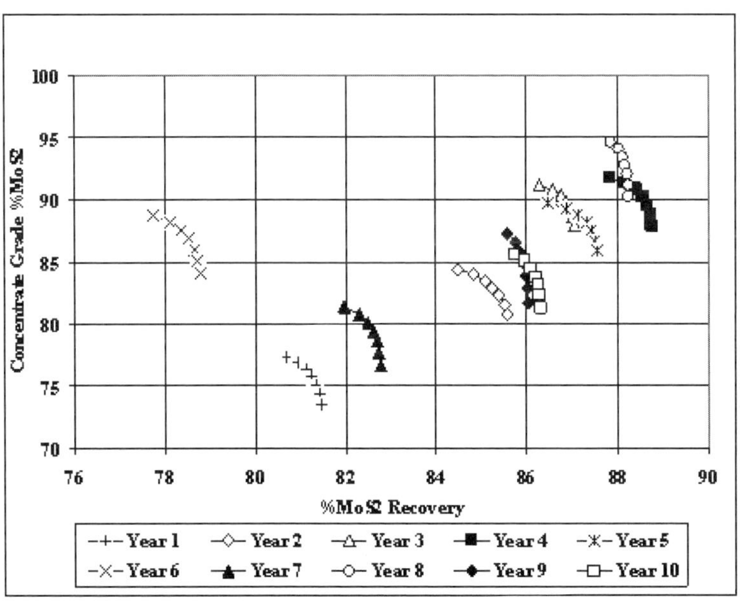


**Figure 12: Simulated average grade/recovery curve by year**

**UNCERTAINTY IN THE ESTIMATES AND RISK**

Potential sources of errors in the forecast flotation plant results arise (Bulled and McInnes, 2005) from:

- Unrepresentative samples of ore
- Insufficient ore sampling and testing
- Flotation testing errors: operation, sampling and assaying
- Inaccuracies in kinetic parameter extraction and FLEET modeling.

Since there is normally a considerable amount of variability within an ore body, the greatest of these errors is expected to arise from unrepresentative and insufficient ore samples. Repeat sample testing and experience in benchmarking FLEET models to operating plants enables the testing and modeling errors to be minimized. All of these errors can be reduced by more sampling and testing but there is always a practical limit in time and cost in producing a design for a feasibility study. Therefore, having an estimate of the statistical error in the projected results is of value in estimating the associated risks of not collecting more data. Obviously, the uncertainty could then be improved (if required) by conducting tests on more samples to reduce the standard error.

Average year by year head grades and flotation parameters were determined for three operating years chosen as examples of years that range from low oxide content, high $R_{max}$ and low uncertainty (year 3) to high oxide content, low $R_{max}$ and high uncertainty (year 13), with year 6 as an intermediate. The data included the best estimates for the average values of those years together with an estimate of the uncertainty (standard error in those estimates). The standard errors were determined for each year in the statistical studies of the data and distribution of data to the blocks. Note that the standard error (uncertainty) for the annual average is lower than the average of the standard errors in the estimates for individual blocks since many more samples are used in calculating an annual value rather than in estimating an individual block. The standard error included in the estimate of the head grades for Cu, Zn and S for each block on the supplied mine plan was estimated at 25% relative (about 5% relative on an annual average) for purposes of the study. These standard error estimates were used to produce a dataset of possible parameters scattered about the mean values in order to conduct a Monte Carlo simulation (Hammersley, 1964) for each of the three years using the FLEET model with standard operating parameters. The possible estimates of results for each of these years when operating under standard conditions are presented as a scatter of MoS2 grade/recovery points for each year in Figure 13; the centre of each scatter represents the best estimate.

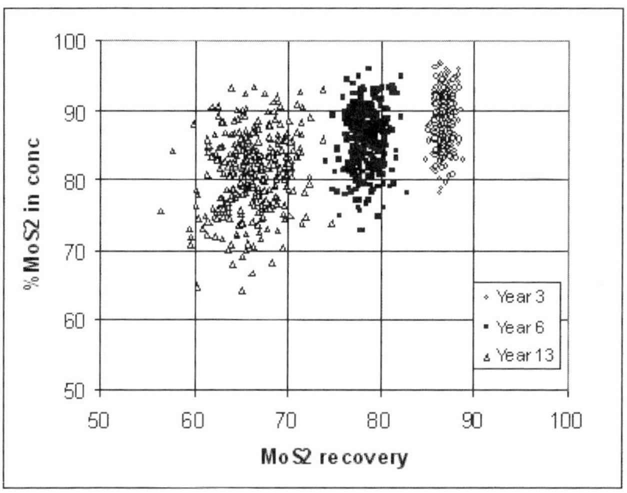


**Figure 13: Possible average results for each of Years 3, 6 and 13**

The results from these 3 years indicate that:

- The uncertainty of the average molybdenite recovery for the year is very similar to the standard error in the $R_{max}$ value; hence there is 90% confidence in the true value being within about 1.5 times that standard error. When the uncertainty is low as in Year 3 with low oxide content, the uncertainty in the estimate of recovery is about +/-1%, but when uncertainty increases with oxide content as in Year 13, it increases to about +/-5%. Testing of more samples from the oxidized area is needed to improve the forecasting precision. The risk of the true recovery being at the low end of the range of uncertainty cannot be significantly compensated for by a change in the plant design since it is a function of the uncertainty in the estimated $R_{max}$.
- The uncertainty in the grade of the final concentrate is high being about +/-5% at best when molybdenite recovery is good as in Year 3, increasing to +/-9% when the oxide content is high and molybdenite recovery is low and the influence of copper recovery becomes more significant as in Year 13. This is more evidence of the need to include chalcopyrite depression in the circuit. Testing of samples from the specific problem areas (high in copper and oxidized ore) is recommended using locked cycle tests and depressant such as Cytec 7260 and Flomin 910 in the cleaners.

## CONCLUSIONS

An ore body has been characterized by using standard flotation tests and the statistical distribution of sample test kinetic data to all of the blocks in the mine plan, and a flotation plant designed by using the FLEET simulation software package. The steps to achieving a design with a quantified error and risk attached to the results were:

- Select eighty-three samples that represent the variability in lithology and feed grade and were geographically spread over the ore body (although concentrated in the first five years of planned operation).
- Conduct standardized rougher flotation tests on each of these samples and extract kinetic parameters for each mineral species (molybdenite, other sulphides and gangue) that could be used in scale-up calculations by the FLEET program.
- Conduct rougher-cleaner flotation tests on composite samples to determine numerical modifiers to kinetic parameters that allowed the effect of cleaner flotation after regrinding to be described in the FLEET program.
- Determine a volume (domain) of the ore body with distinctly different floatation response, i.e. oxidized ore. Allocate a proportion oxide and sulphide to each block depending on its location.
- Determine relationships in the test results between flotation kinetic parameters and head grades for sulphide ore that could be used to estimate the kinetics of each mine block.
- Use the location of the sample points and the geological and mining plans of the ore body to conduct geostatistical analyses that estimated kinetic values for each mine block with a technique that allowed the determination of the precision of the estimates.
- Combine the sulphide estimates from head grade and geostatistics with the oxide estimates as proportions in each block to get a best estimate of kinetics together with an improved precision for the estimate.
- With an average set of kinetic parameters from the first 10 years of planned operation, use FLEET to investigate flotation circuit flowsheets and equipment sizing to achieve an optimum plant design.
- Predict the flotation results for this plant design for every block using the estimated kinetic parameters for that block.
- Combine blocks into production years according to the mine plan and predict a best estimate of the concentrate grade/recovery curve for this plant design in each production year using the average kinetic parameters for that year.
- Use the uncertainty in the average annual kinetics to determine the uncertainty in the predicted average plant results for each year and highlight the risk in having insufficient data for certain areas.

Enlargement of the plant will not compensate for the risk of realizing the lowest estimate of recovery since the uncertainty lies in the estimate of the average value for MoS2 $R_{max}$. The precision could be improved and the risk reduced by conducting tests on further samples to add more information. In this case further sampling and testing was required from the area of oxidized ore and depression of chalcopyrite must be investigated in the cleaner circuit. This recommendation was followed to advance the project later. It may be advantageous to reschedule the mining of blocks to ensure a smooth concentrate production program.

This study demonstrates the power of the FLEET program to handle large amounts of data from an ore body. Statistical techniques were used to distribute flotation kinetics to each block of ore in the mine plan and they were used with FLEET to determine an array of expected flotation plant results across the mine plan.

## ACKNOWLEDGMENTS

The management of SGS-Lakefield Research for permission to write and present this paper.

Michel Dagbert and staff at Geostats International for geostatistical study of the flotation kinetic parameter data and support in understanding the technology

Rachel Bridge, Melanie Cromes and staff at SGS-Lakefield Research for work in flotation testing and FLEET simulations for this project.

## REFERENCES

Bulled, D., McInnes, C., “Flotation plant design and production planning through Geometallurgical modeling”; Centenary of flotation conference, Brisbane, 2005

Bulled, D., “Flotation Circuit Design for Adanac Moly Corp using a Geometallurgical Approach”: Annual Canadian Mineral Processors Conference, Ottawa, Ontario 2007

Dobby, G, Kosick, G and Amelunxen, R, “A focus on Variability within the ore body for improved design of flotation plants”: Annual Canadian Mineral Processors Conference, Ottawa, Ontario 2002

Dobby, G and Savassi O, “An advanced modeling technique for scale-up of batch flotation results to plant metallurgical performance”, *Proc. Centenary of Flotation* (Australasian Institute of Mining and Metallurgy) Melbourne, 2005

Dobby, G. et al., “Geometallurgical Modeling - The New Approach to Plant Design and Production Forecasting/ Planning, and Mine/Mill Optimization”; Annual Canadian Mineral Processors Conference, Ottawa, Ontario 2004

Hammersley, J. M. and Handscomb, D. C., “Monte Carlo Methods”: John Wiley and Sons, 1964.

Mathews J, “Pegasus Gold Inc v Bateman Project Engineering [1999]”; FCA 1734, Federal Court of Australia, 1999

MinnovEX Technologies, 2004: “MinnovEX Flotation Test; Standard Procedure and Parameter extraction”. Available at www.minnovex.com

Preece, R., “Use of point samples to estimate spatial distribution of hardness in the Escondida Porphyry copper deposit, Chile”; SAG 2006, Vancouver, British Columbia

# HIGH RESOLUTION X-RAY MICRO CT (HRXMT)—ADVANCES IN 3D PARTICLE CHARACTERIZATION FOR MINERAL PROCESSING OPERATIONS

J.D. Miller and C.L. Lin

Department of Metallurgical Engineering
College of Mines and Earth Sciences
University of Utah
Salt Lake City, UT 84112

## ABSTRACT

During the past 10 years point-projection X-ray micro CT systems have been used for 3D visualization and quantitative analysis of multiphase particulate systems of interest to the mineral processing community. Although voxel (volume element) resolution has been limited to about 10 microns, valuable information has been gained regarding particle shape, mineral exposure/liberation, particle fracture characteristics, and pore network structures of packed particle beds with limited sample preparation. Analysis of packed particle beds containing as many as 30,000 particles can be accomplished in less than 3 hours with special software to establish the 3D characteristics of each particle in the sample.

Now, due to advances in X-ray optics, the resolution afforded with point-projection micro CT systems has been improved by at least an order of magnitude using a lens-based X-ray micro CT system. With the use of X-ray lens imaging the resolution is determined by the lens rather than by the spot size of the micro-focus X-ray source. One configuration of interest is the high resolution X-ray micro CT (HRXMT) system made by Xradia with a voxel resolution of less than 1 micron. The development of software tools for the analysis of packed particle beds and the use of HRXMT for the characterization of such multiphase particles in 3D is discussed, including the construction of liberation limited grade/recovery curves and particle damage during breakage.

## INTRODUCTION

Cone beam x-ray micro computed tomography (micro CT) systems, introduced commercially a decade ago, are a valuable tool for 3D visualization, characterization and analysis of multiphase systems at a voxel resolution of 10 microns which corresponds to the ability to describe mineral structure and composition of multiphase particles having a size on the order of 100 microns. Micro CT has provided the foundation for advanced research in the characterization of multiphase mineral particle including exposure/liberation analysis, damage and fracture during comminution, pore network structures for LB simulation of fluid flow in porous media (filter cake, column leach, etc.), and other multiphase particle features such as size, shape, texture, etc. As shown in Figure 1, most of the commercial cone beam x-ray micro CT systems are based on the principle of point projection of an x-ray source through the sample onto a detector. In this design, the achievable resolution is both a function of the x-ray source size, and the detector resolution. Commonly the resolution is thought to be driven by the x-ray source spot size.

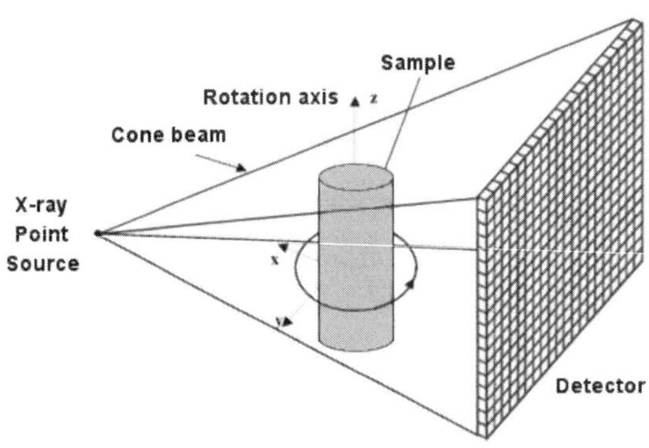


**Figure 1. Conventional cone beam x-ray micro CT system based on point-projection geometry with point source and a flat panel detector.**

Figure 2 illustrates the geometric configuration of a projection type x-ray microscope. High resolution is achieved by magnifying the object onto the detector. Notice, that as the distance between the source and the object gets smaller the magnification increases. However, so does the blur in the detector plane $r_S$ induced by the size of the source spot $S$. This blurring is given by:

$$r_S = S\left(\frac{b}{a}\right) \tag{1}$$

where $S$ is the size of the x-ray spot, $a$ and $b$ are the source to object distance and object to detector distances respectively. To scale the resolution back to the object plane we divide by the geometric magnification $M$ given by

$$M = \frac{a+b}{a} \tag{2}$$

The blurring caused by the size of the x-ray source is commonly referred to as geometric unsharpness or penumbra. In reality, the system resolution is a combination of the source spot size and the resolution of the detector. Therefore, the resolution of a projection based microscopy system can be approximated by a convolution of the detector resolution and the source spot size scaled by the geometric magnification

$$r_{Total} = \frac{\sqrt{r_D^2 + r_S^2}}{M} \tag{3}$$

where $r_D^2$ is the resolution of the detector system, and $r_S^2$ is the blurring due to the geometric unsharpness.

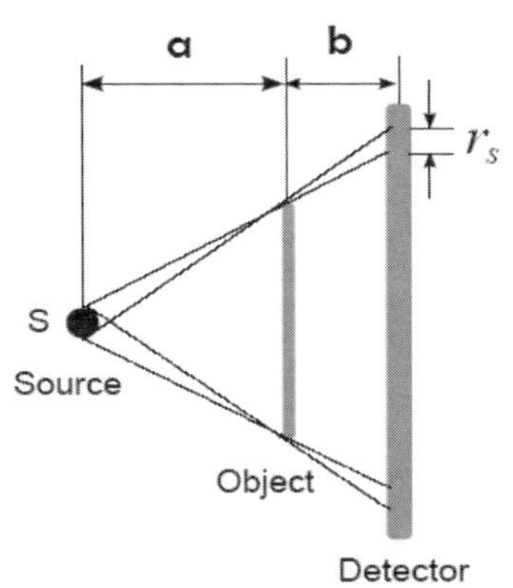


**Figure 2. Schematic of geometrical blurring due to the spot size of x-ray source.**

The conclusion is that for conventional x-ray systems with large detector pixel sizes, a large geometric magnification combined with a small source spot size are required to achieve high resolution. The highest achievable resolution for these systems is limited by the spot size and the closest distance that can be allowed between the sample and the source. As a consequence, traditional x-ray micro CT systems use transmission type sources to achieve a small source spot size combined with a minimal distance between sample and source.

The proximity of the sample to the source imposes a severe restriction on high resolution tomography, since for extended samples the minimum distance between source and sample can be large, leading to a limited resolution according to Equation 3. Inspection of equation 3 shows one possible solution to obtain high resolution is relaxation of the restriction of the sample being close to the source. This solution is to utilize an x-ray detector with high resolution. In fact, by using a high resolution x-ray detector, an imaging resolution better than the x-ray source spot size can be realized for sources with spot sizes larger than the x-ray detector resolution.

Such a system is the High-Resolution Micro CT (HRMXT) from Xradia, which employs an x-ray detector with sub-micron resolution combined with a microfocus x-ray source which corresponds to the ability to describe mineral structure and composition of multiphase particles having a size on the order of 10 microns.. In this system working distances between source, sample and detector are typically around 100 mm, so that full tomography even for larger samples can be achieved. The basic layout of the system is shown in Figure 3.

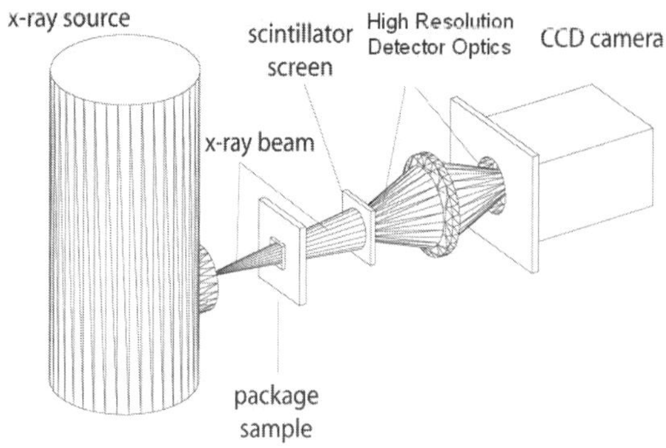


**Figure 3. The Xradia's High-Resolution Micro CT (HRXMT) uses a microfocus x-ray source with 150 kV acceleration voltage. The sample can be translated in x-y-z-theta directions with a high resolution stage. High resolution is achieved with a proprietary x-ray detector with an effective detector pixel size < 1 micron.**

## APPLICATIONS

### Mineral Liberation Analysis

Based on the conventional x-ray cone-beam micro CT, computer tools for 3D segmentation of packed particle bed images and for compositional analysis of individual particles have been developed (Miller and Lin, 2004; Lin and Miller, 2005; Videla, Lin and Miller, 2007). These tools have been shown to be reliable and applicable in various cases. The separation of individual particles in a set allows for further individual compositional analysis of each particle and makes the process of sampling and analysis economically affordable because it allows for acquisition of individual particle information from x-ray micro CT analysis over a set of particles instead of a single particle at a time basis. Analysis of packed particle beds containing as many as 30,000 particles can be accomplished in less than 3 hours using special software to establish the 3D characteristics of each particle in the sample.

Conventional x-ray micro CT technology has a voxel resolution limit on the order of 10 microns. At present, sub-micron resolution is possible with HRXMT as mentioned previously. Comparison of the 3D rendering image of a multiphase ore particle (1.70x0.85 mm) from conventional XMT (20 μm reconstructed resolution) with an image from HRXMT (~2 μm reconstructed resolution) is shown in Figure 4. It is evident that the sharp interfacial features of mineral grains can be located with ease using HRXMT.

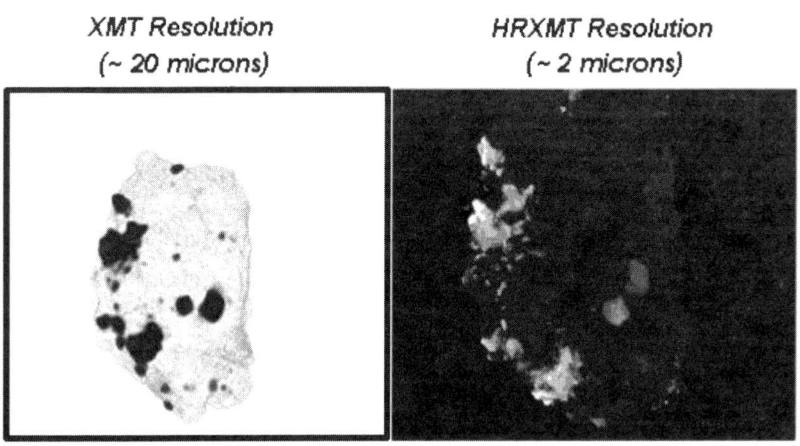


**Figure 4. Comparison of spatial resolution of x-ray micro CT systems for the analysis of a multiphase ore particle (particle size = 1.70x0.85 mm).**

Figure 5 illustrates the ability of the HRXMT system for quantitative analysis to determine the 3D spatial distribution of mineral phases in multiphase particles. Three cross sections along the XY- YZ- and XZ-direction and tri-plane sections are shown as established from the three dimensional reconstruction of a 325x500 mesh ore sample (45x25 μm). The image elements in the reconstruction are cubic, so the spacing between planes equals the voxel resolution which in this case corresponds to 1.09 μm. Here the gray scale levels of the images indicate the relative attenuation coefficient present in the bulk of the sample. Based on the magnitude of the x-ray attenuation coefficient, differentiation of mineral phases within the sample is possible. For example, light gray color identifies pyrite, the bright white color identifies chalcopyrite, and the dark gray identifies silicate minerals.

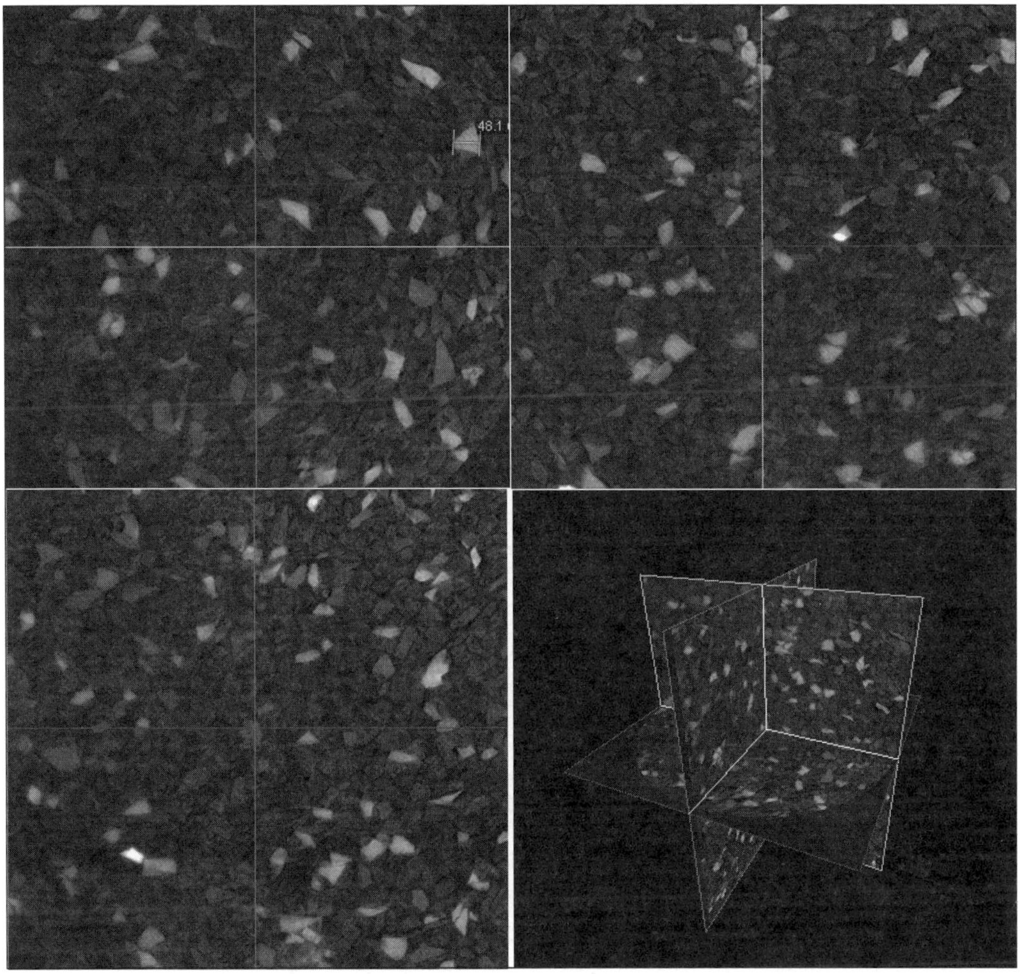

**Figure 5. Cross sectional images along XY-, YZ-, and XZ-directions from the three dimensional HRXMT reconstruction of a packed bed of 325x500 mesh multiphase copper ore particles (45x25 μm) containing chalcopyrite, pyrite and gangue minerals. (Reconstruction resolution of each voxel = 1.09x1.09x1.09 μm).**

Volume rendered images for a subset of three size fractions (+200, 200x325 and 325x500 mesh) from a copper rougher feed sample (512x512x512 voxels) are shown in Figure 6. The width of the cubic image is 0.558 mm. The 3D images of the particle beds are revealed with one quarter of the upper portion opened to show inside sections for each particle size class. Details of the composition/size distribution of different minerals in such a packed bed of multiphase particles can be established based on gray scale value or attenuation coefficient. Differentiation of features is possible because the linear attenuation coefficient at each point depends directly on the electron density, the effective atomic number of the material and the energy of the x-ray beam (McCullough, 1975). In practice, density measurements from XMT data can be done either by calibrating the CT machine with objects of known density or by using dual energy scanning to determine directly the density and atomic number of the material. In summary, it is evident that the utilization of HRXMT not only allows for quantitative analysis of multiphase systems but also allows for textural characterization and the determination of phase continuity at a high resolution.

**Construction of Liberation Limited Grade / Recovery Curves**

The liberation-limited grade/recovery curve can be established from CT liberation analysis of the feed and this curve represents the best separation to be achieved for a particular feed material. The liberation-limited grade/recovery curve is calculated from the liberation spectrum, the histogram showing the amount of mineral component of interest, in each grade class. Typically, twelve grade classes 0%, 5%, 15%, 25%, 35%, 45%, 55%, 65%, 75%, 85%, 95% and 100% by volume are used.

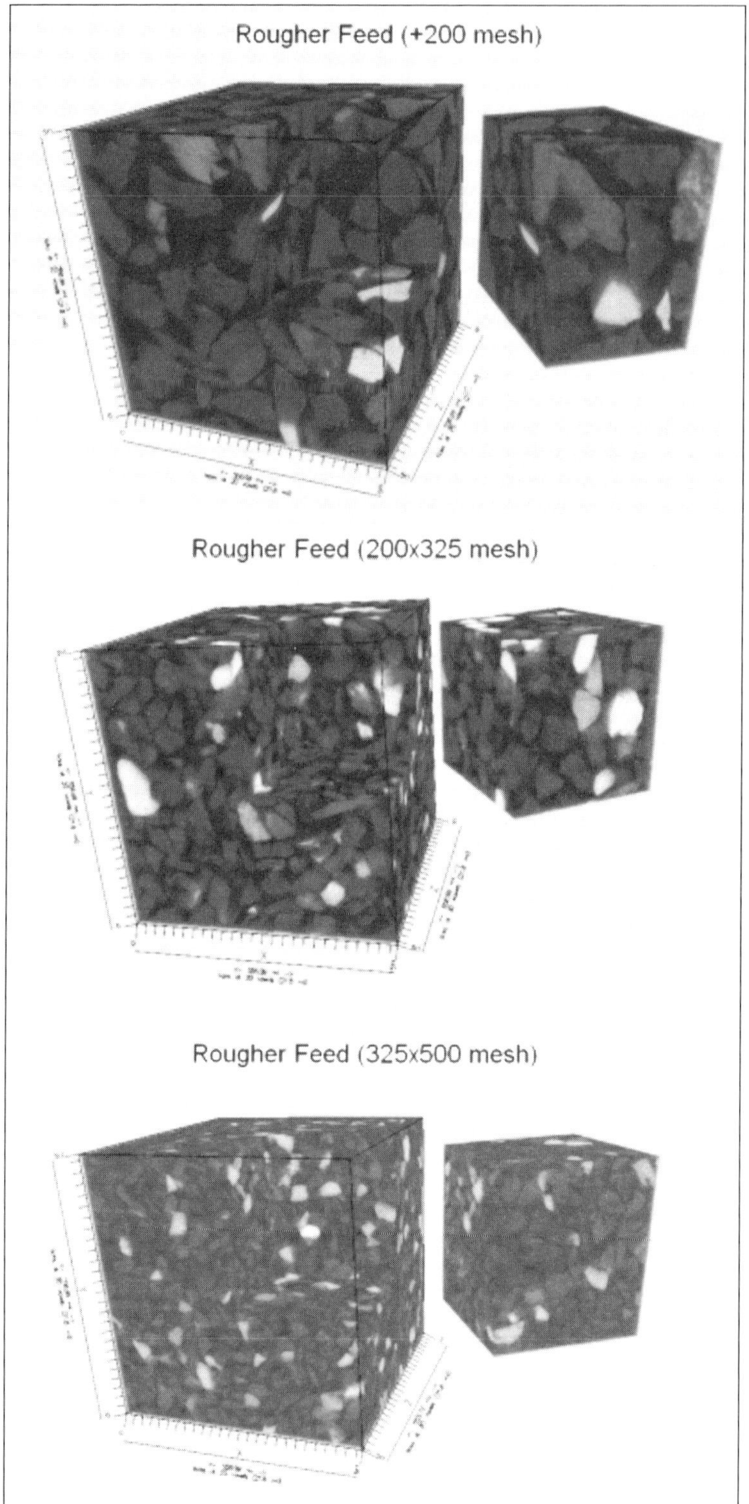


**Figure 6. 3D rendering images using HRXMT reconstruction of packed particle beds sample from three different particle sizes (+ 200mesh, 200x325 mesh and 325x500 mesh) of rougher feed (Reconstruction resolution of each voxel = 1.09x1.09x1.09μm).**

For example, the histograms for liberation analysis (2D and 3D) of a phosphate feed for rougher flotation were constructed from XMT data and are presented in Figure 7. The results indicate that the gangue minerals have good liberation while no phosphate mineral particles are found with complete liberation. Most of the phosphate rich particles contain quartz inclusions. As expected two dimensional liberation analysis increases the apparent liberation of the gangue and phosphate-rich grade classes, when compared to the actual three dimensional liberation analysis.

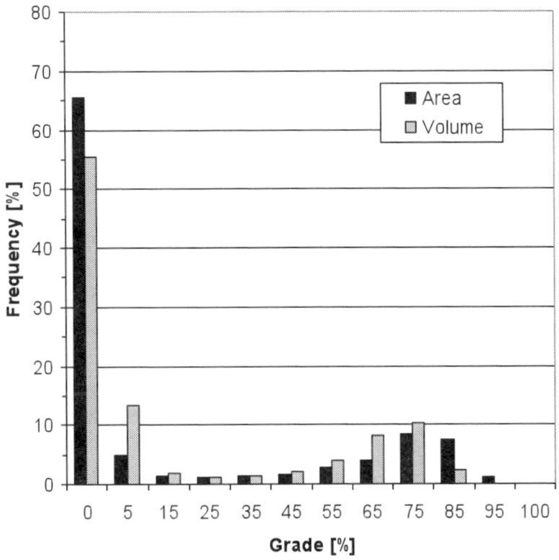


**Figure 7. 2D and 3D liberation spectra of phosphate for the Florida phosphate feed sample.**

The liberation-limited grade/recovery curve represents a boundary for separation efficiency. The grade and recovery for any actual separation can not exceed the limit imposed by this curve. In the best case the actual grade and recovery would fall on the curve and under these circumstances improved separation can only be achieved with further liberation by size reduction. If the grade and recovery for an actual separation falls below the curve then the separation efficiency is limited by other factors (mineral types, surface composition, slime coating, operating conditions, etc.) in addition to liberation limitations (Miller et al, 2009). Basically, the mineral content for all mineral-containing particles in each grade class is calculated and represented as a volume fraction of the total mineral in the feed. The mineral contribution from each grade class beginning with the richest grade class is then accumulated as more and more grade classes are considered until the final grade class, with the least amount of mineral is considered.

The liberation-limited grade/recovery curves constructed from liberation spectra (2D and 3D) of Figure 7 for the Florida phosphate sample are presented in Figure 8. This figure reveals that the liberation limited grade/recovery curve based on the 2D liberation analysis overestimates the expected separation efficiency when compared to the curve based on 3D liberation analysis. The actual grade and recovery from the flotation of this phosphate feed sample is found to lie on the 3D liberation-limited grade/recovery curve. In this regard, phosphate feed sample for this study is characterized by a very good separation and based on the liberation-limited grade/recovery curve the flotation results show that improvement in separation efficiency can not be achieved without further size reduction.

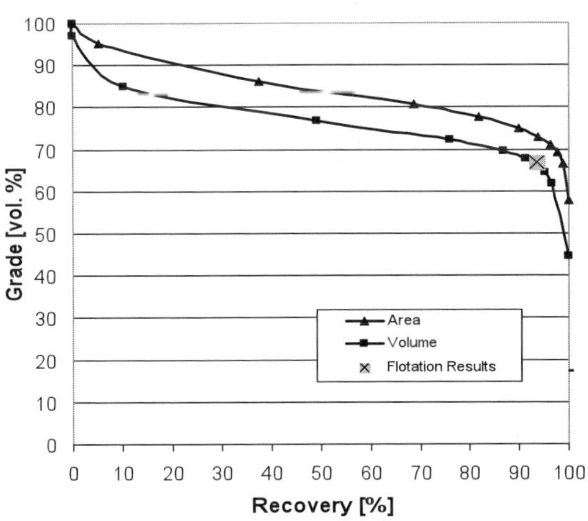


**Figure 8. Liberation-limited grade/recovery curve for the Florida phosphate feed sample.**

**Mineral Association and Texture of Multiphase Particles**

Even though these tools have been found to be effective, no universal procedure can be expected to perform efficiently for all cases due to the incredible variability of rock types and particle shapes from different operations. For instance, classification of mineral phases is based on the x-ray linear attenuation coefficients which are scalar values. In this manner, multiple thresholds combined with a finite mixture method are used to operate on 1D histograms (1D transfer functions) for individual particles to achieve mineral classification. However, the computer tools that have been developed are intended to be of general application and are an excellent starting point for the analysis of new samples of multiphase particles. It will probably be necessary to tune-up the software when new samples are analyzed and further steps may be necessary to improve the analysis in some circumstances. In this regard, multi-dimensional transfer functions allow ore particles to be classified based on a combination of values. In fact, the difference between the associations of minerals within a particle can be accessed with the use of a 2D transfer function (Kniss, Kindlmann and Hansen, 2001).

For application to problems involving mineral associations, the most important issue is locating the interface boundary within a multiphase particle. A two-dimensional transfer function described in the literature (Kniss, Kindlmann and Hansen, 2001) is a most effective method for the classification and analysis of multiphase particles. Besides the 1D histogram of the scalar values (x-ray linear attenuation coefficients), $f$, for each particle, the gradient magnitude (or first derivative), $f'$, is the second variable for the 2D transfer function. The gradient magnitude is useful to discriminate between different mineral phase regions. The gradient magnitude is a scalar quantity which describes the local rate of change in the scalar field, $f$. For example in Figure 9, a 3D rendering image and 3 orthogonal slices of a locked phosphate particle are shown. Figure 10 shows a 1D histogram based of attenuation coefficient data and identifies three materials in the locked phosphate ore particle: air (A), silicate (B), and phosphate (C). It is evident that each mineral has a distribution of attenuation coefficients due to photon sampling characteristics and the submicron variation in composition and porosity.

Figure 11 illustrates a log-scale 2D transfer function histogram of linear attenuation coefficient ( $f$ ) versus gradient magnitude ( $f'$ ) for the multiphase phosphate ore particle shown in Figure 9. The circular regions at the bottom of the histogram represent different phases; air (A), silicates (B) and phosphate (C). The interfacial boundaries between different phases are shown as arches; air and silicates boundary (D), silicates and phosphate boundary (E), and air and phosphate boundary (F). Each of these mineral grains and boundaries can be isolated using the 2D transfer function (Simian, 2001) as shown in Figure 12.

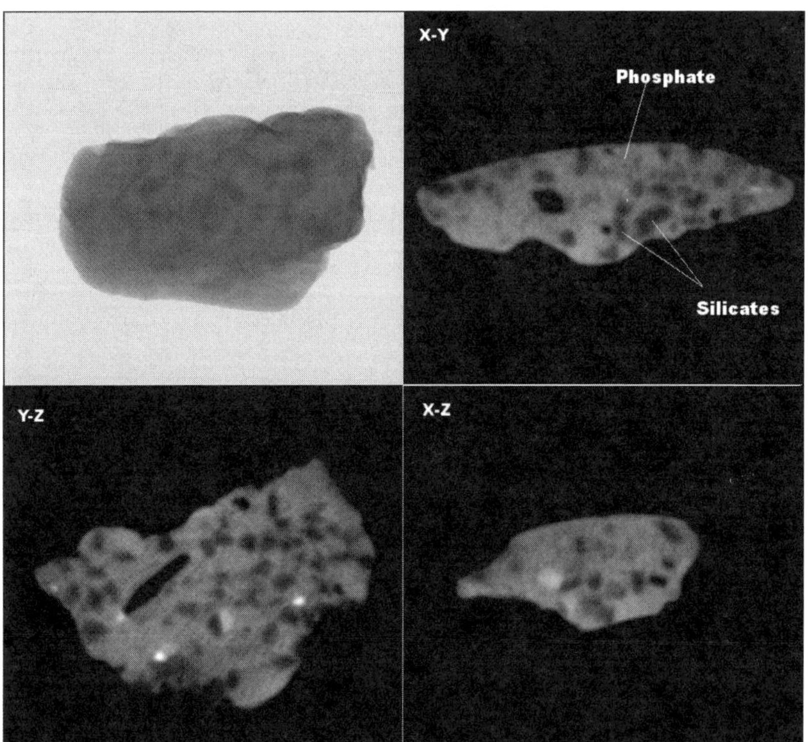


**Figure 9. 3D volume rendering image of a phosphate ore particle with locked silicate mineral grains.**

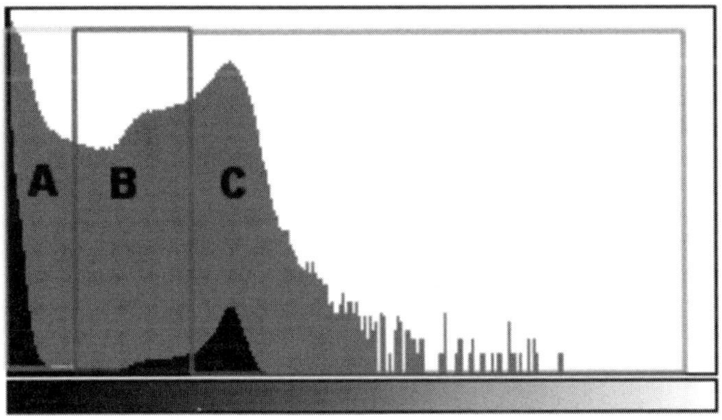


Figure 10. A 1D histogram of a locked phosphate ore particle. The black region represents the data of x-ray linear attenuation coefficient occurrences on a linear scale, the gray is on a log scale. The regions (A, B, C) are identified as air, silicates and phosphate.

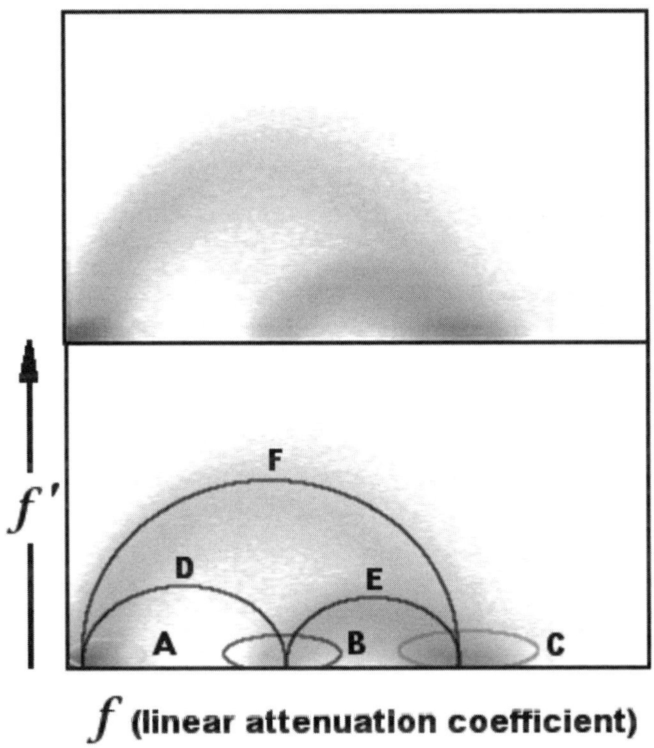


Figure 11. A log-scale 2D histogram (transfer function) of a locked phosphate ore particle. The regions (A, B, C) are identified as air, silicates and phosphate mineral phases. Circular regions (D, E, F) identify the interfacial boundary between phases.

**Figure 12. With the use of a 2D transfer function (linear attenuation coefficient versus gradient magnitude) silicate mineral grains and their interfacial boundary can be uniquely defined fro a locked phosphate particle. (Rendering image generated using Simian, 2001).**

**Damage and Breakage of Multiphase Particles**

Direct determination of intergranular fracture during multiphase particle breakage is a difficult task to achieve. Perhaps the only method of analysis is measurement of interphase area before and after comminution. Conservation of interfacial area after crushing is indicative of transgranular random breakage. On the other hand, if interfacial area is diminished after breakage some degree of preferential grain boundary fracture has occurred. For complete liberation of all grains after crushing the interfacial area goes to zero. Thus the interfacial area criterion is an important metric to assess the significance of preferential grain boundary fracture for different breakage conditions. Recently conditions for grain boundary fracture under slow compression have been established by X-ray micro CT, XMT, (Garcia, Lin and Miller, 2009).

In addition, XMT can be used for the 3-D determination of the internal damage of ore particles which vary in size from 40 mm down to a few hundred microns. For fracture analysis of particles, the crack and internal damage can be easily identified from the evaluation of images obtained from the HRXMT system as shown in right-hand side of Figure 13 with a voxel resolution of 1.5 μm. Data obtained from the conventional micro CT system at low resolution (20 μm) provides insufficient information for detailed analysis as can be appreciated from the image presented on the left-hand side of Figure 13.

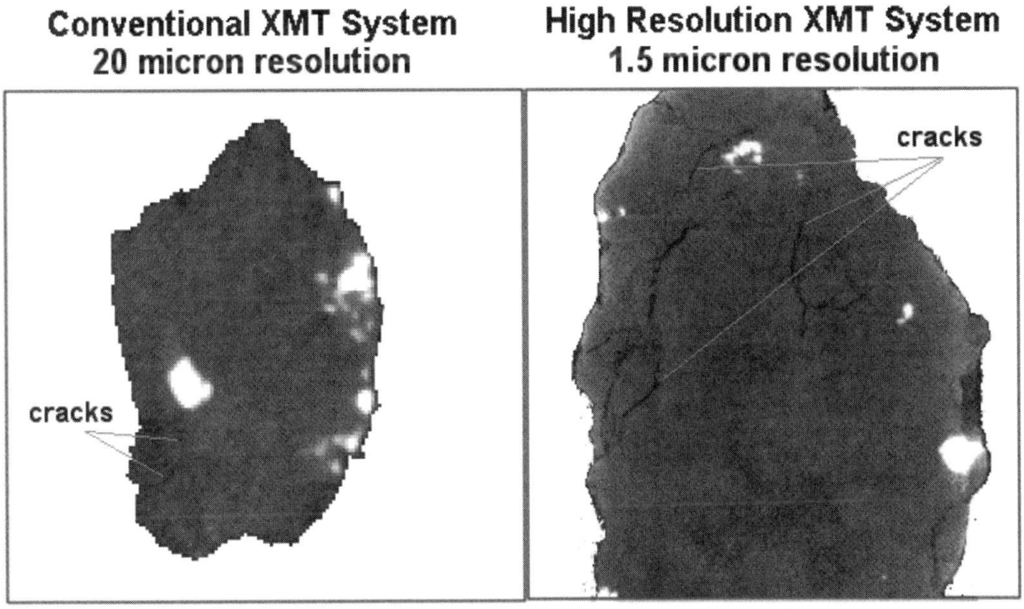


**Figure 13. Comparison of spatial resolution for X-ray micro CT systems, damage analysis of stressed particle (1.70x0.85 mm).**

With the new HRXMT, the voxel resolution will be extended to sub-micron. For example, two sections of the original XMT image, as indicated at z = 142 and 357 are shown on the left-hand side of Figure 14. Sections are spaced at 1.5 microns in this case. The corresponding preliminary isolation of segmented cracks is shown on the right-hand side of Figure 14. It is evident that crack definition is revealed in great detail. A 3D volume rendering image of the crack pattern (in white and brown color) for this particle is shown in Figure 15.

High spatial resolution and the direct processing of raw volumetric data by software that is being developed are the two important benefits offered by this new method. As the CT data get more complex, such as the internal damage state of ore particles, new tools need to be developed in order to extract useful information from the three dimensional data. Quantitative 3D characterization of fracture network at a voxel resolution of one micron includes determination of crack density (number of cracks per particle), individual crack features (length, spacing, orientation, surface area), and crack patterns (percolation of the crack) for each particle. Detailed 3D characterization at this resolution has never been done before.

For a given particle class (size, material type) the distribution of crack densities can be determined using specific software being developed at the University of Utah. Results obtained from the quantitative 3D HRXMT characterization of the damage state (crack density, crack size, shape, surface area, crack network) will determine if different particle breakage methods and loading conditions affect the crack network. With information on the crack network, the fragment size distribution upon particle failure will be predicted from percolation theory. Of equal importance, is the crack surface area created during stress and its relation to stored strain energy. Again different crack surface area distributions are expected for samples of different particle size and material type.

Finally, the initial state of damage will be an important parameter for consideration in data analysis. Minimum energy for breakage data may be rationalized with respect to damage states of parent particles and with respect to the mode of breakage. So, in addition to understanding fragmentation that occurs at the minimum energy condition, damage characterization will facilitate data analysis and normalization. The important issue is that the minimum energy for breakage is related to the damage state of the parent particle. Since the initial state of the parent particle will have some damage with cracks in the range of 1-2 microns in size, high resolution XMT will be necessary to define damage in parent particles and provide the foundation for quantitative minimum energy analysis.

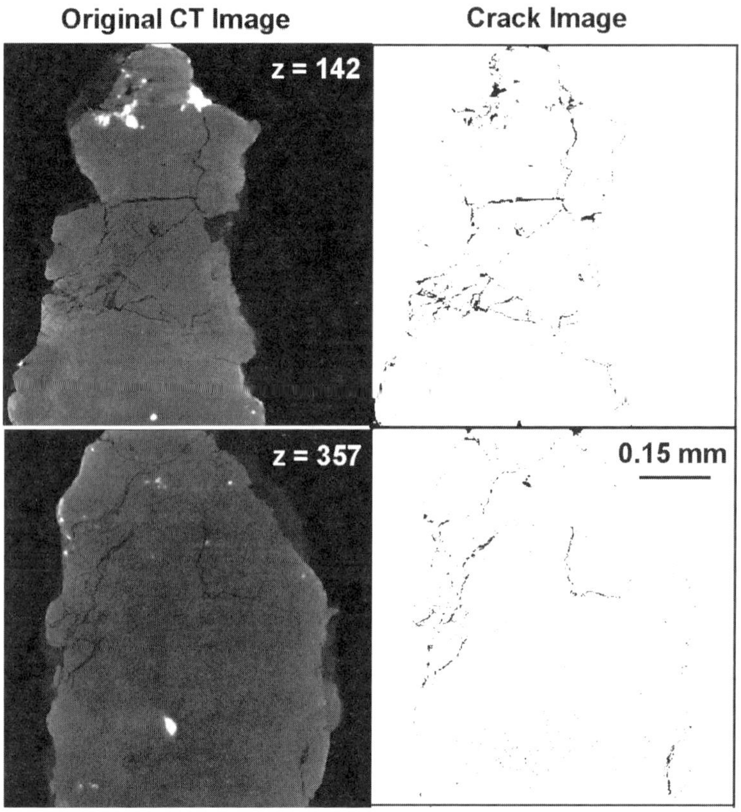


Figure 14. Sections of a damage ore particle (1.70x0.85 mm) obtained from 3D HRXMT image and the corresponding preliminary isolation of segmented cracks.

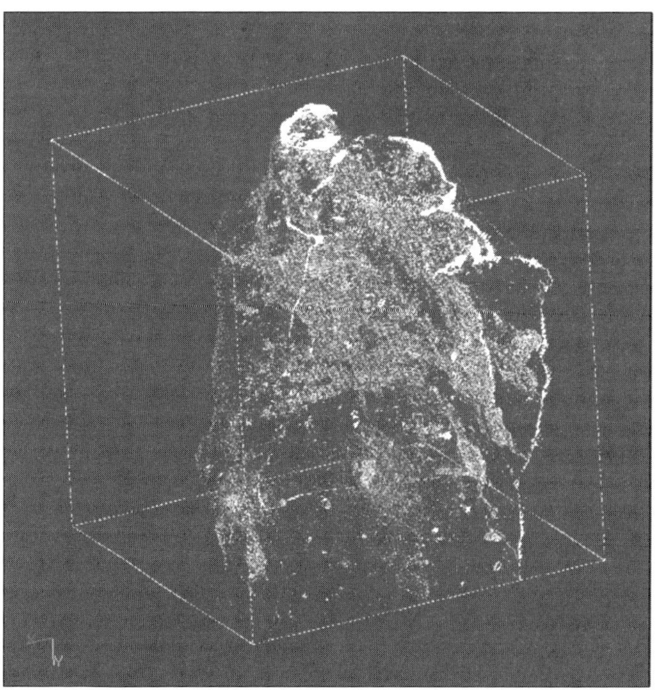

Figure 15. 3D volume rendering of crack pattern (in white and brown color) from HRXMT image of stressed ore particle (1.70x0.85 mm).

**SUMMARY**

High resolution X-ray micro CT (HRXMT) will allow for further advances in the 3D characterization of multiphase particles of interest in mineral processing plant operations. Sample preparation requires only the screening of particle into desired size classes. The sized sample containing thousands of particles can be scanned in a few hours. With a voxel resolution of less than 1 micron, particles as small as 5 microns can be distinguished and described with respect to size, shape, composition, texture, etc. In this way, the HRXMT data can be used for 3D exposure/liberation analysis, the construction of liberation-limited grade/recovery curves, damage and fracture analysis of stressed particles during breakage, and the detailed description of pore network structures of packed particle beds (filter cake, leach heaps, etc.).

**ACKNOWLEDGMENTS**

Some of the financial support for this work was provided by the Florida Institute for Phosphate Research (FIPR). In addition, the algorithm development has been supported by the Center for Advanced Separation Technologies (CAST).

**REFERENCES**

Garcia, D., Lin, C.L. and Miller, J.D., 2009. Quantitative analysis of grain boundary fracture in the breakage of single multiphase particles using X-ray microtomography procedures. *Mineral Engineering*. 22:236.

Kniss, J., Kindlmann, G. and Hansen, C., 2002. Multi-dimensional transfer functions for interactive volume rendering", *IEEE Transactions on Visualization and Computer Graphics*. 8(3):270.

Lin, C.L. and Miller, J.D., 2005. 3D characterization and analysis of particle shape using x-ray microtomography (XMT). *Powder Technology*. 154:61.

McCullough, E.C., 1975. Photon attenuation in computed tomography", *Medical Physics*, 2:307.

Miller, J.D. and Lin, C.L., 2004. Three-dimensional analysis of particulates in mineral processing systems by cone-beam x-ray microtomography. *Minerals and Metallurgical Processing*. 21(3):113.

Miller, J.D., Lin, C.L., Al-Wakee, M.I., and Hupka, L., 2009 Liberation-limited grade/recovery curves from x-ray micro CT Analysis of feed material for the evaluation of separation efficiency. *Preprint 09-025*. 2009 SME Annual Meeting, Denver, CO. Accepted for publication, *International Journal of Mineral Processing (2009)*.

Simian, 2001. http://www.cs.utah.edu/~jmk/simian.

Videla, A., Lin, C.L. and Miller, J.D., 2007. 3D characterization of individual multiphase particles in packed particle beds by x-ray microtomography (XMT). *International Journal of Mineral Processing*. 84:321.

# GAS DISPERSION MEASUREMENTS: USES AND OPPORTUNITIES

Cesar O. Gomez and James A. Finch

McGill University, Department of Mining and Materials Engineering
3610 University, Montreal, Quebec, H3A 2B2, Canada

## ABSTRACT

Measurement of physical as well as chemical variables is required to understand and characterize the performance of flotation units. For example, properties of the air dispersion have profound effects on the flotation process. Recently developed instruments and techniques make possible the simultaneous measurement of gas dispersion properties (gas velocity, gas holdup and bubble size distribution) and concentration of frothing agents. The technology – instruments, installation, and data collection and analysis – are reviewed. The measurements have been used for a variety of purposes: troubleshooting; characterizing cells to define effective operating range; setting gas distribution to banks of cells to optimize performance; selecting among competing sparger systems. These applications are described through a series of case studies.

## INTRODUCTION

Flotation is a process driven by air bubbles that takes place in two zones with widely different hydrodynamics characteristics. The separation of particles results from collection by bubbles in the pulp zone, followed by transport of the bubble-particle aggregates through the froth zone. Although phenomena in these two zones have been described, their quantitative effect on performance remains a subject of research because of the difficulty in decoupling the contributions of the individual zones.

The characteristics of the bubble population generated in flotation machines exert considerable control over the process. We have considered two aspects related to an individual cell for discussion purposes: gas dispersion and gas distribution. Gas dispersion refers to the properties of the bubbles created by breaking-up a mass of air (e.g., bubble size), while gas distribution refers to the spreading of bubbles throughout the cell volume. Design of flotation cells aims to distribute bubbles uniformly, a condition presumably that leads to the best metallurgical performance.

Gas dispersion is a collective term used to represent three parameters that describe the bubble population at a particular location in a cell: superficial gas (air) velocity $J_g$ (or gas velocity, i.e., volumetric air flow rate per cross-sectional area); gas holdup $\varepsilon_g$ (volumetric fraction of gas in the gas-slurry mix); and bubble size $D_b$. Gas dispersion is affected by pulp characteristics and the operating conditions selected for bubble generation, including the type and concentration of the frother in use. Calculation of frother concentration from dosage is unreliable due to factors such as incomplete dissolution and unknown residual frother in recycle waters. Reagents with or containing contaminants with frothing properties (e.g. some xanthates), and commercial frothers that are blends of components with different frothing properties, make prediction of gas dispersion parameters even more difficult.

Sensors and devices to locally measure gas dispersion parameters have been developed (Gorain et al., 1995; 1996; Yianatos et al., 2001; Grau and Heiskanen, 2003). In addition to having compact, robust units for use in industrial installations an important requirement in our case was to have sensors that could be multiplexed and adaptable for on-line measurement (Gomez and Finch, 2002; 2007). These designs have been refined to a point that continuous (gas velocity and gas holdup) or relatively high frequency (bubble size) measurements are possible within the plant environment. A technique for measuring frother concentration in plant pulp samples on-site (Gélinas and Finch, 2005) has been incorporated to establish frother levels during gas dispersion characterization exercises. The experience gathered in many concentrators around the world has demonstrated that the gas dispersion measurement technology is a reliable tool to diagnose the operation of flotation cells and to direct optimization efforts. The technology – instruments, installation and data collection and analysis procedures are reviewed – but emphasis will be on illustrating applications, for example determining cell response to changes in gas velocity used for troubleshooting, testing equipment modifications, and optimizing metallurgical performance.

## MEASUREMENTS USED TO CHARACTERIZE GAS DISPERSION

### Gas Velocity

Gas (air) velocity is measured using a sensor based on the collection of bubbles from the pulp zone into a vertical tube partially immersed in the cell (Figure 1). The collected bubbles rise in the tube until they reach the liquid (pulp) surface and burst. When the tube is closed at the top the gas accumulates, pressure increases, and the pulp (and any froth layer) is pushed down the tube. Monitoring pressure in the tube ($p_1$ curve in Figure 1) tracks the rate of descent of the pulp level which is related to the velocity of gas entering the tube (i.e., the superficial gas velocity). The sensor is operated by manipulation of valve A, which is closed and opened to accumulate and release the gas, respectively. The operation has led to the description "on-off" gas velocity sensor; each closing and opening of the valve constitutes one cycle. Simultaneous and automatic measurement in many cells is accomplished using solenoid valves.

Calculation of the gas velocity from the rate of pressure change (Equation 2) requires the bulk density of the fluid (aerated pulp) which provides the backpressure to the gas accumulating in the tube. By including a second tube maintained permanently full of air (valve B always closed) then, knowing the immersion length difference between the two tubes ($H_{BD}$ in Figure 1), the bulk density can be calculated from the pressure difference measured after the first tube fills with gas (the second tube is referred to as the "bubbler" to distinguish from the original "sensor" tube). The region of interest for the bulk density measurement is that between the end of the sensor and the pulp/froth interface, therefore a shorter tube is used for the bubbler.

A continuous version of the sensor has been developed (Torrealba-Vargas et al., 2003). In this case the sensor includes a calibrated orifice after valve A, which is kept open. The rationale is that a steady state pressure will be reached when the rate of air entering the tube from the cell equals that leaving the tube through the orifice. Gas velocity is calculated from the volumetric flow of air obtained via the orifice calibration divided by the sensor tube cross-sectional area.

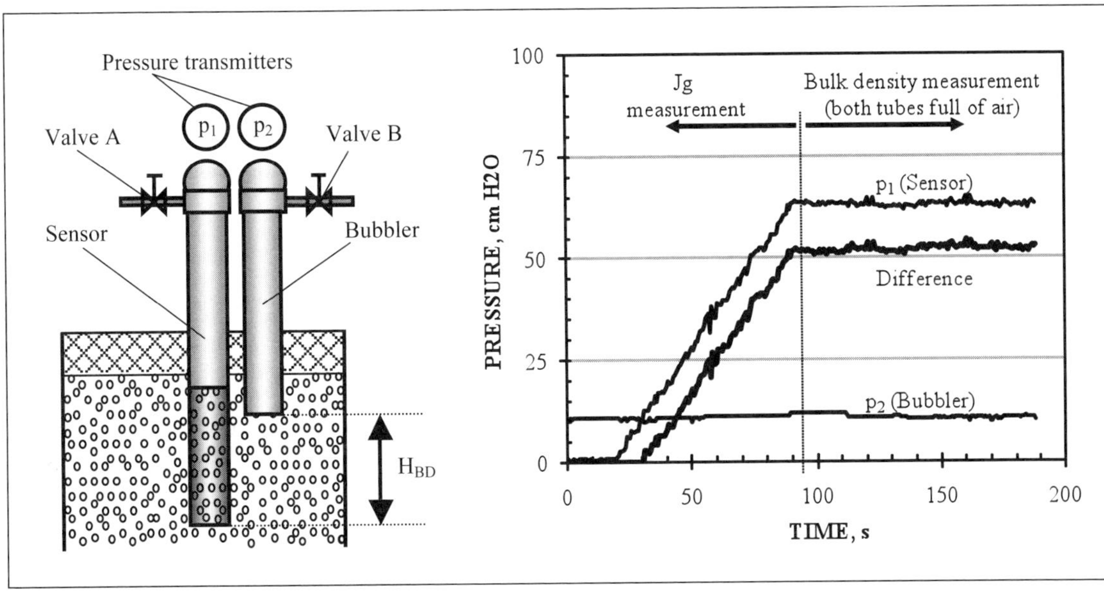


**Figure 1. Schematic of the gas velocity sensor and typical signals**

**Gas Holdup**

Gas holdup represents the volumetric proportion of gas in the slurry. Measurement is based on an equation relating the concentration of a non-conducting dispersed phase (i.e., gas bubbles in our case) to the conductivities of the air-free continuous phase (i.e., slurry) and the dispersion (Maxwell, 1897). A sensor with two conductivity cells designed to work with flowing streams (Tavera et al., 1996) is used (Figure 2), one that measures the conductivity of aerated slurry ("open" cell) while the second, by restricting entrance of bubbles, measures the conductivity of the air-free slurry ("siphon" cell). The conductivity cells have three concentric ring electrodes the two outer ones at the same polarity to constrain the sampling volume to that inside the cell between the rings, to comply with Maxwell's model.

The key for effective operation lies in excluding bubbles from the siphon cell. The conical base hinders entry of rising bubbles. As a consequence, the dispersion inside the cell develops a higher density than that exterior to the cell, generating a siphon action; i.e., a flow out through the bottom opening which completes the exclusion of bubbles with cell contents constantly replenished by pulp entering from the top. To ensure that no bubbles are entrained into the cell from the top, the pulp velocity through the cell must be lower than the terminal velocity of the smallest size bubble present in the system. This condition is met by selecting the bottom-opening size.

The ratio of the two conductivity signals is used to solve Maxwell's equation for the fraction of non-conducting phase (i.e., gas holdup) in the dispersion. This calculation can be done on-line (Figure 2) thus providing a continuous measurement. Fluctuations in pulp conductivity are not a problem since the technique uses the ratio and not absolute values (Gomez et al., 2003a); therefore, pulp conductivity variations affect the open and siphon cell readings in the same proportion.

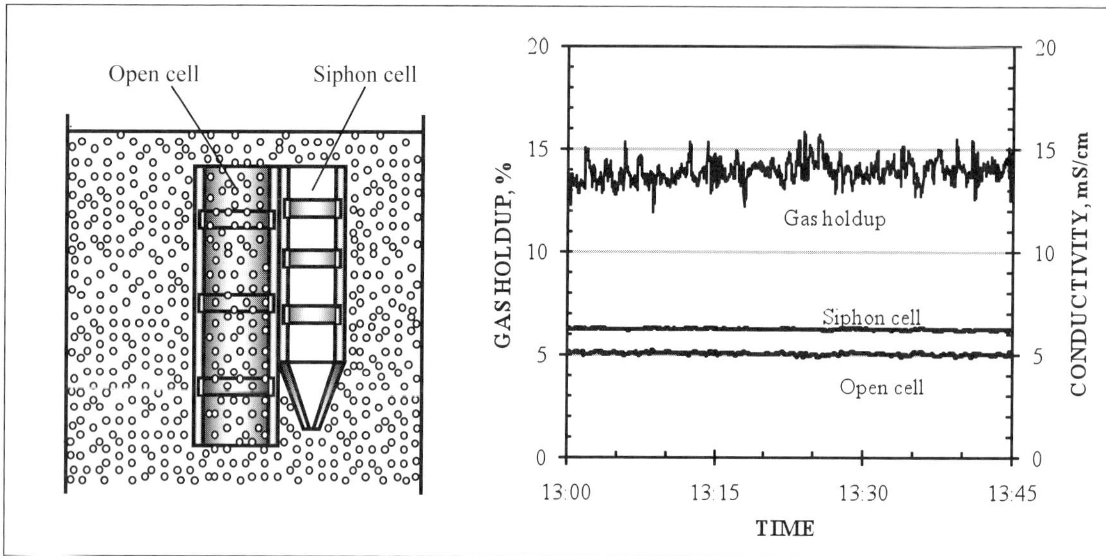


**Figure 2. Schematics of the gas hold-up sensor and typical signals**

**Bubble Size Distributions**

Bubble size ($D_b$) measurement is based on the collection of images which are subsequently processed to determine the size of individual bubbles. The images are collected using a unique device (bubble viewer) designed to sample, expose and image bubbles in the pulp zone (Figure 3). The unit includes an angled (15-degree) viewing chamber, a camera and a light source all mounted on a light-weight aluminum frame to facilitate installation and operation. Bubbles are collected via a sampling tube and directed into the viewing chamber where they are imaged under pre-set backlighting conditions. The chamber remains essentially free of slurry although particles released from bursting bubbles eventually darken the image.

The characteristic that distinguishes the technique is the quality of the images obtained by photographing bubbles against backlighting (i.e., the bubble shadow is imaged) as they slide up the sloped window. By angling, the window spreads the bubbles into a near monolayer, providing both an unambiguous plane of focus and reducing bubble overlap and 'ghost' images of bubbles in the background. The unit is initially filled with water through the top opening (normally using plant process water with frother to prevent coalescence) to provide the liquid column in which bubbles rise. As a consequence of its location above the flotation cell pulp level, the pressure inside the viewing chamber remains below atmospheric during operation.

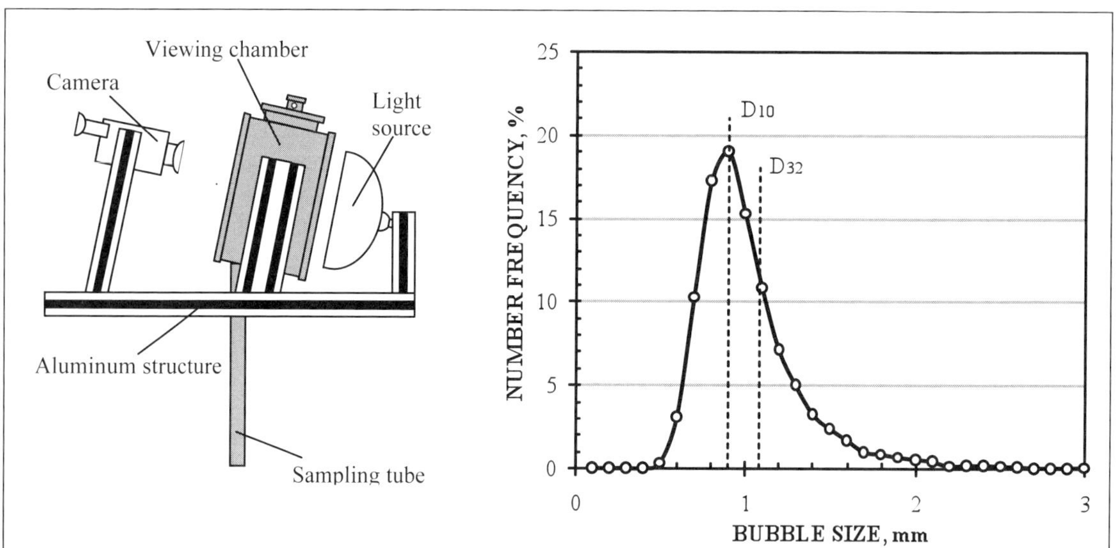


**Figure 3. Schematics of bubble viewer and typical results in the form of size distribution by number and average diameters $D_{10}$ and $D_{32}$**

A crucial aspect of the measurement is the selection of image magnification (pixels per millimeter), which is a combination of the number of pixels in the picture matrix and the area imaged. The smallest arrangement of neighboring pixels with an acceptable circular aspect ratio is taken: in the current version it is 7 pixels. A camera for example with a resolution of 3888 x 2592 pixels (10.1 mega pixels), when positioned to image an area 8 cm wide results in a magnification of 48.6 pixel/mm, and the minimum detectable bubble size will be 0.144 mm (144 μm). Images are automatically processed using a combination of commercial software and Excel applications developed in-house (Gomez et al., 2003b). Parameters to consider during image collection are number of images and collection frequency (between one and two hundred at a rate of one every two seconds to avoid imaging some of the same bubbles twice). Experience has demonstrated that the most effective way of operating the camera is combining a fixed shutter speed in auto-iris mode.

As several thousand bubbles are normally counted, the information generated in a single test is massive (we calculate eighteen properties of each bubble in an Excel file with as many rows as bubbles counted). To facilitate analysis, software was developed to display size distributions. The program retrieves the data and, using the equivalent spherical diameter (diameter of a circle with the same area as that exposed in the image), sorts the bubbles into frequency distributions with bin sizes defined by the user (at arithmetic or geometric intervals). Frequency and cumulative size distributions (Figure 3) are displayed, which can include average diameters such as the number mean and Sauter means ($D_{10}$ and $D_{32}$, respectively).

The accuracy of the measurement ultimately remains with the collection of a representative sample of the bubbles into the sampling tube. It is recognized that sampling is intrinsically compromised by the insertion of the tube if this perturbs existing flow patterns. Once the tube is in place, perfect sampling is only possible when the same hydrodynamic conditions exist around and within the tube, which is not the case (random pulp flow patterns outside vs. a downward flow of water inside). This aspect of the technique continues to be researched. Tests to establish reliability (precision) have demonstrated standard deviations below 1% for number mean and Sauter mean diameters. Establishing validity (accuracy) has been more difficult, as there is no standard alternative method ("gold" standard) against which to compare and no technique for continuously producing bubble swarms of known size distributions. In common with other disciplines where there is no gold standard we have applied techniques such as known groups and dose-response methods (Bailey et al., 2006).

**Frother Analysis**

Frother concentration is determined by a colorimetric method based on the Komarowsky reaction for the analysis of aliphatic alcohols (Coles and Tourney, 1942). The reaction involves the interaction of a frother (specifically the OH group), concentrated sulfuric acid, and an appropriate aldehyde. The color formation is due to the dehydration of the frother by the sulfuric acid, and subsequent reaction of the unsaturated hydrocarbon formed with the aldehyde. A calibration curve of absorbance (at a selected wavelength) vs. concentration is constructed.

The analysis involves four stages:

- Extraction of frother from the sample into chloroform in a separatory funnel with sodium chloride added to aid migration of frother from the aqueous phase, and to reduce formation of foam.
- Extraction of frother from the chloroform into concentrated sulfuric acid in a separatory funnel.
- Komarowsky reaction to produce colored products by adding aldehyde in acetic acid; the mixture is maintained in boiling water for 15 minutes and then immersed in an ice bath to stop the reaction.
- Measurement of absorbance in a UV-Vis spectrophotometer.

The technique has been refined to work with low solubility frothers and to increase the daily analysis rate to about 20 samples. The results of repeat analyses are reproducible and accurate; the relative standard deviation for analysis of pulp samples is about 1%.

## CALCULATION AND CORRECTION OF GAS DISPERSION MEASUREMENTS

Because the volume of a mass of gas depends on pressure and temperature, gas dispersion measurements will be affected by the location and geometry of the sensors. And, as the conditions existing at the measurement points are not the same for the three sensors (and are different from those prevailing in the flotation cell at the bubble collection point), corrections are necessary. In bubble size determination, for example, diameters are measured from images of bubbles in the bubble viewer which is under vacuum; bubbles are larger than their sizes in the cell. A similar situation occurs with the measurement of gas velocity where a volume of gas is accumulated under variable pressure at conditions different from those existing at the location where bubbles enter the tube. Corrections to common conditions are straightforward for bubble size and gas velocity by applying Boyle's law, given that air behaves as an ideal gas under flotation conditions. These corrections require careful documentation of the sensors installation to be effective.

Respecting the inevitable restrictions imposed by local situations, ideally sensors are operated through the same sampling port to measure the three gas dispersion properties at the same location. Similar results are obtained at the same radial distance (Dahlke et al, 2001), but the presence of baffles, direction of the impeller rotation and positions too close to the feed and tailings ports may lead to inconsistent values. In the case of depth, installation must be carefully selected considering that there will be only one position where the three gas dispersion parameters will be known (reference plane in Figure 4 at the depth of the middle ring of the gas-holdup open cell). Our current inability to correct the gas holdup to other locations makes this the reference depth. Integration with frother concentration or with metallurgical results requires that those samples be collected at the same time as the gas dispersion measurements.

To sample the same bubble population all sensors must collect from the same depth (Figure 4). The bubbler tube used to measure the bulk density is installed at about 0.25 m below the pulp/froth interface, with a recommended minimum immersed length difference ($H_{BD}$) of 0.5 m. Calculations and corrections of the measurements require some of the distances indicated in Figure 4:

- Total length of the $J_g$ sensor tube ($H_L$),
- Length of the sensor tube above the cell overflow ($H_0$),
- Immersed length difference between the sensor and bubbler tubes ($H_{BD}$),
- Distance between the middle ring and the bottom end of the open cell in the gas holdup sensor ($H_E$),
- Total length of the bubble viewer collection tube ($H_1$),
- Vertical distance between camera focal point and bottom of bubble viewer ($H_2$),
- Vertical distance between camera focal point and water level in the bubble viewer ($H_3$),
- Froth depth ($H_F$)

Distances H are measured in cm, while pressures p are manometric and P are absolutes, both in cm $H_2O$. There is a small variation (a few cm) in $H_3$ during measurement as air accumulates at the top of the bubble viewer; it is our practice to use the average of the distance before and after images are collected.

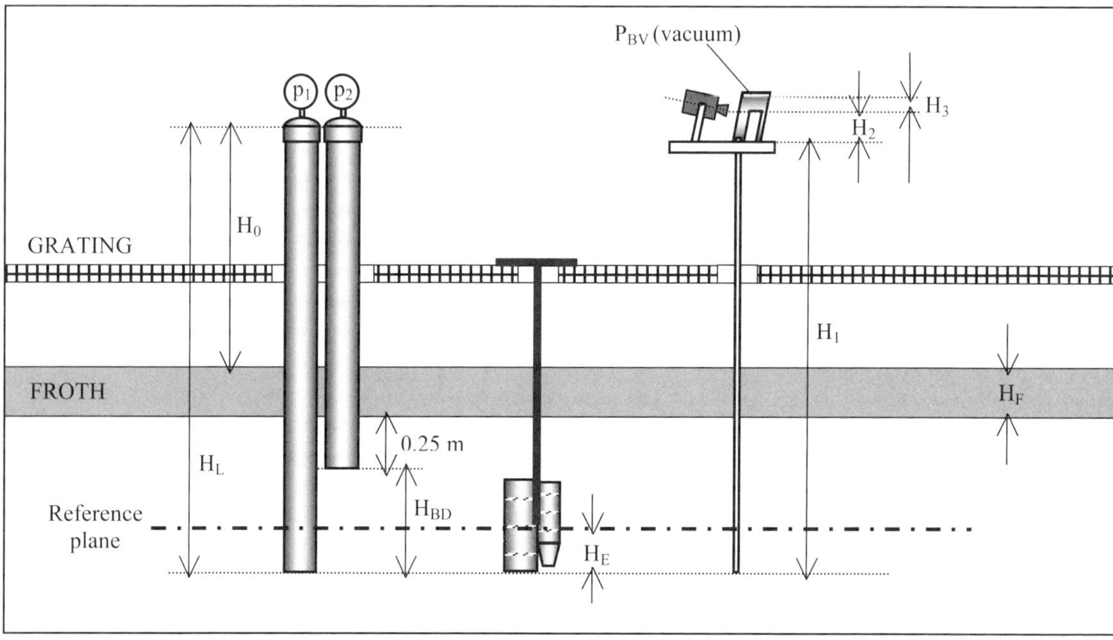


**Figure 4. Installation of sensors and relevant lengths and distances for calculation and correction of gas dispersion parameters**

The calculations performed are described below.

**Gas Velocity**

Operation of the double-tube sensor provides the data for calculating gas velocity. The pressure vs. time variation in both tubes is processed to determine the slope dp/dt (normally an average of five on-off cycles) and then to give the average of pressures $p_1$ and $p_2$ when both tubes are full of air. The bulk density (in g/cm$^3$) is calculated from the pressures $p_1$ and $p_2$, and the immersed length difference between the sensor and bubbler tubes ($H_{BD}$):

$$\rho_{bulk} = \frac{p_1 - p_2}{H_{BD}} \tag{1}$$

The gas velocity is calculated from dp/dt by (Gomez et al., 2003c):

$$J_{g,L} = \frac{P_{atm} + \rho_b H_L}{\rho_{bulk}[P_{atm} + \rho_{bulk}(H_L - H_O)]} \frac{dp}{dt} \tag{2}$$

This equation was obtained applying a mass balance to a tube volume filled with gas (a cylinder with variable volume and pressure), and using the ideal gas equation to relate mass to pressure and volume, and the weight of a column of fluid to relate pressure to volume.

The calculated gas velocity $J_{g.L}$ (in cm/s) corresponds to that entering the sensor tube at a depth ($H_L$–$H_0$). To correct the measured gas velocity to a different location (different hydrostatic pressure) Boyle's law is used. For example, to calculate at the reference plane ($J_{g,ref}$) the following equation is used:

$$J_{g,ref} = J_{g,L}\frac{P_L}{P_{ref}} = J_{g,L}\frac{P_{atm} + p_1}{P_{atm} + p_1 - \rho_{bulk} H_E} \tag{3}$$

Typically, gas velocities at the reference plane are about 3% larger than the values obtained from Equation 2.

**Gas Holdup**

Gas holdup is calculated utilizing the Maxwell model, using averages of the conductivity k over time (typically 5 minutes) for both the open ($k_d$) and siphon ($k_p$) cells, respectively:

$$\varepsilon_g = \frac{1 - \frac{k_d}{k_p}}{1 + 0.5\frac{k_d}{k_p}} \tag{4}$$

**Bubble Size**

Images are collected at the point where the focal plane of the camera meets the sloped window (a vertical distance $H_3$ from the gas-liquid interface in the bubble viewer); at this point the absolute pressure is below atmospheric. To correct bubble size to the cell reference plane, the pressure in the bubble viewer $P_{BV}$ needs to be measured, or in its absence, estimated from equalizing two hydrostatic pressures at the bubble collection point, namely that from the water column inside the bubble viewer and collection tube, and that from the pulp and froth column in the cell:

$$P_{BV} + \rho_{water}(H_1 + H_2 + H_3) = P_{atm} + p_1 \tag{5}$$

The pressure $p_1$ is known (it is measured to estimate the bulk density) and knowing $H_1$, $H_2$, and $H_3$ the pressure in the bubble viewer can be calculated:

$$P_{BV} = P_{atm} + p_1 - \rho_{water}(H_1 + H_2 + H_3) \tag{6}$$

Once $P_{BV}$ is known, measured bubble sizes $D_{b,image}$ can be corrected to reference conditions ($D_{b,ref}$), again using Boyle's law:

$$D_{b,ref} = D_{b,image}\sqrt[3]{\frac{P_{image}}{P_{ref}}} = D_{b,image}\sqrt[3]{\frac{P_{BV} + \rho_{water} H_3}{P_{atm} + p_1 - \rho_{bulk} H_E}} \tag{7}$$

Typically in plant conditions, bubble sizes at the reference plane are about 8% smaller than the values obtained directly from image analysis.

Although the volume of gases is also affected by temperature, no corrections for this are normally applied because temperature variations are usually not large enough to have a significant effect. If large temperature differences are detected between the gas in the pulp and that accumulated during the measurement of $J_g$ or $D_b$, then a temperature correction in Equations 2, 3 and 7 has to be considered. It is recommended that pulp temperatures be measured in case results from cells in different banks, circuits and plants need to be compared or related in the future.

## APPLICATIONS OF GAS DISPERSION MEASUREMENTS

**Cell Diagnosis and Troubleshooting**

As the sensors were under development, plant testing designed to validate working principles and establish measurement reliability found early use to diagnose and troubleshoot cell operation. Gas velocity measurements detected faulty air flow meters, worn dart valves affecting level control, and interactions between individual cells influencing gas rates (Dahlke et al., 2001; Torrealba-Vargas et al., 2003). Gas holdup measurements indicated differences between baffled sections in columns (Tavera et al., 1996) and unreliable gas delivery (Gomez et al., 2003b). Measurement of bubble size demonstrated that there were significant differences between cells down a bank (Chen et al., 2001). These successes gained the attention of operations and gave the technology credibility.

Once the sensors were established as reliable tools they started to be used for specific purposes. One of these applications was to test equipment modifications: for example, bubble size measurements were used to test several operating options of jetting spargers in flotation columns (Gomez et al., 2003a) and compare sparger options (Pyecha et al., 2005), and to establish whether a new rotor/impeller arrangement reduced bubble size (Gomez and Finch, 2002). Sometimes the signals can be put to other uses: for example slurry conductivity has been used to determine whether a new impeller improved particle suspension.

The results of two experiences are illustrated in Figure 5. Measurements of bubble size distribution as a function of air rate were made in the second cell of two parallel banks receiving the same (split) feed. One of these cells was modified by installation of a new rotor/impeller combination. A plot of the number mean and Sauter mean diameters as a function of gas flow rate (Figure 5a) showed, as expected, that bubble size increased with gas flow rate, but the results in the original and modified cells followed the same trend with no difference in bubble size. An advantage of the new configuration was, apparently, improved gas distribution in the cell which allowed operation to higher gas flow rates. In the second illustration, measurement of the slurry conductivity using the gas holdup siphon cell reading was used to compare particle suspension between the installed (original) cells and one modified with a new impeller design. Solids content in the pulp was determined from the slurry conductivity and the slurry water conductivity measured off-line (essentially the calculation for solids holdup is the same as for gas holdup (Gomez et al., 1998)). The results showed that the modified design was more effective in keeping particles uniformly distributed in the cell (Figure 5b).

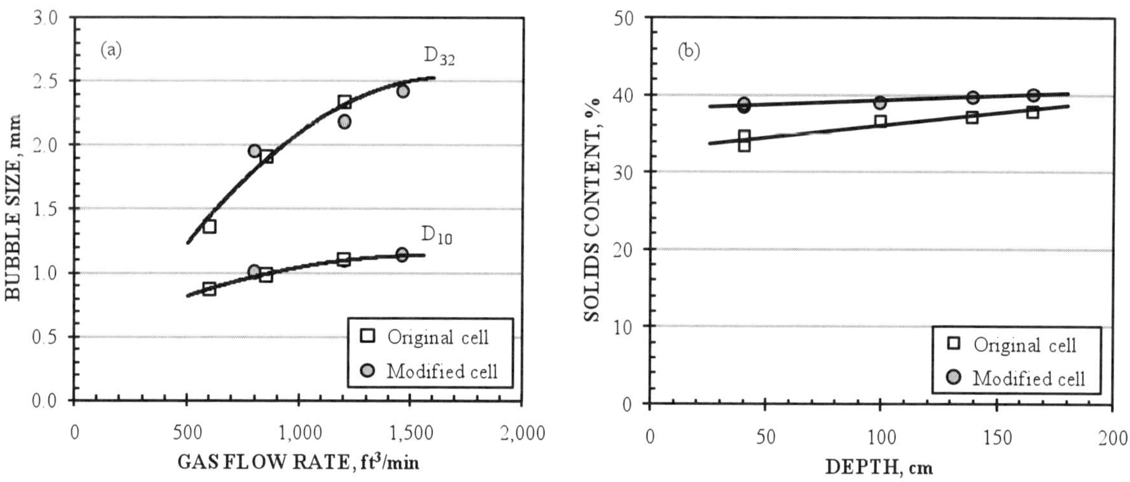


**Figure 5. Testing of equipment modifications using gas dispersion measurements**

**Cell Characterization**

Characterization of gas dispersion in flotation cells is based on the simultaneous use of the three sensors. As noted, installation of the units must be carefully planned and documented to have measurements at a common location in the cell. The purpose is to establish the response of the cell over its full operating range. Once the sensors are positioned the selected conditions are set and, when steady state is reached, the three gas dispersion parameters are recorded. The most important variable to study is probably gas flow rate (or level in the case of self-aspirating cells), but others would be solids content and frother concentration. In any event, frother analysis is necessary to ensure that gas dispersion results are not affected by a varying frother concentration.

There are different flow regimes for bubbles rising in a liquid (Finch and Dobby, 1990). When bubbles are within a relative narrow size range, which is the case of in most flotation machines, they rise at a uniform rate, in a regime called bubbly flow. This regime, which is considered the most effective for flotation, is characterized by a steady increase in gas holdup with increasing gas velocity up to a point where the unit can no longer disperse gas in this manner and the unit moves into a churn-turbulent flow regime and eventually to a flooding condition. Therefore, the gas holdup vs. gas velocity relationship helps to establish the cell operating range. The relationship reflects several aspects of operation including pulp characteristics and bubble loading.

Typical gas holdup vs. gas velocity curves are included in Figure 6 (Dahlke et al., 2005; Nesset et al, 2005). Measurements along a bank of cells (Figure 6a), where pulps with similar characteristics and frother concentrations are expected, showed similar curves with small differences that are probably a consequence of varying mechanical conditions (Cells 3 and 4 in the figure). The results for Cell 6 clearly indicate a major mechanical problem as there was practically no gas holdup response to gas flow rate. Results obtained in cells at different concentrators (Figure 6b) shows curves with different slopes. The different trends probably reflect bubble size differences (the cell at Concentrator 1 seems to generate smaller bubbles than the cell at Concentrator 3), but differences in bubble loading and frother type may also be affecting the results.

A second objective of cell characterization is to define cell response to variables that can be manipulated from the control room, normally gas flow rate. Gas dispersion results have shown, in general, linear dependence of gas velocity and bubble size on gas flow rate (Figure 7a).

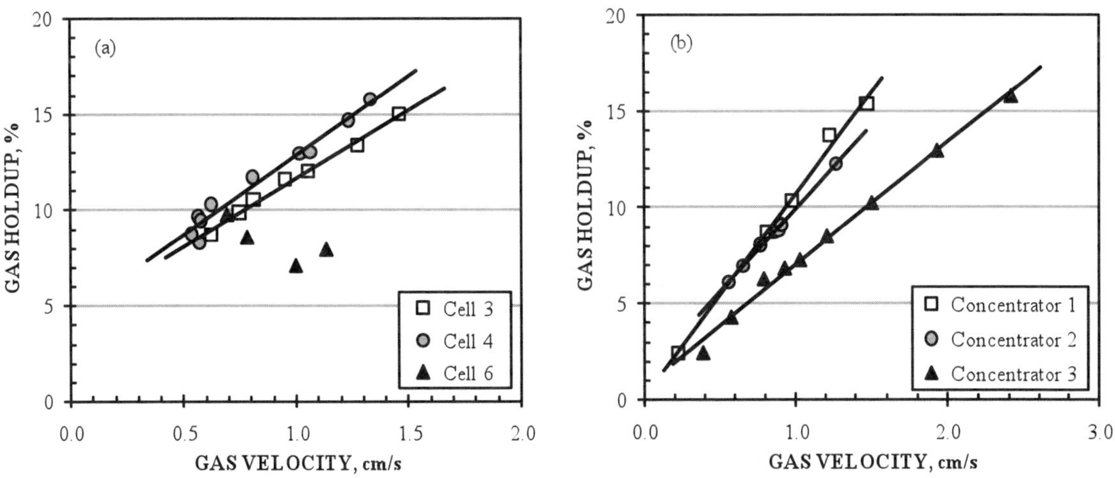


**Figure 6. Typical gas holdup vs. gas velocity curves for mechanical cells**

Once gas velocity and bubble size (Sauter mean diameter) are measured the bubble surface area flux ($S_b$) can be calculated from:

$$S_b = \frac{60 J_g}{D_{32}} \tag{8}$$

with $J_g$ in cm/s and $D_{32}$ in mm (both reported at the same conditions), giving $S_b$ in $s^{-1}$. Deriving $S_b$ from the data in Figure 7a gives Figure 7b which provides the range of bubble surface area flux achievable in the cell, and is important in deciding the target setting to operate at a particular $S_b$.

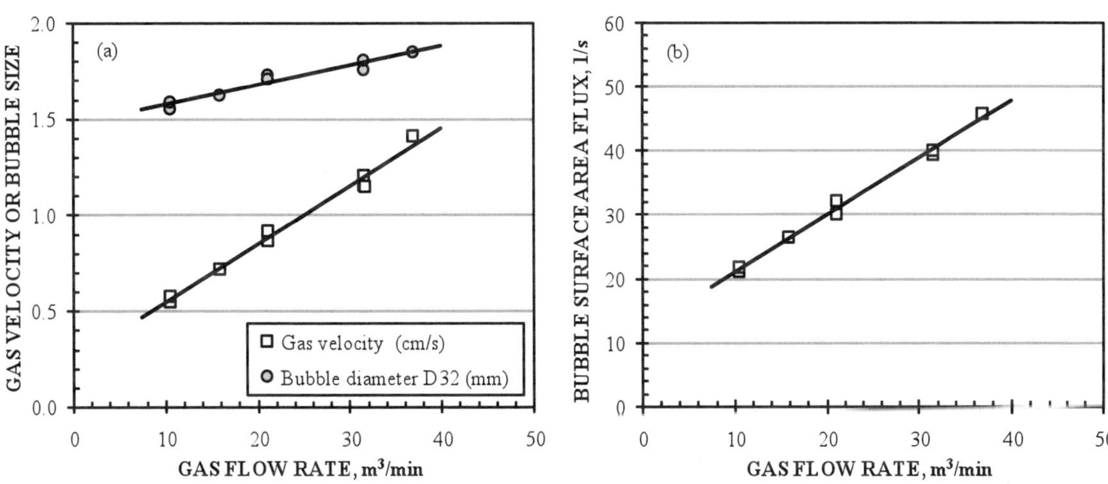


**Figure 7. Cell characterization curves**

**Benchmark Measurements**

We have found that the best approach to cell characterization and optimization work is to first run a benchmark survey. This consists of collecting "as-found" gas dispersion measurements for every cell in the circuit, including frother analysis. The data provide a snapshot of the plant through profiles of gas velocity, gas holdup, bubble size, bulk density, interface level, pulp conductivity and frother concentration.

Some results of a benchmark exercise performed in a circuit with three banks of six cells in a rougher/scavenger/scavenger configuration are presented in Figure 8 (gas velocity and bubble size profiles in Figures 8a and 8b, respectively). The gas delivery strategy at the plant was to have an even distribution down the bank, which was mostly achieved although in every bank there were one or two cells that deviated (cells 2 and 3 in the rougher bank, cell 4 in scavenger 1, and cell 1 in scavenger 2). Bubble size seemed fairly constant (with the exception of cell 2 of the rougher bank, as a consequence of the low gas flow rate delivered to that unit). A small but consistent bubble size difference between the two scavenger banks was noted.

In addition to identifying cells and banks which are running outside specification, benchmark measurements can be used to select units for cell characterization. Cell characterization measurements are time consuming, which restricts the number of cells that can be examined. Selection of representative cells to characterize is important.

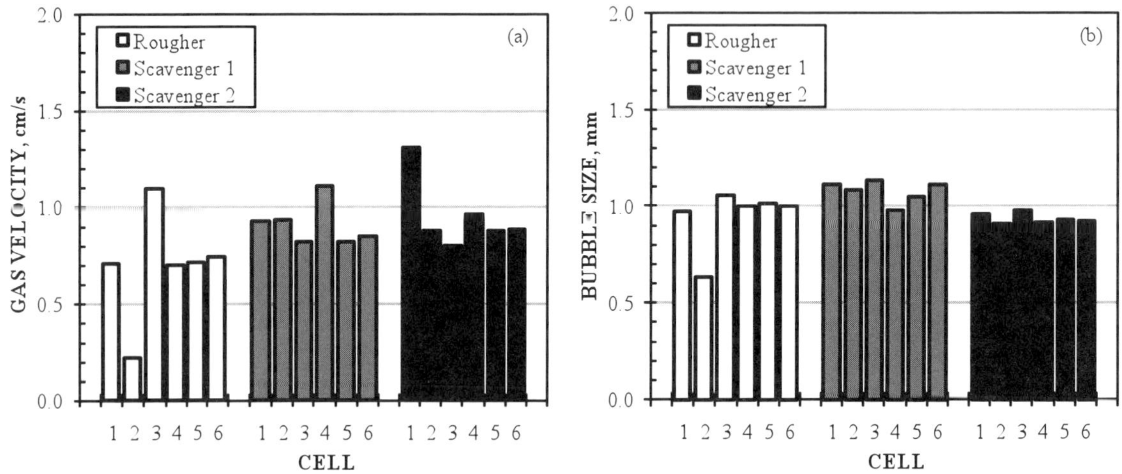


**Figure 8. Typical benchmark measurements results: gas velocity and bubble size**

**Gas Management**

The evidence from many operations around the world has demonstrated that there is rarely any obvious strategy for gas distribution down a bank; i.e., "as-found" gas distribution profiles have no targeted shape (Figure 9), rather they reflect the conditions selected by operators from visual (froth) clues. These profiles demonstrate wide differences between cells in the amount of air delivered which results in differences in bubble size, gas holdup and bubble surface area flux. It is reasonable to expect that different gas distribution profiles down a bank will result in different performance, and that one profile may prove to be the most effective.

Testing the effect of gas distribution profile has been facilitated by a multi-sensor unit developed by putting together electronics and software to drive a battery of identical gas velocity sensors (Doucet at al., 2006). The measurement procedure includes installation of units in every cell at the same radial distance and depth for continuous monitoring, with measurements of bulk density, gas holdup and bubble size once or twice during the survey (by rapidly moving the sensors from cell to cell), and collection of samples for metallurgical appraisal and frother analysis. Operation of the multi-sensor unit provides a permanent gas velocity record in each cell for the duration of the survey which helps to establish whether gas delivery was stable during the exercise.

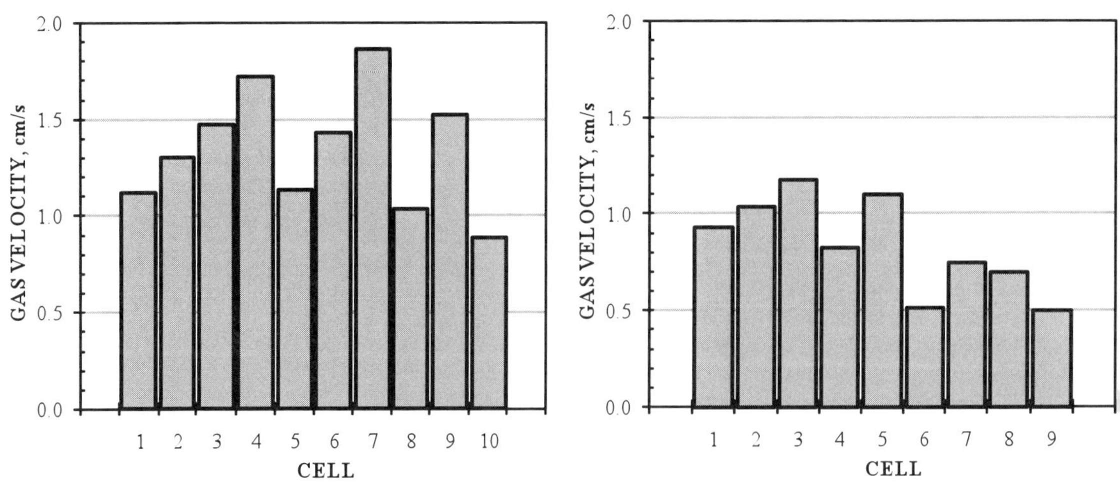


**Figure 9. Examples of "as-found" profiles in bank of cells at different operations**

The first application of gas velocity profiling was on the final Zn cleaner at Brunswick Mine (New Brunswick, Canada) (Cooper at al., 2004). The final Zn cleaner stage consists of two parallel rows of seven Denver DR100 flotation cells. While one of the rows was left "as-found" (i.e., air rates were adjusted by the operators) and used as a reference, the other was operated under a defined profile (the manipulated row). Three gas velocity profiles (increasing, balanced and decreasing for the same total air delivery (Figure 10a)) were tested and each maintained for a week. Under each condition, metallurgical surveys were completed with the down-the-bank grade/recovery established by sampling each cell. The results demonstrated that each profile resulted in a different performance. Figure 10b gives the range (best (closed symbol) to worst (open symbol)) in grade-recovery for each profile type. It was found that the increasing profile produced the best metallurgy, both on average as shown but also on each occasion that the manipulated and reference row were compared. Presently, Brunswick Mine operates all four stages in the Zn cleaning circuit with the increasing profile, which has contributed to an overall improvement in plant performance.

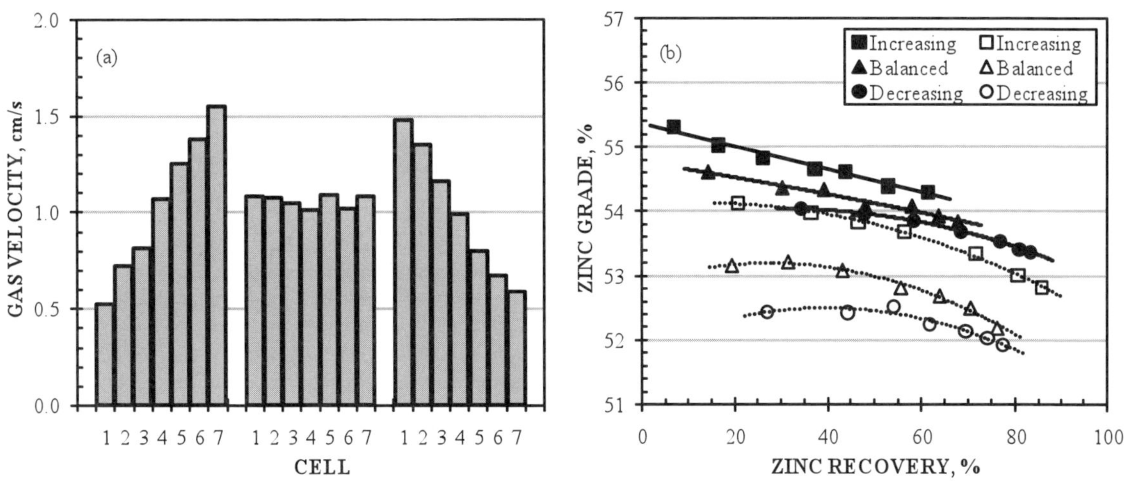


**Figure 10. Gas distribution management: gas profiles and metallurgical performances**

This successful outcome of managing gas distribution has been repeated in other operations (Gorain, 2005). Optimization of the gas distribution profile offers an opportunity for significant metallurgical improvements with no additional investment.

## CONCLUDING REMARKS

Sensors for measuring superficial gas velocity, gas holdup and bubble size distribution in industrial flotation machines are available. The units are robust for plant applications and have demonstrated the ability to collect information continuously or with a relatively high frequency. These sensors can be used to troubleshoot cell operation, to establish the benefit of equipment modifications, to characterize cell operating range and to optimize performance of banks and circuits.

Gas dispersion variables are affected by the pressure and temperature conditions existing at the bubble collection and measurement points. Sensor dimensions and installation geometry must be properly documented to allow for calculation at a reference condition. As no method is available to correct gas holdup, the location of the gas holdup sensor defines the reference plane. The standardization of sensor installation and correction procedures has made possible the effective transfer of the technology to operations. Units and training have been supplied to several groups around the world.

Gas distribution management offers an opportunity for significant metallurgical improvements with no or minimal capital investment. The addition of frother analysis is now included to complement the interpretation of gas dispersion data, and to interpret the impact of frothers on performance.

Our long-term goal continues to be the development of a relationship between gas dispersion measurements, notably bubble surface area flux, and metallurgical performance, which involves expanding our work into the froth zone.

## ACKNOWLEDGMENTS

Funding for sensor development is from two Natural Sciences and Engineering Research Council of Canada (NSERC) Collaborative Research and Development (CRD) grants, one currently sponsored by Vale Inco, Xstrata Process Support, Teck Resources, Shell Canada, Barrick Gold, Agnico-Eagle, COREM, SGS Lakefield Research, and Flottec, and the other by Amira International under the P9N and P9O projects.

## REFERENCES


Bailey, M., Gomez, C.O., and Finch, J.A. 2006. Bubble size analysis in an industrial paper deinking flotation cell. *Progress in Paper Recycling* 15(3): 12.

Chen, F., Gomez, C.O., and Finch, J.A. 2001. Technical note: Bubble size measurement in flotation machines. *Minerals Engineering* 18(4): 427.

Coles, H.W., and Tournay, W.E., 1942. The Komarowsky color reaction for aliphatic alcohols. *Ind. Eng. Chem. Anal. Ed.*, 14(1): 20.

Cooper, M., Scott, D., Dahlke, R., Gomez, C.O., and Finch, J.A. 2004. Impact of air distribution profile on banks in a Zn cleaning circuit. *CIM Bulletin* 97(1083): 761.

Dahlke, R., Scott, D., Leroux, D., Gomez, C.O., and Finch, J.A. 2001, Trouble shooting flotation cell operation using gas velocity measurements. In *Proceeding 33rd Annual Meeting of the Canadian Mineral Processors*, CIM, 359.

Dahlke, R., Gomez, C.O., and Finch, J.A. 2005. Operating range of a flotation cell determined from gas holdup vs. gas rate. *Minerals Engineering* 18: 977

Doucet, J.S., Gomez, C.O., and Finch, J.A. 2006. A first step towards gas dispersion management: Manipulation of gas distribution in flotation circuits. In *Proceedings 38th Annual Meeting of the Canadian Mineral Processors*, CIM, 423.

Finch, J.A., and Dobby, G.S. 1990. *Column Flotation*. Pergamon Press, Oxford.

Gélinas, S., and Finch, J.A. 2005. Colorimetric determination of common industrial frothers. *Minerals Engineering* 18: 263.

Gomez, C.O., Probst, A., Finch, J.A., Moores, N. 1998. Monitoring thickener operation using a conductivity probe. *Minerals and Metallurgical Processing* 15(4):9.

Gomez, C.O., and Finch, J.A. 2002. Gas dispersion measurements in flotation machines. *CIM Bulletin* 95(1066): 73.

Gomez, C.O., Cortes-Lopez, F., and Finch, J.A. 2003a. Industrial testing of a gas holdup sensor for flotation systems. *Minerals Engineering* 3: 1703.

Gomez, C.O., Hernandez, J.R., McSorley, G., Voigt, P., and Finch, J.A., 2003b. Plant experiences in the measurement and interpretation of bubble size distribution in flotation machines. In *Proceedings of Copper 2003*, Eds. C.O. Gomez and C. Barahona, vol. 3, Mineral Processing, 225.

Gomez, C.O., Torrealba-Vargas, J.A., Dahlke, R., and Finch, J.A., 2003c. Measurement of gas velocity in industrial flotation cells", in *Proceedings of the XXII International Mineral Processing Congress*, Eds. L. Lorenzen and D.J. Bradshaw, 1703.

Gomez, C.O., and Finch, J.A. 2007. Gas dispersion measurements in flotation cells. *International Journal of Mineral Processing* 84: 51.

Gorain, B. K. 2005. Optimization of flotation circuits with large flotation cells. In *Centenary of Flotation Symposium*, Ed. G.J. Jameson, 843.

Gorain, B.K., Franzidis, J.P. and Manlapig, E.V. 1995. Studies on impeller type, impeller speed and air flow rate in an industrial scale flotation cell. Part 3: Effect on gas holdup. *Minerals Engineering* 8(12): 1557.

Gorain, B.K., Franzidis, J.P. and Manlapig, E.V. 1996. Studies on impeller type, impeller speed and air flow rate in an industrial scale flotation cell. Part 3: Effect on superficial gas velocity. *Minerals Engineering* 9(6): 639.

Grau, R.A., and Heiskanen, K. 2003. Gas dispersion measurements in a flotation cell. *Minerals Engineering* 16: 1081.

Maxwell, J.J. 1982. *A Treatise of Electricity and Magnetism - 3rd edition*. Oxford University Press, London.

Nesset, J.E., Hernandez-Aguilar, J.R., Acuna, C., Gomez, C.O., and Finch, J.A. 2005. In *Centenary of Flotation Symposium*, Ed. G.J. Jameson, 243.

Pyecha, J., Sims, S., Lacouture, B., Hope, G., Stradling, A., 2005. Evaluation of a Microcel sparger in the Red Dog column flotation cells. In: CD of Proceedings of Centenary of Flotation Symposium, Brisbane, June 2–5, pp. 1017–1025.

Tavera, F.J., Gomez, C.O., and Finch, J.A. 1996. Novel gas holdup probe and application in flotation columns. *Trans. Instn Min. Metall. (Sec. C: Mineral Process. Extr. Metall.)* 105: C99.

Torrealba-Vargas, J.A., Gomez, C.O., and Finch, J.A. 2003. Continuous air rate measurement in flotation cells: a step towards gas distribution management. In *Proceedings of Copper 2003*, Eds. C.O. Gomez and C. Barahona, vol. 3, Mineral Processing, 91.

Yianatos, J., Bergh, L., Condori, P. and Aguilera, J. 2001. Hydrodynamic and metallurgical characterization of industrial flotation banks for control purposes. *Minerals Engineering* 14(9): 1033.

# TAKING MINERAL PROCESSING PLANT SIMULATION TO A NEW LEVEL—INCLUSION OF MINE PLANS AND FINANCIAL PERFORMANCE

John A. Herbst, William T. Pate, and Ming Song

Metso Mining and Construction Technology, 621 S. Sierra Madre-Suite 100, Colorado Springs, CO 80903, USA

## ABSTRACT

Mineral processing plant simulations have been in use in our industry for about twenty-five years. One of the first generalized simulators, ModSim (involving size reduction, classification and separation) was described by Professor R.P. King in 1983. Since that time, several improvements have been made in their ease of use (for example, improved interfaces) and functionality (for example, steady state and dynamic equipment design). Until very recently, little has been done to improve plant simulators' ability to predict performance resulting from planned changes in ores/ore blends coming from the mine and the analysis of the associated financial implications of such changes.

The current paper discusses efforts to bring plant simulators to a new level of usefulness for operations. The resulting production simulator, Metso ProSim, allows fully dynamic (involving time variant phenomenological model) simulations of plant performance for mine plans spanning several years and involving up to twenty ores to occur in a few days time. The simulator tracks plant tonnage, recovery, product grades, operating costs and revenue. Plant operation during simulation observes ore specific equipment limits on throughput and power draw.

An illustrative example of the use of Metso ProSim includes calibration of models, collection and input of cost and equipment limit information, and input of mine plan data. Finally, simulation of plant performance during execution of a one year mine plan is presented.

## INTRODUCTION

Professor R.P. King was a pioneer and visionary in mineral processing flowsheet simulation. Beginning in 1973 Dr. King published his first flowsheet simulator and changed the face of minerals processing plant modeling and simulation forever (King, 1973; King, 1983). Subsequently he and other investigators (King, 1998) continued to develop faster and more powerful process simulators, eventually creating a culture in which plant simulation has become an everyday tool for design and optimization decision making (Herbst, 2000). Notable examples of the latter type of simulators include JK SimMet (Richardson, 1990; Napier-Munn, et al, 1999), a steady state simulator which takes advantage of a significant grinding plant database and MinOOcad (Herbst and Pate, 1998, 2000, 2001; Herbst and Blust, 2000), an object oriented dynamic flowsheet simulator for mine to mill applications including default parameter sets for typical metallic and non-metallic ores.

Within the last few years the mining industry has begun to express a need for simulators which move beyond normal process simulation and into the world of production simulation. Such a need requires tools which allow process flowsheet performance assessment over multiple ore types and economic assessment to determine the value of future projects.

## METHODOLOGY

The starting point for developing the production simulator described here was Metso's existing flowsheet simulation package MinOOcad. MinOOcad is a dynamic simulator; it includes liberation and multi-component separation capabilities (albeit for a single ore type) and it allows the tracking of operating costs in all pieces of equipment, so it provides a good stepping off point for further development.

The first modification made to MinOOcad to turn it into a production simulator was to add multi–ore capability. Because of the object oriented design of MinOOcad (Herbst and Pate, 1998) it was possible to add a property page to the mine bank object (see Figure 1) which allows ore blend change events to be scheduled. It has been anticipated here that the mine may possess a large number of ore types (up to 20 included here) making up each blend and that the blends to the plant may be changed frequently (either on a regular or irregular basis) over a period of several years for evaluation purposes. Randomness within a mining location (e.g., variations within a bench area) was made possible here using a multi-variable Monte Carlo scheme. The properties of each of the ore types involved in the mine plan are captured in a parameter table which can include ore composition, density, crushability, grindability, liberation indices, flotability, abrasivity, etc.

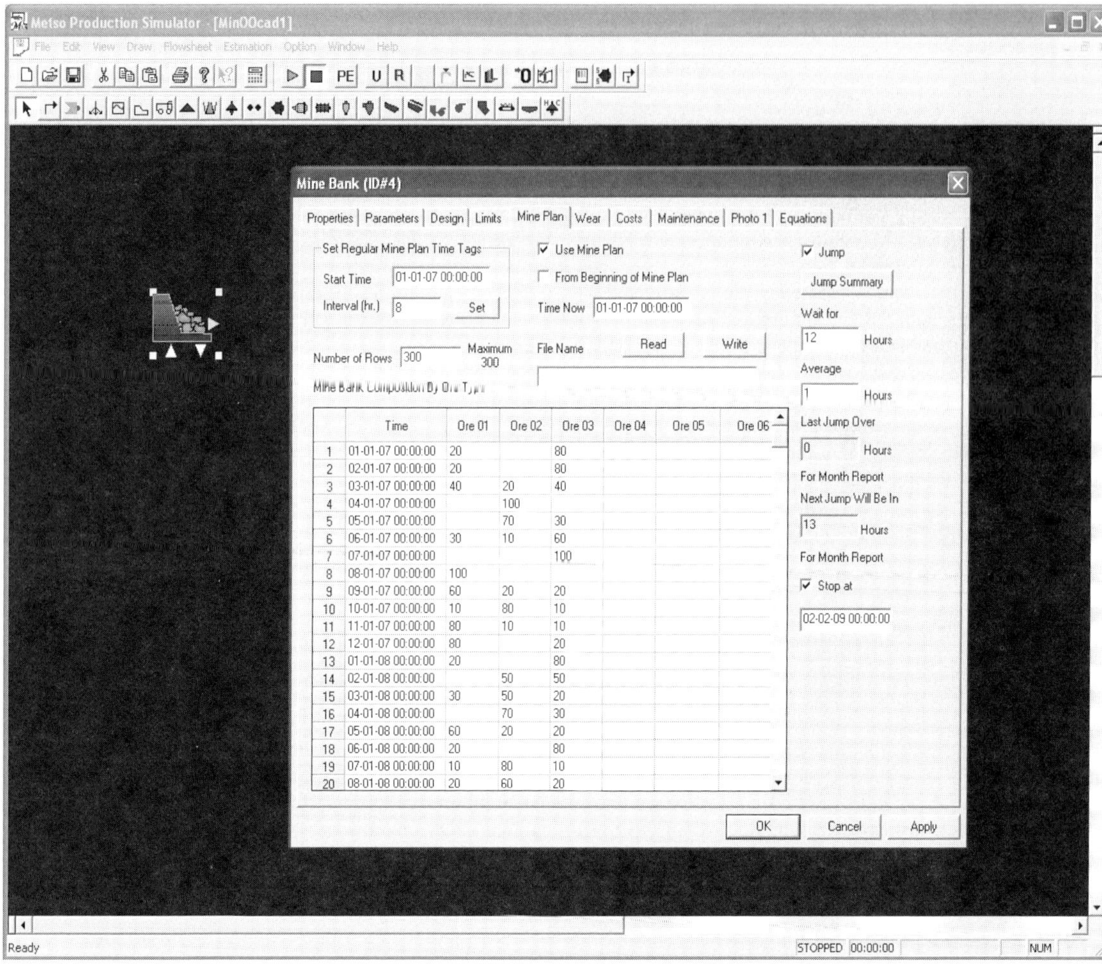

**Figure 1. Simulator mine bank object with mine plan property page opened.**

The challenge associated with adding multiple ore type capability is the increasing number of model equations and the size of the arrays required to describe their co-existence at any given time in each unit operation in any given flowsheet. This adds a huge computational overhead to the simulator. In addition, new auto-calibration features had to be added to allow large numbers of parameters to be estimated easily (see Figure 2). The calibration shown in Figure 2 involved the estimation of 5 crusher parameters and 2 screen parameters. Estimation time was about 2 minutes. The goodness of fit of models is based on the single measure of merit the square root of the sum of squares of error ( model minus measured) divided by the number of points predicted. The best way to think about this single measure of merit is that it tells us the average deviation between measured and predicted values. The root mean squared error for the resulting fit was 2.5% .

The result of adding these features is a new plant simulator front-end which allows the impact of realistic ore conditions to be evaluated in downstream unit operations throughout an entire enterprise.

Different ores and mixtures of ore types experience constraints in different parts of a plant (process or materials handling equipment) which limits the overall processing rate for any given feed. A production simulator must recognize these constraints or limits and make adjustments to the processing rate accordingly. In developing Metso ProSim a Limits Report Object was created with a property page as shown below in Figure 3. In this report the limits for each equipment item in the flowsheet can be examined. In addition the current constraining equipment is identified and the percent of time that each equipment item has been the constraint over a specified period of time is displayed. A basic expert system is available to increment or decrement processing rates to keep the plant "optimally close to" but not significantly exceeding constraints.

A maintenance object has been added to allow scheduled and unscheduled (Monte Carlo) events to occur. The expert system can be made to mimic plant response when downtime occurs. Basically these tools allow plant availability to be assessed.

Several economic analysis objects have been included in the new software. These include operating and capital costs by equipment and by section. Also, production and revenue tables have been introduced (see the production report in Figure 4). Unit costs and fixed costs for energy, wear, chemicals, maintenance and labor are entered as appropriate for

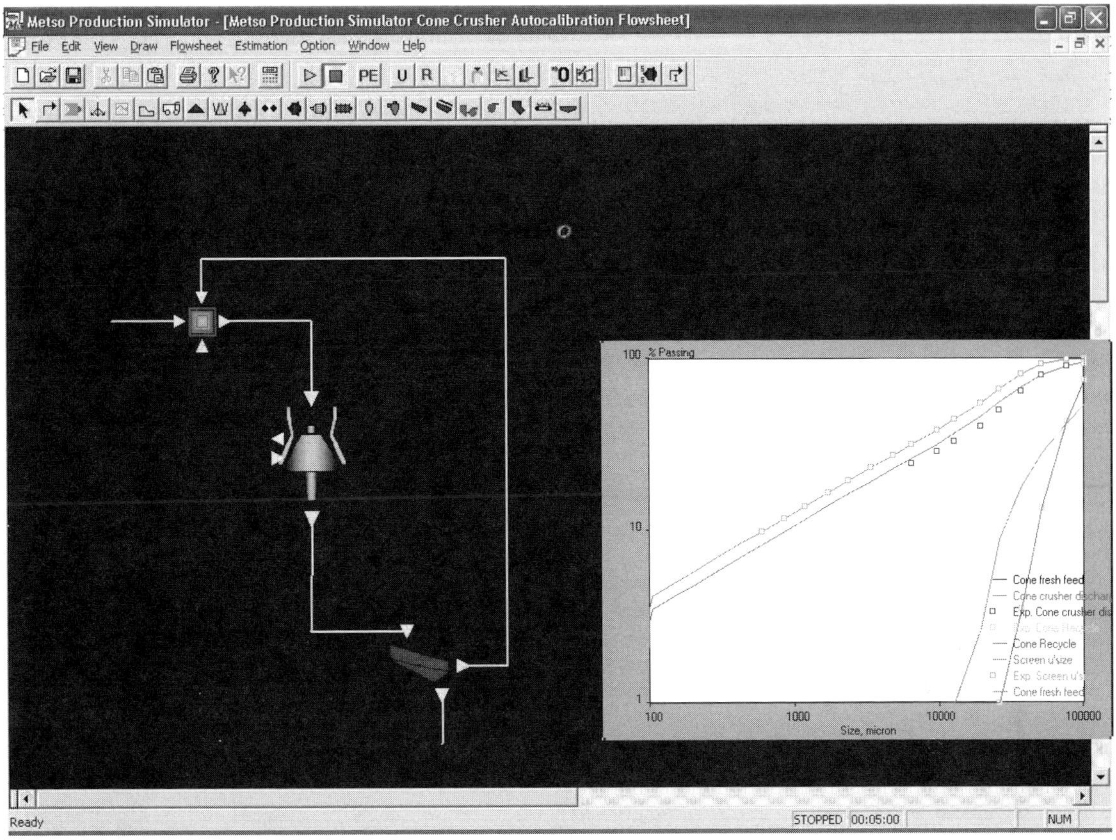


**Figure 2. Auto-calibration for crusher and screen calibration**

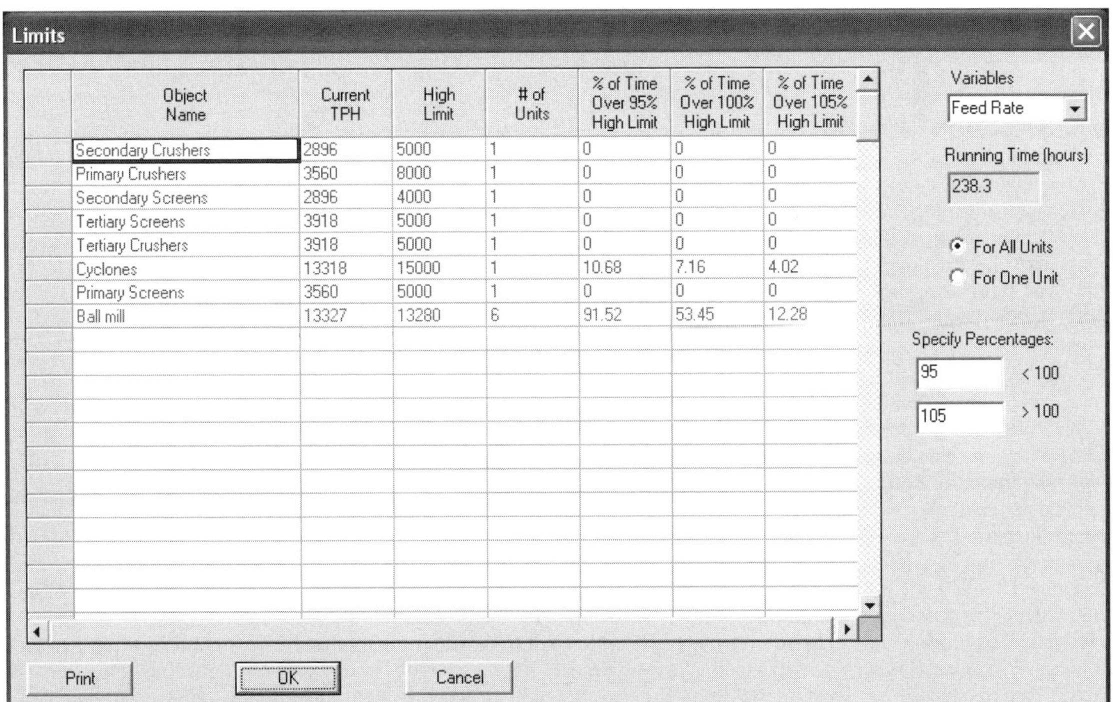


| Object Name | Current TPH | High Limit | # of Units | % of Time Over 95% High Limit | % of Time Over 100% High Limit | % of Time Over 105% High Limit |
|---|---|---|---|---|---|---|
| Secondary Crushers | 2896 | 5000 | 1 | 0 | 0 | 0 |
| Primary Crushers | 3560 | 8000 | 1 | 0 | 0 | 0 |
| Secondary Screens | 2896 | 4000 | 1 | 0 | 0 | 0 |
| Tertiary Screens | 3918 | 5000 | 1 | 0 | 0 | 0 |
| Tertiary Crushers | 3918 | 5000 | 1 | 0 | 0 | 0 |
| Cyclones | 13318 | 15000 | 1 | 10.68 | 7.16 | 4.02 |
| Primary Screens | 3560 | 5000 | 1 | 0 | 0 | 0 |
| Ball mill | 13327 | 13280 | 6 | 91.52 | 53.45 | 12.28 |

**Figure 3. Limits Report Object.**

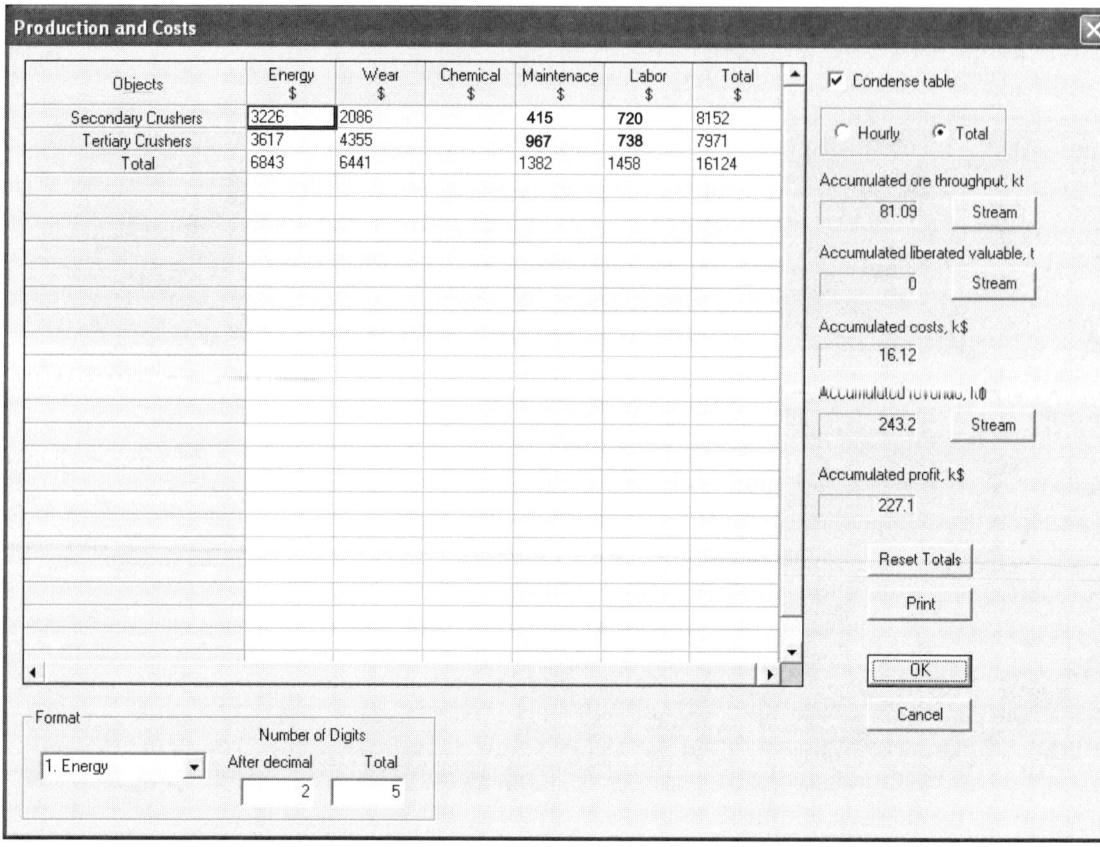


**Figure 4. Production and Costs Report Object.**

It took about twelve man-months to design these objects, code them and thereby create the Metso ProSim software. Another four to six man-months has been invested in testing and debugging. During this period a wide range of applications has been explored. These applications have involved flowsheet design, de-bottlenecking, equipment selection, operator and engineer training and financial evaluations of projects. In the next section two uses are illustrated.

## ILLUSTRATION OF USES

To illustrate some key uses of a production simulator a hypothetical mine-to-mill, (blasting, crushing and grinding) flowsheet was created within Metso ProSim (see Figure 5). Blasting parameters were chosen to yield a typical gyratory feed size distribution. Crushing, grinding and screening equipment dimensions and number of units were chosen to yield a nominal capacity of 85,000 mtph for an average copper ore. Each icon in Figure 5 represents a group of equipment items e.g., the secondary crusher icon represents five cone crushers with a nominal power draw of 300 kW and the ball mill icon represents six 6.91m x 7.92m mills. Typical copper porphyry parameters were used to populate each of the models, along with capital and operating costs.

### Example of Mine Plan Execution—Identification of Bottlenecks

One of the most valuable Metso ProSim features is the ability to simulate the entire mine-to-mill flowsheet involving a mine plan which extends many years into the future. Typically several ore types will be identified as existing in the mine. The mine plan describes the sequence that each ore type will be mined and blended with other ore types over time. The time frame for evaluation typically ranges from one to twenty years into the future. The hardness of each ore type ( defined here based on a Work Index or specific selection function[t/kWh]) must be known to simulate its performance in Metso ProSim. In an ideal world specific selection and breakage functions and other population balance model (PBM) parameters will be available. At a minimum Bond Work Index and/or Drop Weight test parameters for each ore type must be entered to allow the estimate of PBM parameters for objects such as the mine bank, crushers and grinding mills. Once this is done it is possible to simulate many ore types over many years.

In this example a mine plan is run to determine which part of the plant constitutes the main bottleneck for increased tonnage. The mine plan was chosen for illustration purposes to contain only 3 ore types—soft, medium and hard. The expert system adjusts plant feedrate to keep feedrate to every object below its high-high limit( a value determined in the plant or from equipment manufacturer specifications). This strategy will always run the plant at the high limit of one or more pieces of equipment. At the end of the simulation the limits table will show what percentage of the time each piece of equipment was at or above its limits. Identifying the bottleneck, making a change, identifying the new bottleneck is an

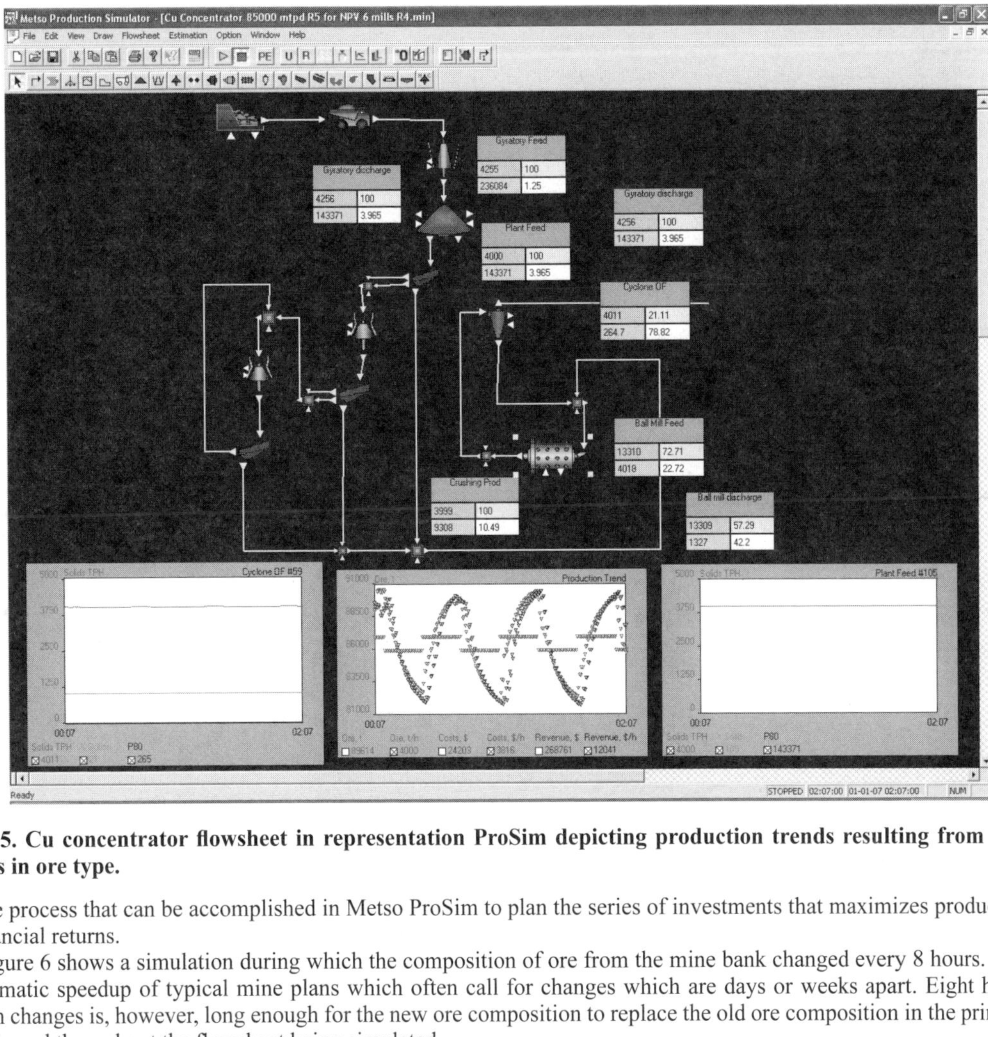


**Figure 5. Cu concentrator flowsheet in representation ProSim depicting production trends resulting from step changes in ore type.**

iterative process that can be accomplished in Metso ProSim to plan the series of investments that maximizes production and financial returns.

Figure 6 shows a simulation during which the composition of ore from the mine bank changed every 8 hours. This is a dramatic speedup of typical mine plans which often call for changes which are days or weeks apart. Eight hours between changes is, however, long enough for the new ore composition to replace the old ore composition in the primary stockpile and throughout the flowsheet being simulated.

Figure 7 shows the high feedrate limits for the major pieces of equipment during the simulation period. The ball mills are seen to operate close to their high limits most of the time. Because the expert system pushes the feedrate up until a high limit is experienced, 91.5% of the time the ball mills are operating at a tonnage which is at a value above 95% of their tonnage high limit, 53.5% of the time the ball mills exceed their high limit because the expert control strategy waits until the high limit is exceeded before it takes corrective action. Once the limit is exceeded the expert strategy reduces the crushing plant feedrate until the ball mill feedrate is below its high limit then starts raising the feedrate again. This strategy results in a limit cycle with the ball mill feedrate above its high limit half the time. Figure 7 shows that 12% of the time the ball mills are actually above 105% of their high feedrate limit.

This and subsequent simulations over longer times and involving a wider range of ore blends suggest that the ball mills are in fact the bottleneck to increasing tonnage to this plant.

### Example of NPV Calculation—Performance from Adding New Equipment

In the previous section, the ball mills were identified as the bottleneck to increasing throughput. In this section the same plant will be simulated with an additional ball mill of the same size as the other six bringing the plant total to seven. To determine if adding a 7th ball mill is a good financial investment, the NPV of adding another mill will be calculated.

Net Present Value is a common means to decide if an investment should be made or not. According to the Wikipedia article on NPV, "Using the NPV method a potential investment project should be undertaken if the present value of all cash inflows minus the present value of all cash outflows (which equals the net present value) is greater than zero." Herbst et al. (2003) show how NPV can be used to decide on the best liner profile for a SAG mill. DEM simulations of the SAG mill with several alternative profiles can be made to determine throughput and liner life. The design influences not only the liner life but also the throughput and power draw as the liner wears. The NPV can then be calculated from the throughput, power draw and liner life. The best design can then be chosen to display a balance between life and throughput. The same NPV concept can now be applied in a much broader investment context within Metso ProSim.

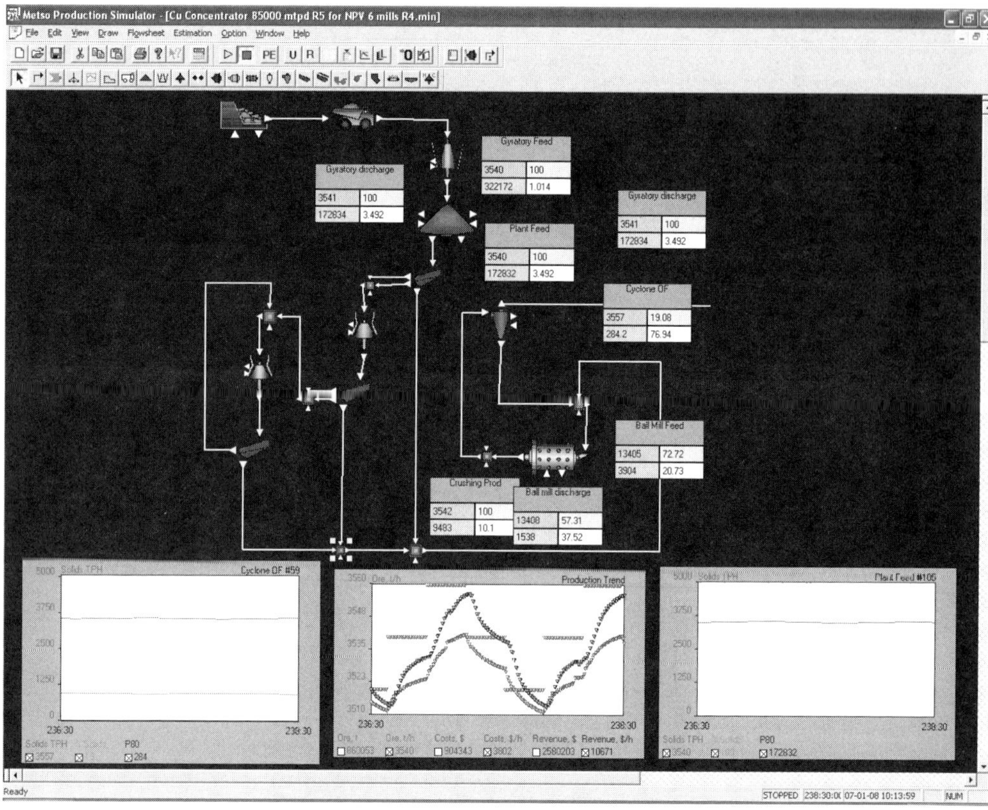


Figure 6. ProSim flowsheet of the copper concentrator showing production trend for the initial part of mine plan execution.

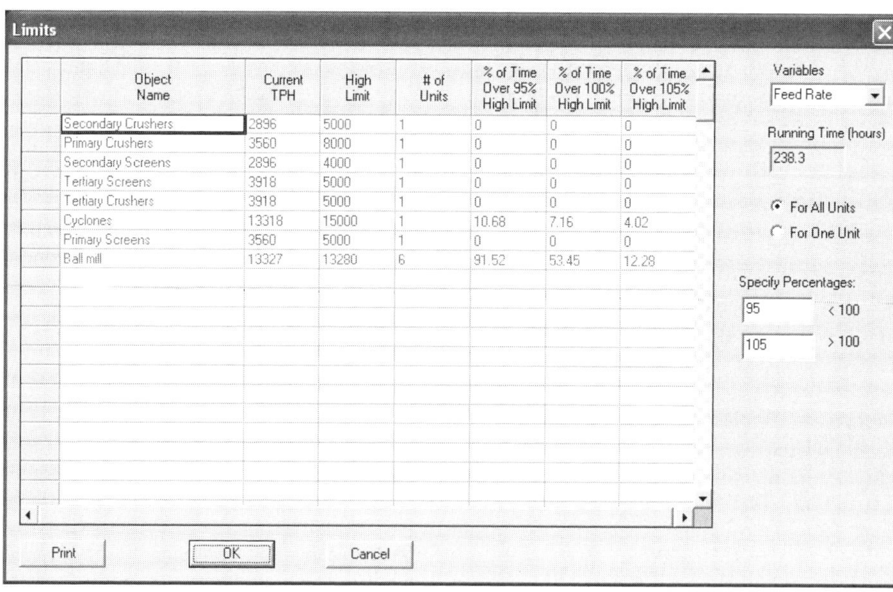


| Object Name | Current TPH | High Limit | # of Units | % of Time Over 95% High Limit | % of Time Over 100% High Limit | % of Time Over 105% High Limit |
|---|---|---|---|---|---|---|
| Secondary Crushers | 2896 | 5000 | 1 | 0 | 0 | 0 |
| Primary Crushers | 3560 | 8000 | 1 | 0 | 0 | 0 |
| Secondary Screens | 2896 | 4000 | 1 | 0 | 0 | 0 |
| Tertiary Screens | 3918 | 5000 | 1 | 0 | 0 | 0 |
| Tertiary Crushers | 3918 | 5000 | 1 | 0 | 0 | 0 |
| Cyclones | 13318 | 15000 | 1 | 10.68 | 7.16 | 4.02 |
| Primary Screens | 3560 | 5000 | 1 | 0 | 0 | 0 |
| Ball mill | 13327 | 13280 | 6 | 91.52 | 53.45 | 12.29 |



Figure 7. High feedrate limits for major pieces of equipment.

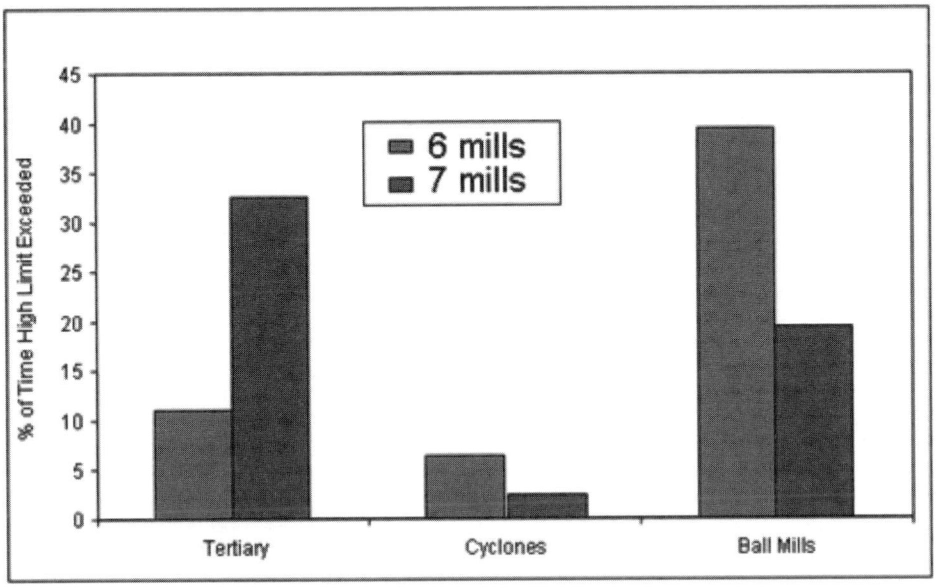


**Figure 8. Migration of constraints to crushing after adding another mill**

The formula for NPV is shown in Equation 1. In this equation, $j$ is the index for the $j^{th}$ day since the equipment investment, $N$ is the total number of days, $R_j$ and $C_j$ are the revenue and costs on the $j^{th}$ day, $i$ is the annual interest rate or corporate hurdle rate corresponding to the minimum rate of return acceptable to undertake a project.

$$NPV = \sum_{j=1}^{N} \frac{(R_j - C_j)}{(1 + i / 365)^j} - \text{Initial Investment} \tag{1}$$

In this example operating costs $C_j$ include crushing and grinding energy and wear parts consumption for liners and balls. The revenue $R_j$ used in the NPV calculation is the value of the copper contained in the liberated copper mineral for 87% recovery. Instead of using just the revenue from the added production of the additional ball mill, the revenue from all mills is used here because the crushing plant has to be able to process the additional tonnage of the 7th ball mill. If the bottleneck shifts from the ball mills to the crushing plant then the NPV with the 7th mill will be lowered.

Since the total plant revenue and costs are used the current NPV will almost always be positive, so a test for improvement must be made using the difference in NPV between alternative investment options. The first or base case NPV involves no additional investment, that is, the plant continues using the current 6 ball mills. The second NPV is calculated with 7 ball mills. In the second case the investment is the cost of a new 6.91 x 7.92m ball mill, plus additional hydrocyclone pack and installation. The value of this investment was determined to be $16 million.

The NPV calculation given here was made to start on January 1, 2007, assumed to be the first day of operation of the ball mill. The investment was made on that day although the NPV could start earlier, e.g., June 1, 2006, the day that the investment was made and the NPV calculated during the last half of 2006 with only 6 mills operating.

The Metso ProSim flowsheet for this plant was run 2 times using the mine plan for 2 years starting January 1, 2007. The first time was to calculate the NPV for 6 mills using 2 years of projected cash flow. This value discounted back to the present was $109,898,000. The second time was to calculate the NPV for 7 mills considering the additional cash flow over the 2 year period. The NPV for this case was $117,660,000. The difference to the share holders from investing in a 7th ball mill is $7,762,000 on January 1, 2007. In other words investing $16 million cash in a mill on January 1, 2007 and discounting the future cash flow over the next 2 years at an annual rate of 10% yields a positive NPV of about $8 million. Most of this benefit accrued from the fact that the seven mill plant yielded an average tonnage which was 10.5% higher than the six mill plant running exactly the same mine plan.

A revisiting of bottlenecks after adding a seventh mill (see Figure 8) shows that the mills are constraining a much smaller percentage of the time and the bottleneck is beginning to shift to the tertiary crushers. The production and financial impact of adding an extra tertiary crusher and or another ball mill could at this point be evaluated and the process continued until the desired plant performance is achieved.

A comment on Metso ProSim speed and accuracy of prediction seems to be appropriate at this point. From the outset of this software development project it was recognized that the simulation time for multi-year mine plan execution could be quite long. At normal MinOOcad speeds (200 times real time) the simulation of one year would require 43.8 hours. Metso ProSim includes smart rules that allow acceleration of calculations whenever possible. The two year mine plan run for this example took less than two hours. Accuracy of simulation predictions depends on many variables including the quality of input data, however, the experience of the authors has been that errors in plant tonnage predictions ( product quality) over extended times is typically between 2 and 3% in plant validation exercises.

## CONCLUSIONS

Driven by mining industry needs, a new production simulator has been developed. The starting point for this development was Metso's MinOOcad simulator. Multi-ore mine plan capabilities, process economics assessment and advanced reporting tools were added to the basic MinOOcad simulator. In addition a new auto-calibration feature has been added to simplify the task of model calibration from plant data. In several instances models have been upgraded to realistically represent process and materials handling performance in the final product, Metso ProSim. The upgraded models and accelerated calculations make it possible to achieve prediction errors on the order of 2 to 3% over several years of mine plan simulation at a speed of thousands of times real time.

This paper ends with two illustrations of the use of this production simulator for plant flowsheet optimization. The first illustration involves using the tool for equipment bottleneck analysis. A typical range of ore types allows the quantitative assessment of equipment which should be de-bottlenecked next in order to increase longer term performance. A second illustration shows how the Net Present Value feature included within ProSim allows the assessment of the current value of flowsheet improvements resulting from capital projects and other changes in order to expedite fiscal decision making about which projects and changes should be implemented.

It is apparent that mineral process plant simulation has in fact moved to a new level with regard to accuracy of prediction and value as a tool for plant decision making.

## REFERENCES

Herbst, J. A. and W. T. Pate, 1998, Dynamic Simulation of Size Reduction Operations from Mine to Mill, *Mine to Mill 1998*, Australasian Institute of Mining and Metallurgy, Brisbane, Oct 11-14, pp. 243-248.

Herbst, J. A. and S. L. Blust, 2000, Video Sampling for Mine-to-Mill Performance Evaluation: Model Calibration and Simulation, *Control 2000 Mineral and Metallurgical Processing*, Ed., J. A. Herbst, SME, Littleton, CO, pp. 157-166

Herbst, J. A. and W. T. Pate, 2000 Dynamic Flowsheet Simulation: A New Tool for Mine Through Mill Optimization, *Proceedings XXI International Mineral Processing Congress, Vol. A*, Rome, Jul 23-27, Ed., P. Massacci, Elsevier, Rome, pp. a3-67 - a3-73.

Herbst, J. A. and W. T. Pate, 2001, Dynamic Modeling and Simulation of SAG/AG Circuits with MinOOcad: Off Line and On Line Applications, *Proc. Intl. Conf. AG/SAG Grinding Technology, Vol. 4*, Vancouver Sep 30-Oct 3, Eds., D. J. Barratt, M. J. Allan and A. L. Mular, University of British Columbia, Vancouver, pp. 58-70.

King, R.P., 1973, "A Computer Program for the Simulation of the Performance of a Flotation Plant," N.I.M. Report 1436, Johannesburg.

King, R.P., 1983, "MODSIM, Modular Method for Design, Balancing and Simulation of Ore Dressing Plant Flowsheets," Report No. G9, Department of Metallurgy, University of Witwatersrand, Johannesburg, South Africa.

King, R.P., 1998, ModSim User Manual, Mineral Technologies, Inc., Salt Lake City, UT.

Napier-Munn, T. J., Morrell, S., Morrison, R.D., Kojovic, T., 2000, JK Mineral *Comminution Circuits: Their Operation and Optimisation*, JKMRC, Brisbane, Australia.

Richardson, J.M., 1990, Computer Simulation and Optimization of Mineral Processing Plants Three Case Studies, Control '90, SME, p. 233.

# IMPACTS OF QUANTITATIVE MINERAL CHARACTERIZATION ON PROCESSING

K. Hoal[1], J. Stammer[1], S. Appleby[1], M. Gregory[2], J. Woodhead[3], and J. Ross[1]

[1] Advanced Mineralogy Research Center, Colorado School of Mines, Golden, Colorado 80401 USA
[2] The Pebble Partnership, PO Box 267, Iliamna, Alaska 99606 USA
[3] 13 Glenn Place, Duncraig, 6023, Western Australia

## ABSTRACT

Quantitative mineral characterization is commonly applied to metallurgical operations, sometimes in the form of diagnostic process mineralogy applied to problem solving, and in other cases in a value-determined geometallurgical (geomet) context. The tools of quantitative mineralogy include automated mineralogical systems such as MLA and QEMSCAN, online analytical methods such as XRD and NIR, and image analysis by microscopy, microprobe analysis, or other techniques. Just as knowledge of critical metallurgical parameters is necessary to identify key process issues, a clear understanding of the mineralogy and locking characteristics of the ore also is required to meaningfully interpret the data. This is why a team-based, integrated approach to geomet is commonly recommended. The key mineralogical attributes that impact processing are: composition, mineral surface characteristics, grain size, texture, and paragenetic relationships. These features influence recovery and the overall value of the project. They are used to identify new ore bodies and mine expansion possibilities, and to characterize ore materials and mine waste during reclamation. Once quantified, assessed, and made digitally available, they are used to determine metallurgical parameters such as locking, liberation, grind size, hardness, breakage, flotability and leachability. They are mapable attributes that define the geomet domains and parameters of a deposit used in block models and mine design.

The impacts of quantitative mineralogy on mineral processing plant design are illustrated by examples from several ongoing geomet projects. The fundamental concern of geomet is to understand ore variability in order to predict and optimize how the ores will behave during processing. Because ore variability is best understood through mineral characterization and geological assessment, it is important to utilize these tools from the start of a project. Examples of concept- and prefeasibility-stage domain definition show alteration and sulfide mineral distribution are important parameters to quantify at an early stage. Some of the impacts of mineral characterization and ore variability quantification on comminution and flotation circuits include the distribution of pyroxene-rich domains in hard kimberlites, potassic alteration and microfracture distribution in porphyry copper ores, Au-bearing arsenian pyrite rims on pyrite in gold ores, and copper activation and rimming of pyrite surfaces in porphyry operations. In addition, the ability to characterize tailing material has an impact not only on reconciliation of operations but also for helping to predict environmental risks.

## QUANTITATIVE MINERAL CHARACTERIZATION

Quantitative mineral characterization is the instrument-based method of high-speed analysis of minerals through quantitative techniques. In particular, we refer to the scanning electron microscopy (SEM)-based systems developed out of the CSIRO in the 1980s (Gottlieb et al. 2000) and which evolved into the Mineral Liberation Analyzer and QEMSCAN analytical instruments. Both systems were developed for the analysis of sulfide minerals and thus utilize an accelerating voltage of 25 keV and the collection of backscatter electrons and x-rays. These instruments and their software applications are now being further developed by FEI, and a new generation of systems is anticipated in the near future. For the examples discussed in this paper, the instrument used was the QEMSCAN E430 located in the Advanced Mineralogy Research Center at Colorado School of Mines, which utilizes a Carl Zeiss EVO50 SEM platform, four Bruker energy dispersive (EDS) detectors, and the iMeasure/iDiscover software for acquiring and quantifying analytical data. The system is capable of 150 mineral determinations/second, and thus is particularly useful for large sample sets and small-sized materials down to about 1 micron. The system relies on 1000-count x-ray spectra over user-defined measurement intervals, and lists of mineral definitions which can be refined for particular projects and material types. In the following studies, the operating conditions, mineral lists, and software applications were modified according to the requirements of the individual projects. The number of installations and users of MLA and QEMSCAN systems have greatly increased in the past two years, such that there are now at least 13 instruments available for public use in North America compared to only one just two years ago. This expansion is due in part to the economic benefits of these systems being able to handle larger, more statistically representative data sets than traditional methods such as optical microscopy. A range of analytical methods is available depending on material type and project goals, and includes field images, particle mineral analysis, trace mineral and specific mineral searches, and bulk mineral analysis. Modal mineralogy is quantified, mineral parageneses and textures are identified, and specific parameters of interest such as mineral association, distribution, grain size, locking and liberation are readily determined. There are many potential applications of the method in a number of industrial and research areas (Hoal et al. 2009). The pitfalls of the method are related to the user: the data are only as good as the mineralogist or geologist performing the evaluation. Because of challenges such as x-ray peak overlaps, boundary phase relationships, the first match rather than best fit allocation, and the need for careful evaluation of mineral list definitions, the method requires mineralogical oversight and evaluation to ensure the results are reliable.

## GEOMET

Geometallurgy (geomet) refers to the interdisciplinary approach of integrating geology, metallurgy and mining in order to understand ore variability and to optimize project value (eg Dobby et al 2004). It is not a new idea: at the turn of the century, the discipline of economic geology developed in part in response to the mining engineering community's desire to better understand the rocks they were processing (Irving 1906). McQuiston and Bechaud (1968) first referenced the term geometallurgy in a textbook of open-pit mining, citing the need for understanding the geology before mine development. Up until the 1970s, many mining personnel were indeed cross trained, with metallurgy students required to take mineralogy and geology courses, and geology students taking classes in ore dressing and metallurgy. However, during a recent course in geomet at Colorado School of Mines, we found students were so unaware of the interdisciplinary impact of their work the first few weeks had to be devoted to establishing a common basis of understanding. Today, the need is for increased communication, efficiencies, and improving value of projects at a time of high costs and low personnel numbers.

The key component in any geomet study is the derivation and statistical application of large data sets. Assays and metallurgical test results are typically quantitative in nature, and they are readily amenable to statistical evaluation and modeling. The geological component has been the most difficult input to quantify, however, and to apply to block models. Nevertheless, the fundamental basis of geomet is geological in nature (Hoal 2008). All geological materials can be reduced quantitatively to two components; their mineralogy and their fabric, or texture. Mineralogy can be quantified in terms of modal abundances, and texture is quantified through a number of parameters such as grain size variability, distribution of fractures and microfractures, preferential alignment of minerals, and locking and liberation characteristics of ore minerals. We are only beginning to develop the textural quantification measures for process parameters that relate to breakage, fragmentation, flotation response, and leachability.

In the end, from exploration throughout the project life to closure and reclamation, geologists must provide improved data sets to engineers for mineral process plant design and development to better understand the variability of the deposit, and to better predict the metallurgical response of the materials. This is best accomplished through geomet evaluations and using information gained through SEM-based quantitative mineralogy.

## IMPACTS ON PROCESSING

Ore variability is a function of the mineralogy and fabric of the ore. It can be expressed as variation in hardness or breakage characteristics over time, or variation in flotation recovery over time. Understanding ore variability is therefore necessary for predicting the behavior of ores in an operational context. The following studies illustrate the impact of quantifying mineralogy and fabric on mineral processing using examples related to materials characterization and domain definition, comminution, flotation, and environmental assessment.

### Characterization and Domain Definition

Quantitative mineral characterization and geomet analysis of a project begin at an early stage, from the beginning of deposit evaluation and collection of samples and data. During mapping and drilling, larges amounts of geological information are generated to define the deposit and to identify the material attributes that eventually may be incorporated into the process flowsheet. Key among these attributes are lithology, distribution of ore minerals, and alteration mineral assemblage. In the Zambian copper belt, for example, Woodhead and Hitzman (2008) identified sedimentological structures and a series of alteration mechanisms in copperbelt shales that revealed not only depositional differences between regions of the belt but also had implications for processing. In particular, the Tenke deposit shows finely laminated, wispy distribution of chalcopyrite in clay-rich shales and carrolite blebs in dolomite, whereas the Lupoto ore materials have a more blocky and brecciated aspect (Figure 1). When quantified, these mineralogical and textural observations become numeric process parameters, e.g., distribution of clay minerals and fine-grained sulfides that are the initial geological data inputs for process plant design.

Geomet-focused core logging requires geologists to be cognizant of the critical parameters that may impact flowsheet design. These parameters may include pyrite:chalcopyrite ratios, degree of silicifation of ores, intensity of sericitic alteration, or distribution of shear zones, for example. Before the widespread use of SEM-based technologies, early geomet studies used geochemical data sets to back-calculate a theoretical modal mineralogy to begin to populate models and define domains. Many examples exist in which a lot of money was spent on testing before the core was examined for mineralogical and textural criteria that impacted grinding, flotation, or leach behavior. In the future, interval sampling of core for quantitative mineralogical analysis will provide that information up front and with greater statistical reliability for identification of geomet parameters and domains in a deposit. Gregory and Lang (2009) applied mineralogical and textural analysis to the definition of geomet domains at the large Cu-Au-Mo porphyry at Pebble, Alaska. The domains were identified on the basis of quantitatively evaluating the effects of two key parameters, alteration mineralogy and sulfide distribution, through the deposit. Three domains are characterized by potassic, pyrophyllitic, and sericitic alteration, which controlled the nature of copper and gold mineralization and redistribution (Figure 2). Identification of the relative proportions and spatial distribution of these geomet domains provides the metallurgical team with a solid framework from which to conduct representative sampling and testing for the process flowsheet and plant design.

### Comminution

For comminution, the distribution of geological elements such as lithology, talc, and fractures, for example, can have a significant impact on breakage, both in blast hole planning as well as during milling and plant operations. Generally, the pertinent parameters that will impact comminution will vary depending on ore type and deposit. From our experience, however, the key factors are most commonly related to alteration mineralogy, and fracture distribution, in particular

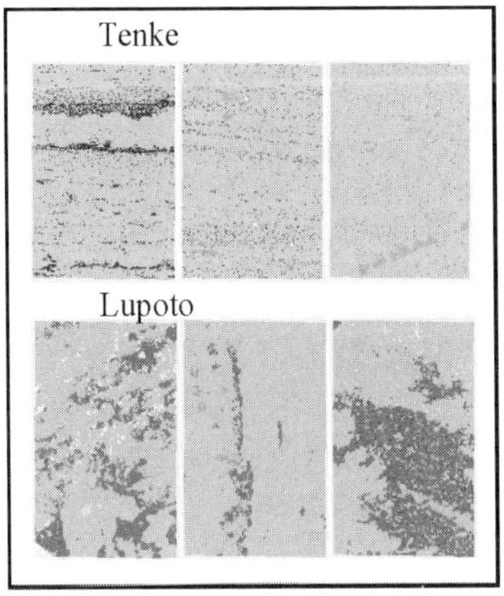


**Figure 1. Sulfide distribution in thin sections of Tenke ore (above) and Lupoto ore (below), Zambian copper belt, showing differences in rock fabric and sulfide grain size that influence sulfide domain definition and extraction. Brown: bornite, yellow: carrolite, orange: chalcopyrite, mustard (lower center image): pyrite (Hitzman and Woodhead, 2008; figures by J. Woodhead).**

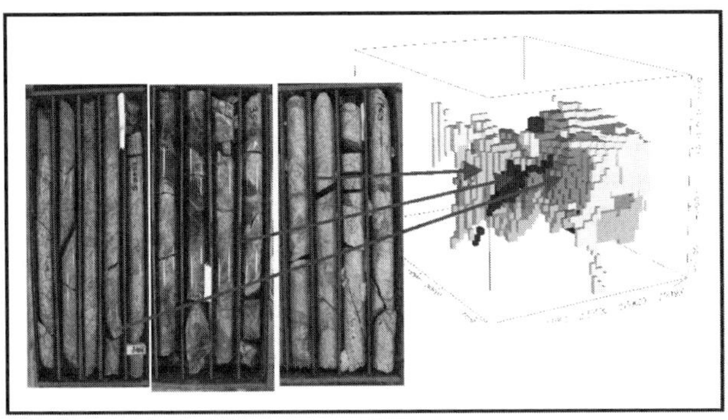

**Figure 2. Example of creating geomet domains on the basis of alteration mineralogy and sulfide deportment data from drill core samples in a Cu-Au-Mo porphyry (Gregory and Lang, 2009; figures by M. Gregory).**

microfractures. Figure 3 shows an example of an ore for which crushing poses a particular challenge. Parts of the material are hard relative to other domains in the deposit, but also hold higher value than the other domains. This is a kimberlite, with localized pyroxene in domains from which diamonds greater than 100 carats in size have been extracted. The mill must consider the optimum grinding protocols to maximize stone size extraction in the hard material. A first step in this analysis is understanding the nature and distribution of the hard material, in order to best approach the processing of it.

For metal ores, description of the ore material in terms of rock hardness, breakage characteristics, and fragmentation enables the mineralogical data to be better integrated into models for crushing and grinding. Because samples from core and blast hole materials are quantifiable in terms of mineralogy and texture, a classification scheme can be developed for these parameters over the three-dimensional extent of a deposit and over the anticipated life of the mine. Such a model should incorporate gangue mineralogy, alteration mineralogy, sulfide mineralogy, liberation and locking statistics for sulfides, degree and nature of alteration, 20- to 40-element geochemical data for assay reconciliation with mineralogy, and the textural parameters of grain size distribution, fabric orientation and extent, degree of shearing, microfractures, mesofractures, phenocrysts and megacrysts. Ross et al. (2009) demonstrated for a porphyry copper deposit that the two primary mineralogically controlled drivers for validating SMC breakage tests are the alteration mineralogy and the microfracture distribution (Figure 4). In an ongoing study, four geomet domains are being studied for breakage prediction, for which potassic alteration appears to have the greatest impact on hardness of the ore, whereas microfracture distribution appears to have the greatest impact on fragmentation of the ore.

**Flotation**

As an example of the impact of mineralogy on flotation characteristics, Stammer et al., (2008) studied the distribution of pyrite, arsenian pyrite and arsenopyrite in Carlin ores from the Chukar deposit, Nevada. Arsenic-rich rims on pyrite were quantified by QEMSCAN methods as a proxy for nano-scale gold mineralization. The results are illustrated in Figure 5. The distinction between arsenian pyrite, pyrite and arsenopyrite in these fine-grained ores using

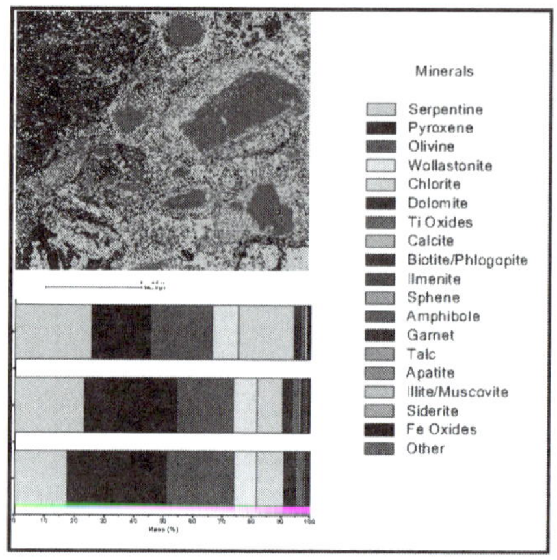


**Figure 3. False-colored image, modal mineralogy and mineral list for kimberlite with challenging grind requirements (after Hoal et al, in press).**

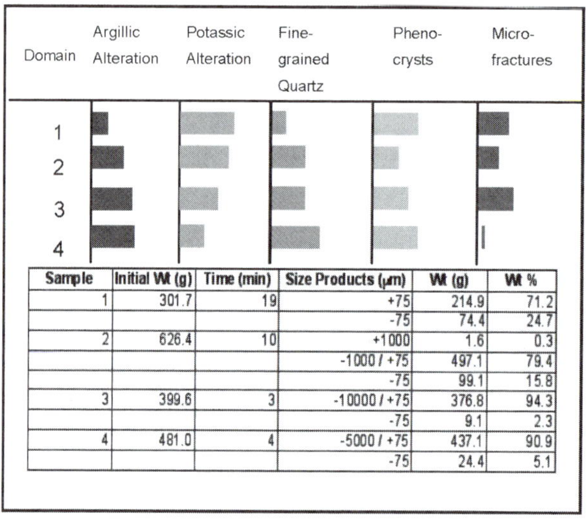


| Sample | Initial Wt (g) | Time (min) | Size Products (μm) | Wt (g) | Wt % |
|---|---|---|---|---|---|
| 1 | 301.7 | 19 | +75 | 214.9 | 71.2 |
| | | | -75 | 74.4 | 24.7 |
| 2 | 626.4 | 10 | +1000 | 1.6 | 0.3 |
| | | | -1000 / +75 | 497.1 | 79.4 |
| | | | -75 | 99.1 | 15.8 |
| 3 | 399.6 | 3 | -10000 / +75 | 376.8 | 94.3 |
| | | | -75 | 9.1 | 2.3 |
| 4 | 481.0 | 4 | -5000 / +75 | 437.1 | 90.9 |
| | | | -75 | 24.4 | 5.1 |

**Figure 4. Textural classification for a porphyry copper deposit, and EPD disaggregation test data for four geomet zones of a deposit. Alteration mineralogy and microtextures are possible key parameters controlling breakage of rocks from these domains (Ross et al., 2009; figure by J. Ross).**

this method provides a new data set for flotation plant operators for understanding the gold deportment with respect to sulfide minerals that was previously not possible. In this deposit, in addition to compositional distinction, a grain size distribution is apparent between pyrite of variable grain size, arsenian pyrite rims on pyrite that are typically <5 micron in size for this deposit, and Au-bearing arsenian pyrite rims that are typically <1 micron in size. This information is useful to the flotation plant to understand As-Au relationships and to predict recoveries of gold, as outlined for pyrite, arsenian pyrite and arsenopyrite by Kappes et al. (2009).

In a study of mineralogy related to flotation plant production, Pichott et al. (2009) demonstrated the economic utility of quantifying the mineralogical variability of concentrates. In one porphyry copper mine, they examined feed, concentrate and tail materials from samples to quantitatively evaluate production efficiencies. These authors demonstrated that the flotation plant was producing 40% pyrite in concentrates because of copper activation of pyrite surfaces, interferences with silicate minerals, and other issues. On the basis of these data, the flotation procedure was modified; the results are shown in Figure 6. Overall, with changes in the plant in response to the mineralogical analysis of feed, concentrate and tailing materials, copper recovery was increased by 6%. The improved recovery is an excellent example of teamwork in operations between geologists and metallurgists, and would not have been possible without the quantitative input into the flotation models of mineralogical information provided to the plant operators by the geomet group.

**Tailing Production**

As a final example, Figure 7 illustrates quantification of sulfide and sulphate minerals in tailing material. This application impacts reconciliation and recovery by further understanding the deportment of minerals and the quantification

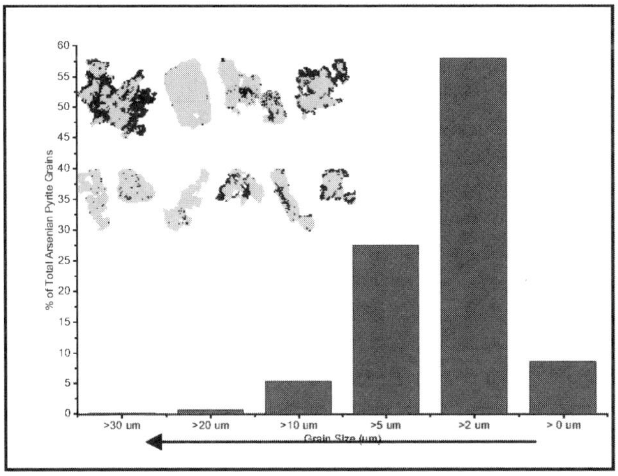


**Figure 5. Graph showing size distribution of pyrite with arsenian rims. 67% of the arsenian pyrite grains are <5 μm in size. The gold-bearing arsenian pyrite rims are <1 μm in size. Inset shows pyrite grains (yellow) with arsenian rims (brown), grains were digitally extracted from rock images. (Figure by J. Stammer).**

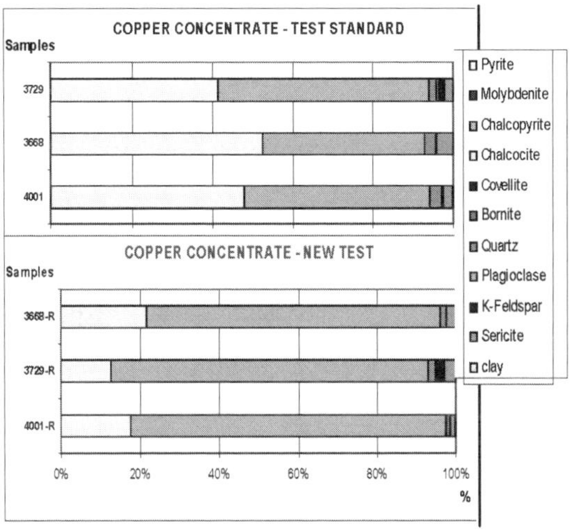


**Figure 6. Modal mineralogy of flotation concentrate product before (above) and after (below) changes were made in the plant; pyrite: yellow, chalcopyrite: orange. Note initially >40% pyrite in the concentrate, and after modifications to the plant less than 20% pyrite, representing a 6% improvement in float recovery (Pichott et al., 2009; figure by S. Pichott).**

of values going to the tailing instead of the concentrate. In addition, the application illustrates how materials are evaluated for environmental impact in much the same way as they are for extraction. This means the same data sets originated during geological and metallurgical evaluation can be utilized to make reasonable predictions regarding environmental uncertainties and risk. In this example, sulfide and sulfate minerals are distributed interstitially to and along the grain boundaries of the gangue silicate minerals, indicating they may be preferentially liberated into the environment at an earlier stage than if they had been locked in silicates. This information is useful in that predictions of element release from sulfides can be made and prepared for, and in this case indicates that long-term release may not be an issue once the interstitial materials have gone into solution.

## CONCLUSIONS

Quantitative mineralogy forms a cornerstone of any geomet study and the information so derived can have significant impact for process plant design and operations. The ability to quantitatively characterize not only mineralogy but also texture, fabric and elemental distribution by SEM-based methods means better understanding of ore variability, and improved prediction capabilities and thus operations. The job of the geological and mineralogical team is to provide the engineering group with the numeric information necessary to develop the process models for the project. For materials characterization, for identification of important parameters and geomet domains, and for impacts on comminution and

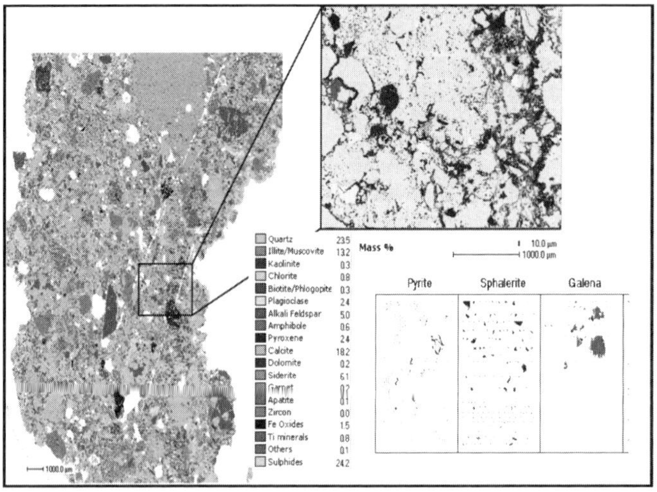


**Figure 7. False-colored images and data quantification for minerals in a tailing pile. The blow-up box shows the distribution of sulfide and sulfate minerals with gangue blocked out (gray). Pyrite, sphalerite and galena are shown in the image grid (lower right) much the same way as they are characterized for processing evaluation.**

flotation, the data derived from quantifying mineralogy and texture should be incorporated into plant modeling and planning. The endpoint is better value for the project by reducing the uncertainty related to materials variability.

## REFERENCES


Dobby, G., Bennett, C., Bulled, D., and Kosick, G., 2004. Geometallurgical modeling: The new approach to plant design and production forecasting/planning and mine/mill optimization. *36th Annual Meeting of the Canadian Mineral Processors, Proceedings,* Ottawa, p. 227–239.

Gottlieb, P., Wilkie, G., Sutherland, D., Ho-Tun, E., Suthers, S., Perer, K., Jenkins, B., Spencer, S., Butcher, A., and Rayner, J., 2000. Using quantitative electron microscopy for process mineralogy applications. *JoM,* April:24–25.

Gregory, M.J., and Lang, J.R., 2009. Hydrothermal controls on copper sulphide speciation, metal distribution and gold deportment in the Pebble porphyry Cu-Au-Mo deposit, southwest Alaska. In: *SGA2009: Smart Science for Explortion and Mining,* Townsville, Australia, 17–20 August.

Hoal, K.O., 2008. Getting the geo into geomet. *SEG Newsletter,* Society of Economic Geologists, 73:1, 11–15.

Hoal, K.O., Stammer, J.G., Appleby, S.K., Botha, J., Ross, J.K., and Botha, P.W., 2009. Research in quantitative mineralogy: Examples from diverse applications. *Minerals Engineering*, 22:402–408.

Hoal, K.O., Appleby, S.K., Stammer, J.G., and Palmer, C., in press. SEM-based quantitative mineralogical analysis of peridotite, kimberlite, and concentrate. Lithos.

Irving, J.D., 1906. University training of engineers in economic geology: Discussion. *Economic Geology*. 1:77–82.

Kappes, R., Brosnahan, D.R., and Gathje, J.C., 2009. Utilizing mineral liberation analysis (MLA) to determine pyrite, arsenopyrite and aresenian pyrite floatabilities for Carlin trend ores. *Preprint 09-117, SME Annual Meeting*, Denver.

McQuiston, F.W., Jr., and Bechaud, L.J., Jr., 1968. Metallurgical sampling and testing. In: Pfleider, E.P., ed., *Surface Mining*. New York: The American Institute of Mining, Metallurgical, and Petroleum Engineers, p. 103–121.

Pichott, S., Godoy, S., and Holmgren, C., 2009. Advanced mineralogical description of the copper feed, concentrate and tailings at Los Bronces Division, Anglo American Chile. *GEOMIN 2009,* Antofagasta, 10-12 June.

Ross, J., Appleby, S.K., Hoal, K.O., and Botha, P., 2009. Quantitative mineralogical study of ore domains at Bingham Canyon, Utah, USA. *Preprint, SME Annual Meeting*, Denver.

Stammer, J.G., Hitzman, M.W., and Highsmith, R.P., 2008. Using automated mineralogy as a research tool to detect sub-micron-sized gold in sediment-hosted Carlin-type deposits. Automated Mineralogy, MEI, Brisbane, August.

Woodhead, J., and Hitzman, M.W., 2008. Quantitative mineralogy of mineralized samples from Tenke, Lupoto, and the ZCB. *AMIRA Project Presentation*, Lusaka.

# BENEFITS OF DYNAMIC PROCESS SIMULATION FOR THE MINERAL INDUSTRY

Prashanth Parthasarathi, Virgil Szaruga and Marian Szatkowski

Andritz Automation Inc., Suite 570, 125 Clairemont Ave., Decatur, GA 30030, USA

## ABSTRACT

Using steady-state computer models for plant design has a long and well established track record in mining and metallurgical industries. In recent years, there has been growing interest in taking process modeling to another level and using dynamic computer models for mineral processing plants (Ergun et al., 2004; Liu and Spencer, 2004; Napier-Munn, 1996; Nikkah and Anderson, 2001; Salazar et al., 2009).

This paper will use an example of a commercial simulation package called IDEAS to show how dynamic simulation can be used throughout the whole life-cycle of a mineral processing plant, i.e., for process design, P&ID verification, control loop tuning and checkout, as well as for operator training. A simple model of a grinding and separation circuit, typical for mineral processing plants, will be used to show how a model of a mineral plant would evolve during the project's life cycle.

## INTRODUCTION

The primary focus of using dynamic simulations in the mineral industry seems to be thus far on the design of process control loops and alternate circuits to improve product quality and/or reduce power consumption (Ergun et al., 2004; Jamsa-Jounela et al., 2003; Liu and Spencer, 2004).

However, dynamic simulation is a powerful tool that can be used for studying a much broader range of applications, such as:

- equipment sizing (tanks, pumps, pipes & valves)
- designing advanced process control strategies
- check-out of Distributed Control System (DCS) and Programmable Logic Controller (PLC) programs
- Hazard and Operability (HAZOP) analyses
- designing and testing start-up and shut-down procedures
- operator training
- de-bottlenecking of operations after the start-up
- energy use optimization

The first six of these eight tasks can be done well ahead of the actual process even being built. After the plant is in operation, the use of simulation assures that running numerous "what if" scenarios, necessary for on-going process improvements, is easy and avoids the risks of damaging process equipment or interrupting production.

The Canadian oil sands industry has accepted dynamic simulation for use in controls checkup and operator training. The nature of the processes for extracting bitumen from oil sands that involves the use of volatile hydrocarbon solvents necessitated a tested process and well trained operator force from day one of any plant operation. In addition, the fast turnover of the workforce in that industry required tools for the continuous training of operators off-line.

In this paper, we will present examples of how a single simulation package can be used in all the applications listed above. For this presentation, a simple model of a grinding and separation circuit, common in mineral processing plants, was simulated using a commercial, steady-state and dynamic simulation package called IDEAS. The reader will be taken through three versions of the same model (steady-state, dynamic and OTS) and will be shown how the model evolves during the project's life cycle.

## IDEAS PROCESS SIMULATOR

The IDEAS simulator is an object-based graphical environment, where the user builds a model by retrieving icon-based "objects" from various libraries and assembling them on a drawing-like worksheet. These objects generally have a one to one correspondence with actual process equipment, i.e., pumps, valves, tanks, transmitters, controllers, etc. The user can therefore create a model by creating a P&ID (or PFD)-like picture of the process. Individual equipment characteristics, such as pipe dimensions or pump curves, and feed characteristics, like composition and PSD, can be specified by filling in dialog box information for each object.

The objects, grouped in libraries, are designed to be interconnected using stream connectors and can thus be used to create complex models. Once objects are connected and the information is entered into the dialog boxes, the simulation is ready to run. No other programming, "subroutine" connections or other software manipulations are needed. The flows in the pipes, as well as the associated pressures throughout the piping network, will automatically be solved, displayed in the dialog boxes, and made available to transmitters and plotters.

IDEAS can perform steady-state mass and energy balances, track components, compounds and element flow and concentration, handle particle size distribution changes, and calculate specific gravity and excess enthalpy. Models can link to spreadsheets for importing multiple case-study scenarios and for exporting material and energy balance data.

Steady-state models created in IDEAS can be modified to incorporate process dynamics by adding objects such as pipes, valves, tanks, PI/PID controllers and other objects. Some unit operation models have two sets of equations already built in to allow switching between steady-state and dynamic modes of operation by simply toggling as switch in their dialog boxes. Other unit operation blocks will need to be replaced in the model with their dynamic version.

Unit operations are modeled in IDEAS using the first principle approach, i.e., the mass and energy conservation and the population balance methodology are the basis for all models of mineral processing operations. In addition to the mass and energy conservation principles used across the whole platform, the mineral processing models use semi-empirical equations and correlations specific from a given operation (hydrocyclones, screens). Furthermore, since we are dealing with natural materials (ores), the composition and the properties of which may vary widely not only from one ore deposit to another but even within the same body of ore, the models also require some basic lab or pilot plant data entered in their dialog boxes to accurately represent actual plant operations. The use of lab test data or results from either a pilot plant or the actual operation allows for verification of unit ops as the model is being built.

IDEAS mineral processing unit operation models have been developed in house based on published literature (King, 2001, Napier-Munn et al., 1996, and Morrell, 2004). The hydrocyclone unit uses two different models: the industry standard Plitt model (Plitt, Finch and Flintoff, 1980; Flintoff, Plitt and Turak, 1987) and the proprietary Krebs model for those applications that use hydrocyclones from this manufacturer.

Not all possible unit operations are available in the package, although the list is being continuously expanded as needed. Users with unique or proprietary models of their unit operations can implement their existing models in the framework of IDEAS in several ways, the easiest being by calling their dll functions(s) from IDEAS objects.

IDEAS has been successfully used in the mining and oil sands industries to develop dynamic simulations of plants by combining process dynamics with first principles models of unit operations. The IDEAS package contains libraries of industry specific models from mine operation to tailings pipelines with all the unit operations in between. It also features a flexible and customizable material properties database with components commonly used in the mineral industry. In addition to steady-state and dynamic modeling capabilities, IDEAS can connect to the plant DCS/PLC control system and be used for operator training via the IDEAS Instructor product. The Instructor product allows the trainer to present operators with a predefined set of training scenarios typical of what would be expected in the plant, and to grade the operators' response according to predefined criteria.

## USE OF SIMULATION THROUGH PROJECT LIFE CYCLE

A grinding circuit was selected to present the various analyses that can be performed using simulations. The process starts with a stockpile. The ore is transported by a belt conveyor, mixed with water and sent to a SAG mill. The pebbles from the SAG mill are sent to a screen and then a pebble crusher. The crusher discharge and the product from the SAG mill are collected in a sump box where they are mixed with additional water. The slurry is pumped to a battery of hydrocyclones, the rejects of which are sent to a ball mill which recycles the ore to the sump box. The product of the hydrocyclone is sent to flotation. The process flow diagram is shown in Figure 1.

### Steady-State Model

Even though the purpose of this paper is to show the benefits of dynamic simulation in the mineral industry, in order to have a complete picture of how a single simulation tool can be used, a brief example of using simulation in the early stage of the plant design, i.e., to develop process flow diagrams and mass and energy balances, will be presented.

In a typical application of steady-state modeling, numerous configurations of the grinding circuit may be produced and analyzed in order to select the best equipment configuration. When the final configuration is selected, the steady-state model may be run for different cases involving different ore types, mineral contents or production rates. Here, a single steady-state model is presented in Figure 1, with the mass and energy balance report generated, Figure 2.

It should be noted that the flow diagram representation of the process has been created by using so-called hierarchical blocks which allow placing a picture of choice on an existing library object (or a group of objects). For example, the SAG square is a hierarchy with the SAG mill set to run in a steady state mode, Figure 3.

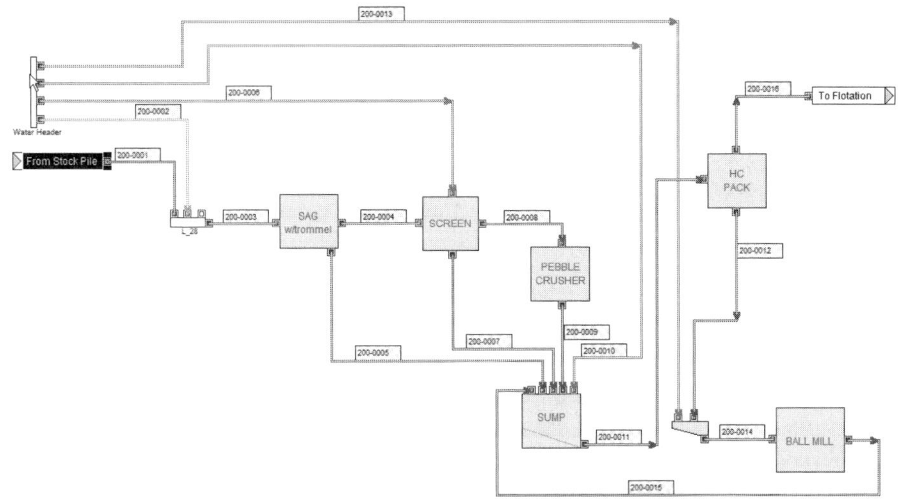


**Figure 1. Steady-State Model of Grinding Circuit**

| | A | B | C | D | E | F | G | H | I | J | K | L | M | N | O | P | Q | R |
|---|---|---|---|---|---|---|---|---|---|---|---|---|---|---|---|---|---|---|
| 1 | **Case Study A** | | | | | | | | | | | | | | | | | |
| 2 | Stream No. | | 200-0001 | 200-0002 | 200-0003 | 200-0004 | 200-0005 | 200-0006 | 200-0007 | 200-0008 | 200-0009 | 200-0010 | 200-0011 | 200-0012 | 200-0013 | 200-0014 | 200-0015 | 200-0016 |
| 3 | Stream Name | | Ore Feed | Water to SAG | SAG Feed | SAG to Screen | SAG Trommel U/F | Screen Sprays | Screen U/F | Pebbles to Crusher | Crusher Discharge | Sump Dillution Water | HC Feed | HC U/F | Water to Ball Mill | Ball Mill Feed | Ball Mill Discharge | Flotation Feed |
| 4 | Stream Flow | t/h | 3,000.0 | 700.0 | 3,700.0 | 2,440.6 | 1,259.4 | 1.0 | 2,113.4 | 328.2 | 328.2 | 2,944.8 | 14,470.2 | 7,527.3 | 297.0 | 7,824.3 | 7,824.3 | 6,942.8 |
| 5 | Stream Flow | m3/h | 1,160.5 | 700.9 | 1,861.9 | 1,218.2 | 649.9 | 1.0 | 1,088.8 | 130.4 | 130.4 | 2,948.6 | 8,511.9 | 3,392.0 | 297.4 | 3,689.7 | 3,692.2 | 5,119.9 |
| 6 | Water Flow Rate | m3/h | 89.9 | 700.9 | 791.3 | 503.9 | 293.6 | 1.0 | 489.8 | 15.1 | 15.1 | 2,948.6 | 5,032.5 | 983.2 | 297.4 | 1,280.9 | 1,283.5 | 4,049.3 |
| 7 | Ores Flow Rate | t/h | 2,910.3 | 0.0 | 2,910.3 | 1,941.7 | 968.7 | 0.0 | 1,628.4 | 313.2 | 313.2 | 0.0 | 9,458.6 | 6,548.2 | 0.0 | 6,548.2 | 6,548.2 | 2,910.3 |
| 8 | Solids | wt % | 97.0 | 0.0 | 78.7 | 79.6 | 76.9 | 0.0 | 77.1 | 95.4 | 95.4 | 0.0 | 65.4 | 87.0 | 0.0 | 83.7 | 83.7 | 41.9 |
| 9 | Temperature | C | 25.0 | 17.0 | 21.1 | 45.2 | 45.2 | 17.0 | 45.1 | 45.1 | 45.1 | 17.0 | 29.3 | 29.3 | 17.0 | 28.0 | 34.3 | 29.3 |
| 10 | Density | kg/m3 | 2,585.1 | 998.7 | 1,987.3 | 2,003.5 | 1,937.7 | 998.7 | 1,941.0 | 2,517.7 | 2,517.7 | 998.7 | 1,700.0 | 2,219.2 | 998.7 | 2,120.6 | 2,119.1 | 1,356.0 |
| 11 | d50 | micron | 49,971.3 | 0.0 | 49,971.3 | 1,695.4 | 1,123.4 | 0.0 | 1,101.2 | 28,190.6 | 11,952.5 | 0.0 | 291.7 | 709.3 | 0.0 | 709.3 | 173.9 | 73.3 |

**Figure 2. Mass Balance for Steady-State Grinding Circuit Model**

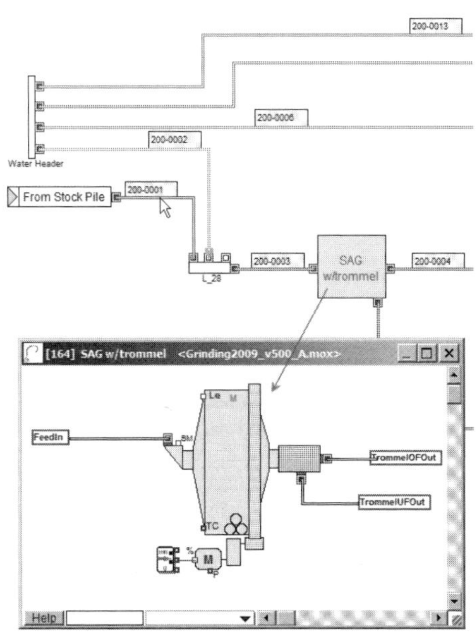


**Figure 3. Hierarchical structure of the SAG Mill unit**

**Dynamic Model – Equipment Sizing and Control Design**

During a traditional plant design process, the design work is divided between multiple disciplines, often between different engineering companies. As each group tends to add a safety factor in their design, this may lead to a significant up-sizing of the process and therefore in increased capital costs. Using a dynamic model of the whole process (a virtual plant), one can check for any design up-sizing of tanks, pumps, pipes, valves and other critical pieces of process equipment resulting in significant capital cost reduction.

Furthermore, the virtual plant allows sizing the equipment not only for the steady-state, design production rates but also for any dynamic changes in the process. Such changes occur as a result of variation in the feed stock or disruptions in the equipment availability. The resulting changes in the amount of material moved through the process during such production swings will be affected by the dynamic and often complex equipment interactions, Having a complete representation of all the dynamic process elements is therefore very important for selecting proper equipment sizes. In addition, complex control strategies, including Advanced Process Control (APC) can be proposed and evaluated against the dynamic model during the design phase, leading to the proper selection of instruments and startup of the process in the most efficient manner.

At the same time, analyses of the virtual plant can reveal design flaws that would not have become obvious until the actual startup, thus resulting in delayed plant commissioning, redesign costs and lost production. Several cases of IDEAS dynamic models revealing mistakes in the design have been documented in oil sands applications [Foulds, Tipman].

Continuing with our example, a dynamic model of the grinding circuit for design and optimization of control loops is presented. The design of any controls in the grinding circuit will have to take into account the complex dynamics of the feed forward mass transfer part (conveyer, SAG, crasher) as well as those of the mass recycle loop (hydrocyclone and ball mill). As an illustration of the complexity of control loop design, a scheme for controlling the particle size (d50) of the product by controlling the density of the feed to the hydrocyclones has been used.

In transitioning from the steady-state to a dynamic model, some of the existing objects have been converted to their dynamic mode by a dialog box switch (mills, hydrocyclones) and additional objects such as a conveyor, tank, pump, valves and PID controllers were added to the model, as shown in Figure 4. The model was also 'equipped' with various transmitters to monitor flows, densities, pressures and PSD of streams.

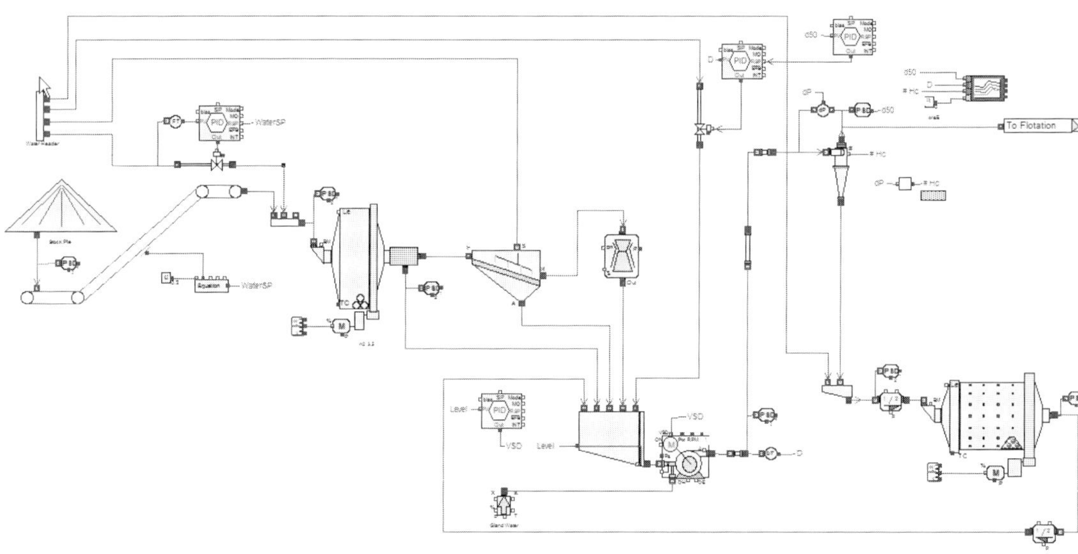


**Figure 4. Dynamic Model of Grinding Circuit**

As can be seen in Figure 4, the control of d50 of the feed to flotation (the product of the grinding circuit) directly involves at least three (3) control loops: the sump level control via the VSD pump, the slurry density control via controlling the dilution water to the sump, and finally, the cascade control of the hydrocyclone separation via adjusting the feed density. The control scheme is further complicated by an independent logic to control the pressure drop in the battery of hydrocyclones, where the number of active cyclones is increased or decreased to maintain the delta P within a predefined range. Without a realistic and complete representation of all process elements, such control loop design would be practically destined to fail.

IDEAS simulator allows tracking and processing particle size distributions of any component selected by the user. The principle of mass conservation with population balance is used in all operations that pass, mix or separate solids according to their sizes. Specific PSD "processing rules" are additionally applied in those unit ops that alter PSD of the processed materials (e.g., comminution, agglomeration, etc). Full distributions of PSD can be tracked in any process stream using PSD transmitters. As shown in Figure 5, the transmitter dialog box allows for customizing the format of the size distribution. The data can be presented in a tabular form or graphically. The transmitter also allows dynamic tracking of d50 (or d20, or d80) of an individual solid (ore) or of the mixture of all solids.

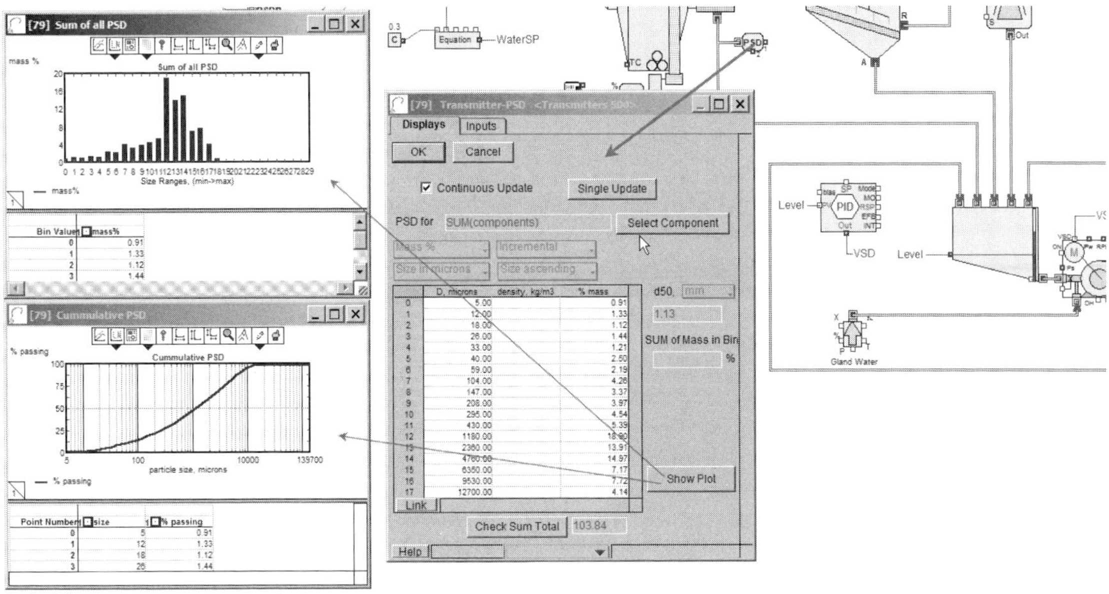


**Figure 5. PSD display using PSD Transmitter**

The simulation studies of this complex control strategy were done using a case of a step change in the ore hardness. The resulting upset in the size distribution of solids throughout the whole circuit resulted in an upset to the product's d50. The changes in the d50 value as the result of a step change from ore A (softer) to ore B (harder), and back, have been shown in Figure 6. The cascade loop to control the d50 by adjusting the feed density was turned off. Only the sump level and the slurry density were automatically controlled at predefined set point values.

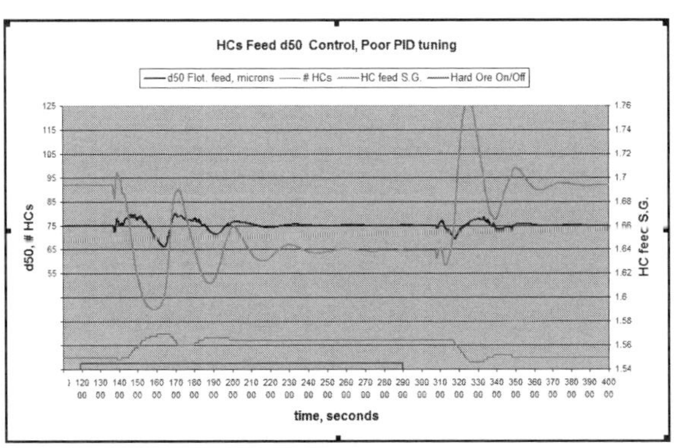


**Figure 6. Dynamic Response of Product's d50 to Change in Ore Hardness without d50 Control Loop**

Figure 7 shows the product's d50 response to the feed type change with the cascade loop for d50 control turned on. A significant improvement in stability of the d50 value was observed by adjusting the feed density. There is still a fair amount of oscillation in the density value. Furthermore, the independent on/off logic controlling the number of active hydrocyclones introduces fast changes to the flow distribution in the battery and these result in small "spikes" in the predicted d50 value.

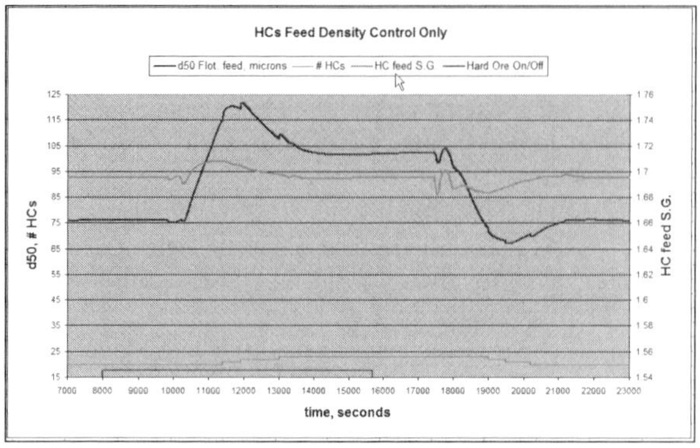


**Figure 7. Dynamic Response of Product's d50 to Change in Ore Hardness with d50 Control Loop**

Once the "proof of concept" study was done and it showed that indeed having this additional control loop will benefit the plant operation in more stable size of the flotation feed, the same model was used to pre-tune all the PID controllers. As shown in Figure 8, the upset in d50 value was reduced even further by proper tuning.

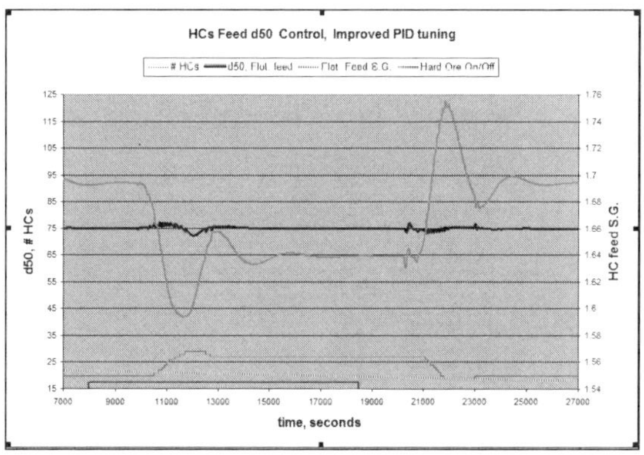


**Figure 8. Dynamic Response of Product's d50 to Change in Ore Hardness with d50 Tuned Control Loop**

**DCS/PLC Staging - Control Logic Verification**

IDEAS simulation package can communicate with a control system (DCS and/or PLC) and therefore any dynamic plant model created for design verification can be used for DCS/PLC staging and for control loop verification. The plant control system to which the model will connect may be the actual control hardware, or the plant controls may be emulated by a software package running on a separate PC or UNIX work station.

When a dynamic process model is first built it will contain only a limited number of local controls loops that are required to run the model. These, typically, will be tank level controls and flow rate control loops as shown in Figure 4. In addition, some extra loops may be added for more in-depth analysis of control strategies. An example of such controls in Figure 4 is the cascade loop for d50 control.

A real plant control system will have all these local controls, and much more, programmed in its DCS/PLC. The exchange of data between the model and the control system is done using two communication objects: Control Input (from DCS to model) and Control Output (from model to DCS). Figure 9 shows the same model used previously but expanded by the addition of I/O objects. The I/O objects can be configured to use the communication protocol that is required by the plant's DCS/PLC hardware.

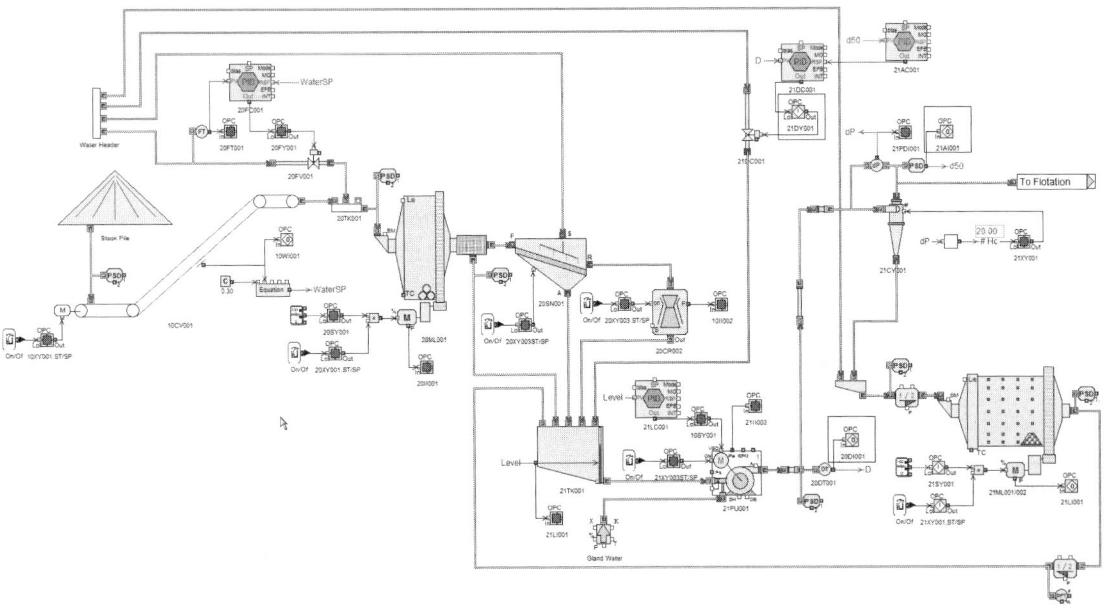


**Figure 9. Dynamic Model with I/O Communication Objects for DCS/PLC Connectivity**

While the example used in this article is simplified and only requires approximately 20 I/O points, real plant simulators will typically have several thousands, if not tens of thousands of I/O points, all communicating with the DCS/PLC in real time.

Early in the controls checkout process, all the I/O points are placed in the model but set to a local mode. This means they do not communicate with the DCS/PLC, and Control Input objects pass values provided by local controllers, switches and sliders to control elements (valves, VSD drives). In Figure 9, these local I/O points are shown with red squares.

In the process of DCS checkout, each I/O point can be switched from "local" to "communication" mode on loop-by-loop basis. This way each communication path and each control algorithm can be checked and debugged individually. In Figure 9, three I/O points related to the cascade control loop are shown to be working in the communication mode (see the highlighted I/O elements without red squares). The two local PID controllers (d50 and Density) are in a stand-by mode as the control is being performed by the DCS.

Through this loop-by-loop testing, the dynamic simulator is used not only to do a standard "loop-back" type of DCS check-out but also for much broader and more in-depth analysis of the control system. Staging the DCS against a detailed process model allows for detecting errors that involve process intent, (such as pressure response, temperature, valve and pump response) in addition to the typical logic errors found during a loop-back type of check out.

The table below shows the benefits of using a dynamic simulator to do DCS check-out.

**Table 1. DCS/PLC Verification Comparison**

| Logic Verification Feature | DCS Loop-Back | IDEAS Model |
|---|---|---|
| I/O and Loop Test | ✓ | ✓ |
| Process Wide Logic Test | × | ✓ |
| Tuning Constants Known Before Startup | × | ✓ |
| Realistic Process Models | × | ✓✓ |
| Remove Control Logic Errors | × | ✓✓ |
| Remove Process Intent Errors | × | ✓✓ |
| Verify Advanced Control Logic | × | ✓ |

Project experience has determined that a DCS checkout done with a dynamic simulator can eliminate over 80% of the errors in control system design before plant commissioning. Using dynamic models for DCS checkout has proven to save time and money for numerous companies. As an example, in one case, a client was able to start up their plant in 16 days instead of 90 days and was above the design production capacity within one month. In another case, a client was able to save three weeks in commissioning time.

**Operator Training Simulators (OTS)**

As a result of DCS/PLC staging process, all the controls are verified and released from the internal IDEAS controls to the DCS/PLC controls, see Figure 10.

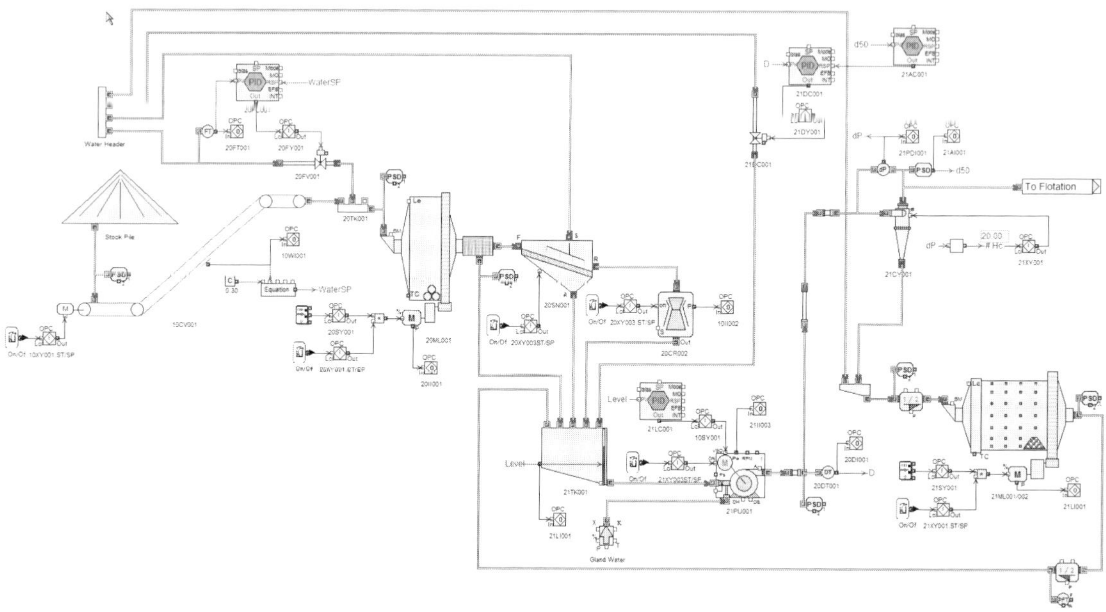


**Figure 10. OTS Model**

At this phase, the model communicates through the actual plant control software. The operators use the actual control system screens and graphics to interact with the process. The dynamic simulator is then ready to be used as a training tool for process operators to:

- learn the startup and shut-down procedures,
- learn to diagnose and correct process upsets and equipment failures in a fail-safe environment,
- learn the emergency procedures,
- learn the real dynamics of the process and develop a feel for process response time ,
- gain understanding of the process and equipment interactions.

The dynamic simulator enhances the learning process by actively involving the operators and providing immediate feedback without risks to production. Also, a well maintained, up to date, training simulator can be used to train newly hired operators off-line and help them quickly become proficient in running a new process.

The operator training is best done using an additional software tool. The IDEAS package has its own, integrated product called IDEAS Instructor with its browser-like interface (Figure 11). The package consolidates the trainees' database including scores and certification, trainers' folders as well as specific training scenarios per each area. IDEAS Instructor is able to measure operator responses to various training scenarios and process malfunctions that are initiated by the software. The Instructor helps to teach and train operators on process upsets and start-up and shutdown procedures before the plant commissioning, thus assuring that the start-up is done with well trained and "seasoned" operators even if the process itself is new and unique.

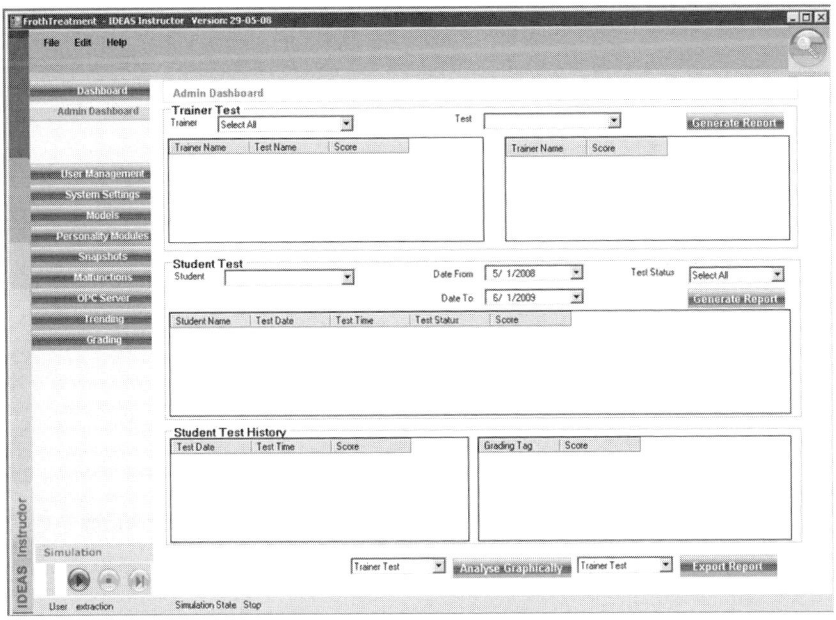

**Figure 11. IDEAS Instructor Interface**

Training using dynamic simulators is known to increase the effectiveness of learning and to reduce the time to train by between 30% and 70% (depending on the study) compared to a class room type of training.

In a copper concentrator in Peru, operators used the simulator to practice start-up, shutdown and emergency sequences in the months prior to start-up. A standardized test with approximately 300 random questions was developed to test operator competency throughout the training process. The test questions were given in three intervals, once before any training, once after class room based training (CBT), and then once again after simulator based training. The results clearly showed that the simulator training made a remarkable improvement in operator competency:

| | |
|---|---|
| Competency before any training | 20.3% |
| Competency after CBT | 26.7% |
| Competency after IDEAS training | 85.0% |

**De-Bottlenecking/Energy Use Optimization**

After the plant is in full operation the dynamic model can be kept in line with the plant modifications or expansions (as-built) at very low cost. The model can be easily decoupled from the OTS and be used as a desk tool for process or control engineers to study potential improvements and troubleshoot any known process shortcomings. New process configurations and improvements in controls can be quickly evaluated, validated and transferred back to the plant control system. This allows for an ongoing process of plant improvements.

**SUMMARY**

The use of steady-state and dynamic simulations for mineral processing plants has been illustrated using the IDEAS simulation package and a simple model of a grinding circuit.

The benefits of using simulation in each phase of a project life-cycle have been presented as:

- Cost-effective evaluation of multiple design or production alternatives,
- Equipment right-sizing; capital cost reduction,
- Controls design integrated with process design and including interactions between equipment,
- Pre start-up verification and optimization of the plant's control system,
- Performing "virtual" startups and shut-downs against the models,
- The most efficient operator training tool,
- Ongoing plant improvements can be tested first on the model before going on line.

## REFERENCES


Ergun, L., Ekmekci, Z., Gulsoy, O. and Benzer, H., 2004, Modelling and simulation of grinding circuit in Madneuli copper concentrator, Physiochemical Problems of Mineral Processing, 38: 231.

Jamsa-Jounela, S.L., Dietrich, M., Halmevaara, K. and Tiili, O., 2003, Control of pulp levels in flotation cells, Control Engineering Practice, 11: 73.

Liu, Y. and Spencer, S., 2004, Dynamic simulation of grinding circuits, Minerals Engineering, 17: 1189.

Napier-Munn, T.J., et al., 1996, Mineral Comminution Circuits: their Operation and Optimisation, JKMRC Monograph Series in Mining and Mineral Processing 2. JKMRC.

Nikkah, K. and Anderson, C., 2001, Role of simulation software in design and operation of metallurgical plants: a case study, Presented at SME Annual Meeting, Denver, Colorado.

Salazar, J.L., Magne, L., Acuna, G. and Cubillos, F. 2009, Dynamic modeling and simulation of semi-autogenous mills. Minerals Engineering. 22: 70

Foulds, G. and Tipman R., 2005, Personal communication.

Plitt, L.R., Finch, J.A. and Flintoff, B.C., 1980, Modeling the hydrocyclone classifier. Euro Symp. Particle Tech, Amsterdam, 790-804

Flintoff, B.C., Plitt, L.R. and Turak, A.A., 1987, Cyclone Modeling: a Review of Present Technology. CIM Bulletin 80 (905), 39-50.

Morrell, S., 2004, A new autogenous and semi-autogenous mill model for scale-up, design and optimization, Minerals Engineering, 17, 437-445.

King, R.P., 2001, Modeling and simulation of mineral processing systems, Butterworth-Heinemann.

# Engineering Commissioning, Start Up, and Case Studies

# EXPANSION OF THE GIBRALTAR MILL

Ken Major[1] and Robert Rotzinger[2]

[1] KWM Consulting, 12576—206 St, Maple Ridge, BC, Canada
[2] Gibraltar Mines Ltd, 10251 Gibraltar Mine Road, McLeese Lake, BC, Canada

## ABSTRACT

The Gibraltar Mine and Mill facility were commissioned in 1972 by Gibraltar Mines Ltd (majority owned by Canex Placer). The mill flowsheet consisted of two stages of crushing followed by a rod mill / ball mill grinding circuit. During operations a number of opportunities were evaluated to expand the mill capacity; however, the economics of the day and the complexity of the project resulted in the project not proceeding. The mine continued to operate until late 1998 when it was shutdown and put on care and maintenance due to low commodity prices. Taseko Mines acquired Gibraltar in 1999 and in 2004, with the improvement in the metal markets, the decision was made to restart operations. Shorty after the restart, a decision was made to re-look at the opportunities for upgrading and expanding the process facilities. Engineering and construction for the expanded facilities was initiated in 2006. The upgraded grinding and flotation facilities were commissioned in 2007. In order to minimize the impact on daily operations, minimize capital expenditures and marry the new facilities into the established site a number of unique opportunities were identified during the design. This paper discusses the flowsheets that were considered and the unique designs that were implemented.

## INTRODUCTION

The Gibraltar Mine operation is located in central British Columbia and was commissioned in 1972 by Placer Development. This was during a period of mine expansion for Placer that included the development of the Endako Mine, also in British Columbia, and the Marcopper Mine in the Philippines. The Gibraltar mill was designed to process 30,000 tons per day and produce a copper flotation concentrate and a by-product molybdenum concentrate.

The mill flowsheet was developed around a conventional rod mill / ball mill grinding circuit. Ore from the mine was delivered to a 54 x 75 gyratory crusher. Primary crusher product was conveyed to a primary screen to scalp rod mill feed sized (screen undersize) ore and convey it directly to the fine ore stockpile. Oversize from the primary screen was conveyed to the coarse ore stockpile and subsequently reclaimed to feed four parallel Allis Chalmers 84" hydrocone crushers. Vibrating screens were located under each of the cone crushers with the screen undersize delivered to the fine ore stockpile and the oversize returned to the coarse ore stockpile. The rod mill feed size was targeted at 80% -19mm.

Ore was delivered to the fine ore stockpile using a tripper conveyor system to distribute the ore over 27 gravity feed points. The grinding circuit was designed with three parallel rod mill / ball mill trains. The fine ore stockpile reclaim systems delivered ore to each of the 3 rod mill feed conveyors via three belt feeders which were each in turn fed from three gravity draw points.

The rod mills were operated in open circuit with the rod mill and ball mill discharge combined and pumped to a set of cyclones to facilitate the closed circuit operation of the ball mills and produce a nominal flotation feed size P80 of 300 microns. The cyclone overflow from each of the ball mill circuits was delivered by gravity to one of three parallel rougher flotation banks. Each bank consisted of 16, Denver 600 cubic foot flotation cells. Rougher concentrate was reground to nominally 60 microns and a bulk concentrate produced via two stage cleaner flotation. The cleaner circuit flotation equipment consisted of 16 Denver 300 cubic foot flotation cells (first cleaner), three 7 foot diameter flotation columns (second cleaner) and 16 Denver 100 cubic foot cells (column scavenger). The molybdenum was then extracted via flotation from the bulk concentrate.

During the life of the mine, a number of open pit resources were identified and processed through the Gibraltar mill. The operating capacity of the mill varied depending on the source of the ore based on the ore hardness and target grind. Over the life of the operation throughput rates ranged from 32,000 tpd to 50,000 tpd and, most recently, the mill had been operating at 36,750 tpd.

In 1998 the mine was placed on care and maintenance due to low metal prices. After acquiring the Gibraltar Mine in 1999, Taseko Mines Limited investigated at a number of process alternatives that had the potential to reduce the cost of production to support re-commissioning the operation including the completion of metallurgical testing and a feasibility looking at the implementation of the CESL hydrometallurgical process for on-site production of copper cathode from the copper concentrate. Gibraltar installed a copper SX/EW facility in 1986 to recover copper from the run of mine waste rock dumps as well as oxide ores and the concentrate treatment process was viewed as an extension of this technology. Late in 2003, with an improvement in copper prices, Taseko Management identified an opportunity to raise the capital required to restart the mine. A re-commissioning plan was generated that identified the equipment repair requirements and flowsheet opportunities that would be incorporated in the capital improvement program during the six month period prior to start-up.

The Gibraltar mine and mill were re-commissioned in October 2004. Following a successful start-up of the existing facilities and with continued improvement in metal prices Taseko re-evaluated Gibraltar's minable resources and determined that there was the potential to increase the mining rate and the production of concentrates. In 2005 Taseko initiated activities to evaluate increasing the mill capacity at Gibraltar.

## EXPANSION CONCEPTS

Increasing the production of the Gibraltar mill has been an ongoing discussion throughout the operating life of the mine. Initially the ore had a very low Bond Ball Mill Work Index of about 10. As the mine plan has progressed, the Bond Ball Mill Work Index has increased to about 15 significantly impacting plant throughput. The operating economics of the site dictate that throughput is the most significant variable to lower unit costs and consequently changes were made in the mill that provided for coarsening of the primary grind feeding flotation to P80 >400 microns. To alleviate bottlenecks in the crushing circuit the rod mill feed size was increased to P80 >25mm and adjusted up or down based on the performance of the rod mills as the ore changed. This enabled the mill to operate consistently above its' design throughput but it was recognized that any additional throughput would require an expansion of the existing mill facilities. The original plant layout allowed for the addition of a 4th rod mill and ball mill train. Over the life of the mine the mill operations department also considered the addition of a parallel single stage ball mill, a tertiary crushing circuit, a parallel water flush crusher and a single stage ball mill circuit, a parallel SAG Mill / ball mill circuit as well as conversion to a SAG Mill / ball mill circuit. For various reasons the different opportunities that were considered did not meet the financial hurdle rate required to justify the capital investment at the time.

During the preparations for the mill re-start in 2004, a number of risk areas were identified that presented potential challenges to maintaining continuous plant operation. Most of the main process equipment in the mill was as per the original installation and some of this equipment was coming to the end of its' serviceable life. The rougher flotation circuit was of primary concern and, when the choices to replace the flotation cells were considered, it was recognized that there was an opportunity to reduce operating costs by replacing the existing cells with larger more efficient units. In the process of the evaluation of the rougher cell replacement it was recognized that the opportunity to expand the plant capacity should be re-assessed.

The primary consideration during these preliminary discussions of the potential expansion options was that for an expansion to be considered it would have to be completed with minimal interruption to the daily operating routine. The concepts discussed included:

- Replacement of the existing cone crushers with larger higher speed units to generate finer product to feed the rod mills. This option was dropped because of preexisting foundation issues with the secondary crushers and the projected high costs and significant production disruption required to upgrade these foundations.
- Installation of a 4th rod mill / ball mill train in the allocated bay. This option was ultimately dropped due to space constraints limiting grinding mill size, the significant upgrades required to the crushing plant and the extensive changes required to accommodate new reclaim facilities under the fine ore stockpile.
- Installation of a SAG Mill and conversion of the existing grinding mills to ball mills.
- Installation of High Pressure Grinding Rolls (HPGR) and conversion of the existing grinding mills to ball mills.

A total of eight flowsheets were identified that fit the main objectives for increasing throughput. Consideration was given to the relative merits of each flowsheet to determine which presented the best opportunities to minimize capital costs, reduce operating and maintenance costs as well as simplify circuit operation. This analysis determined that the best options would be to either install a SAG mill or an HPGR circuit. The main factors in the selection of these two flowsheets were:

1. Minor modifications required to the existing secondary crusher plant.
2. Minor modifications to the fine ore stockpile and reclaim system.
3. Sufficient existing secondary grinding capacity with conversion of rod mills to ball mills for both flowsheet alternatives.

A trade-off study was approved to further assess these alternatives and, to support the efforts of the trade-off study, samples of the mill feed were collected and forwarded to SGS Lakefield for grinding work index testwork and to Polysius for HPGR testwork. The HPGR program included testing to determine the amenability of the ore to HPGR as well as providing estimated power requirements and expected liner wear rates.

## EXPANSION TRADE-OFF STUDY

The expansion trade-off study commenced with the goal to determine which of the technologies would provide the best return on capital to increase mill throughput to 50,000 tpd. The trade off study will be discussed by first reviewing the process descriptions for the grinding circuit options to be evaluated, the impacts on downstream operations, a review of operating costs calculations, estimation of capital costs and project risks.

## PROCESS DESCRIPTIONS

**SAG Mill Option.** In this flowsheet option the ore from the mine would be delivered to the existing gyratory primary crusher. Primary crusher product is conveyed to the existing primary screens located in the secondary crusher building and the panels on the primary screens are modified to provide a 100 mm (4 inch) separation. The screen oversize (+100 mm) would be conveyed to the existing coarse ore stockpile and the screen undersize is conveyed to the fine ore stockpile.

Ore from the coarse ore stockpile is reclaimed and conveyed to the existing cone crushers. The cone crushers are set up for single pass operation to produce –100 mm product and the secondary screens are removed. The cone crusher product is conveyed to the fine ore stockpile. Modifications are made to the fine ore bin feeder reclaim system to accommodate the larger feed size with the existing reclaim feeder conveyors and the existing rod mill feed conveyors are modified to deliver ore to a new SAG mill feed conveyor.

The proposed SAG mill discharges through a combination trommel screen and vibrating screen with the trommel screen designed to remove most of the fine slurry and reduce the energy from the mill discharge in order to provide better distribution to the vibrating screen. The vibrating screen is set up to provide a 12 mm (½ inch) separation with the screen oversize being recycled to the SAG mill feed and the screen undersize being pumped to a pressure distributor to split the feed to the secondary grinding ball mills.

The existing three rod mills would be converted to ball mills to provide six parallel ball mills each operating in closed circuit with a dedicated set of cyclones. These changes would require the addition of three new cyclopacs. The cyclone overflow product at 400 microns would then be delivered to the flotation circuit.

**HPGR Option.** In this flowsheet option, ore from the mine would be delivered to the existing primary crusher. Primary crusher discharge is conveyed to the existing primary screens located in the secondary crusher building and the screens are modified to provide a 60mm (2 3/8 inch) separation. The screen undersize is conveyed directly to the fine ore stockpile and the screen oversize is conveyed to the coarse ore stockpile.

Coarse ore is reclaimed and delivered to the four parallel secondary cone crushers. The crushers would discharge directly onto the existing secondary screens which would be modified to provide a 60 mm separation. Screen oversize is recycled back to the coarse ore stockpile and screen undersize is conveyed to the fine ore stockpile. Fine ore is reclaimed using the existing reclaim feeder conveyor system and the existing rod mill feed conveyors would be modified to deliver ore to the HPGR feed conveyor. The new feed conveyor in turn would deliver the ore to a 600 tonne surge bin. Ore from the surge bin would be reclaimed via variable speed belt feeders that feed two parallel HPGRs.

Product from each of the HPGRs is mixed with water and collected in a sump. The product slurry is pumped to a vibrating screen which provides a 5mm (3/16 inch) separation. The screen oversize is recycled to the HPGR feed and the screen undersize from the two circuits is collected in a common sump and pumped to the ball mill splitter.

As per the SAG mill alternative, the 3 existing rod mills would be converted to ball mills requiring new cyclopacs.

## ANCILLARY PROCESSES

Although the study was focused on determining the optimum flowsheet to expand Gibraltar's grinding capacity, a review of all of the ancillary unit operations was also conducted to determine which would require modification to support the higher plant throughput. The cleaner flotation circuit, concentrate dewatering circuit and molybdenum circuit were quickly determined to have adequate capacity to support operation at 50,000 tpd. The remaining unit operations were identified as requiring more detailed review to determine what modifications would be required to support the expansion. These circuits are briefly reviewed in the following discussion.

### Crushing

The main design limitation for the project was determined to be the existing 54x75 primary gyratory crusher as the project economics (of the day) would not support the capital required to install a new larger crusher. At 50,000 tpd, the existing crusher will need to operate consistently above its' rated capacity; however, this is not an unusual operating requirement and this crushing rate has been achieved at Gibraltar in the past. In addition, several straightforward operating strategies can be employed to manage 24 hour crusher capacity including ore blast designs, crusher open side setting, equipment availability and equipment utilization. For the purposes of the study, both blasting design and crusher utilization (based on the scheduling of ore trucks from the mine) were left as issues to manage in operation. The design of the processing facilities provides significant flexibility to run the crusher at large open side settings to maximize instantaneous crusher throughput without impacting the mill feed size due to the secondary crushing circuit.

**Rougher Flotation.** The original Gibraltar flotation circuit consisted of three parallel banks of 16, 600 cubic foot (17 cubic meter) Denver 600H flotation cells. A project had been initiated by the mill operations department to install one 160 cubic meter flotation cell as a scavenger processing the tailings from the existing three flotation banks. This project was in progress during the trade-off study and was expected to increase metal recovery by increasing the flotation residence time from 17.9 minutes to 21.4 minutes. With the increase to 50,000 tpd, one additional 160 cubic meter flotation cell would be required in order to provide a residence time of 18.3 minutes; roughly equivalent to the original plant retention time. The addition of a further two (3 total) 160 cubic meter flotation cells would be required to provide a residence time of 20.9 minutes.

As an alternative, the option to replace the original flotation cells with two parallel banks of four, 160 cubic meter cells was explored. With the expansion to 50,000 tpd this would provide a combined residence time (9 cells total) of 23.2 minutes and remove the long term issues with continuing to operate the original flotation cells.

**Regrinding.** Gibraltar operates a 450kW regrind ball mill to reduce the rougher concentrate particle size to 60 microns prior to cleaner flotation. The sizing of the regrind mill limits the mass recovery from the rougher flotation circuit to about 2.5%; subsequently impacting the copper flotation recovery.

The estimated regrind power required for a 5% mass pull at 50,000 tpd is 1200 kW. The equivalent vertimill would be a 932kW unit, as the internal recirculating load of the vertimill provides the ability to influence the classification process for the mill.

The study determined that the regrind mill should be treated as a stand-alone project based on its' own economic merit.

**Tailings Pumping.** A review of the tailings pumping and pipeline system was completed to assess the operating limits of the existing system. The Gibraltar tailings system pumps slurry to a highpoint approximately 1,500 feet from the mill and the remainder of the tailings pipeline (11,000 feet) runs downhill to the tailings impoundment. At normal flows, the downhill portion of the pipeline operates at flume flow. As the volume increases, progressively more of the downhill portion operates as full pipe flow and this has a significant impact on the system power and pressure requirements. The maximum system throughput was found to be 52,000 tpd with the system being limited by both the installed pump horsepower and the pipeline pressure rating. The study also identified that the system has some throughput flexibility based on adjusting the flotation feed density (by increasing the density the system is capable of higher throughputs due to the decreased equivalent slurry volume). The system was judged as adequate for the expansion and no modifications to the system were recommended.

## CAPITAL COSTS

Scoping level capital cost estimates were prepared for both of the primary flowsheets (SAG mill and HPGR). These estimates were completed at a ±25% level of accuracy and were developed to support the selection of a preferred flowsheet for further investigation at a basic engineering design level. Independent costs were developed for the replacement of the rougher flotation cells and for upgrading the regrind mill circuit as stand alone projects. These stand alone options could then be analyzed on their own merit or in conjunction with the grinding circuit expansion.

## OPERATING COSTS

Project operating cost schedules were developed for the various unit operations by type. Actual operating expenditure and production data was used where possible to develop fixed and variable costs for each area. Typical variable costs include consumables, wear materials (and labour to replace wear materials) and power. Fixed costs typically include items such as the number of mill operators and operations management personnel, safety and training, assaying, etc. Estimates of expected operating costs were made where there were no analogous unit operations in the existing plant. The SAG mill operating costs were developed based on experience and account for grinding media, liners and maintenance. The HPGR operating costs were estimated from sample data provided by Polysius. This data excluded the costs of wear components which were estimated from the HPGR test program. Manufacturers of HPGRs make the claims for improved grinding circuit efficiency based on the HPGR creating pressure cracks in the ore and this influencing downstream grinding energy requirements; however no value was applied for this in the study operating cost modThe two flowsheet options required very different approaches to costing the operation of the secondary crushing plant. As discussed previously, the existing secondary crushing plant is essential for preparing the HPGR feed and it was estimated that secondary crushers would be required to process 50% of the total mill feed in this case. For the SAG mill option, the secondary crushing plant would mostly be used to convey ore to the stockpile. In this case, it was proposed to use the crushers as single pass units to control the top size to the stockpile and optimize the material flow characteristics to minimize changes to the fine ore bin reclaim feeders. Simulations indicated that less than 17% of the mill feed would require secondary crushing in this case. Consequently the operating costs for the secondary crushing plant for the SAG mill option were reduced to reflect the coarser product size and single pass open circuit operation of the secondary crushers.

## RISKS

An assessment of the risk in each of the comminution options was made to complete the analysis required to decide on the project flowsheet. For the SAG mill option, the process and equipment is well understood and experience identifies the potential risks and, for most risks, the appropriate corrective measures. For example, if the ore is more competent or harder than expected the possible solutions include blasting finer and/or the installation of a SAG reject crushing circuit. Risks for the HPGR circuit were seen as primary process and application related and not equipment specific as HPGRs have been used in a large number of comminution flowsheets (almost exclusively in diamond processing). A significant risk considered for the HPGR circuit was the possibility of the circuit failing to produce a 3 mm product. The design calculations identified that the ball mill feed size ($F_{80}$) needs to be 3 mm or less in order to attain the 50,000 tpd throughput target. As of the Gibraltar project decision date, there had not been any successful installations of HPGRs for copper and gold mines. A couple of projects had been committed to HPGR at the time of the study but, unfortunately, none of these plants were installed and operating in time to influence the Gibraltar expansion flowsheet decision.

## TRADE OFF STUDY FINDINGS

The trade off study concluded that both the HPGR and SAG mill flowsheets were feasible and could meet the overall project requirements. The study found that the SAG mill circuit could be installed for lower capital cost and would have lower operating costs in spite of the HPGRs better energy efficiency. The difference in operating costs can be attributed to the high liner wear rates experienced in the HPGR test work and the low cost hydroelectric power available in central British Columbia. With the SAG mill circuit providing lower capital and operating costs combined with lower technical risk the decision on a preferred flowsheet was obvious. Table 1 lists the advantages and disadvantages of each circuit identified during the study Stage.

## STAGE I EXPANSION DESIGN CONSIDERATIONS AND FLOWSHEET DEVELOPMENT

The results of the 50,000 tpd expansion trade-off study were presented to Taseko management and based on the projected capital costs, operating costs and familiarity with the technology the SAG mill option was confirmed as the preferred alternative. After completion of the trade-off study and identification of the preferred flowsheet, the original plan was to proceed with basic engineering to further identify the equipment requirements and to define a strategy and schedule to complete the brownfield construction activities as well as detail the impacts of these activities on continuing operations.

**Table 1. HPGR vs. SAG for Gibraltar**

| HPGR | SAG |
|---|---|
| **Advantages** | **Advantages** |
| • lower power costs | • well understood technology |
| • no grinding media required | • ability to take advantage of soft ores for incremental throughput improvements |
| • inter particle cracking | • reduce operations and maintenance in secondary crusher plant |
| • narrow range of throughput for ore variability | • lower capital costs |
| • greater online time | • lower labour costs |
| **Disadvantages** | **Disadvantages** |
| • new technology | • does not fit in existing building |
| • does not fit in existing building | • subject to wide ranges in throughput dependent on ore competency |
| • building maintenance crane and building to be designed for 120 ton capacity | • supply of 125mm grinding media |
| • cost of capital spares | • lower equipment availability |
| • handling of coarse particles expected to pass the choke point during startup or shutdown | |
| • elevated ore bin and vibrating screens | |
| • higher capital cost | |

Two significant items arose that changed these plans. First, a review of the long range Gibraltar mine plan identified that the mining fleet could supply a nominal mill throughput of 46,000 tpd without significant capital investment. This resulted in the decision being made to decrease the design capacity of the expansion to match the mine but the decision was also made to continue work based on purchasing the 10.4m diameter SAG mill identified during the trade-off study. This allowed the project to take advantage of the installed grinding power to generate a finer flotation feed, P80 = 300 microns, and improve recovery. Second, with continued improvements in metal market conditions, management decided to immediately embark on the full design and procurement activities for the expansion. The main element identified to expedite the program was the immediate ordering of the SAG mill. Farnell-Thompson was contracted by Taseko to provide engineering and procurement services for the SAG mill and Taseko was able to lock up a 68 week delivery for all of the mill components. (Activities in mining were in general on the rise and within 2 weeks of this order 18 months became the norm for quoted deliveries). The engineering and development of the entire project was then built around the schedule for the mill delivery.

The development of the final project flowsheet for each expansion unit operation as well as a brief discussion of the constraints and considerations that drove the design and the unique features incorporated into each area are reviewed in the following discussion.

## CRUSHING

In general ore delivered to SAG mill grinding circuits is subjected to a single stage of primary crushing and primary crushed ore is reclaimed from a coarse ore stockpile to feed the mill. However, in order to reduce the capital required for the Gibraltar expansion the existing fine ore stockpile was incorporated into the design. The reclaim openings from the stockpile dictated that the maximum ore size delivered to the stockpile was 100mm. To accommodate this, the following changes were considered for the existing secondary crushing plant:

- Removal of one deck from the primary vibrating screen.
- Change the remaining deck to 100mm.
- +100 mm ore remains conveyed to existing coarse ore stockpile.
- Remove secondary screens to effect single pass operation through secondary crushers.

The recommended changes do not eliminate the requirement to operate the crusher circuit but do significantly reduce the amount of ore processed through the secondary crushers (17% of feed at a 1.5x size reduction as opposed to 65% at a 6x size reduction before the changes) and consequently the costs of operating and maintaining the associated equipment. The changes required to the secondary crusher circuit could also be performed with minimum impacts on operations as the circuit has multiple lines which allowed for equipment to be taken offline, modified and then re-commissioned without limiting the ore feed to the concentrator. With the medium work index ore at Gibraltar, the crushing circuit is not required but the existing crushing circuit allowed the project goals to be met with minimum capital and also provides future opportunity to study the effects of precrushing various size components of the mill feed.

**Figure 1. High angle feed conveyor during construction.**

## FINE ORE STOCKPILE AND RECLAIM

Operating as a rod mill / ball mill grinding circuit the fine ore stockpile was designed to feed three parallel circuits. Each rod mill reclaim system consisted of three variable speed belt feeders delivering ore to the rod mill feed conveyor. The feed to each of the belt feeders was reclaimed from the stockpile via three gravity feeders with the amount of feed from each feeder regulated by the depth of ore on the conveyor. To effect the transition of the feeders to deliver -100mm ore for the SAG mill feed, new gravity feeders were designed to incorporate the full opening from the stockpile to the variable speed belt. One gravity feeder was converted per variable speed belt feeder with the other two feeders being deactivated. During plant operations a tunnel was installed adjacent to the mill building and under the rod mill feed conveyors to house a 2.13m wide conveyor to deliver feed to the SAG mill. Each of the rod mill feed conveyors was shortened to deliver ore to this conveyor. The conversion of the stockpile and reclaim system was coordinated on a section by section basis with the conversion of the rod mills to ball mills and consequently could be effected with a minimum impact on operations.

## SAG MILL CIRCUIT DESCRIPTION

The installation of the SAG mill grinding circuit, including the discharge screen and the recycle conveyor systems could be completed as a stand alone green field project with minimal impact on the existing mill operations until the circuit was completed and ready for commissioning. An independent building was erected near the existing mill to house the SAG mill facilities at the same ground floor elevation as the existing mill to effect good access as well as maintaining proper site drainage. The area available to construct the new facilities was constrained by existing infrastructure (main substation), site topography, a creek just north of the plantsite and the desire to keep the distance between the SAG mill and secondary balls mills to a minimum (for operability and to minimize transfer pumping issues). The result of balancing these constraints was that there was insufficient distance between the ore stockpile and the SAG mill feed chute to use a conventional conveying system and the decision was made to incorporate a high angle sandwich conveyor into the ore delivery system (see figure 1).

The selected SAG mill is a fixed speed 10.4m diameter mill with 4850kW twin drives which discharges across a 3.6m x 8.5m single deck horizontal screen (see figure 2). The screen has been mounted directly on top of a concrete pump box to eliminate transfer of vibrations to the building steel. Screen oversize, approximately 12% recirculating load, is transferred using three conveyors back to the SAG feed conveyor. Provision was made in the design and general arrangements for the future installation of a cone crusher if required.

The facility layout resulted in the SAG mill pumpbox being located approximately 150 meters from a central location in the secondary grinding circuit in the existing mill building. Further complicating the slurry transfer is the requirement to be able to distribute the SAG discharge to any or all of the six ball mills that are used for secondary grinding. Gibraltar worked with Derrick Screens to design and size distributors similar to those used for the Derrick multi-start screens (Round Mountain). Derrick recommended operating two distributors for the project flowrate and each of these distributors was designed to split to all of the six ball mills (see figure 3). Three slurry pumps were installed on the SAG mill discharge pumpbox, two operating and one (centre pump) standby, to transfer product to the distributors. The nature of the SAG discharge screen operation results in differences in the particle size distributions along the length of the pump box. With the installation of the two distributors (double splitting the feed) it was anticipated that some of the problems which would result from the mills receiving different size distributions would be alleviated.

**Figure 2. Gibraltar SAG mill and discharge screen prior to commissioning.**

The keys to operating the distributors and minimizing surging through the piping system are ensuring that high line velocities, >6 m/s, are maintained and that the slurry density is maintained between 50% and 60% solids. A secondary benefit of the two operating transfer lines is that a single line can be operated when running low tonnages to allow line velocities to be maintained and eliminate instability and sanding.

## SECONDARY GRINDING

Each rod mill and ball mill circuit operated with both mills discharging into one of two common pump boxes. The rod mills operated open circuit and the ball mills in closed circuit with hydro cyclones. The combined discharge of both mills was directed via dart valves to one of the two common pump boxes, one operating and one standby, each with its' own dedicated cyclone feed pump. To effect the conversion of the mills to their new duty the dart valves were removed, the launders were split and each cyclone feed pump box was set up to operate independently in closed circuit with a cyclopac. The new operating parameters dictated that each of the ball mills be designed to handle 1/6th of the expanded circuit feed versus 1/3rd of the original mill tonnage. The combination of the reduced tonnage per mill and the finer primary grind target necessitated a replacement of all of the cyclopacs.

To achieve the overall tonnage requirements the three rod mills needed to be converted to ball mill operation. The rod mills were identical to the ball mills other than the liners, feed and discharge configurations and the mill operating speeds. New pinions were ordered for the mills to match the existing ball mills, the trommel screens were changed and new discharge trunnion liners were installed. Ball mill liners were also ordered for installation after commissioning when the existing rod mill liners were worn out.

## ROUGHER/SCAVENGER FLOTATION

Prior to the expansion activities, Gibraltar had identified an opportunity to improve flotation performance with the installation of additional flotation capacity and had decided to install an Outotec 160 cubic meter flotation cell prior to the flotation tails pumping system. When the expansion was approved, the decision was made to replace the rougher flotation equipment rather than just installing additional cells to maintain flotation residence time. This decision was based on the condition on the existing flotation machines (after 33 years of service), reduced cell energy requirements as well as anticipated performance and control improvements with modern cells. Based on these factors, negotiations were initiated with Outotec for the purchase of nine additional 160 cubic meter cells to provide two parallel banks of five cells for rougher and scavenger operation.

**Figure 3. Secondary grinding feed distributors.**

The conversion of the rougher/scavenger flotation cells to the big tank cells required the largest interface with existing operations. The installation of the first tank cell outside of the existing flotation area provided the additional flotation residence time to support shutting down and removal of one bank of the existing rougher/scavenger flotation cells. The first bank of four 160 cubic meter cells was installed and commissioned prior to removal of the remaining conventional cells and installation of the second bank of five 160 cubic meter cells.

Each of the new cells was fitted with two pinch valves for level control. A single valve operation is sufficient under normal circumstances; however, during the phased installation of the cells the entire plant throughput was required to be processed through a single flotation bank requiring operation of both valves. This installation also allows significant operating flexibility as plant throughput can be maintained even if a flotation bank requires unscheduled maintenance. To support the new cells upgrades to a number of ancillary systems were required including replacement of the flotation air blowers and overhead crane.

## CLEANER CIRCUIT AND REGRIND

During the period between 1999 and 2004 when the mine was on care and maintenance Gibraltar completed flotation testwork in conjunction with the concentrate hydrometallurgical testwork and flowsheet design activities. This testwork indicated that the copper minerals in the rougher concentrate were present as either coarse mineralization, +100 microns, or fine mineralization, -30 microns, and that each size fraction behaved very differently in the cleaner circuit. To attempt to benefit from these differences, a separate cleaner flotation circuit comprised of five 5 cubic meter flotation cells for treating the fast floating rougher concentrate were designed. Rougher concentrate would be reground using an existing unused 150kW regrind mill prior to flotation and the tails from the new cleaner circuit would be directed to the scavenger cleaner circuit. Concentrate regrind activities were thus split between the 150kW (rougher concentrate) and 450kW (scavenger concentrate) regrind mills. Other possible changes to the regrind circuit, including the installation of the 932kW vertimill identified in the trade-off study, were put on hold until the effects of the revised flowsheet, increasing plant throughput and operating at a finer primary grind could be better determined.

## EXPANSION STAGE 2 (THROUGHPUT OPTIMIZATION)

Prior to completing the Stage 1 expansion activities, Gibraltar initiated an investigation to evaluate the potential to further increase the capacity of the Gibraltar mill by debottlenecking the upstream and downstream unit operations so that the feed rate to the SAG mill purchased for Stage I could be optimized.

Reverting to the previous operating parameters, flotation feed P80 = 400 microns, and optimizing the use of the 9700kW SAG mill the maximum production was estimated at 55,000 tpd, an increase of about 20% over the Stage 1 tonnage. In order to provide assurance that this production rate can be consistently achieved, it was recommended that a cone crusher be installed in the SAG mill recycle system. The decision was made not to include the crusher in the Stage 2 expansion scope but rather to wait until the circuit was commissioned and operated to determine what the limiting production factors were.

In general, the process ancillary equipment (conveyors and pumps) selected in Stage 1 were designed to have adequate capacity to sustain 55,000 tpd. Most of this equipment has sufficient capacity to handle surges of 10% over the Stage 2 tonnage with minor changes to equipment speeds or motor power.

The main processes that were identified as a concern for a Stage 2 expansion were:

- Cleaner flotation.
- Concentrate regrinding.
- Concentrate dewatering.
- Tailings pumping.

## CLEANER FLOTATION

The Stage 2 throughput has a significant impact on the residence time in the cleaner circuit. One of the main efforts of the Stage 1 expansion was to increase rougher flotation residence times in order to optimize copper recovery. The subsequent recommendation was that the existing first cleaner and cleaner scavenger flotation banks be replaced with two banks of five 20 cubic meter flotation cells (See Figure 4) operating in series (1st cleaner followed by cleaner scavenger). To minimize the impact on operating activities, these cells are required to be installed in stages. The initial bank was temporarily installed outside the mill to allow the mill to operate on the temporary cells while the old flotation machines are removed to create sufficient space within the mill building for the permanent installation.

**Figure 4. Engineering model of 20 cubic meter cleaner cell installation.**

## CONCENTRATE REGRIND

An evaluation of the existing regrind mills showed that for regrinding the rougher concentrate and the rougher scavenger concentrates the existing mills were inadequate based on the new design parameters. Based on the expected concentrate mass pulls, it was recommended that the larger (450kW) mill be utilized for the copper rougher concentrate regrind and that a new regrind mill should be installed for the copper rougher scavenger concentrate. The 932 kW vertimill originally identified in the Stage I study was selected for this duty and a location for the vertimill installation inside the existing concentrator was identified (see figure 5).

## CONCENTRATE DEWATERING

At the Stage 1 processing rates, the existing disk filters and rotary kiln dryer can meet production requirements with sufficient time available to adequately maintain this equipment. With the Stage 2 expansion, the operating time required for the existing equipment to meet production requirements combined with the age of this equipment, the poor energy efficiency of thermal drying and the finer concentrate sizing possible with the expanded regrind circuit necessitates replacement of the filters and dryer. Consequently, the Stage 2 expansion scope included installation of a concentrate pressure filter.

**Figure 5. Engineering model for installation of a 932kW Vertimill in the existing concentrator.**

## TAILINGS PUMPING

The existing tailings disposal system includes a Warman 550TUL slurry pump (with 750kW motor) transporting tailings to the impoundment facility via a single 760mm diameter DR21 HDPE pipeline. Two ITT 20x18 SRL-C pumps (with 450kW motor for each pump) in series are installed as a back-up pumps discharging into the same pipeline. The existing system is not adequate to handle the Stage 2 throughput and extensive modifications to the system are included in the project scope. In addition to the existing pipeline, a second 762mm HDPE pipeline will be installed as a back-up to ensure the uninterrupted operation of tailings disposal system. Table 2 presents a summary of the design criteria for the tailings system.

The 762mm diameter pipeline at the Stage 2 nominal and design flow rates will maintain full pipe flow along the entire line profile. The design slurry velocities will be 3.6 and 4.3 m/s for the nominal and design flow rates respectively. Based on operating data, the critical settling velocity of the slurry is estimated to range between 3.8 to 4.5 m/s.

Based on the design criteria it was determined that the existing slurry pumps will require significant modifications and upgrades to handle the increased throughput capacity requirements. The system head is estimated at 80.5m for the design flow rate it was determined that two pumps operating in series would be required. Furthermore, the design head of the system pressure will exceed the maximum pressure rating the existing DR21 HDPE pipe. An analysis of the system hydraulic grade line determined that the first 1,680m of the existing pipe will have to be replaced with a minimum 813mm diameter DR11 (160 psi) HDPE pipe.

The existing tailings pump chamber is not large enough to house the additional slurry pumps and the new tailings pipeline required for the service. In addition, the existing power supply system and MCC's are not adequate for the new duty and will require upgrades. The Stage 2 scope has included installation of a new pump-house with new feed tanks, gland water system, electrical room and an overhead crane for maintenance.

## INSTALLATION AND COMMISSIONING FACTS

- Stage 1—$70m, 21 months—on time and on budget
- Current plant throughput—2,000 tph
- SAG power draw- 7,600 kW, 78 % installed
- Stage 2 budget—$40m—in progress
- Stage 2 Status:
  Vertimill and cleaner cell installation to be completed and commissioned summer 2009
  Tailings pumping system and concentrate filter to be completed summer 2010

**Table 2. Stage 2 Tailings Pumping Design Criteria**

| **Description** | **Unit** | **Data** |
|---|---|---|
| Slurry Data | | |
| Particle Sizing—P80 / D50 | Micron | 420 / 210 |
| Solids Specific Gravity | - | 2.67 |
| Slurry Concentration by Weight | % | 37 |
| Slurry Relative Density | - | 1.30 |
| **Tailings Capacity** | | |
| Solids Transport Capacity—Average | STPD | 55,000 |
| Plant Availability | % | 92 |
| Solids Transport Capacity | STPH | 2,495 |
| Slurry Throughput Capacity (Nominal) | USGPM | 20,718 |
| Flow and Head Design Factor | - | 1.2 / 1.1 |
| Slurry Throughput Capacity (Design) | USGPM | 24,862 |
| **Pipeline Data** | | |
| Pipe Specifications | - | 30" DR21 HDPE, 80 psi |
| Pipe Internal Diameter | Inch | 27 |
| Pipe Length (approximately) | Ft | 12,600 |
| Inlet / Discharge Net Elevation Difference | Ft | (63) |
| Hazen William Factor (Assumed) | - | 110 |

# WET WEATHER DESIGN AND EXPERIENCE FOR MINERAL PROCESSING

Robert Dunne., Jim Orlich, Charles Bucknam, and Dave McLaren

Newmont Mining Corporation, Newmont Technical Facility, 1010 East Dry Creek Road, Englewood, CO80112

## ABSTRACT

Wet climatic conditions especially in high rainfall areas are synonymous with excess surface water issues at mine sites located in this type of location. Proper and adequate water storage and water balance with either sufficient long term evaporation conditions or water treatment and discharge are important consideration in designing a safe and environmentally benign mining operation. Rainfall and evaporation measurements in conjunction with hydrologic modeling provides the basis for the appropriate design criteria for conveyance (channels, pipes, spillways) and retention (collection and sedimentation ponds) structures. Chemical analysis of surface and other waters allows for the proper water treatment strategy to meet all the appropriate regulatory requirements prior to discharge. There are also impacts to mineral processing unit operations (e.g., heap leaching, comminution, flotation and thickening) in high rain fall regions. The issues are discussed and possible solutions are provided.

## INTRODUCTION

At many mine sites, especially those with high rainfall, water management is the most important environmental task facing the mine operators. This paper provides general guidance regarding water management around mine facilities. In addition, details are supplied regarding using hydrologic modeling to determine design flows for hydraulic structures such as channels and spillways and capacity for retention structures. There are also impacts to mineral processing unit operations (e.g., heap leaching, comminution, flotation and thickening) in high rainfall regions. The issues are discussed and possible solutions are provided.

## MINE WATER MANAGEMENT OBJECTIVES

Protecting surface and ground water from contamination (metals, cyanide) is probably the foremost concern to regulatory agencies and the general public. In this regard water at mine sites is classified into two categories, "impacted" and "non-impacted" water. Impacted water includes ground and surface water that has come into contact with mine facilities including rock excavated from the open pits in waste rock, heap leach, or tailing facilities, or with excavated surfaces in the open pits or underground workings. Non-impacted water is surface water that is diverted around the mine facilities (pits, waste rock, heap leach, and tailing facilities); or groundwater that does not emerge into the mine facilities. Any non-impacted water that mixes with impacted water becomes impacted water. The general objectives of water management plans around mine facilities are to:

- Minimize the quantity of impacted water requiring management, sedimentation, or treatment by intercepting surface water with diversion channels before it enters the mine facilities.
- Contain and manage impacted water by conveying the runoff and seepage from the mine facilities to contact water management facilities. Depending on the quality of the contact water, the contact water management facilities may include a sedimentation control facility, a water treatment plant, a process water pond, or a tailing pond.

Typical impacted water management systems for individual mine facilities are:

- Tailing Facilities—Surface water runoff is usually managed within the tailing pond, with the excess water treated, if necessary, before discharge.
- Heap Leach Facilities—Typically the heap leach facility includes an extensive liner, seepage collection, and channel system to collect all water within the facility, which is then sent to the process water ponds, or to the event ponds (i.e., storm water ponds) in the case of large storm events.
- Pits—Surface water is usually managed in the pits by channels that lead to a collection pond or sump. Depending on the water quality, the water may be allowed to infiltrate or sent to a sedimentation control facility or water treatment plant prior to discharge. Where the pit bottom elevation is below the groundwater level, a dewatering system will be installed, which will inhibit the migration of contact water away from the pit.

Surface and groundwater monitoring systems must also be included to detect any migration of impacted water from the facilities.

## PERMITTING REQUIREMENTS

Permitting documents for a mining project require data on rainfall intensity that pertain to the ambient meteorological conditions in the locality of the proposed mine site and support the design of facilities for operations. Rainfall data is needed to:

- Develop a site-wide water balance
- Appropriately size process and diversion structures
- Develop sediment control measures
- Support the development of the Environmental Impact Statement (EIS) document

To support these aspects of data requirement for the general climatic conditions in the region of influence it is important to properly characterize storm events in the area of impact. A review of the local available climate data as well as existing adjacent meteorological stations is the first step in determining what gaps exist. If no rainfall data exists that can be used to assess the site conditions, then a site specific evaluation should be performed to determine if multiple meteorological stations are required to capture site specific climate data variations that may exist due to elevation change, and/or geographic and orographic effects. The appropriate number of meteorological stations should be installed and maintained throughout the life of the project. If sufficient data exists, the data collection requirements on-site can be limited to daily precipitation.

The following parameters are usually monitored:

- Precipitation- 5 minute and daily
- Wind- speed and direction
- Pan Evaporation- daily
- Temperature- hourly
- Barometric Pressure- hourly
- Snowfall Measurements- depth, melt patterns, moisture content

## SURFACE WATER MANAGEMENT

For areas with moderate to heavy precipitation, an in-depth understanding of the surface water conditions (flow rate, TSS concentration and quality) during storm events is required. Surface water flow rate and quality should be collected at locations in all major drainages, springs, seeps, irrigation canals, and major receptor locations adjacent to the ore body as well as at locations that are not anticipated to be impacted by the proposed project. These locations support a before-after-control-impact experimental design for data analysis. Continuous surface water monitoring locations coupled with rainfall monitoring in the contributing catchments is one effective way to collect surface water data during storm events and is required to understand the complex relationships between rainfall, runoff and water quality. These relationships are critical as storm water is a primary migration pathway for environmental impacts from mine operations. Storm water runoff should be analyzed for organic carbon and total metals (Ag, Al, As, Cd, Co, Cr, Cu, Hg, Ni, Mn, Mo, Pb, Sb, Se, Tl, V, and Zn) with additional data on time and flow rate collected over the entire storm and runoff duration. This data should be analyzed to develop predictive tools for rainfall-runoff modeling, and can be used to understand background conditions that could be considered when designing mitigation measures for management of impacted water and sediment control. Investment in continuous monitoring may prove to be invaluable in managing the risk and costs associated with impact mitigation related to surface water.

The assessment program should also consider proposed and alternate locations of project facilities to anticipate collection of up-gradient and down-gradient water quality and quantity. Additionally, the sampling sites should be sited so that they are capable of serving as long term effects monitoring locations in areas that are going to disappear as a result of future project activities and/or construction. Typically 2 years of data is required for development of an EIS and to adequately record the variability in surface water quality and quantity within the specific area of project influence. The first year of monitoring should include monthly sampling. The second year of sampling can be reduced to quarterly sampling if the first year of data suggests limited variability in the quality and quantity on a monthly basis.

Typical surface water monitoring requirements are presented herein. A review of the host country laws is always required to identify any additional local requirements for baseline data collection that are different from the standard requirements land typically include, but are not limited to, the following: field parameters (pH, specific conductivity, temperature, turbidity, dissolved oxygen), general parameters (alkalinity, total dissolved solids, total suspended solids), major ions, nutrients (nitrate, ammonia, phosphate), and dissolved and total metals. Cyanide may also be included for sites that plan to use cyanide in their operations.

## RAINFALL MEASUREMENT

A rain gauge (also known as a udometer or a pluviometer) is an instrument used by meteorologists and hydrologists to gather and measure the amount of liquid precipitation over set period of time. Most rain gauges measure precipitation in millimeters with the level of rainfall reported as inches or centimeters with amounts of 0.25mm or less may be recorded as a trace. Rain gauges have limitations such as reading in a localized area, and high winds, freezing, or snow accumulation can affect results. Gauges are typically placed far enough from structures and trees to minimize interference with readings. The common types of rain gauges are described below:

### Standard Rain Gauge

Developed around the start of the 20th century, the standard rain gauge consists of a funnel attached to a graduated cylinder (see Figure 1). The typical standard rain gauge can measure up to 25mm.

### Weighing Precipitation Gauge

Consists of a storage bin which is weighed to record mass and some models use various types of data recorders. Advantages are that it does not underestimate intense rain, and it can measure other forms of precipitation such as snow or hail. Disadvantages include cost and increased maintenance.

**Figure 1. Standard rain gauge**

**Tipping Bucket Rain Gauge**

The tipping bucket rain gauge consists of a large copper cylinder with a funnel attached to the top to channel precipitation onto one of two small buckets or levers which are balanced in same manner as a scale (see Figure 2). After an amount of precipitation equal to 0.2 mm (0.007 in) falls, the lever tips and an electrical signal is sent to a rotating pen chart or electronic recorder. The major advantage is it provides a measure of the rate of rainfall as well as the total amounts, but is not as accurate as the Standard Rain Gauge (U.S. National Weather Service, Groisman,1994).

**Optical Rain Gauge**

Consist of a row of collection funnels each with a laser diode and phototransistor positioned below. When enough water is collected to make a single drop, it drips from the bottom, falling into the laser beam path which is transmitted to a recorder.

**Radar Measurement of Rainfall**

Radar measurement of rainfall is relatively new technology, it offers particular promise for better understanding storm behavior and rainfall intensity over a complete area of the storm. The methods also include the use of radar to analyze the spatial distribution and intensity of rainfall during storms. The most important advantages of the radar rainfall data are the high spatial and temporal resolution, the large areal coverage, and the ability to obtain and analyze data for locations where no rain gauge data is available (U.S. Geological Survey, 2006a, 2006b).

Meteorological radar is a remote-sensing system that sends an electromagnetic pulse and receives the back-scattered radiation from raindrops in the atmosphere. Radar measurements of the backscattered energy are directly related to a physical quantity called reflectivity, Z, with units of mm6/m3. Radar-reflectivity data are typically obtained in the form of a volume scan that is available every 5–15 minutes and consists of data expressed in polar coordinates. Radar reflectivity and rainfall intensity are functions of the raindrop-size distribution in a sampled volume. Rainfall intensity can be estimated from reflectivity. When rain-gauge data are available then radar calibration or gauge adjustment can be made. It needs to be noted, however, that radar rainfall data are not direct measurements and can, therefore, have relatively high uncertainty. Radar data are sampled from a volume above the ground surface, whereas the gauged rainfall data are collected at the ground surface. Currently, there is no way to truly validate estimates of rainfall depths and intensities from radar. Although rain-gage data can be considered as ground truth, the sampling area (domain) for a gage

**Figure 2. Tipping bucket rain gauge**

is only the size of the gage collector (about 200 cm2), whereas the domain for radar data is several orders of magnitude larger (about 1 km2).This size difference is most significant because of the high spatial variability of rainfall. On this basis, comparisons between radar-sampled rainfall and gauged data need to be used with caution.

## RAINFALL INTENSITY

The intensity of rainfall or the IFD (Intensity-Frequency-Duration) statistics is important from the standpoint of the design of process facilities. In this context, Intensity refers to the rainfall rate, and Frequency is the number of times that a specific limit is exceeded over a given period of time or Duration. The climatic process that governs precipitation is stochastic or completely random at time. This makes it impossible to predict what events are going to occur. Because of this a statistical approach is used. This involves analysis of available information and the development of defined extreme precipitation events for design purposes. To prevent over-conservative designs and to promote a standardized approach the extreme events selected are generally those associated with a given probability of occurrence. For example the pregnant solution pond might be designed with sufficient freeboard (i.e., storage space above the normal maximum operating level) to safely contain a precipitation event which has a one percent probability of being equaled or exceeded in any one year of operation. Another way of describing this precipitation event is to refer to it as the 1 in 100-year event (i.e., obtained by dividing 1 by 0.01). Because of the number of possible combinations of duration and frequency, these statistics are typically reported as a series of intensity-duration graphs at different seven standard frequencies. In some instances government agency requirements will prescribe the 'probability of the event' that the designers must accommodate when the designing some of the project features.

The analysis of IFD's referred to as frequency analysis involving typically data from multiple stations is commonly used in:

- Design and risk assessment of dams and bridges
- Design of roof and storm water drainage systems
- Flood plain management
- Soil conservation studies
- To express the "severity" of a single rainfall event

## HYDROLOGICAL MODELING

The main purpose of hydrologic modeling is to determine design flows for conveyance structures (e.g., channels, pipes, spillways) and retention structure capacity for impoundments (e.g., collection and sediment ponds). The models generally output a hydrograph, which can be used to determine the design flow or required capacity of a retention structure. The precipitation events used in a model should be regularly updated using site rainfall/precipitation measurements. The statistical evaluation of the available on-site and regional precipitation data should be reassessed as additional data is collected to verify the design precipitation depths and precipitation distributions.

Ideally, the model should be calibrated using stream flow measurements during storm events. The calibration effort usually focuses on the calibration of the basin characteristics (e.g., runoff coefficients or curve numbers). For example, if a given storm event resulted in a higher peak flow than that estimated by the model, the basin characteristics used in the model should be modified (i.e., increase the curve number) to better match the measured flow.

Numerous hydrologic modeling programs are available, such as the Hydrologic Engineering Center hydrologic software (HEC-HMS) that is a free and available at http://www.hec.usace.army.mil/software/hec-hms/., the SEDCAD 4 for Windows specifically developed for the design and evaluation of alternative erosion prevention and sediment control systems with a focus on earth-disturbing activities. The program includes hydrology and hydraulics components that can be used to determine design flows and storage capacities (order from SedCAD@cs.com or warnerhofc@aol.com) and Hydrological Simulation Program FORTRAN-(HSPF) (http://www.epa.gov/ceampubl/swater/hspf/).

## MODEL INPUTS

Hydrologic modeling programs generally require the following inputs, which are described in the following sections:

### Precipitation Event

Hydrologic models generally require a design precipitation depth (e.g., 125 mm in 24-hours for the 100-year event). The design precipitation depths should be determined for the site based on a statistical evaluation of the available on-site and regional precipitation data.

The models generally require that the precipitation depth be distributed over the time period (e.g., 24 hours). If on-site storm distributions have not been developed, distributions developed by the US Soil Conservation Service (SCS), now known as the Natural Resources Conservation Service (NRCS), could be analyzed and an appropriate distribution used. Information on these storm distributions can be found in Technical Release 55 (TR-55) by the United States Department of Agriculture, NRCS, Conservation Engineering Division dated June 1986 (available at http://www.wcc.nrcs.usda.gov/hydro/hydro-tools-models-tr55.html).

The antecedent moisture condition, which is the soil moisture conditions at the beginning of a storm, affects the volume of runoff generated by a particular storm event. This condition, which depends on the amount of precipitation prior to the storm event, is discussed in Chapter 10 of Part 630 Hydrology National Engineering Handbook by the United States Department of Agriculture, NRCS, Conservation Engineering Division (available at http://www.wcc.nrcs.usda.gov/hydro/hydro-techref-neh-630.html).

### Basin Characteristics

The size of the basin (i.e., the catchment area) is a critical input to a hydrologic model. This area is usually determined by digitizing the watershed area that will report water to the point of analysis on a topographical map in order to calculate the surface area.

The time of concentration is the time it takes the drop of water falling on the hydraulically most remote point in the watershed to travel through the watershed to the point of analysis. Numerous methods have been developed for determining the time of concentration. Refer http://www.wcc.nrcs.usda.gov/hydro/hydro-tools-models-tr55.html).

Runoff coefficients or curve numbers are specified to determine the amount of precipitation that runs-off by accounting for the amount of precipitation absorbed by the soil.. This involves determining a soil type classification and an appropriate runoff curve number to get the total excess precipitation. While the SCS Runoff Curve Number (CN) method was originally developed for use in agricultural areas in the United States, it is a simple and widely used method that has been successively applied in mining projects. Refer http://www.wcc.nrcs.usda.gov/hydro/hydro-tools-models-tr55.html).

### Hydraulic Structure

Hydraulic structures, such as channels, reservoirs, or ponds, can have an important impact on the results of the hydrologic model because they generally reduce the peak flow (i.e., attenuate the flow). Spillways for tailing facilities or other reservoirs should be designed by routing the flow through the hydraulic structure using appropriate routing calculations. The routing calculations will require the inflow to the reservoir, the stage-storage curve for the reservoir (i.e., how much water is stored in the reservoir for a given elevation), and the stage-discharge curve (i.e., how much flow goes out for a given height of water) for the discharge facility (e.g., the spillway). Referhttp://www.wcc.nrcs.usda.gov/hydro/hydro-tools-models-tr55.html)

## DETERMINING SITE SPECIFIC FLOW CRITERIA

In general, all hydraulic structures should be designed based on the level of risk of failure. For example, if a tailing facility is upstream of a habitable area and a failure of the facility could result in a fatality, a high design event would be warranted. On the other hand, if a diversion channel is being constructed along a minor access road that will be in operation for three years, a low design event may be warranted. For determining the design flow criteria, an evaluation of the cost versus risk (i.e., benefit) should also be used to ensure that an appropriate design flow has been selected.

The probability of having one or more events that exceed the design event during the life of a structure can be calculated using the following equation:

$$P = 1 - (1 - 1/T)^n$$

Where P is the probability that a T-year event will be exceeded one or more times during the n-year life of the structure. For example, if a 25-year design event is selected for a diversion channel that will be in operation for 4 years, the probability of having one or more events that exceed the 25-year design event (i.e., the probability of the channel "failing") is approximately 15%. If the diversion channel is in operation for 25 years, the probability increases to 64%. If the diversion channel is in operation for 100 years, the probability increases to 98%.

## GENERAL DESIGN CRITERIA FOR WATER CHANNELS

At the majority of mine sites, channels are the most common water management structure. This section provides some guidance regarding design criteria for channels.

Freeboard should be included in the channel. As a guide, a minimum freeboard of 0.3 m should be included for flows less than 10 $m^3$/s and 0.6 m for flows greater than 10 $m^3$/s. Additional freeboard may be warranted depending on the flow conditions (i.e., at curves or turbulent reaches).

Channels are generally designed as trapezoidal sections, with side slopes that are stable for the given material (e.g., typically 2H:1V to 3H:1V if not lined with concrete). Where the flow velocities exceed the non-erodible velocities for the native material in which the channels are constructed, erosion protection must be provided. When natural erosion protection such as grasses are not feasible, the preferred erosion protection is often riprap armoring with a geo-textile placed underneath, which has advantages due to its flexible nature compared with rigid systems such as concrete or grouted riprap.

Where the channels cross existing watercourses, inlet structures should be constructed to direct the flow from the watercourse into the channels. Where culverts are required, measures to prevent plugging from sediments should be provided upstream of the inlet and erosion protection should be included at the outlet. An access road should be included along the channel for inspection and maintenance.

## SITE WIDE WATER BALANCE

In order to evaluate the impact of climate conditions, a water recovery model is required to prepare a site wide static water balance. A major factor is the balance of rainfall and evaporative losses as well as water held within the tailings solids.

A water recovery model for tailings facilities is presented in a paper by C Wels & A. MacG. Robertson. The model describes the total water losses ($L_{total}$) consist of four major components:

$$L_{total} = L_{ent} + L_{evap} + L_{rew} + L_{pond}$$

Where:

$L_{ent}$ = Entrainment Losses = void ratio after initial settling x tonnes of tailings / $SG_{tails}$ solids

$L_{evap}$ = Pan Evaporation Rate x Pan Factor x Flooded Area

$L_{rew}$ = Rewetting Losses

$L_{pond}$ = Pan Evaporation Rate x Pan Factor x Pond Area

Details of the components and calculations are presented in the paper. Another example of water recovery calculations from Escondia Mine in Chile are presented in a paper by Chambers, Plewes, Pottie, Murray and Burgess (2003).

## OVERVIEW OF MINE AND METALLURGICAL WATER TREATMENT

Water treatment requirements are generally determined by the mine and metallurgical plant water balances, respectively. In arid climates the metallurgical plants are normally zero discharge facilities, so the plant normally consumes water and treatment is limited to preparing mine, ground or surface water for consumption by the plant which is generally minimal.

Water may need to be treated for discharge to surface facilities from metallurgical plants. In the case of gold leaching, the concentrations of cyanide that are allowable in the discharges should follow guidance provided by the International Cyanide Management Code (http://www.cyanidecode.org/). There are several processes for destruction of cyanide in tailings slurries, the most common being use of Caro's acid and copper catalyzed sulfur dioxide and air. Hydrogen peroxide is also commonly used with copper catalysis for cyanide treatment of clear solutions, particularly in ponds.

In arid environments it is not uncommon to generate excess water from mine dewatering operations which cannot be totally consumed by the dust suppression operations and the water needs of the processing plants. Often trace metals such as arsenic, selenium and antimony need to be removed from the excess mine water prior to discharge; addition of iron solutions followed by solid liquid separation is often used for mine water treatment. When acid rock drainage (ARD) from sulfide oxidation requires treatment, high density sludge lime addition plants are commonly used to neutralize acidity and cationic metals due to their high efficiency of lime usage. Control of mine and process drainage discharges impacted by sulfide oxidation should follow the guidance of the Global ARD (GARD) Guide (www.gardguide.com).

Wet environments with positive water balances requiring containment of both mining and metallurgical processing influenced waters provide the greatest challenge for water treatment as there are a large range of metals and anions which have potential impacts on the downstream receiving water uses. Metallurgical plants in these environments often will need to treat excess water from leaching containment solution ponds and decant water from the tailings facility which may involve a combination of the cyanide destruction and mine water treatment methods identified above. Use of membranes for excess water treatment is gaining wider use in the industry, since they provide a means to contain a wide range of soluble constituents impacting the uses of the water within the processing facility (Groisman, 1994). Technology development in support of water treatment requirements is an industry priority to reduce risks of environmental impacts from metallurgical operations as well as improvement of an important community resource, water.

## CHECKLIST OF KEY PROCESS ATTRIBUTES FOR RAINFALL DESIGN

The following indicators can be used to measure performance:

- Are water management plans developed for the mine facilities?
- Are the design flows and capacities specified for the diversion and retention facilities in accordance with appropriate standards and/or applicable risk-based criteria?
- Are design flows and retention structure capacities being determined using an appropriate hydrologic model?
- Is site specific precipitation data being collected to determine design storm events?
- Is a statistical analysis being used to determine design storm events for different return periods for the site?
- Has a storm distribution been selected for the site?
- Has an appropriate method for calculating time of concentration been determined for the site?
- Have site specific runoff coefficients or curve numbers been selected for the different land cover types?
- Is the hydrologic model calibrated with flow measurements?

## GENERAL DESIGN CRITERIA FOR TAILINGS AND HEAP LEACH FACITILY STORAGE

Tailing storage facilities need to be appropriately designed, constructed, operated, closed and reclaimed to prevent overtopping and for guidance, at a minimum, contain accumulations for fluids resulting from a 25-year, 24- hour storm event.

Tailing storage and Heap Leach Facilities should be appropriately designed, constructed, operated, closed and reclaimed. For guidance these should withstand the run-off from a 100-year, 24-hour storm event.

Runoff generated outside tailing storage and heap leach facilities should be diverted away from the facility. Temporary storm water structures should be designed based on the level of risk of failure. Permanent storm water structures should, at a minimum as guidance, be designed, operated and closed to convey and withstand the 100-year, 24-hour storm event.

Permanent surface water run on diversion structures as guidance should be sized, at a minimum, to convey flow from the 100-year, 24-hour storm event. For facilities up gradient of sensitive regions (e.g., habitations, high-value ecosystems, etc.), additional capacity, or protection, requires consideration based on risk associated with failure.

## MINERAL PROCESSING ASPECTS RELATED TO HIGH RAINFALL ENVIRONMENT

### Gold Heap Leaching

The worst water balance situation occurs in cool, damp climates such as high altitude operations. In such climates, rainfall and snowfall may be significant and evaporation is minimal (Kappes, www.kcareno.com). Generally such heaps can stay in water balance with an aggressive program of summer sprinkling. Arctic heap leaches have been able to stay in water balance because precipitation is lower than the total water requirement needed to saturate the ore.

If the heap and pond systems are properly designed, the active leaching area can be up to 40% of the total area collecting rainfall; it is therefore possible to operate in water balance when rainfall is 2.5 meters/year. For these operations, very large solution surge ponds are required.

Overall evaporative losses include the sprinkler losses, convective loss from air flowing through the heap, and losses due to heating/evaporation from ponds and from other areas not sprinkled. These have been determined at several operations to be up to 20% of total solution pumped in summer months and 10% annually. Thus, direct sprinkler loss accounts for about 60% of the total. Use of drip irrigation can reduce but not eliminate evaporative loss. In tropical climates, noticeable losses occur even during the rainy season. Kappes Cassidy claims that where rainfall is seasonal and up to 2.5 meters per year, the overall annual evaporative loss from all sources, when using wobbler-type sprinklers operated 24 hours/day, is about 7% of solution pumped. Based on a typical heap application rate of 10 l/hr/sq. m, or 88 meters per year (9 inches/day) the evaporative loss of 7% equates to 6.2 meters per year on the areas actually being sprinkled.

Reusable pads (on-off pads) provide significant advantages in areas of high rainfall and this is primarily related to the limited area in which the ore is leached. The size of solution ponds is subject only to the area of the reusable pad. This limits the impact of hydrological cycle and provides a high level of solution control (van Zyl et al 1988).

The collection system/pond outside of the heaps must be designed to accommodate solution from operation as well as storm runoff. The depth and area of heap leach pond(s) are usually chosen to minimize liner costs. However, minimizing evaporation losses and direct precipitation impacts on both the heap water balance are frequent considerations that influence pond depth or shape (van Zyl, et al, 1988.

Open trenches in the heap leach area must be designed to convey high flows from rainfall events without erosion, ripping of synthetic liners or spillage (van Zyl et al, 1988).

**Examples of Gold Heap Leach Projects in High Rainfall Areas**

**Sanu Heap Leach-Ghana.** The Sanu heap leach site that was in operation in 1990's was situated in an area of very high rainfall (Cassiday et al, 1991). Precipitation averages 2.1 meters/year, however in one year of every ten the annual rainfall exceeds 3 meters. Rainfall occurs predominantly in a 5-month rainy season, with occasional one-month rainfall of 0.5 meters.

Rainfall, even in the worst months, was not sufficient to affect heap operations or performance; but was accounted for in solution storage and removal systems. With sufficient pond capacity and a large area of installed sprinklers, it was possible to design a system to stay in complete water balance even in the worst years. However in the design there was allowance for neutralization and discharge in one year of every five.

Six ponds were constructed: three process ponds, a primary overflow pond (that partially fill every rainy season), a secondary overflow pond (that was designed to fill one year out of five), and a neutralization/makeup water pond (used on a batch fill/neutralize/discharge cycle as soon as the secondary pond was filled). The three process and two overflow ponds were identical, each having a capacity of 11600 cubic meters. The neutralization/makeup water pond had a capacity of 1600 cubic meters.

Additionally, evaporation pumps and sprinklers capable of evaporating 20000 cubic meters of solution per month were installed. These were run continuously each rainy season whenever the process ponds were at 75% capacity or more.

**Yanacocha Mine-Peru.** Rainfall at the Yanacocha Mine is high and evaporation rate is low (altitude 3500 meters). Covered side-slopes with plastic has been used to minimize rain collection. Others (Rio Chiquito, Costa Rica - Mallon Minerals Corp) have tried to cover the entire heap during the rainy season, but this has not worked very well because of the mechanical difficulties of moving the cover.

At the Yanacocha Mine the Maqui and Carachugo heaps had skirts or slope covers on one of the slopes to reduce excess water treatment requirements (Esplin, 2009). Liners were thin (40mil) and not much slope preparation was undertaken other than rearranging medium to large rocks and bucket tamping the slope so the liner did not tear during installation or small slope movements. Drip lines were installed beneath them so side slope leaching could continue unabated. The reason this was discontinued was that the environmental group were not convinced the valves directing rainwater off the pad could be left open passively due to the potential risk of a plastic tearing allowing commingling of leach solution at the heap perimeter. In the mid-90's it was proved economically that the cost of water treatment exceeded the installation of side slope liners (Skurski, 2009).

**Flotation**

In areas subject to intense rain fall, some flotation plant designers have incorporated covers to protect the froth zones from collapsing and reducing metal recoveries. There appears to be no consistent criteria applied with regard to when covers should be employed or how to do so leaving it to the discretion of the owners and designers. As an example, the Batu Hijau Project in Indonesia chose to place covers directly on top of each cell which made it necessary to install a number of pieces per cell. The cell covers add to the time required to access and maintain the cells and make it more difficult to observe cell performance and have subsequently been removed. Figure 3 shows another approach recently taken for the Phu Kam Project in Laos. The photo show covers placed over the mechanisms and extending just to the edge of the top of the cells. This design does not impair the ability to observe cell operation and provides some shade for the operations, but might allow some rain to impact the froth zone from the sides. Access to the cell appears to be better and the support columns were used for cable trays and piping support as well. This type of cover would also need to be easily removable to allow access to the mechanisms.

**Thickeners**

For high rainfall areas, the most important design consideration is to 'oversize' the overflow launder, box and pipe, to cater for increased thickener inflow. The calculation for increased overflow capacity can be made from site rainfall data (Brennan,2009).

**Comminution**

The combination of clayey ores and rainfall makes it is a challenge to effectively design and operate a comminution circuit. As a general rule, the favored comminution circuits for this included MMD crushers and or ball mills. The ball mills usually have a low ball charge and operate more as a scrubber. Ball loads vary according to the quantity of "hard' component in the feed. Some circuits pre-screen to remove the hard component and this is crushed prior to grinding.

For fresh rock circuits, covers (e.g., roof) over the crushers are recommended as are appropriate covers on conveyor belts for circuits incorporating High Pressure Grinding Rolls (HPGR). High moisture content in the HPGR feed will result in high roll wear due to erosion of the autogenous wear layer between the studs.

**Figure 3.**

**Gold Circuits**

Tropical climates with abundant vegetation generally have organic contaminants (humates and fumates from the degradation of the vegetation) that will impact the effectiveness of carbon adsorption circuits that remove gold from the cyanide leached slurry. Circuit variations to address this issue include more carbon activation capacity, CCD with Merill-Crowe precipitation and resin adsorption.

## SUMMARY

Wet climatic conditions especially in high rainfall areas are synonymous with excess surface water issues at mine sites located in this type of location. Proper and adequate water storage with either sufficient long term evaporation conditions or water treatment and discharge are important consideration in designing a safe and environmentally benign mining operation. Rainfall and evaporation measurements in conjunction with hydrologic modeling provides the basis for the appropriate design criteria for conveyance (channels, pipes, spillways) and retention (collection and sedimentation ponds) structures. Chemical analysis of surface and other waters allows for the proper water treatment strategy to meet all the appropriate regulatory requirements prior to discharge. There are also impacts on mineral processing unit operations (e.g., heap leaching, comminution, flotation and thickening) in high rain fall regions. The issues and possible solutions are provided.

## ACKNOWLEDGMENTS

The authors acknowledge the following Newmont personnel for their input: Bill Patterson, Brian Simpson, Karen McCaffery.

## REFERENCES

Brennan, S., 2009, Outotec, Perth, Australia, Private Communication

Esplin, T., 2009, Newmont Mining Corporation, Yanacocacha Mine, Private Communication 2009

Cassiday, M.W., Cox, T.J., Domfeh, I. A., Stalker, I. and Jaczkowski B.M., 1991, The Ashanti (Ghana) Heap Leach-Treatment of Decomposed Schist in Tropical Environment, World Gold, Cairns, Australia

Groisman, P.Y.,1994, The accuracy of United States Precipitation Data." Bulletin of the American Meteorological Society 75(2): 215–227.

Kappes, D., Professional Papers, Heap Leach Design and Practice, www.kcareno.com

Skurski, M., 2009 Newmont Mining Corporation, Private Communication

The Tipping Bucket Rain Gauge, U.S. National Weather Service

U.S. Geological Survey., 2006a, Application of Methods for Analysis of Rainfall Intensity in Areas of Israeli, Jordanian, and Palestinian Interest *Compiled by the* U.S. Geological Survey *For the* Executive Action Team, Middle East Water Data Banks Project ISBN 0-16-076155-7

U.S. Geological Survey., 2006b, Application of methods for analysis of rainfall intensity in areas of Israeli, Jordanian, and Palestinian Interest; *Supplement:* Radar analysis of rainfall intensity in the Mujib Basin. *Compiled by the* U.S. Geological Survey *For the* Executive Action Team, Middle East Water Data Banks Project ISBN 0-16-076155-7

van Zyl, D.J.A., Hutchinson, I.P.G. and Kiel, J.E. 1988, Introduction to Evaluation, Design and Operation of Precious Metal Heap Leaching Projects, SME , Littleton, ISB 0-87335-074-X

Wels, C. and MacG Robinson, A., Conceptual model for estimating water recovery in tailings impoundments, Robertson GeoConsultants, Inc., Vancouver, Canada

Chambers, B., Plewes, H., Pottie, J., Murray, L.,, and Burgess, A., 2003, Water Recovery from a Mine in the Atacama Desert, Water in Mining- October, Brisbane, Australia

# ENGINEERING CHALLENGES IN TODAY'S DYNAMIC MARKETPLACE

Kathy Altman, Gordon Shepherd, and Steve Pozder

Samuel Engineering, 8450 E. Crescent Pkwy, Ste 200, Greenwood Village, CO, U.S.A.

## ABSTRACT

Due to the cyclical nature of the mining industry, engineering companies that provide engineering, procurement and construction management (EPCM) services routinely confront two major challenges that are related to available resources. In the good times there is a backlog of work and not enough people to complete the work which oftentimes results in unhappy clients. In the down times there is a deficiency of work and an associated overstaffing that challenges the financial capabilities of the company. This paper discusses methods for addressing both of these situations and addresses the pros and cons of the various decisions.

During the previous year, these swings have occurred in a time period that is very short leading to unprecedented challenges. During the September 2008 Mine Expo in Las Vegas, the outlook was positive and the mood was (overly) optimistic. Base metal prices and precious metal prices had reached unprecedented high prices at the same time. In less than two months, the prices of base metals were falling and anxiety was turning to fear.

This paper also offers some observations regarding the potential benefits that oportunistic operating companies can achieve if they use the economic downtimes as an opportunity to advance their projects and the long-term negative effects of failing to do so.

## INTRODUCTION

### Inflation – Deflation Cycle

The global economy follows the inflation-deflation cycle. The economy, particularly the mining industry, alternates between good times/high metal prices and bad times/low metal prices in a pattern that resembles a sinusoidal curve, as shown in Figure 1.

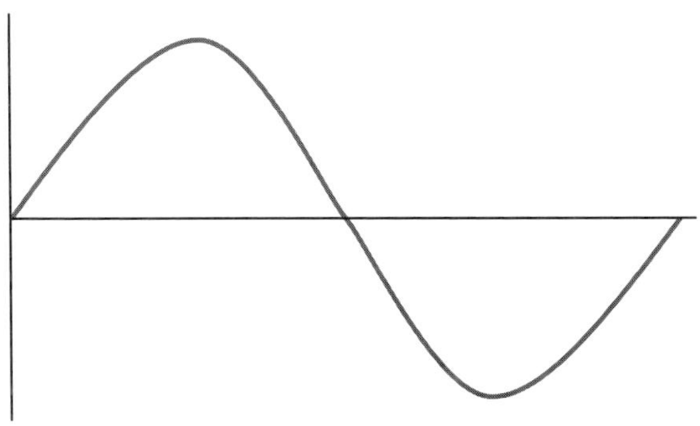

**Figure 1. Inflation – Deflation Cycle**

### Metals Prices

When historical metals prices are evaluated, they follow trends that are similar to the inflation-deflation cycle although they do tend to have a general upward trend over the long term. The graph for the spot copper price over the past five years, as provided by Infomine (2009) is shown in Figure 2.

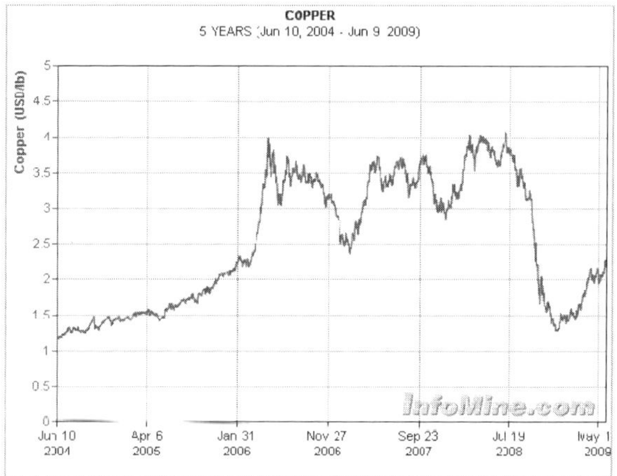


**Figure 2. Historical copper price (2004 to 2009)**

The graph of the historical copper prices shows the precipitous drop in the spot copper price during the second half of 2008. The graph of the Dow Jones Industrial Average (DJI) provided in Figure 3 (Yahoo Finance) shows a similar drop although the DJI started to decline in 2007.

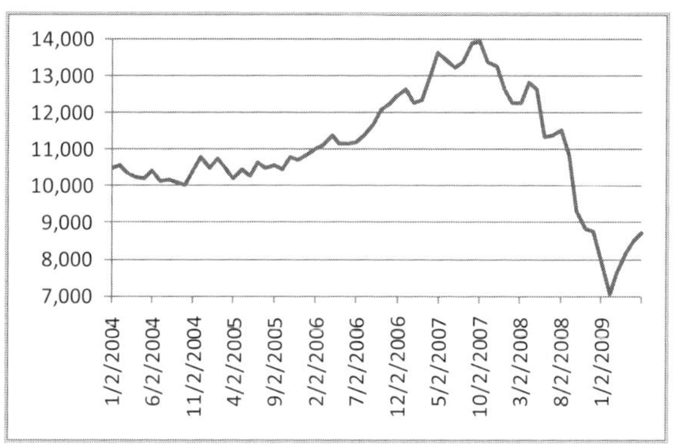


**Figure 3. Historical Dow Jones Industrial Average (2004 to 2009)**

### Additional Challenges and Considerations

In addition to the challenges associated with the cyclical nature of the economy, additional challenges significantly impact the mining industry. Some of the more important challenges include the following.

- Workforce demographics
- Fluctuations in global exchange rates
- Varied business approaches depending on the country
- Mergers and acquisitions

## IMPLICATIONS FOR EPCM SERVICE PROVIDERS

As the economy weakened and metal prices declined, EPCM companies saw a significant decline in the amount of work they were asked to do. Mining companies cancelled or postponed project development due to weakening market conditions and inability to get financing.

### Hours and Availability

Since EPCM companies make their money based on the number of hours billed by their employees, it is common to graph the estimated hours that will be worked versus the number of hours that are available using the staff that is employed by the company. A sample company is used for this demonstration. Assuming there are 100 employees and the normal work week is 40 hours, there are 4,000 billable hours available each week. Some of the staff members are required for support functions such as accounting, human resources, etc. Their work is not typically billed to the clients so it is unlikely that the billability rate will reach 100 percent. A healthy average billability is 65 to 75 percent of the total hours that are worked in a company. In an ideal world the lines on the graph between availability and hours worked are parallel with a uniform gap between the lines, as shown in Figure 4. The "dips" in the graphs corresponds to weeks that have holidays.

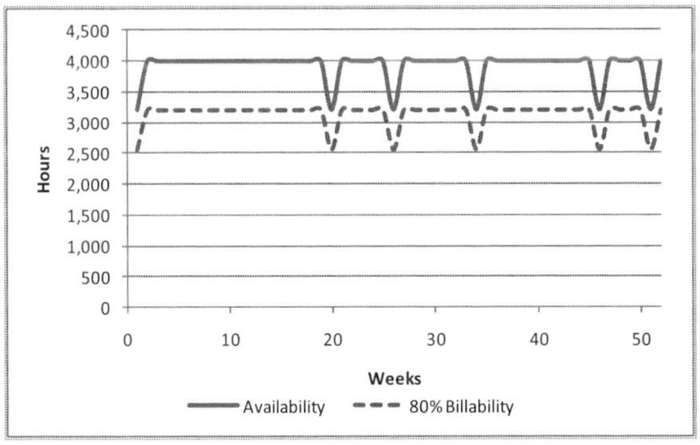


**Figure 4. Ideal EPCM company billability**

In actuality, the estimated number of hours to be worked will decrease over time due to the length and declining manhour nature of projects and the fact that companies do not typically know when they will be starting projects several years in advance. Therefore, the anticipated backlog of work drops off over time, as shown in Figure 5.

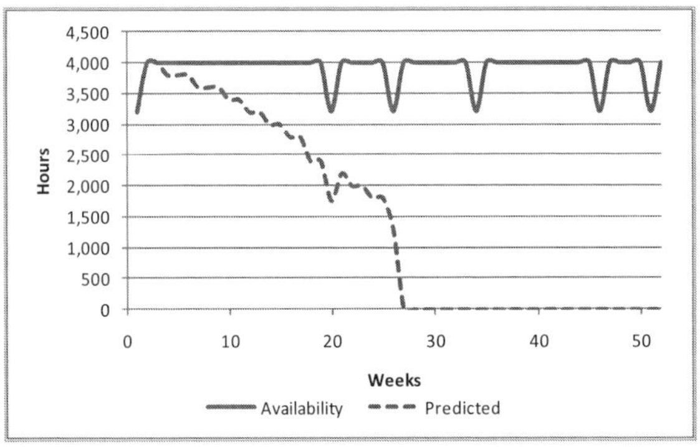


**Figure 5. Typical Backlog of Work**

**Typical Financial Risk Cycle**

Everod Samuel (2009) explained the financial risk cycle for all companies. As all companies follow the cycles, corporate emotions seem as though they are on a roller coaster. As the financial cycle moves toward the peak, emotions move from thrill to peak optimism. It is common for business managers to feel complacency accompanied by invincibility. At this point they reach their point of maximum financial risk. As the cycle starts to fall, anxiety transitions to denial, fear, desperation, panic and capitulation. At the minimum point of the curve, if emotions do not assume control of the actions and logic prevails, a company reaches the point of maximum financial opportunity, as shown in Figure 6 (Prudent Investor). Some of the opportunities will be explained subsequently.

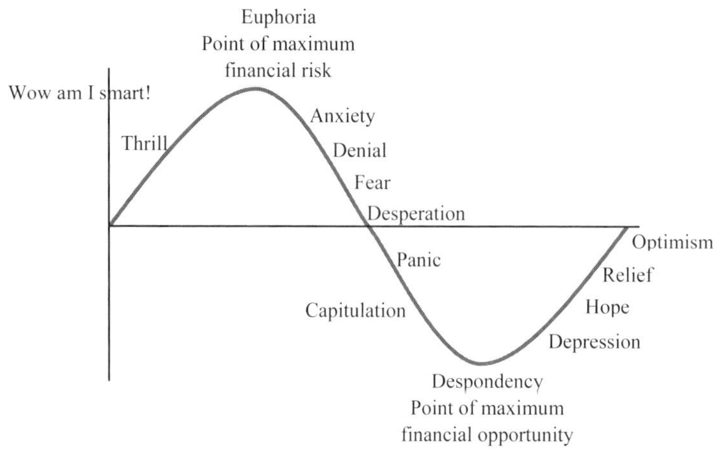


**Figure 6. Financial Risk Cycle**

## IMPLICATIONS FOR MINING COMPANIES

### Typical Inflation – Deflation Cycle for Mining Companies

A similar inflation-deflation cycle as it relates to the mining industry was also described by Samuel (2009). As the price of metals increases, e.g. the upswing of the curve, the majority of mining companies start investing in new projects, plant expansions, etc. This leads to over investment in the industry and excessive debt loads and provides excess capacity for the production of metals. The over-supply of metals, in turn, leads to lower metal prices, weakness in the pricing power and overall devaluation of the mining projects and companies. Owners are forced to sell their products below the cost required to produce them which in turn leads to plant closings, defaulting on debts and the associated reduction in the production of the metals. Due to free market economics the prices for the metals once again increases and the cycle starts to improve. This cycle is depicted graphically in Figure 7 (Prudent Investor).

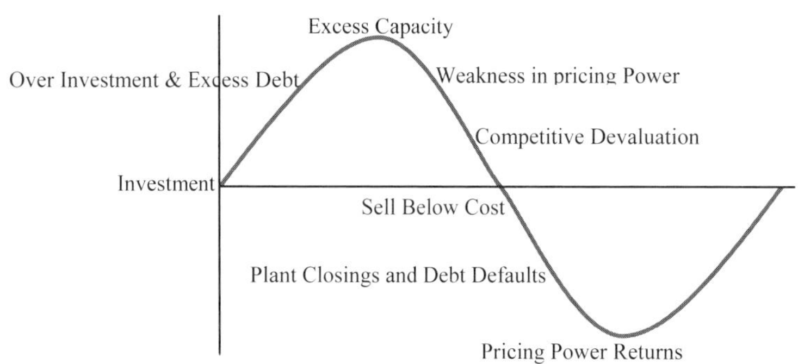


**Figure 7. Typical Inflation – Deflation Cycle for Mining Companies**

## POSSIBLE ACTIONS

### For Mining Companies

As projects are developed, the timing of the development has a large impact on the financial success of the project since the economic results for projects are commonly most sensitive to metals sales prices. Ironically, the company generally has the least amount of control over this variable. This is one reason the mining business is considered high risk. Understanding these relationships, and reacting accordingly, provides the greatest opportunity for mitigating the risks associated with the time a project should be developed. However, the following explanation shows that the best decisions are counter intuitive to traditional behavior.

**Typical project timing**. Since logic does not always prevail, companies often fail to take advantage of financial opportunities. Companies perceive that they should not start new projects unless the demand for metals and the metal price is high or well on the way to a high price. Figure 8 shows why this approach may not be the optimal approach for project development.

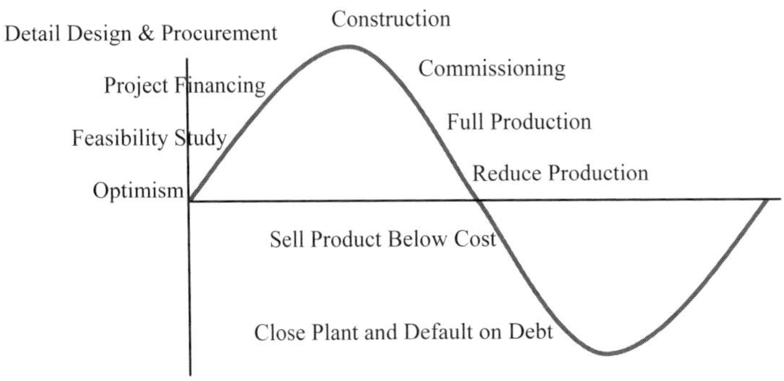


**Figure 8. Bad project timing**

As shown graphically, waiting to start a project until the price of metals increases often results in commissioning after the peak metal price has been reached and the price of the metal is declining due to the oversupply.

**Good project timing.** An alternative to the typical project timing can be achieved by companies that are structured to maximize their financial opportunities and have managers who not afraid to make those decisions. Figure 9 depicts this scenario graphically.

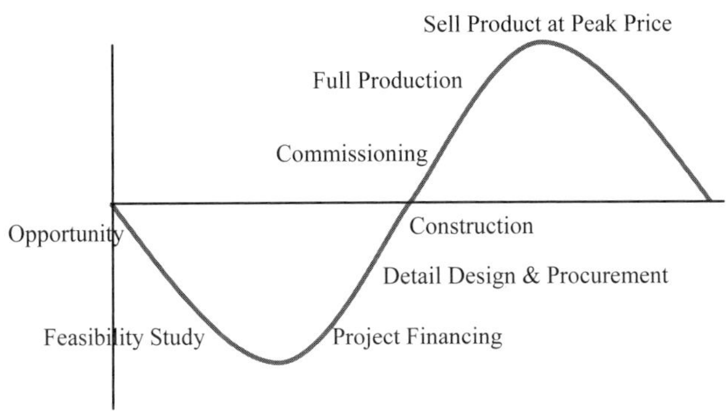


**Figure 9. Optimum project timing**

Since the mining industry is cyclical, if a company is in a position to take advantage of the opportunities that are associated with developing a project while metal prices are weak, they will be in a position to commission the operation and begin production at a time when the competition from other new projects is minimal and be in a position to sell their products when prices are at their peak which, in turn, allows them to repay the debt on the project quickly and be in a position to potentially weather the downturn unencumbered by debt when it does come.

There are a number of factors that contribute to the opportunities associated with developing a project under poor or optimal conditions. As discussed below, these factors include:

- Availability of resources
- Equipment pricing
- Delivery schedules

***Availability of resources.*** Perhaps the most important resource needed to develop a project is the availability of qualified personnel to complete the design, manage the project from both the owner's side and the EPCM side, commission and operate the mine. Due to the limited number of people that are trained to work in the mining industry and experienced at doing so, qualified people are at a premium during the upswing and peak in the economic cycle. It is difficult to entice the most qualified people to work on new projects and the salaries that are required to do so are high. At the same time, it is difficult to get the attention of the large engineering consulting companies. The largest companies will only accept large projects that require lengthy periods of time to design, procure the equipment, construct, commission and begin operating the facilities.

Alternatively, during the deflationary period, many of the most qualified people are available and most EPCM companies are willing to submit proposals to develop projects. In short, the companies that do take advantage of the opportunities that are available during the deflationary period get the "A team" as opposed to the "C or D team."

***Commodity and equipment pricing.*** Historical data collected by Samuel Engineering also confirms that the pricing of equipment and commodities falls as the deflationary cycle starts and increases during inflation when there is a strong demand for mining equipment and the price of the metals used to fabricate the equipment is high.

The historical price of oil is shown in Figure 10 (Energy Information Administration). The price peaked at approximately $147 per barrel in the second half of 2008 and dropped to just under $37 per barrel in the first half of 2009 before beginning another increase in price.

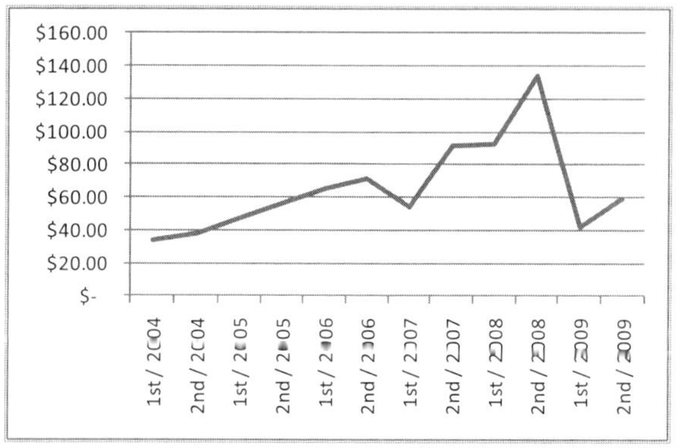


**Figure 10. Historical oil prices (2004 to 2009)**

The historical prices for fabricated steel are shown in Figure 11. Once again, the graph shows that prices peaked in 2008 and fell in 2009.

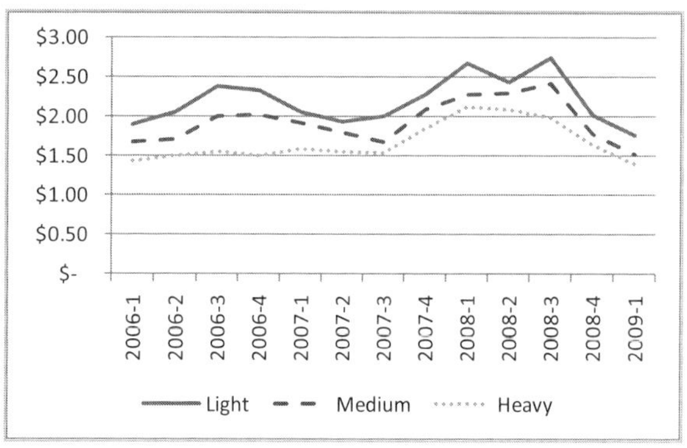


**Figure 11. Historical fabricated steel prices (2004 to 2009)**

The prices for gearless mill drives, crushers, and grinding mills are shown graphically in Figures 12, 13, and 14.

Figure 12 shows that the price of gearless mill drives steadily rose until it reached a peak that was nearly twice as much as the 2004 prices during the second half of 2008.

Figure 13 again shows a peak in the price of crushers during second half of 2008 with prices falling by more than 30 percent in 2009. The same type of trend is shown for grinding mill prices in Figure 14.

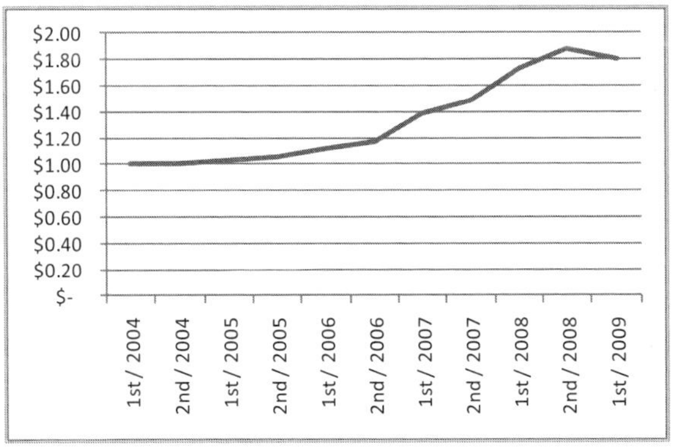


**Figure 12. Gearless mill drive pricing**

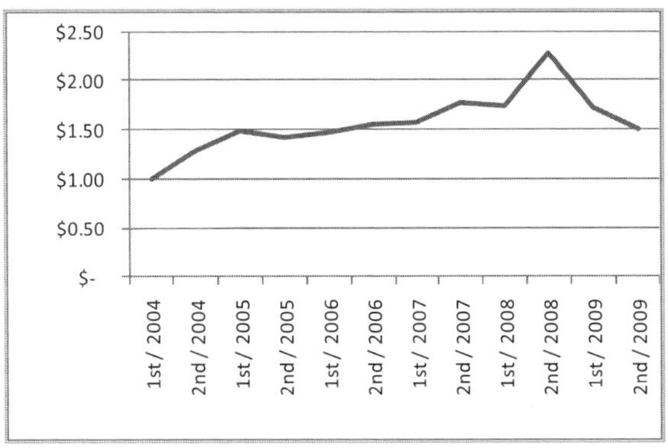


**Figure 13. Crusher pricing**

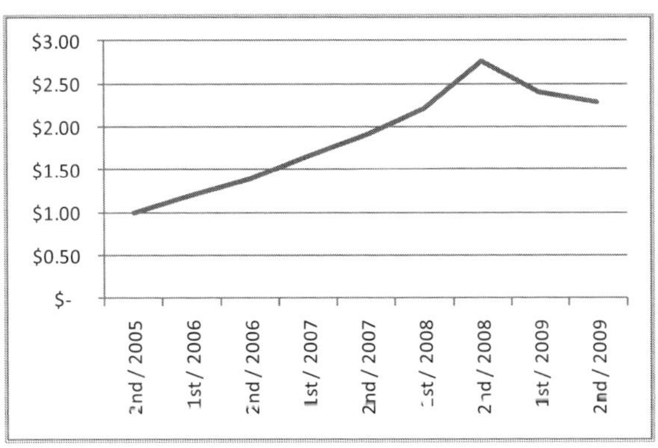


**Figure 14. SAG and ball mill pricing**

*Delivery schedule.* In addition to lower prices during the deflationary periods, the delivery times for major mining equipment are also reduced due to fewer orders and greater availability of fabrication resources. The historical delivery schedules for gearless mill drives, crushers, and grinding mills are shown in Figures 15, 16 and 17, respectively. Since these large pieces of equipment are commonly on the critical path of the schedule for mining projects, reduced delivery times may have a significant positive effect for the execution of mining projects.

Figure 15 shows that delivery times for gearless mill drives fell from a peak of 110 weeks in the first half of 2008 to 90 weeks in the first half of 2009.

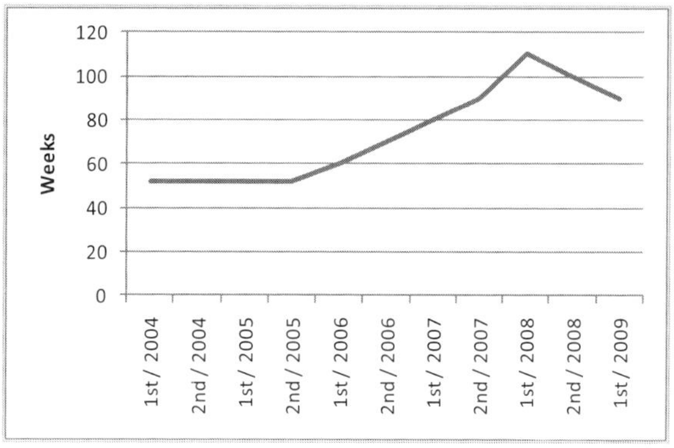


**Figure 15. Gearless mill drive delivery**

The delivery time for crushers peaked at 78 weeks during 2007 and the first half of 2008 before dropping to 48 weeks in the second half of 2009, as shown in Figure 16.

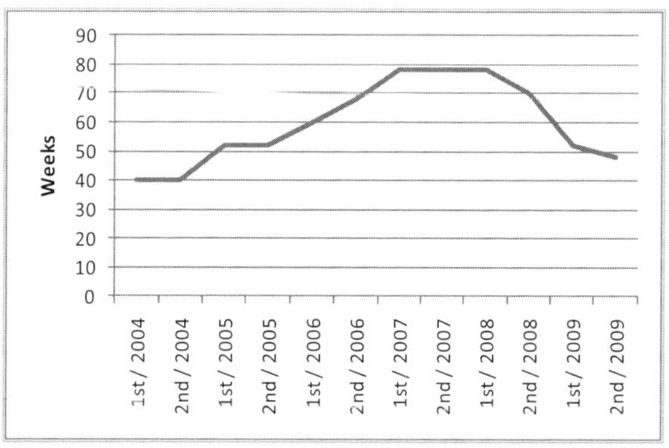


**Figure 16. Crushing equipment delivery time**

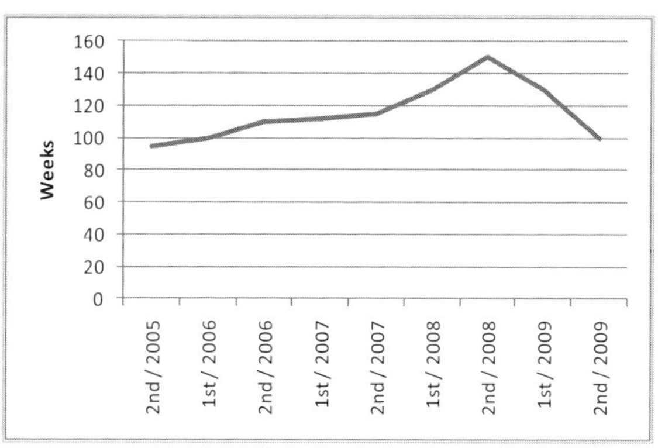


**Figure 17. SAG and ball mill delivery time**

The graph provided in Figure 17 shows that there has been a reduction of almost one year for the delivery schedules of semi-autogenous grinding (SAG) and ball mills. The duration dropped from about 150 weeks in the second half of 2008 to about 100 weeks in the second half of 2009.

**For EPCM Companies**

The most significant management challenge for EPCM companies is the management of personnel. During inflationary periods, there is typically an excess of work and not enough qualified people to do the work. Then, during the deflation there is a lack of work and it is not possible to provide reimbursable work for all of the personnel on the staff. Managing these ups and downs so there are enough people when they are needed is a challenge. There are several ways to manage the situations.

**Hire and layoff.** The most obvious actions for managing the need for more or less personnel is to hire people when they are needed and lay them off when they are not needed. The downside of this option is that it is generally expensive and difficult to hire people when there is a peak demand. Then, if people are laid off when they are no longer needed, the company may get the reputation that they do not provide a good work environment and the loyalty of employees will echo the loyalty showed to them by the company.

Two other options with regard to hiring and laying off employees are available. First, it is possible to bring in experts who are specifically qualified to work on the projects as contractors instead of hiring full time employees. These people are available and they do not have the expectation that the employer will retain their services when the project is completed and they are no longer needed.

Secondly, when the EPCM companies react to the deflation logically and seize opportunities as they become available, it is possible to hire the most qualified people when other companies let them go. This means spending money as an investment to improve the quality of resources when there is a lack of work to support them in anticipation of future needs.

**Unilateral actions.** In large companies, unilateral actions are common. An example is across the board salary cuts such as a 20 percent reduction in pay for all employees or laying off the project staff when the project ends. Or it may be laying off a percentage of the people employed in each department. This type of action requires little thought and seldom results in an optimal solution to the deflation.

**Employment preservation initiatives.** Alternatively, an effective alternative to unilateral action is considering each individual circumstance and taking actions that are most appropriate for each circumstance. Downturns do provide a unique opportunity to terminate unproductive or disruptive employees. Other opportunities include salary reductions, temporary layoffs and reduced work weeks. The benefit to maintaining employees is that it is easy to quickly respond to the changes in the workload. While there is discomfort for the employees that have reduced income it is oftentimes a better alternative than a layoff. In the long term this approach should result in loyal employees who understand what they can expect from their employer as good times come and go.

The objective of thoughtful choices is that the availability of human resources can be adjusted as the work load comes and goes. This is shown graphically in Figure 18.

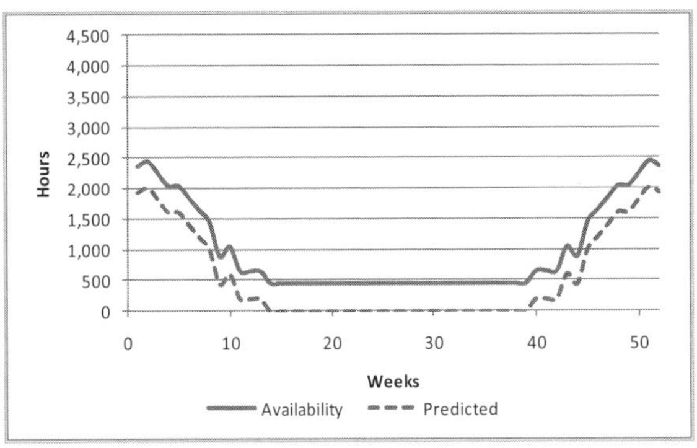


**Figure 18. Ideal relationship between availability and work load**

## CHALLENGES FOR ALL

### Workforce Demographics

A challenge facing the United States is an aging workforce. Using data from the U.S. Bureau of Labor Statistics,as shown in Figure 19, over the period of time from 1980 to 2008, the number of workers in the age range between 25 and 54 reached about 100 million in the late 1990s and has remained at that level since then. On the other hand, the number of workers over the age of 55 who are employed full time has shown a steep increase since the mid 1990s. They now number nearly 50 million out of a total workforce of a total workforce of approximately 125 million fulltime workers.

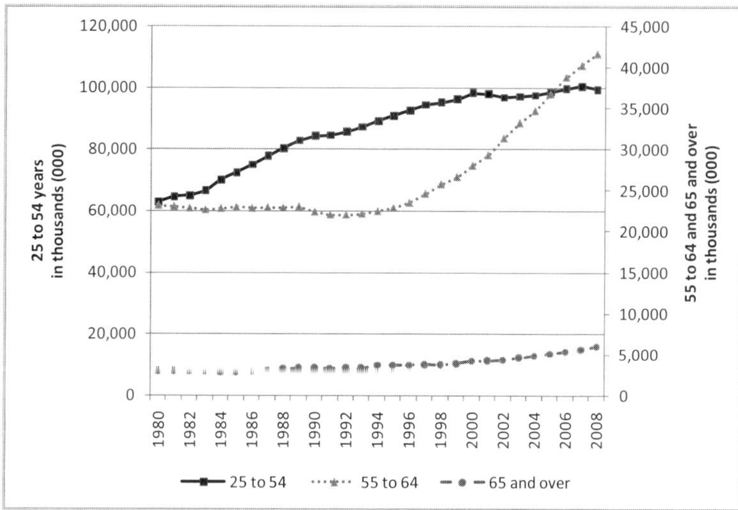


**Figure 19. Workforce demographics in the United States**

Since the total workforce that is employed fulltime has increased since 1980, the impact of this data may be shown more clearly by plotting the percentage of the total workforce in these age ranges. This data is plotted in Figure 20.

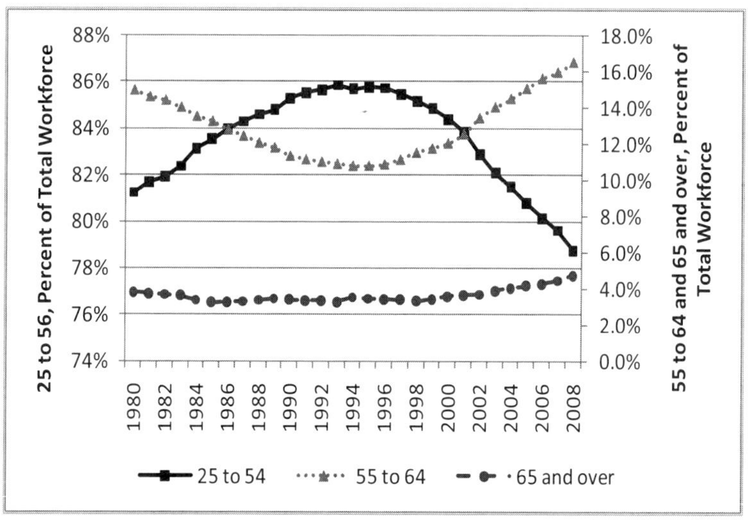


**Figure 20. Percentage of workforce in the United States**

The percentage of fulltime workers between the ages of 25 and 54 in the United States increased from about 81 percent in 1980 to approximately 86 percent in the mid 1990s. Since that time the percentage of fulltime workers between the ages of 25 and 54 has declined steeply to a current level of about 79 percent of the total fulltime workforce. During this same time, the percentage of fulltime workers between the ages of 55 and 64 declined from approximately 15 percent in 1980 to about 10 percent in the mid 1990s and, then, increased to about 16.5 percent in 2008. At the same time, the percentage of fulltime workers over the age of 65 was between 3.2 and 3.6 percent between 1980 and 2002. Since 2002 the percentage of fulltime workers over the age of 65 increased dramatically reaching 4.7 percent in 2008. Due to concern about the number of qualified workers available to work in the mining industry the Society for Mining, Metallurgy and Exploration (SME) hosted the Mining Leadership Forum in 2006. Freeman presented data that showed, based on the demographics of engineers available to work in the industry, a state of crisis has been reached (2006).

**Global Currency Exchange Rates**

Due to the global nature of the mining industry, currency exchange rates may play a critical role in the development of mining projects. A North American company may develop projects in South America, Africa, Asia, or South America. The EPCM services may be awarded to a company in Australia, Canada, or Europe as well as the United States or other countries. The role of currency exchange rates has the potential to have a major impact on the development of the project because it will impact the cost of the project and, therefore, may be a determining factor in the choice of an EPCM company and equipment suppliers. Currency exchange rates have been volatile just as equipment pricing has varied. Figures 21 and 22 show the currency exchange rates for the Canadian dollar and the Australian dollar versus the United States dollar over the past five years. (Yahoo Finance, 2009)

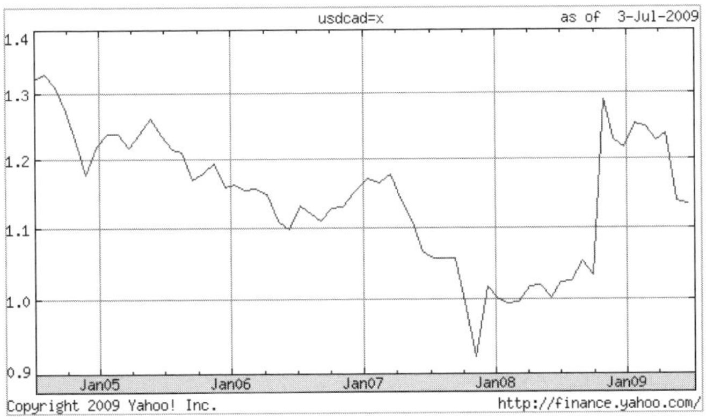


**Figure 21. Currency exchange rate between Canadian dollar and U.S. dollar**

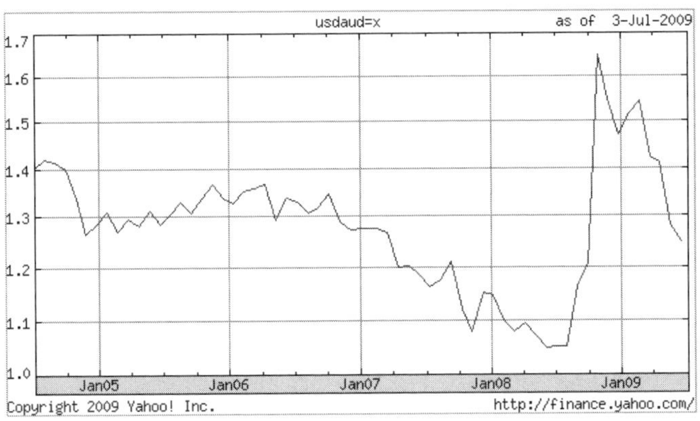


**Figure 22. Currency exchange rate between Australian dollar and U.S. dollar**

Based on these graphs, a constant value of $US 100,000 varied between about $CAD 91,000 and $CAD 132,000 and between $AUS 105,000 and $AUS 165,000. As the exchange rates fluctuate, it may be more cost effective to complete an engineering design in different countries.

**EPCM Approach**

The "typical" approach to project execution used by EPCM companies in the United States and Canada is quite different from the one used in Australia. In North America EPCM companies generally hire full time employees who stay with the company over the execution of a number of projects. However, in Australia, projects completed by EPCM companies generally rely on hiring contract employees. These contract employees tend to move from one engineering company to the company of a competitor as projects are awarded to different companies. It is not uncommon for employees to move to a different company based a small, incremental increase in salary or wages. Loyalty to the companies is not a significant consideration when this business approach is used.

This approach has the potential to improve or to detract from the success of the project execution. Large projects that can afford to pay higher contract rates may be able to attract the most qualified people to work on the project. Concurrently, there is a risk that key personnel assigned to a project may move to a different project that has been awarded to a different company during the project execution.

Although there are no guarantees in the U.S. or Canada, the team assigned to a project based on the proposal is likely to be available to follow through with the execution of the overall project. Another advantage to the availability of fulltime, long-term employees is that interdisciplinary teams working for an EPCM company in North America commonly develop camaraderie and knowledgeable working relationships during the execution of multiple projects that are executed by their employers.

**Mergers and Acquisitions**

Over the past several years, there have been a number of mergers and acquisitions of companies with previously diversified businesses. For example, some of the large equipment suppliers have acquired engineering consulting companies. These acquisitions have given the supply companies the capability to complete engineering projects internally without relying on outside EPCM companies. Also, a number of smaller companies have merged to form larger companies and a number of larger engineering or consulting companies have purchased smaller companies that may have previously done business in a related industry. All of these changes will likely result in changes to the industry although it has not been determined what the changes and what the ramifications may be.

**SUMMARY**

Companies that manage their finances successfully and seek non-traditional and creative approaches to changing situations are most capable of seizing opportunities when they present themselves. For example, if company owners and managers maintain a healthy cash reserve during the good times, they will be in a position to upgrade staff and offer discounts and deferred payment plans to mining companies that want to develop projects during the downtimes. On the other hand, companies that spend the proceeds as they become available and fail to keep a cash reserve are unable to offer more creative solutions to their clients during the down turns.

**A Number of Options**

Everyone, companies and individuals alike, are confronted with a plethora of choices. The choices that are made directly impact the opportunities and challenges that will confront the decision makers in the future. Figure 23 shows growth curves for three different types of companies. Many (average) companies survive for a limited period of time and, then, fail. Successful companies grow to a certain, manageable size and continue on a steady course for an extended period of time. The growth curves for excellent companies show that these companies go through very turbulent times as they transition from one mode of operation to the next. During these times, the managers are able to change corporate cultures and make decisions that prepare them for future growth. When the company emerges from the turbulence caused by changes, another period of rapid growth follows. This scenario is not possible for managers who are unable or unwilling to adjust course, make difficult decisions and implement creative and dynamic actions.

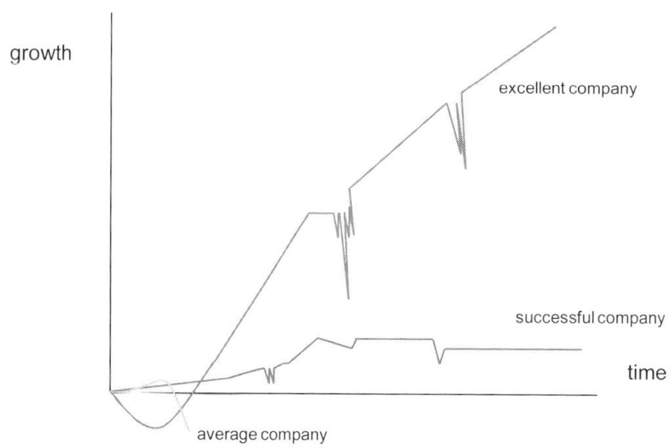


**Figure 23. Representative company growth curves**

**Thoughtful or Reactionary Choices**

**No decisions.** Both EPCM and mining companies that make no conscious decisions and take no action during a downturn are likely to fail and, at best, will not thrive. A large percentage of all companies that are started in the U.S. do fail.

**Typical decisions.** Companies that make typical or traditional decisions, such as hiring full-time employees when they need them and laying off employees when they are not needed, are much more likely to be average over the long term. However, the lack of ingenuity and intellectual creativity is not likely to result in an excellent, highly successful company.

**Excellent decisions.** Companies that have managers who are able to resist emotional responses ranging between euphoria and panic and are able to make logical, intellectually thoughtful decisions, are much more likely to see sustainable high growth rates over the long term. This type of decision-making ability will allow a mining company to earn and retain sufficient capital during the economic good times to allow them to proceed with development of new projects during the economic down turns. The competitive advantage of this approach is that competition and availability of resources allows the project development to be completed more efficiently and at lower costs. The company also has greater potential to market the products, i.e. metals, at higher sales prices which in turn maximizes corporate profits.

EPCM companies that have managers who make the best decisions will benefit from an ability to recruit and retain the most qualified staff. This benefit alone will perpetuate the ability to make the best decisions, seek the most creative solutions to problems that are encountered, recognize and pursue available opportunities and offer value-added services to their clients. Of course, offering the highest value to clients increases the success of the EPCM company and concurrently allows the client companies to be more successful. It is a cycle of success that improves the outcome for everyone involved.

**Additional Considerations**

The dynamics of the past few years have added additional considerations that do impact both mining companies and EPCM providers. Some key considerations include the relationships between international currency exchange rates and the differences in how business is conducted in different countries as well as the trend toward fewer, larger companies to execute the engineering and construction management of new projects. All of these considerations, along with the main considerations discussed in this paper, provide additional incentive for managers and investors to make logical, informed decisions and to attempt to avoid decisions based on emotions or reactions.

**Future Success is in the Balance**

The changes that occur in the marketplace provide both financial risks and financial opportunities for all companies, including mining companies and EPCM companies. The extreme dynamics during the past year have made these risks and opportunities more apparent because the expected cycle between inflation and deflation was shortened and managers had to make quick decisions in order to mitigate the risks and take advantage of the opportunities. If they followed some of the best practices explained in this paper, their companies should be well on the way to weathering the storm and preparing for a successful future accompanied by high, sustained growth. If not, the companies may survive but their future will not be nearly as bright.

## REFERENCES

Bureau of Labor Statistics, http://data.bls.gov/PDQ/outside.jsp?survey=ln, accessed July 5, 2009

Energy Information Administration, http://tonto.eia.doe.gov/dnav/pet/hist/rwtcM.htm, accessed June 10, 2009

Freeman, Leigh, *SME Mining Leadership Forum Overview*, SME Mining Leadership Forum, Park City, Utah, 2006

Infomine Dynamic Charting Tool, http://www.infomine.com/investment/charts.aspx?mv=1&f=f&r=5y&c=ccopper.xusd.ulb#chart, accessed June 10, 2009

Prudent Investor, http://prudentinvestor.blogspot.com/2005/07/chart-of-day-deflation-cycle.html, accessed June 10, 2009

Samuel, Everod, 2009, *The Recession as Seen by the Engineer*, SME Mineral Processing Division Annual Meeting, Colorado Springs, Colorado.

Yahoo Finance, http://finance.yahoo.com/q/hp?s=%5EDJI&a=00&b=1&c=2004&d=05&e=1&f=2009&g=m. accessed June 10, 2009

Yahoo Finance, http://finance.yahoo.com/currency-converter#from=USD;to=EUR;amt=1, accessed July 5, 2009

# GIBRALTAR MINE EXPANSION—DESIGN CONSIDERATION FOR A "FAST-TRACK" BROWN-FIELD PROJECT

Rob Rotzinger[1] and Max Bellisomo[2]

[1]Gibraltar Mines Ltd, 10251 Gibraltar Mine Road, McLeese Lake, BC, Canada
[2]Wardrop Engineering, Suite 800—555 West Hastings Street, Vancouver, BC, Canada
(Formerly, Hatch, 1066 West Hastings Street, Vancouver, BC, Canada)

## ABSTRACT

The Gibraltar Mine expansion and modernization is being carried out in two parallel and overlapping stages. Stage one was completed on budget and on schedule in March 2008, while stage two is scheduled for completion in 2010.

The financial benefits forecasted from early completion of the expansion and modernization program motivated an expedited schedule and a "fast-track" approach to the project. Engineering activities therefore did not follow the accepted project development sequence through feasibility studies and basic engineering but progressed directly from a conceptual level to detailed engineering. This coupled with the fact that the majority of the project areas were located in a brown-field environment and within an operating plant meant that construction activities had to be carefully planned to meet both schedule and budget requirements as well as minimizing production losses due to plant stoppages.

This paper presents an overview of the engineering strategies and management processes utilized to successfully undertake this project.

## INTRODUCTION

The Gibraltar Mine is a large scale open pit copper-molybdenum porphyry operation located in central British Columbia, 362 km north of Vancouver (see Figure 1). The site is located 45 km north of the City of Williams Lake and is accessed via 20 kilometers of paved road from Highway 97.

The mine and related process facilities were originally commissioned in 1972 by Gibraltar Mines Ltd. (majority owned by Canex Placer Limited) and the mine operated until December 1998 when it was placed on care and maintenance due to low copper and molybdenum prices. Taseko Mines Limited acquired Gibraltar in 1999 and the mine and processing facilities were re-commissioned in October 2004.

The modernization and expansion of the Gibraltar Mine began with a trade-off study started in the fall of 2005. The study evaluated the opportunity to increase the ore processing capacity of the existing concentrator and implement modern process technologies and equipment in order to reduce unit costs. In March 2006, the Stage I expansion project was approved based on the results of this trade-off study and the project proceeded directly to detailed engineering. The stage one expansion was designed to increase the plant throughput from 36,750 to 46,000 stpd.

**Figure 1 – Site Location**

The first stage expansion comprised the addition of a new stand-alone SAG mill facility housing a 34' SAG Mill fed by a high angle conveyor, discharge vibrating screen, associated re-circulating pebble conveyors and slurry pumping system. The remainder of the stage one scope consisted of brown field modifications to the existing plant including modifications to the existing ore storage and reclaim system, conversion of the existing three rod / ball mill circuits to six individual ball mill circuits operating in parallel and the replacement of the existing rougher flotation system with new 160 $m^3$ tank cells. The existing substation was also expanded and the site power system reinforced to support the additional load required by the increased plant throughput. Figure 2 shows the overall process facility site layout as well as the location of the new facilities.

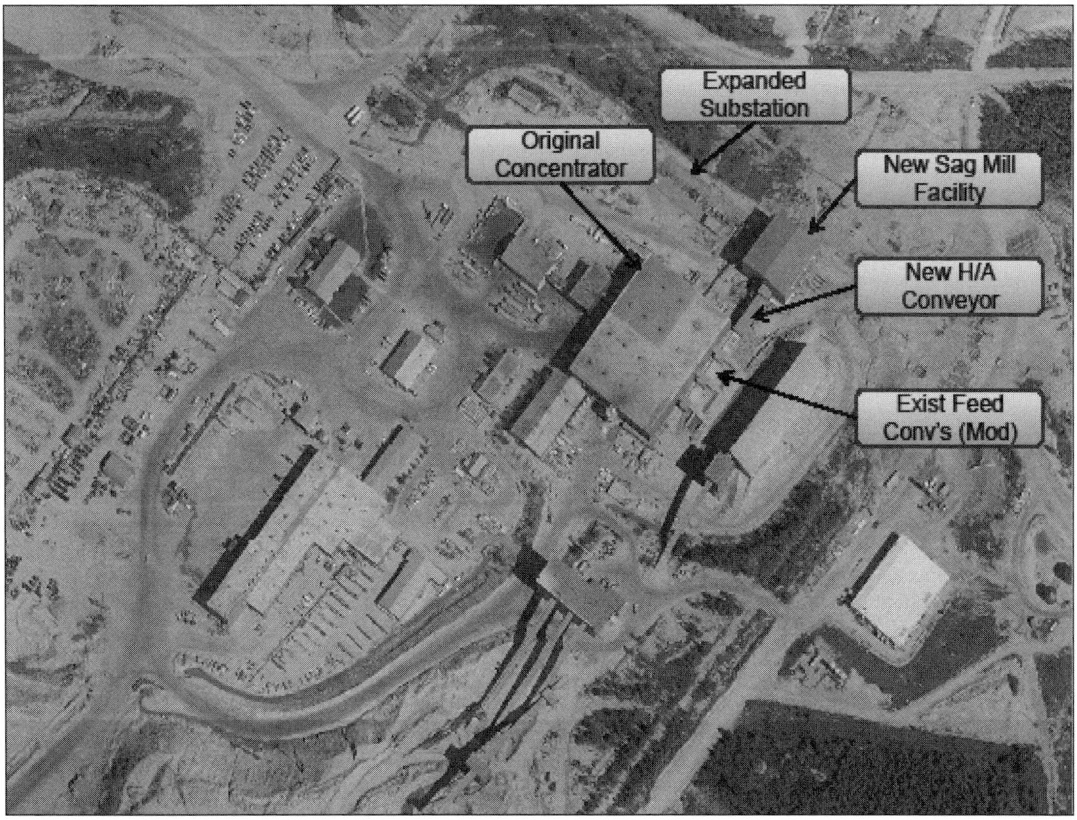


**Figure 2. Gibraltar Mine site plan indicating expansion areas**

A process de-bottlenecking scoping study was initiated in parallel with the stage one expansion detailed engineering and was completed during the second quarter of 2007. The study provided an outline of how the process plant throughput could be further increased to 55,000 stpd and formed the basis of the second stage expansion. The stage two expansion project was approved in May 2007 and detailed engineering was initiated as engineering for the stage one expansion was reaching completion.

The stage two expansion consists of various brown-field modifications to the existing concentrator and includes the addition of a Metso Vertimill to provide tertiary regrind capacity, replacement of the existing Denver cleaner flotation cells with new tank cells, replacement of the concentrate filtering and drying equipment and replacement of the tailings pumping facility as well as other ancillary system upgrades. A new in-pit crushing facility and overland conveying system was also undertaken at site, although managed as a stand-alone entity.

The combined goal of the two stage expansion project was to transform the existing mill into a modern 55,000 stpd concentrator with minimum interruption to copper and molybdenum production. The stage one expansion was successfully completed in March 2008, on time and on budget. Engineering for the stage two expansion was completed in late 2008 and construction is scheduled to be completed in 2010.

## ENGINEERING

The Gibraltar expansion projects proceeded directly from scoping level conceptual studies to detailed engineering on a "fast track" schedule. The Stage one expansion proceeded to detailed engineering based on the results of a trade off study originally intended to identify only the most cost effective comminution alternative to expand the mill capacity. The study investigated and developed the processing requirements for the expansion but did not investigate many of the downstream supporting unit operations in detail as they were common to the various grinding circuit options being analyzed. This resulted in the project commencing with the process requirements being well established but very limited strategies for achieving these requirements.

As we all well know, the execution of any engineering design is best achieved by careful pre-planning, detailed field investigations and a full development of engineering concepts to ensure that scope and schedule are fully defined prior to detail design taking place. This is doubly true when the work is to be performed inside an operating plant. However, in the copper and molybdenum market of the day, the additional project certainty that completion of a feasibility study and basic engineering would have provided to Gibraltar were grossly overshadowed by the potential lost revenue of the associated delay in increasing plant production.

Proceeding directly from concept to detailed engineering presents several challenges to established project practices, but even more so in a brown-field, operating plant environment. At the conceptual stage the project constraints are only identified at a very high level, ancillary equipment and services costs are based on factored estimates (and these historical factors are notoriously unreliable for brown field projects) and project demolition requirements for existing structures and equipment have not been thoroughly detailed. In short, both projects began with a generalized scope, a tight timeline and no road map to guide and direct the execution process.

The lack of scope definition made it all but impossible to estimate engineering resource requirements, schedules and costs at the onset of the engineering process. The iterative nature of the work in a brown field environment (as concepts are developed and field checked issues are uncovered which force re-work of the original concepts) makes it necessary for the continuous updating of engineering strategies.

The initial step in the project development process, therefore, was to establish a robust but flexible work break down structure which identified all the project expansion areas and provided a high level definition of the project scope. The work breakdown structure provided the framework for managing and trending subsequent engineering, procurement, and construction activities independently until the final scope for each project area had been fully defined. Preliminary engineering and procurement manpower budgets were allocated to each work area and formed the baseline to be used for regular hours and cost reports. Progress monitoring could not follow conventional "earned value" methodologies due to the lack of scope definition but focused rather on trending spent plus forecast efforts to project completion against the original budget estimates. Such a forward focused approach ensured that all the latest area scope definitions were incorporated into the engineering estimates at any point in time in order to provide a "best total estimate" of costs at project completion.

The brown field nature of the Gibraltar site forced significant restrictions and constraints on the design of the new facilities. The site had limited footprint on which the facilities could be constructed and the presence of existing infrastructure within most areas available for the expansion further restricted design. In addition, constant consideration had to be given to minimizing disruptions to ongoing operations.

Early gathering and confirmation of existing site data played an important role in setting the foundation for detailed engineering activities. Most of the engineering drawings for the original Gibraltar plant were available on site; however, 30+ years of modifications to the plant had occurred without rigorous documentation. Regular site visits were therefore conducted to visually verify ongoing layouts versus the existing site infrastructure and key measurements confirmed by hand wherever possible. Surveys were initiated for the more critical areas and where hand measurements did not prove feasible.

As the engineering process proceeded for each project area, the constraints were identified in detail followed by integrated design sessions with engineering, operations and construction personnel to develop a high level execution plan for the area.

The key project constraints can be generalized as follows:

- complete the project within the allotted budget
- complete construction as soon as possible
- minimize plant downtime and production disruptions
- maximize use of existing infrastructure
- maximize value by re-use of existing equipment if appropriate and by replacing equipment if warranted
- manage engineering to minimize weather impacts on construction
- produce a functional/operable circuit
- produce a circuit that is maintenance friendly
- build for low operating costs and minimize capital costs
- build process facilities that when complete are logical and not ad hoc

After each area plan was set, detailed engineering proceeded with regular design reviews and referenced back to the project goals and constraints. As issues were identified that impacted on the key constraints the design work would revert back to earlier stages, sometimes all the way back to the concept stage, to identify possible solutions or find new methods.

Area layouts were developed in a 2D environment to establish the early concepts and design opportunities identified to address the area specific constraints. The new SAG Mill facility for example was identified as being a "stand alone" green field facility with relatively few constraints once the construction area had been cleared of existing services and bulk earthworks had taken place. Other project areas within the existing plant however required multi-discipline modification that effected plant operations and, in many cases, necessitated a "phased" approach in order to minimize downtime. Engineering deliverables for the relocation of existing site buried services and overhead power lines, required to expedite earth-moving activities, were thus identified as the priority in the Sag Mill facility area while the development of detailed erection methodologies and modeling of existing infrastructure components were the priorities in operational brown-field plant areas, where the early concepts developed by the 2D layouts required additional definition to ensure that the proposed equipment and services could be constructed within the selected plan envelopes.

Process flowsheet were finalized in parallel to the early layout activities and long lead items identified in order to prioritize the development of related technical equipment specifications and set the procurement plan to be followed by the project. Procurement activities were then initiated and vendor drawings for critical equipment expedited to confirm the integrity of the early layout concepts.

The layouts also dictated where specific types of equipment configurations were necessary to suit the site physical constraints. Some examples are provided as follows:

- The limited distance between the existing fine ore storage facility and associated conveyors and the new proposed sag mill facility (location constrained by a fish bearing creek and the desire to minimize the pumping distance from the SAG mill to the secondary ball mills) necessitated the selection of a high angle conveyor.
- The footprint of the replacement rougher flotation cells in the existing concentrator had to be specifically defined to fit within the constraints placed by the existing structure, while the tails discharge minimum height required for gravity feed to the tailings pump box and the existing crane elevation required that split shaft agitators be specified. New equipment component weights further determined that a new larger crane had to be installed on the existing crane rails for erection purposes.

P&IDs and 3D models were developed for all project areas and for each of the identified phases once the 2D layouts and process flowsheets were reviewed and their integrity confirmed. These P&ID's and 3D models were continuously updated by all disciplines as concepts were finalized and new vendor information received and constituted the key engineering control deliverables for the project. The use of images taken from the 3D models also proved invaluable as a tool to integrate and communicate with construction, operations and maintenance personnel in the plant.

Project envelopes required for subsequent construction activities were clearly defined during the engineering process for each of the brown-field project areas. Existing services relocations within the identified project envelopes was managed directly by site personnel as "up-front" work packages; a strategy that allowed the engineering effort to be minimized, the construction team to become familiarized with the operating plant and the work to proceed while taking advantage of operational opportunities and avoiding seasonal weather issues.

New building enclosures and modifications to existing external building structures and related foundations were prioritized wherever possible in order to erect and clad the various facilities in the summer months and allow for the installation of equipment and related services indoors during the winter months. Overhead cranes and accesses to such facilities were therefore sized to facilitate major components erection after the buildings had been erected. In very congested areas, such as in the case of the modification to the fine ore stockpile discharge conveyors, the new building structures themselves were designed such that they could be erected over the existing structures in order to maintain the operational integrity of the existing system while facilitating the demolition and modifications of existing equipment in an indoor environment.

Internal foundation designs were developed to allow construction activities to take place outside of the critical path and without disrupting operations; the design being such that it allowed for the concrete to be poured above the existing slabs and, in some cases, also embedding existing columns. Possible cracking to the existing slabs caused by the additional imposed loads was prevented by providing saw–cuts around the new foundations, while the design itself facilitated the subsequent demolition of embedded steel during subsequent construction phases.

Structural steel design prioritized reuse and "in-situ" strengthening (if necessary) rather than replacement of existing steel structures wherever possible. This approach minimized demolition requirements and facilitated further non-critical path construction activities. It however necessitated that the structural integrity of the existing structures be checked for the additional imposed loads.

Shutdown requirements were further minimized by the early identification of piping and electrical tie-ins within the project areas, which allowed for modification to take place "on the fly" during available plant maintenance downtime and prior to construction itself taking place.

A critical stage of the engineering process for each individual project area was a formal design review. This review was undertaken as soon as the major design elements for the particular area had crystallized. The purpose of this review was to ensure that the engineered product achieved the project goals and to ask several key questions. The key questions can be generalized as follows:

- Can the facility be built?
- Have all safety considerations been identified and addressed?
- Does the proposed design and construction phasing minimize interferences to existing plant operations?
- Is the facility operable without significant inconveniences?
- Is the facility maintenance friendly?
- Will the unit operation feel like it was designed / built with significant constraints? If so, how should the facility be re-structured?

Usually the final review process resulted in only minor changes to the design. Occasionally, however, the review resulted in engineering for the area being returned all the way back to the concept stage.

## PROJECT MANAGEMENT

One would be somewhat remiss to discuss the engineering process without at least touching on some of the main project management challenges that had to be addressed to facilitate this projects' execution. A dynamic design, procurement and construction process with a short fuse timeline and with a limited information base to start from can lead to many potential pitfalls. The project required a management team that could take quick action and make decisions based on limited information. The key elements to making good decisions in this environment are a broad understanding of the projects goals and a keen understanding of how to minimize the risk associated with each decision. There is no fixed management structure that suitably addresses a process with so many variables and so much uncertainty. This necessitated the development of a multi-skilled project team able to understand and take into consideration the sites' operating goals and capable of balancing the requirements of the engineering and procurement process with ever present constructability and operational challenges. The building of such a flexible and integrated team capable of steering the project through all of its ambiguities was the key to the success of this project. The selected team was multidisciplinary and was comprised of members representing all stakeholders. The team was continually challenged to look ahead, to the path from where the project was at any particular point in time to completion, rather than looking backwards to where some efforts may have been misdirected in the past. This type of a process requires trust, open communications and the strong support from senior management towards all team members. It requires all parties to focus on achieving the projects' goals rather than the often disparate goals of their functional group or individual company.

## A CASE STUDY: MODERNIZATION OF THE ROUGHER FLOTATION CIRCUIT

The modernization of the rougher flotation circuit at the Gibraltar Mine is an excellent illustration of the concepts presented earlier in this paper. As with the overall project, the flotation circuit was fast-tracked with basic engineering commencing in May 2006 and a deadline for system commissioning of July 2007. The area scope consisted of replacing the existing 3 banks of sixteen 600 $ft^3$ Denver flotation cells, which had reached the end of their useful service life, with new modern flotation cells. The design goals for the area were to double the flotation residence time to support the higher mill tonnages as well as improve metal recovery and to install modern equipment which would support operations for a planned 25 year mine life. Ten 160 $m^3$ Outotec tank cells complete with all ancillary equipment necessary for a complete system were selected for the duty. The mine had previously purchased a single 160 $m^3$ Outotec flotation cell which had not yet been installed and was to be incorporated into the project plan.

Engineering commenced with the development of a high level execution strategy. The option to construct a standalone rougher flotation facility was quickly discarded as the cost for the installation would be too high and site restrictions would have forced excessive separation between process unit operations, impacting operability. The only remaining option was to utilize the same physical space as the existing rougher flotation cells. A photograph of the old flotation circuit is presented in Figure 3.

**Figure 3. Existing flotation circuit prior to the changeover**

Three possible strategies were identified to accomplish the change over:

| | |
|---|---|
| Option 1 | Shutdown plant to perform the changeover |
| Option 2 | Install a "sacrificial" bank of outdoor flotation cells to maintain production during the changeover |
| Option 3 | Perform a phased decommissioning of the old flotation cells in parallel to a similarly phased installation of the new flotation cells |

Obviously, Option 1 was quickly dismissed since one of the project goals was to minimize the impact that construction of the new circuit would have on day-to-day operations and considering that, given the price of copper and molybdenum in mid-2007, each day that the concentrator was offline would result in a loss of revenue of in excess of $500,000.

Option 2 required the installation of a bank of flotation cells outside of the main concentrator building to be used temporarily while the existing flotation cells were demolished and new permanent flotation cells installed in their place. While this option was a workable solution, the costs associated with a temporary installation of a bank of 160 $m^3$ cells plus either relocating the cells in their entirety or purchasing a second set of tanks (so that only the bridges and mechanisms would require relocating) led to this option being rejected in favor of the third alternative.

Option 3 entailed a phased decommissioning of the existing flotation cells and phased installation and commissioning of the new flotation cells. Furthermore, the installation of the single 160 $m^3$ cell purchased by the mine could be completed as planned (as a rougher scavenger cell) and integrated within the phased construction sequence to form part of the final installation. Although this option did not present the most straightforward construction sequence, it was selected based on a lower installation cost.

Once the preferred option was selected, the installation of the first 160 $m^3$ flotation cell commenced and was completed in August 2006. The cell was installed as a scavenger processing the tailings from the three existing rougher flotation banks. The installation of this first cell provided sufficient flotation retention time to remove one of the Denver rougher banks and continue to operate the mill with minimal impact on recovery. This left nine 160 $m^3$ flotation cells to be installed in the space occupied by the existing Denver flotation cells.

Basic engineering commenced with preliminary 2D layouts to confirm that two banks (one of 4 x 160 $m^3$ cells and the second of 5 x 160 $m^3$ cells) could be located within the footprint of the existing 48 cells. The rougher scavenger cell would ultimately become the 5$^{th}$ cell in the first operating bank. As construction would occur in an area of the concentrator that was vital to plant operations; careful planning and constructability reviews were required to ensure that production disruptions were minimized. In addition, working within the existing building and infrastructure envelope presented significant layout constraints which became key issues when completing the design and construction plan.

The construction and operating sequence to be followed during the transition was further defined as follows:

- Build and commission a pumping system to allow the three existing grinding lines to discharge to a new common pump-box, which would in turn be used to feed two of the existing flotation banks.
- Decommission and demolish one of the three banks of existing flotation cells while running full throughput through the remaining two banks and the new scavenger cell.
- Construct one bank of four new flotation cells in the space made available by the removal of the first bank of existing flotation cells and modify the flotation feed pumping system to feed the new bank of cells.
- Commission the new bank of flotation cells and switch plant throughput through the new flotation cells including the scavenger cell.
- Decommission and demolish the remaining two banks of existing flotation cells while running full throughput through the new bank of flotation cells.
- Construct the second bank of five new flotation cells in the space made available by the demolition of the last two banks of existing cells and complete the flotation feed pumping system to split the feed from the grinding lines between the two flotation banks.
- Commission the second bank of new flotation cells.

This detailed construction sequence provided the framework for how the project could proceed within the imposed constraints. The next step was to develop the engineering deliverables necessary to support the agreed strategy and further optimize the design in each discipline group.

The engineering team, construction management team and operations team provided continuous joint input into a detailed engineering schedule to support the construction methodology for the area, which was broken down into three distinct phases, as follows:

Phase 1

1) Construct all concrete foundation work. The concrete matt foundations for both new banks of flotation cells to be poured above the existing slab and under the operating flotation cells, encompassing the existing flotation cells support steelwork which would be demolished later.
2) Install a new heavier duty crane on the existing crane rails, as required to erect the new, larger flotation cell components.
3) Install the new flotation feed pump-box and associated pumps to temporarily feed two banks only of existing flotation cells. The flotation feed pump-box would later be reused in the permanent installation.
4) Install temporary flotation feed piping, electrical services and instrumentation to defined tie-in points to support the operation of two banks of existing cells fed from the new pumping system
5) Shutdown the concentrator and tie-in the pumping system to the existing flotation banks.

Phase 2

6) Decommission and demolish one bank of 16 existing flotation cells and support steelwork
7) Install the new steel support structure on the previously poured matt foundation and one bank of four replacement flotation cells in the space thus made available.
8) Install ancillary systems and cold commission the new bank of cells.
9) Shutdown the concentrator and tie-in the new bank of flotation cells. Commission the new flotation cells. Figure 8 provides a picture of the plant upon completion of phase 2.

Phase 3

10) Decommission and demolish the remaining two banks of old flotation cells and associated steelwork.
11) Erect the new steel support structure on the previously constructed matt foundation and install the second bank of five new flotation cells in the space thus made available.
12) Install ancillary piping, electrical and instrumentation systems and cold commission the new bank of cells.
13) Perform final tie-ins and commission the final circuit, consisting of both parallel banks of replacement cells, into operation. Figure 9 illustrates the system on completion.

Having defined the construction sequencing, engineering activities and design optimization could then proceed to support the defined construction sequence. The 2D layouts produced to prove the early concepts were developed into 3D models for each of the phases. The models incorporated the existing structures and were populated with preliminary equipment information where no actual data was yet available. These models were continuously updated as certified equipment information was received and design concepts finalized during the detailed engineering process. Erection sequences were further developed from the model for each phase. Sequencing models for Phase 3 are illustrated in figures 4 to 7.

**Figure 4. Model indicating the first set of four new cells operating with the remaining old cells in the foreground on completion of Phase 2**

**Figure 5. Model indicating the existing floor steel in place after the remaining two banks of old cells had been decommissioned and removed**

**Figure 6. Model indicating the existing floor steel demolished and the new support steel for the second bank of cells erected on the new matt foundation**

**Figure 7. Model indicating all nine cells and services operating in the final configuration on completion of Phase 3. The tenth cell can barely be seen in the bottom left corner**

The models clearly indicated new equipment, services tie-in requirements, items to be removed or modified and provided the basis for the identification of any interferences and conflicts. Snapshots from the models were forwarded to site at regular intervals, to be reviewed with the construction management team and representatives from the operations group prior to finalizing detailed deliverables and bulk material purchases. These reviews proved essential to the success of the project, as several interferences were identified during this process which allowed corrective actions to be taken before additional downtime was incurred.

Detailed scope of work documents and material take offs were subsequently produced by each of the disciplines and for each of the identified phases, to support the "staggered" construction strategies and associated bulk purchases made necessary by the "fast track" approach.

**Figure 8. First set of Outotec flotation cells in operation at completion of Phase 2. The remaining two banks of old Denver cells can be seen in the foreground.**

**Figure 9. The complete new Outotec flotation cells circuit at completion of Phase 3.**

**Case Study Highlights:**

Careful planning and the development of detailed construction phasing plans were critical to the success of the rougher flotation cell replacement.

Only two plant shutdowns, one for 24-hours and one for 8-hours, were required to complete the project work and operations was able to maintain full plant throughput during the remainder of the construction period. Furthermore, the plant shutdowns were scheduled together with maintenance activities to further minimize the impacts on production.

The same general methodology that was used on the rougher flotation circuit was successfully applied to all other brown-field operating areas of the project, including the secondary grinding circuit upgrade, the cleaner flotation circuit modernization and the tailings disposal area upgrade.

## CONCLUSION

The fast track approach implemented for the Gibraltar Expansion Projects combined with the brown-field operating plant setting presented many potential pitfalls to the projects' design and execution. Limited engineering had been performed prior to project approval and the level of engineering detail did not facilitate a detailed scope and cost estimate. Furthermore, the brown-field nature of the project resulted in numerous design iterations and the operating plant environment meant that construction activities were secondary to operational issues.

The success of the project can be attributed to a combination of the following factors:

- The risk / reward balance was recognized by all parties at project inception and reasonable contingency factors were included in the project.
- Construction management and operations personnel were involved in the entire engineering process, from concept development through to completion.
- Early and continued involvement by the engineering team in the project execution process.
- Clear separation of design activities best performed by the construction/operations personnel "in the field" and those activities best performed by the engineering team "in the office".
- Development and maintenance of an innovative, forward looking and flexible project team involving all stakeholders and engineering disciplines.
- Clear and open communication between all of the functional groups.
- Strong upper management support for this process and the project team.

## ACKNOWLEDGMENTS

Acknowledgement must be extended to the team of dedicated personnel from both the Owner's team and Hatch Ltd. The site operational knowledge an engineering expertise which they brought to the planning and execution of the expansion project, combined with their personal commitment, proved invaluable and ensured the success of it's execution.

Special mention to Jason Kolba from Hatch who provided some of the input in the case study presented.

# Sampling, Bench, and Pilot Plant Test Work

# THE COMBINED PRESSURE ACID LEACH (CPAL) PROCESS, BATCH AND PILOTING TESTWORK

J. Quinn, J. Turner, and D. van der Meulen

SGS Lakefield Oretest Pty Ltd, 431 Victoria Road, Malaga, WA, 6090, Australia

## ABSTRACT

World nickel resources are discussed followed by a summary of the main pyrometallurgical and hydrometallurgical nickel processing technologies including flowsheets for both laterite and sulphide ores. Highlighted are Western Australian nickel resources and processing technologies suited to the ore types of the region.

Poseidon Nickel Ltd is presently evaluating the use of Combined Pressure Acid Leaching (CPAL) to process various ore types such as nickel sulphides, nickel laterites and oxidized sulphide ores mined at operations in the Goldfields region of Western Australia. The CPAL process involves milling a combined nickel sulphide ore or concentrate and limonitic laterite ore in pregnant leach solution (PLS), from a heap leach or atmospheric leach circuit. The addition of sulphides to limonitic ores and the use of PLS to slurry the feed results in improved rheological properties (i.e., reduced slurry viscosity). In the subsequent high pressure oxidative leach, the sulphide component oxidises to sulphate releasing enough energy to autogenously heat the system. The oxidation of sulphide minerals and the hydrolysis of the ferric iron and aluminium (from the PLS) release acid that is consumed by the laterite portion of the ore. Under the correct operating conditions and feed combination, the CPAL process is operated with little or no sulphuric acid addition and no supplementary steam heating.

The CPAL process generates solutions with high autoclave discharge solution nickel tenors >10 g/L Ni due to the consecutive leaching stages. This factor combined with the reduced dilution from the injection of direct steam impacts on the downstream circuit volume and equipment size. The CPAL process is robust allowing the processing of difficult ore types including arsenic-bearing feeds and ores with low iron to magnesia ratios which are typically problematic for smelter operations.

SGS Lakefield Oretest Pty Ltd, Perth, Australia, operated hydrometallurgical batch and continuous pilot plant test work for Poseidon Nickel Ltd in 2008 to evaluate the CPAL process. Batch CPAL test work results showed nickel and cobalt extractions in excess of 95%, at an operating temperature of 220°C and an 900 kPa oxygen overpressure, after 90 minutes of leaching with no supplementary acid addition. Solution nickel tenors reached 13 g/L.

A pilot-scale four-stage (pot) vertical cascade autoclave was used for the combined pressure acid leaching of a combined feed (70:30%w/w nickel laterite / sulphide). The blended feed was milled in a typical PLS and fed to the autoclave. Autoclave operating temperature and pressure were 220°C and 900-1100 kPa oxygen overpressure, respectively, with supplementary acid addition at a rate of 150 kg per tonne of blended ore. CPAL stage nickel and cobalt extractions reached 92% and 79% (the cobalt extraction is significantly higher when the PLS-milling extraction is taken into account), respectively, with relatively low iron and aluminium dissolution.

## INTRODUCTION

Nickel is a transition metal with an atomic number of 28 and the most common oxidation state is +2. It is the 5$^{th}$ or 6$^{th}$ (depending on source) most abundant terrestrial element after iron, oxygen, silicon and magnesium but 24th in terms of crustal composition. The International Nickel Study Group (INSG) reported in 2008 that primary nickel production in 2007 was 1.43 kt (36% Europe, 26% Asia, 23% America) while primary consumption was 1.32 kt (52% Asia, 32% Europe, 12% America) (The Nickel Institute 2008). Approximately 2/3$^{rds}$ of all primary nickel is used for stainless steel production, up from about 40% in 1980. While nickel contributes some corrosion resistance to stainless steel – chromium is added for that purpose, nickel is added to stabilise the austenite structure (face centred cubic lattice) at lower temperatures. The austenite structure results in high ductility, low yield stress and relatively high tensile strength (i.e. it can be formed without cracking). Other uses include rechargeable batteries and superalloys (USGS 2009).

It is well documented (Dalvi et al.(1) 2004, USGS 2009, Nice 2004) that land-based nickel laterite resources (60-70%) exceed nickel sulphide resources (30-40%), though historically, commercial nickel production from sulphide minerals has been pre-dominant (~70%). The most common processing route is flotation followed by smelting, converting and refining operations (for sulphide minerals). Over the coming decades sulphide deposits will begin to deplete and even in the event of another discovery of a major nickel sulphide deposit it is unlikely new smelting operations will be established. The last greenfield sulphide smelter built was in Kalgoorlie by WMC in 1972. Most new sulphide deposits are either being fed through existing smelters or opting for a hydro-processing route.

In 2004, it was estimated that Australian nickel resources (on a contained nickel basis) accounted for 14% of all nickel sulphide and 13% of nickel laterite (world land-based) resources (Dalvi et al.(2) 2004, Krause and Kerfoot 1999). Australia is one of few countries in the world to have significant resources of both nickel ore types, commonly located in close proximity to one another or even present in the same ore body. There is a potential opportunity to use the resources to the advantage of process chemistry, industry economics and the environment. Laterite processing technologies which enable more efficient processes and which limit reagent and energy consumption are vital to process viability.

Although in recent years there have been multiple examples of cost overruns, commissioning / start-up issues and typically high operating costs associated with most hydrometallurgical laterite processing plants there have been clear advances in process technology, flowsheet design and selection. Hydrometallurgical flowsheets eliminate the production of sulphur dioxide and acid, show improved nickel and cobalt recoveries and typically allow for the processing of "dirty" ore types.

Prior to 1998, no laterite projects in Australia were in operation. In the late 1990's three WA projects came online: Cawse, Bulong and Murrin Murrin. Ravensthorpe began operation in 2007 and utilised the Enhanced Pressure Acid Leaching (EPAL) flowsheet to process both low and high magnesium ore types. Bulong, Cawse and Ravensthorpe have since ceased operation, leaving Murin Murrin as the only hydrometallurgical laterite processing plant still in operation in Western Australia.

The 2008-2009 global economic crisis and the crash in metal prices highlighted the need for an economically viable process option. The need to reduce costs and the environmental footprint of a hydrometallurgical process are paramount.

Poseidon Nickel Limited has several nickel resource development projects in the Goldfields region of Western Australia. The nickel resources include sulphide and oxide deposits as well as oxidised dump material from previously mined ore. Mt. Windarra was operated by WMC from 1974-1991 and shutdown due to low nickel prices. Poseidon is investigating various processing routes for the different ore types. Much of the oxides and low grade sulphide material seem amenable to heap leaching.

In 2008, SGS Lakefield Oretest Pty Ltd undertook an extensive batch and pilot plant testwork programme on behalf of Poseidon Nickel Ltd to investigate the Combined Pressure Acid (and oxidative) Leaching (CPAL) process. The CPAL process entails milling nickel sulphide (ore or concentrate) and limonitic laterite ore in pregnant leach solution (PLS) from a heap leach or atmospheric leach circuit. The addition of sulphides to limonitic ores and the use of PLS to slurry the feed result in improved rheological properties (i.e., reduced slurry viscosity). In the subsequent high pressure oxidative leach, the sulphide component oxidises to sulphate releasing enough energy to autogenously heat the system. The sulphide oxidation coupled with the hydrolysis of the ferric iron and aluminium (present in the PLS) produce acid to leach the laterite portion of the ore. Under the correct operating conditions and feed combination, the CPAL process operates with no sulphuric acid addition and no supplementary steam heating. In the event sulphuric acid is required for the CPAL process, pyrite from the ore deposit could be roasted to form a metallurgical gas which would be cooled, conditioned and sent to the wet end of a sulphuric acid plant, providing reagent for the process.

The CPAL process also enables relatively high autoclave discharge solution nickel tenors to be achieved (>10 g/L Ni) due to the consecutive leaching stages. This factor combined with no live steam addition (dilution) results in reduced downstream circuit volumes and equipment size. In existing plants, the consecutive leaching stages and autogenous heating could allow for increased throughput capacity and possible de-bottlenecking.

## NICKEL EXTRACTION TECHNOLOGIES (PROCESSING ROUTES)

### Nickel Pyrometallurgy

The following section briefly summarizes pyrometallurgical methods for the processing of nickel sulphides and laterites, limitations of the processes, cost drivers and environmental aspects related to production.

### Pyrometallurgical Processing of Sulphide Ores

Nickel sulphide ores are typically beneficiated and processed through a flotation concentrator prior to smelting. Electric or flash smelting is typically employed with flash smelting being the most common practice. Electric furnaces are typically used for the processing of complex ores / materials. Electric smelting requires an ore roasting step prior to smelting to reduce its sulphur content and volatiles (WBG 1998).

Flash smelters are fed with dry sulphide concentrate (typically, <1% moisture), pre-heated air and oxygen-enriched air (30-40% oxygen), or pure oxygen. Iron and sulphur oxidize, creating heat through the exothermic release of energy which is adequate to smelt the feed concentrate and produce a liquid matte (up to 45% nickel) and a fluid slag. Furnace matte still contains iron and sulphur that are oxidized in the converting step to sulphur dioxide and iron oxide by injecting air or oxygen into the molten bath. Oxides form a slag which is skimmed off and typically further processed in an electric furnace for the recovery of residual nickel before discard. Process gases are cooled and processed for dust removal and sulphur dioxide is converted to sulphuric acid in an acid plant.

A major refining practice is electro-refining which dissolves impure nickel anodes (produced from the converting stage) in an electrolytic bath where pure (99.99+%) nickel is plated at the cathode (Vale Inco Thompson, Jinchaun). Several operations perform matte leaching via sulphate-oxidative (Rustenburg, Impala), chlorine/chloride (Sumitomo, Exstrata) or ammonia (Sherritt Gordon, BHP Billiton) leaching followed by downstream processing (Krause and Kerfoot 1999). Another refining route is the carbonyl process (employed by Vale Inco at Sudbury and the Clydach Refinery) which produces high purity nickel pellets and specialty products. The process uses carbon monoxide to produce a nickel carbonyl gas which under the conditions precipitates nickel (metal) on the surface of pre-formed pellets. By-products of the process include copper and precious metal residues which require separate treatment. Another option is fluid bed roasting followed by a chlorine-hydrogen reduction which produces high grade nickel oxides (95+% nickel).

Smelting involves high capital costs associated with major process equipment including the flash furnace, convertors, slag treatment, power generators and an acid plant to capture sulphur dioxide. Smelting operations are coming under ever-increasing scrutiny around the world as green house gas emission controls become more stringent.

Relative to hydrometallurgical operations, smelter extractions are low, especially in the case of cobalt which is largely lost to the slag phase. Ore types which contain PGM's have typically been processed through smelters although recent pressure hydrometallurgical technologies have proven the ability to recover PGM's. Major smelting costs include oxygen and fuel. The process produces large amounts of dust and sulphur dioxide which must be captured and necessitate dust and gas handling systems (and an acid plant). Sulphuric acid must also be marketed as a by-product of sulphide smelting.

**Pyrometallurgical Processing of Laterite Ores**

The general laterite pyrometallurgical flowsheet includes: drying, calcining, reduction and (electric furnace) smelting which produces matte or ferronickel. Typically, saprolitic or garnieritic ores are treated using pyrometallurgical methods. These ore types generally have lower iron content when compared to the limonitic ores typically processed via hydrometallurgical routes. Ore composition and moisture content play key roles in determining whether a laterite is amenable to smelting. Unlike sulphide smelting, which is able to unlock significant amounts of energy through the exothermic reaction of sulphide oxidation, laterites necessitate significant energy input to heat the system (typically a combination of electrical energy and fuel). Coal is added to the kilns to produce coke which controls iron reduction and sulphur is also added to the kiln for matte production. Most of the iron reports to the slag as iron oxide and is removed in the skimming process.

Ferronickel production has been practiced in Noumea, New Caledonia since 1879. The Doniambo operation (Eramet Group) has been in operation since 1910 and produces both ferronickel and matte. Other major ferronickel producers include Cerro Matoso (BHP-Billiton) in Columbia, Falcondo (Exstrata) in Dominican Republic, Pomalaa (ANTAM) in Indonesia and Codemin (Anglo) in Brazil. Vale Inco is also in the process of building a new ferronickel processing plant at the Onca Puma operation in Brazil. One of the few other matte producers is PT Inco in Indonesia.

Laterite smelting uses large amounts of energy (electrical and fuels) during drying and smelting. Cobalt is generally not recovered in laterite smelting operations and thus, there are no cobalt by-product credits (Dalvi et al.(1) 2004). Ore chemistry plays a critical role in determining whether laterites can be smelted.

**Nickel Hydrometallurgy**

On a commercial scale, the use of hydrometallurgical nickel processing methods for laterite and sulphide ores has been limited. Over the past decades, the use of pressure oxidative (POX) leaching methods has been practiced commercially for zinc, copper and refractory gold ores with limited use for nickel concentrates. Recent developmental testwork has shown the potential for new POX processes in the nickel industry.

For the processing of laterites, both the Caron and PAL processes have been demonstrated on a commercial scale. Recent development projects, in the past 10 years, have had limited success. The following section briefly describes selected hydrometallurgical processes for both sulphide and laterite ore types, cost drivers and environmental issues related to production.

**Hydrometallurgical Processing of Sulphides**

The hydrometallurgical processing of zinc sulphide concentrates via pressure oxidative (POX) leaching methods has been practiced commercially since the 1970's and the early 1980's (Kerfoot et al. 2002, Muir 2005). The process generally produces hematite and elemental sulphur with small amounts of sulphate produced. The production of elemental sulphur was a major step forward from the roasting process which produced acid that had to be marketed or processed and discarded as a stable compound (typically, gypsum). The experiences of zinc producers and more recent applications for refractory gold processing have aided the application of POX technology to copper and nickel sulphides. One of the major issues related to high temperature POX operation is the production of molten sulphur which coats sulphide particles and limits extraction. Copper minerals seem to be most effected by the coating and chemical additives are effective in limiting the effect of the molten sulphur. Many projects have attempted to extend the POX methods to copper and nickel concentrates but few have been commercialized. The processes typically utilise a sulphate or sulphate/chloride-based systems, possibly integrated with fine grinding. A major issue related to copper leaching processes lies in the difficulties in treating primary and secondary copper sulphides via a single process (Miller et al. 2008).

Several copper sulphide ores have been treated commercially by pressure oxidative leaching. Inco operated an ammonia leaching process on a roasted low-grade pentlandite-pyrrhotite concentrates in the late 1950's but the process was generally unsuccessful and only operated for three years (Habashi 2004). The process was similar in principle to the Caron process (described later, in the laterite processing section). Sunshine Mining and Refining in Idaho used nitrogen species catalysed pressure oxidation to process a copper/silver deposit from the early 1980s to the mid-1990s. More recently, the Mt Gordon operation in Queensland, Australia, utilised a low pressure oxidative ferric leach to process a chalcocite ore. Freeport McMoran's Bagdad operation has been processing a chalcocite/chalcopyrite ores via a pressure oxidative leach process since 2003. The Sepon process was commercialized is Laos in 2005 and entails atmospheric and pressure acid leaching. The process is a ferric assisted atmospheric leach in which the atmospheric leach residue is floated to recover pyrite. Pyrite is then treated through a high pressure oxidative leach to produce ferric, heat and acid to be used in the atmospheric leach (ATL) circuit.

Nickel processing via a POX flowsheet is yet to be successfully commercialised although many processing routes have been investigated. Two notable development project include Norilsk Nickel's Tati project which utilises Activox technology (project is currently on hold), and Vale Inco's Voisey's Bay operation which employs a high pressure sulphate/chloride based leach that is scheduled to start up in 2011.

The Activox process was developed by WMT (Western Minerals Technology, now Norilsk) in Perth, Australia. The process involves the ultra-fine grinding of sulphide ores, typically copper and nickel minerals, followed by low temperature (100°C) oxidative sulphuric acid pressure leaching. The ultra-fine grinding process is energy intensive and little energy from the process is re-used. Extensive testwork has been completed on implementing the Activox process at Norilsk Nickel's Tati Nickel operation in Botswana but the project is currently on hold because of the current economic climate. The use of ultra-fine grinding, for the activation of minerals prior to leaching, was also outlined by Gerlach (1973) and Corrans and Angove (1992).

Inco (now Vale Inco) patented a leaching process which utilises chlorine/chloride for the Voisey's Bay (Newfoundland, Canada) flowsheet. The flotation concentrate undergoes an atmospheric chlorine leach prior to a high pressure oxidative sulphuric acid leach. The flowsheet involves solvent extraction followed by nickel electrowinning. Oxygen and chlorine gas, evolved at the anode, are captured and recycled to the atmospheric leach stage. Anolyte is recycled to the leaching stages in the flowsheet. Extensive mini-plant and demonstration plant operation has been completed and the commercial operation is scheduled to begin in 2011.

The CESL (Cominco Engineering Services Limited, now Teck Cominco) process employs a moderate temperature (150°C) chloride-enhanced oxidative sulphuric acid pressure leach (Jones 1999). The process is typically used to process nickel and copper sulphides but the patent also covers the leaching of lateritic ores. The process has undergone extensive pilot plant testwork and Vale's Salobo operation in Brazil began the operation of a small-scale plant utilising CESL technology in late 2008 (Defreyne 2009).

Two notable atmospheric leaching processes have been developed to treat sulphide concentrates: the Albion and Galvanox Processes.

The Albion process was developed in 1993 by MIM Holdings (now Xstrata Plc) to treat base metal (typically copper/zinc possibly containing platinum group metals - PGM's) sulphide concentrates (Pease 2005). The process utilises an Isamill to produce a finely-ground concentrate followed by an atmospheric leach. The atmospheric leach is auto-thermally heated by sulphide oxidation. Temperature is controlled by oxygen addition rate and feed slurry density. The ultra-fine grind is said to "activate" the mineral, increasing leaching rates due to the higher mineral surface area which limit mineral passivation. The process can be implemented with an alkaline and cyanide leach for refractory gold ores. Process licenses have been issued to Dominicana Gold Proprietary Ltd. to process gold and silver containing tailings and to European Goldfields Limited for processing gold ore. Xstrata has performed extensive piloting testwork on zinc and copper concentrates, refractory gold concentrates and smelter dusts with contain high levels of arsenic (Pease 2005, Hourne et al. 2006).

The Galvinox process was developed by researchers at the University of British Columbia and patented in 2005. The process is a sulphate based (typically, oxidative) atmospheric leach which operates at about 80°C. The process utilizes the galvanic interaction between chalcopyrite and pyrite. Ferric reduction is promoted at the surface of pyrite eliminating passivation of the chalcopyrite surface. By controlling the pulp potential the pyrite is not oxidised. Oxygen requirements for the process should be relatively low.

Typically, base metal hydrometallurgical processes have been limited to ore types with low PGM content. The Platsol process was developed (piloted) at SGS Lakefield, Canada in 2005 for Polymet Mining. The process treats copper/nickel/cobalt concentrates that contain PGM's by exposing the minerals to a pressure oxidative leach in the presence of 10-20 g/L sodium chloride. The sodium chloride promotes the dissolution of PGM's and gold. Piloting testwork products included copper cathode and PGM precipitates while nickel and cobalt were either processed through to individual hydroxide products or a mixed hydroxide product. The Platsol process is one of the few hydrometallurgical processes able to effectively extract both base metals and PGM's in a single leach stage.

Major advantages of the POX processes versus smelting operations include the elimination of sulphur dioxide (and its associated capture and acid production), the production of elemental sulphur, increased cobalt recoveries (typically doubled via hydrometallurgical processing), stable hematite production in the autoclave, lower energy input and the ability to process problematic ore types (arsenic-bearing minerals). Disadvantages include limited industrial applications, tailings management issues and possibly water resource requirements.

**Hydrometallurgical Processing of Laterites**

The two major hydrometallurgical processing routes for laterite ores are the Caron or PAL processes. Both processes have been commercially applied for over 40 years but with only a limited number of plants in operation. Both processes were initially operated by Freeport Sulfur Company operations in Cuba (Dalvi et al.(1) 2004). Heap leaching and atmospheric leaching have only recently been applied at a commercial scale and have recently garnered much attention for low grade deposits.

The Caron process was first developed in the 1920's and commercialized in 1944 by the Nicaro Nickel Co. in Cuba (Dalvi et al.(1) 2004). The process includes drying and milling the ore followed by roasting (calcining at ~700°C) under a reductive atmosphere which converts nickel and cobalt into its metallic form. The ore then undergoes an ammonia / ammonium carbonate selective atmospheric leach. Ammonia stripping results in nickel carbonate precipitation. Subsequent calcining yields a nickel oxide product. The pyrometallurgical front end of the Caron process is energy intensive. The process has been labelled as inefficient and outdated (Dalvi et al.(1) 2004, Reid 1996) primarily due to high energy usage and lower metal recoveries. Nickel and cobalt recoveries are also generally low when compared to smelting and PAL processes. Ammonium sulphate is produced as a by-product and must be marketed. Downstream processes are also reagent intensive. One advantage is the ability to process high magnesium ores. The BHP Billiton Yabulu operation in Queensland, Australia, utilises a modified Caron process flowsheet which includes solvent extraction to separate nickel and cobalt and produces two final products.

The first commercial PAL plant in Moa Bay, Cuba started operation in 1959. The plant is still in operation and the nickel/cobalt sulphide product has been refined in Fort Saskatchewan Alberta since 1994. The early 1990's saw many mining companies consider laterite resources and begin to develop hydrometallurgical processing routes. Advances in horizontal autoclaves, developments in solvent extraction and nickel/cobalt separation technologies and relatively low sulphur and acid prices (at the time) enhanced the viability of the processes.

Over the past 10-15 years several PAL plants have come online, generally, with limited success. Four laterite PAL plants have operated in Western Australia with the first operation, Cawse, coming online in September 1998, soon followed by Bulong and Murrin Murrin, both in December 1998 (Taylor 2007). Bulong and Cawse have since ceased operation. Ravensthorpe came online in 2007 and closed in early 2009 due to the sharp decline in metal prices. Murrin Murrin is the only laterite processing plant still in operation in Western Australia.

Cawse was the only one of the original three operations which was able to incorporate a physical upgrading (beneficiation) stage prior to leaching. There are generally limited opportunities for upgrading laterite ores, unlike sulphides which are concentrated by flotation prior to leaching. The coarse silica fraction was removed increasing the grade of the ore by 30-50% while retaining 70-80% of the metal value (Kyle and Furfaro 1997). The Cawse flowsheet (piloted at SGS Lakefield Oretest) incorporated pressure acid leaching operating at a temperature 250°C and system pressure of 3800 kPa. The leach residence time was 75-105 minutes. The downstream circuits produced a mixed hydroxide product (MHP) followed by ammonia re-leach, nickel solvent extraction/ electrowinning and cobalt ammonium sulphide precipitation.

Bulong operated the PAL process at 250°C, 4000 kPa and 75 minute retention time. It was the only direct solvent extraction (DSX) flowsheet with cobalt and nickel solvent extraction followed by nickel electrowinning. The cobalt stream underwent cobalt sulphide precipitation and redissolution, impurity removal and cobalt electrowinning. Bulong ceased operation in 2003, mainly due to its inability to source a cheap supply of sulphuric acid (Taylor 2007).

According to Motteram et al. (1997), Murrin Murrin was the most conservative process flowsheet as identified by technical, start-up and financial risk analysis. Although Murrin Murrin experienced setbacks (as did all the PAL plants) at start-up (Taylor 2007, Nice 2005) the relatively conservative nature of project may be the reason that Murrin Murrin is the only Western Australian PAL plant still in operation. The process flowsheet involves pressure acid leaching at 255°C and a system pressure of 4300 kPa for 90 minutes. The downstream processes produce a mixed sulphide product (MSP) which is subsequently dissolved in a pressure oxidative leach, followed by iron, copper and zinc removal, zinc and cobalt solvent extraction, hydrogen reduction (ammonia). The final products are nickel and cobalt briquettes and ammonium sulphate.

Ravensthorpe utilised the enhanced pressure acid leaching (EPAL) process developed (piloted at SGS Lakefield Oretest) and patented by Liu et al. in 2006. The process consisted of the pressure acid leaching of the limonite ore followed by atmospheric leaching of pre-treated saprolite. The process is similar in concept to that of the "High Temperature Neutralization of Laterite Leach Slurry" patented by Amax Inc (Queneau and Chou 1975).

In recent years, Ravensthorpe and Coral Bay (Philippines) are the only two hydrometallurgical operations to come on-line but there are currently many projects in the development stages (Goro, Vermelho, Ambatovy, Weda Bay - Halmehera, Ramu and Gladstone, to name a few).

Heap and Atmospheric leaching of nickel laterites has recently received much attention. Minara, BHP Billiton and Vale have patents related to heap leaching of nickel/cobalt laterites. Minara, Heron and Metallica have undertaken extensive heap leach testwork. There are many new atmospheric leach processes which are currently in the development stages. BHP Billiton's Gag Island project is investigating the use of atmospheric sequential leaching of limonite and saprolite ores.

Wedderburn (2005) estimated similar operating expenses for the three process (PAL, heap and atmospheric leaching) options but capital expenditure for the PAL operations were significantly higher. Atmospheric Leaching circuits have lower energy requirements, use lower cost materials and require lower capital expenditures though metal recoveries are generally lower when compared to PAL processes (Wedderburn 2005).

## THE HYDROMETALLURGICAL PROCESSING OF MIXED SULPHIDE/LATERITE ORES

Several leaching processes have been developed to take advantage of the energy and acid produced when sulphur-bearing minerals are oxidized. The following section outlines several processes which have been developed that blend sulphur-bearing materials and oxide ores under an oxidative environment to allow for autogenous heating and acid production which drives the leaching of valuable metals.

Seidel and Fitzhugh (1968) outlined the hydrothermal sulphidization oxidation (HSO) process to treat oxidized nickel ores with sulphur-bearing compounds. The process outlined a sulphidization stage (235°C) in which elemental sulphur or pyrite was used to form nickel and iron sulphide and sulphate. The sulphidization is followed by an oxidation stage (200°C) in which the metal sulphides are oxidized and dissolve while ferric sulphate is hydrolysed to form hematite.

Inco was awarded a patent in 1974 which outlined the high pressure oxidative leaching of a blend of laterite (limonite, saprolite) and sulphur-bearing compounds. The high temperature oxidation of the sulphur bearing minerals leads to the production of heat and acid which drove the leaching of the nickel and cobalt.

Salinovich and Blandy (WMC Resources) patented a process to hydrometallurgically treat a mixture of oxide and sulphide material in which the sulphide component is oxidized to produce sulphuric acid which drives the leaching process and the associated heat released by the oxidation reaction heats the system. The sulphide components outlined in the invention include sulphur, sulphur matte or sulphur-bearing concentrates. Energy expenditure for process heating can be minimized or eliminated.

A thorough review of the hydrometallurgical co-treatment of limonitic ore and sulphur-bearing materials was undertaken by Ferron and Fleming (2004). The paper outlined the implications of the addition of sulphur-bearing materials to the autoclave in both HPAL (laterite processing) and Platsol (ores containing platinum group metals) flowsheets. Two of the major operating costs involved in HPAL circuits are energy and acid consumption. The sulphur bearing materials added to high pressure oxidative leaching circuits are an alternative to acid injection and steam heating. The additions of elemental sulphur, pyrrhotite and Cu-Ni low grade bulk concentrate to limonitic ores were investigated in batch leaching testwork.

The energy released through the oxidation of the sulphur-bearing minerals allows a laterite leaching process (which typically operates with a negative heat balance) to operate without supplemental heating. Ferron and Fleming also noted, qualitative, that the addition of sulphur-bearing minerals to the limonitic ore had no visual effect on rheological properties of the slurry. If rheology became an issue, the ore types could be added to the autoclave individually. Once at high temperatures rheology improves significantly. The impact of the type of sulphur-bearing material added on discharge solids density was discussed. The addition of elemental sulphur does not affect the discharge as it is completely oxidized. The addition of pyrite will lead to hematite precipitation and increased solids content in the leach discharge. Any metal values in the sulphur-bearing material would also be recovered. The autogenous process also has an impact on downstream process requirements due to limited or no steam dilution in the autoclave. The major cost comparison between processes lies in the cost associated with oxygen production versus acid and traditional heating requirements (Ferron and Fleming, 2004).

High pressure acid leaching (HPAL) processing plants in Western Australia have, in the past, added sulphides ("transition" sulphide) or non-smeltable concentrates to the process (Salinovich and Blandy 1996). The reducing nature of the sulphides can limit the applicability of sulphides to a HPAL autoclave due to issues with autoclave lining deterioration. This issue is eliminated by the oxidizing environment in the CPAL autoclave.

## SERIAL LEACHING PROCESSES

Several processes have utilised the basic concept of serial (consecutive) leaching in hydrometallurgical applications. The sequential leach can minimize reagent usage, concentrate valuable metals in solution and limit the necessity for separate processing circuits.

As discussed earlier, the EPAL process is a serial leaching process which entails pressure acid leaching of the limonitic ore and a subsequent atmospheric leach/neutralization stage in which saprolite ore is added, valuable metals present in the saprolite dissolve while free acid is neutralized by magnesia present in the ore.

The Combined (referring to combined processes) Leaching Process was patented by Costa and Mendes (CVRD) in 2008. The patent outlines the use of serial leaching stages, notably atmospheric leaching followed solid liquid separation with the solution stream passing to a high pressure acid leaching. Residual acid from the first stage leach aids in leaching during the second stage. The dissolved iron and aluminium from the atmospheric leach stage hydrolyse to regenerate acid in the subsequent HPAL stage. The patent claims a 40% reduction in the addition of acid to the high pressure leach due to the acid re-generation from the hydrolysis of dissolved iron and aluminium from the atmospheric leach stage.

Consecutive or Simultaneous Leaching of Nickel and Cobalt was patented by Liu et al. (BHP Billiton SSM Technology) in 2006. The process involves a first leaching stage in which a laterite or partially oxidized sulfide ore is leached utilizing acid to produce a PLS which is subsequently used in the leaching of a sulfide ore or concentrate. The ferric iron present in the PLS is used to maintain the oxidation / reduction potential of the sulfide leach slurry high enough to assist in the leaching process. The patent also outlines the same process in a combined leaching process (laterite ore and or partially oxidized ore with a sulphide ore or concentrate).

The Murrin Murrin Operations (Minara Resources) patented a process in 2007 to recover nickel and cobalt values from lateritic ores (Rodriguez and Wedderburn 2007). The process entailed beneficiation, with the fine size fraction undergoing a pressure leaching stage. Subsequent to pressure leaching the discharge slurry undergoes a solid/liquid separation stage with a portion of the discharge solution being used as a leachant in a heap leach of the coarse fraction.

## THE CPAL PROCESS

The Combined Pressure Acid Leach (CPAL) process comprises a method for the recovery of base metals from a combination of nickel sulphide (ore or concentrate) and nickel oxide ores (laterites). The process combines the two feed types and a clarified and heated pregnant leach solution (PLS) from a heap leach or atmospheric leach circuit (containing nickel, cobalt, ferric iron and aluminium) in a milling process. The use of PLS to slurry the combined feed results in improved rheological characteristics (reduced viscosity) which allows for higher density feed to be processed. The subsequent pressure leach is carried out using an oxidant, typically air or oxygen (but also possibly ferric iron or manganese dioxide present in the feed ore). The oxidant allows for the sulphur / sulphide material to readily oxidize to sulphate releasing heat and producing acid. The sulphide oxidation coupled with the hydrolysis of ferric iron and aluminium (present in the PLS) produce acid which drives the leaching process. The CPAL process has the potential to operate without sulphuric acid addition or supplementary heating. The CPAL process also includes the possible roasting of pyrite ore in conjunction with an acid plant to produce acid for use in the process, if necessary. Typically, the heap or atmospheric leach feed would process saprolite/smectite ore while the CPAL feed would be a limonitic ore type.

The CPAL process also enables relatively high autoclave discharge solution nickel tenors to be achieved (>10 g/L Ni) due to the consecutive leaching stages. This factor combined with the fact that live steam (dilution) is not required to heat the slurry can have a significant impact on downstream circuit volume and equipment size.

The CPAL process enables the processing of ores which are typically not amenable to smelting processes. Material with low Fe:MgO ratios tend to affect slag viscosity and cannot be processed through a smelter. High MgO content necessitates increased reactor temperatures to lower slag viscosity. The CPAL method is able to process arsenic-bearing ores which typically require blending to manage health and safety and ensure a safe working environment. The CPAL process is tolerant of feed containing arsenic and materials with low Fe:MgO ratios.

## HYDROMETALLURGICAL TESTWORK PROGRAMME

### CPAL Batch Test Work

Selected results from an extensive batch test work programme are presented in this section. In the batch test work laterite and sulphide ore samples were combined at prescribed ratios and were slurried in PLS in a 5 L Titanium autoclave. The autoclave head was secured and the agitator speed was adjusted to 350 rpm. Autoclave temperature was controlled to approximately 5°C less than the required set point. Once the desired autoclave temperature was achieved, oxygen was injected into the autoclave. The off-gas from the headspace of the autoclave was removed via a condenser and the water vapour collected and returned to the reactor vessel. Samples were collected at selected times throughout the 250 minute test. The samples were immediately filtered and the solids were washed. Solids and solution samples were analysed by ICP-OES. Metal recoveries were calculated based on the solids assays (calculated head grade). Results of two batch tests are outlined.

Both tests were operated with a 70:30%w/w laterite/sulphide blend. The CPAL process was operated at 220°C with an oxygen overpressure of 800 kPa. The tests were operated at a feed density of 28% w/w and with no supplementary acid.

**Table 1. CPAL batch testwork parameters**

| CPAL Batch Test Work Parameters | | | | | |
|---|---|---|---|---|---|
| **Tests** | **Feed % Solids** | **Temp. °C** | **$O_2$ Overpressure kPa** | **Retention min** | **Acid : Ore kg/t** |
| 1 and 2 | 28.0 | 220 | 900 | 250 | 0 |

Table 2 presents the composition of the blended feed used for both tests. Nickel and cobalt in the feed were 1.24% and 0.055%, respectively. Total sulphur content was 3.99% and sulphide content was 3.20%. Assays of the PLS solution used in each test are shown in Table 3. The PLS used for test 2 was slightly more dilute than the PLS solution used in test 1.

**Table 2. Batch testwork feed assays**

| **Assay** | **Solids Assay ($\%_{w/w}$)** | | | | | | |
|---|---|---|---|---|---|---|---|
| **Sample Type** | **Ni** | **Co** | **Fe** | **Al** | **Mg** | **S (t)** | **$S^{2-}$** |
| Test 1 70:30 Blended Feed (Test 1 and 2) | 1.24 | 0.055 | 12.2 | 4.74 | 5.46 | 3.99 | 3.20 |

**Table 3. Batch testwork PLS assays**

| **Assay** | **Solution Assay (mg/L)** | | | | | | | | **[$H_2SO_4$] (g/L)** |
|---|---|---|---|---|---|---|---|---|---|
| **Sample Type** | **Ni** | **Co** | **Fe(t) / $Fe^{2+}$** | **Al** | **Mg** | **Ca** | **Cr** | **S(t)** | |
| PLS Test 1 | 4347 | 326 | 41490 / 1787 | 8548 | 21350 | 371 | 834 | 79500 | 15.3 |
| Diluted PLS Test 2 | 4122 | 309 | 39119 / 1694 | 8105 | 20243 | 352 | 791 | 75400 | 14.5 |

The results from test 1 and test 2 are shown in Table 4 and Table 5, respectively. The results show greater than 90% extraction of nickel and cobalt after 90 minutes of leaching for both tests. Effective iron and aluminium precipitation can be seen in the solution assays. Solution iron concentration decreased from about 40 g/L to 3-5 g/L at the conclusion of the tests. Solution aluminium concentration decreased from roughly 8 g/L to roughly 2 g/L. In excess of 90% of the sulphide content was oxidized after just 30 minutes of leaching. Acid production through iron and aluminium precipitation and sulphide oxidation is evident in the increased sulphuric acid concentrations throughout both tests.

**Table 4. Batch test 1 results**

| Time (min) | Extraction (%) | | Discharge Solution Assay (mg/L) | | | | $[H_2SO_4]$ (g/L) | $S^{2-}$ ($\%_{w/w}$) |
|---|---|---|---|---|---|---|---|---|
| | Ni | Co | Ni | Co | Fe | Al | | |
| 0 | 0 | 0 | 4347 | 326 | 41490 | 8548 | 15.3 | 3.2 |
| 30 | 73.8 | 89.9 | 8011 | 579 | 14275 | 1392 | 48.1 | 0.11 |
| 60 | 90.1 | 93.6 | 11151 | 604 | 11749 | 1501 | 52.6 | 0.08 |
| 120 | 94.6 | 95.4 | 11832 | 623 | 8095 | 2251 | 53.6 | 0.08 |
| 190 | 96.5 | 96 | 12580 | 646 | 6493 | 2615 | 55.3 | 0.05 |
| 250 | 97.5 | 96.4 | 12997 | 669 | 5621 | 2507 | 56.2 | 0.06 |

**Table 5. Batch test 2 results**

| Time (min) | Extraction (%) | | Discharge Solution Assay (mg/L) | | | | $[H_2SO_4]$ (g/L) | $S^{2-}$ ($\%_{w/w}$) |
|---|---|---|---|---|---|---|---|---|
| | Ni | Co | Ni | Co | Fe | Al | | |
| 0 | 0 | 0 | 4122 | 309 | 39339 | 8105 | 14.5 | 3.2 |
| 30 | 81.7 | 89.6 | 9488 | 514 | 10337 | 1199 | 44.5 | 0.13 |
| 60 | 88.8 | 92.7 | 10087 | 527 | 7908 | 1357 | 48.5 | 0.11 |
| 90 | 92 | 94.2 | 10487 | 545 | 6263 | 1694 | 49.4 | 0.08 |
| 120 | 93.9 | 94.7 | 10993 | 545 | 4879 | 1946 | 50.6 | 0.08 |
| 190 | 94.9 | 95.1 | 11270 | 565 | 4361 | 2011 | 50.5 | 0.06 |
| 250 | 96.7 | 95.5 | 11019 | 549 | 3614 | 1771 | 49 | 0.06 |

Figure 1 shows nickel and cobalt concentrations throughout the testwork. Cobalt dissolution occurs slightly faster than nickel dissolution. The graph also shows the increasing nickel solution concentration throughout the tests.

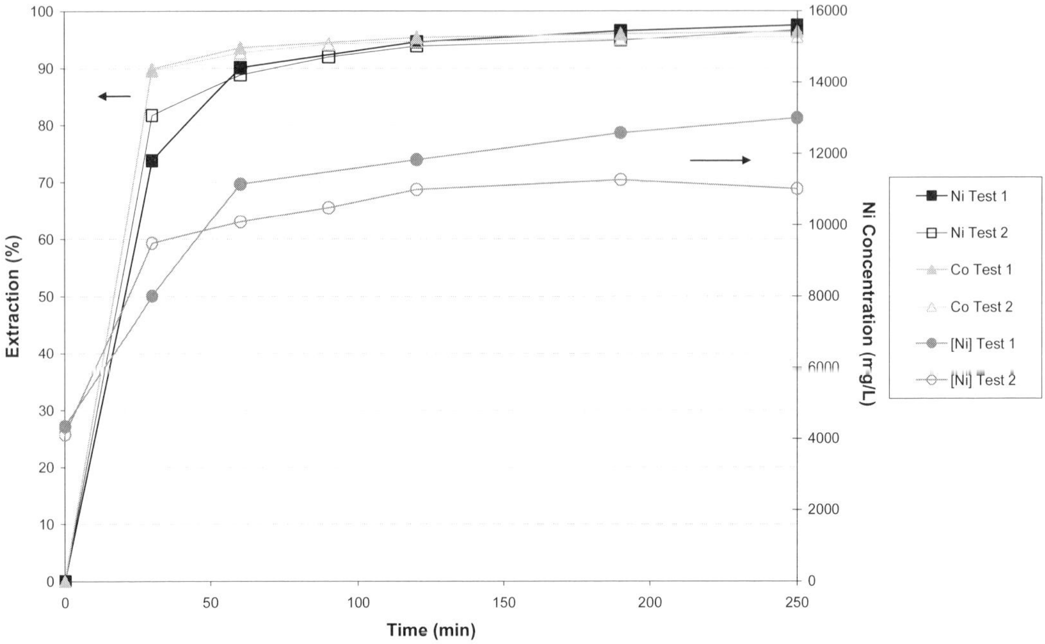


**Figure 1. Nickel and cobalt extractions and nickel solution assays for test 1 and 2**

Figure 2 shows decreasing aluminium and iron concentrations resulting from hydrolysis (for both tests). As the iron and aluminium concentrations decreased the free acidity in the leach slurry increased.

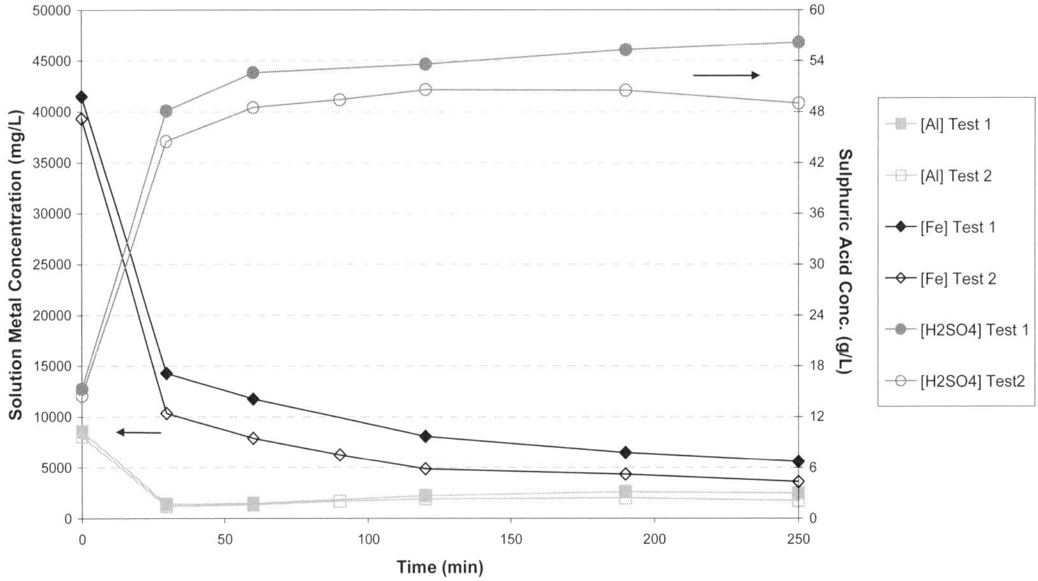


**Figure 2. Aluminium and iron solution assays and free acidity for test 1 and 2**

Figure 3 shows the decreasing sulphide concentration in the solids. As the sulphide component of the ore oxidised it released acid, as can be seen in the increased sulphuric acid concentrations as the leach progressed.

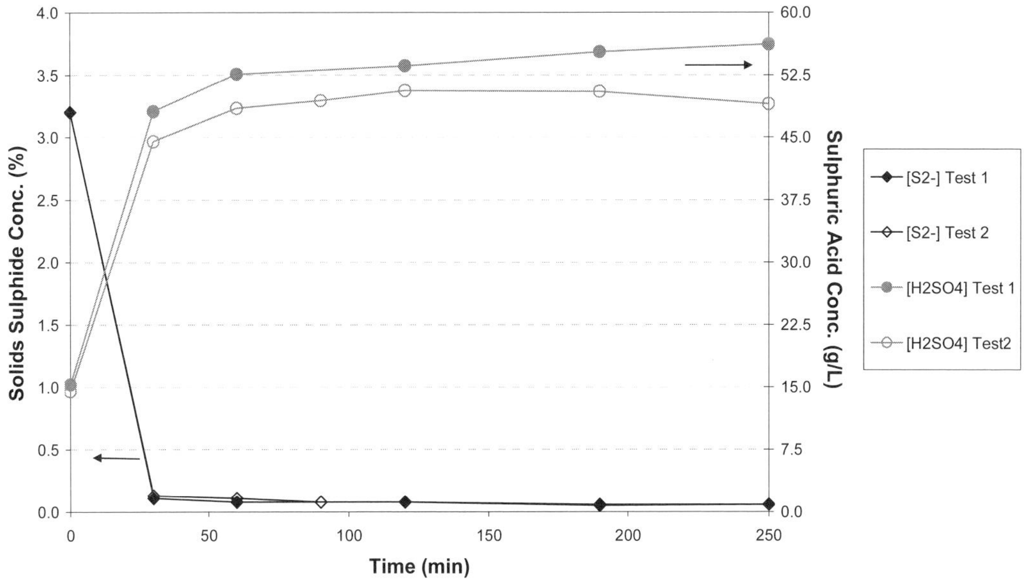


**Figure 3. Solids sulphide content and free acidity for tests 1 and 2**

The batch testwork results showed greater than 90% nickel and cobalt dissolution after 90 minutes of combined pressure acid leaching with no supplementary acid addition. The tests showed effective iron and aluminium precipitation and sulphide oxidation which produced acid to drive the leaching process.

**CPAL Piloting Test work**

An eleven day piloting campaign was undertaken at SGS Lakefield Oretest for Poseidon Nickel in September 2008. The results presented were taken from twelve hour mass balance (MB) test periods and profile samples taken during the piloting campaign.

A blended feed comprising 70%w/w bulk nickel laterite and 30%w/w bulk sulphide was prepared (-6 mm) prior to the commencement of piloting. The composition of the blended feed sample is shown in Table 6.

**Table 6. Individual assays of the three feed types used to produce the two blended feed samples**

| Assay | Solids Assay ($\%_{w/w}$) | | | | | |
|---|---|---|---|---|---|---|
| Sample Type | Ni | Co | Fe | Al | Mg | S (t) |
| 70:30%w/w Blend | 1.63 | 0.052 | 15.90 | 3.99 | 5.21 | 4.63 |

The blended feed was milled in a typical PLS and fed to the autoclave. Table 7 shows the composition of the PLS (prior to milling). PLS nickel and cobalt concentrations were 3821 mg/L and 314 mg/L, respectively. The total iron (Fe(t)) concentration was 37548 mg/L while aluminium was 7993 mg/L. Sodium levels were high at 20103 mg/L due to the addition of sodium hydroxide to the original PLS to target a sulphuric acid concentration of 25 g/L. In the high pressure acid leaching of laterite ores high solution sodium concentrations have resulted in low levels of aluminium dissolution (Kyle 1999) due to natro-alunite formation. The high sodium levels have also been seen to retard jarosite conversion to hematite which can have a significant effect on acid consumption (Johnson et al. 2002, Whittington 2003). Chloride was also present in the heap leach solution at 1241 mg/L.

**Table 7. Typical heap leach solution assays**

| Assay | Solution Assay (mg/L) | | | | | | | | $[H_2SO_4]$ (g/L) |
|---|---|---|---|---|---|---|---|---|---|
| Sample Type | Ni | Co | Fe(t) / $Fe^{2+}$ | Al | Mg | Na | S(t) | Cl | |
| PLS | 3821 | 314 | 37548 / 1542 | 7993 | 15927 | 20103 | 84663 | 1241 | 24.64 |

A pilot-scale four-stage (pot) vertical cascade autoclave was used for the combined pressure acid leaching of the PLS-milled blended feed. Autoclave operating temperature and system pressure ranged from 220-240°C and 3250-4350 kPa (900-1100 kPa oxygen overpressure), respectively, with supplementary acid addition for certain test periods. The PLS-milled CPAL feed blend was fed into 100 L batching tanks where ferric sulphate was added to a target solution iron concentration. Table 8 shows the CPAL feed solution total iron and aluminium concentrations and the CPAL feed solids sulphide concentration.

CPAL acid availability was estimated based upon supplementary acid addition rates, the assumption that 98% of the iron in solution formed hematite, that 95% of the aluminium in solution formed alunite and that 90% of the sulphide sulphur was attributed to pyrite which fully oxidized. The acid availability for test 2 is significantly higher (452 kg/t) than test 1 (271 kg/t) due to elevated iron levels in the feed solution and supplementary acid addition.

**Table 8. CPAL feed solution and solids assays used to estimate acid availability**

| MB Period | Acid Addition Rate kg/t | Feed Solution Assays mg/L | | Solids Assay $\%_{w/w}$ | Acid Availability kg/t |
|---|---|---|---|---|---|
| | | Fe(t) | Al | $S_2$ | |
| 1 | 0 | 38280 | 9791 | 2.49 | 271 |
| 2 | 150 | 42620 | 8913 | 2.56 | 452 |

Based on the CPAL feed solids assays, stoichiometric ratios of acid consumption, a target discharge free acid concentration (40 g/L) and typical extraction values (shown in Table 9) acid requirement were estimated for each mass balance period. The calculation assumes 100% aluminium dissolution (of which 95% precipitates as natro-alunite). The CPAL mass balance feed solids assays (post-milling) and estimated acid requirements are shown in Table 10. The feed solids assays for tests 1 and 2 are of similar composition and thus the acid requirement values are similar (385 and 384 kg/t).

**Table 9. Typical extraction values used to estimate acid requirement**

| | Ni | Co | Fe | Al | Mn | Mg | Ca |
|---|---|---|---|---|---|---|---|
| **Extraction (%)** | 93.0 | 94.0 | 2.0 | 5.0 | 92.0 | 90.0 | 95.0 |

The autoclave feed solids compositions and estimated acid requirements are shown in Table 10. The feed solids nickel and cobalt concentrations were 1.4% and 0.02%, respectively.

**Table 10. CPAL feed solids assays and estimated acid requirement**

| MB Period | Solids Assays $\%_{w/w}$ | | | | | | | | Acid Requirement kg/t |
|---|---|---|---|---|---|---|---|---|---|
| | Ni | Co | Fe | Al | Mn | Mg | Na | Ca | |
| 1 | 1.42 | 0.020 | 14.9 | 3.86 | 0.065 | 4.54 | 0.145 | 0.962 | 385 |
| 2 | 1.41 | 0.020 | 15.4 | 3.72 | 0.072 | 4.54 | 0.314 | 1.030 | 384 |

Test conditions for the two mass balance periods as well as the acid surplus (or deficit) values (based on the acid availability and requirement calculations) are shown in Table 11.

**Table 11. CPAL process variable targets and actual values**

| MB Period | CPAL MB Targets / Actual Values | | | | | | Acid Surplus (Deficit) kg/t |
|---|---|---|---|---|---|---|---|
| | Feed % Solids | Temp. °C | Pressure kPa | Retention min | Acid : Ore kg/t | Feed [Fe] g/l | |
| 1 | 33.0 / 30.9 | 220 / 220 | 3350 / 3343 | 120 / 108 | 0 / 0 | 40 / 38.3 | (114) |
| 2 | 33.0 / 32.7 | 220 / 220 | 3150 / 3153 | 90 / 103 | 150 / 162 | 40 / 42.6 | 68 |

Mass balance period results are shown in Table 12. Mass balance period 1 was operated with no acid addition to pot 1. The CPAL operating targets were 220°C and 3350 kPa (total pressure) with an increased retention time of 120 minutes. The absence of acid addition resulted in low nickel and cobalt extractions (71.1% and 47.4%, respectively). Discharge iron and aluminium solution concentrations were relatively low revealing significant levels of precipitation. In the absence of acid addition metallurgical performance was poor, though the results clearly show the acid generating potential of the process. The test was operated with no supplementary acid addition and the discharge acid concentration was 21.1 g/L.

Nickel and cobalt extractions for CPAL mass balance period 2 reached 92% and 79%, respectively. Overall cobalt extractions were significantly higher when the PLS-milling stage extractions are taken into account. The test showed significant levels of iron and aluminium precipitation evident in the low solution iron and aluminium concentrations. The relatively low sodium levels in the discharge solution suggest significant natro-alunite precipitation. About two thirds of the total iron in the discharge solution was ferric.

**Table 12. CPAL mass balance period results**

| MB Period | Extraction (%) | | CPAL Discharge Liquor (mg/L) | | | | | $[H_2SO_4]$ g/L | Slurry ORP mV |
|---|---|---|---|---|---|---|---|---|---|
| | Ni | Co | Ni | Co | Fe(t) / $Fe^{2+}$ | Al | Na | | |
| 1 | 71.1 | 47.4 | 6628 | 373 | 9680 / 7931 | 477 | 6580 | 21.1 | 414 |
| 2 | 91.9 | 78.8 | 8701 | 394 | 6714 / 2279 | 968 | 880 | 46.7 | 488 |

Figure 4 shows results of profile (kinetics) sample results for mass balance period 1 and 2. A profile samples was taken from each CPAL stage. Nickel extraction (for both tests) increases with retention time and increased free acidity in the leach.

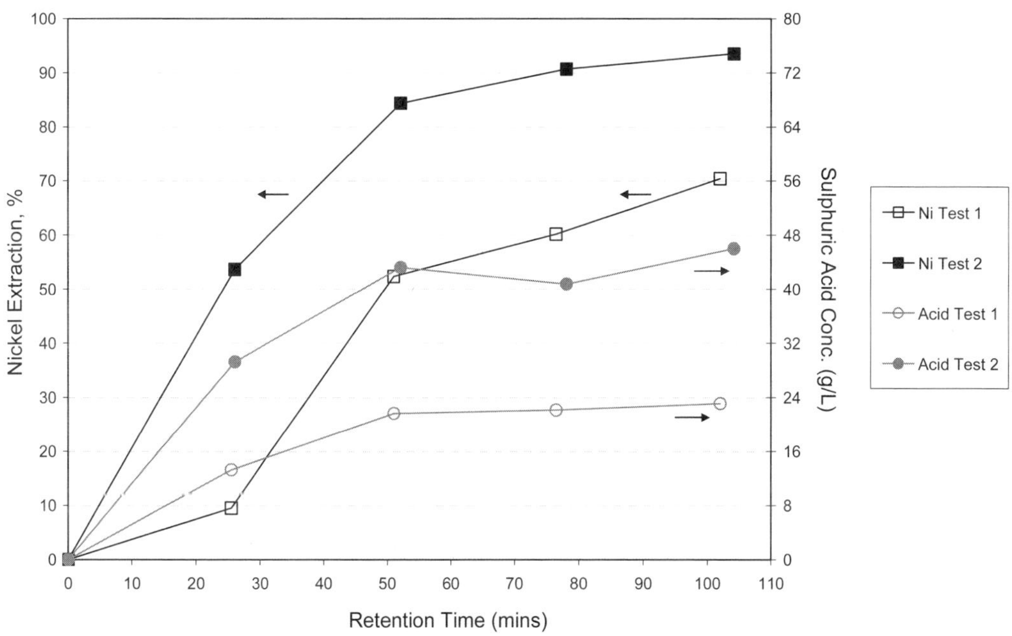


**Figure 4. Pilot plant CPAL nickel extraction and sulphuric acid concentration for test 1 and 2**

The CPAL pilot plant acid to ore ratio (from the pilot plant test work) of 150 kg/t is relatively low when compared to (past and present) Western Australian PAL operations which ranged from 375 to 518 kg/t (Mayze 1999). Sherritt International's Ambatovy project in Madagascar whose lateritic deposits have a relatively low magnesium content estimates acid consumption at 200 kg/t (MRA 2005). Optimisation of the feed laterite to sulphide ratio and the PLS composition could have the added advantage of lowering the acid requirement. The feed laterite to sulphide ratio (extent of the sulphide oxidation reaction) also dictates whether supplementary heating is required.

The piloting testwork results showed greater than 90% nickel and approx' 80% cobalt dissolution at an acid addition rate of 150 kg/t of ore. Results showed effective iron and aluminium precipitation. High feed and low discharge solution sodium concentrations indicated significant levels of natro-alunite/jarosite formation which may have limited hematite production and associated acid production. Estimated acid requirements were good indicators of metallurgical performance.

## THE FUTURE

Poseidon Nickel Limited plans to commission further research and preliminary engineering to identify the optimum configuration for the combination of nickel sulphide and laterite ores and concentrate. Following the successful demonstration of the technology Poseidon will move to commercialise the process either at new or existing facilities.

## CONCLUSIONS

The world has historically relied upon sulphide deposits as its major source of nickel. It is widely acknowledged that the production of nickel from laterite ores will become more significant as sulphide deposits are depleted. As no new nickel smelters are likely to be built, new projects will rely upon hydrometallurgical processing options.

The recent downturn in the global economy coupled with the financial and operational difficulties of almost all major nickel laterite hydrometallurgical projects have highlighted the need for technology which reduces costs through minimization of reagent and energy consumption. Reduction in energy requirements is linked to reducing the environmental footprint. The resource industry is under increasing pressure to adopt a sustainable development approach.

Poseidon Nickel Limited has several nickel resource development projects in the Goldfields region of Western Australia. The nickel resources include sulphide and oxide deposits as well as oxidised dump material from previously mined ore. Poseidon has patented the combined pressure acid leach (CPAL) method to process various ore types through a single operation. Pressure acid leach plants have commercially treated low grade nickel sulphides in Western Australia. The processes have been restricted due to the reducing nature of the sulphide ores which impact PAL autoclave materials of construction. Various oxidative leaching methods (HSO, Inco and WMC) have been investigated to process mixed laterite/sulphide feeds which utilise the heat and acid released through sulphide oxidation to lower energy and reagent requirements.

The CPAL process comprises a method for the recovery of base metals from a combined nickel sulphide (ore or concentrate) and nickel oxide laterite. The process combines the two feed types and a PLS in a milling process. The use of PLS with the combined feed results in improved rheological characteristics (reduced viscosity) which can allow for higher density feed to be processed. The subsequent oxidative pressure leach allows for sulphide material to readily oxidize to sulphate releasing heat and producing acid to heat the system and leach the laterite portion of the ore. The hydrolysis of ferric iron and aluminium, from the PLS, produces acid which also aides in leaching. The CPAL process has the potential to operate without supplementary sulphuric acid addition or heating. The CPAL patent also includes the possible roasting of pyrite ore in conjunction with an acid plant to produce supplemental acid for use in the process, if necessary. Typically, the heap or atmospheric leach feed would process saprolite/smectite ore while the CPAL feed would be a limonitic ore type.

The CPAL process also enables relatively high autoclave discharge solution nickel tenors to be achieved (>10 g/L Ni) due to the consecutive leaching stages. This factor combined with the fact that live steam (dilution) is not required to heat the slurry has a significant influence on downstream circuit volume and equipment size. The use of the CPAL process and the ability to concentrate nickel in solution through sequential leaching processes and the absence of steam dilution could enable substantial increase in throughput capacity and possibly allow for de-bottlenecking of current operations.

The CPAL process enables the processing of ores which are typically not amenable to smelting practices. Material with low Fe:MgO ratios tend to affect slag viscosity and cannot be processed through a smelter. High MgO content necessitates increased reactor temperatures to lower slag viscosity. The presence of arsenic in the smelter feed can have an effect on managing health and safety and blending practices may be essential to a safe working environment. The CPAL process is tolerant of feed containing arsenic and materials with low Fe:MgO ratios.

SGS Lakefield Oretest Pty Ltd, Perth, Australia, operated hydrometallurgical batch and continuous piloting of the CPAL process in 2008. Batch CPAL test work results showed nickel and cobalt extractions in excess of 95%, at an operating temperature of 220°C and 900 kPa oxygen overpressure, after 90 minutes of combined pressure acid leaching with no supplementary acid addition. Solution nickel tenors reached in excess of 13 g/L.

A four-stage vertical cascade autoclave was used for the Combined Pressure Acid (and oxidative) Leaching (CPAL) of a 70:30%w/w nickel laterite / sulphide feed blend. The blended feed was milled in a typical pregnant heap leach solution and fed to the autoclave. Autoclave operating temperature and pressure ranged from 220-240°C and 3000-4250 kPa (800-1000 kPa oxygen overpressure), respectively. Estimated acid requirement / availability were good indicators of metallurgical performance. The results clearly demonstrate the potential of the CPAL process to produce acid through iron and aluminium hydrolyses and sulphide oxidation. The highest nickel and cobalt extractions obtained during the CPAL piloting testwork were 92% and 79%, respectively. Results showed effective iron and aluminium (hematite / jarosite / alunite) precipitation resulting in acid production, resulting in an effective base metal leaching process.

## BIBLIOGRAPHY

Berezowski, R. April 22, 2004. "Nickel Extraction Technology Developments". *Presentation for MEMS 13th Annual Conference.*

Corrans, I.J., and Angove, J.E. 1992. "Activation of Mineral Species". US Patent 5232491.

Costa, R.dS., Mendes, F.D. (CVRD). 2007. "Combined Leaching Process". International Patent WO/2007/053919.

Dalvi, A.D., Bacon, W.G., and Osborne, R.C. (1), 2004. The Past and the Future of Nickel Laterites, *PDAC 2004 znternational Convention, Tradeshow and Investors Exchange..*

Dalvi, A.D., Bacon, W.G., and Osborne, R.C. (2), 2004. "Past and Future of Nickel Laterites Projects". In Imrie, W.P., Lane, D.M. (eds.) *International Laterite Nickel Symposium TMS 2004* – 133rd Annual Meeting and Exhibition. Charlotte, North Carolina, pp 1-27.

Defreyne, J., 2009. "Early Copper Production Results from Vale's Hydrometallurgical CESL Refinery". *Alta 2009 Copper Technical Proceedings.*

Dixon, D.G., and Tshilombo, A.L., Dec. 8 2005. "Leaching Process for Copper Concentrates", US Patent 2005/0269,208.

Ferron, C.J., and Fleming, C.A., 2004. Co-treatment of Limonitic Laterites and Sulphur-bearing Minerals as an alternative to the HiPAL process, *International Laterite Nickel Symposium – 2004 Pressure Acid Leaching*. p. 245-261.

Gerlach, J. K. ; Gock, E. D. ; Ghosh, S. K. Gerlach et al., 1973. "Activation and Leaching of Chalcopyrite Concentrates in Dilute Sulfuric Acid", *International Symposium on Hydrometallurgy -AIME*, 1973, chapter 16.

Habashi, F., 2004. "The origins of pressure hydrometallurgy". *Pressure Hydrometallurgy 2004*. pp. 3-20.

Hatch, 2006, "Skye Resources Inc. – Fenix Feasibility Study: Final report". Section 5.

Hore, M, Robinson, J, and Lonsdale, G., April 8, 2008, "Combined Pressure Pachuca Acid Leaching of Oxide-Sulphide Ore Blends - 10298", SGS Final Report for Poseidon Nickel Ltd.

Hourne, M., MacDonald, C.A., Rohner, P., and Woodall, P., 2006. "Albion Process and Leaching High Arsenic Materials at Mount ISA". *Alta Copper 2006,* Perth, Australia.

Johnson, J.A., McDonald, B.I., Whittingdon, Quan, L.P., and Muir, D.M. 2002. Process Water Salinity Effects in the Pressure Leaching of Cawse Nickel Laterite Ores. *Chloride Metallurgy 2002. Practice and Theory of Chloride / Metal Interaction.*

Johnson, G.D., and Streltsova, N., June 29, 1999. "Method for the Processing of Copper Minerals", US Patent 5917116.

Jones, D.L. (Cominco Engineering Service Ltd). 1999. "Chloride Assisted Hydrometallurgical Extraction of Nickel and Cobalt from Sulphide and Laterite Ores". International Patent WO/1999/009226.

Kerfoot, D.G.E., Krause, E., Love, B.J., Singhal, A. (Inco Ltd), 2002. "Recovery of Nickel and Cobalt Values from a Sulfidic Flotation Concentrate by Chloride Assisted Oxidative Pressure Leaching in Sulfuric Acid", International Patent WO/2002/024966.

Kyle, J.H., 1999. Water Quality in Nickel Laterite Processing. *Australian Minerals Foundation Conference – Water Management in Metallurgical Operations.*

Kyle,J.H., 2003. Jarosite / Alunite in Nickel Laterite Leaching – Friend or Foe?. *Alta 2003 Nickel / Cobalt – 9 Technical Proceedings,* Perth, Australia.

Kyle, J.H, and Furfaro, D., 1997. "the Cawse nickel/cobalt laterite project metallurgical process development". Cobalt/Nickel '97. Nickel/cobalt '97, vol. 1, eds. Cooper, W.C., and Mihaylov, I., Canadian institute of Mining, Metallurgy and Petroleum. Montreal, 379-389.

Krause, E., and Kerfoot, D., October 1999. "Nickel Processing Technologies and the Next Millenium". Presentation at 1999 Parker Center Hydrometallurgy Conference.

Liu, H., Duarte, H., and Meihack, W., (BHP Billiton SSM Technology PTY LTD). May 26, 2006. "Consecutive or Simultaneous Leaching of Nickel and Cobalt". International Patent WO/2006/053376.

Mayze R. (Bateman Brown & Root). 1999. "An Engineering Comparison of the Three Treatment Flowsheet in WA Nickel Laterite Projects". Alta 1999: Nickel/Cobalt Pressure Leaching & Hydrometallurgy Forum.

Mezei, A., Verbaan, N., Wang, N., Todd, I., and Kyle, J.H., Effect of Salinity on Nickel Laterite Processing, *Chloride Metallurgy 2002. Practice and Theory of Chloride / Metal Interaction.*

Miller, K.J., Sylwestrzak, L.A., and Baxter, K.G. 2008. "Treatment of Copper Sulphide Deposits: Evaluation of a Galvanox Versus Sepon Circuit Configuration". ALTA 2008 Nickel-Cobalt, Copper & Uranium Conference

Motteram, G., Ryan, M., Weizenbach, R., 1997. "Application of the pressure acid leach process to Western Australian nickel/cobalt laterites". Nickel/cobalt '97, vol. 1, eds. Cooper, W.C., and Mihaylov, I., Canadian institute of Mining, Metallurgy and Petroleum. Montreal, pp. 391-407.

MRA, 2005. "Ambatovy Deal a Boost for Madagascar's Mining Sector". Mining Review Africa (MRA), Issue 4.

Muir, D., April 2005. "Overview of Current Cevelopments in POX". Presentation at Pressure Oxidation: Modeling, Reactor Design and Process – CSIRO Minerals.

Neudorff, D., Jan. 5, 2006. "Method for nickel and Cobalt recovery from Laterite Ores by Reaction with Concentrated Acid and Water leaching". US Patent 2006/0002835 A1.

Nice, R.W., 2004. "The Western Australian Nickel Project – What Have We Learned". Pressure Hydrometallurgy 2004. pp. 155-179.

Opratko, V., Bell, J.A.E., and Ferrajuolo, V., May 7, 1974. "Acid Leaching of Laerite Ore", United States Patent Office, 3,809,549.

Palmer, C.M., and Johnston, G.D., July 2005. "The Activox Process: Growing Significance in the Nickel Industry". JOM.

Poseidon Nickel Limited, *Poseidon Nickel Annual Report 2008.*

Queneau, P.B., and Chou, E.C. (Amax Inc.). 1975. "High Temperature Neutralization of Laterite Leach Slurry". US Patent 3991159.

Pease, J., August 2005."Complex Leaching Becomes Much Simpler". *Australian Mining Article*. pp. 26-32.

Reid, J.G. 1996. "Laterite Ores – Nickel and Cobalt Resources for the Future". *Nickel '96 Mineral to Market*, Eds. Grimsey, E.J. and Neuss, I., The Australian Institute of Mining and Metallurgy, Melbourne, pp. 11-16.

RM Research Pty Ltd, September 29, 2006. Niagara Mining Limited: Exploration intensifies at Windarra…multi-target program.

Rodriguez, M., (Poseidon Nickel Limited), 2008. Australian Innovation Patent 2008100563 "Method for the Recovery of Nickel from Ores". Australian Government, Australia IP.

Rodriguez, M., Wedderburn, B.J., 2007 (Murrin Murrin Operations Pty Ltd). 2007. "Hydrometallurgical Method for the Extraction of Nickel and Cobalt from Laterite Ores". International Patent WO/2007/016737.

Roorda, H.J., and Hermans, J.A., 1981. "Energy Constraints in the extraction of nickel from oxide ores (I)". *Erzmetal*. Vol. 34, no. 2, pp 82-88.

Salinovich, T., and Blandy, C.W.D., (WMC Resources Ltd.) Australian Patent Application number: 199530448 B2, Patent number: 709751. 1990. March 21, 1996.

Seidel, D.C., and Fitzhugh, J.R., April 1968. A Hydrothermal Process for Oxidized Nickel Ores, *Mining Engineering*. P. 80-86.

Taylor, A., 2007. "Nickel Processing Technology 10 Years On From Cawse, Bulong And Murrin Murrin". *Alta Nickel/Cobalt 2007 Proceedings*, Perth Australia.

Taylor, A., & Jansen, M.L., 2000. "Future Trends in PAL Plant Design for Ni/Co Laterites", *Ni/Co 2000, ALTA Metallurgical Services*, Melbourne, pp 1-13.

Taylor, A., & Jansen, M.L., May 15-17, 2000. "Hydrometallurgical Treatment of Copper Sulphides: Are We on the Brink?", *ALTA Nickel/Cobalt 2000 Proceedings*, Perth, Australia.

The Nickel Institute, website: http://www.nickelinstitute.org, May 2008.

US Geological Survey, January 2009. Mineral Commodities Survey (Nickel).

Wedderburn, B., September 2005. "Recent Trends in Nickel Laterite Technology". Presentation at AJM Conference – Sydney.

Whittington, B.I., McDonald, R.G., Johnson, J.A., and Muir, D.M., 2002. "Pressure acid leaching of arid-region nickel laterite Part I: effect of water quality." *Hydrometallurgy* 70 (2003) 31-46.

World Bank Group (WBG). 1998. "Nickel Smelting and Refining". Pollution Prevention and Abatement Handbook. Pp 349-352.

# SAMPLING FOR ON-STREAM ANALYSIS AND COMPOSITE SAMPLES

Christian von Alfthan and Matti Kongas

Outotec Minerals Oy, Riihitontuntie 7 D, 02200 Espoo, Finland

## ABSTRACT

It is not practical to feed a full mineral processing plant process flow through an analyzer system in most cases. A representative sample is much easier to handle and analyze accurately. Typical mineral processing plant slurries can be considered to behave almost like a liquid, but because the slurry contains particles and trash, sampling system design must also take these into account.

A primary sample is taken from process stream in one or more steps. The flow rate of the primary sample has to be high enough to allow reliable transfer to an analyzer or composite sampler by pumped or gravity flows. Often the primary sample has to be re-sampled as the sample presentation system to the analyzer uses a lower secondary sample flow rate than the primary sample.

Primary samplers can be classified into two groups: samplers for process flows under pressure and samplers for gravity flows. The present trend to use high capacity big flotation cells has increased the process flows. Thus frequently two stage or even three stage primary sampling is needed to get suitable primary sample flow. Some of the most common sampler types are presented.

The sample flow must have a minimum speed to keep the solids in suspension. There is frequently trash in the process stream to be sampled. The engineering and installation of sample lines requires more attention than is generally realized. The paper gives simple rules, which can be used in addition to common sense to implement dependable sampling systems for on-stream analysis and composite sampling.

## INTRODUCTION

Slurry flows in mineral processing plants are always turbulent to keep the solids in suspension. The turbulence makes the flow well mixed in many process locations. Slurries with particle sizes < 100 micron are practically always well mixed. The terminal fall velocity in water by gravity for these small particles even with high specific gravity is small compared to the mixing velocities provided by the turbulence for flows with a speed of more than 1.5 m/s. In normal slurry the small particles reduces additionally the terminal velocity of big particles. Thus there are many process locations were a simple fixed sampler could give a representative sample.

Secondly the flow into the sampling cutter or nozzle should be ideally the same as the surrounding process flow to avoid segregation. A too high flow into the nozzle will collect fine particles from a large region but heavy big particles tend to continue their path more straightly and only a small part goes into the sample. This isokinetic principle is important but for usual particle sizes not very critical. For instance for 100 micron particles the allowed mismatch in velocities is about 1 m/s. For this information and a more detailed theoretical discussion of fixed samplers see (Hughes 1983).

In practice fixed samplers due to their reliability and low price have been the prevalent solution for on-line analyzer sampling for decades. A small bias in the sample can be allowed for process control applications, for which trends are most important, but not for samples used for material balance calculations. The composite sample increments collected for material balance calculations must not only be representative of the concentrations but also of the mass flow. In principle only correctly installed gravity samplers can be representative of a changing mass flow.

## FIXED SAMPLERS

The fixed samplers can be divided into three groups. The samplers for pressure flows and gravity flows are already traditional samplers used for decades. The pressure samplers can easily be designed also for large flows but for gravity flows it is not possible due to the vertical cutters becoming too narrow. For this reason for flows larger than about 1000 $m^3/h$ two-stage samplers are preferred. The second stages of the two-stage samplers are actually pressure samplers so the sample pipe must be calculated as a pressure pipe to get the right sample flow. Only the one-stage gravity sampler pipes have gravity flow and thus give a sample flow relative to the process flow useful for mass balance calculations. Only the main types will be presented in the following.

### Pressure Samplers

Pressure samplers generally use a round nozzle preferably in the middle of a vertically rising stream. The assumption is that the sample is mixed in front of the nozzle and there is no horizontal segregation. For horizontal, gravity or pressure streams one solution is to make a U in the pipe forcing the process stream first down and then up and insert the nozzle in the up-going part of the U. At the bottom of the U are often however collected stones, so in many cases a dump-valve has to be installed.

The pressure pipe sampler PSA is as a rule installed in a vertical orientation for good representativity, generally immediately above a process pump. A good alternative is after a long (>10 x diameter of pipe), straight vertical stretch of the pipe. In a horizontal pipe there is a risk of segregation in the vertical direction.

From flotation cell end or intermediate boxes proper sampling is difficult to get. A simple thief sampler also called suction pipe sampler SPA is most often used. Ideally it is installed so that the isokinetic principle is applied as well as possible.

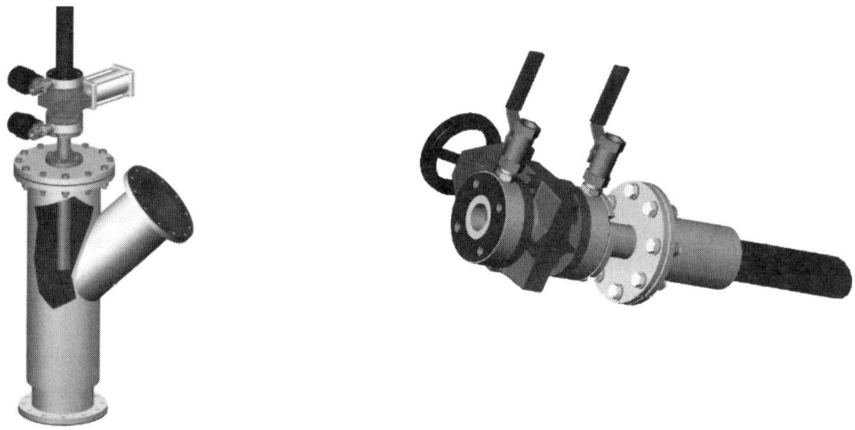

**Figure 1. Pressure sampler PSA with automatic valves and suction pipe sampler SPA with manual valves**

**One-Stage Gravity Samplers**

In gravity flows there is generally some segregation in the vertical direction. For this reason vertical cutters are used whenever possible. If the gravity flow is in a pipe the flow is made into a rectangular shape in a box. In a rectangular launder only a cutter needs to be inserted. Flow should be smaller than about 1000 $m^3/h$ for these samplers.

**Figure 2 LSA and SKA samplers**

**Two-Stage Gravity Samplers**

***The CPS Codelco primary sampler (patented)***

The sampler takes a primary sample with a "boot shaped " sampler made of a round tube. It takes with its open "toe" a sample of the gravity flow launder stream. The sample flow is turned into the vertical direction in the boot and returned back into the process flow through an opening in the back. The secondary sample is taken from the upper part of the vertical flow by a round nozzle. See Figure 3. The sampler hangs from above the launder and can be turned around a hinge out of the stream for inspection or repair. The sampler has been tested and found good for well-mixed flows without froth.

***Two stage cutter sampler (patented)***

The sampler is generally suitable for sampling of near horizontal non-pressurized pipes or launders with flows higher than 420 m3/h. The primary adjustable cutter in the center of the stream cuts a representative sample from the process flow. The sample is reshaped in the cutter to a broad low cross-section. This makes it possible to use a second cutter to reduce the sample flow to the required size. Best sampling result are obtained for turbulent flows with a reasonably constant flow rate. See figure 4. This sampler allows segregation in the vertical direction and froth.

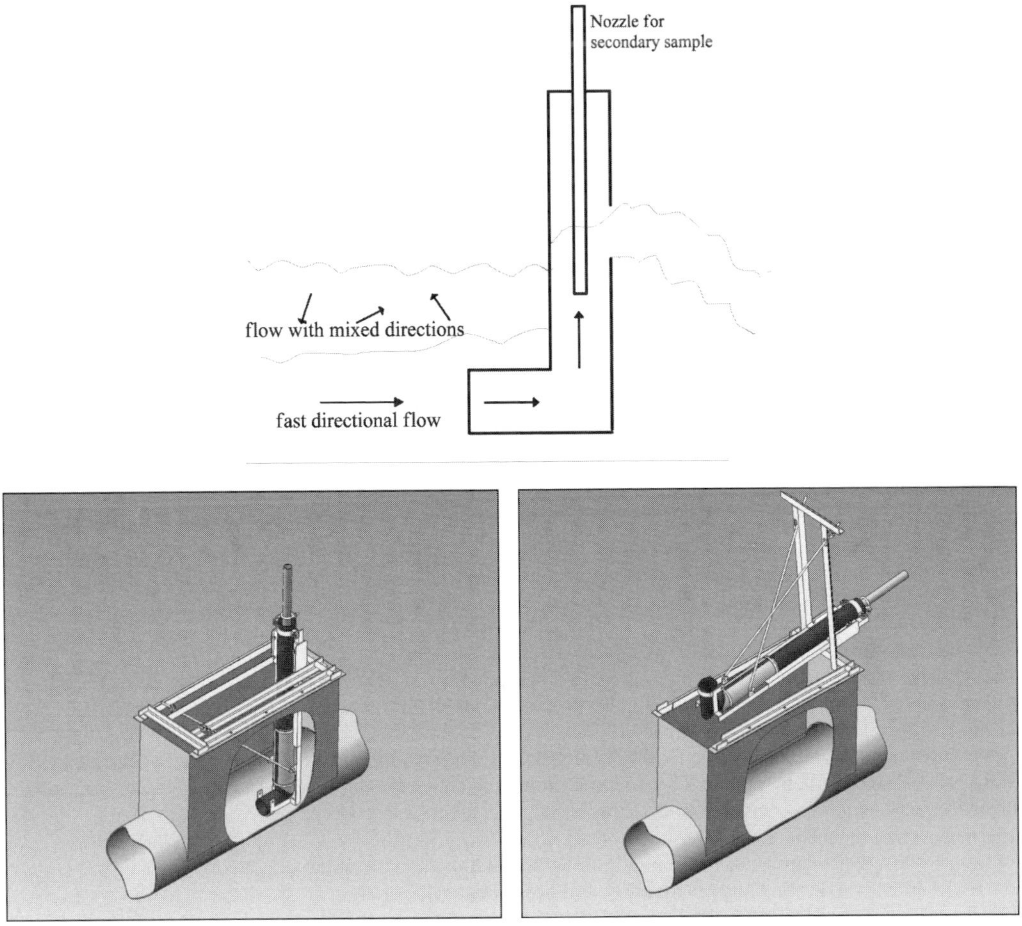


**Figure 3. The Codelco primary sampler working principle and as installed in a gravity pipe**

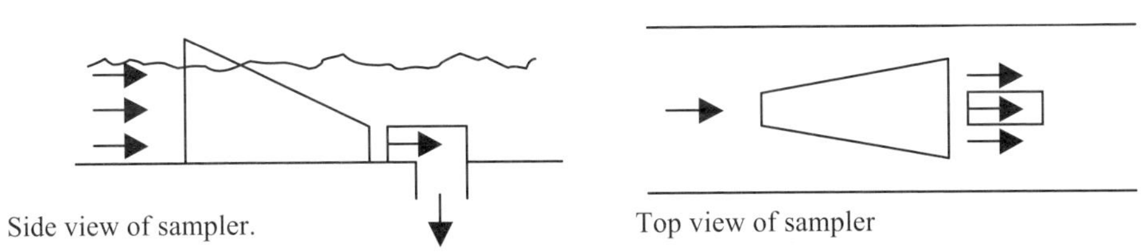


**Figure 4. The two-stage cutter sampler working principle**

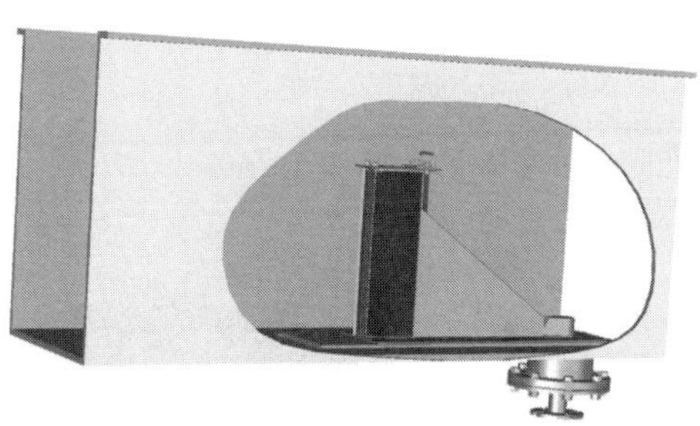

**Figure 5. The two-stage cutter sampler as installed**

## METALLURGICAL SAMPLING

According to the textbooks accurate sampling can only be done with a correctly installed moving cutter sampler, which transverses the whole stream. For most streams a composite sample cannot be obtained with one stage but usually at least two or more stages are required. This results in a large sampling system with a big head loss especially for the intermediate samples, which have to be returned to the main stream. See figure 6.

**Figure 6. First stage of classical composite sampling system usually followed by a vezin sampler to take the composite sample**

The classical solution with only moving cutters is generally a high maintenance item. When it is operated continuously or semi-continuously to get a suitable sample for an on-line analyzer, reliability has been poor. The head loss for modern flat plants is also a problem.

To get a practical solution a new concept was introduced. The sampling system uses mixing of the slurry followed by several parallel fixed cutters replacing the traditional moving cutter. For big streams several similar stages following each other can be used until a transversing cutter takes the final composite sample.

The newest and most advanced sampling system using this concept is the MSA multi-stage metallurgical sampler. It provides a truly representative slurry sample both for metallurgical accounting and on-stream analysis. The sampler operates automatically and offers high availability with low maintenance.

The modular design offers flexibility in meeting a wide range of process flows while maintaining the simplicity and compactness of the sampler, finding space for good installation is easy.

Operation of the sampler is based on a combination of static cutters followed by moving sample cutters. Static multi-cutters and moving multi-cutters reduce the volumetric flow to a suitable size for final metallurgical sampling. Additional sampling stages are added, as necessary, to meet the requirements of even very high process flow rates. See figure 7.

Moving cutter to the left taking no sample, in the middle taking full stream and to the right taking no sample

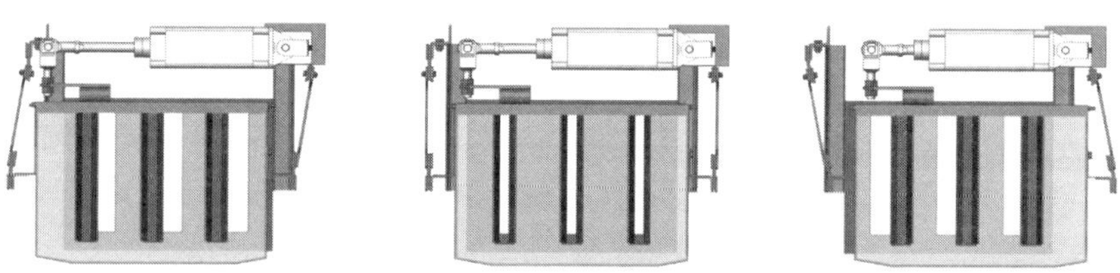

Moving cutter to the left taking no sample, in the middle taking full stream and to the right taking no sample

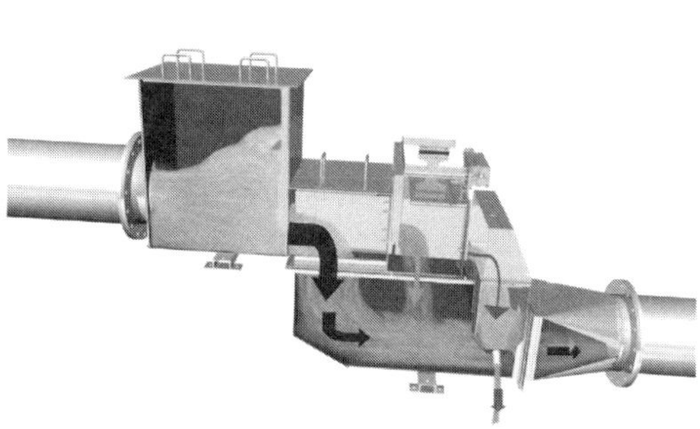

**Figure 7. MSA and the moving cutters principle**

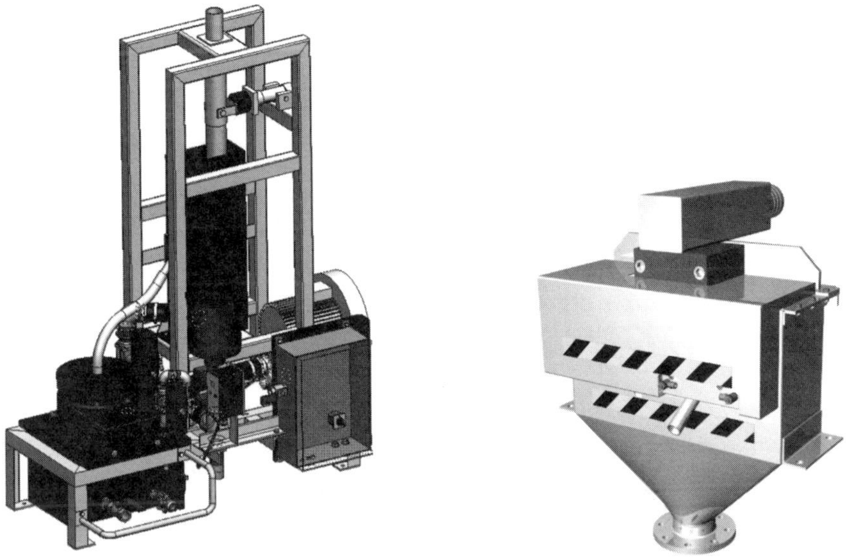

**Figure 8. SSA with sample filter and LMC composite sampler**

The sample from the crosscut sampler is a continuous flow of about 150 l/min suitable for an on-line analyzer. Single cuts at intervals are also possible. A flexible hose sampler takes the final composite sample from this flow in the on-stream analyzer multiplexer or with a separate SSA or LMC standalone composite sampler unit. See figure 8.

The MSA sampler is designed to require minimal maintenance: The multi-cutters can be easily cleaned/replaced without stopping the process flow. The process flow goes through the down sloping sampler without sanding.

- No risk for sanding
- Large, easily accessible openings to prevent trash from plugging the multi-cutters
- A patented new type of moving cutter gives good durability for continuous on-line analyzer sampling. No rails or wheels just rubber hinges. See figure 6.

The MSA sampler is based on modules each with several parallel static or moving sample cutters. The process flow is reduced by up to two static multi-cutter modules and a final moving multi-cutter module to a suitable flow for final metallurgical sampling. The slurry is mixed before each module and sampled in 1:4...15 ratios with static multi-cutters first and moving cross cutters at the last sampling stage. See table 1.

**Table 1. Metallurgical sampler range**

| Model | Nominal capacity $m^3/h$ | Stages | Number of static cutters | Main launder width mm |
|---|---|---|---|---|
| MLC80 | 18-60 | 1 | None | DN80 pipe |
| MSA 1/40 | 60-180 | 1 | None | 400 |
| MSA 2/50 | 180-600 | 2 | 3 | 500 |
| MSA 2/80 | 600-1800 | 2 | 4 | 800 |
| MSA 3/140 | 1800-6000 | 3 | 5/3 | 1400 |
| MSA 3/200 | 6000-18000 | 3 | 5/4 | 2000 |

## SAMPLER INSTALLATION REQUIREMENTS

The samplers with fixed cutters and nozzles must be located in a part of the process stream, where the slurry is well mixed, not immediately after two joining streams for example. Gravity samplers with vertical cutters should not be placed close to a horizontal bend, which causes the flow to follow the outer edge of the pipe or launder maybe completely missing the cutter.

The process flow out of a gravity sampler must be ascertained so the flow does not fill up the sampler. Outflow pipe inclination and size must be sufficient.

Space must be reserved for sample pipe bends and valves.

Service platforms are needed.

## SAMPLE PIPE

The engineering and installation of sample piping is more difficult than for process piping but often it gets much less attention and time. In a plant project first the plant and process piping of course must be planned, but then often the time is up and hastily sample piping is sketched in without further thought with disastrous results for the sampling. In a fairly recent example only two out of twelve sample lines worked. Two gave flows flooding the analyzer and eight did not give a drop.

Sample lines must be properly installed and flows calculated. Sample lines have a small diameter usually 32-60 mm and this means that friction losses are about 10 times higher than in typical process tubes. Head requirement is typically 10% or more of the sample pipe length. See figure 9.

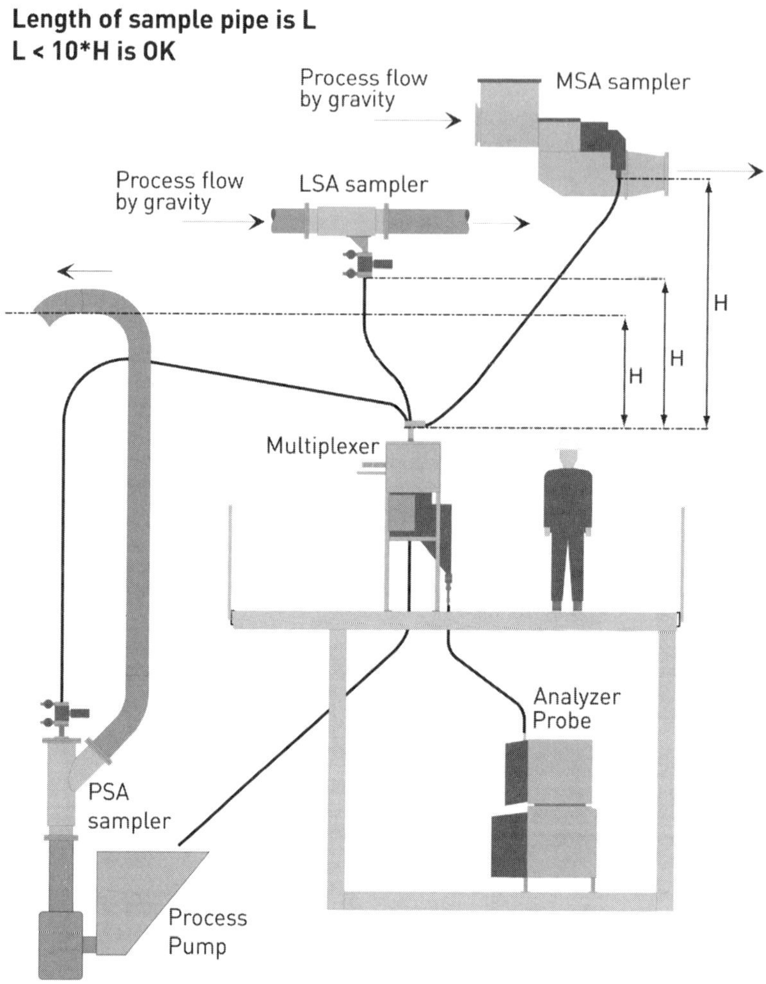


**Figure 9. Example of sampling installations with head requirements**

Sanding and blocking is always a danger. Pipes should be straight and flow empty if flow is stopped. Near horizontal plastic pipes must thus be well supported. The inner surface of the pipe should be smooth so trash cannot be caught at steps. Butt-welding of plastic pipes is not good. See figure 10.

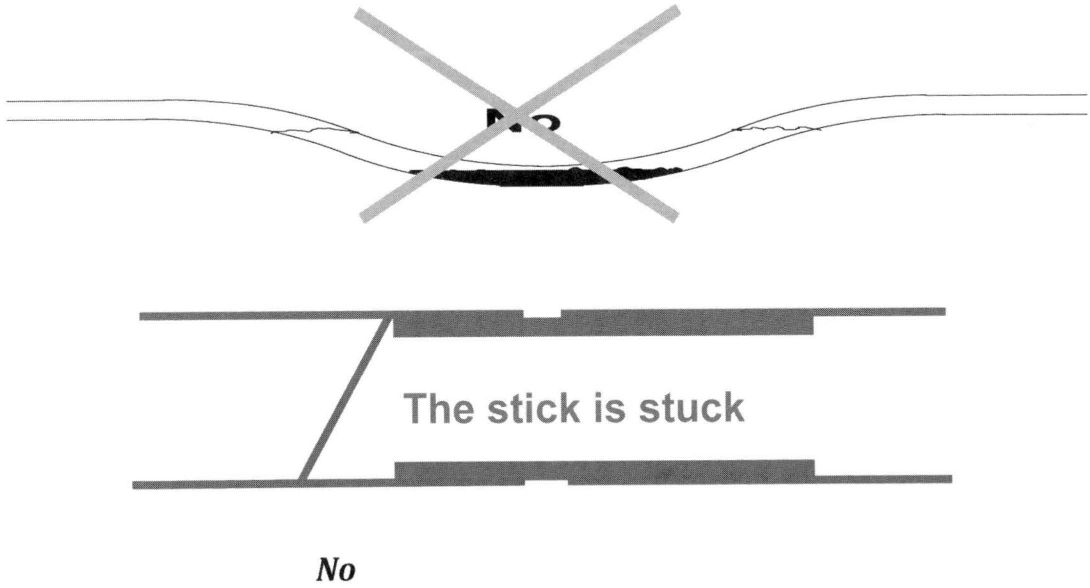


**Figure 10. Not allowed solutions in sample piping**

Bends should have a radius of about 500 mm to avoid wear. Commercial 90° Bends tend to wear out fast and give a high flow resistance.

Gravity sample lines must be engineered to be gravity lines also in practice. Often at the analyzer end there is a flow brake or deaccelerator, which brings high-speed flow to slow down to a full pipe flow into the analyzer. If the upper end of the pipe is not open to air the pipe flow will turn into pressure flow and often the head and the size of the pipe is such that we will get an unwanted high flow. If for example the cutter of the LSA is submerged in slurry this effect will occur. The flow resistance of the sampler will raise the level in the sampler, thus the process pipe feeding a LSA sampler should be only about half full. The flow should on the other hand not be a too fast having a low level, as after a steep decline in the process pipe.

**CONTROLLED SAMPLING**

On-stream analyzers have traditionally used continuously flowing primary samples. Controlled sampling is a built-in feature in new generation Courier analyzers. The analyzer can have an active role in controlling the primary sample acquisition and delivery to the analyzer. Remote controlled sample valves are required and the desired sampling sequence has to be configured. A combination of continuous and controlled sample lines can be used in the same analyzer system.

For on-demand sampling, analyzer scheduler starts the primary sample flow some time before the next measurement and/or composite sampling. After measurement the sample flow is stopped and the sample cutter and sample line are flushed with water. This radically improves the availability of sample lines and reduces the volume of returned samples.

The primary sample usually flows continuously, but is interrupted at programmed intervals to flush the sampler nozzle or cutter and the sample line. The flushing is scheduled not to interfere with normal measurements and composite sampling cycle. See figure 11.

**RESAMPLING FOR ANALYSIS OR COMPOSITE SAMPLING**

The sample flow coming to the analyzer still contains trash and possibly entrained air. These must generally be removed before the sample is presented to the analyzer. Stable flow conditions independent of process flow or sample flow is also generally required. Also one analyzer measures generally several samples in sequence.

A convenient solution is to use simple hose samplers in parallel feeding a sample tank with level control gravity feeding a suitable stable flow to the analyzer. This unit is generally called multiplexer. The slurry entering the tank goes through a sieve eliminating trash and air is escaping the slurry in the tank if necessary aided by froth sprays. If three position cylinders are used one position feeds the sample tank the middle position bypasses the analyzer and going to the third position the sample can pass over a cutter taking a composite sample. See figure 12.

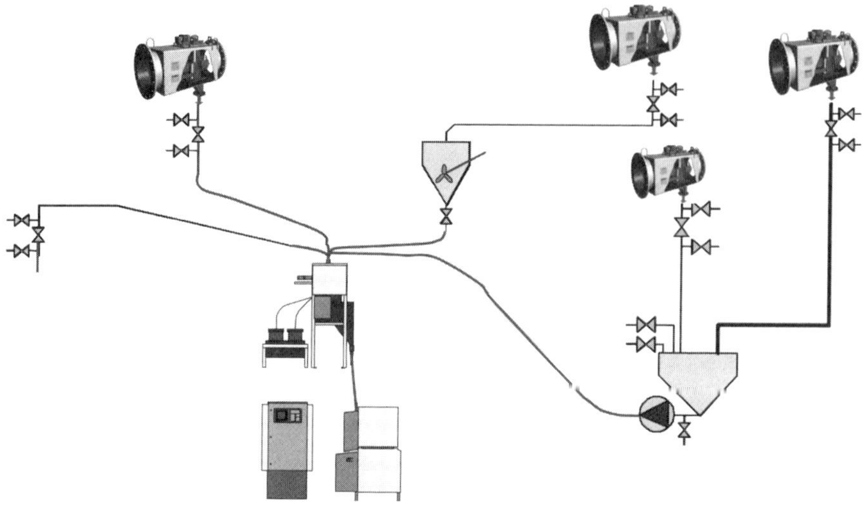

**Figure 11. Typical controlled sampling system with automatically controlled valves**

In online XRF analyzers stable flow conditions in the sample cell to get a representative sample into a thin layer at a window film is of great importance. See (Braden&al 2002). The characteristic x-rays of most metals emanate from a layer, which is less than one millimeter. A representative sample should be in that region. The window should stay clean. A correctly designed cell like the Jet-cell in the Courier analyzers will keep the window clean by uniform abrasion of the solids in the slurry. The sample in the thin layer is always to some extent biased. The trick is to keep the bias stable over time. All this the Jet-cell and the resampling system of the Couriers achieve.

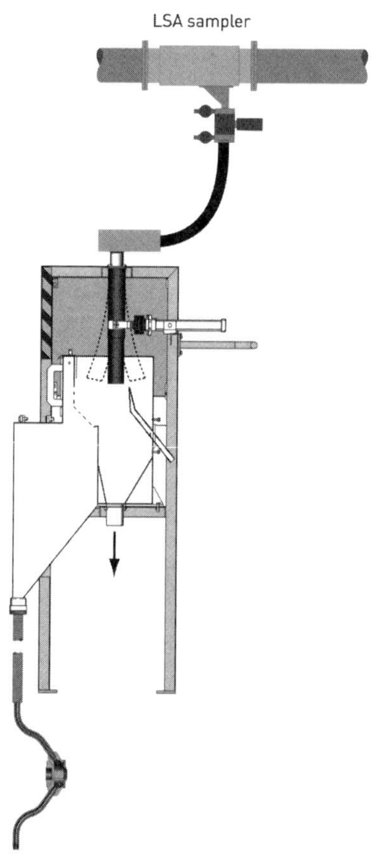


**Figure 12. A typical multiplexer showing the 3-position hose sampler**

For particle size analysis using the automatic caliper the resampled flow is not so critical but should still be stable and free of trash. The primary sampling on the other hand is more critical because segregation will influence the result with full impact.

For particle size analysis using laser diffraction the laser beam must pass through a thin diluted slurry layer. To achieve that only a few composite sample cut like samples are added to a big water volume. In this case the hose sampler has a bypass position in the middle and on both sides cutters the hose can travel over.

## REFERENCES

T.F.Braden, M.Kongas, K.Saloheimo 2002. On-line composition analysis of mineral slurries. Mineral Processing Plant Design, Practice and Control Proceedings volume 2

D. V. Hughes 1983. Sampling systems for on-stream x-ray analysers in ore-dressing plants. Trans Inst M C Vol 5, No 4.

# THE DESWIK PILOT STUDY REGRIND MILL—BENEFITS OF AN ONSITE SMALL SCALE FINE GRIND TEST STUDY


Stephen Massey

Deswik Milling


## ABSTRACT/ INTRODUCTION


The advantage of an onsite pilot study is that it can address many of the operational issues that make each site unique. Examples of these issues include: material driven effects causing equipment wear rates and power consumption for a given throughput and grind size which ultimately leads to determining the point where grinding for recovery plateaus and costs begin to outweigh yield. Other site unique factors that give cause for a pilot study may be plasticity of ore, slurry density, and even temperature of slurry feed for example in very hot or cold climates.


## INTRODUCTION

Deswik Milling now offers the convenience of a Deswik25 pilot mill to enable onsite fine grinding trials as part of plant design and feasibility studies.

The Deswik25 is a compact vertical mill that is driven by a hydraulic motor in the same manner as its larger industrial equivalents (Deswik250, Deswik1000 and Deswik2000). The hydraulic drive offers a high tip speed (energy intensity) at a lower power input than equivalent electric drives. The Deswik25 is capable of throughputs of around 500 dry kg per hour. It weighs about 2 tonnes and fits an area approximately 8 $m^2$. It requires a level pad, 3 phase power, low pressure water and air supply.

## RECENT PILOT FINE GRIND STUDY FINDINGS

Grinding trials in the Deswik Mill have recently been conducted on Copper-Lead-Zinc tailings in Australia with considerable success. The study has enabled metallurgists to quantify the effect of particle size on metal recovery during the retreatment of their tailings material. The grinding data also provided scale-up information which allowed the design of primary and concentrate regrind circuits for the tailings retreatment plant.

Additional information gained from the grinding study confirmed that the intense grinding action in the Deswik prohibited the accumulation of dense material in the mill (Galena). These results were in line with previous industrial experience at Kroondal (South Africa) where there has been no Chromite tramp material found in the Deswik1000 after nearly 20 months of operation. Galena build however is an example of a condition to watch out for when fine grinding and can best be monitored over a consistent pilot study fine grind.

## MILL FEED

Slurry feed can be unique from site to site for a variety of reasons. In remote parts of Western Australia for example the ambient air temperature can be as high as 50 degrees centigrade. One could assume that because a lab grind study indicates a mill grinds ore for a given F80 to a given P80 in a given timeframe without problems this will translate well in the field. Did the lab test simulate hot ambient conditions? What happens if this same grind in the field leads to the mill running dangerously hot causing premature wear and obscure results?

## WHY FINE GRIND AND WHY PILOT STUDY FINE GRIND

Logical steps toward deciding on an Ultrafine Grind Mill Study:

1. Are valuable minerals being trapped in problematic/low grade ore?
2. If yes, will UFG liberate minerals cost effectively? Not sure?
3. Send ore to a lab grind test to establish a grind curve.
4. Different sampled sizes from the grind test should then be put through a series of post metallurgical treatments to see which size marries up with which downstream process to achieve optimum recovery. The steps back from this "optimum" should also be recorded because they ultimately could be the most cost effective solutions. For example grinding to 5 microns may lead to 95% recovery but throughput could be very slow. Grinding to 10 or 12 microns may occur quite rapidly and plateau thereafter where even one extra micron reduction may reduce throughput in dramatically.
5. From the lab scale grind test: throughput can be determined for a given P80 target; full Scale Mill solutions can be estimated; wear and opex can also be extrapolated.
6. Is everything outlined in point (5) above certain? Consider on site Pilot Mill Study particularly if feed material tends to vary, climate is very hot (or even very cold).

## MATTERS TO CONSIDER BEFORE EMPLOYING PILOT MILL

Issues to consider before decision to carry out onsite Pilot mill Study:

1. Do you have a continuous feed for the Pilot mill—up to 500kg per hour of solid material This should be done as continuous feed not batch.
2. How do you intend measuring ultra fine recovery? Many mines may not possess a lazer size analyzer
3. If pilot regrind circuit is not yet established where does the recovered material go?
4. Power, water and air requirements. 415volt, 25 amp CB
5. Area approximately 8sqm

## OVERGRINDING

In many cases where an ultrafine grind is called for such tailings retreatment or refractory gold bearing ores, the finer the grind the greater the recovery however it is recognized that in some cases overgrinding can be an issue and needs to be controlled. In such circumstances a Pilot Mill Study is an ideal approach because it gives the time and opportunity to "tune" the Mill to suit the target.

The Deswik can be tuned in a number of ways such as:

1. Impellor spacing
2. Number of impellors
3. Media quantity and size distribution of beads
4. Shaft speed

## WHAT IF TIME OR RESOURCES DO NOT ALLOW FOR PILOT STUDY?

Do the trial in a lab mill which is a small scale replica of the intended purchase Mill but if you have reason to believe climate, feed density, or ore consistency may be problematic close simulation will be difficult. If you have no reason to suspect these items will be of concern then lab Grind tests should be adequate.

## CONCLUSION

Deswik is pleased to announce that contracts have now been secured to commence Pilot Studies in the Deswik25 for Norton Goldfields Mt Morgan and a confidential Major client. During rare moments of downtime the mill will be housed in an industrial facility south of Brisbane when in house development can occur.

Pilot study fine grinding can add considerable weight to a feasibility study and Deswik now has the right solution to enable simple, cost effective, onsite Pilot Scale Fine Grinding to take place. Conversely, if the mine is not set up with appropriate measuring/analytical equipment or is unsure of the post-met treatment process which should occur on the "fines" then lab testing should be the more appropriate way forward.

A Pilot Mill study should remove all uncertainty which may otherwise exist if ore, climatic conditions or slurry density are issues. Additionally in these problematic cases the Pilot Study will enable tuning of the Mill parameters to suit the conditions.

# Flotation

# DESIGN AND OPERATION OF MECHANICAL FLOTATION MACHINES

Michael G. Nelson[1], Dariusz Lelinski[2], and Sami Gronstrand[3]

[1]University of Utah Mining Engineering Department, 135 South 1460 East, Room 313, Salt Lake City, Utah, U.S.A.
[2]FLSmidth Minerals, 2850 South Decker Lake Drive, Salt Lake City, Utah, U.S.A.
[3]Sami Grönstrand, Outotec Minerals Processing, P.O. Box 86, FI-02200 Espoo, Finland

## ABSTRACT

Mechanical flotation cells continue to increase in size. Recent introduction of 300 m³ cells has required careful analysis using conventional hydrodynamics and computational fluid dynamics (CFD), followed by extensive testing, first with water and then with slurry. This chapter provides current information regarding the sizes and designs of mechanical flotation machines from major suppliers, hydrodynamic data, and the installation bases for machines of various sizes and types. Also included are recommended operating procedures for large machines, including mechanical and electrical monitoring criteria, and procedures for machine shutdown and reconditioning.

## INTRODUCTION

Mechanical flotation machines have been designed in a wide variety of configurations. Machines used for most mineral flotation applications may be designated as either mechanical or pneumatic. Pneumatic machines include flotation columns, Jameson cells, and others, and are not considered in this chapter. Mechanical machines are of three types:

1. Self-aerating tanks, with the rotor near the top of the tank;
2. Externally-aerated tanks, with the rotor near the center of the tank; or
3. Externally-aerated tanks, with the rotor near the bottom of the tank.

Self-aerating machines are manufactured only by FLSmidth Minerals, under the names WEMCO® Smartcell™ and WEMCO® 1 + 1®. Externally-aerated machines are manufactured by FLSmidth Minerals, under the names Dorr-Oliver (bottom rotor), Agitair (bottom rotor), and *EXCELL* (center rotor); by Metso Minerals; and by Outotec OY. Many of these machines have been described previously (Nelson, et al. 2002). This chapter provides current information on machine design and sizes available, describes new machines and new features on machines described previously, and provides detailed information on design, installation, startup, and operation of mechanical flotation machines of all makes.

## AVAILABLE EQUIPMENT

All the information in this section was provided by representatives of the respective manufacturers, or taken from manufacturers' websites and printed literature. Machine descriptions and benefits are given as provided by each manufacturer, and readers are left to form their own opinions regarding the material presented. The respective manufacturers are presented in alphabetical order, with no implication as to which the authors believe to be superior.

### FLSmidth

FLSmidth Minerals provides five types of flotation machines, the Agitair, the Dorr-Oliver®, the WEMCO® 1+1®, the WEMCO® SmartCell™, and the *XCELL*. Both the WEMCO® machines are self-aerated; the other three are externally aerated. FLSmidth Minerals has delivered more 53,000 flotation cells to operations worldwide.

The first Agitair machine was designed in 1934, and an expanded product line, with three different mechanism designs, was sold by the Galigher Equipment Company for many years (Dreyer 1976). Figure 1 shows a typical Agitair machine; FLSmidth still offers Agitair machines, but reports than none have been sold for many years.

Figure 2 shows a typical Dorr-Oliver® machine. FLSmidth lists the advantages of these cells as follows:

- Low power consumption—streamlined pump-action vortex profile rotor and overhung type stator are more efficient than other externally aerated units;
- Non-clogging, low maintenance design rotor;
- High air dispersion capability—high shear rotor-stator combination gives a wider range of airflow adjustment than competitive systems and results in greater control of the flotation process;
- Superior metallurgical performance—intense recirculation in a well-defined mixing zone multiplies the chances of contact between mineral particles and air bubbles, providing for greater mineral recoveries and higher concentrate grades;
- Easier restarting mechanism—overhung stator design and vortex rotor profile keeps pumping channels free of settled solids, providing easier restarts after unplanned shutdowns; and
- Low reagent costs—air is a natural reagent in the flotation process, and a wide air dispersion capability permits tuning the flotation plant to the optimum configuration.

Figure 1. Large Agitair Machine

Figure 2. FLSmidth Dorr-Oliver machine

The WEMCO® 1+1® machine was patented in 1969. FLSmidth literature describes the 1+1® as follows: In direct comparisons by major mining companies, Wemco 1+1 cells have repeatedly delivered higher recovery and grade with easier start-up, simpler operation, lower reagent consumption, longer mechanism life and less required maintenance. At the heart of each WEMCO® 1+1® cell is a patented rotor-disperser that delivers intense mixing and aeration. Ambient air is drawn into the cell and uniformly distributed throughout the pulp, providing optimum air/particle contact. In larger cells, a false bottom and draft tube channel slurry flow, ensuring high recirculation and eliminating sanding. The combination of efficient aeration and optimum solids suspension give WEMCO® 1+1® cells the highest species recovery and concentrate grade performance available, and reduced reagent consumption. Features include:

- Self-induced aeration,
- Elevated rotor position,
- Large design clearances,
- 13 cell sizes,
- Reliable, automatic level control,
- Durable rotor-disperser materials,
- Rugged mechanism assembly,
- Reversible operation, and
- Flexible, modular tank and box design.

The WEMCO® SmartCell™, shown in Figure 3, was developed in 1995, by installing a redesigned 1+1® mechanism in a cylindrical tank. FLSmidth literature describes the Smartcell™ as follows: The WEMCO® SmartCell™ flotation series retains the rugged WEMCO® 1+1® aeration mechanism that has been proven in thousands of installations worldwide. Its massive cast-iron bearing housing maintains accurate shaft alignment under all loads and moments, ensuring long service life. Induced air flow provides sufficient aeration, mechanical simplicity and capital economy. Like the 1 + 1®, the SmartCell™ can be serviced online and restarted easily under full load. Features include:

- Cylindrical tanks,
- Froth crowders,
- Automatic air control,
- Ultrasonic pulp level sensing,
- Froth flow measurement, and
- Sizes from 5 to 300 $m^3$.

Benefits include:

- Cylindrical tank design, which improves mixing efficiency and air dispersion, provides better surface stability, less pulp turbulence, lower capital costs, and reduced power consumption;
- Hybrid draft tube and beveled tank bottom, which improve hydrodynamic mixing and coarse particle recovery and increase solids suspension; and
- Radial launders and mixing baffles, which increase froth mobility, decrease froth residence time, increase recovery, and enhance froth stability.

**Figure 3. WEMCO® SmartCell™ machine**

The *XCELL* was introduced in 2003 by FFE Minerals USA, a subsidiary of FLSmidth. When FLSmidth acquired Dorr-Oliver EIMCO in 2007, the *XCELL* became part of the FLSmidth flotation product line. Figure 4 shows an *XCELL* machine.

According to FLSmidth literature, the *XCELL* was designed to improved metallurgical performance by improving fine *and* coarse particle flotation and enhancing froth recovery.

The symmetrical, cylindrical tank design improves both hydrodynamics and froth recovery, resulting in improved metallurgical performance. The *XCELL* machine enhances the performance of the cylindrical tank by placing the concentrate launders on the tank exterior. The *XCELL* machine is unique in that external launders are offered on all cell sizes from the smallest the largest (1.5 to 350 $m^3$).

The *XCELL* rotor has eight blades and is designed for radial flow. The rotor is constructed with a mild steel core and a long-wearing, cast urethane cover. Air injection ports are located under the rotor top plate between each pair of blades. The rotor is in the center of the tank to separate the energy input region around the rotor from the froth surface, allowing for a quiescent froth-pulp interface but maintaining a short transport path from the contract region in the stator volume to the froth. This allows the *XCELL* to operate with shallow froths, if required, and reduces bubble-particle disengagement by keeping the transport path relatively short. Centering the rotor in the tank also provides the opportunity to control the air rate. The ability to control the rate of air input enhances the operator's ability to control and optimize metallurgical performance. The unique position of the *XCELL* rotor provides a savings of up to 35% in overall power usage, relative to other, similarly-sized flotation machines. In spite of the fact that the *XCELL* machines use less power than other machines, measurements have indicated that slurry pumping rates are similar to both self- and externally-aerated designs, providing similar contact frequency numbers at significantly lower energy consumption or

higher contact frequency numbers at equal power consumption.

**Latest developments.** FLSmidth Minerals and Rio Tinto recently commissioned the two SuperCells®, the world largest production cells. One is 300 $m^3$, the other 330-350 $m^3$. The cells are installed in the bulk copper/ molybdenum flotation circuit at Rio Tinto's Kennecott Copper. The SuperCells® use a universal tank that can be fitted with any of the three types of mechanisms manufactured by FLSmidth Minerals. The installation is shown in Figure 5.

Prior to installation, a rigorous development program was completed, which integrated dimensionless scale-up parameters with state-of-the-art CFD models of the machines. This rationalization process allowed FLSmidth to utilize operational information from over 100 installed 257-$m^3$ SmartCells™, along with the fundamental models developed for the design of the WEMCO® SuperCell®. Similar investigations were included in the design phase for the use of the *XCELL* and Dorr-Oliver® flotation mechanisms in the SuperCell®. Examples of the CFD models of all three machines are presented in Figure 6. This fundamental understanding of equipment design allowed the construction and installation of the machines to be fast-tracked, and completed in 110 days from start to finish.

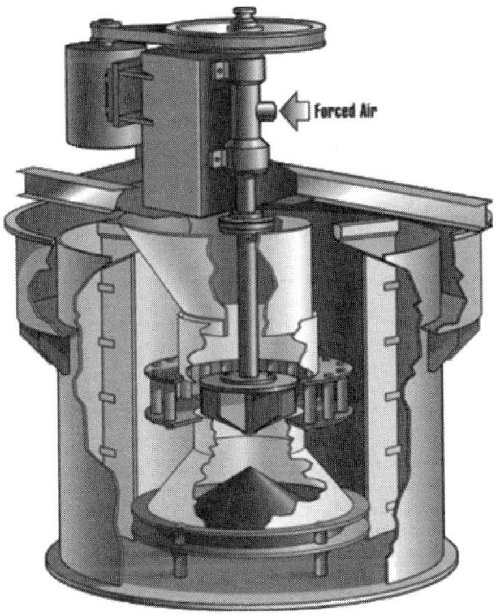


Figure 4. FLSmidth *XCELL* machine

Figure 5. WEMCO® SuperCell®

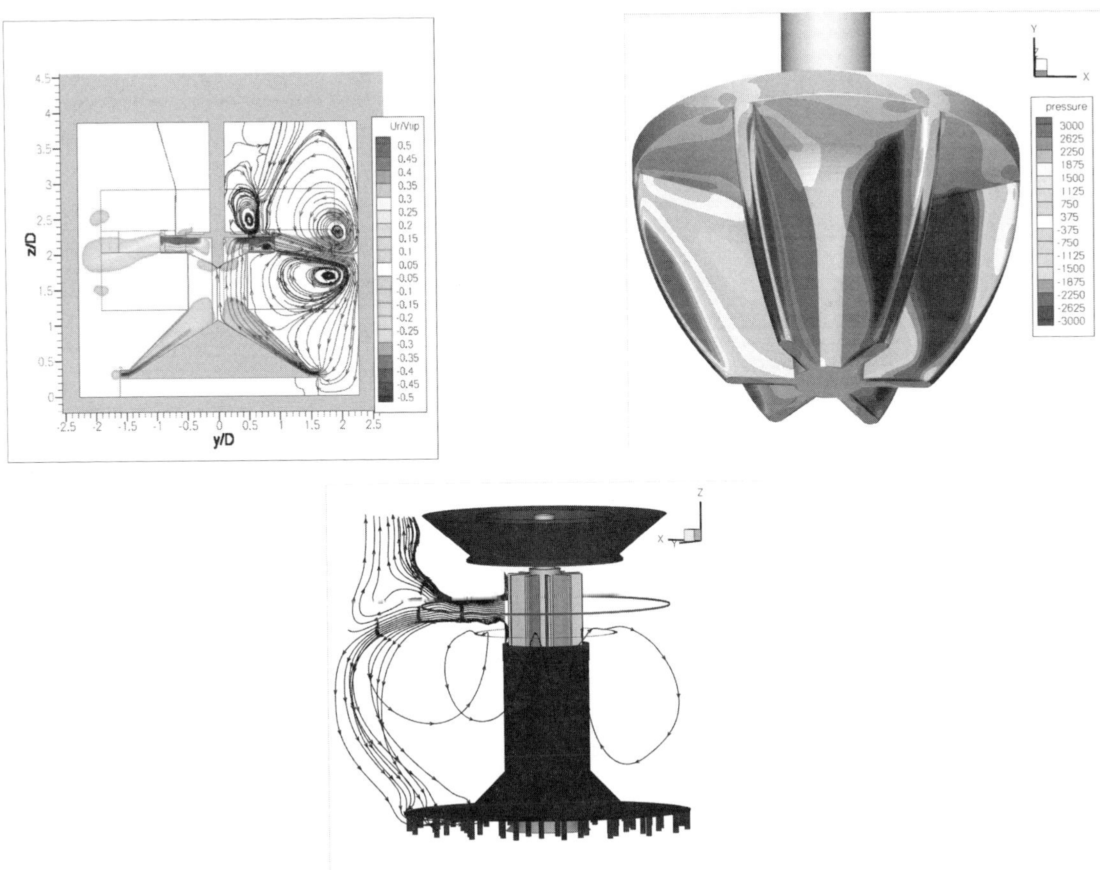


Figure 6. CFD models of the *XCELL*, Dorr-Oliver®, and WEMCO® mechanisms

**Current machines offered.** Machines offered by FLSmidth are shown in Tables 1-4. Dorr-Oliver® machines are offered in three configurations: rectangular (R), rectangular with U-shaped bottom (U), and cylindrical (RT).

**Table 1. Dorr-Oliver® machines**

| Cylindrical Tanks | | | Rectangular and U-bottom Tanks | | |
|---|---|---|---|---|---|
| Model | Effective Cell Volume, *m*³ | Motor, *kW* | Model | Effective Cell Volume, *m*³ | Motor, *kW* |
| DO-1.5 RT (pilot) | 1.5 | 7.5 | DO-1 R | 0.02 | 0.6 |
| DO-5 RT | 5 | 7.5 | DO-10 R | 0.24 | 1.1 |
| DO-10 RT | 10 | 14.9 | DO-25 R | 0.60 | 2.2 |
| DO-20 RT | 20 | 29.8 | DO-50 R | 1.2 | 3.7 |
| DO-30 RT | 30 | 37.3 | DO-100 R | 2.4 | 5.6 |
| DO-40 RT | 40 | 44.8 | DO-300 UT | 7.2 | 11.2 |
| DO-50 RT | 50 | 56 | DO-600 UT | 14.3 | 22.4 |
| DO-70 RT | 70 | 74.6 | DO-1000 UT | 23.8 | 29.8 |
| DO-60 RT | 60 | 74.6 | DO-1350 UT | 32.2 | 37.3 |
| DO-100 RT | 100 | 111.9 | DO-1550 UT | 36.9 | 44.8 |
| DO-130 RT | 130 | 149.2 | DO-1550 UT | 36.9 | 44.8 |
| DO-160 RT | 160 | 149.2 | | | |
| DO-200 RT | 200 | 186.5 | | | |
| DO-330 RT | 330 | 410 | | | |

**Table 2. WEMCO® 1+1® machines**

| Model | Effective Cell Volume, *m*³ | Tank Length, *m* | Tank Width, *m* | Typical Lip Length, *m* | Typical Froth Area, *m*² | Maximum Air Flow, *m*³*/min* | Motor, *kW* |
|---|---|---|---|---|---|---|---|
| 18 | 0.028 | 0.31 | 0.25 | 0.61 | | | 0.37 |
| 28 | 0.085 | 0.46 | 0.34 | 0.91 | | | 0.75 |
| 36 | 0.31 | 0.91 | 0.48 | 1.83 | | | 2.24 |
| 44 | 0.59 | 1.12 | 0.62 | 2.24 | | | 3.73 |
| 56 | 1.16 | 1.42 | 0.73 | 2.84 | | | 5.60 |
| 66 | 1.73 | 1.68 | 0.80 | 3.35 | | | 7.46 |
| 66D | 2.83 | 1.52 | 0.80 | 3.05 | | | 11.2 |
| 84 | 4.25 | 1.60 | 1.12 | 3.20 | | | 11.2 |
| 120 | 8.50 | 2.29 | 1.57 | 4.57 | | | 18.7 |
| 144 | 14.16 | 2.74 | 1.96 | 5.49 | 9.20 | 6.80 | 22.4 |
| 164 | 28.32 | 3.02 | 2.18 | 6.05 | 11.3 | 10.5 | 44.8 |
| 190 | 42.48 | 3.58 | 2.51 | 7.16 | 15.6 | 15.9 | 74.6 |
| 225 | 84.96 | 4.17 | 2.91 | 8.33 | 20.8 | 24.6 | 149.2 |

**FLSmidth Minerals Mixed Rows.** In the past, mineral producers had to choose between self-aerated and externally-aerated mechanical flotation systems. FLSmidth has introduced a mixed flotation circuit that uses WEMCO® self-aerated *and* Dorr-Oliver® externally-aerated machines. Combining flotation cells with markedly different processing actions into a single process bank increases recovery rates. The WEMCO® design is better for coarse particle recovery; the Dorr-Oliver® cell is better for recovering fines and increasing concentrate grade. Large cells use a "Universal Tank" that can accommodate either mechanism. Both cells feature beveled bottoms, and a common radial launder design. Cell-to-cell connectivity is enhanced by patented, hinged dart valves, eliminating the need for a junction box and saving floor space. FLSmidth provides mixed rows using the combination of WEMCO®, Dorr-Oliver®, and *XCELL* machines that is best suited for the application in question. Figure 7 shows a typical mixed- row installation.

**Table 3. WEMCO® SmartCell™ machines**

| Model | Effective Cell Volume, $m^3$ | Installed Power, *kW* |
|---|---|---|
| 1.5 (pilot) | 1.5 | 7.5 |
| 5 | 5 | 30 |
| 10 | 10 | 37 |
| 20 | 20 | 50 |
| 30 | 30 | 75 |
| 40 | 40 | 90 |
| 50 | 90 | 90 |
| 60 | 60 | 150 |
| 70 | 70 | 150 |
| 100 | 100 | 150 |
| 130 | 130 | 185 |
| 160 | 160 | 185 |
| 200 | 200 | 250 |
| 250 | 250 | 315 |
| 300 | 300 | 373 |
| 350 | 350 | 373 |

**Table 4. *XCELL* Machines**

| Effective Cell Volume, $m^3$ | Tank Diameter, *m* | Tank Height, *m* | Weir Lip Height, *m* | Launder Width, *m* |
|---|---|---|---|---|
| 1.5 (pilot) | 1.36 | 1.25 | 1.13 | 0.15 |
| 3 | 1.72 | 1.83 | 1.58 | 0.20 |
| 5 | 2.03 | 2.01 | 1.81 | 0.20 |
| 10 | 2.62 | 2.47 | 2.17 | 0.20 |
| 20 | 3.30 | 2.95 | 2.75 | 0.20 |
| 30 | 3.80 | 3.38 | 3.13 | 0.20 |
| 50 | 4.50 | 4.00 | 3.85 | 0.25 |
| 70 | 4.96 | 4.47 | 4.22 | 0.25 |
| 100 | 5.55 | 4.88 | 4.58 | 0.30 |
| 130 | 6.05 | 5.30 | 5.00 | 0.30 |
| 160 | 6.33 | 6.00 | 5.70 | 0.30 |
| 350 | 8.64 | 6.36 | 6.07 | 0.50 |
| | | | | |
| | | | | |
| | | | | |
| | | | | |

**Figure 7. Row of Dorr-Oliver® cells (left) and mixed row of Dorr-Oliver® and WEMCO® SmartCells™ (right)**

**Metso Minerals**

Metso Minerals offers two types of machines. The RCS machine, shown in Figure 8, was developed in the mid-1990s. The primary design goal was to create the two classic zones within a flotation cell, an active lower zone to ensure effective particle suspension and transportation and a relatively quiescent upper zone to minimize bubble-particle separation. A circular cell design was adopted as being the most suitable tank shape to provide symmetrical hydraulic flow patterns with minimum upper-zone turbulence.

The DR machine, shown in Figure 9, is based on the DR design developed by the former Denver Equipment Company in 1968 (Daman, 1968). It features rectangular tanks, connected in the open, "hog-trough" configuration.

Figure 8. Metso Minerals RCS™ machine

Figure 9. Metso Minerals DR machine

Features of the RCS™ machine include:

- *Maximum particle-bubble contact within the mechanism and the flotation tank,*
- Effective solids suspension during operation and re-suspension after shutdown,
- Effective air dispersion and distribution throughout the complete cell volume,
- Flotation air provided by a separate air blower,
- Aeration rate manually or automatically controlled at each mechanism,
- V-belt drive standard up to 70 $m^3$ cell volume, and
- Gearbox drive with extended output shaft bearings and drywell construction for cell volumes over 70 $m^3$.

Benefits of the RCS™ machines include:

- Mechanism minimize local high velocity zones within the impeller and diffuser to extend wear life,
- Impellers and diffusers supplied in high abrasion resistant elastomers or molded polyurethane,
- Impeller profile designed to minimize absorbed power,
- The DV mechanism fully suspended from the cell superstructure, and can be removed as a complete unit for routine maintenance, and
- Wear parts are replaceable within the flotation machine without removal of the mechanism.

Features of the DR machines include:

- Incorporates vertical circulation of pulp by combining a "recirculation well" with a top-feed impeller; provides positive vertical circulation of pulp;
- Cell-to-cell type machines minimize sanding, as the pulp is 100% mixed in every cell as it passes through the feed pipe and impeller; and
- DR principle of vertical recirculation of pulp minimizes stratification and sanding.

Benefits of the DR machines include:

- Increases effective aeration through supercharging,
- Maintains solids in suspension by vertical recirculation of pulp,
- Impeller ejects pulp-air mixture over entire bottom of cell, lifts and suspends solids, and
- Draws large volumes of pulp from upper zone to break up any sanding at the bottom of the cell.

**Current machines offered.** Tables 5 and 6 show available models of the RCS™ and DR machines, respectively.

**Outotec Oyj**

All Outotec flotation cells are externally aerated, from the smallest OK-0.5-R to the largest TankCell-300. Outotec believes it is important to maintain the maximum flexibility in terms of manipulating the gas flow rate and number of bubbles. The first Outotec flotation cells were the U-shaped OK-16 and OK-38, introduced in the 1970s and 80s. The cylindrical cell, Outotec's TankCell® was introduce in the 1990's, and is now sold in sizes up to 300 $m^3$.

All Outotec cells are designed using hydrodynamic analysis and modeling with computational fluid dynamics (CFD). The balance between high intensity micro-turbulence and macro scale laminar flow velocities must be found to suit the particle sizes being floated.

The bottom part of the cell is considered the contact zone, and is designed for suspending the solids and contacting the air bubbles with the particles by its turbulent conditions. Air is fed to the rotor through a hollow shaft. The flotation air is uniformly dispersed into the slurry through the slots of the rotor. Particles are collected by air bubbles forming aggregates in the contact zone, and particle-bubbles aggregates rise by buoyancy towards the froth zone. The laminar flow field in the quiescent zone is dedicated for selectivity; it allows separation of valuable particles from unwanted particles. Particle-bubble aggregates form a froth phase, which acts as a cleaning zone, which further rejects unwanted solids and water, upgrading the froth. The froth flows over the concentrate lip into the concentrate collection launder. Hydrophilic solid particles are carried by the flow fields in the cell out of the cell through the valves at the bottom of the cell tank. Figure 9 shows the conceptual operation of the Outotec TankCell®.

**Table 5. Current offerings of Metso Minerals RCS™ machines**

| Model | Effective Cell Volume, $m^3$ | Motor, *kW* (per cell; slurry @ 1.35 $g/cm^3$) | Air Flow, $m^3/min$ | Air Pressure, *kPa* |
|---|---|---|---|---|
| RCS 5 | 5 | 15 | 3 | 17 |
| RCS 10 | 10 | 22 | 5 | 22 |
| RCS 15 | 15 | 30 | 7 | 25 |
| RCS 20 | 20 | 37 | 8 | 27 |
| RCS 30 | 30 | 45 | 10 | 31 |
| RCS 40 | 40 | 55 | 12 | 34 |
| RCS 50 | 50 | 75 | 15 | 38 |
| RCS 70 | 70 | 90 | 18 | 42 |
| RCS 100 | 100 | 110 | 22 | 47 |
| RCS 130 | 130 | 132 | 27 | 51 |
| RCS 160 | 160 | 160 | 30 | 54 |
| RCS 200 | 200 | 200 | 35 | 58 |

**Table 6. Current offerings of Metso Minerals DR machines**

| Model | Effective Cell Volume, $m^3$ | Slurry flow, $m^3/hr$ | Motor, *kW* (per cell; slurry @ 1.35 $g/cm^3$) | Air Flow, $m^3/min$ | Air Pressure, *kPa* | Maximum cells per section |
|---|---|---|---|---|---|---|
| DR 15 | 0.34 | 25 | 1.1 | 0.67 | 7 | 15 |
| DR 18sp | 0.71 | 55 | 3.0 | 1.33 | 8 | 12 |
| DR 24 | 1.4 | 110 | 5.0 | 2.5 | 10 | 9 |
| DR 100 | 2.8 | 215 | 7.5–11 | 3.8 | 10 | 7 |
| DR 180 | 5.1 | 415 | 11.– 15 | 4.0 | 14 | 6 |
| DR300 | 8.5 | 580 | 18–22 | 7.7 | 18 | 5 |
| DR 500 | 14.2 | 760 | 25–30 | 11.3 | 18 | 4 |
| DR 1500 | 42.5 | 1,780 | 55–75 | 19.8 | 23 | 2 |

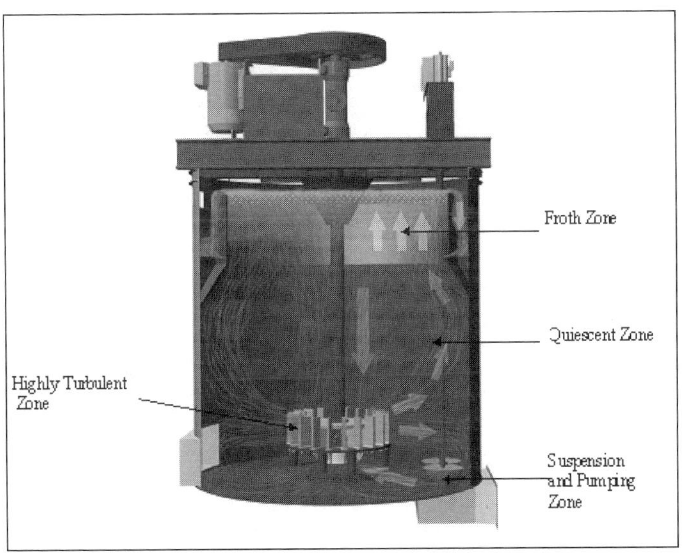


**Figure 9. Outotec TankCell®, conceptual operation**

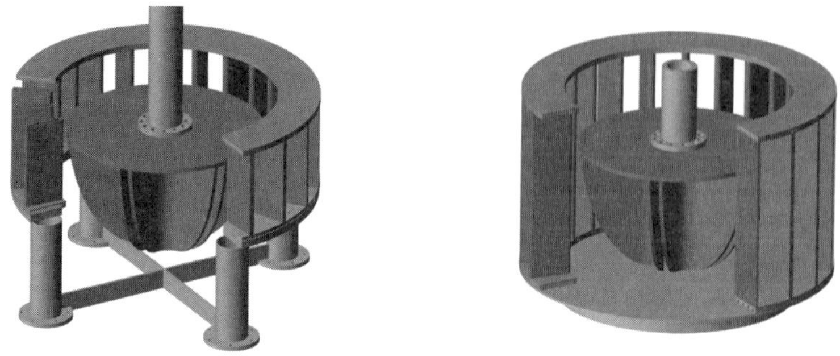

Figure 10. Conventional OK mechanisms. FreeFlow (left); MultiMix (right)

**Conventional Mechanisms.** Each mechanism is designed to maintain solids suspension, disperse the bulk gas flow from the shaft into small bubbles, and provide the acceleration required for the particle-bubble attachments to attach to the bubbles. These mechanisms are described in detail in Nelson, et al. (2002).

In the conventional mechanisms the rotor is mounted on the lower shaft by a bolted flange joint. The two conventional mechanisms, shown in Figure 10, were developed for different duties.

Generally, for fine and middling particles (< 100 *microns*) a MultiMix design is recommended. For coarser particles, a FreeFlow design is more suitable. Both the rotor and the stator are lined with thick elastomer (typically polyurethane or natural rubber) and the rotor is statically balanced for use. A new mechanism design is now available, but vast majority installed Outotec cells use one of the conventional mechanisms.

**The FloatForce® Mechanism.** It is known that ingested air deteriorates the performance of pump impellers. Air occupies the space that should be filled with slurry, and in the extreme case of excess air addition, pumping can stop completely. In the FloatForce® mechanism, shown in Figure 11, the air is introduced to a dedicated area of the impeller. The core of the rotor is used only for slurry pumping, with minimal disturbance by the air. Thus the mixing capacity remains high even when a high air feed rate is used.

**Figure 11. Outotec FloatForce® mechanism**

Figure 12 shows a comparison of the power draws of the conventional OK and the FloatForce® mechanisms. Even at high aeration rates, the FloatForce® mechanism maintains a higher power draw, indicating that it is still pumping slurry.

The FloatForce® mechanism was designed to provide improved metallurgical performance. It also provides decreased wear. Because impact velocity affects the component wear exponentially, peripheral tip velocity of the FloatForce® rotor is kept as low as possible, while still maintaining adequate turbulence to disperse air into small bubbles. The FloatForce® mechanism has wide channels for pumping and air dispersion to eliminate clogging of the impeller.

Stator blades up to a certain size are individually bolted and light enough for one person to handle without a lifting device, which simplifies change-out. For larger sizes, blade sets are provided for one-quarter of the stator. This makes replacement of damaged blades quicker and less expensive than for the conventional mechanisms, where all stators were changed one-half at a time. Size, shape, and lining thickness of the stator blades have been optimized using CFD provide maximum wear life, as shown in Figure 13.

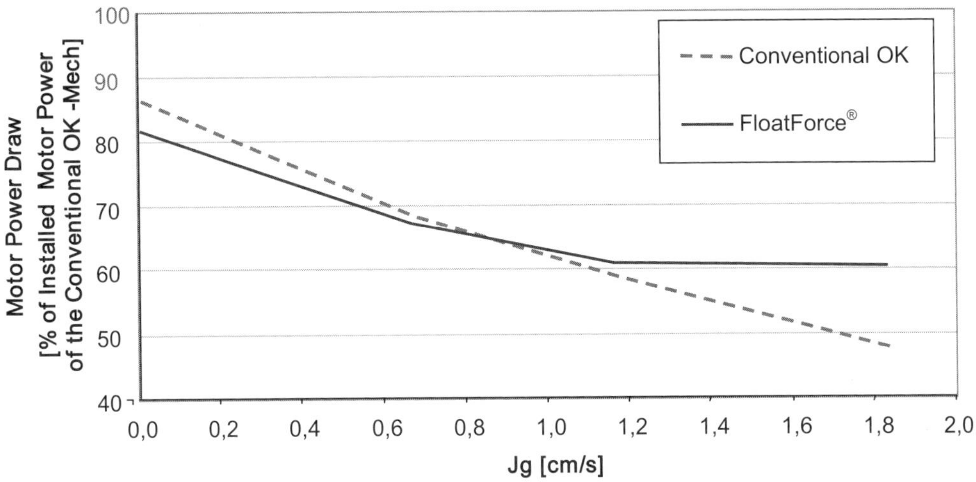


**Figure 12. Power draw vs. superficial gas velocity ($J_g$), for two mechanism types at constant impeller speed**

The FloatForce® mechanism can be retrofitted to existing Outokumpu/Outotec flotation cells. It is available in the same sizes, configurations, and connecting dimensions as all previous mechanisms. For new projects, the FloatForce® mechanism does not use different stator designs or bottom clearances, as in the conventional mechanisms. Instead, the adaptation for particle size is done by selecting the size and speed of the mechanism. For finer particles and easier duties, a mechanism one size smaller than nominal run at higher speed is selected. For coarse particles, a larger mechanism at lower speed will provide sufficient mixing the high turbulence that breaks the connection between a large particle and a bubble. The benefits reported by for the FloatForce® mechanism are summarized in Table 7.

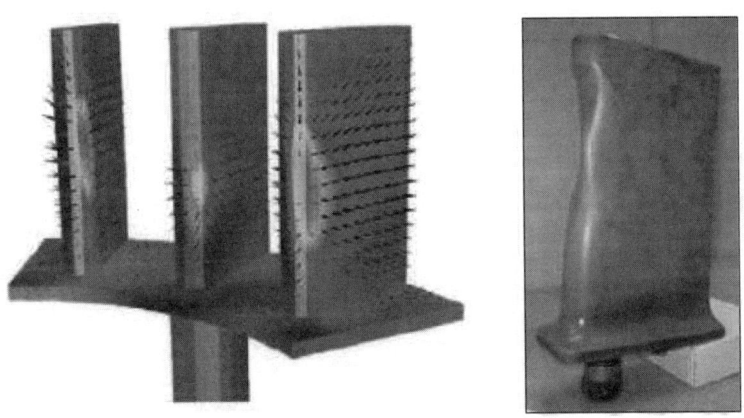

**Figure 13. High-wear areas of an Outotec stator: CFD pressure distribution (left); Actual worn blade (right).**

**Other Design Improvements.** Outotec uses CFD modeling to test design modifications that may be needed to address changing operating conditions and constraints. For example, increased energy costs may lead to coarser grinding and require flotation of composite (semi-liberated) coarse particles. In this case, design of the cell must avoid excess turbulence, but provide decisive, upward carrying flow to shorten the travel time of the bubble-particle aggregates while avoiding excessive turbulence in the froth phase. Using CFD modeling, Outotec designed the patented FlowBooster, an additional impeller installed above the primary mechanism. As shown in the CFD graphics in Figure 14, the rotation of the second impeller drives the slurry down towards the primary mechanism, amplifying flow in the turbulent, mixing portion of the cell. It also enhances quiescent, laminar circulation in the top part of the tank, aiding the upward flow of collected material exiting the rotor-stator.

**Table 7. Reported benefits of the Outotec FloatForce® mechanism**

| Key Feature | Effect | Potential result |
|---|---|---|
| *Increased mixing* at the same aeration rate | Increases bubble-particle collisions | Higher recovery |
| | Increases the suspension of coarse solids | Higher coarse recover<br>Coarser grind size (if liberation allows) |
| | Enables the use of higher slurry density | Added solids retention time<br>More throughput |
| | Pumps more slurry | Less sanding |
| Maintained mixing at a *higher dispersed aeration* rate | Increases bubble surface area flux ($S_b$) | Higher recovery |
| Maintained mixing *and* aeration rate (when selecting lower speed for new equipment) | Reduces power draw in no air / startup situation | Reduced energy costs<br>Reduced capital expense (motor and cables)<br>Lower spare parts cost |
| Individual lightweight stator wear parts | Enables easier and safer stator maintenance | Faster maintenance<br>Less time spent in confined space |
| | Enables testing of various wear materials | Longer wear life, higher availability |
| | Enables changing of individual stator blades in case of impact of foreign objects | Reduced wear part costs |

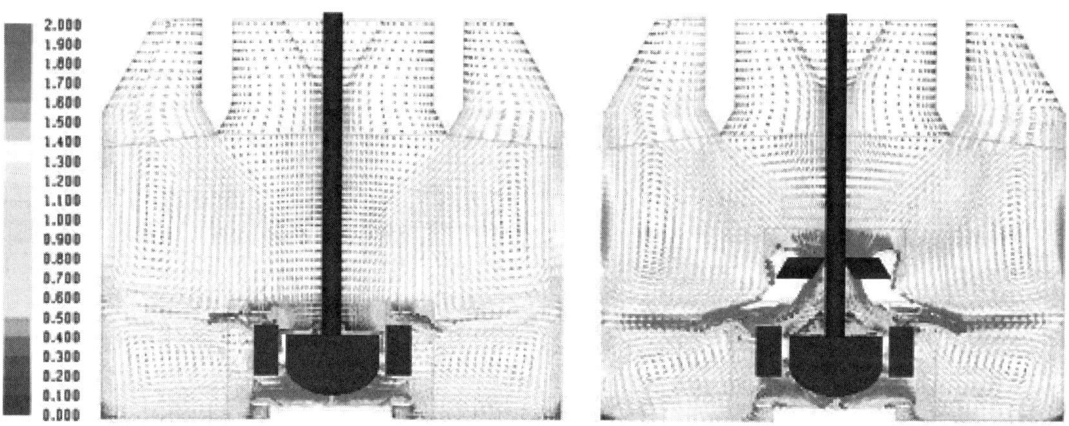


**Figure 14. Flow velocities in an Outotec cell: Conventional mechanism (left); FlowBooster impeller added (right)**

**Scale-up.** For externally aerated machines, hydrodynamic scale-up is relatively easy, because the air feed is provided by an external source, and the mechanism's only function is pumping slurry. In a cylindrical vessel, when the volume is tripled, the surface area only doubles, and the diameter increases by a factor of only 1.3. Flotation residence time is naturally related to volume, and gas flow rate through froth removal is related to surface area. Thus as cells become larger, concentrate flow rate does not increase proportionally to cell volume. Nonetheless, as lower-grade ore bodies are mine, more tonnage has to be treated to yield the same amount of concentrate, and larger cells better suited to this task. As is discussed below, in standard cell designs, power density decreases as cell volume increases. However, this does not necessarily imply a decrease in flotation performance.

**Current machines offered.** Table 8 shows rectangular (R) and U-shaped (U) machines; Table 9 shows cylindrical machines. The variables used in the tables are:

- Nominal Volume = minimum active volume or volume available for slurry and air.
- Air Feed Rate = air feed, when superficial gas velocity $J_g$ in the tank is 1.5 *cm/s*ec.
- Installed Mechanism Motor = Nominal electric motor power for typical slurry (specific gravity <1.4 g/cm$^3$)
- Specific Operating Power = mechanism motor power *plus* blower power, when operating with typical slurry and typical air feed rate. Higher air feed rate reduces motor power and increases blower power, and lower air feed rate does the opposite.

**Table 8. Outotec OK-R and OK-U Cells**

| Model | Effective Cell Volume, $m^3$ | Tank Length, *m* | Tank Width, *m* | Typical Lip Length, *m* | Typical Froth Area, $m^2$ | Maximum Air Flow, $m^3/min$ | Motor, *kW* |
|---|---|---|---|---|---|---|---|
| OK-0.5-2R | 0.5 | 1.9* | 1.1 | 1.9* | 0.8 | 1.0** | 3.75 |
| OK-1.5-2R | 1.5 | 2.8* | 1.6 | 2.8* | 2.0 | 2.0** | 7.5 |
| OK-3-2R | 2 | 3.4* | 2.0 | 3.4* | 3.0 | 3.0** | 15 |
| OK-8-U | 8 | 2.3 | 2.3 | 1.5 | 5.1 | 4.5 | 22 |
| OK-38-U | 18 | 2.8 | 2.8 | 5.6 | 7.8 | 7.0 | 37 |
| OK-16-U | 38 | 3.7 | 3.7 | 12.7 | 12.3 | 12.0 | 75 |

* Values for two connected cells, one motor on two shafts. **Values for one cell, individual air control in every shaft

**Table 9. Outotec TankCells®**

| Model | Effective Cell Volume, $m^3$ | Tank Diameter, *m* | Typical Lip Length, *m* | Typical Froth Area, $m^2$ | Maximum Air Flow, $m^3/min$ | Motor, *kW* |
|---|---|---|---|---|---|---|
| TankCell-5 | 5 | 2.0 | 4.7 | 3.0 | 2.8 | 11 |
| TankCell-10 | 10 | 2.5 | 6.3 | 4.7 | 4.3 | 18.5 |
| TankCell-20 | 20 | 3.1 | 7.9 | 7.1 | 6.7 | 37 |
| TankCell-30 | 30 | 3.6 | 9.4 | 9.7 | 9.0 | 45 |
| TankCell-40 | 40 | 3.8 | 10.0 | 10.6 | 10.0 | 55 |
| TankCell-50 | 50 | 4.7 | 12.7 | 16.5 | 15.0 | 110 |
| TankCell-70 | 70 | 5.3 | 14.1 | 21.3 | 19.0 | 110 |
| TankCell-100 | 100 | 3.0 | 16.3 | 27.5 | 25.0 | 132 |
| TankCell-130 | 130 | 6.4 | 17.6 | 31.0 | 28.0 | 160 |
| TankCell-160 | 160 | 6.8 | 19.0 | 35.2 | 32.0 | 185 |
| TankCell-200 | 200 | 7.2 | 20.1 | 40.0 | 36.0 | 250 |
| TankCell-300 | 300 | 8.0 | 22.0 | 49.0 | 45.0 | 300 |

**Installed base.** There are over 4,000 Outotec flotation cells operating in the world. Of those, over 2,000 are of the cylindrical TankCells®. Applications include sulfide metals, gold, platinum group metals, iron ore (reverse flotation of silica), coarse phosphates and other industrial minerals, coal, and potash. Table 9 shows some important, recent installations.

## SPECIFICATION OF NEW MACHINES

Several parameters must be specified for new machines. Each is discussed briefly below.

### Residence Time

The following discussion is summarized from Nelson, et al. (2002). Residence time is the time that a unit volume of slurry takes to travel from through a process or machine. If there is no short-circuiting, residence time in *min* is simply process volume in $m^3$ divided by the flow rate to the process, in $m^3/min$.

The required residence time for a given application is determined from laboratory or pilot-plant testing. The laboratory data are then used to calculate the parameters for a kinetic model, such as the one proposed by Klimpel [1980], where recovery, R, is given by

$$R = R_o \left[ 1 - (1-e^{-Kt})/Kt \right] \qquad (1)$$

where

$R_o$ = recovery at time zero,
t = time, and
K = flotation rate constant.

When the required residence time is known, the flotation capacity is determined by a simple calculation [Poling, 1980]. This procedure may be summarized in the following equation:

$$C = Q_{pulp} \times R \qquad (2)$$

where
C = total required flotation capacity, $m^3$, and
R = required residence time, *min*.

In this calculation, the solids feed rate to the mill, the solids specific gravity, and the pulp density are used to calculate the volumetric flow rate of the pulp.

When the required total tank volume is determined, the number of machines of a given size is found from Equation 10:

$$\tau = V_{eff}/Q_{pulp}, \text{ where} \qquad (3)$$

$\tau$ = theoretical residence time, *min*,
$V_{eff}$ = effective cell volume, $m^3$, and
$Q_{pulp}$ = volumetric flow rate of pulp, $m^3/min$.

The effective tank volume must subtract the volumes of the froth layer and the internal tank components. It must also account for the air holdup, which is the amount of air suspended in the pulp during machine operation.

Table 13, provided by Metso Minerals, shows relationships between test-bench and plant retention times for various materials. It should be used as a *guide* where no other information is available. It is clear that bench times need to be increased to obtain correct plant design retention times for plant design. Outotec reports using scale-up factors somewhat higher than those shown. FLSmidth further caution that care should be taken during laboratory testing to ensure that the froth removal rate results in the same concentrate grade as that anticipated in the plant design.

In the past, when rectangular cells were used in "hog-trough" configurations, there was extensive discussion of connection patterns for those cells. It was typical to connect two to four cells in "open" configuration, and connect one group to the next through a step-down junction box, with level control. There was much discussion of the best way to configure rows of flotation cells, to achieve the best mixing and minimize short circuiting.

With the advent of large, cylindrical machines, the grouping of cells became less common. Large machines are now often connected to one another individually, through connection boxes and level control valves. This seems to pose no problems regarding retention time or short circuiting, but it does limit the operator's options in dividing the froth for selective retreatment. Further, the minimum number of cells required may not be the usual five or six per row in an open circuit. As the residence time in a cell increases, the size of the perfectly-mixed portion will also increase, so fewer cells are needed. For closed circuits (typically used in cleaning stages), where the tails are circulated back to other stages, only one cell per row may be needed, as the tailings will have a chance to be recovered elsewhere.

Outotec's procedure for selecting the machine size, number of machines, and launder configuration for a given application goes beyond laboratory measurement of retention times and the application of empirical scale-up factors. Figure 15 is a simplified flowsheet showing the selection process for a typical sulfide application.

The numbers are of course case dependent. For instance the Carry Rate (CR) limitation depends on the Enrichment Ratio (ER). For a higher the ER, the CR should be lower. This relationship is based on a database of Outotec installations. Other factors also affect cell design, like the Air Efficiency, which is the mass of floatable material to be lifted by the air, and particle size, which affects the Transport Distance. Performing design tests to determine these factors allows a reasonable selection of the size and number of cells, as well as the desirable launder configuration.

**Drive**

Large flotation machines use AC motors. The motors for large (>250 $m^3$) machines may require medium-voltage power, which changes power-supply requirements and motor control configurations for the entire mill. The relative costs and advantages of medium-voltage power must be carefully considered for each site. Variable frequency drives are often used for development and testing, but their high installed cost usually precludes use in operations.

For machines up to 200 $m^3$, multiple V-belt connections are more reliable and easier to maintain. For larger sizes, gear-box drives are used by some suppliers. However, for externally-aerated machines, additional cooling of the drive may be required to compensate for the heat added by the air being blown through the shaft. When belt drives are used for machines larger than 200 $m^3$, a re-design of the drive is required to transmit the large amount of torque required by the mechanism. Absorbed power is based measurement of *actual kilowatts* (not just amperage and power factor) during routine operation. Thus the complex power triangle be measured to accurately characterize cell performance.

**Power Density**

Power density, usually measured in $kW/m^3$, characterizes the intensity of mixing or agitation in a flotation cell. Again, the important consideration is *absorbed* power density, as contrasted with *installed* power density. Installed power is based on the nameplate ratings of the drive motors, which in every case are oversized to account for atypical or upset conditions. Absorbed power is determined by measuring voltage, current, and power factor, so the complex power relationships can be calculated.

In all cases, power density decreases a cell volume increases. In externally-aerated cells, the tank diameter and to some extent the tank's horizontal, cross-sectional area determine the required mixing intensity. Power is not consumed in moving pulp vertically because the gas bubbles carry the floatable material to the top of the cell. Sanding can be a problem in large cells, when scale-up is not done properly. In Outotec cells, the size and speed of the mechanism are

**Table 13. Residence time guidelines for various applications**

| Material[1] | Typical rougher solids concentrations %[2] | Typical rougher residence times, min[3] | Typical lab flotation times, min | Scale-up factor |
|---|---|---|---|---|
| Copper | 32-42 | 13-16 | 6-8 | 2.1 |
| Lead | 25-35 | 6-8 | 3-5 | 2.0 |
| Molybdenum | 35-45 | 14-20 | 6-7 | 2.6 |
| Nickel | 28-32 | 10-14 | 6-7 | 1.8 |
| Tungsten | 25-32 | 8-12 | 5-6 | 1.8 |
| Zinc | 25-32 | 8-12 | 5-6 | 1.8 |
| Barite | 30-40 | 8-10 | 4-5 | 2.0 |
| Coal | 4-8 | 3-5 | 2-3 | 1.6 |
| Feldspar | 25-35 | 8-10 | 3-4 | 2.6 |
| Fluorspar | 25-32 | 8-10 | 4-5 | 2.0 |
| Phosphate | 30-35 | 4-6 | 2-3 | 2.0 |
| Potash | 25-35 | 4-6 | 2-3 | 2.0 |
| Sand (Impurity Float) | 30-40 | 7-9 | 3-4 | 2.3 |
| Silica (Iron Ore) | 40-50 | 8-10 | 3-5 | 2.6 |
| Silica (Phosphate) | 30-35 | 4-6 | 2-3 | 2.0 |
| Effluents | As received | 7-12 | 4-5 | 2.0 |
| Oil | As received | 4-6 | 2-3 | 2.0 |

Notes: 1. Material must be in floatable form.
2. For cleaning applications, use 60% of the solids concentration for roughing.
3. For cleaning applications, required residence time is approximately 65% of that required for roughing.

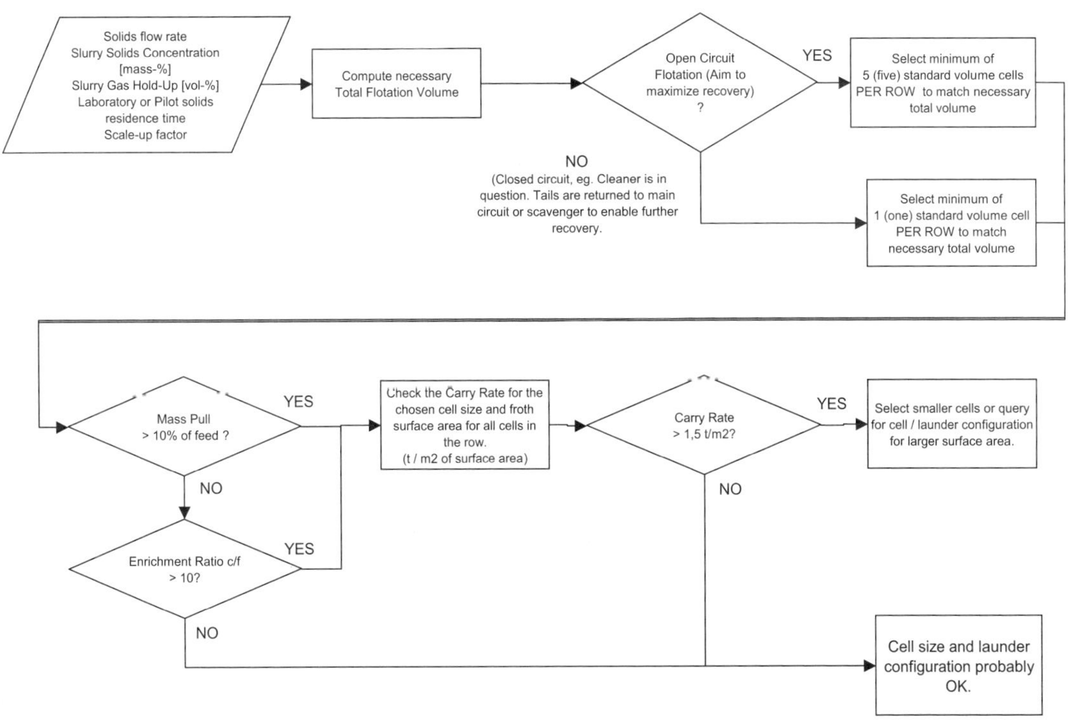


**Figure 15. Outotec procedure for selecting numbers and sizes of cells**

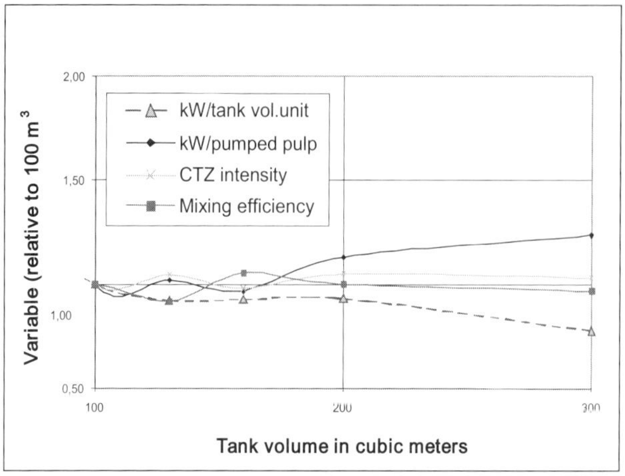


**Figure 16. Power and mixing parameters for Outotec TankCells®**

selected so that the mixing power and efficiency in proportion to the tank area remain the same in all cell sizes. In other words the volume of the pulp that the mechanism pumps is kept constant in relation to the cross-sectional area of the cell. Figure 16 shows various power and mixing parameters, measured for Outotec machines.

Reduced power density does not mean the flotation performance is equally reduced. In proper hydrodynamic scale-up, the intensity of turbulence in the vicinity of the mechanism is maintained throughout the size range. Outotec refers to this intensely turbulent region around the mechanism as the core turbulent zone, or CTZ. Outotec cells are designed so that every cell in the product line has the same kinetic flotation rate. This was shown in recent tests at Codelco's Chuquicamata Concentrator, where the first TankCell®-300 machine was tested against two TankCell®-160 machines, connected in parallel. The larger machine showed the same or improved metallurgical performance at lower energy consumption.

Experience has shown that higher density ores respond better to flotation in cells with higher power density. For example, in platinum flotation, 3 $kW/m^3$ is a standard value, while in copper flotation, the values range from 0.7 to 1.2 $kW/m^3$. This factor must be carefully analyzed for each application; the optimum power density is often not intuitively obvious. For example, in platinum cleaners, power density is greater than in platinum roughers; in copper cleaners, power density is lower than in copper roughers.

## Mechanism

Mechanism designs vary widely among manufacturers. In all cases, it is important that the lower portion of the mechanism (shaft, and rotor) be statically and dynamically balanced, *after* the installation of liners and *before* use. Careful balancing is especially important in the case of large cells, where small imbalances in the mechanism can cause large vibrations in the supporting structure. Vibration measurements should be made on each mechanism immediately after installation, and operating vibration should be checked against the baseline on at least a quarterly basis, so that corrections can be made as needed.

Mechanical loading forces in the mechanism are easily calculated using first-order, theoretical mechanics. Unfortunately, vibration characteristics are much more difficult to predict, and a systems approach is required. In mechanical flotation cells, besides mechanical properties (dimensions, masses, forces, etc.), the rheology of the slurry, the effect of air dispersion, generic and local flow patterns, and wear must be taken into account.

There are no standards for specifying allowable vibration levels in mechanical flotation cells. The closest applicable ISO standard is for to rotating machines with excitation frequencies between 2 and 2,500 *Hz*, or 120 to 15,000 *rpm* (ISO, 2009), but the majority of the large flotation cells operate below 2 *Hz*. Nonetheless, Outotec has found that the guidelines for response to vibration given in ISO 10816-3:2009, shown in Table 14, to be useful for flotation machines. Table 15 shows vibration data for several Outotec TankCells®. Outotec and other manufacturers often measure vibration in the top support bearing of the shaft, as shown in Figure 17, using 3-axis accelerometer and a frequency band of 0.5 to 15 *Hz*.

**Table 14. ISO 10816-3:2009 guidelines for response to vibrations**

| Vibration level | 0.0–1.4 *mm/sec* | 1.4–2.8 *mm/sec* | 2.8–4.5 *mm/sec* | > 4.5 *mm/sec* |
|---|---|---|---|---|
| Condition and suggested response | Newly commissioned machine | Ok for normal, continuous operation | Maintenance needed on next shutdown | Maintenance required immediately |

**Table 15. Vibration data for Outotec TankCells®**

| | TankCell®-100 | TankCell®-200 | TankCell®-300 |
|---|---|---|---|
| Year of Startup | 1998 | 2002 | 2007 |
| Tank diameter × height | 6.0 × 4.3 *m* | 6.8 × 5.8 *m* | 8.0 × 7.0 *m* |
| Installed power | 132 *kW* | 220 *kW* | 315 *kW* |
| Measured Vibration Velocity RMS | 4.3 *mm/sec* | 2.9 *mm/sec* | 2.2 *mm/sec* |
| Measured Vibration Peak Displacement | 0.16 *mm* | 0.05 *mm* | 0.03 *mm* |

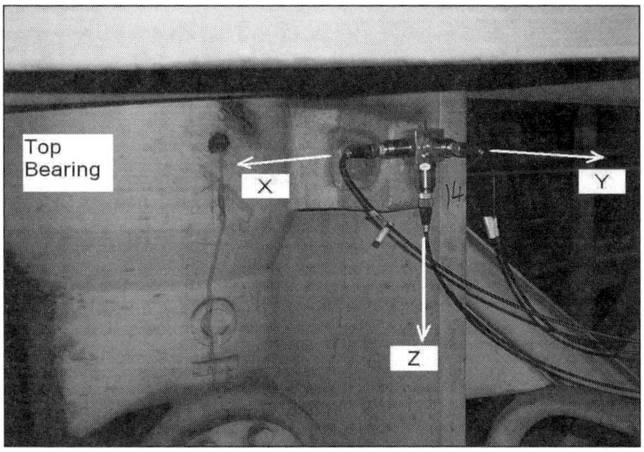


**Figure 17. Installation of 3-axis accelerometer on an Outotec shaft bearing housing**

**Dart Valves**

Dart valves are the standard device for controlling flow between cells, and thus for regulating the pulp level in a cell. As shown in Figure 18, dart valves use a conical closing mechanism, or plug, that fits against a circular seat. As the conical mechanism is raised by the valve rod, the opening gets larger and flow through the valve increases. This characteristic is not necessarily linear, and should be analyzed in combination with the characteristics of the controller to ensure that flow control can be maintained in the desired region. Plugs with a specially-designed taper provide a more nearly linear response characteristic. On large cells, two valves are often used—one with high gain, for large changes, and one with lower gain, for fine control.

Dart valves can be installed in an external connection or discharge box, or inside the cell. They can also be installed in "down-flow" or "up-flow" configuration. Figure 18 shows an external valve in down-flow configuration, on the left, and an internal valve installed in up-flow configuration, on the right. The benefits of internal dart valves include a reduced footprint, lower equipment cost, easier access to valve plug and seat wear parts, and no air and froth accumulation in the connection or discharge box. Up-flow dart valves save hydrostatic head, and this design is typically preferred. However, in up-flow valves, oversized material tends to accumulate in the space just below the dart valves, and a flushing nozzle with an amply-sized rock port are recommended the allow removal of potential accumulation. If there is no restriction on the height of the discharge box, a down-flow configuration is preferred, to avoid the problems of solids accumulation. However, the down-flow valves do require some back pressure, or at least zero gage pressure for satisfactory operation. For this reason, a breather pipe is required directly after the valve. An excellent and detailed description of dart valve design and operation has been prepared by Bourke (2006).

**Tankage**

The tankage associated with a machine includes feed boxes, the cell itself, connection boxes, and discharge boxes.

Feed boxes provide a constant head for feed to a machine or row of machines. A feed box is usually designed to admit feed at the top and pass it out to the flotation process at the bottom, through an open connection. The box should be designed to accommodate process upsets, and should fit as closely as possible to the machine, minimizing required floor space. Feed boxes typically have a sloping bottom to enhance flow into the flotation cell. The size of the opening that connects the feed box to the cell is based on the minimum linear velocity required to maintain slurry suspension. Of course, high velocities can cause high wear, so connections should be lined with wear-resistant materials. In concentrators with SAG mills in the grinding circuit, large fluctuations in grinding output are common. In such installations, large feed boxes can provide valuable surge capacity in the flotation circuits.

Flotation cells come in two basic shapes, rectangular and cylindrical. Rectangular cells are usually of the "hog-trough" design, to provide longer retention time, and often have a U-shaped or horseshoe bottom, to conform to

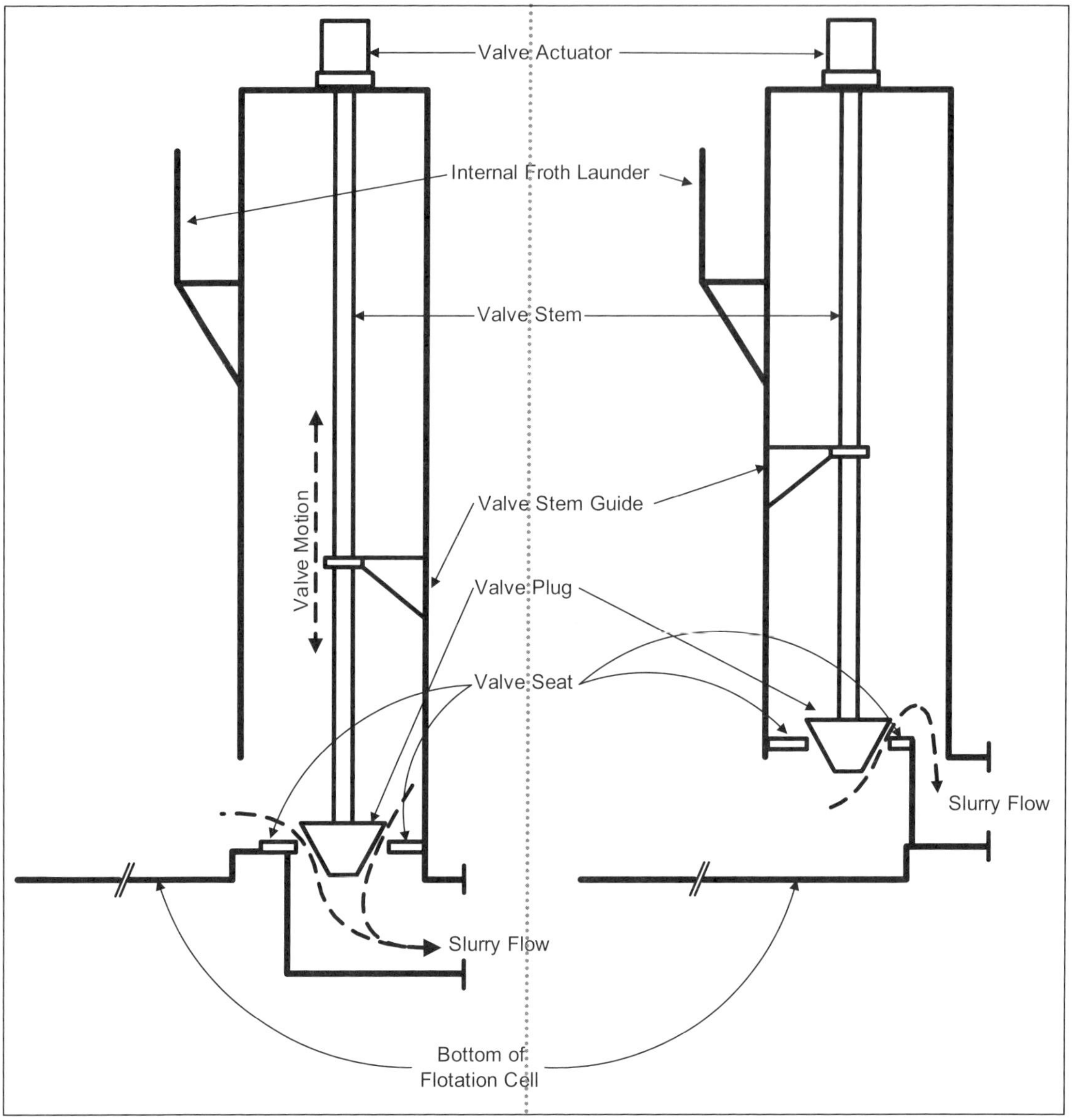


**Figure 18. Dart valve designs: External down-flow (left) and Internal up-flow (right)**

hydrodynamic flow patterns, and to minimize wear and sanding. The largest installed rectangular machines are 85-*m3*, WEMCO 1 + 1 machines. Rectangular cells are still supplied for some portions of flotation circuits in smaller-volume plants. Cylindrical machines are symmetrical, and thus provide better hydrodynamics. They eliminate stagnant areas found in the corners of rectangular cells, and the turbulence resulting from corner effects. Plant layout using cylindrical cells requires careful consideration to minimize required floor space. However, some advantages can be achieved by configuring the layout so that connections can be modified, allowing one cell to be taken out of service for maintenance without shutting down an entire section.

Connection and discharge boxes are practically the same. Both use dart valves to control the flow through the box. Connection boxes, installed between adjacent cells, have a lower hydrostatic head, usually just enough to maintain flow through the bank of cells and accommodate changes in throughput. Discharge boxes have a higher hydrostatic head, to provide the required flow to the tailings circuit. Both boxes should be designed based on the minimum linear velocity required to maintain slurry suspension. Again, high velocities can cause high wear, so connections should be lined with wear-resistant materials.

There are four designs for connection and discharge boxes used with cylindrical tanks. The first, and conventional design, is based on the design still used with rectangular cells. These are rectangular boxes, attached to the outside of the associated cell or cells. The second design, which saves floor space and provides easier access, is a small, cylindrical box, again attached to the outside. The third design places the dart valves inside the cell, with uses a smaller box, below the floor level. The advantages of this design are reduced floor space, as shown for Outotec machines in Figure 19, lower

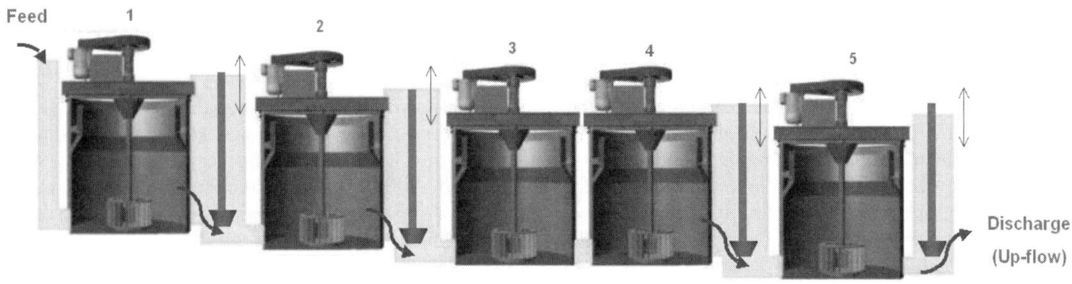

**Figure 19. Floor space saved by internal dart valves (bottom row), as compared with external dart valves (top row)**

equipment cost, easier access to the valve and seat wear parts, and no accumulation of air and froth in separate box. In the three cases just described, the dart valve moves vertically. In the fourth design, the dart valve is inside the tank, but the tanks are connected directly on their adjoining walls. The valve plug is connected to the valve stem by a linkage that allows the valve stem's vertical motion to swing the valve plug in and out of an opening connecting the tanks. This fourth design saves both vertical and floor space, but requires a more complex dart valve mechanism, which is more subject to binding and malfunction.

### Launders and Froth Removal

Froth removal is critical to satisfactory function of any flotation circuit. Recent theoretical analyses have considered kinetics for *two* processes in flotation, particle-bubble attachment and transport to the froth interface, and froth recovery. Froth recovery is the fraction of particles reaching the froth interface that finally report to the froth launder. In some cases, this has been reported to be as low as 10. It is important that the entire froth handling system have capacity adequate to handle the highest anticipated volume of froth produced. This includes launders, piping, and pumps used to transport froth product to holding tanks or thickeners.

Circuit design should include an estimation of the froth volume produced from each cell in a given time, and a calculation of the launder "lip length" required to allow timely removal of that froth. As cells have become larger, the peripheral lip length and cell cross-sectional area, relative to cell volume, have decreased. Thus internal launders of various configurations have been incorporated in cell designs. Some suppliers use radial launders, others a series of concentric rings. Internal launders have two functions. First, they provide the required lip length for froth removal; second, they ensure a minimum travel distance to the launder from any point in the froth mass.

Froth crowders are sometimes used to decrease the surface area at the top of the cell, thus forcing the froth to move more quickly to the launders. This can be important in scavenger cells, where the froths are relatively thin. Note however that, in large cells, the decreasing ratio of surface area to cell volume can result in a condition of "over-crowding" in the froth, where particles that have attached to bubbles cannot travel through the froth for final recovery to the concentrate.

### Liners

Liners are discussed in some detail elsewhere in this volume. Liners are usually installed in two locations in flotation cells, the tank and the launders, and the mechanism.

The tank and launders are lined to prevent corrosion and abrasion of the steel, so the liner should be robust, but not too thick. The most common liners used are hand-laid natural rubber and sprayed-on polyurethane, with a suitable, corrosion-inhibiting primer. Lining of larger tanks, and of tanks that are fabricated and shipped in segments for assembly on site, require special surface preparation and installation skills, and should be done only by qualified contractors. In locations where large daily temperature fluctuations are expected, the lining system should be selected to withstand the "cold wall effect," which is caused by a temperature differential across the interface between the liner and the substrate.

### Lifting and Transportation

These considerations, often overlooked, are key to smooth and successful installation and maintenance. With large cells, manufacturing location and methods for welding, applying liners, and other manufacturing functions must be analyzed very carefully in relation to the size of the tank or tank sections.

Lifting lugs should be installed before cell linings, so that welding of lugs does not compromise lining integrity. Lugs or other lifting aids should be installed on all components that may need to be separately removed, including belt guards, mechanism components, launders, dart valves, and tank sections. Belt guards and other components should be designed so they can be separated into pieces that are easily handled. All bolt holes and connection points should be checked for alignment and compatibility before cell components are shipped to the plant site. For components that are lined with elastomer or other protective coatings, this check should be made before the lining is installed.

Similarly, with large cells, method of shipment, shipment route, and shipping schedule must be carefully considered. Large cells must often be manufactured in sections, to facilitate transport. In some cases, special permits for over-the-road shipping must be obtained, and the large sizes may make shipment be rail impossible. Where large cells are installed in existing plants, designers should carefully consider the sizes of doorways, overhead clearances, bridge crane capacities, and other factors, to ensure that the manufactured sections can be put into place as required.

## INSTALLATION

### Foundations

Foundations for large cells should be designed by a competent structural engineer. Large cell mechanisms generate high torque. For cells that have rotors near the bottom, this torque can cause flexing in of the tank bottom, if the support given by the foundation is inadequate. This can in turn lead to failure of welds in the tank bottom or the parts of the mechanism attached to the bottom.

Almost all concentrators eventually increase capacity above the original design level, so that sumps, launders and pipelines can become undersized. It is preferable to oversize sumps and other utilitarian structures, in anticipation of the inevitable. Regardless of design, all sumps overflow at some point. It is the important to have a design that facilitates easy washdown and cleanup. One excellent configuration is to install the flotation cells above the sump level, and provide the sump level with a sloping floor, frequently placed hose connections, and a good system of drainage channels, connected to the sump.

The dynamic requirements for cell foundations are related to the excitation frequencies of the equipment installed, and the nominal or natural frequency of the entire system. Here, the system means the foundation structure supporting the cells, and cells *holding slurry*, and the cell mechanisms provide the main excitation frequency. The system must be supported so that the nominal frequency is higher than the lowest excitation frequency by a safe margin. This margin is defined in ISO 10816-3:2009. Support is considered stiff if the nominal frequency of the entire system is 25% higher than the main excitation frequency. Standard practice for most flotation cell installations fall into this category, but special care must be taken where the underlying earth is unstable, due to moisture, permafrost, or some other anomalous condition. Thus, the cost of foundations for a given cell increases if the rotational speed of the flotation cell increases.

### Power Connections

As previously mentioned, the motors required for larger cells may require medium-voltage power and motor control centers. This possibility should be carefully considered in combination with the cost of power and the cost of installation at each site. If medium-voltage power is not available, the cost of installing transmission lines and substations may prove prohibitive.

All machines should be provided with instruments to indicate real power consumption to the control operator. Power consumption, especially when compared with feed rate, can to a large degree indicate what is happening in the cell—whether it is sanded, whether the air intake is plugged, etc.

### Sampling Connections

Sampling connections have two purposes. First, they must provide a representative sample of a given flow stream; second, they must be capable of connecting to an on-line analysis system. Each connection should be configured so that, even if it is connected to an on-line system, a separate sample can be manually removed at any time, for independent laboratory analysis. Sampling connections are easily installed during construction, and connections for samplers should be installed so that every stream can be sampled, if necessary. This is true even if there is no plan to sample a given stream when the operation is started. Sampling ports should be designed to be easily cleaned of blockages, and for easy attachment of connections to the sampling device or analyzer.

### Walkways, Platforms, Stairways, and Railings

Walkways should be installed to provide access to all locations that may require attention from operating or maintenance personnel. This includes sampling ports; motor connections; drive belts; instrument mountings; manual and automatic control valves for slurry, air, or wash water; and drainage ports. Of course, all structures designed for human foot traffic must comply with applicable health and safety codes.

Access to the interior of the cells is frequently necessary for inspection and maintenance. In smaller cells, permanent ladders are often installed. In larger cells, access is through a manhole near the bottom of the cell. In both cases, when operators enter the cell, the cell must be locked and tagged out, and operators must wear safety harnesses for recovery, as required by local health and safety codes.

## STARTUP

All procedures must follow local safety procedures for lockout and tagout, personal protective equipment, etc. Certain simple procedures during startup can obviate serious problems during operation, first with the cells empty, then with water-filled cells, then with water-filled, fully-aerated cells. All checkout procedures are made *before* slurry is introduced to the cells. The checkout and startup of the systems for reagent addition and control are not covered in this chapter.

### Empty Cells

**Electrical systems.** At the beginning of the startup, a preliminary check of electrical components should be made. Bump the cell drive motors, blower motors (where used) and all pump motors, and check each for correct rotation direction. If motors start and rotate correctly when bumped, run each one long enough to confirm that its power draw is in the expected range. Finally, confirm correct operation of any variable-frequency drives.

**Mechanical systems.** Check each cell mechanism first by hand rotation, to make sure it rotates freely and bearings are not binding. Make measurements to confirm that the rotor is concentric within the stator or draft tube. Make a final inspection of welds in the structure and and cell lining.

Next, while the cells are still empty, power up each cell mechanism and check for vibration, visually and with a vibration sensor. After the mechanism has run for about 30 minutes, shut it down and again check the concentricity of the mechanism and the welds on the mechanism structure.

**Control systems.** First, check the dart valves. Confirm that the stroke, function, and calibration of all the dart valves are all correct. Make sure that valve shafts are moving freely, not binding in the valve guides. Confirm that the valve plugs are concentric in the seats, and that the plugs seat properly. Blow down the air lines for the valve control systems, check the instrument air pressure and make sure that it functions over the correct pressure range. Then check control loop calibration and function, and confirm that control loop parameters are set as specified by the supplier.

Next, where blowers are used, confirm the correct function of the blower control system. Cycle the blower through its prescribed range, and make sure the blower and the flow sensors respond correctly. Also, confirm that any control valves or dampers in the flow control system function correctly

Finally, if the cell mechanisms or pumps are fitted with variable-frequency drives (VFDs), confirm that each VFD is functioning correctly. Cycle the motor through its design speed range, and confirm that the VFD and all associated instruments are functioning correctly.

### Water Test

In the water tests, the electrical, mechanical, and control systems are tested simultaneously. Fill the cell with water. (Note that initially air is not introduced to the machine. Self-aerated machines will draw in some air when the mechanism rotates, but this should be minimized by keeping the air control ports at the top of the draft tube closed.) Make a careful visual inspection to locate any static leaks in the feed box, cell, connection and discharge boxes, auxiliary tanks, and piping. Any leaks located should of course be repaired before proceeding. When it is confirmed the water-filled system has no static leaks, start the mechanism and make another careful inspection for leaks throughout the system.

Now start the blower or open the air control ports at the top of the draft tube. When a blower is used, once again check flow control function and confirm correct operation of flow control valves or dampers in the air supply system. Run the fully-aerated, water-filled machine for the time suggested by the manufacturer, and check the temperatures of the gearbox, the shaft bearings, the air supply piping (in systems with blowers), and the temperatures of all pump motors, couplings, and gear boxes. Finally, confirm the correct function of the level control system.

### Slurry Test

After all the preceding checks have been made and everything is confirmed to be satisfactory, begin to introduce slurry to feed box. Allow slurry to replace water in the system at its natural flow rate. As the slurry replaces the water, start metallurgical operation and again check system functions: Vary the air flow, vary the level, and check dart valve response. During this time, continue to monitor all temperatures and power draws, and frequently walk around each cell checking for leaks, vibration, or any other unusual circumstances.

## OPERATION

Successful and optimized flotation plant operation relies on human understanding, observation, and intervention.

These control points in particular should be monitored: pulp level, froth thickness, froth pull, air flow, and reagent addition. All are important in determining flotation response, but in every operation, *regardless of the machines being used*, they are closely interrelated, and no one variable can be changed without influencing the others.

### Pulp Level

In self-aerated machines, pulp level is the variable most commonly changed by operators to influence the process outcome. If the pulp level is raised, the cell is said to be "pulling hard." The thickness of the froth layer will decrease, and the process will move towards increased recovery on the grade-recovery curve. If the pulp level is lowered, the opposite occurs. If the pulp level is raised too high, the pulp will overflow into the froth launders, and the machine will not function properly. This is especially true if the froth/pulp interface is unstable.

### Air Flow

In externally-aerated machines, it is more common to change air flow than pulp level. Increasing the air flow will make the machine pull harder, and move again towards increased recovery on the grade-recovery curve. Decreasing air flow will have the opposite effect. Of course, there is a limit to the amount of air that can be distributed in the pulp. If this limit is exceeded, "burping" or "geysering" will occur on the surface of the cell, and stable operation will be disrupted. In both types of machines, there is for any given feed conditions a combination of pulp level and air flow that will provide the best recovery. Only experienced operators can consistently operate at the optimum point.

### Reagent Addition

Reagent addition is not usually changed in routine operations. It is usually set based on experience and laboratory testing for the various ore types that are known to exist in the mine. Often it is impossible to know in advance when the ore type in the mill feed changes. Thus changes in the reagent addition scheme are usually made based on visual observations of the froth structure and the tailings, and on the online tailings analysis, after discussion among operations and engineering personnel.

## SAMPLING

The basis for evaluation of metallurgical performance of any process is a mass balance around the concentrate. Thus representative and regular samples, and accurate and timely analyses, are necessary. Each plant will have its own specific sampling schedule.

### Routine Sampling

In large mills, it is typical to sample concentrates and tailings for online analysis two to four times per hour. Once or twice per hour, separate samples, taken by hand or with an automatic device, are also taken and composited into shift or daily samples. It is typical in many large operations to splits the composite samples, with one split being analyzed for elemental composition and the other analyzed for mineralogy, using an MLA™ or QEMSCAN™ machine.

### Special Sampling

Routine sampling may not indicate long-term changes in the feed, the process, or the process equipment. Special sampling, or metallurgical auditing, should be undertaken on at least an annual basis to re-establish performance baselines for the plant. Special sampling includes down-the-row sampling, grind-release or size-by-size sampling, and residence-time-distribution analysis.

**Down-the-row sampling.** Down-the-row sampling is preferably conducted on two or three rows of each section (rougher, cleaner, scavenger, etc.) in a plant. Feed, concentrate, and tailings samples are taken from each bank in a given row. (A bank is a group of cells with common level control.) Note that the tailings from one bank are the feed to the next. In the evaluation, mass balances are calculated for each bank and each row. Down-the-row sampling is an arduous and time-consuming process. It requires preparation and careful forethought, especially in determining how and where to take representative samples. A carefully-conducted, down-the-row sampling campaign can indicate long-term, gradual changes in the mechanical conditions of the cells, the properties of the ore, the performance of the grinding circuit, and other variables. By comparing results of down-the-row sampling with the results of grind-release studies in the lab, down-the-row sampling can give the metallurgist a good indication of where the plant is operating, in relation to the ideal grade-recovery curve. This in turn can provide can provide worthwhile ideas for improving plant performance.

**Grind-release sampling.** Grind-release sampling can be conducted in the plant or in the laboratory. In-plant sampling, perhaps the more important, includes sampling feed, concentrate, and tailings for the entire plant or any portion of the plant, screening the samples into appropriate size fractions, and analyzing each size fraction for content of the constituent of interest. These analyses can readily indicate, for example, that all the chalcopyrite larger than 200 mesh is reporting to the tailings. The results of grind-release sampling are especially valuable when used in conjunction with analysis by an MLA™ or QEMSCAN™ machine.

**Residence-time-distribution analysis.** Residence-time analysis has been described in detail previously (Nelson et al. 2002). Residence time analyses are usually conducted by the manufacturer on new flotation machine designs. It is important to realize that similar analyses can be conducted in the plant, using salt tracers, dyes, or radioactive tracers (where allowed). Ideally, the tracer is *not* added at the head of a flotation row, but in the feed distributor for a number of rows. Samples are taken from the feed to each row, to determine how well the feed distributor is working. Tailings samples are taken at the end of each row, and the residence time for each row calculated by the standard method. The tailings analyses can indicate gradual changes in the conditions of machines, connection and discharge boxes, and dart valves, and allow ready comparison of individual rows with one another, based on the baseline residence-time distribution measurements, performed on new machines.

## MAINTENANCE

The importance of good maintenance is obvious, and cannot be overstated. Modern instruments and control systems make routine monitoring of important parameters, such as power draw, vibration, and bearing temperatures, relatively easy. However, nothing can replace regular and thorough visual inspection during shutdowns, by experienced operators and mechanics.

**Drive Assembly**

Temperatures of the motor, bearings, and gear drive (if used) should be checked regularly. Statistical quality control methods may be used to detect important changes or long-term trends in readings.

**Mechanism**

Regular lubrication of shaft bearings is important, and bearing temperatures should be checked manually on a regular basis. The linings on the shaft, rotor, and stator should be inspected in each shutdown for wear, cracks, and chunk-type failure. Components with severe damage should be removed and replaced. Concentricity of the rotor inside the draft tube or stator should be checked.

Rotation should be reversed every shutdown, to achieve even wear on both sides of the blades. If rotation is not reversed regularly, symmetry will not be maintained, and then, when rotation is reversed, the motor may trip out due to excess current draw. Vibration should be checked against baseline values at least quarterly.

In self-aerated machines, the air openings should be checked to insure they are clear. Similarly, in forced-air machines, the air intake openings in the rotor should be cleared of any residual slurry or debris.

**Liners**

The integrity of tank liners should be inspected during every shutdown. Deterioration of liners can lead to liner failure over small or large areas, which in turn leads to increased corrosion and wear of the steel substrate, causing eventual failure of the tank component. Liner integrity is especially important at the flow-through points between various tankage components—feed boxes and cells, or cells and connection or discharge boxes.

**Dart Valves**

Darts and seats should be carefully inspected for wear, proper alignment, and concentricity. Movement of the valve shaft in its guides should be checked to make sure the shaft is not bent, and there no binding or misalignment. The air lines and moisture traps on the dart valve controllers should be checked and blown down weekly. The flotation circuit need not be shut down to complete this operation.

**Launders**

Cell overflow lips and froth launders should checked and washed down regularly, as needed to prevent accumulation of concentrate and assure equal flow along the entire length of the lip. In large cells, connection points between the various components of interior launders may experience high wear, and should be inspected on each shutdown.

**Cleanout**

Regular cleanout of the cell and its mechanism is important. Occasional upsets in the grinding circuit cannot be avoided, and oversize material will almost always accumulate in flotation cells. There is a limit to the amount of material that can be accumulated in any cell before performance is adversely affected. To assure satisfactory operation, this material must be removed periodically.

The easiest way to remove oversize material is with water. In some cases, much of the oversize can be removed by pumping water at a high flow through the circuit. In some plants, operators insert hand-held, high-pressure water pipes into the cell, to assist in moving oversize material into the flow. However, this practice can damage the lining on the cell and mechanism. At points where the lining is damaged or detached, the high-pressure water tends to raise the lining and increase the damage and detachment. When oversize material cannot be removed with water, the remaining alternative is the use of shovels and buckets.

## REFERENCES

Bourke, P. 2006. Dart plug valves—the modern slurry control tool. http://www.outotec.com/34892.epibrw (accessed May 2009).

Daman, Jr., A.C. 1966. Development and Theory of the D-R* Denver Flotation Machine (Denver Bulletin No. F10-B121). *Deco* Trefoil, Nov-Dec-9-16.

Dreyer, J.P. 1976. The Development of Agitair Flotation Machines. *J. S. Aft. Inst. Min. Metall.* June: 445-7.

ISO. 2009. Standard10816-3:2009, *Mechanical vibration -- Evaluation of machine vibration by measurements on non-rotating parts -- Part 3: Industrial machines with nominal power above 15 kW and nominal speeds between 120 r/min and 15,000 r/min when measured in situ.* Geneva, Switzerland: ISO.

Klimpel, R. R. 1980. Selection of Chemical Reagents for Flotation. In *Mineral Processing Plant Design*, A. L. Mular and R. B. Bhappu, editors. New York: SME-AIME.

Nelson, M. G., F.P. Traczyk, and D. Lelinski. 2002. Design of Mechanical Flotation Cells. In Mineral Processing Plant Design—Operating Practice and Control, Halbe, D. R., editor, Littleton, Colorado: SME.

Poling, G. A. 1980. Selection and Sizing of Flotation Machines. In *Mineral Processing Plant Design*, A. L. Mular and R. B. Bhappu, editors. New York: SME-AIME.

# THE APPLICATION OF FROTH FLOTATION FOR GOLD RECOVERY AT NEWMONT MINING CORPORATION

Jim Orlich, John Gathje, and Ronel Kappes

Newmont Mining Corporation, 10101 E. Dry Creek Rd., Englewood, CO 80112

## ABSTRACT

The application of froth flotation technology for gold recovery is strongly influenced by the mineralogical association, as well as particle size and gold/silver content of the gold. The paper reviews some of the ore types that can be encountered in gold ores and discusses their effect on flowsheet selection, plant design and operation. Examples are discussed for a number of gold recovery circuits utilized at properties operated by Newmont Mining Corporation.

## INTRODUCTION

In most mineral processing design considerations, the mineralogy of an ore plays a key role. This is particularly applicable to gold ores, where a myriad of different gold-bearing minerals have been identified. These minerals include (Marsden and House, 2006):

- Placers
- Free-milling ores
- Oxidized ores
- Silver-rich ores
- Iron sulfides
- Arsenic sulfides
- Copper sulfides
- Antimony sulfides
- Tellurides
- Carbonaceous ores

Newmont Mining Corporation owns and operates gold mines on five continents and as such has encountered most of these gold-bearing minerals in one form or another. Frequently, froth flotation is used to upgrade/concentrate these minerals for a variety of reasons including:

- Production of a saleable concentrate
- Pre-concentration of sulfides for subsequent refractory ore treatment (pressure oxidation, roasting or ultra-fine grinding)
- Production of a concentrate for subsequent cyanidation

This paper provides an overview of the typical ore mineralogy and gold associations encountered at Newmont Mining Corporation's current operations and projects, particularly where the application of froth flotation has played a crucial role. Included in the overview are considerations of typical flotation strategies which may be employed in the recovery of these minerals. In addition to this overview, several case studies of current/future Newmont flotation operations are reviewed.

## ORE MINERALOGY AND GOLD ASSOCIATION

### Gold Minerals

***Free Gold (Native Gold/Electrum).*** Gold grains that exist as readily identifiable particles rather than gold existing on a molecular level within the matrices of host minerals, are typically referred to as "free, or free-milling gold." Free gold grains can be further categorized depending on the gold/silver ratio, as native gold if the particle contains between 15-5% silver or as electrum if the silver content exceeds 15%. Free gold particles greater than 200um are usually readily recoverable by gravity separation but are too large for effective flotation recovery. Flash flotation (high %solids within the grinding circuit) can be used to enhance the flotation of coarse free gold particles and "normal" flotation can be used very effectively to recover free gold at smaller particle sizes (<200 μm), although the selectivity for free gold flotation decreases rapidly at finer particle sizes (<20μm) (Marsden and House, 2006). It is generally found that the flotation properties of electrum are greater than native gold and increase with the proportion of silver contained in the gold grains.

Because gold is so ductile, it is usually not readily reduced to fine particle sizes and can therefore build up in grinding circuits due to its high density (Woodcock et al., 2007). In the process the gold surface can be damaged and pick up surface coatings, which may reduce its floatability. As a consequence, it is usually recommended to have a gravity (and/or flash flotation) circuit prior to flotation where coarse native gold is known to be present in the feed (Woodcock et al., 2007).

There are many flotation collectors that have been used successfully for the recovery of free gold, including the workhorse of the flotation industry, xanthate. Commonly, the monothiophosphate, dithiophosphate or dithiophosphinate based collectors are used when considering free gold flotation. The selection of collectors will usually in part be determined by the objective of the flowsheet, including bulk flotation (recovery of all gold-bearing minerals including gold minerals associated with sulfide minerals) or a more selective flotation approach (recovery of gold minerals and rejection of the bulk of the sulfide minerals).

***Gold Tellurides.*** Generally, gold-telluride minerals are considered to be fairly hydrophobic and in many instances can be floated without the addition of a collector. However, as is the case for native gold flotation, the type of collector addition selected is in many instances driven by the purpose behind the flowsheet and further downstream processing objectives.

### Iron and Arsenic Gold-Bearing Minerals

Iron and arsenic gold-bearing minerals tend to fall into two categories, i.e., free-milling gold ores and refractory gold ores. Free-milling sulfide gold ores may contain fine free gold and gold associated with mainly pyrite and arsenopyrite. For refractory sulfide gold ores, the gold is commonly associated with pyrite, arsenopyrite and arsenian pyrite (pyrite enriched in arsenic) as is the case for example for a Carlin-trend type ore deposit.

***Pyrite and Arsenopyrite—Free Milling Sulfide Gold Ores.*** Flotation of the fine free gold and auriferous sulfide minerals is commonly applied to sulfide-containing free-milling ores (Woodcock et al., 2007). In the instances where free gold is present in the feed, a gravity separation circuit is usually included prior to the flotation circuit. These auriferous sulfides minerals are usually pyrite and/or arsenopyrite and are easily recovered using either a xanthate and/or a dithiophosphate type collector. High flotation gold recoveries are usually possible.

***Pyrite, Arsenian Pyrite and Arsenopyrite—Refractory Sulfide Gold Ores.*** In the case of refractory sulfide gold ores, gold is commonly associated with pyrite, arsenopyrite and arsenian pyrite (pyrite enriched in arsenic). Additionally, the pyrites in these types of ores (such as often found at Newmont's operations in Carlin, Nevada), can be divided into four distinct morphological types (Figure 1). Typically the gold content of these ores are strongly related to their arsenic content (Figure 2). Thus, it is typically the finely disseminated pyrite (that contains most of the gold) that needs to be recovered to ensure good gold recoveries are achieved. Due to the finely-locked nature of the sulfides, it is quite a difficult flotation separation to achieve. Although finer grinding will improve the sulfide mineral liberation to some extent, it is well documented that this will enhance surface oxidation of the arsenian and arsenopyrite leading to reduced floatabilities of these minerals (Miller et al., 2000). Additionally, finer grinding may not be economically feasible due to the higher milling costs associated with it. Recovery of the middling particles then becomes crucial to the success of the flotation process. Currently the only collector capable of ensuring good bulk sulfide flotation in this type of scenario is xanthate, particularly potassium amyl xanthate (PAX).

The $N_2$TEC™ (Gathje and Simmons, 1997) technology, which also utilizes PAX as the collector, has been found to be very successful in improving the floatability of the reactive auriferous sulfide ores. Under $N_2$TEC™ flotation conditions, air is replaced by nitrogen as the flotation gas, which limits surface oxidation of the reactive pyrite species and improves their floatability, ultimately improving gold recovery significantly.

### Copper Sulfide Gold-Bearing Minerals

Gold is commonly recovered as a by-product of copper production. Auriferous copper ores may contain chalcopyrite and other copper sulfides, but also pyrite (Woodcock et al., 2007). The gold then can occur as native gold, gold associated with the copper sulfides and gold associated with pyrite. Again, common practice would dictate the use of a gravity concentration (and/or flash flotation) circuit prior to flotation to recover gravity recoverable native gold. Conventional collectors used may include xanthate, dithiophosphate and dithionocarbamate. Often a combination of collectors may be utilized, e.g., xanthate/dithiophosphate, dithionocarbamate/dithiophosphate or xanthate/ dithionocarbamate. To ensure a selective flotation separation is achieved, a flotation pH > 9.5-10 or even higher is usually utilized to depress most of the pyrite during rougher flotation.

## GANGUE DEPRESSION

Gangue minerals, or minerals that provide no economic value to the process, occur in two broad categories. The first category is sulfide minerals with pyrite and pyrrhotite being the most prevalent, although there are a host of other sulfide minerals that may either add no economic value to the process or result in higher operating costs. Examples are minerals that contain elements such as bismuth, antimony, selenium or mercury and which will incur penalty costs if their concentration in a copper concentrate is too high. The second major category is non-sulfide gangue (NSG), that can be further sub-divided into silica-bearing minerals such as quartz and a host of silicates (talc, various clays such as sericite and kaolinite), and carbon bearing minerals such as organic carbon and a variety of carbonate minerals.

The term "gangue depression" refers to the intentional management of the flotation system so as to impede or eliminate the flotation of the gangue during recovery of the value minerals. Gangue depression in the treatment of gold ores is important for several reasons:

1. It is imperative to depress (reject) both sulfide and NSG gangue during flotation of copper/gold ores in order to produce a final copper concentrate that is of suitable grade for marketing or subsequent processing. Typical NSG present in gold ores can include quartz, talc, clays, etc., whereas the most prevalent sulfide gangue is typically pyrite. Both aspects of gangue depression are present at the Phoenix Mine.
2. When producing concentrates for subsequent hydrometallurgical treatment the depression of gangue can help reduce the amount of species that might, for example, act as acid consumers or that might impede liquid/solids separation. For example, in the pressure oxidation of a gold bearing sulfide concentrate a high concentration of silica (in various forms) can lead to slurries that are difficult to flocculate and thicken. This was of concern during recent process development of feed materials from Newmont's Yanacocha District.

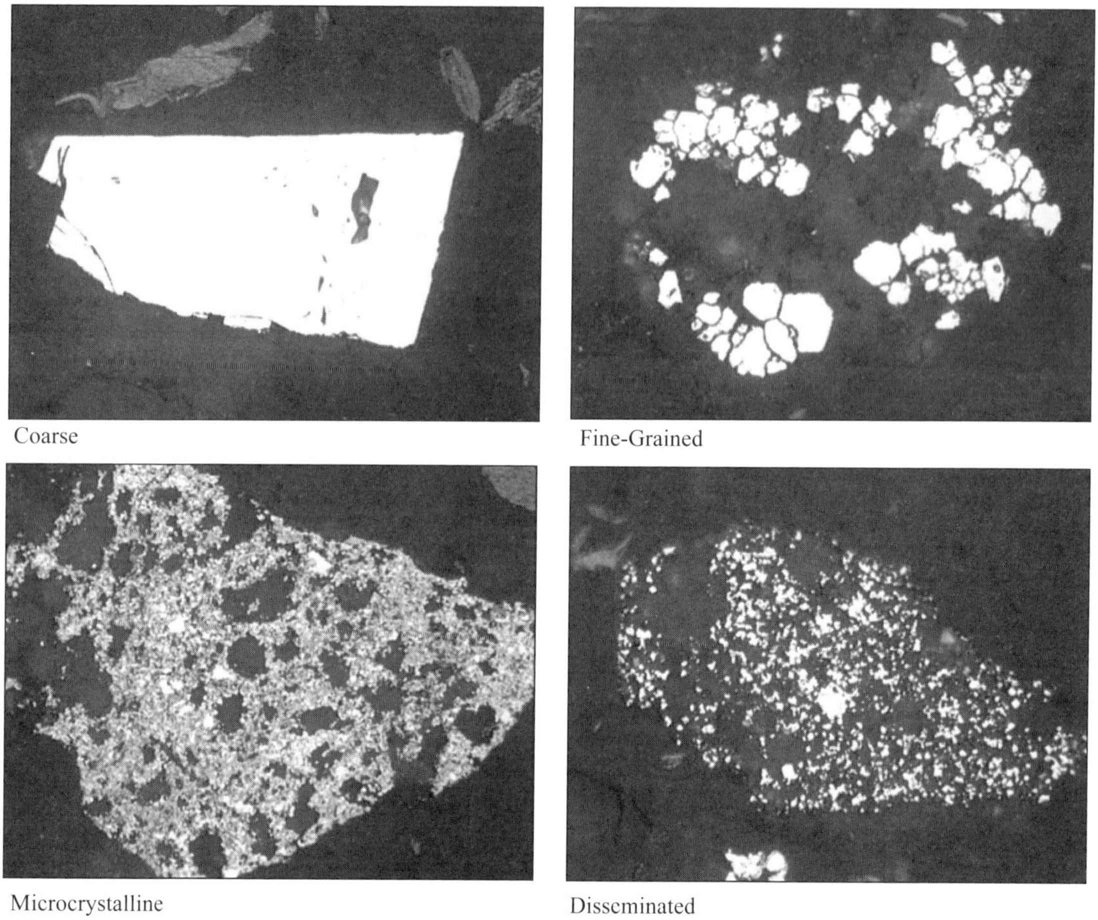


**Figure 1. Morphological Types of Pyrite (AMTEL Report, 2007)**

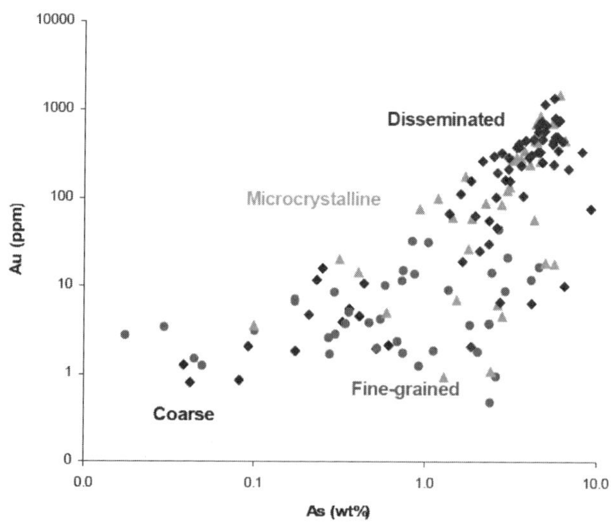


**Figure 2. Typical Gold (in ppm) and Arsenic Concentration (in wt.%) in Carlin-trend Pyrite (AMTEL Report, 2007).**

3. The depression of gangue cannot be ignored even when producing concentrates to be used for pyrometallurgical treatment. For example, the rejection of gangue can provide a roaster feed that has better materials handling characteristics. An example is Mill 5 at Newmont's Carlin operation where better rejection of gangue would help improve downstream processing prior to roasting.

**Reagent Options**

The options that are available for the depression of gangue during the treatment of gold ores can be quite varied but generally fall within one of the following broad groups of reagents:

- Sodium silicate is a well known reagent which is typically used for the depression of quartz (or other siliceous minerals) and is available in many forms and grades. It can act as either a depressant or as a dispersant. In the present discussion no differentiation is made between these two types of responses since they ultimately help to achieve the same goal. Dosages rates can vary from low (<250 gpt) to high (>1000 gpt). Overdosing of this reagent will lead to brittle froth which can, in turn lead to reduced recoveries.
- Carboxymethylcellulose (CMC) is an anionic water soluble polymer which is a derivative of cellulose. The functional properties of CMC depend on the degree of substitution of the cellulose structure and on the chain length of the cellulose backbone structure. The types of CMC products available to the flotation engineer can range in grade (typically measured as viscosity) and quality. The CMC products are used for the depression of talc, serpentine, chlorite, pyrophyllite and other minerals with similar flotation properties.
- Starches offer the most diversity and probably the greatest degree of complexity in their selection for use in flotation. They can be derived from a multitude of sources but the ones most commonly used are derived from corn. Their applications can vary significantly but in the case of gold ores they are useful for the depression of organic carbon and some NSG.
- Synthetic polymeric depressants have come into use in recent years and offer a new and sometimes very useful tool for the depression of gangue. These products can be applied to the depression of both NSG and sulfide gangue depending on their functional groups. They are typically more restricted in their applications than the other depressants and for this reason their testing and utilization can be difficult.
- The most common depressant for sulfide gangue (pyrite) depression is lime. It is generally very effective by itself but it is not unusual to use it in conjunction with other depressants such as cyanide, sodium sulfite or other reagents.
- A "reagent" that is often overlooked and little appreciated is water. High slurry densities are beneficial for minimizing plant equipment sizes but the addition of water (lowering slurry densities) can be extremely useful for controlling the rejection of gangue. Process development and ultimate plant design needs to account for this option as it is often the most cost effective option.

**Application of Gangue Depression**

When selecting the "right" depressant the flotation engineer should have a full understanding of not only the mineralogy but also the dynamics within the particular flotation process.

-
- For example, the presence of talc or pyrophyllite will undoubtedly lead to the selection of CMC or possibly a starch to achieve their depression. Like most depressants the CMC's are susceptible to overdosing, sometimes within a very small range of addition rates. The overdosing can, for example, be manifested by the generation of a brittle froth which can severely impact the flotation process. Overdosing may also lead to the depression of the sulfide minerals of interest (i.e., copper sulfides) and too much starch or CMC has been found to depress the flotation of free gold. Present plant testing at Newmont's Phoenix mine has indicated that a change by as little as 20 to 30 gpt of CMC can determine whether or not its desired effect is achieved, and whether adverse effects are set into motion.
- The selection and application of a given depressant can be impacted by the system's pH although in most cases the working range is generally rather forgiving. A case in point is the use of lime as a depressant for pyrite. At the Phoenix mine the use of lime produces pH 11.5 in the cleaner circuit but pyrite depression can still be a problem. By comparison, the pyrite depression in sulfide ores from Newmont's Yanacocha District can be very sensitive to lime additions and at pH 10.5 changes of as little as 0.5 pH units can significantly change the pyrite's flotation response. It should be noted that the final copper concentrate at the Phoenix mine contain gold values that are predominantly free gold, so the use of high lime has not been noted to depress the free gold.
- Conditioning time is typically short for depressants such as CMC or sodium silicate but some conditioning is always beneficial. The depressing action of some reagents can diminish with time and stage additions are the most effective for optimum metallurgy.
- Newmont has achieved very interesting results using of the synthetic polymers and has found that they offer applications that are "outside the box." However, they must be used with caution since small changes in the dosage rates can have very deleterious effects on the metal recoveries.
- The locking of values (typically as sulfides) at very fine sizes (<20μm) with gangue is not untypical for many of Newmont's ores. In such cases an aggressive system for depression of the gangue will result in significant values losses. As a result there needs to be a balance in the operating system to achieve optimum metallurgy. An example is Newmont's Mill 5 operation at Carlin where a significant portion of the gold is contained in very fine arsenian pyrite (<10μm) that is locked with the gangue (siliceous and carbonate). It is not practical to liberate

these sulfide values so their recovery as middlings is imperative. Therefore, a very aggressive depression of the gangue would lead to value losses.

- It is not unusual for a depressant to have an effect downstream of the flotation process. For example, depending on the depressant, the liquid/solids separation characteristics of the concentrate and/or tails can be affected either in a good or a bad way. An example is flocculation which is typically aided by the use of lime but impeded if the pH is modified by caustic or soda ash.
- In the selection of any depressant it is important to use a product from a vendor who will supply a consistent and reliable product. Trying to resolve problems that are due to changes in reagent quality is not an efficient use of the scarce resources often present in a plant operation.
- The selection of the reagent cannot always be based on costs. A low grade and less expensive reagent can often require higher dosages and be less cost effective than a more expensive grade with a higher activity that can be used at lower dosages.

## CASE STUDIES—NEWMONT FLOTATION MILLS/PROJECTS

### Callie Deeps

The Callie Deeps deposit is located in the Tanami Desert in the Northern Territory approximately 500 km north west of Alice Springs, Australia. The Tanami underground mine is the largest underground gold mining operation in Australia. The deposit contains coarse gravity-recoverable gold and the so-called "nugget-effect" is very prevalent. Currently, ore at the mine is being processed using gravity separation followed by cyanidation. The project is currently undergoing a review to establish various processing opportunities that may exist by including flotation of the gravity tail.

The flotation approach utilizes two collectors, PAX and a mercaptobenzothiazole/dithiophosphate blend at natural pH. The objective in flotation is to recover all the native gold (together with any free milling sulfide gold-bearing minerals) that was not recovered during the gravity concentration stage. The resulting flotation concentrate will ultimately be leached with cyanide.

### KCGM

Kalgoorlie Consolidated Gold Mines (KCGM) is a joint venture between Barrick Gold and Newmont Mining Corporation, located in Kalgoorlie, Western Australia. Gold is primarily associated with pyrite, but there is also some fine native gold and gold tellurides. The gold concentrate produced is dried and roasted prior to cyanidation. The flotation tailings are also leached with cyanide using CIP (carbon-in-pulp). Recently, ultrafine grinding of the flotation concentrate was introduced, followed by agitated cyanide leaching of the ground product, minimizing the quantity of sulfide concentrates that need to be roasted (Marsden and House, 2006).

Since most of the gold in these ores is associated with pyrite, a fairly strong flotation reagent scheme is utilized. Potassium amyl xanthate is used together with copper sulfate at natural pH (~8.0-8.5) to ensure adequate bulk sulfide flotation is achieved.

### Hope Bay

Newmont Mining Corporation recently acquired the Hope Bay project located in Nunavut, Canada. Gold in this orebody is primarily native and associated with sulfide minerals. In most instances the gold-bearing sulfides is considered free-milling (amenable to cyanidation) although there appears to be some refractory gold-bearing sulfides present in certain areas in the deposit.

Since the majority of these gold-bearing sulfides are free-milling, the main objective in the flotation process is to recover the majority of the native gold and gold-bearing minerals. The current process flowsheet also includes a gravity concentration stage prior to flotation, since the gravity recoverable gold content of these ores can be quite significant. Rougher flotation is carried out at natural pH (~8.0) and a combination of PAX and diethyl dithiophosphate/di-sec-butyl monothiophosphate is used as collectors. The purpose of the xanthate is to ensure that adequate bulk sulfide flotation is achieved, while the diethyl dithiophosphate/di-sec-butyl monothiophosphate blended collector is added to assist native gold recovery.

### Lone Tree

Gold at Lone Tree occurs primarily in the form of arsenian pyrite that results in poor flotation response by conventional air flotation. The Lone Tree Flotation Mill, located near Battle Mountain, Nevada, was operated from 1997 through 2007 using the $N_2$TEC™ flotation technology to produce a sulfide concentrate for subsequent oxidation and gold extraction. The mill was designed to grind 4,085 mt/d of ore to on the order of 80% passing 270 mesh and produce 494 tpd of gold bearing sulfide concentrate which was subsequently treated by pressure oxidation on site, or was shipped to other Newmont autoclave or roasting facilities in Nevada. Remaining cyanide soluble gold present in the flotation tailing was treated by CIL cyanidation. The $N_2$TEC™ technology was implemented at Lone Tree by covering the flotation cells and recycling the captured gases along with additional nitrogen obtained from the on-site oxygen plant. Primary reagents were potassium amyl xanthate and lead nitrate, and sulfuric acid addition to a pH of 5.5.

**Carlin**

The Carlin-trend ores have a similar mineralogy to the Lone Tree ores, with gold commonly associated with arsenian pyrite (pyrite enriched in arsenic), as well as arsenopyrite, and pyrite. A sulfide flotation circuit was retrofit into Carlin Mill #5 in 2005, designed to treat blends of whole ore bio-oxidation residue, oxide and sulfide ores. Bio-oxidation was used to pre-treat the readily oxidized arsenian pyrite, while gold occurring with other sulfides was recovered by sulfide flotation with PAX as the primary collector and using lead nitrate as an activator. The circuit consists of rougher and cleaner flotation at a pH of 5.5-6.0 (maintained with sulfuric acid addition). Flotation tailings are subsequently cyanide leached to recover gold associated with the oxidized pyrites. A later change in the reagent scheme was been made to include an auxiliary collector designated as Cytec-blended reagent 6205.

One of the biggest challenges of flotation at Carlin is dealing with the varieties of current and future ore types and mineralogies. In late 2008, the bio-oxidation pads were shut-down resulting in increased acid consumption and higher mass recovery of sulfides. The ore mineralogy is further complicated by increasing amounts of carbonate minerals (calcite/dolomite) which impacts the ability of the mill operations to control the pH at adequate levels (<6.0) when present in high quantities (~15.0 wt.%). Additionally, the high carbonate-containing ore types seem to contain a greater proportion of the refractory gold-bearing sulfides as finely disseminated material, requiring larger quantities of flotation concentrate mass recovery, which impacts (negatively) downstream processing (thickening and filtration). It is also considered that the carbonate containing minerals generate slimes which may also negatively impact the flotation kinetics of the gold-bearing sulfide minerals. Research into ways of improving recovery with these high carbonate ores includes the possible use of a CMC depressant or a dispersant such as sodium silicate, which might mitigate the impact of slimes during flotation. The use of the $N_2$TEC™ flotation technology has also shown a benefit of between 6-8% overall gold recovery depending on the ore type.

**Batu Hijau**

Batu-Hijau is a classic porphyry complex with bornite and chalcopyrite as the dominant copper minerals and host of the gold mineralization (De Mull et al., 2002). Mill operation began in 2001 with a flotation circuit consisting of rougher flotation and three stages of conventional cleaners producing copper concentrates grading from 26%-33%Cu. Rougher flotation is carried out at an alkaline pH (>9.5) to minimize pyrite flotation using a combination of PAX and a blended collector containing dithiophosphate/mercaptobenzothiazole. The flowsheet is unique in that the process water is a blend of sea water and site run-off water. Batu Hijau has also implemented controlled potential sulfidization with sodium hydrosulfide to improve recoveries when treating ores that have oxidized on stockpiles. Annual recoveries of copper and gold have ranged from 80.6%-89.2%Cu and 75.2-81.9%Au from ore grading 0.47%-0.75% Cu and 0.32-0.62 g/t Au.

**Boddington**

The Boddington copper-gold deposit is located near Perth in Western Australia and is currently scheduled for plant start-up in mid-2009. In this deposit gold occurs as native gold, gold associated with copper minerals (chalcopyrite, bornite and cubanite) and free-milling gold-bearing sulfide ores. A gravity concentration stage is implemented prior to flotation. In flotation the circuit consists of rougher flotation utilizing a dithionocarbamate collector under alkaline conditions with lime (pH ~10) to selectively produce a copper concentrate. Subsequently, copper sulfate is added (activator) together with PAX to produce a scavenger concentrate. The cleaner scavenger tail is leached with cyanide for free-milling gold recovery.

**Phoenix**

The ores treated at the Phoenix Mine, located near Battle Mountain, Nevada, are primarily Cu/Au skarns with relatively low average ore grades (0.15%Cu, 7.5 g/t Ag and 0.7 g/t Au). Phoenix ore types are primarily chalcopyrite dominant sulfides; however, there are significant quantities of chalcocite/covellite in some ores as well as oxide copper minerals in transition and oxide hosted ores. Gold at Phoenix occurs in association with iron and copper sulfides with significant and variable quantities of gravity recoverable gold. The particle size of the gold is also highly variable ranging from <20um to as coarse as +150um.

The Phoenix flowsheet was designed to accommodate highly variable ore mineralogy and gold associations and includes gravity, flotation and leaching of the flotation tailings. The gravity circuit consisting of Knelson centrifugal gravity concentrators produces a gravity concentrate that is subsequently leached using a high intensity cyanidation (HIC) process. Because of the significant amounts of relatively coarse free gold, a flash flotation cell was added to float the Knelson gravity tailing with the concentrate pumped to cleaner flotation. The flotation circuit consists of rougher flotation at pH 10.5 with lime followed by regrinding and two-stage cleaning at a pH of 11.5. The flotation reagents are generally selective copper collectors to produce a copper concentrate containing from 15%-25%Cu. The flotation tailings are subsequently cycloned to remove cyanide soluble copper, and treated by cyanidation.

Flotation performance is significantly affected by varying gangue mineralogy, most notably pyrrhotite, talc, and sericite. The use of CMC as a gangue depressant for talc and possibly sericite is being evaluated. Varying pyrrhotite content has been noted to adversely affect flotation recovery and froth quality, and the addition of the primary collector to the SAG mill appears to somewhat mitigate this effect.

## REFERENCES


AMTEL Report, 2007, "Gold Deportment in Tailings of the Bio-Oxidation/Flotation/CIL Carlin Plant: Issues & Opportunities," *AMTEL Report 07/07*.

De Mull, T.J., Spenceley, J. and Hickey, P., 2002, "Batu Hijau: From Discovery to Production," Mining Engineering.

Gathje, J.C. and Simmons, G.L., 1997, U.S. Patent 5,653,945.

Marsden, J. O., House, C.I., 2006, "The Chemistry of Gold Extraction," SME Inc. 651 pp.

Miller, J.D., Du Plessis, R., Kotlyar, D.G., Zhu, X. and Simmons, G.L., 2000, "The Low Potential Hydrophobis State of Pyrite in Amyl Xanthate Flotation with Nitrogen," *SME Preprint* 00-113.

Simmons, G.L. and Gathje, J.C., 1998, U.S. Patent 5,837,210.

Woodcock, J.T., Sparrow, G.J. and Brickard, W.J., 2007, "Flotation of Precious Metals and Their Minerals," *Froth Flotation: A Century of Innovation*, SME Inc. 891 pp.

# DESIGNING MODERN FLOTATION CIRCUITS USING JKFIT AND JKSIMFLOAT

D.A. Collins, S. Schwarz, and D.J. Alexander

JKTech Pty Ltd, 40 Isles Rd, Indooroopilly, Queensland, Australia

## ABSTRACT

Determining the flotation response of ore types from small scale flotation tests has traditionally been the "holy grail" of the flotation industry. Modeling methodologies proposed by AMIRA P9 and commercialized through JKTech is widely accepted as the best flotation optimization technology available but is limited to existing operating plants. The floatability of ore types prior to processing cannot be characterized using this method unless a pilot campaign is undertaken.

After revision of large floatability data sets from existing industrial operations, a new methodology was proposed that involves a laboratory based flotation test procedure along with parameters obtained from a floatability database or a plant survey campaign to characterize the floatability of new ore types. The full methodology is labeled the floatability index test procedure (JKFIT).

This paper outlines the JKFIT procedure and the results of applying this methodology to a design study for a silver/lead/zinc ore. JKFIT was applied to many different future feed types forecast for processing in the expansion of an existing operation. The results were used to predict the different performance in various circuit designs and configurations.

The JKFIT test has the potential to become the standard test of flotation because it allows full ore body characterization and prediction of key grades and recoveries before the ore enters the plant. When used in conjunction with JKSimFloat, the JKFIT test will provide a powerful tool for plant design and improve the confidence in flowsheet layout before final decision on project specifications is reached.

## INTRODUCTION

The JK approach to simulating circuits has always been to decouple machine characteristics from ore characteristics and has spawned many very successful computer simulation software packages; most notably JKSimMet (McKee & Napier-Munn 1990) and JKSimFloat (Schwarz and Alexander 2006a; Schwarz et al 2006; Schwarz and Alexander 2007). Machine characteristics are easily measureable and readily available in the database of parameters on record with JKTech (Schwarz and Alexander 2006b). Ore characteristics need to be determined for individual ore bodies and breakage tests have been available some time, from the batch wise original Bond Work Index tests to the new multi-particle test (Shi et al 2008).

JKSimMet has been used very successfully in designing comminution circuits using this type of methodology (Morrison and Richardson 2002). The problem in applying this approach to flotation design has always been the shortfall in flotation techniques available to determine ore floatability. In the past it has been thought sufficient to run bench scale float tests and apply the heuristic method of multiplying residence times by two and a half (or a similar factor) (Lane et al 2005). Those with further time and resources may have completed locked-cycle tests and those with the luxury of large amounts of sample material may have undertaken the costly practice of running a pilot plant. The flow sheets evolving out of these techniques were limited to the experience of the engineer commissioning the work. Rarely these techniques have utilized the flexibility of computer simulation available today through packages such as JKSimFloat. The only problem remaining is how to determine the floatability parameters.

JKTech has been undertaking campaigns for many years to determine the floatability of ores from existing plants. These floatability modeling campaigns have been conducted on many different ore bodies and ore types on various circuits in many countries around the world (Alexander et al 2000; Alexander and Wigley 2003; Schwarz and Kilgariff 2005; Alexander et al 2005; Coleman et al 2006; Coleman et al 2007; Schwarz et al 2008). In conjunction with this growing database of knowledge, a test, and more importantly, a methodology was conceived to allow the floatability parameters to be established for new ore types yet to pass through these float plants (Alexander and Collins 2008). It stands to reason that if parameters can be obtained from a database of measured machine parameters (Schwarz and Alexander 2006), then the flotation response of any ore can be simulated if the floatability parameters are determined by this new technique.

## FLOATABILITY

Floatability is defined as the amenability of attachment between mineral particles and air bubbles. It is important to note that it is not a flotation rate constant, however, can be used to estimate the rate of the flotation process by using the following equation (Alexander et al 2000):

$$k = P.\ S_b.\ R_f \tag{1}$$

where: $k$ = overall flotation rate constant (min$^{-1}$)
$P$ = ore floatability (-)
$S_b$ = bubble surface area flux (min$^{-1}$)
$R_f$ = froth recovery factor (fractional but often expressed as a percentage)

Although, there is a great deal of research being undertaken in the area of surface chemistry with the ability to measure hydrophobicity and induction times, these measurements alone are not providing a full understanding of floatability. Unfortunately, there is no technique at present for direct physical measurement of the floatability of particles.

There is, however, a technique to estimate the feed stream floatability indirectly. In the method developed by the AMIRA[1] P9 project and described by Harris (1997), floatability component flotation tests are undertaken on as many streams as possible around a circuit and the floatability distribution is regressed from these and survey results using a combination of linear (Alexander and Morrison 1998) and non-linear optimization procedures (Harris 1997). This methodology assumes that the floatability components in the cell, bank or circuit are conserved. In other words, the total mass of material with floatability $P_i$ in the concentrate and tail of a flotation cell, bank or circuit must be equal to the total mass of material with floatability $P_i$ in the feed. This assumption has been shown to be applicable in many plants, (Runge et al 1997; Alexander et al 2000; Alexander and Wigley 2003; Schwarz and Kilgariff 2005).

The current industrial approach to characterizing the flotation response of greenfield ore types where there is no existing plant in production is by one of these three methods:

1. Laboratory Rate Test
2. Laboratory Locked Cycle Test
3. Pilot Plant Test

The list is in order of increasing reliability and confidence but also increasing cost and time requirements. The laboratory rate test is the most common method to characterize new ore types as access to reliable pilot plants and the large mass of sample required to perform these, and the locked-cycle tests, often makes them prohibitive. The rate test involves taking timed concentrate samples and analyzing the elemental content of the concentrates and the remaining tail. From this data, elemental recovery versus time curves can be generated (see Figure 1).

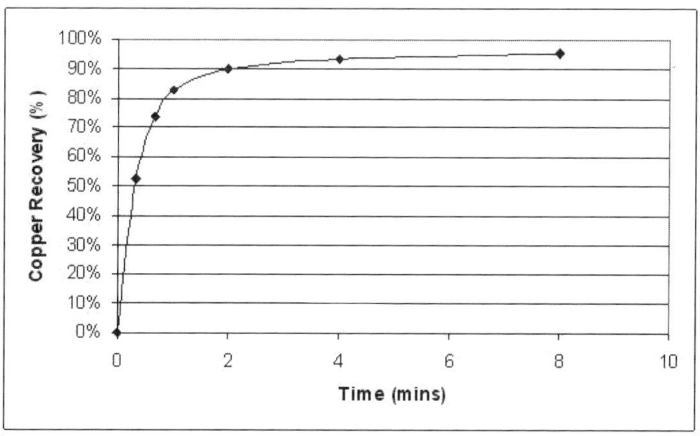


**Figure 1. A typical elemental recovery time profile generated from a laboratory rate test**

To determine full scale prediction from this test, a design engineer will use various "rules of thumb" (Lane et al 2005). The most common rule of thumb is to determine the required plant recovery (for example 95% in Figure 1) and multiply the time at this recovery by a general factor (typically 2.5) to determine the required rougher bank residence time (about 20 minutes in the case of Figure 1). The methodology has the advantage that it is simple and requires low sample mass. However it does not provide any information on the likely cleaner performance, especially re-circulating loads, and is often inaccurate when predicting the recovery of the other elements in the system, particularly non-sulphide gangue materials and therefore inaccurate at predicting grades. Hence, a new test was required which was also simple but permitted better prediction of the full metallurgical response.

**The JK Floatability Index Test (JKFIT)**

The JKFIT methodology includes calibration of the bench scale floatability parameters with known plant scale parameters which allows these values to be used in the JKSimFloat steady state simulator. The JKFIT methodology incorporates the following features:

- the bubble surface area flux scale-up analysis (Gorain et al 1997);
- the current industrial scale floatability data-base, including the floatability proportionality constant values
- the assumption that material can be lumped into three flotation classes namely fast, slow and non floating
- the floatability index calibration;
- size and mineralogical liberation analysis.

[1] Australian Mineral Industries Research Association

Gorain et al (1997) proposed that the flotation rate constant was linearly related to the bubble surface area flux (see Figure 2 below). It was also proposed that the slope of this relationship is related to the ore floatability. Furthermore, Gorain et al (1997) also showed that scale up from small scale cells (250 liter) to larger scale cells (3 m$^3$) using a constant ore floatability was possible. This was confirmed in a separate study at BHP Billiton Mount Keith Nickel Operations by Alexander et al (2000) where scale up of the flotation performance from a 60 liter cell to a 100 m$^3$ was shown.

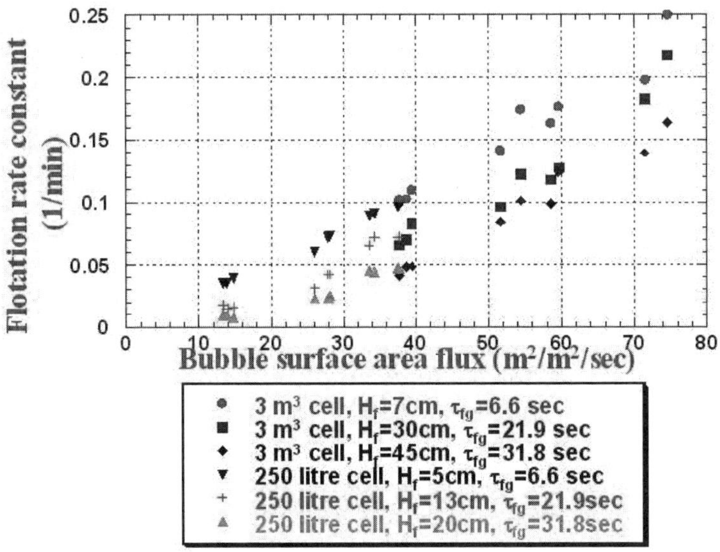


**Figure 2. Flotation Rate Constant versus Bubble Surface Area Flux (after Gorain et al, 1997)**

In small laboratory cells, varying the impeller speed can greatly increase the kinetics of the flotation system, much more than the response of the increase in bubble surface area flux in the system. In the JKFIT method, impeller speed is kept constant and only air rate is changed to alter the bubble surface area flux. This produces a linear relationship between flotation rate and bubble surface area flux.

Concentrates are collected for each air rate to determine recovery over time. From these samples, elemental assay, size by elemental assay and MLA analysis (if available) is analyzed. Recovery at each time can then be determined for critical minerals and the flotation rate calculated using the following plug flow equation:

$$R = 1 - e^{-k\tau} \qquad (2)$$

where: $R$ = overall mineral recovery
$k$ = flotation rate constant
$\tau$ = concentrate time (min)

An estimate of the floatability can then be attained for each component (fast, slow and non-floating), mineral, size class and liberation class (where data is available) by least squares regression fitting of Equation 1, see Figure 2. It is not necessary that the resulting floatability values accurately represent the real floatability of the system but that they are reproducible and can be calibrated with known floatability parameters of actual plants. The floatability parameters can then be compared to existing values in the database to determine plant scale floatability values.

Once full scale floatability parameters are determined, JKSimFloat can then be used to estimate the flotation response of the full scale ore. This simulation can be enhanced with information from grinding ore characterization tests (e.g., drop weight test) and JKSimMet. In addition to prediction of the flotation response, the engineer also has the ability in JKSimFloat to manipulate the flowsheet and optimize the performance of the flotation circuit before any final decision on project specifications is reached.

**Definition of JK Floatability Index (JKFI)**

Floatability is the probability of a particle reporting to the froth phase of the flotation process and is affected by the particles amenability to attachment and the possibility of detachment at any time in the pulp phase. In the definition of the JKFI, particles with similar floatability are grouped together, consistent with the floatability component model. The average floatability of all particles in the component group defines the floatability of the group as a whole.

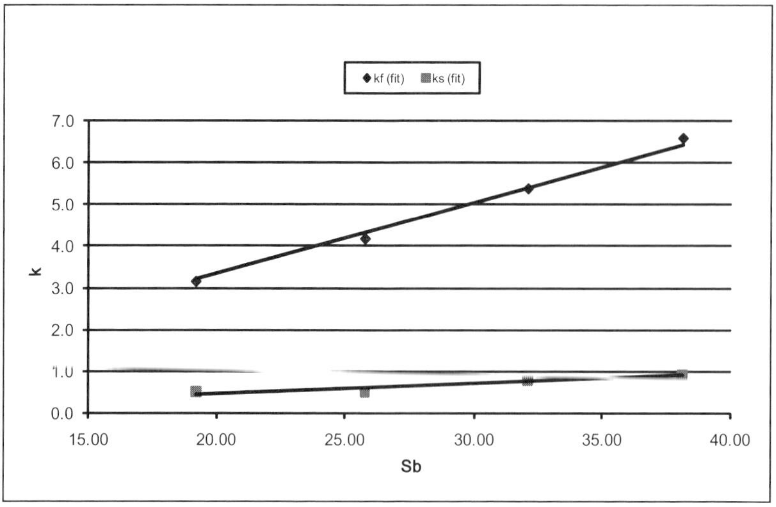


**Figure 3. Typical k vs. Sb regression**

The JKFI is defined as:

$$JKFI = (m_f + m_s).100000 \cdot \sum_{i=1}^{n} (m_i \cdot P_i) \tag{3}$$

The JKFI calculation takes into account the amount of non-floating material ($m_{nf}$) in the ore as well as the contributions of both the $P$'s and $m$'s of the floatable material. The total sum is multiplied by a factor of 100 000 to create whole numbers from decimal numbers for easier reading.

**Harmonic Mean**

When the floatability of the fast floating component, $P_f$, is of orders of magnitude larger than that of the slow floating component, $P_s$, it has an overriding influence on the value of JKFI. Large changes in $P_s$ or $m_s$ make very small changes to the index. The reason for using a harmonic mean to represent the ores floatability characteristics is its higher dependence on the lower value $P$, that is, the slow floating fraction (Collins 2007).

$$JKFI^* = 100000 \cdot \frac{(m_f + m_s)}{\sum_{i=1}^{n} ({}^{m_i}/_{P_i})} \tag{4}$$

The JKFI values can enable plant personnel to rank the floatability of various ore types, reagent regimes, grind sizes as well as benchmark against other ore types around the world.

## CASE STUDY

A study was conducted at a silver/lead/zinc mine to enable the response of different ore types to flotation to be determined. A full survey campaign was performed across all lead and zinc circuits to develop the floatability component model by size according to the standard JKTech flotation methodology, including measurements of cell characteristics such as bubble size, superficial gas velocity, gas hold-up, froth recovery, entrainment, residence time and water recovery. A sample of the SAG mill feed was also obtained at the same time as the survey campaign and sent to the JK Centre in Brisbane, Australia for JKFI testing. The floatability parameters obtained for the plant-scale results were compared with the floatability parameters obtained from the JKFI test and calibrated accordingly. A summary of the calibration results is given in Table 1.

In general, the JKFI resulting from the plant is higher than in the laboratory. As stated above, however, the actual numbers from the laboratory do not necessarily have to represent the floatability in the plant as long as they are reproducible and can be calibrated to the plant. The calibration involves determining the laboratory-scale floatability values ($P$) from the plant-scale floatability values, maintaining the proportion of each floatability component as determined in the laboratory. This gives a single scale-up number for each mineral, which can then be used to scale the laboratory floatability parameters for other ore types to be used in the calibrated JKSimFloat simulator.

**Table 1. JK Floatability Index results for plant and laboratory scale test**

| | Lead circuit | | Zinc circuit | |
|---|---|---|---|---|
| | JKFI (plant) | JKFI (lab) | JKFI (plant) | JKFI (lab) |
| Galena | 863 | 218 | 41 | 15 |
| Sphalerite | 2355 | 127 | 304 | 33 |
| Iron sulphides | 61 | 3 | 2 | 0.08 |
| Silver minerals | 1497 | 484 | 43 | 6 |
| Gangue | 5 | 0.6 | 3 | 0.04 |

**Table 2. Results from lead circuit simulations for different ore types**

| | JKFI* | Galena Grade (%) | Pb circuit Recovery (%) |
|---|---|---|---|
| Calibration | 545 | 80.7 | 69.6 |
| A | 839 | 66.4 | 73.7 |
| B | 602 | 63.8 | 71.6 |
| C | 858 | 90.4 | 75.6 |
| D | 324 | 85.1 | 66.0 |
| E | 1143 | 92.2 | 74.5 |
| F | 873 | 85.1 | 66.0 |
| G | 1271 | 87.3 | 75.4 |
| H | 438 | 82.5 | 57.9 |
| I | 435 | 82.5 | 65.0 |
| J | 792 | 76.3 | 71.9 |

**Results From Ore Variability Tests**

JKFI tests were conducted on 10 different ore types, with the laboratory scale floatability values scaled to plant-scale according to the method as described above for both the lead and zinc circuits. The plant-scale floatability values were used in the JKSimFloat simulator to predict the performance of both lead and zinc circuits when processing the new ore types. The metallurgical results for galena from the lead flotation circuit simulations are given in Table 2. Note that the JKFI values presented in Table 2 are the harmonic mean scaled to the plant, as opposed to the values presented in Table 1. The slow floating fraction can be accounted for more accurately using the JKFI* values, as this will significantly impact the material re-circulating through the circuit.

In general, a higher JKFI* resulted in better metallurgical performance, as evidenced by ores C, E and G with higher galena recovery across the lead circuit. The opposite is true also, with lower JKFI* values resulting in lower recoveries, e.g., ores D, H and I. A closer investigation of the results from the laboratory JKFI test indicated these ores also contained more non-floating galena, which restricted the overall recovery of galena within this circuit.

While the mineral recovery can be relatively compared with the JKFI* value, the resulting concentrate grade is also impacted by the floatability of the gangue species (outside the scope of this paper). This is shown clearly with ores A, B, and J, where the JKFI* values were higher than the calibration sample and resulted in higher recoveries but lower concentrate grades. It was also noted that ores C and E produced higher JKFI* values and higher concentrate galena grades and recoveries.

Similarly, the zinc circuit results are shown in Table 3.

As discussed above for the lead circuit, a higher JKFI* value generally resulted in higher zinc circuit recoveries of sphalerite, as shown by ores E, G and I. For the ores with low JKFI * values and low circuit performance (A, B, D, F and J), they would be targeted for further investigation and methods to transfer the majority of slow floating sphalerite into faster floating material.

**Recommendations from JKFI Tests**

The results displayed above enable plant personnel to identify ores that may result in poorer metallurgical performance through the flotation circuit, and identify methods to improve the flotation performance. With the flexibility of the simulator in JKSimFloat, metallurgists can change the circuit configuration or operating conditions to best optimize the metallurgical performance across the flotation circuit based on the JKFI results, such as shutting some cells down for low grade ores, or increasing air rates/froth depths as necessary, to achieve the target grade specifications.

**Table 3. Results from zinc circuit simulations for different ore types**

| | JKFI* | Sphalerite Grade (%) | Zn circuit Recovery (%) |
|---|---|---|---|
| Calibration | 102 | 85.5 | 78.5 |
| A | 91 | 90.0 | 49.0 |
| B | 54 | 88.3 | 69.2 |
| C | 128 | 90.9 | 56.4 |
| D | 22 | 75.8 | 47.2 |
| E | 442 | 84.4 | 81.0 |
| F | 53 | 70.6 | 12.7 |
| G | 1665 | 96.1 | 85.7 |
| H | 348 | 74.0 | 70.4 |
| I | 947 | 80.4 | 83.0 |
| J | 37 | 90.7 | 62.6 |

From the results above, it is evident that some ores will be very amenable to flotation through the circuit, while some ores will need to be further evaluated as to the economic viability of treating. This could include mineralogical analysis to determine liberation characteristics of the desired minerals at various grind sizes as well as associations with gangue minerals. Other options would be to blend various ore types, aiming to take advantage of the higher performance in different areas. As blending is known to not follow a simple mathematical relationship in terms of flotation response, JKFI tests can be performed on various blends and compositions to generate the floatability parameters for use in further JKSimFloat simulations.

## CONCLUSION

The current methodology to determine the floatability parameters for a particular ore type is to survey an existing plant, measure cell characteristics and perform batch tests on streams within the circuit. The JK Floatability Index (JKFI) test has been developed to determine the response of new ore types to flotation prior to processing. The recovery vs time results from the simple laboratory flotation test are combined with scale-up parameters (either from existing plants or the JKTech database) to determine the floatability parameters in a plant-scale environment. Simulating the planned circuits in JKSimFloat enables optimization of operating conditions and circuit configuration for new ore types, reagent schemes and grind sizes.

This test is fast becoming the new standard in designing flotation circuits from drill-core and other samples, with positive comparative results from validation studies conducted around the world.

## REFERENCES


Alexander, D.J., and Collins, D.A. 2008. A methodology for measuring the floatability of ores: the Floatability Index Test. In *Proceedings of MetPlant 2008*, Perth, WA, Australia, August. The Australasian Institute of Mining and Metallurgy.

Alexander, D.J. and Wigley, P. 2003. Flotation Circuit Analysis at WMC Ltd. Olympic Dam Operation. In *Proceedings of the AusIMM 8th Mill Operators Conference*, Townsville, Australia, August. The Australasian Institute of Mining and Metallurgy.

Alexander, D.J., and Morrison, R.D. 1998. Rapid estimation of floatability components in industrial flotation plants. *Minerals Engineering*, 11, 133-143.

Alexander, D.J., Bilney, T., and Schwarz, S. 2005. Flotation Performance Improvement at Placer Dome Kanowna Belle Gold Mine. In *Proceedings of the 37th Annual Canadian Mill Operators Conference*, Ottawa, Canada, January.

Alexander, D.J., Runge, K.C., Franzidis, J.P. and Manlapig, E.V. 2000. The application of multi-component floatability models to full scale flotation circuits. In *Proceedings of the AusIMM 7th Mill Operators Conference*, Kalgoorlie, Australia, October. 167-178.

Coleman, R.G., Urtubia, H.E., Alexander, D.J. and Bailey, C.W. 2007. Integrated grinding and flotation simulations for the improvement of circuit performance at BHP-Billiton's Minera Escondida. In *Proceedings of the 39th Annual Canadian Mill Operators Conference*, Ottawa, Canada, January.

Coleman, R.G., Urtubia, H.E., and Alexander, D.J. 2006. A comparison of BHP-Billiton's Minera Escondida Flotation Concentrators. In *Proceedings of the 38th Annual Canadian Mill Operators Conference*, Ottawa, Canada, January.

Collins, D. A. 2007. Validation of the JK Floatability Index Test. *MSc Thesis*. The University of Queensland.

Gorain, B.K., Franzidis, J.P. and Manlapig, E.V. 1997. Studies on impeller type, impeller speed and air flow rate in an industrial flotation cell—part 4: effect of bubble surface area flux on flotation performance. *Minerals Engineering* 10, 367-379.

Harris, M.C. 1997. A practical framework for flotation circuit modeling and simulation. Paper presented at *SAIChE '97, the 8th National Meeting of the South African Institute of Chemical Engineers*, Cape Town, South Africa, 16-18 April.

Lane, G., Brindley, S, Green, S. and McLeod, D. 2005. Design and Engineering of Flotation Circuits (in Australia). In *Proceedings of the Centenary of Flotation Symposium*, Brisbane, Australia, The Australasian Institute of Mining and Metallurgy, 127-140.

McKee, D.J. and Napier-Munn, T.J. 1990. The status of comminution simulation in Australia. *Minerals Engineering*, 3(1-2), 7-21.

Morrison, R.D. and Richardson, J.M. 2002. JKSimMet: A simulator for analysis, optimization and design of comminution circuits. In *Mineral Processing Plant Design, Practice, and Control*, SME, Vol 1, 442-460

Runge, K.C., Frew, J., Harris, M.C. and Manlapig, E.V. 1997. Floatability of streams around the Cominco Red Dog Lead Cleaning Circuit. In *Proceedings of the AusIMM 6th Mill Operators Conference*, Australasian Institute of Mining and Metallurgy, Madang.

Schwarz, S. and Alexander, D. 2006a. Gas dispersion measurements in industrial flotation cells. *Minerals Engineering*, Vol 19, 554-560.

Schwarz, S. and Alexander, D. 2006b. JKSimFloat V6.1PLUS: Improving flotation circuit performance by simulation. In *Proceedings of the Mineral Process Modelling, Simulation and Control Conference*, Laurentian University, Sudbury, 35-47.

Schwarz, S. and Alexander, D. 2007. Optimisation of flotation circuits through simulation using JKSimFloat. In *Proceedings of the 33rd International Symposium on Application of Computers and Operations Research in the Mineral Industry (APCOM)*, Santiago, Chile, 461-466.

Schwarz, S. and Kilgariff, B. 2005. Flotation optimization studies of the Perilya Broken Hill lead circuit. In *Proceedings of the Centenary of Flotation Symposium*, Brisbane, Australia, The Australasian Institute of Mining and Metallurgy.

Schwarz, S., Alexander, D., Whiten, W.J., Franzidis, J.P. and Harris, M.C. 2006. JKSimFloat V6: Improving flotation circuit performance and understanding. In *Proceedings of the XXIII International Mineral Processing Congress*, Istanbul, Turkey, September 2006, 1717-1722.

Schwarz, S., Crnkovic, I. and Alexander, D. 2008. Flotation circuit analysis at Zinifex Century Mine. In *Proceedings of the 40th Annual Canadian Mill Operators Conference*, Ottawa, Canada, January.

Shi, F., Kojovic, T., Larbi-Bram, S. and Manlapig, E. 2008. Development of a new particle breakage characterization device—the JKRBT. In *Proceedings of the Comminution 2008 Conference*, Falmouth, Cornwall, June

# PURPOSEFUL INSTRUMENTATION FOR EFFECTIVE FLOTATION CONTROL

Jeff McKay[1], Olivier Guyot[2], and Alain Broussaud[2]

[1]Metso Cisa North America, 2513 W 11370 S South Jordan, UT 84095 USA
[2]Metso Cisa 3 Ave Claude Guillemin BP 6009 45060 Orleans France

## ABSTRACT

Advances made in instrumentation technology in recent years have opened the window to advanced control of flotation processes. However, in order for new instrumentation to be beneficial for process control, and thus provide economic improvement for the end user, the measurements provided need to result in control actions. Without this link, any technology falls short of its ultimate purpose and has no definable process control payback. This paper discusses some of the advanced flotation control sensing technology including vision, ultrasonics, laser, bubble size and air hold up measurements, on-stream analyzers, etc and relates these to advantageous process control uses or the lack thereof. Examples are given of currently used techniques and discussion of potential uses in the future.

## INTRODUCTION

Technology has moved forward at a rapid pace in recent years, leading to the development of many new types of flotation sensors or improvements on well proven types of sensor. Table 1 is provided as an introduction to what parameters we might wish to measure, why and what technology is available.

## SELECTING FLOTATION INSTRUMENTATION FOR PROCESS CONTROL

The most effective way to ensure that flotation instrumentation will be effectively utilized for process control is to determine how and why a measurement will be used for control. In essence, one must conceptually develop a control strategy.

The steps outlined below are key to making a good decision.

### Step 1 Determine what you CAN control first

By determining what it is you CAN adjust or control, you will greatly simplify your approach to control. Many process control problems appear daunting at first glance due to complexity of equipment or shear number of process variables but are reduced to simple logic upon review of what can really be done to change the process.

Next, ask yourself the following two questions:

- What would make you increase this variable? and
- What would make you decrease this variable?

Even if you CAN adjust something, does not mean you should regularly change it. For example, pH or eH (oxidation/ reduction potential); both of these play a critical role in flotation, and, indeed are controlled by addition of chemicals. However, there is generally no good reason to adjust the pH or eH setpoint as there is a defined and fixed optimum at which the process functions best for flotation separation. This is usually a function of ore type and without a strong indication that ore type has changed (which is generally not available without offline confirmation), there is no reason to change such a control parameter. Also, if you have a reason to increase or decrease something, you must also have a reason to do the opposite. If not, this tells us that we should just place the setpoint for this parameter at either its minimum or maximum and let it go. For example, if you make a decision to increase collector when mineral recovery is high, you must have logic that reduces collector under a different set of circumstances or all you have concluded is to run at a high fixed level.

### Step 2 Ask what you need to know to manipulate this variable

This is where the connection between valuable sensors and actions that bring benefit to the client is made. If a sensor is not necessary in this analysis, then it ultimately adds no benefit to process control. This may be because the necessary information is already available using another sensor and so the information is redundant. Sometimes, use of two sources of data may provide a more reliable control. For example, use of sump level and pump amps will be a more certain measurement of the circulating load in a flotation circuit than either one by itself. Once you have bridged the gap between measurement and control action, you have the basis for justifying a purposeful flotation sensor and getting benefit from its use. Without this, the sensor may provide some auxiliary use but its value will be tenuous at best.

If we apply this approach to several key flotation measurements, we quickly see those with the greatest value for process control.

**Table 1**

| Parameter | Need | Technique | Strength | Weakness |
|---|---|---|---|---|
| Level | | | | |
| Cell | Primary control variable for froth/pulp separation and pulling rate in self aspirating cells | 1. Bubble tube<br>2. Pressure or DP cell<br>3. Ultrasonic direct<br>4. Ultrasonic w/ target<br>5. Sonar/Radar | 1. Inexpensive, simple<br>2. Mechanically simple<br>3. Non-contact<br>4. Direct interface measurement<br>5. Non-contact/direct | 1. Density dependent<br>2. Density dependent<br>3. Affected by froth<br>4. Float shaft may stick<br>5. Interference from froth walls, more expensive |
| Sump/Tank | Critical indicator of downstream constraint | | | |
| Flowrate | | | | |
| (Gas-Air/N2) | Primary control of pulling rate in non-self aspirating cells | 1. Float/Variable area<br>2. Thermal mass<br>3. Vortex | 1. Inexpensive, simple<br>2. True mass press independent<br>3. Simple | 1. Pressure dependent<br>2. Sensitive to moisture<br>3. Limited turn down range |
| Slurry Tailings | Useful in defining dynamic mass balance and circuit physical constraints | 1. Doppler<br>2. Ultrasonic channel<br>3. Coriolis<br>4. Vibration/Cidra | 1. Simple—noninvasive<br>2. Non-contact<br>3. Can measure smaller flowrate<br>4. Non-Invasive, can handle aerated slurries | 1. Less reliable—affected by pipe type<br>2. Depends of geometry of channel and inferred velocity<br>3. Expensive, complex, plugging<br>4. More expensive, more complex |
| Slurry Concentrate | Useful in defining dynamic mass balance and circuit physical constraints | | | |
| Froth Velocity | | | | |
| Pulling Rate | Critical for optimization/control of plant concentrate production | Camera/Vision system | Simple technique, insensitive to lighting and conditions, very little maintenance | Several thousand / camera, placement and/or number important for success |
| Froth Height | | | | |
| | Useful when used with froth velocity in defining froth volume leaving cell | Ultrasonic sensor | Non-contact, inexpensive | Separate sensor |
| | | Laser coupled with vision | Non-contact, simple extension of vision system | Modification of camera mount required |
| Valve Position | | | | |
| Cell Dart | Useful in defining dynamic mass balance, feed-forward and circuit physical constraints | 1. Output signal<br>2. Proximity switch | 1. Very simple<br>2. Simple real indication | 1. Inferred measurement—can be wrong<br>2. Susceptible to alignment issues |
| Column Pinch | Useful in defining dynamic mass balance and circuit physical constraints | | | |
| Distributor Dart | Useful in defining dynamic mass balance and circuit physical constraints | | | |
| Amps | | | | |
| Pump Motor | Useful in defining dynamic mass balance and circuit physical constraints | Amp meter | Very simple and reliable | Inferred indication of load |

**Table 1 (Continued)**

| Parameter | Need | Technique | Strength | Weakness |
|---|---|---|---|---|
| Grade, % metal analysis | | | | |
| Conc | | | | |
| Final | • Critical in determining if we are making minimum(optimum) for contract profitability<br>• Useful in dynamic mass balance modeling | 1. On-stream analyzers<br>2. Model estimates<br>3. Vision based<br>a. Texture<br>b. Color | 1. Direct measurement<br>2. Less expensive to set up<br>3. Less expensive to set up | 1. Sensitive to sampling Complex devices –require frequent maintenance<br>2. Not direct measurement so not as trustworthy hard to maintain<br>3. Site specific and sensitive to light, etc. |
| Rougher | • Critical in determining if we are making minimum necessary to make acceptable final conc in cleaners<br>• Useful in feed forward control<br>• Critical in ratio control<br>• Useful in dynamic mass balance modeling | | | |
| Feed | | | | |
| | • Useful in feed forward control<br>• Critical in ratio control<br>• Useful in dynamic mass balance modeling | | | |
| Tailings | | | | |
| Rougher | • Not useful for process control as we always want lower | | | |
| Cleaner | • Useful in determining if cleaners are overloaded and need to pull harder | | | |
| Bubble Size | | | | |
| Cell | Useful in determining there is sufficient frother and/or bubbles are loaded. | 1. Camera/Vision system<br>2. Insitu vision based instruments | 1. Very simple, not very sensitive to lighting or conditions to get a good relative indication<br>2. Direct visual measurement | 1. Not sufficient by itself. Use with level, froth velocity, etc.<br>2. Not currently designed for continuous measurement. Designed for circuit analysis and process evaluation |
| Column | Useful in determining there is sufficient frother and/or bubbles are loaded. | | | |
| Air Hold Up | | | | |
| | Currently no value in process control as air hold up is a function of bubble size (which is a strong function of equipment design and therefore not adjustable and air rate which is already measured | Insitu vision based instruments | Direct visual measurement | Not currently designed for continuous measurement. Designed for circuit analysis and process evaluation |
| Froth Accoustics (JKMRC) C. Vanegas, P. Holtham / Minerals Engineering 21 (2008) 883–888 | | | | |
| | Currently no process control application as other sensors already provide similar data or no connection to a control action has been determined | Ultra sonic sensor based | Unique approach with promise | Not yet industrially proven, in research and beta testing |
| Bubble Load Measurement (JKMRC) D.R. Seaman et al. / Int. J. Miner. Process. 74 (2004) 1–13 | | | | |
| | Currently no process control application as other sensors already provide necessary data or no connection to a control action has been determined | Direct weight per volume sampling technique | Unique direct weight per volume sampling technique | Not continuous sensor. Tedious for real time control |

As background to better understand Table 2, the following explanation of function of each control parameter and the typical effect of increasing or decreasing each is provided.

1. Cell Dart Valve Position—Decreasing a flotation dart valve is generally understood to mean closing that valve. As the valve is progressively closed, it creates a restriction to the discharge flow from the cell and causes the pulp level to rise (froth depth decrease) and vice versa. This is the principle way a plant operator controls the operation of a flotation circuit. Typically this is done through control loops which automatically adjust the dart valve(s) further open or closed as necessary to maintain a desired pulp level or froth depth in the flotation cell.
2. Cell Aeration Rate—is generally adjustable through either manual valves which restrict air flow into the shaft of the flotation cell or through air flow control loops where an air flow SP is entered by the operator and the control loop manipulates the control valve to achieve the desired air flowrate. Aspirated flotation cells generally do not need or use control of aeration rate. Generally, up to a point of saturation, higher air flowrates translate to a higher bubble surface area flux and greater extraction rate of floatable material over the lip of the cell and vice versa. Beyond a maximum limit for any given cell, increasing aeration rate creates extremely large bubbles and the flotation cell begins to "burp." This is a very inefficient area of function for the flotation cell and generally produces inferior separations so experienced plant operators do not run their cells under such conditions.
3. pH or eH SP—Most sulfide flotation is run under basic pH conditions. The higher the pH, the more depressing the flotation environment becomes. Higher mineral recoveries are generally achieved at lower pH values, but the flotation is less selective. For most mineral/ore types, there exists an optimal pH which is generally the lowest pH where a minimally acceptable selective depression of gangue minerals is achieved. This is optimal from both a mineral recovery and reagent cost perspective. Likewise, eH or redox potential is used in some flotation processes such as recovery of MoS2 values from bulk copper concentrates in order to depress the other sulfide species. Again, the optimal eH is the minimum level where an acceptably clean concentrate can be produced. Increasing pH or running more negative eH generally depresses minerals more and reducing pH (in the basic range) or running a less negative eH generally depresses minerals less.
4. Frother SP—The role and function of frother is to produce a stable froth for the flotation separation process. This is achieved by reducing bubble size and stabilizing the formed bubbles. Often, frother effectiveness is a function of concentration in solution and one way to stabilize its effect is to control the parts per million of frother in the mineral slurry. In general, increasing a frother dosage produces finer bubbles and more stable froths while reducing frother dosage creates bigger bubbles and less stable froths. Some collectors have frothing capabilities and thus reduce the need for frother. Also, some frothers have collecting properties for some minerals. Different frothers also tend to linger or dissipate at much different rates so care should be taken when adjusting frother dosages to avoid long term problems in the flotation process.
5. Collector Dosage—Collectors work in a variety of ways including either chemical or physical adsorption. In general, once a sufficient amount has been added to achieve the necessary degree of hydrophobicity for the mineral to be floated, adding more does not achieve any benefit. One effective way to control this is to ratio the collector according to either bulk tonnage feeding the flotation process or tonnage of the target mineral or metal value feeding the flotation process. Up to the point of maximum hydrophobicity, increasing collector has a positive effect on mineral recovery. Beyond this point, the effect is either neutral or can cause a decrease in recovery if it adversely affects froth stability. Adjusting of collector dosage is definitely more complex than level, air or even frother dosage adjustments.
6. Cell Agitator Speed—Most all flotation cell agitators are run as fixed speed devices. The role of the agitator is to suspend the particles in slurry, aid in bubble/particle collision and to generate bubbles by shearing off air as it is introduced through the impeller. Lower agitator speeds are conducive to more quiescent froth phase conditions and thus higher concentrate grades, but generally achieve lower mineral recoveries. Higher agitator speeds result in greater energy applied to the formation of bubbles, more energetic bubble/particle contact and better suspension of particles. Users with either coarse material or harder to float material generally prefer faster agitator speeds and more energy. Users with finer material or simpler floating material generally prefer slower agitator speeds and less energetic flotation cells. Agitator speed can general be changed only by changing belts unless the motor is equipped with a variable frequency drive or VFD.

## APPLICATION TO VARIOUS CONTROL PARAMETERS

Table 2 shows the results of application of this approach to a number of key flotation control parameters.

**Table 2.**

| Control Parameter | Pertinent data to make control decision | Reason to Decrease | Reason to Increase |
|---|---|---|---|
| *Cell Dart Valve Position* | 1. Pulp level/Froth depth<br>2. Valve position<br>3. Downstream constraints<br>a. Sump levels<br>b. Pump amps<br>c. Slurry flowrates (if available)<br>d. Conc grade values<br>e. Froth velocity/ Pulling rate | 1. Cell level not high and<br>2. Cell not pulling pulp and<br>3. Valve not at minimum<br>4. Downstream constraints not a limit (roughers)<br>5. Upstream constraints at a limit (cleaners) | 1. Cell level not high or<br>2. Cell pulling pulp or<br>3. Downstream constraints at a limit (roughers)<br>4. Upstream constraints not at a limit |
| *Cell Aeration Rate* | 1. Pulp level/Froth depth<br>2. Control valve position<br>3. Air flowrate<br>4. Downstream constraints<br>a. Sump levels<br>b. Pump amps<br>c. Slurry flowrates (if available)<br>d. Conc grade values<br>e. Froth velocity/ Pulling rate | 1. Cell level not high and<br>2. Cell not pulling pulp and<br>3. Valve not at minimum<br>4. Downstream constraints not a limit (roughers)<br>5. Upstream constraints at a limit (cleaners) | 1. Cell level not high or<br>2. Cell pulling pulp or<br>3. Downstream constraints at a limit (roughers)<br>4. Upstream constraints not at a limit |
| *pH or Eh SP* | 1. Ore type | 1. Ore has significantly changed else maintain | 1. Ore has significantly changed else maintain |
| *Frother SP* | 1. Bubble size<br>2. Froth velocity<br>3. Feedrate/feed flowrate (feed forward action)<br>4. Downstream constraints<br>f. Sump levels<br>g. Pump amps<br>h. Slurry flowrates (if available) | 1. Bubble size coarse and<br>2. Downstream equipment not overloaded with froth<br>3. Froth depth shallow compared to froth velocity<br>4. Conc grade not too low | 1. Bubble size not fine and<br>2. Downstream equipment overloaded with froth<br>3. Froth depth not shallow compared to froth velocity |
| *Collector SP* | 1. Feedrate (feed forward control)<br>2. Feed grade (feed forward control)<br>3. Collector flowrate<br>4. Mineral recovery (grade) | 1. Ore type change<br>2. Feedrate increase<br>3. Feed grade increase<br>4. Very low recovery | 1. Ore type change<br>2. Feedrate decrease<br>3. Feed grade decrease<br>4. Very high recovery |
| *Cell Agitator Speed* | 1. Ore type<br>2. Mineral recovery(grades) | 1. Flotation needs more energy to achieve recovery | 1. Mineral floating well (high recovery) and therefore can work with less energy (generally not know) |

## EXAMPLES

### PTFI—Freeport Indonesia

Field trials instrumenting one row of flotation cells at Freeport Indonesia in 2002 indicated that froth velocity control of the flotation cell level setpoints (using a froth velocity profile to pull according to first order flotation kinetics down the row) resulted in significant metallurgical improvements. Results during the trial were a 2.6% to 3.9% Cu and 3.7% Au increase in recovery coupled with a 25% reduction in frother dosage. Based on these results, PTFI installed flotation cameras/froth velocity control on all 180 flotation cells at site. The froth velocity control resulted in a 51 day payback conservatively based on 1.5% improvement in copper recovery.

### BHP Escondida

A 102 camera VisioFroth™ system was installed at the BHP Escondida Phase IV concentrator near Antofagasta Chile in 2003. Expert control logic optimized the rougher and cleaner flotation circuits in addition to regrind and dewatering processes. A modified froth velocity profile was employed. Bubble size along with froth velocity and froth depth were used to control frother addition increases and concentrate sump loading and bubble size were used to determine when to reduce frother addition. Feed forward control was programmed into the DCS. A plantwide dynamic mass balance model was also created and used for estimating values pertinent to advanced control. Results from an official on/off statistical evaluation indicated that OCS© vision based expert control achieved an average increase in copper recovery of 1.6% with a net decrease in final concentrate (based on Escondida's desired way to control the cleaners) of 0.5%. Adjusting to bring concentrate grade to the same value yielded a net improvement of 1.0% in copper recovery.

### Northgate Kemess

The Northgate Kemess concentrator in Northern British Columbia, Canada, uses froth velocity from 22 wireless froth cameras to control a concentrate flowrate, measured by Cidra flowmeters. The OCS© expert system adjusts cell levels to pull the flotation cells in ratio to one another in order to maintain the desired total flowrate of copper concentrate.

### Teck Pogo (SME 2009)

Teck Pogo uses vision based froth velocity to control cell aeration and level setpoints in order to maximize cell pull to fill a froth product tank. Upstream pH indication is also used as a feed forward indication of "frothiness" which adjusts aerate rate limits to low values when frothy material is indicated and higher values when more brittle froth character is indicated. This approach has greatly reduced the occurrence of sunken flotation cells from 9.3% to only 1.5% and resulted in tighter control of concentrate storage tank level as shown below in Figure 4.

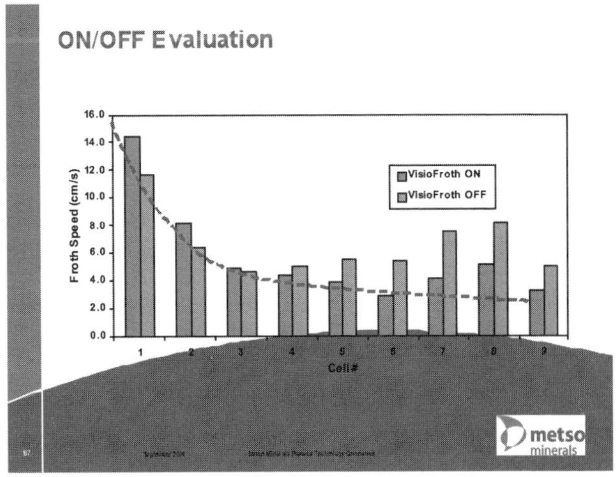


**Figure 1. Froth Velocity Comparison with and without OCS© VisioFroth Expert Control**

**Figure 2. Escondida Laguna Seca Rougher Flotation Circuit**

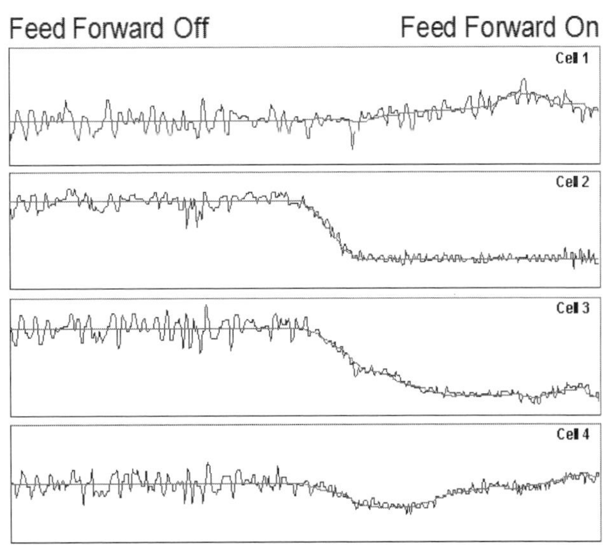


**Figure 3. Affect of Feed Forward Control on Rougher Flotation Cell Level at Laguna Seca Concentrator**

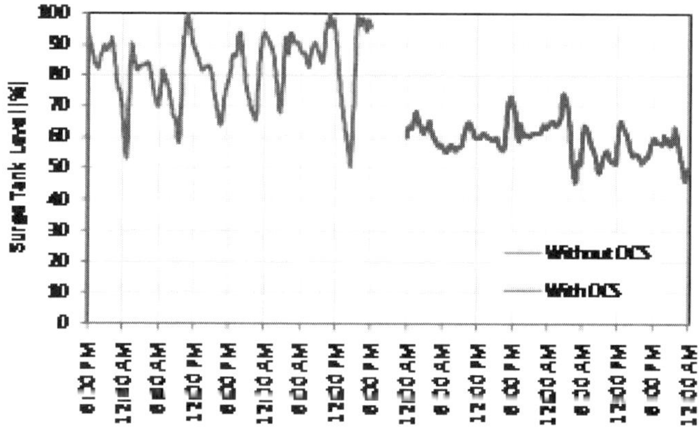


**Figure 4. Effect of Froth Velocity Control on Surge Tank Level Stability**

**Calcium Carbonate Reverse Flotation**

A North American Calcium Carbonate producer is using vision based froth texture as input to a neural network model to estimate insol content for use in expert control of collector for reverse flotation of Insol from calcium carbonate product. Manual samples are drawn and lab assays entered every 4 hours. The Vision Texture/NN model based insol estimate provides estimated assays during the interim between lab samples and allows the control to make collector dosage adjustments (both up and down) faster. This keeps product within specification better and reduces collector consumption when it is not necessary.

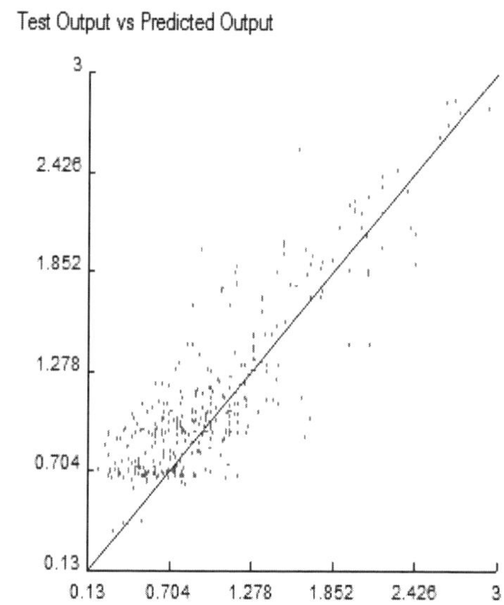


**Figure 5. NN Predicted Feed Insol (froth texture based) vs actual**

**CONCLUSION**

Instrumentation is critical to both regulatory and advanced process control of the flotation process. However, this importance does not translate to more instrumentation = better control. Reliable real sensors are always better than virtual sensors or model estimates and model estimates are better than no sensor at all. Frequently, sensor reliability forces the need for virtual sensors and the combination provides a sufficiently reliable data source for process control. Location of sensors is also a critical issue that bears consideration in the use of instrumentation for control.

Instrumentation for process control should be selected based on what value it brings in making a control decision. By looking from the control action backward, flotation personnel will find greater justification for plant instrumentation and more definable payback for their investment.

**BIBLIOGRAPHY**

C. Vanegas, P. Holtham / Minerals Engineering 21 (2008) 883–888 *On-line froth acoustic emission measurements in industrial sites*

D.R. Seaman et al. / Int. J. Miner. Process. 74 (2004) 1–13 *Bubble load measurement in the pulp zone of industrial flotation machines—a new device for determining the froth recovery of attached particles*

Chan,X, Sawada,M. and McKay, J. SME 2009. *Implementation of Advance Control on Rougher Flotation at Teck Pogo Mine*

# BENCHMARKING OF FLOTATION PLANTS: A KEY ELEMENT OF FLOTATION MODELLING

James Connolly and Glenn Dobby

SGS Minerals Services, 1140 Sheppard Ave West, Toronto, Canada M3K2A2

## ABSTRACT

The ultimate success of any mineral process model hinges on reliable calibration to operations at existing sites. This is especially so with froth flotation, where the chemistry and flowsheet varies significantly from plant to plant. The model calibration process is referred to as benchmarking, which consists of collecting plant operating and metallurgical data and simultaneous measurement of flotation kinetics of the minerals in the feed stream. The subsequent model tuning, repeated at multiple plant sites, provides important relationships on key model parameters such as froth recovery and entrainment. This paper reviews the approach taken with the FLEET flotation modelling package by providing a detailed description of the flotation benchmarking process.

## INTRODUCTION

The modelling of froth flotation systems has advanced considerably over the past 10-15 years, with significant activity and research at many research institutions and companies around the world. One of these is the FLEET modelling system developed by MinnovEX Technologies Inc., now part of SGS Minerals Systems, with the financial and technical support of seven mining companies. FLEET methodology has been successfully applied to flowsheet design for future ore types, plant optimization and reagent studies, but the driving force was geometallurgy (Bulled, 2007a; Bulled and Connolly, 2009). The objective is to model complete flotation circuits from kinetic flotation tests on drill core, and to link the mineral kinetic parameters (such as rate constants and ultimate recovery) with quantitative mineralogical data. These kinetic values are distributed across the mine block model and become the feed information for a flotation model that predicts flotation performance for all blocks in the block model. The kinetic test developed for this program is now referred to as an MFT, and the flotation model is FLEET.

The benchmarking methodology that is applied by the SGS group contains traditional as well as non-traditional data collection techniques. This methodology has been developed and enhanced through benchmarking programs at approximately 25 operating plants to date. Some of the larger plants that have been benchmarked using this methodology include: Minera Escondida (both concentrators), Olympic Dam, Collahausi, Cerro Verde, Bulyanhulu, Samarco (iron ore), Perilya (zinc) and BMA's coal plants in eastern Australia.

The approach to benchmarking has six distinct phases, illustrated in Figure 1, which will be discussed in the sections to come.

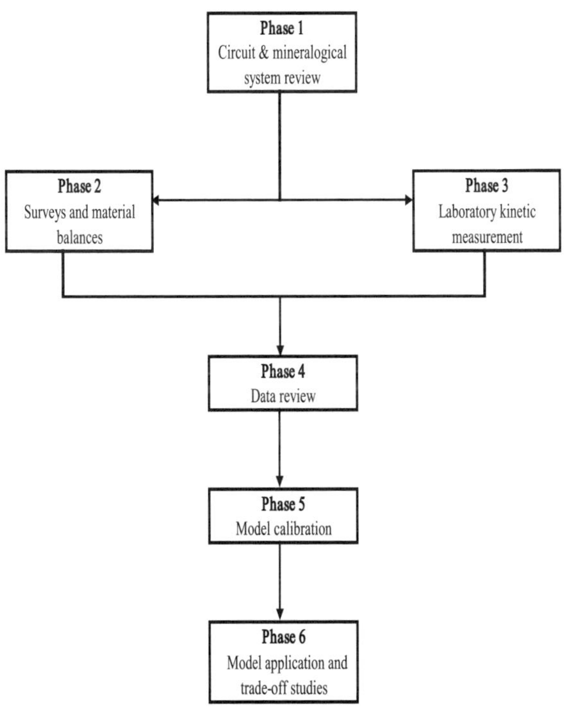


**Figure 1. The six steps of a flotation benchmarking process with FLEET.**

## PHASE 1. CIRCUIT AND MINERALOGICAL SYSTEM REVIEW

Phase 1 is a review of the circuit and mineralogical system. This is conducted before sampling begins and includes: defining specific site objectives, sourcing recent plant historical data, reviewing sampling points with plant operations staff and deciding on the assay and mineral system that will be employed.

Assigning kinetic parameters to metals does not have physical meaning for sulfide mineral systems when modelling the separation of ore particles by flotation—the analysis needs to be on a mineral basis. A key step in every project is to decide on the mineral system that will be employed, and assay accordingly.

Consider, for example a copper sulfide ore containing pyrite as a sulfide gangue mineral. If the ore contains only chalcopyrite, Cu and S assays can be easily converted, using stoichiometric formulae, to chalcopyrite, pyrite and nonsulfide gangue (NSG), where the NSG is everything but chalcopyrite and pyrite. Trace quantity valuable or penalty elements, such as gold, platinum or arsenic are balanced individually, but don't make a significant contribution to the mass balance. If the ore contains chalcocite, covalite and chalcopyrite in varying proportions throughout the orebody, a mineralogical analysis of the feed is required, where the objective is to quantify the copper mineral speciation and therefore determine the average copper concentration in the copper sulfides.

This assumes that the individual copper species have a similar rate of flotation ($k_{avg}$) and maximum recovery by true flotation ($R_{max}$). It is recognized that this may not be the case, and the material balance derived from the assumption of similar flotation rate will be somewhat in error. However, this is usually a fairly small error, and one that is taken into consideration in the modelling. The alternative option, of mineralogical sample work on each stream, is generally uneconomic.

The conversion factors determined on the feed sample are applied in the MFT and to all streams in the flowsheet. This approach is sufficiently flexible that it can be applied to all base metal systems (lead, zinc tin, nickel, molybdenum, etc.), precious metals (platinum and palladium, gold, uranium) and coal or reverse flotation.

## PHASE 2. CIRCUIT SURVEYS AND MATERIAL BALANCES

Phase 2 is conducting circuit surveys and material balances. Automated sampling equipment is used where possible, and an important element is establishing as good a water balance as possible. Depending on the objectives, it is common not to sample all streams within multiple cleaning loops, recognizing that some circulating streams rarely come to steady state while the material balance and model assume steady state. Size-by-size analysis of plant streams is conducted periodically; however, while providing very useful information, the additional expense of collecting this data cannot always be justified.

The intent of the survey campaign is to adequately capture stable operating conditions reflective of 'normal' operation. The survey campaign includes capturing all major equipment specifications, equipment operating parameters, and relevant stream characteristics.

Steady state operation of the plant, with exceptions outlined above, must be ensured prior to surveying. Operator performance indicators such as flotation feed conditions, circulating loads and circuit densities should be monitored prior to starting a survey. Consultation with the operators prior to surveying is valuable as process upsets and abnormalities are more readily assessed by those who have intimate knowledge of normal operating conditions.

Owing to normal variations in plant operating conditions it is impractical to attempt to survey a circuit by considering circuit residence times. Rather it is beneficial to view the survey as a 'snap shot' of the operation with the survey sample times and sampling frequency being defined by stream variance and manpower constraints. A typical flotation survey campaign incorporates 5 to 8 surveys over a reasonable time period. The intent being to capture, within reason, a range of exogenous variables including: mineralogy, liberation and hardness in terms of both throughput rates and associated feed particle size distributions. Typically, full circuit surveys are conducted over a period 2 to 3 times the rougher residence time. Sample frequency of ½ the rougher residence time normally accommodates stream and sampling variance.

Sampling points for stream characterization are normally selected with the intent to over-define the final mass balance. It is normal practice to define a stage as a collection of flotation cells operating in series or parallel which have a common feed, concentrate and tailings (example: roughers, scavengers, cleaners, cleaner scavengers and column cells). Stream characterization should include mass percent solids and assaying (in accordance with mineralogical and trace element requirements) on all sampling points. Additionally, size reduction nodes (primary milling, secondary milling and or regrind milling) are identified.

All previously defined stages should fall within an area of common sizing (example: the regrind mill area of common sizing is normally common to a cleaner, cleaner scavenger and column). All flotation streams entering or leaving an area of common sizing (or sizing stage) should undergo size fraction analysis. Four size classes adequately define most minerals systems if chosen well, namely coarse, intermediate coarse, intermediate fine and fine. Best practice would be to conduct a full size fraction analysis of the concentrates of a sizing stage, and then define the coarse screen size as the screen size where 5% to 10% of the material is retained (on a cumulative mass basis). The fine screen size is normally selected as the finest screen at which it is practical to wet screen. The intermediate screen would be the resultant geometric mean of the coarse and fine screens. In circuits where streams are common to two or more sizing stages then care should be taken in the selection of screen sizes and commonality in screen selection should be ensured.

At a minimum, the feed stream to the flotation circuit is identified for laboratory flotation kinetic testwork. Laboratory kinetic testwork is sometimes conducted on those streams where mineral floatability is substantially changed or modified [example: regrind overflow, significant changes to pulp chemistry (Eh or pH)]. From both a practical and modeling point of view sample aging should be quantified prior to the start of any campaign. All streams identified as requiring laboratory kinetic testwork should initially be tested at aging 0 circuit residence time, 1 residence time, and 3 residence times. Sample aging within circuit residence times seldom impacts circuit modeling and calibration but if sample aging is pronounced, its impact on both the sampling of streams for laboratory flotation testwork and modeling must be taken into consideration.

During the survey period plant operating conditions (including: stream mass flows, stream volume flows, stream densities, flotation cell air, flotation cell level, and flotation equipment motor power) are to be recorded. During the campaign period it is advisable to record: plant equipment specifications (including: flotation equipment dimensions, lip lengths, flotation equipment motor installed power); plant configuration; and water recovery profiles down each flotation stage. For a more fundamental study both bubble size and air hold-up measurements may be considered.

For a well defined survey reconciliation, excess data is recommended. It is good practice to reconcile the excess data as the surveying proceeds on a mass basis (solids mass flows, water mass flows). Once assays are completed the balance is performed on a component basis (unsized assay, sized assay, and by size) and material and component balances are reported per flotation stage on a sized and unsized basis.

## PHASE 3. LABORATORY KINETIC TESTING AND MINERALOGICAL WORK

Phase 3 is measurement of kinetic parameters in parallel with the plant sampling. Typically, tests are conducted on samples of feed slurry only, taken directly from the plant during the survey exercise. Occasionally, samples of reground material (such as rougher concentrate) are also tested.

The key information that characterizes mineral behavior in a flotation circuit is the suite of kinetic parameters for the constituent minerals. In the FLEET modelling system, this information is collected via a specifically designed kinetic test named the MFT (for Minnovex Modified Test). To date more than 1,000 MFT's have been conducted on approximately 60 mineral deposits, including base metals, precious metals, iron ores and coal (Dobby and Savassi, 2005; Bulled and McInnes, 2005). These tests are normally conducted on drill core, where the ground ore is prepared in a laboratory mill. For the purposes of benchmarking however, the feed to the MFT usually is plant flotation feed slurry, already ground and having most or all of the reagents already added.

Typically, a sample of flotation feed pulp is obtained from an on-line sampler shortly after the start of a survey, and the MFT is immediately conducted in the plant laboratory. Occasionally, duplicate MFT's are conducted. This section will review the concept and methodology of the MFT, and the derivation of the kinetic parameters.

The basis for the MFT scale up approach is that the kinetics of bubble-particle collection is the primary source of flotation selectivity in both lab and industrial cells. The main objective of the test is to generate parameters that represent the true flotation mechanism in the collection zone of the batch cell, rather than trying to replicate the performance of the industrial circuit in the lab.

The standard MFT cell is equipped with a froth crowder to prevent the formation of a dead zone behind the impeller shaft. The compressed air line must have a main valve ahead of the flowmeter, so that the air flowrate can be set prior to the testwork. A T-shaped paddle allows uniform scraping that is independent of the operator. Upon scraping, the paddle stands at 10 mm above the pulp level mark, so that the operator can only scrape the froth that is above a minimum froth layer. If the froth layer is thinner than that, there will be no concentrate collected.

In order to measure the true flotation kinetics in the collection zone of the batch cell it is necessary to maximize the mass transfer through the froth, so that the particles that become attached to the bubbles in the pulp will end up in the concentrate, without any significant drop back from the froth. This situation, which is modelled as one of full froth recovery of attached particles (Savassi, 2005) is the reason for the froth removal rate during the MFT being relatively high: one scrape every 2 seconds. Obviously the high scraping rate will also increase the degree of entrainment through the froth (Johnson, 1972, Trahar, 1981, Dobby and Finch, 1990). This problem is solved by means of a phenomenological model that allows true flotation and entrainment recoveries to be decoupled based upon the behavior of the gangue. The model also takes into account the differences in the flow regime (batch / perfect mixing) from the lab to the collection zone of the industrial cells.

In summary, the MFT aims at determining kinetic parameters, rather than trying to replicate plant performance in the lab. The standard experimental procedure has been designed to prevent any significant interference from the operator. Froth removal rate is relatively high in order to ensure that all particles that become attached to the bubbles in the pulp will report to the concentrate. Recovery by true flotation and entrainment is decoupled by means of a phenomenological model. The upcoming sections provide more details about the MFT methodology; further details can be found in Savassi et al., 2007.

### Experimental Procedure

After noting the wet weight of the pulp at the start of the test, the MFT cell is placed under the rotor, the impeller is switched on (with the air still off) and a feed sub-sample is siphoned out for assaying and size analysis. At the end of the test, another sub-sample is siphoned out from the tail and the pulp wet weight noted again.

A key aspect in batch flotation testwork is the definition of the zero time, that is, the instant at which the test 'starts'. In the MFT procedure the operator must wait for a moment after switching on the air until the froth structure is fully developed. Zero time is defined as the instant the paddle first passes over the cell lip.

As with many standard kinetic tests, four individual concentrates are collected at consecutive time intervals. Water addition rate and slurry weights are measured, so that water recovery per time interval can be calculated Feed, combined concentrate (weighted per mass of each of the four concentrates) and tailings are then screened into four size intervals (it is not necessary to conduct size analysis on individual concentrates).

Operating MFT conditions (chemical environment, airflow, impeller speed, etc) are determined by preliminary testwork. Time settings must be such that 40 to 60 % of the valuable mineral is recovered in the first concentrate and about 70% of the remaining in the second. The third and fourth concentrates are collected by the same length of time, with the valuable mineral reaching a plateau toward the end of the test. The top screen opening must be selected to retain around 5% of the mass of the combined concentrate. The fine screen opening often is 38 microns.

As previously discussed, assigning kinetic parameters to metals does not usually have physical meaning in terms of the separation of ore particles by flotation, and the elemental assays are converted to the same mineral system as employed for the plant surveys—using a combination of minerals that sum to 100% plus any required trace assays.

### Material Balance

Raw data from the MFT, including solids and water masses, are run through a material balance program (usually BILMAT) to obtain a coherent and size-by-size material balance. Feed assay is derived from (a) an assay on the feed sample, (b) back-calculated from the feed sample size fractions, and (c) back-calculated from the five product streams (four concentrates and one tailings). Since the test is carefully controlled and the data is rigorously handled, the agreement between these three values is usually very good.

### Parameter Extraction

**The k-distribution** The principal objective of the MFT parameter extraction is to determine the k-distribution of the minerals in the ore. The k-distribution (Figure 2) is based on classes of particle flotation rate, similar to the way in which the Rosin-Rammler distribution is based on classes of particle size. Each class in the k-distribution, however, contains particles of different size and liberation. Ultrafines and poorly liberated particles belong to the slowest flotation class, while fully liberated particles of intermediate size to the fastest one (King 1976).

The k-distribution of a given mineral is described by three kinetic parameters:

- $R_{max}$: the ultimate recovery the mineral can achieve by true flotation.
- $k_{avg}$: the weighted average flotation rate, excluding unfloatable particles.
- Alpha: a measurement of the spread in the k-distribution.

As shown in Figure 2, the percentage of unfloatable particles (zero flotation rate) corresponds to $100 - R_{max}$, while $k_{avg}$ is the flotation rate at the mid point of the floatable range. Alpha is usually between 1 and 5 for hydrophobic minerals and up to 10 for non-sulfide gangue. Clearly, a sharp difference in the k-distributions of gangue and valuable minerals is essential for effective separation by flotation in both lab and plant cells.

Note that the resulting average rate constant for non-floating gangue is a very different value from that derived from traditional kinetic tests. In traditional tests an average rate constant is quoted for all of the gangue, which consequently will numerically be very small. In the MFT approach, the majority of the gangue ($100 - R_{max}$) is recognized to have a zero value for rate constant, being recovered only by entrainment, while the $k_{avg}$ refers only to the gangue that has a non-zero flotation rate; this is usually due to locking of the gangue with the faster floating minerals. Hence, the MFT-derived $k_{avg}$ value for gangue often has a value that is 20-30 % of the $k_{avg}$ for the fast floating mineral.

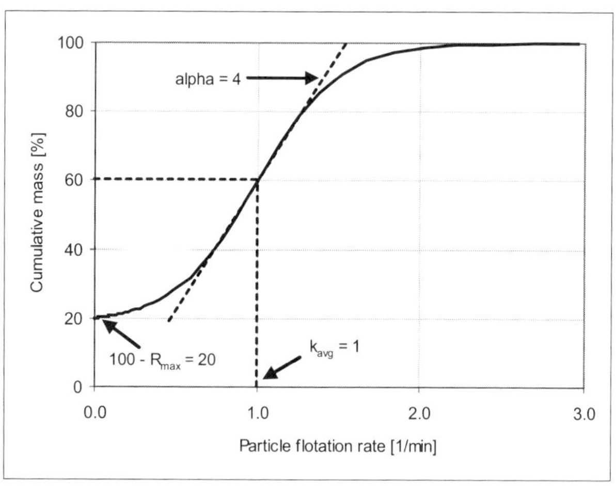


**Figure 2. The k-distribution of an example mineral that has a maximum recovery of 80%, average flotation rate constant of 1 and alpha value of 4: the slope of the tangent to the curve at $k_{avg}$ is determined by the alpha value**

**Decoupling true flotation and entrainment** The overall recovery in flotation derives from both true flotation and entrainment mechanisms. To be representative, however, the kinetic parameters must refer to true flotation exclusively. In the MFT parameter extraction the nonfloating gangue (referred to in the remainder of this paper as NSG, for nonsulfide gangue) is used as a hydrophilic tracer to decouple true flotation and entrainment (Savassi, 2005). A partition curve is used to describe the effect of particle size upon the degree of entrainment. The entrainment parameters of the NSG are numerically adjusted to meet the following criteria:

- most, if not all, of the NSG in the fourth concentrate is recovered by entrainment only
- most, if not all, of the NSG in the finest size interval is recovered by entrainment only.

The entrainment parameters of the other mineral species in the system (either hydrophobic or hydrophilic) are estimated from those of the NSG and the specific gravity of the mineral using a conversion model. Given the overall recovery of the mineral and the recovery by entrainment, the true flotation recovery is estimated by difference.

**Quantifying the impact of changes in the grind** When starting the MFT from dry ore (for instance, a drill core sample in a production planning project), the grind that will be attained in the industrial plant is not usually known at the time the MFT is conducted. Therefore, in order to reconcile lab and plant data, it is necessary to quantify the impact of changes in the grind on the true flotation recovery. The traditional approach to obtain such information is to conduct a series of tests at different grinds; a procedure that is costly, time consuming, and requires significant quantities of drill core.

A significant feature of the MFT is the ability to estimate the impact of the grind on flotation kinetics (particularly $R_{max}$) from a single test. This is based on the assumption of consistent recovery per size class, independent of the mass distribution per size class. In other words, each particle interval behaves independently of the others, allowing the overall $R_{max}$ and $k_{avg}$ to be estimated as a weighted average of the size intervals. The validity of this approach has been confirmed several times, and example of which is shown in Figure 3.

**Outputs from the MFT**

The MFT is a rigorous test, and as such provides a great deal of information. This includes:

- reconciled material balance for both minerals and metals
- size distribution for each mineral component, as well as the fitted Rosin-Rammler values for P80 and slope (m)
- kinetic parameters ($R_{max}$, $k_{avg}$ and alpha) on a size-by-size basis for each mineral component
- kinetic parameters for each mineral component forecast at user selected P80 and m values, and parameters to calculate $R_{max}$ and $k_{avg}$ at other grinds

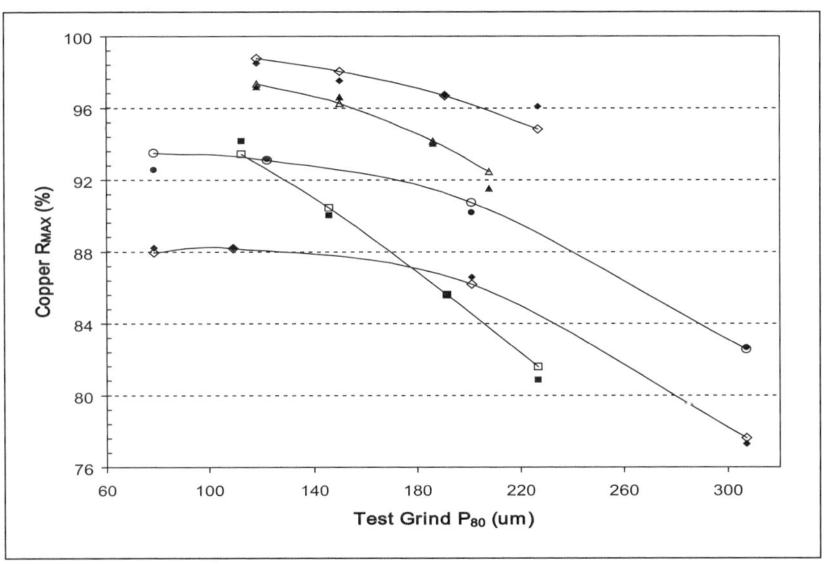


**Figure 3. Effect of the grind on flotation kinetics for five porphyry ore samples. The closed symbols are $R_{max}$ values obtained from four individual grinds per ore sample. The open symbols are forecast $R_{max}$ values from one MFT per ore sample.**

## PHASE 4. DATA REVIEW AND COMPARISON WITH HISTORICAL DATA

Phase 4 is a review of the benchmark data. The focus is two fold, the first of which is comparing plant survey results with recent historical data to ensure that the surveys are reasonably representative of plant performance. The second is to examine the nature of the plant results in conjunction with the simultaneous kinetic measurements.

The first step in evaluating the plant survey results is to conduct a reasonably simple (yet important) overview of how the data compares with recent historical plant performance. For example: 'Do the feed grade and recovery seem reasonable and typical?' A natural approach to this is to plot frequency curves of key plant operating performance and mark the location of the survey values. An example is shown in Figure 4, showing rougher concentrate assay from five surveys versus plant recent historical values (a cumulative frequency plot is shown; cumulative plots are much simpler to work with than straight frequency curves). In this example, all of the values are on the upper side of rougher concentrate grade, but one is at the 99th percentile—this would raise a question regarding the validity of the survey.

Another copper example is shown in Figure 5. In this example, five plant survey results are plotted against recent historical data. The upper graph is rougher copper recovery versus feed Cu:Fe ratio, and the lower graph is cleaner circuit upgrading ratio versus cleaner circuit recovery. In both cases, one survey clearly falls outside of the historical data range, again raising a concern regarding the results of those two surveys.

The MFT data also requires a thorough review. Obvious checks are on feed grade, as discussed earlier. Feed size should be examined for both F80 (again, compared with recent historical data) and distribution. The MFT analysis extracts a fit to the Rosin-Rammler equation, for the whole ore and for the individual mineral species. The slope of the Rosin-Rammler curve (referred to here as the 'm' value) provides an insight to the classification efficiency. This is achieved by plotting 'm' versus P80 (for the ore and for the primary mineral species) and comparing this with the SGS historical database.

MFT $R_{max}$ versus rougher recovery for the primary floating mineral are immediate values to examine. For ores that liberate well, we usually observe an $R_{max}$ value of the primary mineral that is reasonably similar to the rougher recovery. However, for ores with complex interlocking, such as an altered copper sulfide ore with fine chalcocite and pyrite, it is not unusual for the plant rougher recovery to be substantially lower than the MFT Rmax. This occurs either because the plant circuit does not have sufficient rougher residence time and or cannot be pulled as hard as required to achieve the $R_{max}$ level and at the same time make an acceptable concentrate grade (without invoking an unrealistically large cleaner circuit). Essentially, a finer grind or more rougher residence time is required.

An example of how $R_{max}$ values may compare with rougher recovery values is shown in Figure 6, obtained from a benchmark campaign on a copper porphyry operation. In this figure, the y-axis is the ratio of rougher recovery to Rmax and the x-axis is average particle size of four size classes. The ratio is shown for copper sulfide, pyrite and nonfloating gangue. The copper sulfide curve clearly shows that the ratio is close to 1.0 for the finer fractions, but decreases as particle size increases. Modelling work indicates that the reduction in coarse particle recovery in going from the MFT (with a very thin froth layer) to the plant roughers is due to froth dropback in the full scale circuit. At the same time, the ratio of recoveries for gangue decreases dramatically as particle size increases. This again attributed to froth drop back of coarse gangue particles carrying small amounts of copper sulfide.

The $R_{max}$ value of the primary floating mineral is usually closely related to the quantity of the mineral that has been ground into an acceptable size range. An example, shown in Figure 7, is for copper flotation from several benchmark plants, covering a range of grind sizes. $R_{max}$, as measured by the MFT on rougher feed slurry, is plotted on the y-axis. The x-axis notes the quantity of 'displaced' copper, where we define 'displaced' (somewhat arbitrarily) as the percentage of copper mineral in the feed that is less than 2 um and greater than 200 um—the -2 um material will have very low kinetics of particle collision and the +200 um particles are generally small grains of copper sulfide locked to a gangue particle. Even though there are different copper sulfide systems examined and different sets of reagents, the trend is clear –copper recovery is maximized by preventing the copper minerals from reporting to very coarse or very fine size classes.

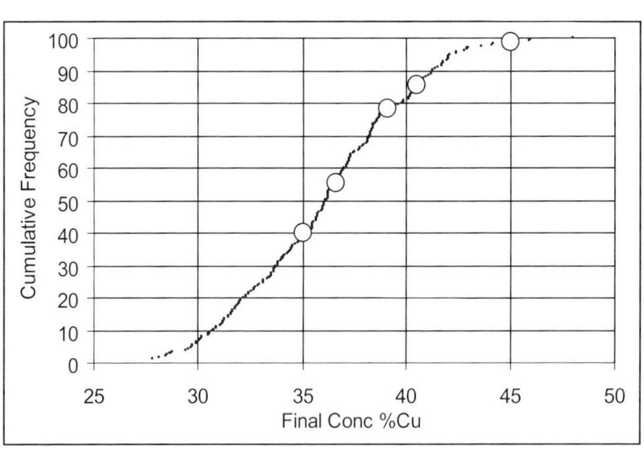


**Figure 4. An example of plant survey results (open circles) in comparison with a frequency distribution of recent historical plant performance for daily composites.**

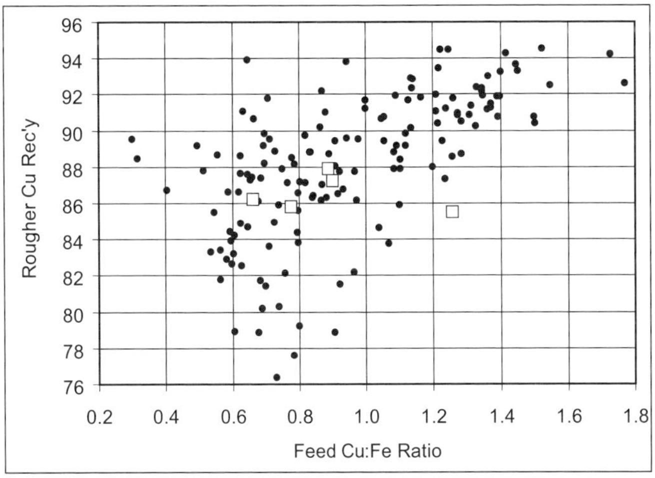


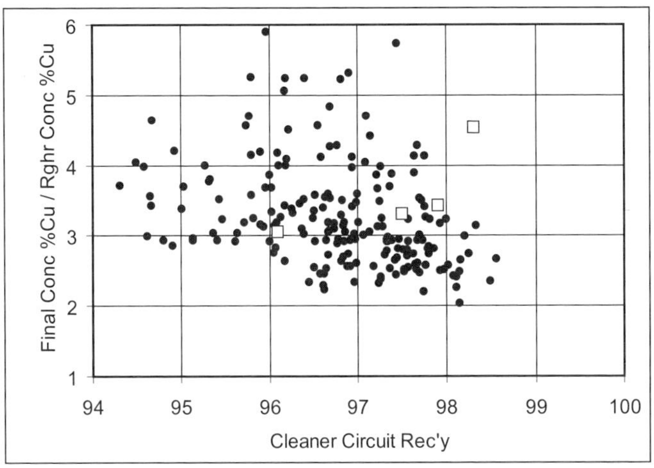


**Figure 5. Examples of comparing plant survey results (open squares) with characteristic performance of the plant. Upper graph: a copper operation where rougher copper recovery is loosely tied to feed Cu:Fe ratio. Lower graph: a copper operation where cleaner circuit upgrading ratio is loosely tied to cleaner circuit recovery.**

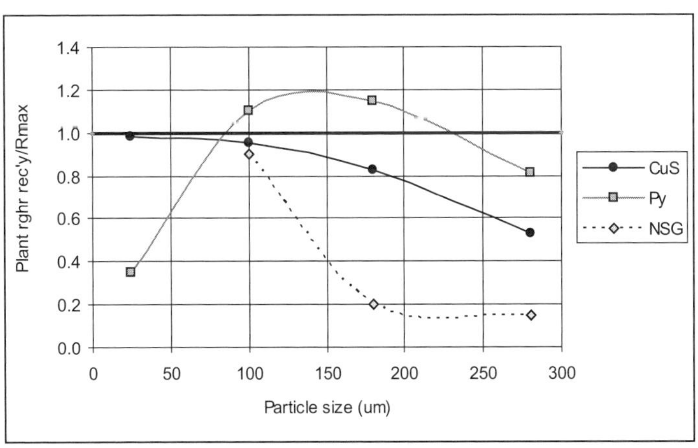


**Figure 6. The ratio plant recovery to MFT Rmax value is plotted versus particle size class for copper sulfide, pyrite and nonfloating gangue, as measured during a benchmarking campaign at a copper porphyry concentrator.**

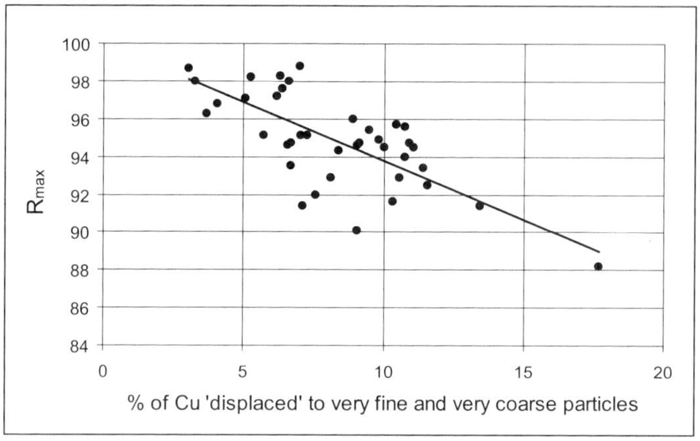


**Figure 7. Rmax versus percentage of copper displaced to the very fine (-2 um in this case) and very coarse (+200 um) size fractions, for a series of copper ores as measured on benchmarking campaigns.**

## PHASE 5. FLEET MODEL CALIBRATION

Phase 5 is model calibration. The primary tuning factors are froth recovery ($R_f$) per stage, entrainment factor (ENT) per stage, and kinetic modifiers that result from significant mid-circuit reagent addition or from regrind. The measured kinetics are inputs to the FLEET model and the model tuning factors are adjusted to provide a simulated circuit performance similar to the survey result. This is a challenging exercise, and requires the expertise of a process engineer well versed in flotation systems and the model structure.

The primary objective of the model calibration is to adequately quantify a number of froth phase parameters, including: the degree of entrainment (ENT), the froth phase recovery factor ($R_f$), and water recoveries across the flotation stages. Additionally, the quantification of the change to mineral floatabilities (or modifiers) is required to adequately describe circuit changes owing to size reduction (or regrind), pulp chemistry (Eh of pH), and depressant or activator addition points. To this end, flotation model calibration requires the fitting of the surveyed mineral stage recovery data to the model predicted data through the manipulation of ENT, $R_f$, water recoveries and modifiers.

The calibration process is informed by knowledge of the mineral system and information regarding the froth such as stickyness drawn from images, video and/or onsite observation. The cell operating conditions are further constrained within reasonable bounds, where known particle size will affect the degree of entrainment accepted within the calibration process (see for example Figure 8). This process also draws on the SGS database of past projects to define 'reasonable' flotation performance estimates.

Similarly, measured mineral floatabilities constrain the modifiers during the calibration process, and water recoveries make reference to the mass balanced stage water recoveries and measured water recovery profiles. All significant surveys are used during the calibration process to ensure robust calibration.

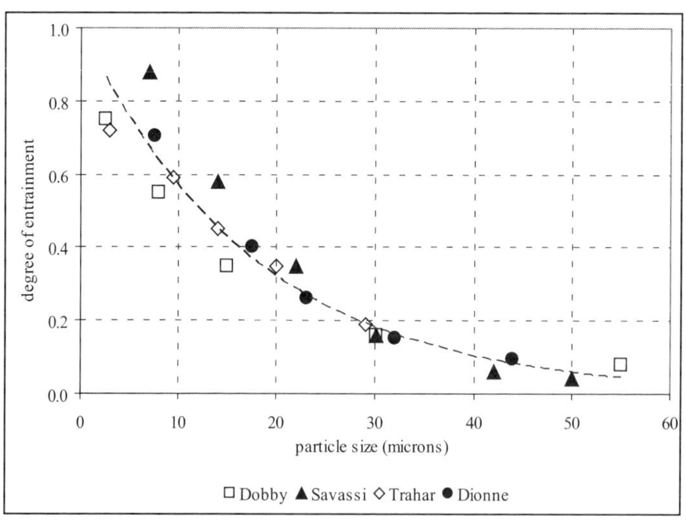


**Figure 8. Effect of particle size on the degree of entrainment of solids.**

## PHASE 6. MODEL APPLICATIONS AND TRADE-OFF STUDIES

Phase 6 is application of the model to address site specific issues. The combination of a rigorous kinetic test (the MFT) and a calibrated FLEET model for the plant presents several opportunities for the process engineer. Some of these include the following:

- Future ore performance can be accurately forecast via MFT testing on drill core followed by circuit modelling, using the new kinetic parameters and the calibrated FLEET model. This can be extended to a complete geometallurgy-based production forecasting program.
- Froth pull rates at individual stages, especially within the cleaning circuit, can be optimized.
- The effect of additional rougher or scavenger capacity can be modelled.
- Changes to the circuit flowsheet can be modelled, not only on current ore but also for future ores (in conjunction with MFT results on drill core of future ores).
- When linked with a comminution model such as CEET (Bulled, 2007b), a trade-off analysis of throughput versus recovery (and grade) can be made.
- The combination of the MFT and circuit model provide a powerful tool for reagent studies. This can be in two stages: (a) first, as an initial evaluation of expected plant performance, by using MFT tests on different collectors followed by plant modelling with the calibrated FLEET model; and (b) enabling a rigorous analysis of in-plant reagent performance, where MFT's are conducted on plant feed and the model is re-tuned on plant surveys conducted with the different reagents and dosages.
- Key process units that are under-performing or that present process bottlenecks can be identified for further investigation. Optimal mass pull rates can be found per stage to select process air and level setpoints. This information allows management to set achievable metallurgical targets for operations and enables metallurgical staff to deliver to that level.

## CONCLUDING REMARKS

A standard and rigorous approach to flotation benchmarking has been described. The extent to which the exercise is applied can be tailored to each plant. For example, with a smaller operation and a tighter budget, plant surveying can be restricted to two or four-hourly samples from the automated sample cutters, supplemented with one or two key streams that are hand cut. Alternatively, for larger operations, surveys can include detailed sampling as well as size-by-size analysis of plant streams. In either case, the modelling is achieved by conducting MFT's on only the feed stream.

Future development will focus on linking quantitative mineralogical data with mineral kinetics. When fully applied, this will be a powerful approach for geometallurgy-based production forecasting and for a more thorough analysis of regrind performance.

## REFERENCES


Bulled, D. 2007a. Flotation circuit design for Adanac Moly Corp using a geometallurgical approach, *Annual Canadian Mineral Processors Conference*, Ottawa

Bulled, D. 2007b, Grinding circuit design for Adanac Moly Corp using a geometallurgical approach, *Annual Canadian Mineral Processors Conference*, Ottawa

Bulled, D. and Connolly, J., 2009. A case study in the flotation characterization of an ore body and its use in circuit design and risk assessment, *Mineral processing Plant design 2009*, SME, Tucson

Bulled, D. and McInnes, C., 2005. Flotation plant design and production planning through geometallurgical modelling, *Centenary of Flotation*, Aus. IMM, Brisbane

Dobby, G. and Finch, J., 1987. Particle size dependence in flotation derived from a fundamental model of the capture process. *Int J. of Mineral Processing*, vol 21, pp 241-260.

Dobby, G. and Savassi, O., 2005. An advanced Modelling Technique for Scale-Up of Batch Flotation Results to Plant Metallurgical performance. *Centenary of Flotation*, Aus. IMM, Brisbane

Johnson, N.W., 1972. The Flotation Behaviour of Some Chalcopyrite Ores, PhD Thesis, University of Queensland.

King, P., 1976. The use of simulation in the design and modification of flotation plants. Flotation, A.M. Gaudin Memorial Volume, Fuesternau, M.C. (ed.) AIME, vol 2, pp 937-962.

MinnovEX Technologies, 2004. MinnovEX Flotation Test: Standard Procedure and Parameter Extraction. Available at www.minnovex.com.

Savassi, O., 2005. A compartment model for the mass transfer inside a conventional flotation cell. *Int. J. of Mineral Processing*, vol 77, pp 65-79.

Savassi, O., Dobby. G. and McInnes, C., 2007. The MFT: A Standardized Methodology for Scale-up of Batch Flotation Results, *Southern Hemisphere Minerals Conference, XXII ENTMME / VII MSHMT*, Ouro Preto, Brazil

Trahar, W.J., 1981. A Rational Interpretation of the Role of Particle Size in Flotation. *Int. J. of Mineral Processing*, vol 8, pp 289-327

J. N. Kohmuench[1], M. J. Mankosa[1], Harold Wyslouzil[2], and G. H. Luttrell[3]

[1] Eriez Manufacturing, 2200 Asbury Road, Eriez, Pennsylvania, USA
[2] Canadian Process Technologies, 7168 Honeyman Street, Delta, British Columbia, Canada
[3] Virginia Tech, 100 Holden Hall, Blacksburg, Virginia USA

## ABSTRACT

Conventional, mechanical flotation has long been considered the workhorse of the flotation industry. However, column and other non-traditional flotation approaches continue to provide benefit where traditional processes fall short. This benefit is best demonstrated in applications treating ultrafine material where the use of smaller bubbles and froth washing provide increased capacity and a high-grade product. Several technologies now exist which incorporate this approach. At the other end of the particle size distribution, flotation in a fluidized-bed offers the opportunity to improve the recovery of particles in excess of 3 mm. This technology is well suited for applications where liberation is achieved at relatively coarse sizes, which is common for the potash, phosphate and diamond industries. This approach can also significantly reduce the quantity of material that requires grinding or regrinding given that this technology can provide throwaway reject. This chapter reviews the design and guidelines for flotation of fines with column flotation as well as provides up-to-date information on the flotation of ultracoarse particles using new technology. Operating guidelines for these high-efficiency flotation technologies have also been provided.

## INTRODUCTION

Froth flotation is widely regarded as one of the most efficient and economical processes for separating fine particles. In the mining industry, most industrial installations make use of mechanical (conventional) flotation machines. These machines consist of a series of agitated tanks (called banks) through which dilute feed slurry is passed. The agitators keep particles suspended and disperse air that enters through the rotating impeller assembly. Depending on the application, these machines may be arranged in single banks or in complex circuits that may include rougher, scavenger and cleaner banks. In theory, flotation is a highly versatile process that can be applied to any system containing particles with different surface wettabilities, either natural or induced via the addition of suitable reagents. On the other hand, the effectiveness of traditional flotation processes is limited to a relatively narrow range of particle sizes (Figure 1). Very large particles do not respond well to flotation due to factors such as low buoyancy, high detachment and poor suspension. At the other extreme, ultrafine particles are difficult to recover by flotation due to hydrodynamic limitations that reduce the frequency of bubble-particle collisions. The flotation of ultrafines can also suffer from poor selectivity due to the unwanted entrainment of fine gangue particles that are hydraulically carried into the concentrate product with the froth water. These upper and lower particles size barriers have been a longstanding problem for flotation practitioners.

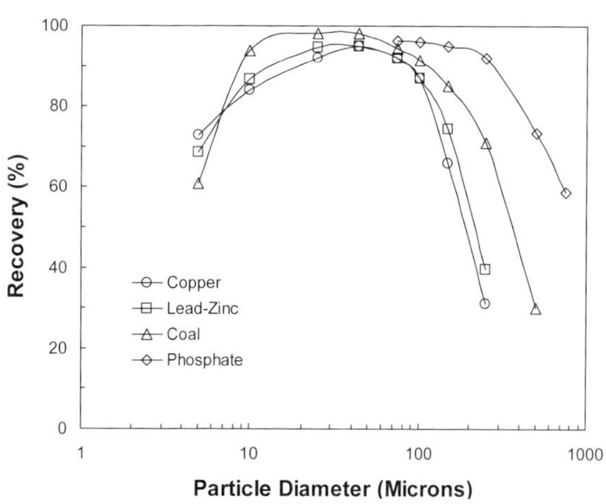


**Figure 1. Range of particles sizes effectively treated by traditional flotation machines.**

Several types of next-generation flotation machines have been developed during the past several decades to effectively deal with the problems of treating very fine or very small particles. Perhaps the most noteworthy example of this technological evolution is the development of column flotation. Column cells have become a popular choice throughout the mineral processing industry for the selective flotation of fine particles (i.e., <100 micron). Columns equipped with well-designed froth washing systems enhance the overall quality of the froth concentrate by minimizing the carryover of fine impurities into the floated product. In some cases, the wash water can also partially offset the need for depressants, which lowers chemical dosage rates. The efficiency of column cells is such that a single stage of processing can often replace two or three stages of cleaning in mechanical cells. In addition, columns avoid the generation of new fine slimes by eliminating the intense agitation that is required by conventional machines equipped with a rotor/stator. At the opposite end of the particle size spectra, new flotation machines are being developed and applied commercially for the treatment of very coarse particles (>250 micron). One of the most successful of these is a fluidized-bed flotation device known as the HydroFloat™ separator. This technology combines the capacity advantages of a density separator with the flexibility of a flotation process. These new flotation technologies have made it possible for plant operators to improve the performance of both their fine and coarse flotation circuits. In many cases, these systems often reduce production costs, lower power demand and cut reagent consumption. The following sections provide a brief overview of the working features of these advanced flotation technologies. The results obtained from two case studies are also provided to illustrate the potential benefits of these systems for mineral beneficiation.

## COLUMN TECHNOLOGY

Column flotation cells are ideally suited for the flotation of fine particles (<100 microns). As the name implies, column cells are typically configured with a height to diameter ratio that is greater than unity and occasionally as great as 10 or more (Figure 2). Unlike conventional machines, feed slurry enters in the upper section of a flotation column and is distributed over the cross-section using a piping network. The feed moves counter-current to a rising swarm of fine bubbles that are generated by a suitable air sparging device. Particles that collide with and attach to the bubbles are carried up the column and are discharged to the product launder. Particles that do not attach settle down through the column and report to the underflow tailings stream. In practice, fresh or clarified wash water is introduced over or into the froth to minimize the nonselective carryover of fine gangue into the froth product. Sufficient water should be added to fully displace the feed water that would otherwise report to the floated product with the froth water. A relatively deep froth must also be maintained in order to provide additional upgrading and to properly distribute the wash water. When properly configured, these features allow a flotation column to provide a higher grade product than is attainable using conventional technologies.

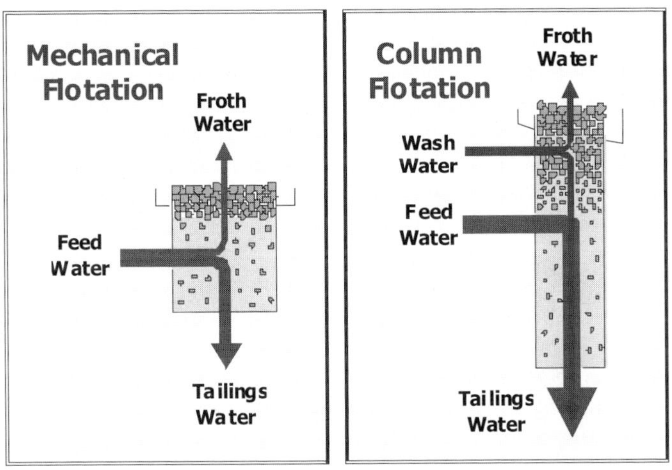


**Figure 2. Schematic illustration of mechanical and column flotation machine designs.**

### Column Types

Several different types of column flotation cells are offered commercially. The major differences in these various technologies are the design of the gas sparging system. The spargers, which are the most important component in a column cell, must provide the maximum rate of bubble surface area through the column without burping or surging. For this reason, older poorly performing static spargers (e.g., porous bubblers, filter cloth or punched rubber) have been replaced at many industrial sites by efficient dynamic spargers that utilize high energy dissipation to more completely disperse air. While details related to the specific design features of the various sparging technologies have been presented in the literature (McKay et al. 1988; Davis et al. 1995; Finch 1995), three of the most popular dynamic sparging systems are described below.

**SlamJet™.** One of the first dynamic sparging systems for column flotation cells was developed by the U.S. Bureau of Mines. This design was essentially made up of a series of small orifices (1-3 mm) that were placed along the length of a pipe inserted through the column wall. This approach worked satisfactorily, but was plagued with maintenance problems with regard to excess wear and plugging. Canadian Process Technologies (CPT) solved these problems with the introduction of their SparJet™ aeration system. This system uses a series of removable air-lances which includes a single orifice located at the end of the sparger. High velocity air is injected into the column cell to create and disperse fine bubbles. This technology was further refined by incorporating a self-closing mechanism that eliminates any backflow of slurry into the aeration system (Figure 3). This technology is now commercially distributed as the SlamJet™ sparging system.

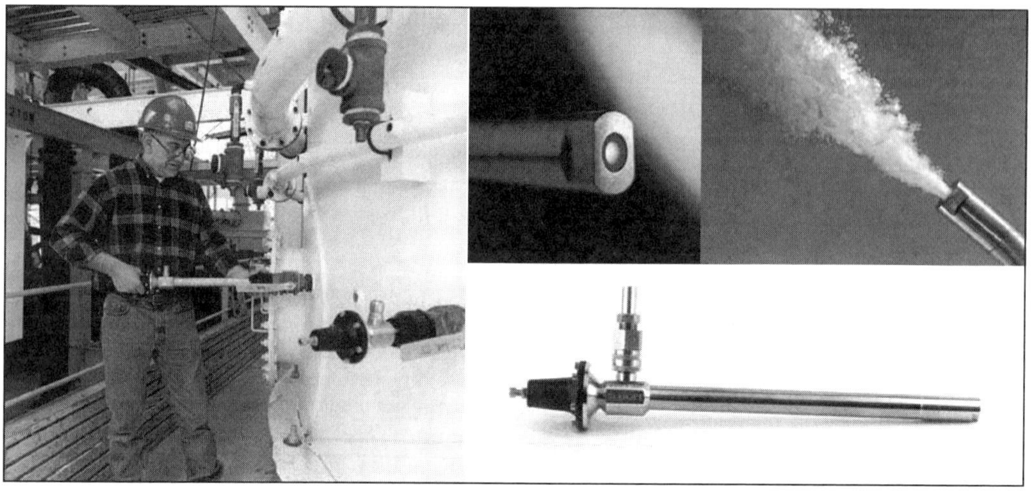

**Figure 3. SlamJet™ sparging system with self-closing nozzle.**

**Microcel™.** Another popular dynamic sparger for column cells is the Microcel™ technology. This system was conceived after several years of fundamental research that showed that the rate of flotation could be substantially enhanced through the use of smaller air bubbles (Yoon et al., 1992). In this technology, small (<0.8 mm) microbubbles are generated by circulating slurry from the lower section of the column through parallel in-line static mixers into which compressed air is injected (Figure 4). The sparging system has the ability to efficiently generate large amounts of very small bubbles at high gas rates. Studies indicate that the combination of small bubbles and high gas rates improve flotation kinetics by greatly enhancing the frequency of bubble-particle collisions. As such, the Microcel™ system offers a high production capacity and/or improved concentrate recoveries when compared to older static sparging systems. The systcm is also easy to maintain since the spargers are located external to the column.

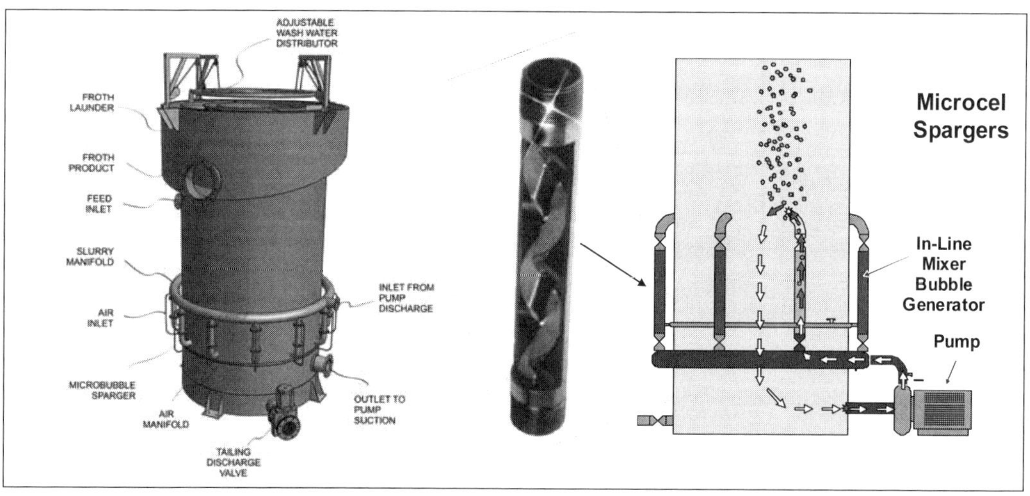


**Figure 4. Microcel sparging system incorporating in-line mixers and pump.**

**Cavitation-Tube™.** Another effective sparging system that can meet the rigorous performance demands of industrial installations is the Cavitation-Tube™ sparger (Figure 5). This sparger consists of an hour-glass shaped tube constructed of wear resistant material. During operation, a mixture of flotation pulp and compressed gas is circulated at very high velocity through the device using a centrifugal pump. The high velocity and throat geometry creates cavitation in the slurry that is beneficial for the flotation of ultrafine particles (Zhou et al., 1994; Zhou et al., 1997). The gas nuclei help to activate flotation by promoting the attachment of larger bubbles. In other words, the tiny bubbles act as secondary collectors for particles, which can reduce flotation collector dosage and enhance the probability of particle attachment. This leads to substantially improved flotation recovery of poorly floating fine particles and reduced reagent cost, which is often the largest single operating cost in flotation plants.

**Figure 5. Cavitation-Tube™ sparging system.**

**Operating and Design Variables**

Column cell performance is strongly dependent on a variety of operating and design variables. Some of the most important of these include carrying capacity, retention time, aeration rate, wash water rate and chemical dosage. Additionally, tracer studies have indicated that mixing is also an important consideration. The proper application of these design and operating parameters is essential to ensure a successful column cell installation.

**Wash Water.** The use of wash water in column flotation is important and necessary in order to provide the optimum separation efficiency. In addition, a froth depth of 0.6-1.2 m is typically required to ensure good distribution of the wash water and to prevent short-circuiting of flotation pulp to the froth product. The flow of wash water should exceed the volumetric flow of water reporting to the froth product to prevent entrainment. In most cases, less than about 1% of the feed water will report to the froth product if the wash water is properly controlled. The amount of water carried by the froth can be calculated from:

$$Q_w = M_c\,[(100\text{-}P)/P] \qquad (1)$$

where $Q_w$ is the water flow demand per unit cell area ($m^3/hr/m^2$), $M_c$ is the dry mass rate of froth concentrate per unit cell area ($t/hr/m^2$) and P is the froth percent solids content. For example, a column cell producing 1.2 $t/hr/m^2$ of dry froth concentrate at 18% solids will carry about 5.5 $m^3/h/m^2$ of water from the pulp into the froth [i.e., 1.2 (100/18-1) = 5.5]. Entrainment should theoretically be eliminated when the number of dilution washes (defined as the froth water demand divided by the wash water addition rate) reaches a value of one. However, froth mixing usually requires that 1.25-1.50 dilution washes be used to fully suppress hydraulic entrainment. Higher gas and frother addition rates will typically increase the froth water demand and, as a result, the amount of wash water required. Excessive wash water flows should be avoided since the extra wash water passing downward through a column will create an undesirable reduction in the slurry retention time and, hence, a potential reduction in recovery. Very high water additions may also destabilize the froth by stripping frother from the bubble surfaces. High water rates may also have a detrimental impact on product grade by increasing axial froth mixing, thereby reducing the wash water effectiveness (Yianatos et al. 1988).

The design of the wash water distributor can also significantly affect column performance. In some cases, the distribution piping is intentionally submerged below the cell lip so that a drained froth can form above the distributor. This arrangement allows the depth of the drained froth and the extent of froth drainage to be varied by raising or lowering the distributor. Changes to the vertical position of the distributor can be used to control the split of water between the froth concentrate and tailings streams. In some cases, multi-level concentric distribution rings may also be used to overcome problems associated with poor froth mobility. The inner rings are typically located above the outer rings to reduce drainage and improve the fluidity of the froth in the center of the column. More recently, one or more internal launders have been employed in lieu of tiered wash-water rings. And in some cases, the water distributor may be located just above the top of the froth. This arrangement does not allow the froth mobility to be controlled by adjusting the distributor location, but it does make it easier to identify and correct plugging problems that may severely impact the performance of the distribution network.

**Aeration.** The primary advantage of an efficient and commercially available sparging system is the ability to generate large amounts of very small bubbles. This capability is commonly reported in terms of the superficial bubble surface area rate ($S_b$), which is defined as the total bubble surface area per unit of time passing through a given column cross-sectional area. This value can be calculated from the superficial gas velocity ($J_g$) and bubble diameter ($D_b$) using:

$$S_b = 6\,J_g/D_b \qquad (2)$$

The impact of $S_b$ on flotation recovery is illustrated by the test results given in Figure 6 (Kohmuench et al. 2004). These data show that recovery increases sharply as $S_b$ increases above 50 $sec^{-1}$ and eventually reaches a plateau at 100-150 $sec^{-1}$. The data also indicate that the relationship between recovery and $S_b$ is generally independent of sparger type.

The problem is that many commercial sparging systems, particularly static bubblers, simply cannot produce high $S_b$ values. Field tests also suggest that operation at $S_b$ values above 100-150 $sec^{-1}$ often produces poor separations because of "runaway" froths that cannot be effectively washed. A superficial gas rate in the range of 70-95 $m^3/h/m^2$ is often found to be suitable for most industrial applications. These values correspond to total aeration rates of approximately 990-1350 $m^3/hr$ of air for a 4.25 m diameter column. The gas rates at the lower end of the range would generally be used for spargers that generate smaller bubbles, while the higher gas rates are typically needed for less efficient spargers. A proper combination of gas rate and bubble size will generally provide a gas holdup in the flotation pulp of 15-18% or more.

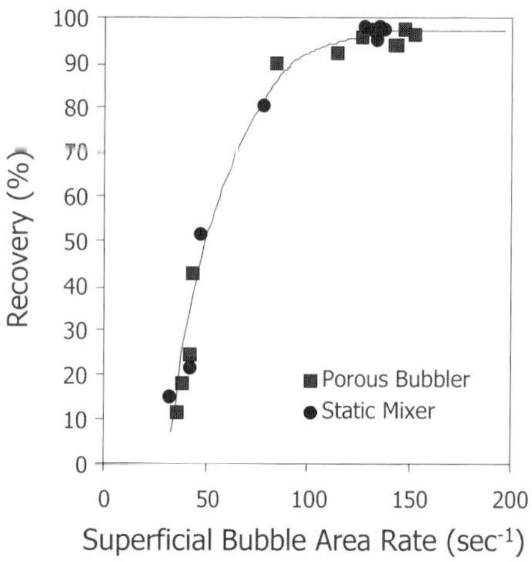


**Figure 6. Effect of $S_b$ on flotation recovery for different gas sparging systems.**

Caution should be used during the metering of gas flow rates. A properly designed system should be equipped with a flow meter that is calibrated to read correctly at a specified operating pressure. The operating pressure should be held constant by placing a pressure regulator ahead of the flow meter so that the meter will always operate at its design pressure. If the flow meter is placed after the control valve, then the operating pressure is unknown and the true gas flow rate cannot be determined. Improper metering of the gas flow rate can be a particularly serious problem when laboratory and pilot-scale tests are conducted for the purpose of collecting scale-up information. In addition, a great deal of confusion exists regarding the specification of compressors for column applications. Much of this problem is related to improper use of gas flow terminology (Sullair Bulletin, 1992). For example, column manufacturers normally report gas flow rates as a standard volumetric flow per time. This value is only valid at 1 atm of pressure and 20°C of dry air. The actual flow rate specified by compressor manufacturers is typically reported in terms of inlet conditions or free air. Although this amount of air enters the compressor, it is not necessarily the amount of air delivered to the column due to compressor seal leakage. As a result, the actual flow may be only 95% of the inlet flow. Furthermore, corrections to the gas flow rate must be made to account for differences in elevation (atmospheric pressure) and humidity. Air temperature generally has little impact on the capacity of an oil-flooded screw compressor, but may affect the performance of an air-cooled compressor. These complications generally require that professionals be consulted to ensure that the compressor is properly sized for the specified air requirements.

**Residence Time.** Like all flotation processes, column flotation cells must be designed to provide sufficient slurry residence time to achieve the target recovery of floatable material. The mean residence time can be estimated by dividing the volume of the flotation pulp (V) by the volumetric flow rate of tailings (Q), i.e.:

$$\tau = V(1-\varepsilon)/Q \qquad (3)$$

Note that the volume of the cell must be corrected for the fractional gas holdup ($\varepsilon$). A detailed procedure for estimating the column residence time required to achieve a given recovery has been presented in the literature (Dobby and Finch 1986) and is well beyond the scope of this article. However, as a rough rule-of-thumb, a column cell will typically require at least twice the residence time of a bank of conventional cells and three or more times the residence time of a batch laboratory flotation cell. One reason for the long residence time is to minimize the bypass of floatable particles to the tailings stream. Bypass can be caused by axial mixing that allows some material within the cell to report to tails without being influenced by the bubble swarm produced by the sparging system. Typically, as the residence time is increased, the probability of bypass is reduced. Bypass is often not an issue for conventional flotation cells due to their cell-to-cell arrangement. In contrast, column cells are often arranged in parallel and can be severely impacted by bypass. Consequently, it is not unusual for a column cell to be designed with a very long residence time when compared to a bank of conventional machines arranged in series. When calculating residence time, it is also important to note that the active volume for columns is defined as the volume of pulp contained between the bottom of the froth and the top of the air spargers. This volume is substantially less than the total volume of the tank structure. The calculated volume must also be reduced to account for the volume occupied by the air bubbles (i.e., air holdup).

**Froth Capacity.** The carrying capacity for a column is defined as the mass rate of flotation concentrate per unit time per cell cross-sectional area. While carrying capacity restrictions can often be ignored for conventional flotation circuits, it is of great importance in the design of column cells. This is largely due to the fact that the specific surface area of the cell (ratio of the cross-sectional area to the volume) is much higher for conventional cells. Maximum carrying capacity is achieved when all available bubble surface area has been consumed. When operating above this critical limit, any additional floatable solids within the system will be lost to the tailings. Therefore, it is important to design column cells with sufficient cross-sectional area to ensure a high recovery of floatable solids can be maintained. While this factor may not be an issue for some applications where the expected float yield is quite low, this constraint is often the limiting factor on recovery and capacity for high yield situations such as cleaner circuits. Maximum carrying capacity has been well defined by numerous researchers with extensive data available for coal and mineral applications. These studies indicate that carrying capacity is linearly related to the size and density of particles in the froth (Sastri, 1996). However, carrying capacity is best estimated from laboratory and pilot-scale flotation testing. During these evaluations, the feed rate (dry solids) to the column cell is sequentially increased. The product rate will increase in direct proportion to the feed rate until the maximum froth capacity is reached and all of the available bubble surface area has been consumed. Up to this point, a linear relationship is realized with a slope defined as the product yield (Figure 7). Once the maximum carrying capacity is reached, the product rate will remain constant or may decline due to heavy froth loadings. The required column cross-sectional area can be determined by dividing the expected froth tonnage (t/hr) by the carrying capacity (t/hr/m$^2$).

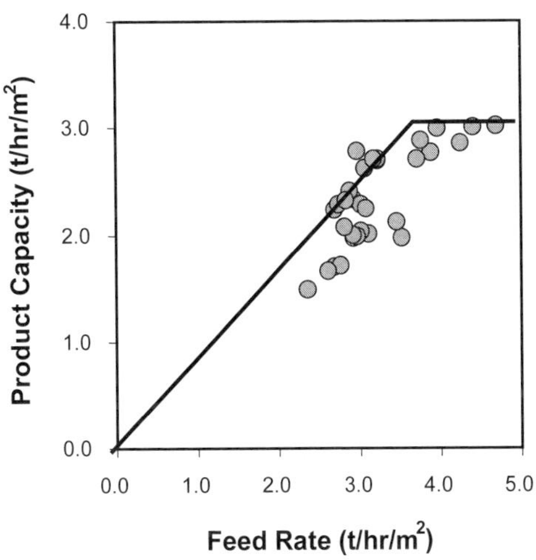


**Figure 7. Example of a carrying capacity plot for a pilot-scale test (Kohmuench et al., 2006).**

**Froth Handling.** Handling of the product can be an issue for column cells due to the large volumes of froth generated by this technology (Davis et al., 1995). In particular, concentrates containing large amounts of hydrophobic particles can become excessively stable, creating serious operating problems related to backup in launders and other downstream unit operations. Attempts to overcome this problem by selecting weaker frothers or reducing frother dosage can be successful, but may lead to lower column recoveries. Therefore, several circuit modifications have been developed to deal with the froth stability problem. For example, column launders need to be considerably oversized with steep slopes to reduce backup. Horizontal froth travel distances must be kept as short as possible and adequate vertical head must be provided between downstream operations and column launders. In addition, many installations have resorted to using defoaming agents or high-pressure launder sprays to deal with the stability problem. Newer column installations avoid this problem by including a deaeration tank to permit time for the froth to collapse (Figure 8). Special provisions may also be required to ensure that downstream dewatering units can accept the large froth volumes.

**Reagents:** The addition of flotation reagents is typically required in nearly all conventional flotation applications. Column flotation is no different and also requires the addition of a frother that will sufficiently lower the surface tension of the pulp to allow for the creation of a deep and stable froth. Many types and blends of frothers are available; however, the typical dosage of each type is dependent on many factors that include frother strength, site water chemistry, flotation circuit configuration, dewatering techniques, and whether the plant maintains an open or closed water circuit. Regardless, sufficient frother must be added to allow for maximum flotation yield. The ultimate chemical dosage may be lower if the reagent has the potential to cause other problems in the plant (i.e., froth build up). As such, the choice and dosage of these chemicals should be determined through consultation with professional chemical vendors.

**Figure 8. Use of high-pressure launder sprays and deaeration tank to improve froth handling.**

**Application Example**

Table 1 provides an example of the potential benefits that can be realized using column technology for the flotation of fine particles. The data compares the results obtained after replacing mechanical cells with columns at a major phosphate plant in Brazil. The upgrade reduced the number of cells in the plant from 66 mechanically agitated machines down to just six columns. Despite using less equipment, the recoveries of coarse apatite and barite were improved by 1% and 2%, respectively. However, much larger gains in recovery were observed for the fine circuits. The recovery of fine apatite was improved by 56% for the natural fines in the feed and 18% for the generated fines produced by the grinding circuit. The retrofit also reduced power costs by 6% and lowered collector demand for barite and apatite by 70% and 86%, respectively. As such, the column cells have proven to be extremely effective for upgrading fine material into salable products.

**Table 1. Comparison of mechanical and column flotation cells for the processing of fine phosphate ore.**

| | Mechanical Cell | Column Cell | Change % |
|---|---|---|---|
| Circuitry Data: | | | |
| -Number Cells | 66 | 6 | -91 |
| -Power (kW-hr/t) | 108 | 100 | -7 |
| Collector (g/t): | | | |
| -Barite Circuit | 567 | 170 | -70 |
| -Apatite Circuit | 2803 | 404 | -86 |
| Barite Recovery: | 95 | 97 | 2 |
| Apatite Recovery: | | | |
| -Natural Fines | 50 | 78 | 56 |
| -Generated Fines | 65 | 77 | 18 |
| -Coarse/Regrind | 71 | 72 | 1 |
| Fine Product: | | | |
| -$P_2O_5$ | 33.5 | 34.7 | 4 |
| -$Fe_2O_3$ | 6.1 | 5.8 | -5 |
| Coarse Product: | | | |
| -$P_2O_5$ | 36.2 | 35.9 | -1 |
| -$Fe_2O_3$ | 3.1 | 3.1 | 0 |

## HYDROFLOAT TECHNOLOGY

The HydroFloat™ separator was specifically designed to overcome the limitations of traditional flotation equipment in recovering coarse particles (Kohmuench et al., 2005). The HydroFloat cell, which is capable of treating particles in the 0.25-3 mm size range, combines the flexibility of a froth flotation process with the high capacity and low cost of a density-based separator. As shown in Figure 9, the HydroFloat consists of a tank subdivided into an upper separation chamber and a lower dewatering compartment. During operation, the unit operates much like a traditional hindered-bed (teeter-bed) density separator with the feed particles settling against an upward flow of fluidization water provided via a network of distribution pipes that extend horizontally across the bottom of the separation chamber. The novel feature of this device is an external aeration system that injects small air bubbles into the hindered-bed. The air bubbles are created by injecting compressed air and water through a high-shear mixer. The air-water mixture is then passed through the distribution network at the bottom of the cell to act as the fluidization medium. The air bubbles, which are typically 0.8-1.2 mm in diameter, selectively attach to particles that are naturally hydrophobic or have been made hydrophobic through the addition of appropriate collectors. Unlike flotation, the resulting bubble-particle aggregate does not need to have sufficient buoyancy to freely rise in the flotation pulp. Instead, the teetering effects of the hindered bed forces the low-density agglomerates to rise to the top of the bed and overflow into the product launder. Buoyancy calculations show that even a single attached bubble is usually sufficient to create the density difference needed to recover even large particles from a high-density teeter-bed. Denser minerals can also be floated in preference to lighter minerals, although this situation may limit the topsize of particles that can be effectively recovered. Hydrophilic particles continue to move down through the teeter-bed and eventually settle into the dewatering chamber. These particles are discharged as a high solids stream (e.g., 75% solids) through a control valve at the bottom of the separator. The valve is actuated in response to a control signal provided by a pressure transducer mounted to the side of the separation chamber. This configuration allows a constant effective density to be maintained within the teeter bed, which is important for stable operation.

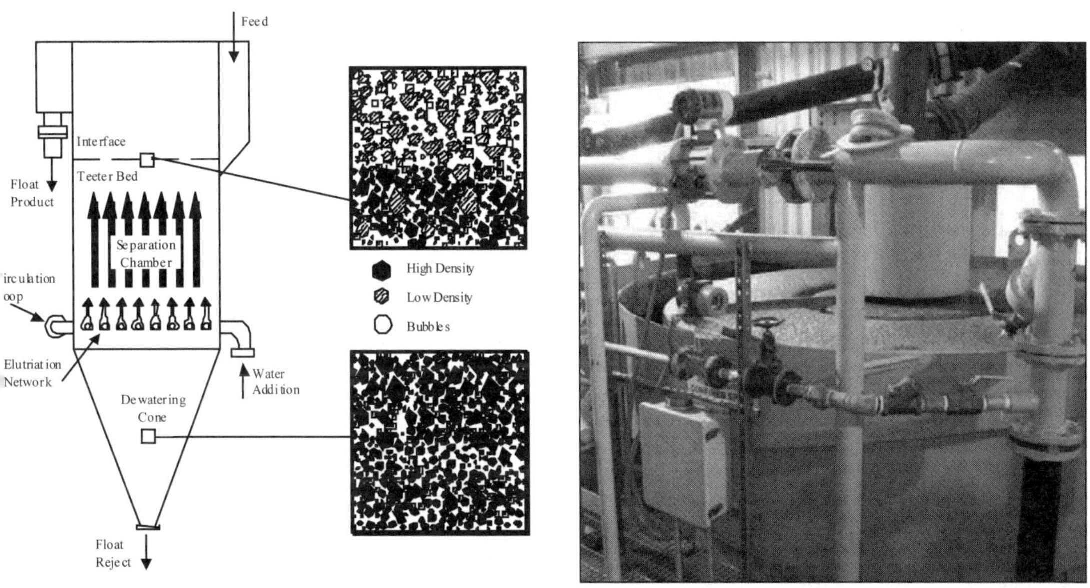


**Figure 9. Operating principle of the HydroFloat™ coarse particle separator.**

### Operating and Design Variables

Like any flotation process, the performance of the HydroFloat separator is dependent on a variety of operating and design variables. Some of the most important of these include feed rate, aeration, fluidization water, bed depth and reagent dosing. The effect of these variables on the separation performance of this technology is described in the following sections.

**Feed Rate.** One of the significant advantages of the HydroFloat™ separator is high unit capacity in a small footprint. In most cases, the relationship between specific capacity and dry solids feed rate is approximately linear up to a critical overload value at which recovery declines sharply (Figure 10). However, due to the high solids density in the unit, this critical limit is at least 2-3 times higher or more than attainable using conventional flotation technology. In addition, this high level of capacity can be maintained while achieving superior product recoveries and comparable grades (Table 2). The percent solids of the feed introduced to the HydroFloat can also impact on flotation recovery. In general, the recovery improves as the feed solids content increases. This improvement is attributed to several factors including a reduction in hydraulic loading and an improvement in bubble-particle contacting as the feed solids content increases. The large holdup and long residence time of solid particles within the teeter bed allows the unit to operate at a slurry residence time that is only 10-25% of that typically required in traditional flotation processes.

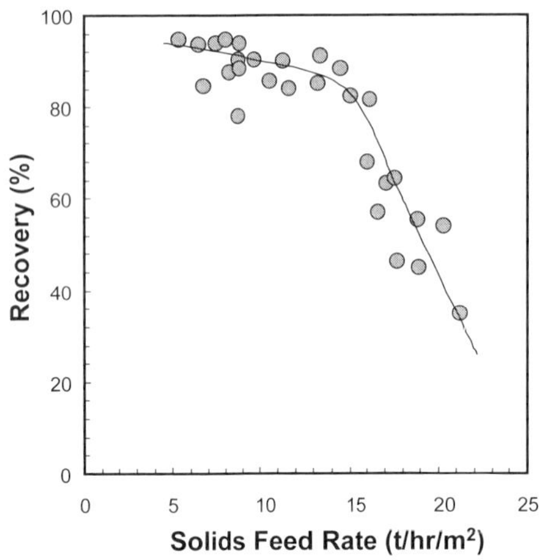


**Figure 10. Effect of solids feed rate on concentrate recovery.**

**Table 2. Comparison of conventional flotation and HydroFloat™ performance.**

| Application | Feed Rate ($t/hr/m^2$) | | Recovery (%) | | Grade (%) | |
|---|---|---|---|---|---|---|
| | Existing | HydroFloat | Existing | HydroFloat | Existing | HydroFloat |
| Phosphate | 10 | 25 | 91 | 98 | 4-6(1) | 4-6(1) |
| Potash | 7 | 16 | 50 | 90 | 60(2) | 57(2) |
| Coal | 3 | 20 | 50 | 60 | 8-10(3) | 8-10(3) |

(1) Percent Insols; (2) % $K_2O$; (3) % Ash

**Aeration:** Flotation in a fluidized teeter-bed is similar to other flotation processes in that an increase in aeration rate tends to increase recovery and decrease concentrate quality. For example, Figure 11 shows the effect of aeration rate on the recovery and grade of phosphate concentrate generated by a pilot-scale unit (0.37 $m^2$). The higher recovery should be expected since an increase in gas flow rate results in a greater probability that floatable solids will encounter air bubbles. Likewise, the decrease in product grade with increasing air rate can be attributed to several factors including increased hydraulic entrainment, greater turbulence in the particle bed, and the recovery of more composite particles as middlings. It should be noted that, due to its unique design, the separator requires substantially less air that traditional flotation processes. For most applications, aeration rates of less than 10-20 $m^3/hr/m^2$ are required to maintain good recovery values. This compares favorably to conventional flotation machines which typically demand at least 50 $m^3/hr/m^2$ or more gas to operate effectively. In fact, high aeration rates must be avoided in fluidized-bed separators since the increased resistance of the particle bed to the passage of large volumes of gas bubbles can result in high gas holdups and bed collapse when operating under this condition.

**Fluidization Water.** The flow rate of fluidization water is another important operating variable for the fluidized-bed flotation separators. This influence can be effectively demonstrated in Figure 12, which shows the effect of increasing fluidization water rate on product quality for upgrading of a coarse phosphate ore. The data show that the amount of unwanted acid insoluble material (insols) in the floated product increased linearly from approximately 5 to 10% as the fluidization water rate increased. The higher water velocity through the teeter-bed helps lift the large bubble-particle aggregates out of the bed and, hence, provides a slight increase in recovery. Unfortunately, the increased water velocity also tends to carry free gangue (silica sand) into the overflow product, thereby reducing the quality of the concentrate. The number of injection pipes and the spacing between the water injection holes area also a very important design parameter. For example, the data collected from commercial installations show that the performance of the cell can be improved by reducing the number of water injection pipes (and hence number of water injection holes) to improve water velocity and distribution. When using too many holes, the system pressure can be too low, thus allowing the air bubbles to coalesce before entering the teeter-bed. The turbulence created by the poorly dispersed air and water is very detrimental to overall performance in terms of both grade and recovery. The optimum number and spacing of holes in the distribution piping network is case specific and depends on the sizes and densities of particles in the feed stream.

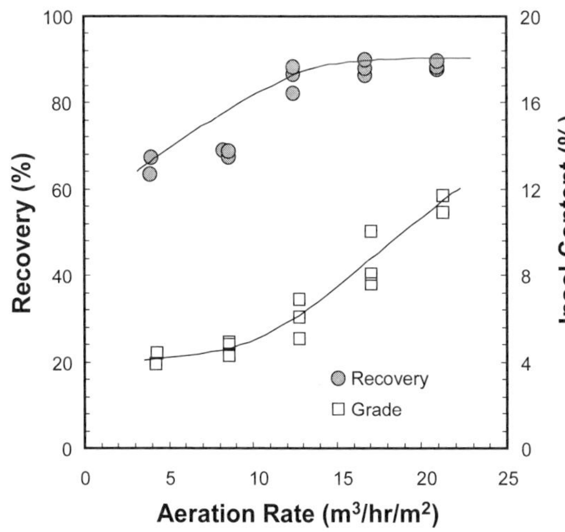


**Figure 11. Effect of aeration rate on concentrate recovery and grade.**

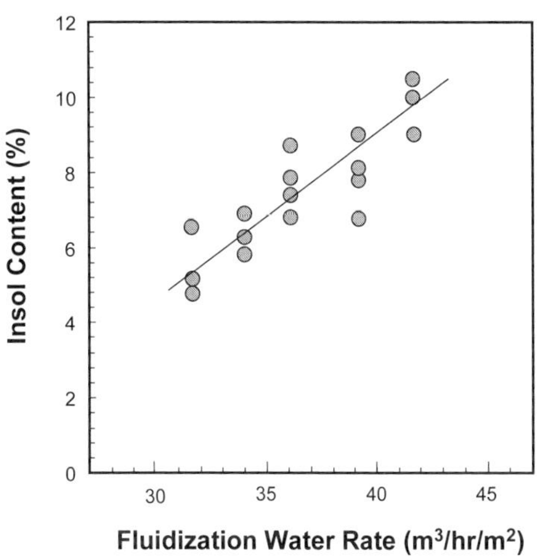


**Figure 12. Effect of fluidization water rate feed rate on concentrate grade.**

**Bed Depth.** The bed depth and feed injection points are important to the operation of the fluidized-bed flotation cell due to their respective impacts on product quality and concentrate recovery. The bed depth is defined as the vertical distance between the top of the teetering bed of solids and the top of the fluidization water injection tubes. While the location of the injection point varies from system to system, the feed inlet is typically placed one third of the distance down from the top of the cell. In most cases, a shallower bed depth decreases product grade while having little influence on product recovery. The decline in concentrate quality can be explained by the larger amount of nonfloatable gangue that is short-circuited into the overflow product as the bed level approaches the overflow lip. The depth of the feed injection point is defined as the distance between the feed injection point and the top of the teeter bed. The feed injection point tends to have a stronger effect on recovery of floatable solids than on concentrate quality. For most industrial separations, the recovery of valuable mineral increases as the feed inlet moves closer to the overflow weir and decreases as the feed introduction point is lowered. This trend suggests that particles introduced deeper into the teeter-bed are more difficult to recover since they have to travel through the high solids teeter-bed without becoming detached or misplaced.

**Reagents:** Reagents similar to those used in conventional flotation systems are required in the fluidized-bed flotation technology. The floated particles must be naturally hydrophobic or made hydrophobic via the addition of suitable collectors. Since the quiescent separation zone within the fluidized bed promotes the recovery of weakly floating particles, some reductions in collector demand (up to 35% or more) have been noted when using this technology in commercial applications. A lower collector dosage should be expected due to the reduced surface area of the coarser feeds that are typically upgraded by the separator. The device also requires the addition of a suitable frothing agent, despite the fact that little froth actually forms atop the cell. The frother is required to promote the generation of bubbles and to minimize bubble coalescence. At lower dosage levels (<5-10 ppm), an increase in frother dosage typically improves cell performance as a result of improved recoveries resulting from enhanced bubble generation and dispersion. However, excessive frother dosages (above 5-10 ppm) are detrimental to the operation of the separator. A reduction in recovery is often observed above a critical frother dosage and, in some cases, a complete collapse of the teetering bed can occur under this operating regime. Very small bubbles (<0.5 mm) generated at high frother dosages tend to create bubble-particle aggregates with less buoyancy than larger bubbles (~1 mm) formed at lower frother dosages. As such, the bubble-particle aggregates formed with larger bubbles have a lower apparent density and are recovered more readily in the separator.

**Application Example**

In order to demonstrate the potential benefits of the HydroFloat™ technology, full-scale prototypes were installed to upgrade coarse feed streams at an industrial phosphate beneficiation plant. These retrofits took advantage of existing tank structures that were already in place and being utilized as pneumatic flotation cells. A single 2.5-meter diameter HydroFloat was used to replace rougher-scavenger pneumatic cells. The results obtained from the commissioning of the circuit are plotted in Figure 13. The single HydroFloat™ rougher cell also provided metallurgical results that were superior to those provided by the pneumatic rougher flotation cell installed on a parallel coarse feed stream. The single-stage HydroFloat™ achieved recoveries approaching and exceeding 90%, while the existing cells struggled to maintain 80% recovery in a single stage. Moreover, the single-stage HydroFloat™ cell provided results close to those obtained using the combined two-stage rougher-scavenger pneumatic cells operating in the plant. The test data also showed that an impressive feed rate of 20 t/hr/m$^2$ was achievable using the HydroFloat™ cell. Based on the good results obtained from the initial coarse circuit tests, all three of the remaining pneumatic cells in the coarse circuit were converted to HydroFloat units. This allowed the coarse circuit feed to be increased by over 56% (i.e., from 145 to 227 t/hr). In addition to the improved metallurgy, the HydroFloat retrofit also provided benefits in terms of reduced reagent consumption. The fatty acid and fuel oil blend was reduced by 9.8% and the fuel oil dosage as reduced by 9.6%. A very large reduction of 56.7% in surfactant dosage was also possible due to the lower water addition rates used with the HydroFloat™ technology.

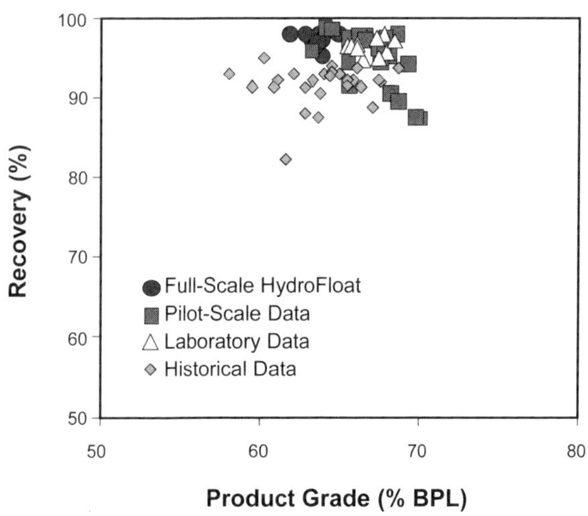


**Figure 13. Comparison of historical pneumatic cell data with data from different sizes of HydroFloat units.**

## SUMMARY

The industrial deployment of advanced flotation technologies now provide plant operators with the opportunity to recover greater amounts of valuable solids at a lower cost than previously attainable. For fine particles, column cells equipped with dynamic spargers, such as the Microcel™ and Cavitation-Tube™ technologies, provide high levels of energy dissipation that improves flotation recovery by generating smaller bubbles and promoting bubble-particle contacting. When combined with froth washing, these technologies enhance production while simultaneously improving flotation selectivity and reducing reagent consumption. For coarse particles, next-generation flotation technologies such as the HydroFloat™ cell have made it possible to selectively separate coarse particles once considered to be unfloatable. This compact flotation device combines the high capacity of a density separator with the operational flexibility of a flotation process. The flotation of particles suspended in a teeter bed provides an ideal environment for separating coarse particles at a high level of efficiency and very low processing cost. These two technologies now provide an effective means for extending the range of particle sizes that can be treated by flotation.

## ACKNOWLEDGEMENTS

The authors would like to acknowledge the tremendous contributions of the industrial participants involved in this work. These contributions, in terms of time, manpower, expertise, and willingness to install and test new technology are gratefully acknowledged.

**REFERENCES**


Davis V.L., Bethell P.J., Stanley F.L., and Luttrell G.H., 1995. "Plant Practices in Fine Coal Column Flotation," *High Efficiency Coal Preparation: An International Symposium*, Chapter 21, S.K. Kawatra, Ed., SME, Inc., Littleton, CO, pp. 237-246.

Dobby G.S. and Finch J.A., 1986. "Flotation Column Scale-Up and Modeling," *CIM Bulletin*, Vol. 79 (89), pp. 89-96.

Finch J.A., 1995. "Column Flotation: A Selected Review – Part IV: Novel Flotation Devices," *Minerals Engineering,* Vol. 8 (6), pp. 587-602.

Kohmuench, J.N., Davy, M.S., Ingram, W.S., Brake, I.R., and Luttrell, G.H., 2004. "Benefits of Column Flotation Using the Eriez Microcel," *Proceedings of the Tenth Australian Coal Preparation Conference*, Polkolbin, New South Wales, Australia, October 17-21, 2004, pp. 272-284

Kohmuench, J.N., Mankosa, M.J., Westerfield, T.C. and Luttrell, G.H., 2005. The HydroFloat Separator - A New Technology for Recovering Coarse Phosphate Sands, *Proceedings of the Centenary of Flotation Symposium*, June 6-9, Brisbane, Queensland, Australia, 1063 pp.

McKay J.D., Foot D.G., and Shirts M.D., 1988. "Column Flotation and Bubble Generation Studies at the Bureau of Mines," *Column Flotation' 88*, SME, Littleton, Colorado, pp. 173-186.

Sastri S.R.S., 1996. "Technical Note: Carrying Capacity in Flotation Columns," *Minerals Engineering*, Vol. 9 (4), pp. 465-468.

Sullair Corporation, *Sales Bulletin*, Industrial Compressors – Compressor Capacity, Bulletin E-78, April 1992, 3 pp.

Yianatos J.B., Finch J.A., and LaPlante A.R., 1988. "Selectivity in Column Flotation Froths," *International Journal of Mineral Processing*, Vol. 23, pp. 279-292.

Yoon, R.-H., Luttrell, G.H., Adel, G.T., and Mankosa, M.J., 1992. "The Application of Microcel Column Flotation to Fine Coal Cleaning," *Coal Preparation – A Multinational Journal*, Vol. 10, pp. 177-188.

Zhou, Z.A., Xu, Z. and Finch, J.A., 1994. On the Role of Cavitation in Particle Collection During Flotation-A Critical Review, *Minerals Engineering*, Vol. 7, pp. 1073-1084.

Zhou, Z.A., Xu, Z., Finch, J.A., Hu, H. and Rao, R., 1997. Role of Hydrodynamic Cavitation in Fine Particle Flotation, *International Journal of Mineral Processing*, Vol. 51, No. 1-4, pp. 139-149.

# FLOTATION OF FINE OIL DROPLETS IN PETROLEUM PRODUCTION CIRCUITS

C. Hank Rawlins

The Center for Advanced Mineral and Metallurgical Processing, Montana Tech of The University of Montana, 1300 West Park Street, ELC Building, Butte, MT, USA

## ABSTRACT

The oil and gas industry produces 3-5 barrels of water for each barrel of oil, which must be separated from the hydrocarbon fluids and cleaned for disposal or reinjection. Land based oil field can use mechanical flotation cells that have been developed for traditional mineral processing circuits, however offshore production installations requires smaller, lighter, and motion insensitive flotation cells that have been developed utilizing hydraulic principles only. Hydraulic induced gas flotation (IGF) in the oil industry must remove very fine oil droplets (10-50 microns) from hypersaline water using methane as the bubble gas. This paper will discuss droplet breakup and coalescence, along with droplet-bubble interaction with regards to oil and water fluid properties and hydraulic principles. Also discussed will be hydraulic IGF design with respect to orientation, residence time, gas and solids handling, flow pattern, oil skimming, and gas bubble generation.

## OILFIELD MINERAL PROCESSING

World crude oil production averaged 73.8 million barrels per day (BPD) (3.1 billion gallons per day) in 2008 (DOE/EIA), however, the total amount of water produced along with the crude oil fluids totaled over 194 million BPD (8.15 billion gallons per day) (Veil 2009), resulting in approximately 3 barrels of water produced along with 1 barrel of crude oil. This water may be termed connate, interstitial, or formation water, but is commonly referred to as produced water. Initial well production is designed to minimize or eliminate water production, as this liquid occupies production volume and requires expense for treatment and disposal. However all crude oil wells will eventually generate produced water due to water breakthrough or enhanced oil recovery (EOR) activities, thus a water treatment system is incorporated in most all production systems. The volume of produced water is often the limiting factor for sizing the entire production system.

Produced water treatment is governed by the U.S. Environmental Protection Agency (EPA) through the Clean Water Act. Article 40 CFR 435 of the National Pollutant Discharge Elimination System (NPDES) is entitled "National Effluent Limitations Guidelines for the Oil and Gas Extraction Point Source Category" and sets technology based limits for (overboard) discharge of produced water based on location. Oil and grease discharge limits (in produced water) for territorial seas (0-3 miles) and the outer continental shelf (>3 miles) are 29 mg/l monthly average and 42 mg/l daily maximum. However, petroleum production in onshore or coastal zones has zero discharge of produced water, except under specific instances for agricultural use or stripper wells (<10 BPD). In addition to oil and grease limits, regulations are set on toxicity, hypoxia, as well as other specific components (benzene, lead, thallium, phenol, etc.). The general trend worldwide regarding produced water is to reach zero discharge, thus the primary disposal route will be reinjection into the production formation or a separate zone.

Drilling a disposal well is a sunk cost that adds no fiscal benefit to the operating company, thus maximizing the lift of the disposal well is critical to minimize overall disposal costs. The damaging mechanisms for disposal wells include near well-bore formation plugging, invasive plugging, perforation plugging, or wellbore fill-up which can all be caused by contaminants in the produced water (Barkman and Davidson 1972). These contaminants include suspended silts, clays, scale, oil droplets, and bacteria. The solid and liquid contaminants are removed in a series of steps that utilize many traditional mineral processing unit operations specifically designed for oilfield use. Recovery of the oil from produced water, in addition to protecting the disposal well, also recoups some of the financial burden of treating the produced water. One hundred parts per million by volume (ppmv) of oil has a $0.21 per barrel value (based on $50 per barrel crude spot price) which can cover a significant part of the processing cost.

Using the definition "any naturally occurring homogeneous substance or compound that results from the processes of nature and obtained usually from the ground," crude oil or petroleum can loosely be classified as a mineral. Furthermore using the American Geologic Institute definition for ore as "a mineral, or aggregate of minerals, more or less mixed with gangue, which from the standpoint of a metallurgist can be treated at a profit" treatment of crude oil can be considered *oilfield mineral processing*. Crude oil is mixed with gangue materials, principally water, formation sand and clay, scale solids, salts, and other precipitates, and must be liberated to produce a crude oil "concentrate." Crude oil in itself is not an end product, much like a sulfide flotation concentrate from a zinc, copper, or lead circuit, but must be refined to produce a final saleable commodity. The offshore platform that receives raw crude and performs the upgrading duty to produce a clean crude concentrate is analogous to a mineral processing plant, and the crude oil refinery which receives the clean crude concentrate to produce commodity hydrocarbon products is analogous to a smelter.

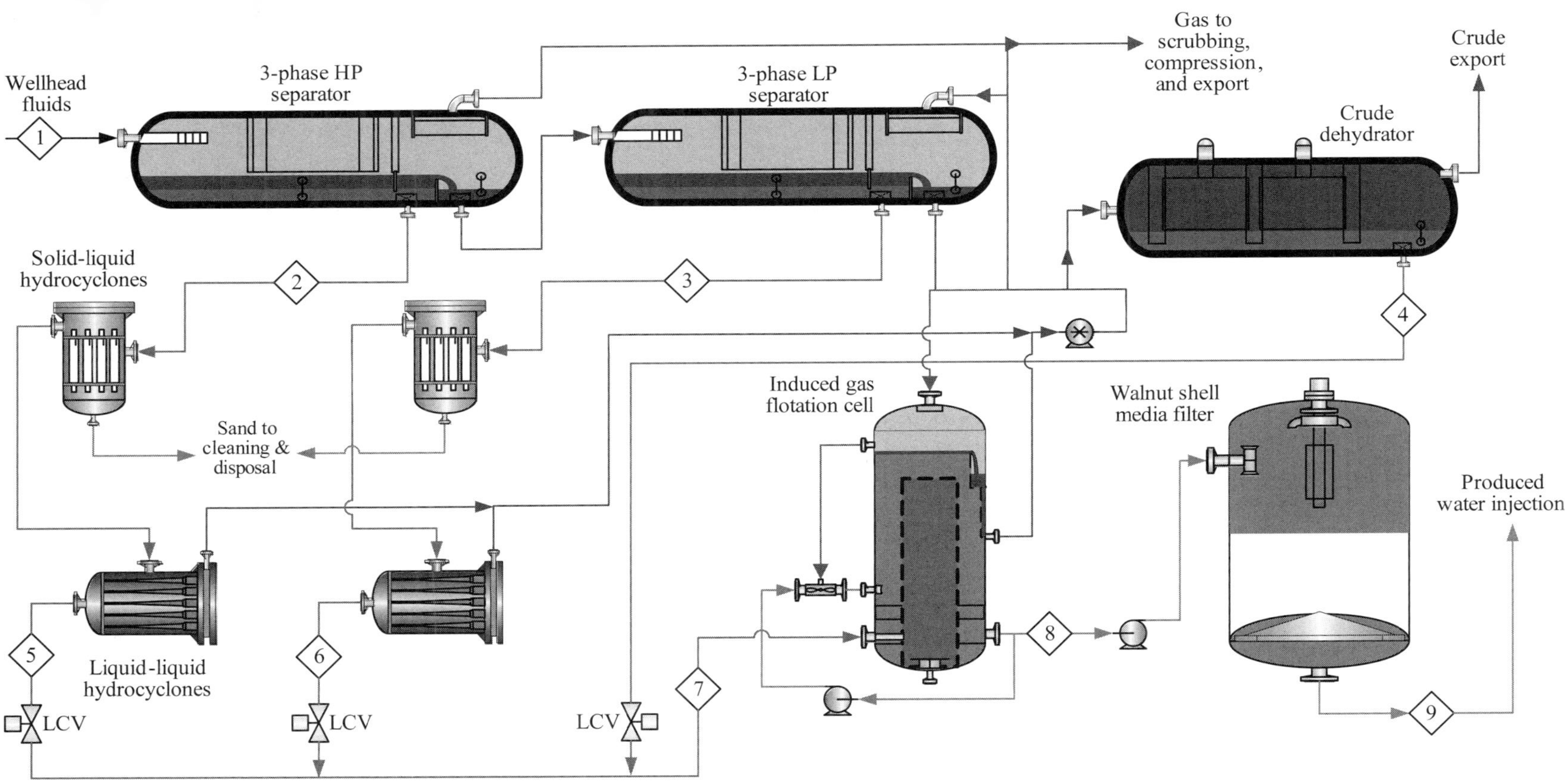


| Stream Description & No. | | Wellhead Fluids 1 | HP Separator Outlet 2 | LP Separator Outlet 3 | Dehydrator Outlet 4 | HP Deoiler Outlet 5 | LP Deoiler Outlet 6 | Flotation Inlet 7 | Flotation Outlet 8 | Filter Outlet 9 |
|---|---|---|---|---|---|---|---|---|---|---|
| Flow Condition | Units | | | | | | | | | |
| Pressure | (psig) | <1350 | 675-1350 | <260 | 50-150 | 100 | 100 | 50 | 45 | 25 |
| Temperature | (°F) | 120-180 | 120-180 | 120-180 | 150-300 | 120-180 | 120-180 | 150-200 | 100-150 | 100-150 |
| Oil Concentration | ppm(v) | 5-99% | 500-2000 | 500-2000 | 500-2000 | <50 | <50 | <50 | <25 | <5 |
| Mean Oil Drop Size | microns | Continuous | 100-300 | 100-300 | 100-300 | <50 | <50 | <40 | <10 | <5 |

**Figure 1. Process flow diagram for the raw crude and produced water treating systems for an offshore Gulf of Mexico based crude oil process system.**

**Offshore Produced Water Treatment**

Figure 1 shows a process flow diagram for an offshore Gulf of Mexico based crude oil process system, similar to what would be found on a floating or fixed based platform structure, with emphasis on the produced water treatment system. Wellhead fluids (raw crude) are received through a choke valve into a series of three phase separators. The production separators are large, horizontal, cylindrical pressure vessels with dimensions from 10-15 foot diameter by 30-50 foot in length. Three phase production separators operate on the principles of pressure let-down and gravity settling. Pressure reduction allows gas dissolved in the crude and water to effervesce, while gravity settling results in the formation of three distinct layers with gas on top, crude in the middle, and water (with solids) at the bottom. The high pressure (HP) and low pressure (LP) separators operate in series with a sequence of pressure let-down to capture natural gas in stages. The recovered gas reports to scrubbing and compression for export to a pipeline. The crude oil goes through an electrostatic dehydrator as a polishing stage to meet the export specification of 0.5% BS&W (bottom sediment and water).

The produced water treatment system is designed to capture and combine all separator water streams and prepare them for common disposal, which in this case is injection into a subsurface formation. Produced water from the HP and LP separators follow parallel paths operating at different pressures. Water from the HP separator exits at 675-1350 psig and 120-180°F while the water from the LP separator exits at <260 psig and similar temperature. Oil in the separator produced water stream is in the form of droplets that have not risen due to buoyancy under Stokesian principles. A concentration range is from 500-2000 ppmv oil-in-water, with a median oil droplet diameter of 100-300 microns. Water discharging from the dehydrator will be at a lower pressure (50-150 psig) and higher temperature (150-300°F), but with similar oil droplet concentration and median droplet diameter. However, the water flow rate from the dehydrator comprises <5% of the total produced water treatment throughput.

Produced water from the separators is first treated with a closed underflow multi-liner solid-liquid hydrocyclone vessel termed a desander. Inorganic solid particles (mainly sand and scale) greater than 15 microns are separated and periodically discharged for cleaning and disposal. The desander operates with a nominal 15-25 psi pressure drop and details of its design and operation are given elsewhere (Rawlins and Wang 2000). The produced water stream is then treated with a multi-liner liquid-liquid hydrocyclone vessel termed a deoiler. Dispersed oil droplets greater than 50-75 microns are captured to form a continuous oil concentrate (<10% oil) reject stream which is sent to the dehydrator or a slop tank for oil recovery. Details of the design and operation of a deoiler system are given elsewhere (Ditria and Hoyack 1994). Produced water exiting the deoiler contains up to 50 ppmv of oil (thus 90-98% removal efficiency by the deoiler) with a median droplet size <75 microns. This stream passes through a level control valve (LCV) which is used to control the liquid level in the production separators. The turbulence created by the pressure drop through a control valve may decrease the median droplet diameter by up to 90%, so the LCV is placed after the deoiler to optimize operation of that unit process. However, droplets remaining in the produced water stream after the deoiler are subject to shearing, but since the overall concentration of oil is much less the net effect is minimized. Oil droplets that are not captured by the deoiler and/or sheared by the LCV are termed fine oil droplets. Under quiescent or laminar conditions fine oil droplets in high concentration will interact and coalesce to increase the median droplet size. However after the deoiler and LCV the total oil-in-water concentration is very low, thus the probability of growth by coalescence is diminished. As a point of reference for this paper the term *fine oil droplets* applies to a median droplet size <50 microns in a very dilute system (<50 ppmv).

A flotation cell is the primary unit operation used to recover fine oil droplets from produced water systems. Both mechanical induced gas flotation (IGF) and hydraulic dissolved gas flotation (DGF) cells have been used for residual oil removal from produced water since the late 1960's (Sport 1970; Bassett 1971). Mechanical IGF cells developed from the WEMCO® multi-cell rectangular-shaped mechanically-driven rotor/stator design. The primary difference from a mineral processing flotation cell is that the flotation chamber is fully encapsulated to prevent hydrocarbon vapor release into the atmosphere. Mechanical IGF cells have found significant use in applications that are not space or motion sensitive such as land-based or fixed-platform fields. However deepwater floating platforms are not suitable for mechanical IGF use as the footprint requirements are too strict and motion of the platform does not allow a fixed skimming level for the oil froth resulting in re-entrainment of the oil or splashing of the produced water. The latest generation of offshore flotation cells are primarily hydraulic IGF or DGF designs with a vertical orientation and no moving parts. The vertical orientation both minimizes footprint, but also allows a thicker oil skim pad which is much less sensitive to platform motion. Vertical hydraulic IGF cells are similar in concept to a column flotation cell, however the length to diameter ratio is normally 3:1, whereas a column cell may be 10:1. These flotation cells rely partly on dissolved gas effervescing from solution (due to pressure letdown), but the primary bubble source is hydraulic entrainment of gas through an eductor, sparger, or shear pump. In all cases the flotation gas is natural gas sources from the facility. Air is never used as the flotation gas, as oxygen in the system can lead to adverse chemical reactions, corrosion, or biological growth. Figure 2 shows a four-stage mechanical IGF cell installed at a land-based oil facility and a vertical IGF cell installed on an offshore fixed-platform. A flotation cell will generally reduced the oil concentration to <25 ppmv and capture oil droplets down to a median size of <10 microns. Currently eight companies provide over twenty different models of flotation cells specifically designed for produced water filtration (Rawlins 2009).

The final stage of produced water treatment before injection is normally deep bed media filtration. These units are designed to remove the residual oil droplets and particulate solids to prevent formation damage. The discharge oil and solids concentration is normally just a few ppmv. Several types of media are used such as garnet or silica, but the preferred bed media is walnut or pecan shells. Typically sized at 12/20 or 20/30 mesh, this bio-media has an excellent combination of oil adhesion and resistance to attrition during the backwash cycle.

**Figure 2. (left) Four-stage mechanical IGF cell installed at a land-based oil facility (courtesy of Whittier Filtration, Inc.) and (right) a vertical IGF cell (gray vertical vessel near center) installed on an offshore fixed-platform (courtesy of FLSmidth gMAX Systems).**

## PROPERTIES OF THE FLUIDS IN PETROLEUM CIRCUITS

In flotation of fine oil droplets in a petroleum production water treating circuit, the continuous phase is produced water. Produced water differs significantly from the fresh/recycled water used in mineral processing circuits in the amount of dissolved solids. All formation (produced) water contains substantial concentrations of dissolved solids, primarily sodium chloride. The concentration and distribution of ions differs greatly from seawater giving produced water unique properties. Total dissolved solids (TDS) concentrations have been reported from 200 ppm to saturation, which is near 300,000 ppm (McCain 1990). Typical values lie in the range of 80,000-120,000 ppm, which is 3-4 times the TDS value of seawater. The most common cations are $Na^+$, $Ca^{2+}$, and $Mg^{2+}$, and the most common anions are $Cl^-$, $SO_4^{2-}$, and $HCO_3^-$.

### Properties of Produced Water

In modeling the mechanisms of fine oil droplet flotation in produced water, the properties of the continuous fluid are required. The density and viscosity are needed for a simple Stokesian buoyancy calculation, while further calculations on bubble-droplet interaction require the interfacial tensions (water-oil and water-gas). These thermophysical properties are a function of pressure, temperature, and water composition. Of these factors, temperature and composition are determining while the pressure normally exhibited in a flotation cell (<100 psig) has negligible effect. Figure 3 shows the effect of temperature and salinity on water density and viscosity at atmospheric pressure.

Temperature and salinity have competing effects on both density and viscosity. An increase in temperature decreases both properties, while an increase in dissolved solids has a direct proportional effect. Produced water containing 15% NaCl at 50°C has density of 1099 kg/m$^3$ and viscosity of 0.74 cP. Temperature has a stronger effect on viscosity and dissolved solids content a stronger effect on density. In terms of Stokes' Law, both density and viscosity have unity exponent however the density term is a relation between oil and water ($\rho_{water}$-$\rho_{oil}$) while viscosity is a direct factor. Temperature has a slightly stronger effect on viscosity than salinity. For a given droplet diameter, doubling the temperature (at a fixed salinity) increases the rise velocity by ~50% and doubling the salinity (at a fixed temperature) decreases the rise velocity by ~10%, therefore temperature is the prevailing parameter. While this discussion has primarily focused on the rise of oil droplets in produced water a similar relationship exists with the rise of gas bubbles in produced water. The density difference between water and gas is several orders of magnitude higher than that between water and oil, but the viscosity effect of the water phase is the same. The net effect is that gas bubbles rise about one order of magnitude higher velocity compared to a similar diameter oil droplet.

The interfacial tension (IFT) between gas-water ($\gamma_{gw}$) and oil-water ($\gamma_{ow}$) is a key factor in determining droplet/bubble stability. Figure 4 shows the effect of temperature and salinity on these parameters at ambient pressure. The air-water IFT curve is given as a point of reference and shows that increasing temperature decreases IFT. Methane has a slightly lower IFT (3-7%) to fresh water compared to air showing that methane bubbles in water are slightly less stable than air bubbles and the driving force for bubble coalescence is enhanced (Jennings and Newman 1971). The effect of dissolved solids is to increase surface tension. The data shown for methane-water IFT at various salinities is estimated from data for air-water IFT with NaCl molality given at ambient pressure and temperature (Argaud 1992). Increasing the salinity to 10 wt.% solids (NaCl) brings the methane-water IFT to near the same values as air-water. Salinity has the effect to stabilize gas (methane) bubbles and work against bubble coalescence. The magnitude of temperature and salinity effect on methane-water IFT is approximately the same, but working in opposite directions.

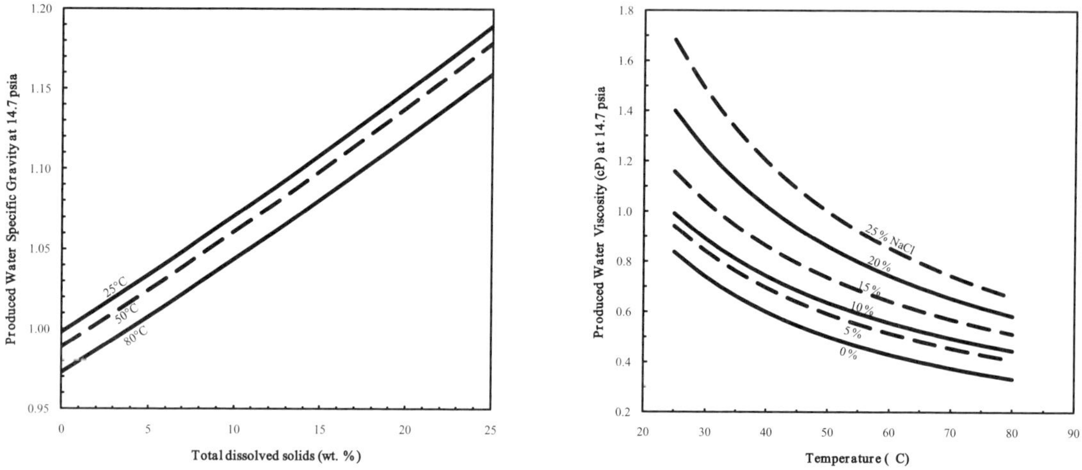


**Figure 3. (left) The effect of dissolved solids (as salinity) and temperature on the specific gravity of produced water (data from McCain 1990 and Street et al. 1996) and (right) the effect of salinity and temperature on the viscosity of produced water (data from McCain 1990).**

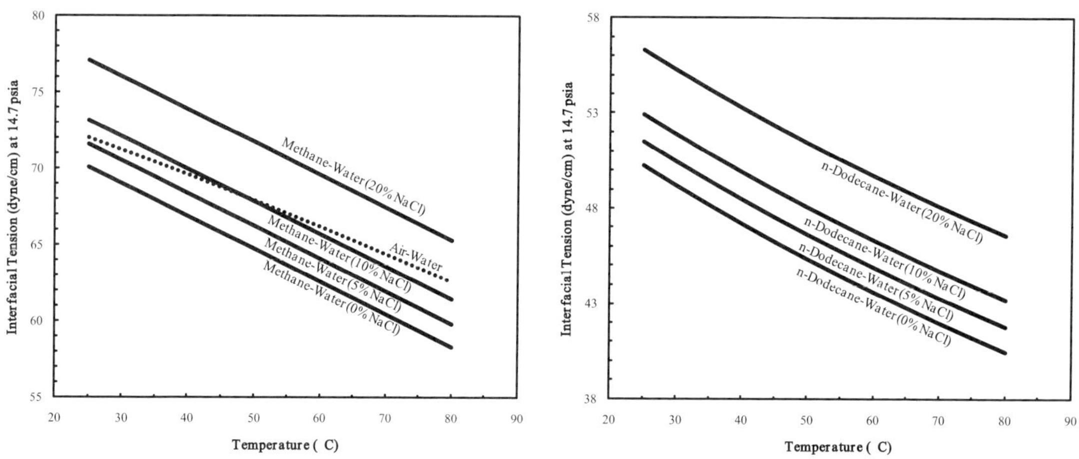


**Figure 4. (left) The effect of dissolved solids (as salinity) and temperature on the interfacial tension between methane and water at ambient pressure (data from Jennings and Newman 1971; Street et al. 1996; Argaud 1992) and (right) the effect of salinity and temperature on the interfacial tension between n-dodecane and water at ambient pressure (data from Street et al. 1996; Argaud 1992).**

Figure 4 also shows the interfacial tension between oil and water. While methane is a single component gas and is the most prevalent component of flotation gas used in hydraulic IGF cells, crude oil has a widely varying composition. As a reference, data is presented for n-dodecane ($C_{12}H_{26}$) which has the characteristics of a light crude-oil condensate. It has a specific gravity of 0.80 (46 API) and is a neutral (non-polar) oil. The IFT range for n-dodecane and pure water is about 30% less than the IFT for methane pure water. Oil droplets are less stable and have a stronger driving force for coalescence compared to bubble-bubble coalescence. The effect of temperature is to decrease IFT, while salinity increases IFT. Compared to methane-water, the magnitude of temperature and salinity on IFT are very similar.

As both methane and n-dodecane are neutral hydrocarbons, the IFT between these compounds is lower than that between either of them and water. Figure 5 shows the IFT between methane gas and n-dodecane liquid across the temperature range 25-80°C (ambient pressure). The IFT value is about 50% less than the IFT between n-dodecane and water. Temperature has a small decreasing effect on IFT over the observed range, reducing the value about 9%. These values disregard the effect of polar hydrocarbons that may be present at the liquid-gas interface, and the effect of dissolved or dispersed solids in the water. All of these factors will increase the IFT between methane and n-dodecane. Decreasing the carbon content of the alkane ($C_n$) decreases the oil-gas IFT ($\gamma_{og}$). Liquid i-Pentane ($C_5H_{12}$) has approximately half the IFT with methane compared to liquid n-dodecane. Raw crude oil is a mixture of at least 30-40 hydrocarbons ($C_1$, $C_2$,... $C_n$) and the real $\gamma_{og}$ is often lower than that shown in Figure 5.

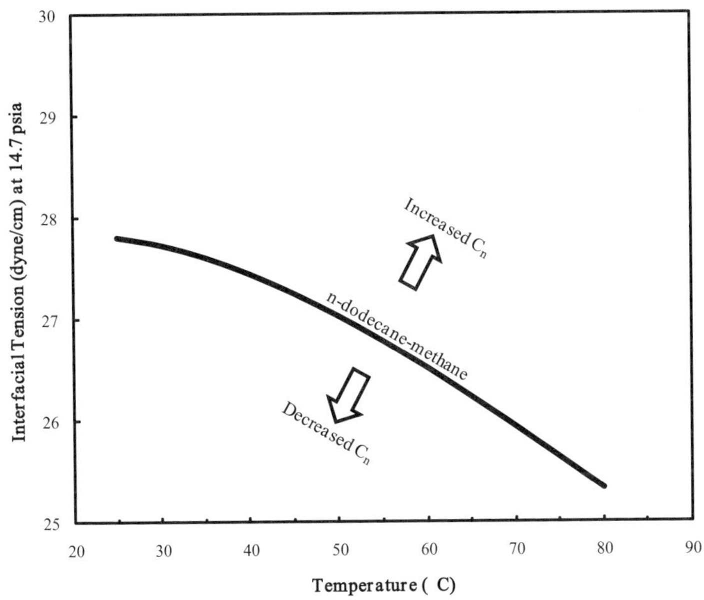


**Figure 5. The effect of temperature on the interfacial tension between methane and n-dodecane (data from Mc-Cain 1990).**

**Spreading Coefficient and Free Energy**

One method to estimate the efficacy of oil droplet flotation by methane bubbles is to evaluate the spreading coefficient. The spreading coefficient of a fluid (S) is the imbalance between the interfacial tensions acting along the contact line between the fluid phases (Grattoni et al 2003). For the water-oil-gas system, the oil spreading coefficient at the water-gas interface ($S_o$) is given by Equation 1.

$$S_o = \gamma_{wg} - \gamma_{ow} - \gamma_{og} \tag{1}$$

If ($\gamma_{ow} + \gamma_{og}$) is less than $\gamma_{wg}$, then $S_o$ will be positive which indicates spreading of the oil layer to form a boundary between the gas and water. A negative $S_o$ indicates a non-spreading condition, in which the oil will form a finite contact angle with the gas and water interface. Spreading of the oil droplet along the gas-water interface (i.e., formation of an oil coating on the gas bubble) will only happen if the balance of interfacial tensions yields a positive spreading coefficient, therefore gas flotation is more effective with a positive $S_o$.

Figure 6 shows the effect of temperature and salinity on the spreading coefficient of the dodecane-methane-water system. In pure water a dodecane droplet has a $S_o$ of -7.93 dyne/cm at 25°C. Temperature has a small increasing effect on $S_o$, raising the value by ~4% when the temperature doubles to 50°C. Salinity has a slightly greater effect increasing $S_o$ by 7% as the salinity doubles from 10% to 20%. In all cases shown $S_o$ is negative indicating that n-dodecane will not spread (or wet) the interface between methane and water, but will form a definite contact angle. Decreasing the carbon content of the alkane ($C_n$) will increase the spreading coefficient such that light oils (i.e., propane and ethane) or raw crude oils that contain a high content of these light hydrocarbons will actively coat bubbles of methane in a saline-water system. Chemical additions are not commonly used in oil droplet flotation in order to minimize treatment cost, however a surfactant that lowers the oil-water interfacial tension ($\gamma_{ow}$) will increase the spreading coefficient to enhance bubble coating by the oil droplet.

Further analysis with the spreading coefficient is to evaluate the overall change in system energy involved with oil coating a gas bubble. The free energy (G) of the interface between a gas bubble and water or an oil droplet and water can be determined from the interfacial tension ($\gamma$) and bubble/droplet surface area (A), as shown in Equation 2.

$$\gamma = \left( \frac{\partial G}{\partial A} \right)_{P,T,n} \tag{2}$$

For constant pressure, temperature, and composition then G is a product of $\gamma$ and A. If the system energy is reduced ($\Delta G<0$) by converting a discrete oil drop and a discrete gas bubble into an aggregate, then the oil will coat the gas bubble. The creation of an aggregate from the discrete particles can be represented by Equation 3.

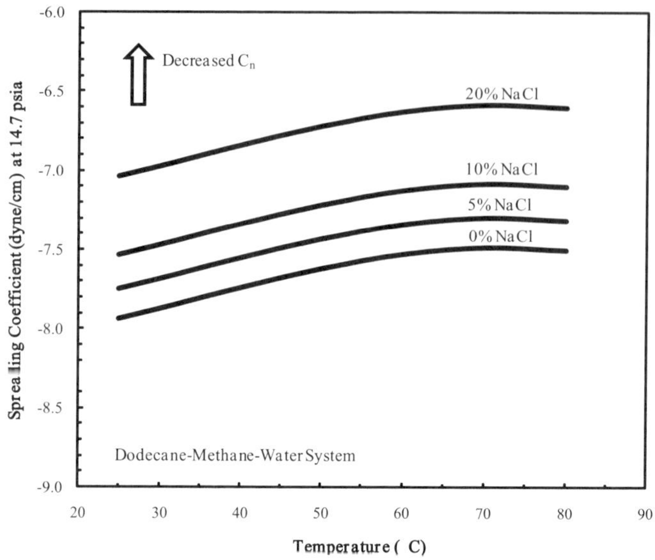


**Figure 6. The effect of temperature and salinity on the spreading coefficient of the dodecane-methane-water system.**

$$\left(\gamma_{og} A_{bubble} + \gamma_{ow} A_{oilfilm}\right) - \left(\gamma_{ow} A_{drop} + \gamma_{wg} A_{bubble}\right) < 0 \tag{3}$$

### Critical Diameter Ratio

If the energy of the discrete oil droplet ($\gamma_{ow}A_{drop}$) plus the discrete gas bubble ($\gamma_{wg}A_{bubble}$) is greater than the energy of an oil coated bubble, then the droplet will coat the bubble. The oil acts to replace the water-gas interface with an oil-gas interface plus oil-water interface with a net reduction in free energy. The reduction in energy is both interfacial tension and bubble/droplet size dependent. As the surface area of the bubble is in both terms, changing the bubble size alone has less effect than changing the droplet:bubble ratio. Holding the interfacial tension values constant, a droplet/bubble diameter ratio can be determined below which the droplet will not coat the bubble, but attach through a different mechanism. This diameter is called the critical diameter ratio ($D_{cr}$). Figure 7 shows the net energy curve for 50 μm, 100 μm, and 200 μm diameter methane bubbles in water (0% salinity) at ambient temperature and pressure in contact with a n-dodecane droplet of varying diameter.

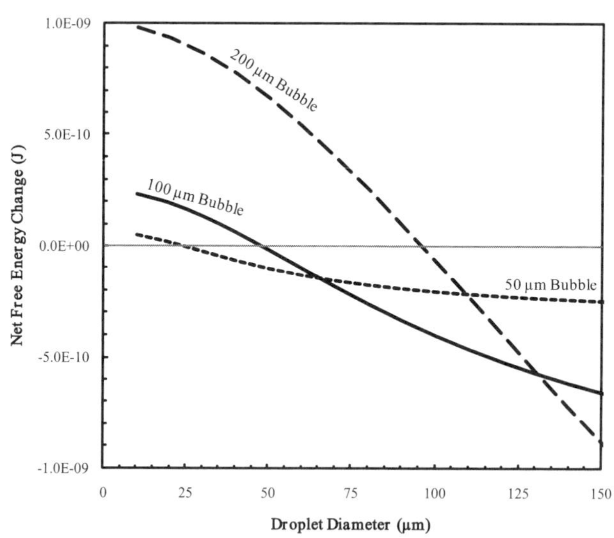


**Figure 7. Net surface energy change for coalescence of oil droplet (n-dodecane) of varying diameter onto gas bubble (methane) in water (0% salinity) at ambient pressure and temperature.**

The critical diameter ratio is determined for each bubble size at the point where the net energy change goes below zero. At $\Delta G<0$ the reduction in energy drives the coating of the gas bubble with the oil droplet. For the scenario depicted in Figure 7, $D_{cr}$ has a constant value of 0.48 independent of bubble diameter. A 100 micron methane bubble will capture (by coating) n-dodecane droplets 48 microns and larger. Droplets smaller than this diameter may be captured by a different mechanism.

The value of $D_{cr}$ will change with water salinity and temperature. Figure 8 shows the effect of salinity (0-20%) at a constant temperature and pressure and the effect of temperature (25-80°C) at a constant salinity and pressure for the dodecane-methane-water system. Salinity has a moderate decreasing effect on $D_{cr}$ reducing the value by 0.065 (~13.6%) across the range of 0-20% NaCl content. Temperature has a slight increasing effect on $D_{cr}$, raising the value by 0.052 (~10.9%) as the temperature is increased from 25° to 80°C.

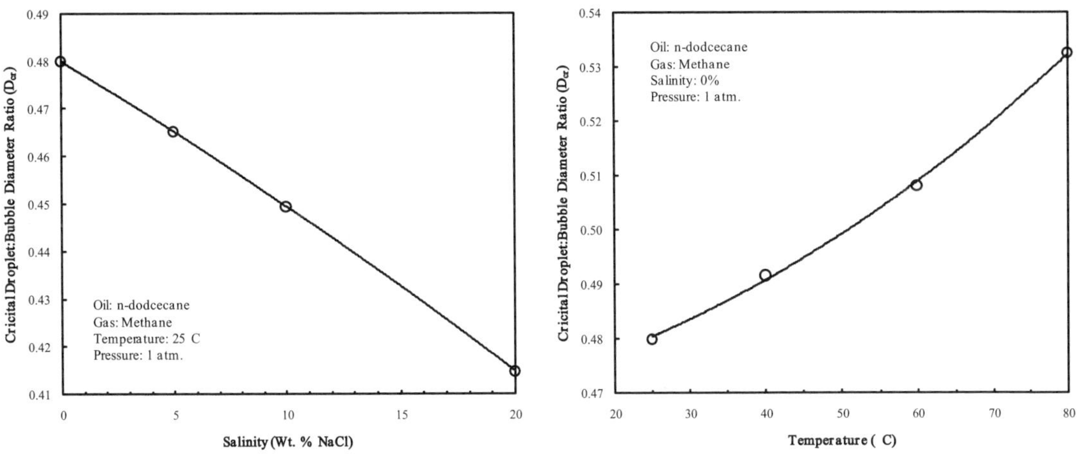


**Figure 8. Effect of salinity (left) and temperature (right) on critical droplet:bubble diameter ratio ($D_{cr}$) for the dodecane-methane-water system at ambient pressure.**

Capture of a distribution of oil droplets by coalescence (coating) of methane bubbles is promoted by increased salinity and lower temperature. Minimization of $D_{cr}$ allows capture of smaller droplets with the gas bubbles. Typically these factors are not adjustable within the system, however changes to the IFT, bubble diameter, or droplet diameter can be undertaken to improve oil droplet capture. The most important factor is to decrease methane bubble size or increase the oil droplet size, as droplet capture is a ratio of the bubble diameter. To capture a 10 micron droplet, an approximate 20 micron gas bubble is the largest that can be used. Another important factor is to maximize the methane-water interfacial tension. The magnitude of $\gamma_{wg}$ is larger than $\gamma_{og}$ or $\gamma_{ow}$. Impurities in methane such as higher hydrocarbons, oxygen, or carbon dioxide will all decrease $\gamma_{wg}$ thus the methane used should be kept clean and pure.

## GAS BUBBLE AND OIL DROPLET BUOYANCY

Small gas bubbles and oil droplets in an aqueous system can be assumed to have a spherical shape, thus their buoyant rise velocity can be approximated using Stokes' law. The common form of Stokes' relationship is given in Equation 4, where $u_\infty$ is the terminal velocity (cm/s), $g$ is gravitational acceleration (981 cm/s²), $d_p$ is the bubble/droplet diameter (cm), $\rho_l$ is the liquid density (g/cm³), $\rho_p$ is the bubble/droplet density (g/cm³), and $\mu_l$ is the liquid viscosity (g/cm·s).

$$u_\infty = \frac{g d_p^2 \left(\rho_l - \rho_p\right)}{18\mu_l} \tag{4}$$

Stokes' law is valid for dilute suspensions and where the rising droplet or bubble has a Reynolds number (Re) <0.1. For Reynolds numbers up to unity Stokes' law has a 9% error, thus it can be termed approximately valid for Re<1.0. The Reynolds number for a rising spherical particle in a fluid can be determined from Equation 5.

$$\mathrm{Re} = \frac{\rho_l u_\infty d_p}{\mu_l} \tag{5}$$

Produced water treated by flotation normally has <100 ppmv oil content thus the suspension can be approximated as dilute. The oil droplets will be distributed amongst a size distribution that is dynamic due to continual droplet breakup and coalescence, but the average droplet size is <40 microns. As a population of gas bubbles is introduced into the system they will interact with the distributed oil droplets to capture and coalesce forming an aggregate particle. Assuming a coating-capture mechanism where the oil forms a film on the gas bubble, the aggregate will take the shape of the gas bubble, thus the aggregate will have a spherical shape and the resultant rise velocity be approximated by Stokes' law.

Figure 9 shows the calculated rise velocity for methane gas bubbles, n-dodecane oil droplets, and methane-dodecane aggregates in water (0% salinity) at ambient pressure and temperature. The velocities were calculated using Equation 4, and the aggregates were assumed to be methane bubbles coated with an even film of n-dodecane. At the given conditions, methane bubbles rise at 3.8 times the rate of dodecane droplets. The valid Stokes' region for methane bubbles is <120 μm and for n-dodecane droplets is <190 μm. An aggregate made from the coalescence of a 100 μm methane bubble and 100 μm n-dodecane droplet has a diameter of 126 μm, with an intermediate velocity and is within the Stokes' region (<150 μm). Attachment of the gas bubble increased the droplet velocity from 13.6 cm/min to 32.7 cm/min, which represents a factor of 2.4. For an n-dodecane droplet at the $D_{cr}$ size (48 μm), attachment of a 100 μm bubble will result in a velocity increase factor of 16.4. Maximum flotation effect will be achieved by selecting a bubble size such that the oil droplet mean size is the corresponding $D_{cr}$. If the bubble size has a $D_{cr}$ larger than the oil droplet size then the oil droplets may not be captured. If the bubble size has a $D_{cr}$ smaller than the oil droplet size the oil droplets will be captured, however the rise velocity of the aggregate will be less than the optimum value.

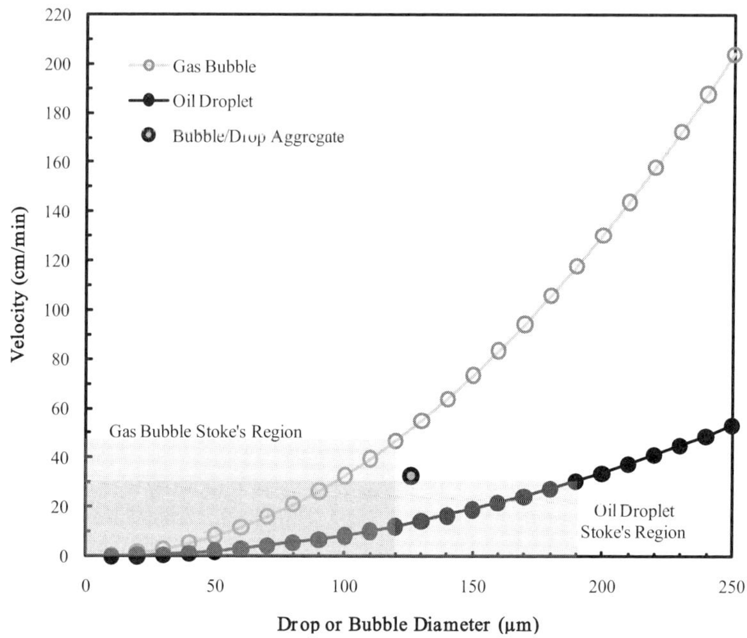


**Figure 9. Stokes' buoyant rise velocity of methane gas bubble, n-dodecane droplet, and gas-bubble aggregate in water (0% salinity) at ambient pressure and temperature.**

Increasing the produced water salinity (higher density) and mean oil droplet size aid flotation, while increasing the water viscosity, oil density or gas density hinder flotation. Pressure and temperature effects on flotation are reflected in the water, oil, or gas density and viscosity.

The primary purpose of oil droplet flotation is to reduce residence time. Whereas mineral flotation seeks to separate and concentrate a specific mineral in a mixture of other minerals, oil flotation seeks to separate and concentrate the oil from a bulk liquid phase. Most oil will naturally float by itself (heavy oils and bitumen may have a specific gravity >1), and given enough residence time in a gravity separating vessel the produced water will be cleaned of the dispersed oil phase. A very fine oil droplet however may have such a low velocity that the required vessel residence time may be impractical. A 25 μm droplet of n-dodecane in fresh water has a rise velocity of 0.55 cm/min. In a flotation vessel with a 2 meter reaction zone, the droplet will take 6.1 hours to rise that distance in quiescent conditions. Attaching that droplet to a 50 μm gas bubble increases the velocity to 8.2 cm/min, which reduces the rise time to 24.4 minutes. If the droplet can be attached to a 100 μm bubble, then the rise time is further reduced to 6.1 minutes. The bubble rise velocity must be matched with the direction and velocity of the bulk flow in the vessel. Downward bulk (plug) flow will act counter to the rising droplet or bubble, and generate a cutoff size below which the bubble or droplet will not separate, but travel with the bulk fluid flow. Upward plug flow will assist oil droplet and gas bubble flotation, however there are challenges to removal of the clean water.

The reduction in residence time results in a corresponding reduction in vessel volume, and for offshore platform based produced water treatment vessel size and weight are limited by platform space and load capability. A vertical vessel with a 2 meter quiescent reaction zone treating 25,000 BPD (730 gallons per minute) will have a 17.9 m diameter for a 25 μm n-dodecane oil droplet, 4.6 m diameter for a 25 μm n-dodecane droplet with 50 μm methane gas bubble, and 2.3 m diameter for a 25 μm n-dodecane droplet with 100 μm methane gas bubble.

## ATTACHMENT OF OIL DROPLET TO GAS BUBBLE

Effective flotation relies on attachment of the introduced gas bubble to the dispersed oil droplet. There are numerous factors involved in successful collision and attachment including bubble/droplet size ratio, bubble swarm density, mean bubble size, and bubble distribution, produce water salinity, oil viscosity, spreading coefficient, fluid velocity, and turbulence. Of these factors gas bubble size, distribution, and dispersion are the most important for flotation efficiency. Oil droplets can attach to gas bubbles by four mechanisms, which are depicted graphically in Figure 10.

The primary mechanism for oil droplet flotation is full or partial encapsulation of the gas bubble by the oil droplet as shown in Figure 10a (Strickland 1980; Oliveira et al. 1999; Grattoni et al. 2003; Moosai and Dawe 2003; Niewiadomski et al. 2007). In this mechanism both the oil droplet and gas bubble exhibit buoyant rise in the fluid in the opposite direction of gravity. The velocity of the gas bubble is sufficiently higher than that of the oil droplet due to diameter and density, thus from the gas bubble inertial frame of reference the oil droplet is settling downward as if under gravitational influence. Because the oil droplet has lower inertia than the surrounding water it will not directly impact the gas bubble, and water streamlines formed from the rising bubble tend to direct the approaching droplet around the bubble. As the droplet passes around the bubble the trajectory in the rising wake pulls the droplet towards the bubble, and if the droplet has sufficient diameter the droplet will touch the gas bubble surface. Once the droplet has touched the bubble surface the intervening water film thins then breaks to allow the hydrocarbon liquid and gas to come in contact (Niewiadomski et al. 2007). Attachment of the oil droplet to the gas bubble is governed by the spreading coefficient and free energy minimization ($D_{cr}$) as detailed previously. Full encapsulation, where the oil completely covers the gas bubble surface provides the strongest bond between the hydrocarbon liquid and gas thus prevents removal of the oil by shear forces. In addition, an oil coated bubble may more easily attract a free oil droplet thus multiplying the flotation effectiveness (Strickland 1980). Should the droplet size be insufficient for full encapsulation, the oil droplet may form a lens at the aft of the bubble to minimize contact with the aqueous phase, or if the spreading coefficient is negative the oil droplet may form a point contact with the gas bubble. Lens formation is a stronger bond than point contact.

A second mechanism was shown (Sylvester and Byeseda 1980) when gas bubbles 200-700 microns diameter were observed to capture oil drops 1-15 microns diameter in the hydrodynamic wake of the rising bubble, instead of chemical adhesion to the surface. This mechanism is depicted in Figure 10b. The "falling" oil droplet follows the water streamline around the rising bubble. The droplet has insufficient diameter to break the water layer at the bubble surface, however the droplet is small enough to become trapped in the turbulent wake at the aft-end of the rising bubble thus follow the bubble in its upward trajectory. This mechanism does not create a strong bond between the hydrocarbon gas and liquid however this does provide a mechanism by which large bubbles can capture droplets smaller than the $D_{cr}$.

A third mechanism postulated is shown in Figure 10c, whereby bubble clusters form a buoyant "mat" to assist in droplet entrapment along with hydrophobic binding (Rodrigues and Rubio 2007). The formation of bubble clusters has been observed in mineral flotation, in which case the bubbles are held together by bridging particles that are simultaneously attached to two or more particles (Fuerstenau et al 2007). In oil flotation, the oil droplets, especially in the case of point attachment, or suspended solids, can act as a bridging structure to form a cluster or mat of bubbles. The bubble mat sweeps upward with a wider diameter than single particles and is able to more effectively sweep the water phase. Bubble coalescence is retarded by high ionic content of the water, plus stabilizing chemicals in the water, thus the bubbles attach to particles (liquid or solid), but do not form a larger bubble. Bubble cluster formation is promoted by high density of small diameter bubbles and a high density of finely disseminated hydrophobic particles.

The fourth mechanism, shown in Figure 10d, is a product of the dissolved gas flotation (DGF) mechanism in that gas bubbles nucleate on particle/droplet surfaces, then grow to sufficient size to aid in flotation as the gas comes out of solution (Oliveira et al. 1999; Rodrigues and Rubio 2007). For this mechanism to take place, sufficient gas must effervesce from solution, which usually coincides with a fairly large pressure drop at the inlet of the flotation cell. Unlike a DGF cell, a separate mechanism for dissolving gas is not utilized, but the gas is forced into solution while the fluids are in contact in the reservoir. Gas will flash from solution at all stages of pressure let-down including well-bore inflow, pipe transport, choke let-down, and inside three-phase separators. The solubility of methane, and other hydrocarbon gases, in water is several orders of magnitude lower than the solubility in hydrocarbon liquid, therefore most of the gas effervescing from solution will come from the oil droplets. The gas bubbles will nucleate within the droplets or at the droplet surface, and continue to grow in diameter as long as the partial pressure is lower in the gas phase compared to the liquid phase. Eventually the nucleating bubbles will contact, and may coalesce to form a larger discrete bubble. The coalesced bubble will form full or partial encapsulation or point contact depending on the volume of the bubble and the spreading coefficient, which is similar to the mechanism in Figure 10a. The amount of gas effervescing depends primarily on the pressure and temperature difference between the upstream piping and the operating conditions of the flotation cell, and the composition of the hydrocarbon fluid. These factors determine the bubble-point pressure.

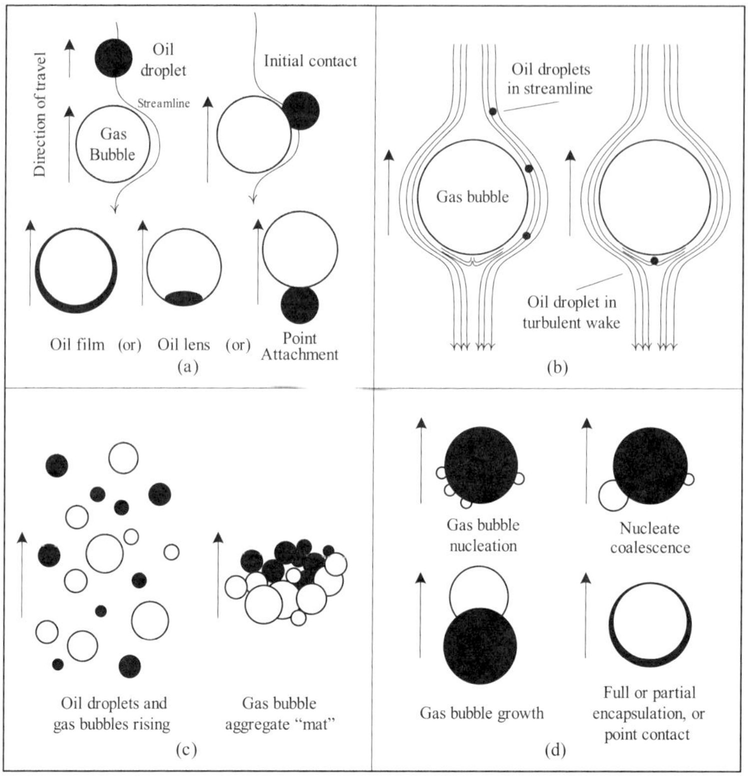


**Figure 10. Mechanisms for gas bubble capture of oil droplets suspended in water: (a) direct impingement with full or partial encapsulation by chemical adhesion, (b) hydrodynamic capture of oil droplets in the wake of a rising gas bubble, (c) clustering of gas bubbles to form a buoyant mat, and (d) gas bubble nucleation, coalescence, and growth on the surface of an oil droplet to result in full or partial encapsulation.**

## OIL FLOTATION UNIT PROCESS

The design of oil flotation in produced water systems as a unit process is governed by location and discharge regulations. Geographic location, primarily onshore or offshore, dictate the size and weight restrictions, while regulations govern the discharge method and oil content limits. Onshore produced water treatment, due to lack of space and weight constraints, may use a gravity settling tank (e.g., API tank) with long residence time, but due to injection requirements will often use deep bed media filtration to reduce the oil content to very low levels. Flotation cells are not used frequently for onshore produced water treatment. Offshore produced water treatment will commonly use liquid-liquid hydrocyclones followed by a vertical flotation cell to meet discharge specifications, as these two unit processes have high throughput to footprint ratio. A gravity settling tank or deep bed media filter are rarely used offshore. An offshore flotation cell must be designed for minimal footprint, minimal weight, high single pass efficiency, gas and oil slugs, suspended solids, motion sensitivity, high uptime, and simplistic operation. These requirements have led to the evolution of the compact flotation unit (CFU), which is a vertical hydraulic single cell, as the primary choice for fine oil droplet removal from produced water in offshore operations. This design is similar in operation to a column flotation unit used in mineral and coal processing. Aspects of this design are discussed as follows.

***Orientation.*** The preferred orientation for the primary reactor in a flotation cell should be vertical, as opposed to horizontal. Vertical orientation minimizes footprint for offshore use, and minimizes liquid level surface area subjected to motion from the facility or gas slugs. Distribution of gas bubbles in a radial pattern to flow vertically in a cylindrical shell is more homogenous compared to distributing bubbles along the length of a horizontal partially filled cylindrical vessel (Smith et al 1996; Lee and Frankiewicz 2004). Plug flow pattern and velocity can also be maintained more homogenous to prevent short circuiting, and skimming of oil from a cylindrical cross section is more efficient than along the length of a rectangular section (Owens and Lee 2007). A vertical orientation allows introduction of the feed water to form a swirling flow pattern, which cannot be accomplished in a horizontal flotation cell. Unlike a mineral or coal column flotation cell which may have a 10:1 or 15:1 height to diameter ratio, the produced water flotation cell most commonly has a 2:1 to 3:1 height to diameter ratio.

***Cell Type.*** The preferred method of flotation action in the flotation cell reaction zone should be induced/dispersed gas, as opposed to mechanical or cyclonic. Mechanical and cyclonic flotation cells introduce high shear forces that can break apart the existing oil droplets, or break apart the droplet/bubble aggregate (Sylvester and Byeseda 1980; Van der Zande et al. 2001; Tavera and Escudero 2003; Welz et al. 2007; Niewiadomski et al. 2007). An induced/dispersed gas flotation cell (including dissolved gas type) has much more flexibility in customizing the bubble distribution and size, as well as introduction location for the gas stream, without creating high shear forces that are detrimental to the oil droplet or droplet/bubble aggregate. Cyclonic flotation cells, while having the smallest footprint, lack the residence time necessary to handle process upsets.

***Residence Time.*** The residence time, or hold-up volume, of the flotation cell should be designed to handle process upsets. These upsets could include oil carry-under from upstream separation equipment due to improper operation, or due to introduction of non-normal fluids into the produced water stream. In some oil and gas facilities the flotation cell may be viewed as the "catch all" for liquids drained from other vessels. Under these abnormal conditions the residence time should be sufficient to maintain the appropriate outlet oil concentration, except in extreme cases. Column flotation cells have low residence time due to small hold-up volume, while the more common oil and gas vertical flotation cells have at most a 3-4 minute residence time. The residence time will determine the height of the vertical flotation cell.

***Gas Slug Handling.*** Gas breakout due to pressure or temperature changes upstream of the flotation cell will result in the presence of free gas as part of the inlet flow stream entering the flotation cell. Small amounts of gas may be beneficial in providing introductory flotation action, but large slugs may lead to high inlet velocity or churning flow, which can lead to oil droplet shearing and/or large disturbances in the oil/water interface, both of which lead to oil droplet carry under, or disturbances in the oil froth collection zone, which will lead to water carryover (Lee and Frankiewicz 2004). Gas slugs of 25-50% gas void fraction (GVF) are common. One method used to accommodate gas slugs is an inlet gas/liquid cyclone. A gas-liquid cyclone requires introduction of the feed (contaminated) stream at or near the circumferential center of the vertical flotation cell, which may preclude introduction of the bulk fluid flow in a tangential or swirling pattern.

***Flow Pattern.*** The flow pattern can be counter-current, co-current, or swirling, as illustrated in Figure 11. In a counter-current flow pattern the contaminated feed water is introduced at or near the top of the flotation cell reaction zone, thus the oil droplets are introduced near their exit point. The bulk water flow is thus forced down in the direction of gravity, which goes against the natural direction of travel for the oil droplets. The gas bubbles rise against the bulk water flow leading to a natural "filtering" action of the water by the rising bubble swarm due to increase probability of collision between the droplet and bubble (Tavera and Escudero 2003). Free gas in the feed fluids introduced near the top of the flotation cell will have less disruptive effect on the reaction zone, if a gas/liquid cyclone or slugging knock-out device is not used. In addition, free solids settling through the reaction zone may be cleaned on their travel downward. Downward bulk flow does limit the size of gas bubbles and oil droplets that will report to the froth layer, as the rising bubbles/droplets must reach a terminal velocity higher than the downward plug flow velocity (Reay and Ratcliff 1973; Gutkowski et al. 1981; Lee and Frankiewicz 2004).

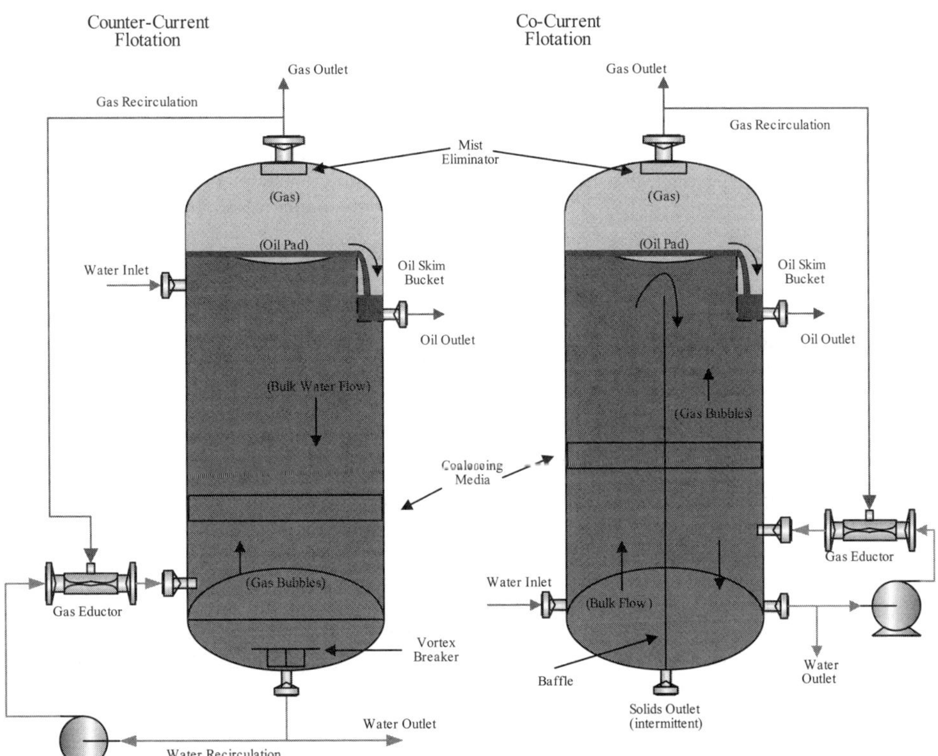


**Figure 11. Vertical compact flotation unit for oil flotation in produced water with counter-current flow pattern (left) and co-current flow pattern (right).**

A co-current flow pattern, where the feed fluids are introduced at or near the bubble source, allows longer time of contact between the concentrated contaminated stream and the bubble swarm, and forces all flow upward (initially) towards the froth layer such that the bulk flow and natural inertial flow of the droplets coincide. However, the resulting clean water phase must then be reversed to eventually exit near the bottom of the reaction zone and shown in Figure 11. The internals required for this clean water reversal must be designed to prevent disruption of the oil/water interface, and high downward velocities that may entrain small oil droplets (if part of the vessel cross sectional area is taken up by upward feed flow, the returning downward clean water will have a smaller area through which to flow compared to a counter-current design, hence the downward velocities will be higher for the same diameter vessel).

Swirling flow is a regime that is added to the co-current or counter-current flow pattern. An advantage of swirling flow is that it forces the oil droplets and gas bubbles towards the axial center of the vessel for easier collection, and allows longer contact time between the oil droplets and gas bubbles, compared to straight axial flow. In addition, a low velocity swirl introduces a slight turbulence into the bulk flow that may prove advantageous to oil droplet and gas bubble collision efficiency (Strickland 1980; Smith et al. 1996; Grattoni et al. 2003) by imparting inertia to the oil droplet and gives more time for the droplet/bubble to interact due to turbulent flow eddies (Sylvester and Byeseda 1980; Niewiadomski et al. 2007). The swirling flow pattern must be kept near laminar to prevent gas vortexing and droplet/bubble breakup. In a swirling flow pattern collection of the oil froth layer is best accomplished at the center of the swirl. A swirl pattern can be induced by internal nozzles or tangential inlet.

***Solids Handling.*** Produced water entering the flotation cell will contain a small amount of suspended solids. These solids may be free flowing in the water phase, or encapsulated in the oil or oil/water boundary. During the flotation reaction, some of the solids will become liberated and settle into the bottom of the reaction chamber. Small solids may stay encapsulated with the oil phase and report to the oil froth. The settling solids must be accounted for in the flotation cell design, as their build up in the bottom of the vessel can be detrimental to the flotation cell. Settled solids may lead to corrosion or reduction in vessel residence time. These settled solids must be periodically or continuously removed to prevent adverse effects. An axial flow exit of the clean water (i.e., center of the bottom dished head) allows the solids to continuously exit the flotation cell with the discharge water and prevents accumulation. However these solids have a coating of oil which may lead to higher oil-in-water values in the discharge stream. In addition an axial flow discharge must have a vortex breaker (especially with swirling inlet flow) to prevent gas and oil ingress into the outlet stream and the vortex breaker must be designed to prevent solids build-up. A non-axial discharge (i.e., side of the vessel) prevents oily solids from discharge, but then requires a second discharge point for intermittent solids removal.

***Oil Skim Method.*** Oil and gas bubbles report to the surface of the water phase to form a froth layer or oil pad. The method of removal depends partly on the inlet flow stream design. An inlet gas/liquid cyclone prevents a central oil bucket, and may require a circumferential or chord shaped weir. A tangential/involute inlet allows a central oil bucket, which requires more complicated piping for oil removal. In all cases a non-mechanical froth removal system is desired to reduce complexity and cost. A non-mechanical typically involves an overflow weir system that relies on the overall liquid height and thickness of the oil layer. The thickness of the oil pad is part of the design methodology for oil skimming. A thin pad skims the oil continuously to prevent re-entrainment, but may allow excess water into the discharge oil stream. A thick pad minimizes water carryover, but in upset conditions may allow oil droplets to be re-entrained in the water phase. Thick pad methodology is the preferred design for column flotation and allows better responsiveness to active interface control.

***Gas Introduction Location.*** Flotation gas may be introduced external or internal to the flotation cell. An external design introduces the bubble gas into the inlet piping of the flotation cell then passes the mixture through a mixing zone, such as an in-line auger or static mixing element. The claim is to provide intimate bubble/droplet mixing, however the mixing device is probably detrimental to the inlet stream by promoting droplet breakup (Sylvester and Byeseda1980; Van der Zande et al. 2001; Tavera and Escudero 2003; Welz et al. 2007; Niewiadomski et al. 2007). Introduction of the gas bubbles into the bottom of the flotation cell reactor (internal) allows generation of controllable bubble size distribution and development of a full dispersion of these bubbles across the area of the reaction chamber. In addition the (near) laminar flow in the reaction zone aids in directing the bubbles towards the most advantageous path for interaction with the oil droplets.

***Gas Introduction Method.*** The method of introducing gas flow into the bulk fluid stream to generate flotation bubbles has undergone the most investigation. Methods investigated include porous frit plate, sintered tubes, sparge nozzle, shear orifice, shear pump, dissolved air, electrolysis, and mechanical impeller. A porous glass/metal device (plate, tube, sparger, etc.) has been used successfully in the mineral flotation industry, but shown to have many operational difficulties in produced water flotation. These difficulties include plugging/fouling by contaminants in the gas stream and build up of precipitated scale on the pore throat (Lee and Frankiewicz 2004). While these devices are simple to operate and relatively inexpensive, they are limited on the bubble size produced and gas throughput (Rubio et al. 2002). A gas eductor is the primary method used for produced water flotation due to low price, simplistic operation, and relative efficiency at bubble generation. This mechanism, usually in combination with a shear plate or orifice, is capable of generating bubbles down to 200-300 micron median size (macro-bubbles). For micro-bubble generation (<100 microns) devices include electroflotation (10-20 microns), cavitation tube sparger (20-30 microns), and dissolved air flotation (40-50 microns). In addition a system comprised of an eductor, shear nozzle, and Stokes settling chamber is capable of generating bubbles down to 5 micron diameter (Owens and Lee 2007). The device or system used must be able to accommodate methane gas with possible liquid contamination and water with very high total dissolved solids and small amounts of suspended solids.

***Gas Recirculation.*** Recirculation of the bubble gas is often undertaken to minimize the amount of field gas consumed by the flotation cell. This gas is taken off the top of the flotation cell, above the froth zone, and recaptured for introduction as the bubble gas. An eductor device accomplishes this task well due to the vacuum created in the eductor throat. Relatively little mist carryover is observed as the gas exit velocities are very low. The primary concern with gas recirculation is possible contamination by hydrogen sulfide stripped from the water and saturation of the recycle gas by water however no detrimental performance on the flotation cell is documented (Hubred et al. 1998). From a mechanical view a recycle line provides minor added complexity of design or cost, and the major concern is the supply of field gas from the end user.

***Coalescing Media.*** Most of the second and third generation vertical flotation cell designs contain a section of coalescing media in the flotation reaction zone. These may be placed in the inlet stream (co-current flow), or across the mid to lower section of the main reaction chamber (counter-current flow). Coalescing media type include balls, saddles, parallel corrugated plate, and cross-flow corrugated plate. In a counter-flow design the oil droplets are introduced in the upper part of the reaction chamber and flow downward to impact the coalescing media, while the gas bubbles are introduced below the media to flow upward. The coalescing media helps capture very small descending oil droplets and allows them to grow to a larger size that are more easily floated.

## SUMMARY

Treatment of oil well fluids requires the same approach methodology as traditional mineral processing, and uses much of the same unit processes. Approaching the treatment of crude oil as oilfield mineral processing allows technology transfer with the goal of finding optimum solutions for both industries based on two viewpoints. An example is water treatment in the petroleum which requires the removal of small (<100 μm diameter) oil droplets, for which induced gas flotation is commonly employed. Oil flotation differs from mineral flotation in that oil droplets are naturally hydrophobic, have a lower density than water, and have a spherical shape. Unlike mineral particles, oil droplets have a very low energy of cohesion (35-56 erg/cm$^2$ for oil vs. 2000-5000 erg/cm$^2$ for minerals) and are subject to breakup from water shear forces, or may coalesce upon contact with other oil droplets (Gutkowski et al. 1981). Produced water has increased density, viscosity, and IFT with hydrocarbon fluids due to the high content of dissolved solids (mainly NaCl). The IFT between the three phases drives the spreading coefficient and critical bubble:droplet diameter ratio to determine how oil and gas bubbles will interact. Once a oil droplet contacts a methane gas bubble it will attach through one of four mechanisms, of which complete encapsulation provides the strongest bond. Complete encapsulation occurs when the droplet diameter is at least 0.4-0.5 times the diameter of the gas bubble. The droplet-bubble aggregate then rises faster than the oil droplet alone, which provides the impetus for using gas flotation. The rise velocity of oil droplets or gas bubbles is governed by Stokes relationship, and because of the several of orders of magnitude increase in rise velocity, the residence time and size of the flotation cell can be significantly reduced compared to gravity separation. Offshore produced water treatment uses primarily a vertical hydraulic single cell flotation unit that is similar in concept to a mineral flotation cell, but must be designed for minimal footprint, minimal weight, high single pass efficiency, gas and oil slugs, suspended solids, motion sensitivity, high uptime, and simplistic operation.

## REFERENCES


Argaud, M.J. 1992. Predicting the interfacial tension of brine/gas (or condensates) system. Advances in Core Evaluation III Reservoir Management, Reviewed Proceedings of the Society of Core Analysts Third European Core Analysis Symposium in Paris, France. 14-16 September, 1992. 147-174.

Barkman, J.H. and Davidson, D. H. 1972. Measuring water quality and predicting well impairment. *Jour. Pet. Tech.* 24:865-873.

Bassett, M.G. 1971. WEMCO Depurator™ system. SPE paper 3349 presented at the Rocky Mountain Regional Meeting of the Society of Petroleum Engineers of AIME held in Billings, MT, 2-5 June 1971.

Department of Energy, Energy Information Administration (DOE/EIA). 2009. March 2009 International Petroleum Monthly, April 13, 2009. http://www.eia.doe.gov/ipm/supply.html

Ditria, J. C. and Hoyack, M. E. 1994. The separation of solids and liquid with hydrocyclone-based technology for water treatment and crude processing. SPE paper 28815 presented at the SPE Asia Pacific Oil & Gas Conference held in Melbourne, Australia, 7-10 November 1994.

Fuerstenau, M. C., Jameson, G., and Yoon, R.-H. 2007. Froth Flotation-A Century of Innovation. Society for Mining, Metallurgy, and Exploration, Inc., Littleton, CO. 357-358.

Grattoni, C., Moosai, R., and Dawe, R.A. 2003. Photographic observations showing spreading and non-spreading of oil on gas bubbles of relevance to gas flotation for oily wastewater cleanup. *Colloids and Surfaces A: Physiochem. Eng. Aspects* 214: 151-155.

Gutkowski, B., Hupka, J., and Miller, J.D. 1981. Flotation of oil droplets from water. In: Interfacial Phenomena in Mineral Processing: Proceedings of the Engineering Foundation Conference; Franklin Pierce College, Rindge, New Hampshire, August 2-7, 1981. 287-301.

Hubred, G., Meekel, W.A.C., and Wyslouzil, H.E. 1998. Flotation Method and Apparatus. U.S. Patent 5,814,228.

Jennings, H.Y. and Newman, G.H. 1971. The effect of temperature and pressure on the interfacial tension of water against methane-normal decane mixtures. *Trans. AIME* 251:171-175.

Lee, C.-M., and Frankiewicz, T. 2004. Developing vertical column induced gas flotation for floating platforms using computational fluid dynamics. SPE Paper 90201 Presented at the SPE Annual Technical Conference and Exhibition, Houston, TX, 26-29 September 2004.

McCain, W.D. 1990. The Properties of Petroleum Fluids, 2nd edition. Pennwell Publishing Company, Tulsa, OK, 1990.

Moosai, R., and Dawe, R.A. 2003. Gas attachment of oil droplets for gas flotation for oily wastewater cleanup. *Sep. Purif. Tech.* 33: 303-314.

Niewiadomski, M., Nguyen, A.V., Hupka, J., Nalaskowski, J., and Miller, J.D. 2007. Air bubble and oil droplet interactions in centrifugal fields during air-sparged hydrocyclone flotation. *Intl. J. of Environ. Pollution* 30.2: 313-331.

Oliveira, R.C.G., Gonzales, G., and Oliveira, J.F. 1999. Interfacial studies on dissolved gas flotation of oil droplets for water purification. *Colloids and Surfaces* 154: 127-135.

Owens, N., and Lee, D.W. 2007. The use of micro bubble flotation technology in secondary & tertiary produced water treatment—a technical comparison with other separation technologies. TUV NEL 5th Produced Water Workshop, Aberdeen, Scotland, 30-31 May 2007.

Rawlins, C.H. and Wang, I.I. 2000. Design and installation of a sand separation and handling system for a Gulf of Mexico oil production facility. SPE paper 63041 presented at the 2000 SPE Annual Technical Conference and Exhibition held in Dallas, TX, 1–4 October 2000.

Rawlins, C.H. 2009. Flotation cell design for produced water treatment: Phase 1—analysis of intellectual property. Internal report, 20 January 2009.

Reay, D., and Ratcliff, G.A. 1973. Removal of fine particles from water by dispersed air flotation: effects of bubble size and particle size on collection efficiency. *Canad. J. Chem. Eng.* 51: 178-185.

Rodrigues, R.T. and Rubio, J. 2007. DAF-dissolved air flotation: potential applications in the mining and mineral processing industry. *Intl. J. Min. Proc.* 82: 1-13.

Rubio, J., Souza, M.L., and Smith, R.W. 2002. Overview of flotation as a wastewater treatment technique. *Minerals Engineering* 15: 139-155.

Smith, J.S., Burns, L.F., Valsaraj, K.T., and Thibodeaux, L.J. 1996. Bubble column reactors for wastewater treatment. 2. The effect of sparger design on sublation column hydrodynamics in the homogeneous flow regime. *Ind. & Engr. Chem. Res.* 35: 1700-1710.

Sport, M.C. 1970. Design and operation of dissolved-gas flotation equipment for the treatment of oilfield produced brines. SPE paper 2717 presented at the First Annual Offshore Technology Conference held in Houston, TX, 18-21 May 1969.

Street, R.L., Watters, G.Z., and Vennard, J.K. 1996. Elementary Fluid Mechanics, 7th edition. John Wiley & Sons, Inc., 1996.

Strickland, W.T. 1979. Laboratory results of cleaning produced water by gas flotation. SPE Paper 7805 Presented at the SPE 1979 Production Operators Symposium, Oklahoma City, OK, 25-27 February, 1979.

Sylvester, N.D. and Byeseda, J.J. 1979. Oil/water separation by induced-air flotation. SPE 7886 Presented at the SPE 1979 International Symposium on Oilfield and Geothermal Chemistry, Houston, TX, 22-24 January, 1979.

Tavera, F.J., and Escudero, R. 2003. De-oiling of industrial water effluents using column flotation. In: Proceedings of the Yazawa International Symposium, Metallurgical and Materials Processing: Principles and Technologies, Volume III: Aqueous and Electrochemical Processing. Editors: F. Kongoli, K. Itagaki, C. Yamauchi, H.Y. Sohn. The Minerals, Metals, & Materials Society. 315-326.

Van der Zande, M.J., Janssen, P.H., and van den Broek, W.M.G.T. 2001. Size of oil droplets under high-water-cut conditions. SPE Paper 67250 Presented at the SPE Production and Operations Symposium, Oklahoma City, OK, 24-27 March, 2001.

Veil, J.A. 2009. The inextricable linkage of water and energy. 19th Produced Water Seminar, January 14-16, 2009.

Welz, M.L.S., Baloyi, N., and Deglon, D.A. 2007. Oil removal from industrial wastewater using flotation in a mechanically agitated flotation cell. *Water S.A.* 33.4: 453-458.

# Extractive Metallurgy

# NEW CHLORIDE BASED PROCESS FOR THE RECOVERY OF BISMUTH FROM POLYMETALLIC CONCENTRATES—NICO DEPOSIT, NWT, CANADA

M. Canizares[1], M. Gladkovas[1], A. Mezei[1], M. Samuels[2], and R.E. Goad[2]

[1]SGS Mineral Services, 185 Concession Street, Lakefield, Ontario, Canada, K0L 2H0
[2]Fortune Minerals Limited, 140 Fullarton St., Suite 1902, London, Ontario, N6A 5P2

## ABSTRACT

The paper provides an overview of the NICO deposit process development with emphasis on the recovery of bismuth.

The NICO "IOCG" (iron oxide – copper - gold) type deposit is located in the Mazenod Lake district near the Snare Hydro electric complex, approximately 160 km northwest of Yellowknife, North Territories, Canada.

The mine development is planned based on a reserve of 21.8 million metric tons of ore averaging 0.13% Co, 0.16% Bi and 1.08 g/ t Au. Pilot plant results predicted that selective sulphide flotation would recover 70% of the bismuth in a concentrate averaging 45% Bi, 0.71% Co and 150 g/t Au.

Gold is recovered by cyanidation of the bismuth concentrate. The cyanidation residue is treated by a novel process involving chloride assisted oxidative leaching followed by high purity metal recovery with concomitant lixiviant regeneration that does not involve the use of gaseous chlorine. The process occurs with selective rejection of arsenic and iron into the tailings stream. The process relies entirely on commercially proven unit operations and equipment.

## INTRODUCTION

The NICO cobalt-gold-bismuth-copper deposit mine development is planned based on a reserve of 21.8 million metric tons of ore averaging 0.13% Co, 0.16% Bi, and 1.08 g/ t Au that will be mined using a combination of open pit and underground methods within a lower grade overall resource. These reserves, which support a minimum 15-year mine life at a production rate of 4,000 tons per day, include Underground Proven and Probable Mineral Reserves of 1,204,000 tons, grading 5.07 g/t gold, 0.14% cobalt and 0.19% bismuth and Open pit Proven and Probable Mineral Reserves of 20,613,000 tons, grading 0.85 g/t gold, 0.13% cobalt and 0.16% bismuth. The economic metals occur throughout the NICO deposit in concordant and discordant sulphide-rich fractures and disseminations. Petrographic and electron microprobe analyses indicate that the cobalt is contained predominantly within a solid solution series between arsenopyrite and cobaltite, with minor cobaltian loellingite (Goad et. al, 1998 and 2000). Bismuth occurs as native bismuth, bismuthinite, and bismuth tellurides, while copper is contained within chalcopyrite (Ferron et. al, 1999, Mezei et. al, 2000). Extensive bench and pilot scale flotation and hydrometallurgical testwork carried out during the past decade and on various samples led to a conceptual flowsheet that recovers cobalt, bismuth and gold into selective bismuth and cobalt concentrates which can in tune be processed for the recovery of cobalt, bismuth and gold. Recent large scale pilot exercises (2007 and 2008) confirmed the technical feasibility of the process and produced commercial plant design criteria (Mezei et al., 2008). This paper provides a description of the Chloride Leach Electro-Winning process for the extraction of bismuth contained in typical NICO pilot plant concentrates. A summary of key experimental data is also reported. This process replaces the previously published ferric-chloride leach-cementation process as it consistently produces higher grade bismuth cathodes (99.5% Bi compared to 80-90% Bi) and it regenerates the lixiviant without relying on gaseous chlorine.

## PROCESS CHEMISTRY CONSIDERATIONS

Previous leaching testwork on NICO selective bismuth concentrates using ferric ion assisted oxidative hydrochloric leaching system (Equation 1) reported extractions in excess of 99% system (Mezei et. al, 2000). The possibility of recovering bismuth as intermediate grade (80-90% Bi) cement was also reported in the same paper.

$$Bi_2S_3 + 6FeCl_3 = 2BiCl_3 + 6FeCl_2 + 3S^\circ \qquad (1)$$

Subsequent work highlighted certain disadvantages of the cementation process such as inconsistent cement grade, iron and arsenic contamination and excessive iron levels in the barren solution. In addition, the regeneration of the lixiviant required the use of gaseous chlorine. Extensive large scale piloting of the process indicated the possibility of arsine gas generation during the cementation process. At the same time, comprehensive testwork on bismuth concentrates originating from various areas of the NICO deposit confirmed the possibility of effective leaching of bismuth and, most importantly, within a wide range of bismuth grades of the concentrate-feed. Furthermore, carrying out the leaching in a counter-current fashion allowed for near stoichiometric ferric consumption which resulted in the production of a pregnant leach solution (PLS) suitable for bismuth electrowinning (EW).

The main cathodic electrochemical reaction for bismuth electro-deposition is described by Equation 2, whilst Equation 3 describes the concomitant main anodic electrochemical reaction that regenerates the ferric ions. At the same time, Equation 3 suggests that the pregnant leach solution (electrowinning feed) should be free of ferric ions as to prevent the occurrence of the reverse reaction which would render practically the iron regeneration insignificant whilst severely impairing the electrochemical efficiency of Bi electrodeposition. This can be achieved by counter-current leaching. Therefore, the basic process chemistry and electrochemistry considerations presented so far allowed to predict that by carrying out the bismuth leaching in counter-current, the resulting PLS will contain only ferrous ions, hence allowing for the (i) judicious use of the oxidizing agent, (ii) its regeneration and (iii) concomitant bismuth electro-recovery.

Finally, Equation 4 illustrates the theoretical possibility of a third anodic reaction that could occur due to accidental over potential, which could generate gaseous chlorine. In practice however, the possibility of that reaction is quite remote since its standard potential is significantly higher compared to the reaction of ferrous to ferric oxidation (Equation 3). Generally, in a well-balanced system the oxidation of iron will occur favorably versus the evolution of chlorine gas at the anode.

$$Bi^{3+} + 3e^{-} = Bi \qquad Evs^{SHE}{}_{at\ 50°C} = 0.335\ V \tag{2}$$

$$3Fe^{2+} - 3e^{-} = 3\ Fe^{3+} \qquad Evs^{SHE}{}_{at\ 50°C} = -0.799\ V \tag{3}$$

$$2Cl^{-} - 2e- = Cl_2(g) \qquad Evs^{SHE}{}_{at\ 50°C} = -1.329\ V \tag{4}$$

*Note: $Evs^{SHE}$ stands for electrochemical potential (in Volts) versus the Standard Hydrogen Electrode.*

## CONCEPTUAL FLOWSHEET

The conceptual process flowsheet is illustrated in Figure 1. Accordingly, the bismuth is leached in a two stage counter-current sequence whereby the fresh feed is contacted with a semi-spent pregnant leach solution (Stage 2 PLS) generated in stage two leaching stage that is carried out previously on partly leached feed (using freshly regenerated lixiviant as detailed further below). Both leaching stages involve liquid - solid separation, yet only the second stage residue requires filtration and washing. Stage two final residue is filtered hot and washed using acidified wash water during the first wash to avoid precipitation of the bismuth, then neutral water for the subsequent washes. Stage one PLS is subjected to the primary bismuth electrorecovery (EW1) with simultaneous regeneration of the ferric ions (as chloride). The ER1 barren solution discharge is split into two streams. The main stream of the EW1 barren solution discharge containing regenerated ferric ions is blended with wash water to adjust its volume, then it is subjected to a final adjustment of the free acidity to 20 g/L HCl with $H_2SO_4$ and sodium chloride to a concentration of 120 g/L free NaCl. The ferric-regenerated, chloride and acidity-adjusted new lixiviant solution is directed to stage two leaching. The second stream of the ER1 discharge barren solution ("bleed") is directed to stage two electrowinning (EW2) aiming to remove the residual bismuth in a "liberator" type cell. The resulting bismuth-free EW2 discharge stream produced is further subjected to iron removal rendering it amenable for disposal as a final barren, or for subsequent recycling into the process as make-up water and/or wash water. The process can be carried out continuously in its entirety, or, semi-continuously, with the two leaching stages conducted on a batch basis and the two electrowinning stages continuously.

The continuous version of the flowsheet planned for commercial implementation showing details of the reaction tanks, thickeners, surge capacities, etc. is illustrated in Figure 2. The process parameters were determined by testwork – a summary of the results is reported further below in this paper.

## TESTWORK RESULTS SUMMARY

Extensive base line and subsequent optimization testworks were carried out on bismuth concentrates produced by a pilot plant exercise carried out at SGS Minerals Lakefield, Ontario facility.

Typical feed compositions are summarized in Table 1.

The process parameters were determined by bench scale testwork carried out on each individual unit operation, with emphasis on leaching, electrowinning/lixiviant regeneration and iron removal. The optimum conditions were fine-tuned under simulated steady-state conditions by recycling regenerated lixiviant.

The final confirmation was carried out through a series of locked-cycle leaching-electrorecovery-iron removal tests operated in a fully integrated mode. The results presented below discuss each unit operation individually. An example of overall steady-state balance on key unit operations is also provided.

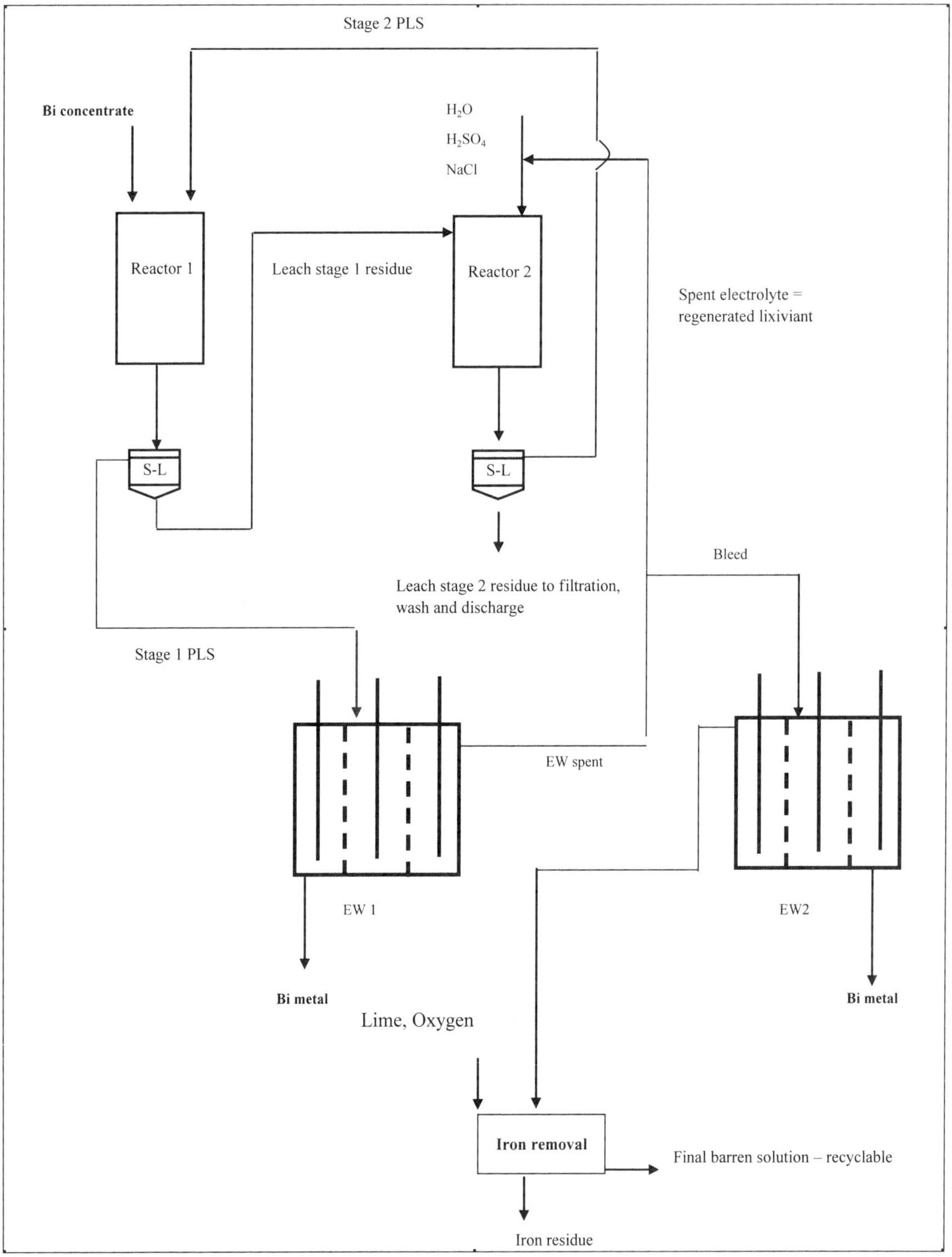


**Figure 1. Conceptual flowsheet – Chloride Leach – Electrowinning Process**

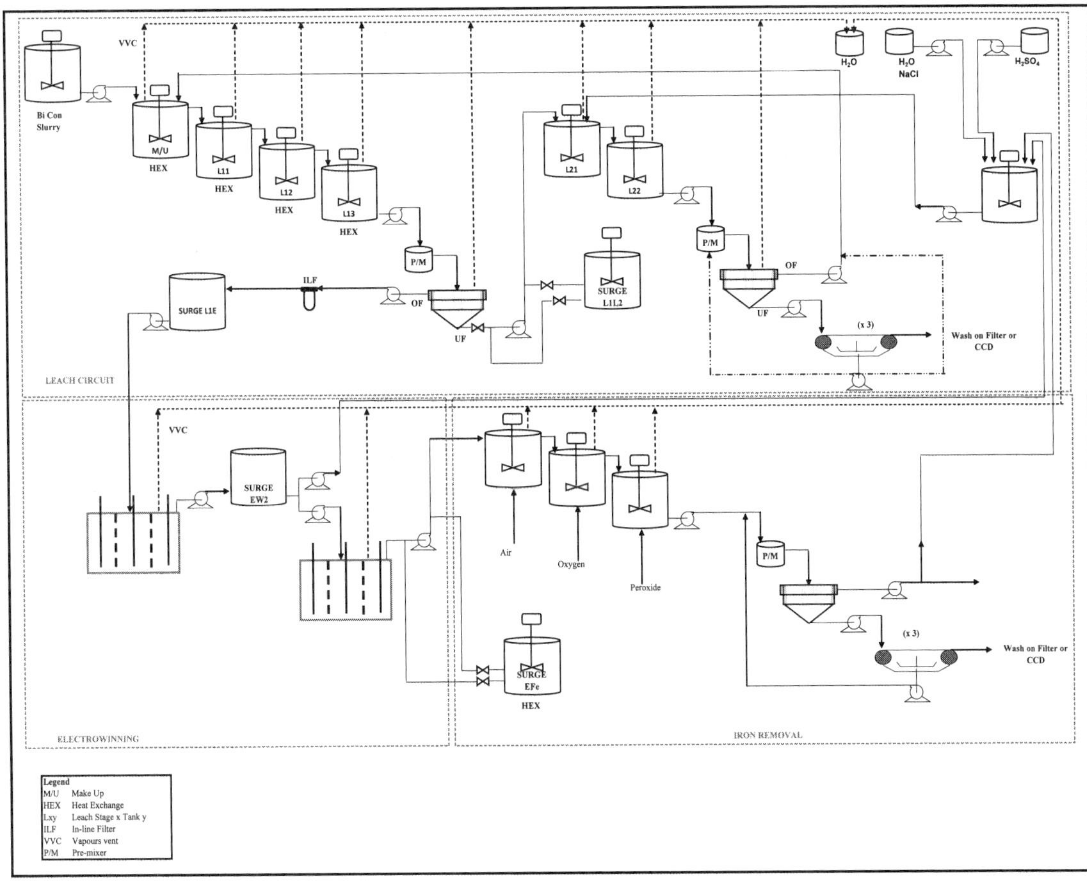


**Figure 2. Continuous flowsheet – Chloride Leach – Electrowinning Process**

## Counter Current Leaching

The locked-cycle leaching sequences were carried out in batch mode using concentrate charges weighing about 1500 g equivalent dry (determined exactly) for each individual test. The feed was in form of filter cakes of known moisture content, produced by the pilot plant (of compositions shown in Table 1).

The leaching agent (lixiviant) consisted of regenerated spent electrolyte having its acid and chloride levels adjusted, as part of the regeneration process. The composition of the lixiviant was determined by chemical analyses before advancing it to the leaching stage, as well as between the stages. The leaching apparatus consisted of a standard bench scale agitated mixing set-up fitted with temperature control, pH/emf measurement devices and sampling ancillaries. The work was carried out in ventilated hood connected to an alkaline scrubber.

## Parameters Summary

Leach stage I:

- Residence time: 180 minutes;
- Temperature: 95 - 100°C;
- Solid content: as resulted by previous leach and weight loss: ~5-7% solids;
- Initial ferric concentration: 40 g/L $Fe^{3+}$;
- Overall ferric stoichiometry ratio of 3 to 1 of $Fe^{3+}$ to Bi contained in the fresh feed;
- NaCl concentration: 120 g/L;
- HCl initial concentration: 20 g/L made from NaCl and $H_2SO_4$.

Leach stage II:

- Residence time: 60 minutes;
- Temperature: 95 - 100°C;
- Solid content: as required by overall stoichiometry, generally about 10-15% also correlated to the electrowinning (EW) stage parameters.

**Table 1. Typical bismuth concentrates samples tested**

| Sample ID | Bi % | Co % | Cu % | Au g/t | Ag g/t | Ni g/t | Zn g/t | As % |
|---|---|---|---|---|---|---|---|---|
| 1 | 40.2 | 0.87 | 2.89 | 339 | 24.6 | 340 | 540 | 3.66 |
| 2 | 36.9 | 0.98 | 2.99 | 355 | 30.1 | 330 | 550 | 4.17 |
| 3 | 37 | 0.81 | 3.05 | 313 | 16.2 | 280 | 595 | 3.15 |
| 4 | 42.3 | 0.64 | 3.13 | 307 | - | - | - | 2.38 |
| 5 | 39.8 | 0.91 | 2.89 | 349 | - | - | - | 3.94 |
| 6 | 41.5 | 1.12 | 2.73 | 370 | - | - | - | 4.88 |
| 7 | 38.4 | 0.10 | 0.31 | 424 | - | - | - | 4.01 |
| 8 | 36.1 | 1.15 | 2.78 | 405 | - | - | - | 5.25 |
| 9 | 37.1 | 0.81 | 3.08 | 324 | - | - | - | 3.28 |
| **Low gold grade bismuth concentrate samples** | | | | | | | | |
| 10 | 49 | 0.4 | | 119 | | | | |
| 11 | 51.5 | 0.4 | | 126 | | | | |
| 12 | 47.2 | 0.5 | | 114 | | | | |
| 13 | 50.45 | 0.5 | | 117 | | | | |
| **Sample ID** | **Fe %** | **S %** | **$SO_4^{2-}$** | **$S^{2-}$** | **S°** | **Cd g/t** | **Pb g/t** | **Al g/t** |
| **High gold grade bismuth concentrate samples** | | | | | | | | |
| 1 | 14.2 | 21.0 | <0.4 | 21.0 | 0.19 | <20 | 920 | 3800 |
| 2 | 14.9 | 21.5 | <0.4 | 21.4 | 0.13 | <10 | 900 | 4000 |
| 3 | 14.6 | 21.5 | <0.4 | 21.3 | <0.5 | <10 | 930 | 4000 |
| 4 | 13.5 | 21.4 | - | - | - | - | - | - |
| 5 | 14.5 | 21.5 | - | - | - | - | - | - |
| 6 | 15.2 | 20.9 | - | - | - | - | - | - |
| 7 | 12 | 21.4 | - | - | - | - | - | - |
| 8 | 15.1 | 22.7 | - | - | - | - | - | - |
| 9 | 14.5 | 22.3 | - | - | - | - | - | - |
| **Low gold grade bismuth concentrate samples** | | | | | | | | |
| 10 | - | - | - | 21.0 | - | - | - | - |
| 11 | - | - | - | 20.7 | - | - | - | - |
| 12 | - | - | - | 20.6 | - | - | - | - |
| 13 | - | - | - | 20.4 | - | - | - | - |
| **Sample ID** | **Mg g/t** | **Ca g/t** | **Mn g/t** | **Cr g/t** | | | | |
| **High gold grade bismuth concentrate samples** | | | | | | | | |
| 1 | 9300 | 8700 | 240 | 96 | | | | |
| 2 | 9000 | 8500 | 250 | 59 | | | | |
| 3 | 10000 | 9000 | 260 | - | | | | |

**Results**

The results of the locked cycle indicated that:

- Overall (dual-stage) Bi extraction efficiencies were about 99% leaving residues assaying about 1% Bi after the second contact;
- Iron co-extraction efficiencies were below 2% whilst arsenic co-extractions were insignificant, indicating very good selectivity versus these deleterious elements – it was noted that arsenic displayed an extremely favorable behaviour during the second contact whereby whilst in the first contact the arsenic concentration reached 80-100 mg/L after the second contact the highest concentration (after a few cycles) was consistently within the 10 – 20 mg/L range;
- After the first contact, around 1 g/L of Fe as $Fe^{3+}$ was present in the resulting PLS, requiring a second contact to reduce the $Fe^{3+}$ to $Fe^{2+}$, i.e. counter-current leach;

**Table 2. Typical counter-current leach efficiencies**

| Test ID | Leach Efficiency | Initial $Fe^{3+}$ | Final $Fe^{3+}$ | Final Bi |
|---|---|---|---|---|
| | % | g/l | g/L | g/L |
| Locked cycle leach 1A | 99.4 | 37.6 | 1.2 | 40000 |
| Locked cycle leach 1A-2 | 9.6 | 1.2 | ~0 | 46900 |
| Locked cycle leach 1B | 98.9 | 37.6 | 4.8 | 34900 |
| Locked cycle leach 1B-2 | 14.2 | 4.8 | ~0 | 44100 |
| Locked cycle leach 2 | 98.6 | 35.8 | 1.25 | 85432 |
| Locked cycle leach 2-2 | 13.3 | 1.25 | ~0 | 90600 |
| Locked cycle leach 3 | 98.5 | 37.4 | 1.32 | 77800 |
| Locked cycle leach 3-2 | 64.5 | 2.32 | ~0 | 87600 |

Note: x-2 indicates the second stage leaching

Settling thickening tests results on the final leach discharge produced underflow solids densities in excess of 48% wt. in presence of 30 g/t cationic flocculant CIBA Magnafloc 455. The thickener underflow unit area requirement was 0.04093 $m^2$/t/day, and the thickener hydraulic unit area requirement was 0.00975 $m^2$/t/day. Vacuum filtration tests results indicated that the underflow could be further dewatered to 18% moisture at a test-predicted belt-vacuum filtration rate of 1800 kg/$m^2$/hour at 5 mm cake thickness.

**Bismuth Electrowinning**

Base line and optimization tests revealed that counter-current leach was beneficial not only in that it ensured judicious (i.e. total) use of the ferric iron, but also by ensuring cathode substrate integrity (by preventing corrosion) during the subsequent electrowinning (EW) stage. In addition, the absence of ferric ions allowed maximizing the current efficiency of Bi electrodeposition.

Preliminary and extensive optimization testworks carried out using a Hull cell, and subsequently conventional electrowinning cells confirmed the possibility of producing good quality bismuth cathodes. Current efficiency was low however, and the ferric regeneration efficiency lower than expected, suggesting that the ferric ions generated during the anodic process were mostly reduced due to the reversal of the reaction illustrated by Equation 3. This prompted the need for separating the anodic and cathodic sections of the electrowinning cell by a porous medium (membrane, diaphragm or equivalent). Two porous media were tested and the results pointed to Lanxess Ionac® MA-7500, a heterogeneous anionic ion exchange membrane containing a quaternary amine characterized by strong ionization capability which renders high perm-selectivity. Separating the anodic and cathodic sections allowed for the electrodeposition of bismuth on the cathode and concomitant re-oxidizing of the iron at the anode to produce a solution that was recycled to the leach stage to recover the Bi from the concentrate, as detailed further below.

**Primary Bi EW Conditions.** The feed to the primary bismuth electrowinning produced by the locked-cycle counter-current leaching sequences contained approximately 90 g/L bismuth and 40 g/L of Fe as $Fe^{2+}$ in its entirety. Electrowinning was carried continuously in a cell with a working volume of 5100 mL for the cathodic section and 4200 mL in the anodic section. Brass cathodes and graphite anodes were used. A photograph of the electrowinning set-up is depicted in Figure 3.

Key EW operating parameters included:

- Cathodic current density ~ 100 A/$m^2$;
- Operating current based on confirmed 90% current efficiency, consistent with the optimized values;
- Cell temperature from 48 to 50°C;

The operation was carried out with frequent analytical determinations from the catholyte and anolyte solutions for control and evaluation of the EW performance.

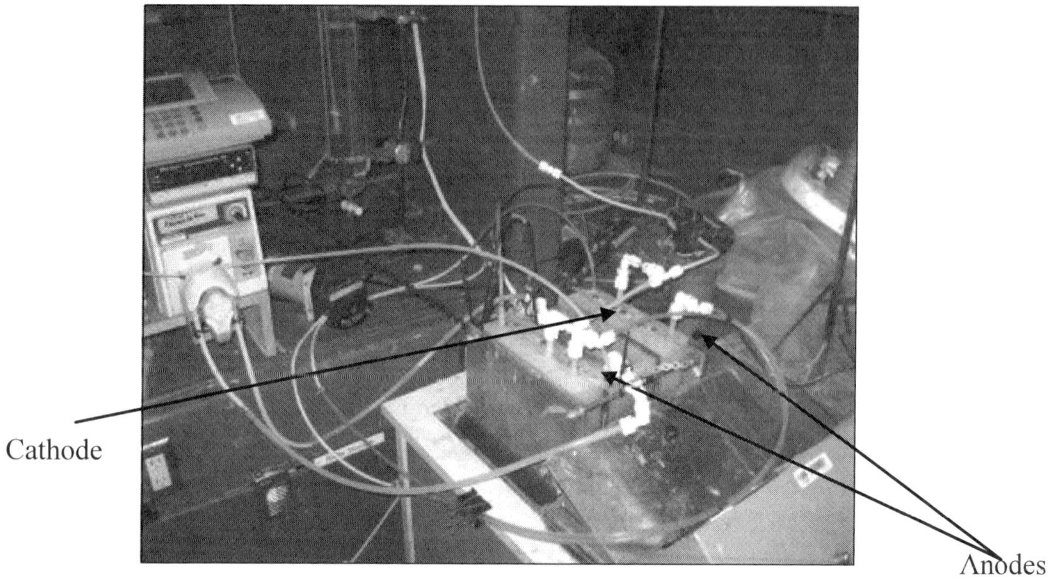


**Figure 3. Bismuth electrowinning set up**

**Primary Bi EW Results.** Details of the cell operation conditions and results are summarized in Table 3. The bismuth deposit displayed a pronounced dendritic formation tendency. In addition, occasional powdery metal drop-off occurrences were observed. These problem self-corrected as the operation progressed, and without the need of using additives. Whilst the deposition current efficiency oscillated between 80 and 90% during the initial cycles, it stabilized above 90% towards the final cycles. Typical assays are provided in the next section (Table 5).

**Table 3. Primary Bi electrowinning conditions and results**

| | Day 1 | Day 2 | Day 3 | Day 4 | Day 5 | Day 6 | Day 7 | Day 8 | Day 9 | Day 10 |
|---|---|---|---|---|---|---|---|---|---|---|
| Cathodic Current Density ($A/m^2$) | 200 | 100 | 100 | 100 | 100 | 100 | 100 | 100 | 100 | 100 |
| Average Voltage, V | 4.3 | 3 | 2.8 | 4.3 | 2.9 | 2.9 | 2.8 | 2.7 | 3 | 2.9 |
| Plating time, h | 3 | 5 | 2 | 8 | 11 | 8.25 | 8 | 4.25 | 5 | 5 |
| Feed Bi, g/L | 90.6 | 60 | 60 | 87 | 87 | 87 | 92.6 | 92.6 | 96 | 100.4 |
| Spent Bi, g/L | 39.7 | 48.8 | 49.1 | 59.2 | 49.45 | 45.8 | 45.5 | 45.7 | 49 | 44.8 |
| Feed $Fe^{2+}$, g/L | 37.6 | - | - | 37.6 | 37.6 | 37.6 | 40 | 40 | 41.25 | 42.31 |
| Spent $Fe^{2+}$, g/L | 2.7 | - | - | 0.33 | 0.37 | 0 | 0 | 1.55 | 0.39 | 0 |
| Total Metal plated, g. | 56.12 | 61.32 | 20.82 | 91.65 | 121.68 | 103 | 97.5 | 52.2 | 65.96 | 68.6 |
| Current Efficiency,% | 61.5 | 92.5 | 94.1 | 86.4 | 83.5 | 91.6 | 92 | 94.5 | 90.6 | 94.3 |

**Secondary Bi EW.** The secondary ("liberator") cell was constructed identically to the primary cell and it was operated similarly except that the cathodic current density was maintained relatively high, at about 200 $A/m^2$ through the entire operation. This allowed for relatively fast residual Bi depletion kinetics however at the cost of lower current efficiency, ranging from 62 to 76% (Table 4). Interestingly, the lower current efficiency was produced when the cathode was covered with a polypropylene bag, which was tested to better contain the Bi powder within the cathodic section. This was necessary because the hydrogen bubbles that formed under the "terminator" conditions tended to detach the finer grains of Bi metal thus making them susceptible to be carried over to the anodic section. Although the physical appearance of the metal deposit was not as attractive compared to the deposit produced by the primary Bi EW, the chemical analyses were quite comparable (Table 5).

**Table 4. Secondary Bi electrowinning conditions and results**

| | Day 1 | Day 2 | Day 3 |
|---|---|---|---|
| Cathodic Current Density (A/m$^2$) | 200 | 200 | 200 |
| Average Voltage, V | 2.5 | 2.6 | 2.8 |
| Plating time, h | 5.9 | 4 | 4 |
| Feed Bi, g/L | 45 | 45 | 45 |
| Spent Bi, g/L | 1.63 | 3.43 | 3.76 |
| Feed $Fe^{2+}$, g/L | 40.25 | 40.25 | 40.25 |
| Spent $Fe^{2+}$, g/L | 0.11 | 1.27 | 0.63 |
| Total Metal plated, g. | 32.28 | 22.36 | 17.97 |
| Current Efficiency, % | 75.36 | 77.09 | 61.95 |

**Cathode assays.** The overall-average cathode grade exceeded 99.5% Bi, with antimony as main impurity ranging from 0.1 to 0.4% Sb, and averaging about 0.3536% Sb (Table 5). Other impurities such as Ti, Zn, Cu, etc. were mostly product of the cathode substrate and the sum of their assays did not exceed 0.05%.

**Table 5. Typical Bismuth cathode assays**

| Sample ID | Bi % | As % | Cu % | Fe % | Mn % | Sb % | Zn % | Ni % | Pb % | Ti % |
|---|---|---|---|---|---|---|---|---|---|---|
| Main Cell Day 1 | 99.5956 | 0.0570 | 0.0290 | 0.0063 | 0.0001 | 0.3100 | 0.0020 | - | - | - |
| Main Cell Day 2 | 99.5782 | 0.0390 | 0.0045 | 0.0170 | 0.0003 | 0.3600 | 0.0010 | - | - | - |
| Main Cell Day 3 | 99.5782 | 0.0390 | 0.0045 | 0.0170 | 0.0003 | 0.3600 | 0.0010 | - | - | - |
| Main Cell Day 4 | 99.5809 | 0.0390 | 0.0060 | 0.0130 | 0.0001 | 0.3600 | 0.0010 | - | - | - |
| Main Cell Day 5 | 99.4809 | 0.0390 | 0.0060 | 0.0130 | 0.0001 | 0.4600 | 0.0010 | - | - | - |
| Main Cell Day 6 | 99.5461 | 0.0240 | 0.0075 | 0.0093 | 0.0001 | 0.4100 | 0.0030 | - | - | - |
| Main Cell Day 7 | 99.5461 | 0.0240 | 0.0075 | 0.0093 | 0.0001 | 0.4100 | 0.0030 | - | - | - |
| Main Cell Day 8 | 99.5285 | 0.0352 | 0.0132 | 0.0102 | 0.0001 | 0.4100 | 0.0028 | - | - | - |
| Main Cell Day 9 | 99.6285 | 0.0450 | 0.0170 | 0.0078 | 0.0000 | 03000 | 0.0010 | - | - | 0.0007 |
| Main Cell Day 10 | 99.5795 | 0.0900 | 0.0076 | 0.0081 | 0.0001 | 0.3100 | 0.0010 | - | - | 0.0038 |
| Lib Day 1 | 99.1100 | 0.0500 | 0.0043 | 0.0360 | 0.0001 | 0.3900 | - | 0.0036 | 0.0060 | 0.4000 |
| Lib Day 2 | 99.6597 | 0.0230 | 0.0031 | 0.0660 | 0.0002 | 0.1800 | - | 0.0050 | 0.0300 | 0.0330 |
| Lib Day 3 | 99.6793 | 0.0170 | 0.0090 | 0.0099 | 0.0001 | 0.1900 | - | 0.0150 | 0.0720 | 0.0077 |

**Overall leach-electrowinning balance.** The overall balance example illustrated in Iron removal is necessary to control the build up of impuurities such as iron, arsenic and possibly traces of sulphate. The bleed ratio was determined based on the amount of iron leached. The test was designed and executed based on preliminary experience with iron-arsenic removal systems, whereby provisions for oxygen addition and final hydrogen peroxide "polish" were made.

The results indicated though that successful iron removal was achieved without the need for an oxidizing agent (since the iron was already oxidized in the liberator cell). Nevertheless, provisions for possible/occasional need for oxidants in the commercial process were considered, as indicated in the process description.

The results (Table 7) indicated that two hours residence time at 60°C ensured iron removal to levels below 2 ppm Fe. To be noted that the kinetic samples indicated the possibility of removing the iron and arsenic selectively, while leaving most of the Bi in solution.

Iron removal is necessary to control the build up of impurities such as iron, arsenic and possibly traces of sulphate. The bleed ratio was determined based on the amount of iron leached. The test was designed and executed based on preliminary experience with iron-arsenic removal systems, whereby provisions for oxygen addition and final hydrogen peroxide "polish" were made.

The results indicated though that successful iron removal was achieved without the need for an oxidizing agent (since the iron was already oxidized in the liberator cell). Nevertheless, provisions for possible/occasional need for oxidants in the commercial process were considered, as indicated in the process description.

The results (Table 7) indicated that two hours residence time at 60°C ensured iron removal to levels below 2 ppm Fe. To be noted that the kinetic samples indicated the possibility of removing the iron and arsenic selectively, while leaving most of the Bi in solution.

Table 6 shows that practically, the ferrous to ferric oxidation during the electrowinning was complete, and, on the other hand, the reverse process of ferric to ferrous reduction occurring during the counter-current leach was also quantitative.

The favourable behaviour of the arsenic is also highlighted whereby whilst limited arsenic leaching indeed occurred during the first contact (~108 ppm As tenor), most of it reprecipitated during the second contact (to ~ 20 ppm As tenor level). This selectivity of the leaching process versus arsenic is important not only from a point of view of the cathode quality produced, but also with respect to the treatment of the bleed stream as detailed below.

The balance also reflects the fact that the bleed solution stream was retrieved from the cathodic section of the main EW cell instead from the combined discharge stream. This was necessary because the iron in the anodic section was in its oxidized form ($Fe^{3+}$) and diverting that stream to the cathodic section of the liberator cell would have caused EW cell corrosion and poor electrochemical efficiency.

**Iron Removal**

Iron removal is necessary to control the build up of impurities such as iron, arsenic and possibly traces of sulphate. The bleed ratio was determined based on the amount of iron leached. The test was designed and executed based on preliminary experience with iron-arsenic removal systems, whereby provisions for oxygen addition and final hydrogen peroxide "polish" were made.

The results indicated though that successful iron removal was achieved without the need for an oxidizing agent (since the iron was already oxidized in the liberator cell). Nevertheless, provisions for possible/occasional need for oxidants in the commercial process were considered, as indicated in the process description.

The results (Table 7) indicated that two hours residence time at 60°C ensured iron removal to levels below 2 ppm Fe. To be noted that the kinetic samples indicated the possibility of removing the iron and arsenic selectively, while leaving most of the Bi in solution.

**Table 6. Example of overall balance leach – electrowinning**

| Leach | | | | |
|---|---|---|---|---|
| | Bi | Fe | $Fe^{3+}$ | As |
| **Solutions** | mg/L | | | |
| Spend from previous cycle | 44000 | 35800 | 35800 | 50 |
| First contact PLS | 85432 | 36000 | 1250 | 108 |
| First contact wash | 13392 | 5700 | 0 | 20 |
| Second contact PLS | 90600 | 37400 | 0 | 9 |
| **Solids** | % | | | |
| Feed | 43.68 | 12.73 | n/a | 4 |
| First contact residue | 0.84 | 22.8 | n/a | 7 |
| Second contact residue | 43.48 | 13.54 | n/a | 4 |
| **Distribution and balance** | % | | | |
| Overall recoveries | 98.6 | <10 | n/a | <1 |
| First contact weight loss | 44 | | | |
| Second contact weight loss | 17 | | | |
| **EW** | | | | |
| | **Bi** | **Fe** | **$Fe^{3+}$** | **As** |
| | **mg/L** | | | |
| Feed | 90600 | 37400 | 0 | 9 |
| Spent | 39700 | 37400 | 37400 | 9 |
| **Liberator** | | | | |
| | **Bi** | **Fe** | **$Fe^{3+}$** | **As** |
| | **mg/L** | | | |
| Feed | 39700 | 37400 | 0 | 9 |
| Spent | 1630 | 37400 | 37400 | 9 |

**Table 7. Iron removal conditions and results**

| Time (Minutes) | Reactor | | | Sample | | |
|---|---|---|---|---|---|---|
| | Temp | Slurry | | Fe mg/L | As mg/L | Bi mg/L |
| | °C | pH | Emf,mV* | | | |
| 0 | 36.70 | -0.10 | 617.20 | 42200 | 2 | 3270 |
| 20 | 80.40 | -0.23 | 705.70 | - | - | - |
| 40 | 84.60 | 0.53 | 724.60 | - | - | - |
| 60 | 83.40 | 0.96 | 709.50 | 19000 | <3 | 3100 |
| 80 | 80.30 | 1.04 | 689.00 | - | - | - |
| 100 | 63.40 | 1.33 | 612.50 | - | - | - |
| 120 | 66.40 | 4.65 | 561.20 | <2 | <3 | <2 |
| 140 | 64.20 | 4.63 | 507.90 | - | - | - |
| 160 | 61.90 | 4.62 | 446.30 | - | - | - |
| 180 | 61.20 | 4.62 | 438.90 | <2 | <3 | <2 |
| 240 | 60.60 | 4.62 | 490.70 | <2 | <3 | <2 |

## CONCLUSIONS

Bismuth contained in representative NICO deposit bismuth concentrates was successfully recovered by a process involving a two stage counter-current leaching sequence followed by two-stage electrorecovery with simultaneous regeneration of lixiviant and a final iron removal stage. High grade cathodic deposits, averaging in excess of 99.5% Bi were produced at overall recoveries of about 98%. The lixiviant regeneration occurred without chorine generation and the barren discharge bleed-stream contained less than 3 ppm Fe and less than 2 ppm As, rendering it suitable for recycle into the process.

## ABOUT FORTUNE MINERALS

Fortune Minerals (TSX-FT) is a diversified natural resource company with several mineral deposits and a number of exploration projects, all located in Canada. They include the Mount Klappan anthracite coal deposits in British Columbia, and the NICO cobalt-gold-bismuth deposit, the Sue-Dianne copper-silver deposit and other base and precious metals exploration projects in the Northwest Territories.

## REFERENCES

Goad, R. E., 1998, "The NICO Proterozoic Iron Oxide-Hosted Polymetallic Deposit, Southern Great Bear Magmatic Zone, Northwest Territories, Canada. Paper presented at the 100th CIM Annual General Meeting, Montreal, Quebec, Canada.

Goad, R. E., Mumin, A.H., Duke, N.A., Neale, K.L. and Mulligan, D.L. 2000, Geology of the Proterozoic Iron Oxide-Hosted, NICO Cobalt-Gold-Bismuth, and Sue-Dianne Copper-Silver Deposits, Southern Great Bear Magmatic Zone, Northwest Territories, Canada. In Porter, T.M. (Ed.) *Hydrothermal Iron Oxide Copper-Gold & Related Deposits: A Global Perspective*, Australian Mineral Foundation, Adelaide, 249-267.

Ferron C. J., Mezei, A., and Goad R.E., 1999, Process Development: Recovery of Cobalt, Gold and Bismuth from Polymetallic Concentrates - NICO Deposit, NWT, Canada. Part I: Leaching. Paper presented at TMS – EPD'99 Congress, San Diego, California.

Mezei, A., Ferron C.J., Krstic S. and R. E. Goad, Recovery of Cobalt, Gold and Bismuth from the NICO Deposit, NWT, Canada - Part II: Recovery of Bismuth. Paper presented at the SME 2000 Congress, Minor Elements 2000, Salt Lake City, Utah.

Mezei, A, Fleming C.A., Canizares, M., Ashbury, M., Brown J., Todd, I, Imeson, D., Jackman, R. Goad, R.E, Samuels, M., Ferron, C.J. Hayden, A., Molnar, R., Konigsmann, K, King, D., Marshall. D., Ngai, E., Recovery of Cobalt, Gold and Bismuth from the Polymetallic Concentrates - NICO Deposit, NWT, Canada – Pilot Plant Results. Paper presented at the ALTA 2008 Congress, Nickel and Cobalt, Perth, WA, May, 2008.

# PROCESS DEVELOPMENT TESTING FOR CONTINUOUS ION EXCHANGE APPLICATIONS

Gordon J. Rossiter

Ionex Separations, 448 West 19th Street, Suite 346, Houston TX 77008 USA

## ABSTRACT

Ion Exchange and specifically **C**ontinuous **I**on e**X**change (**CIX**) is a Cyclic Separation Process in which species are transferred between a solid phase and a liquid phase. In hydrometallurgy the chemistry of the solution phase can be quite complex requiring a pragmatic approach to ion exchange process development. A practical phased approach to industrial process design is discussed keeping in mind the cost impact of each phase; phases include the initial selection and evaluation of adsorbent candidates; the development of a practical cyclic process; the laboratory verification of the process; the initial development of a process control strategy.

## INTRODUCTION

**C**ontinuous **I**on e**X**change (**CIX**) has recently been considered for several new metal recovery projects. In all cases it is driven by an increased appreciation of the benefits of state of the art **CIX** systems compared to previous attempts to utilize ion exchange technology in a conventional (fixed bed) or in a pseudo continuous system (NIMCIX, Porter systems). Amongst the projects currently under review are new uranium plants employing both acid and alkaline leaching; several Ni/Co plants to remove excess nickel from the cobalt electrolyte; Ni laterite leachates followed by direct **CIX** recovery of nickel has been successfully piloted; copper recovery from low grade solutions is ready to be commercialized.

**CIX** is now firmly implanted in such processes as: corn syrup purification, recovery of amino acids, sugar cane juice purification, softening of sugar beet juice, anti-biotic recovery and purification, potassium nitrate production, water purification. All mentioned processes previously employed conventional fixed bed ion exchange.

The development steps required to assess the viability of **CIX** are not particularly difficult or costly but **data interpretation requires persons of experience**. This paper illustrates the simple laboratory test procedures, using examples of how it is done, and the rationale behind the various steps. The reader will find that from the beginning of the investigation the data being produced are constantly being evaluated with a view to providing good design data.

## CIX PROCESS BENEFITS

**CIX** has matured greatly in recent years due to the introduction of new equipment technology together with more sophisticated process techniques. The technology offers many benefits that are to date little understood by those familiar only with conventional fixed bed ion exchange. These new system designs provide the hydrometallurgical design engineer a new alternative in his approach to metal recovery using ion exchange. Benefits to be realized are:

1. Reduced resin charge (50%-80%). Resin charge reduction implies a smaller plant and reduced cost.
2. Higher purity products (by a factor of 3-4)
3. Reduced chemical consumption (eluents) (40%-60%)
4. Reduced water consumption (40%-60%)
5. Reduced waste; in some cases zero waste discharge
6. Extends resin life thus reducing operating cost.
7. Permits the use of a pre-elution stage and/or post adsorption scrubbing stage to remove co-loading impurities and produce a purified eluate often greatly simplifying the downstream product recovery.

The reader is referred to reference 1 for an explanation of the basis of the above claims. In general it can be stated that **CIX** utilizes the concept of counter-current contacting in all steps of the ion exchange cycle and consequently is always superior to a conventional (fixed bed) batch process design. The economic driver for **CIX** comes not only from these process efficiencies but also from benefits achieved downstream on account of the more concentrated and purer product solutions. As an aside, the new state-of-the-art CIX systems are mechanically simple and utilize a single moving part and have been proven to be reliable in industrial operations.

## PROCESS DEVELOPMENT STEPS

Fast and effective testing on a small scale will provide the engineer with enough information to go to vendors and develop a reasonable handle on the capital and operating costs; thereby minimizing development expense. The major steps in the development can be summarized as:

1. Laboratory Test work
2. Pilot Plant Test work
3. Demonstration Plant

Laboratory test work is relatively inexpensive and produces data enabling an initial commercial design. This is important to assess the viability of **CIX** as an alternative to other separation and purification techniques. Pilot and demonstration plants are more expensive and may be necessary in certain cases. This paper discusses only the laboratory development steps.

Prior to embarking on these initial studies the engineer is advised to contact *both the supplier of resins used in mineral and metal processing and also the suppliers of CIX systems*. Ionex Separations, as a CIX system supplier, highly recommends that preliminary advice be sought from these sources as a way to reducing the work involved in resin selection, test work design and data evaluation.

**LABORATORY TEST WORK**

A summary of the laboratory steps needed to generate design data for a pilot or commercial **CIX** system are:

1. Resin selection
2. Resin candidate evaluation
   a. Capacity, Kinetics & Selectivity Checks
   b. Isotherm Checks
3. Single Column Tests – Process Cycle Development
4. Cycle Tests – Process Confirmation

**Stage 1 Resin Selection**

When considering the application of **CIX** to a specific recovery or separation one normally takes a typical feed stream and conceptualizes the process. The first step is to understand the solution chemistry and then propose suitable resins for experimentation. From knowledge of the species of interest and the accompanying elements one can usually select the type of resin[1] and subsequently specify the candidate resins. One might select 6-8 candidate resins of the type selected from different resin suppliers as each supplier's resin will have different characteristics (selectivity and kinetic properties) when tested on the specific solution. There is much art in the design of ion exchange resins and it is often not only the functional group that requires attention but also the polymer bead size, size distribution and structure. The latter can have an impact on the kinetics and resin life.

Table I below is a short list of typical separations where resins types have been chosen based on discussions with persons of experience.

**Table 1 Resin Candidates – Typical Separations**

| Metal of Interest | Solution pH & Impurities | Resin Type |
|---|---|---|
| Cu PLS | pH 1-3, Fe, sulfate | IDA, HPPA |
| Ni PLS | pH 1-3, Fe, sulfate or Cl | BPA |
| Co electrolyte | pH 2-6, Ni, sulfate | BPA |
| U PLS | pH 10-11, carbonate, V, Mo, P, | SBA |
| U PLS | pH 1-2, sulfate, V, Mo, Si, Fe | SBA |
| Sb, Bi | pH < 1, sulfate, | AMP |
| Mo | pH < 1, As, SO4 | WBA |
| Re | pH < 1, As, SO4, Se | SBA & WBA |
| Au | pH > 2 Cu, Zn, | SBA |
| Fe | pH < 1 | DPP |
| Zn | pH > 3 | DEHPA, IDA |
| IDA = Iminodiacetic acid, BPA = Bis-picolylamine, SBA = Strong Base Anion | | |
| AMP = Amino Phosphonic, WBA = Weak Base Anion, HPPA = Hydroxyproylpicolylamine | | |
| DPP = DiPhosPhonic, DEHPA = Di - 2 EthylHexylPhosphate (impregnated adsorbent) | | |

The characteristics of a resin candidate to evaluate are:

1. **Capacity** to load the species of interest – kg_species/m$^3$_resin
2. **Kinetics** of the mass transfer steps of adsorption and elution
3. **Selectivity**

The above parameters are those to be determined in the initial phase of development and when established they provide the basis for a CIX design. The productivity of a system ([kg/hr]/m$^3$_resin) relates directly to both adsorption kinetics and elution kinetics. The mass transfer rates in the adsorption and elution steps, set, in great part, the resin charge which in turn sets the CIX system size. Running the adsorption tests at different velocities provides data to optimize the design of this step. Experimenting with elution velocities will in turn provide data for elution design.

Selectivity in the loading step is important as a first purification step. **Selectivity in scrubbing and pre-elution is a powerful aspect of CIX which when combined with natural resin selectivity can produce product purities previously unknown to conventional IX practitioners.**

The adsorption-scrub-pre-elution combination properly designed can produce an eluate product of suitable purity for direct downstream recovery. Typical downstream metal recovery steps include electro-winning, precipitation, crystallization, cementation amongst others. In some cases CIX will eliminate the necessity for an intermediate SX concentration and purification step.

[1] Types of resin refers to the functional group immobilized on the polymer resin structure; e.g., Cation or Anion, Strong or Weak, Chelating, etc.

**Stage 2a Resin Candidate Evaluation**

The first tests, at this stage, simply employ laboratory beakers and give information on kinetics and selectivity. The method used here requires laboratory beakers and hot plate stirrers. Resin aliquots are placed in beakers and the feed solution is added. The resin solution mix is stirred at constant speed and timed samples then are taken[2]. Samples need to be analyzed for all impurities in addition to species of interest.

The choice of solution and resin volumes is dependent on the specific conditions of the target separation. Factors to take into account are: likely resin loading, expected variation in the feed metal concentrations. These tests are important and the data interpretation is critical so that a suitable resin is found and more important is not to eliminate an overall good resin. In these tests one has to remember that normal adsorption flow rates in **CIX** systems are between 5-30 BV/hr. This translates into **solution** contact times ranging from 2-12 minutes. (Note: Typical fixed bed adsorption flow rates are typically 2-5 BV/hr). This difference in fluid contact times is an important factor in the interpretation of the initial test data for **CIX** systems. **Resin** contact times in **CIX** systems range from ½ → 6 hours. In general, more concentrated feed solutions extend fluid residence times and shorten resin contact times.

**Examples**

The examples below are made to illustrate some typical observations that can be obtained from beaker tests. They are not exhaustive and it is recommended that the development engineer consult with persons of experience to fully understand the results.

**Example 1: Molybdenum Extraction from Acidic Solution**

In this case we assumed that Mo would exist as an anionic oxy anion species. Several resins with anionic functionalities were chosen and a time/contact test was run for each resin. This consists of stirring a small aliquot of resin in a beaker with the feed solution – one beaker test for each resin. Ensure that all tests utilize a fast stirring rate that is equal in intensity for all beakers. The results of a typical test are shown in Figure 1.

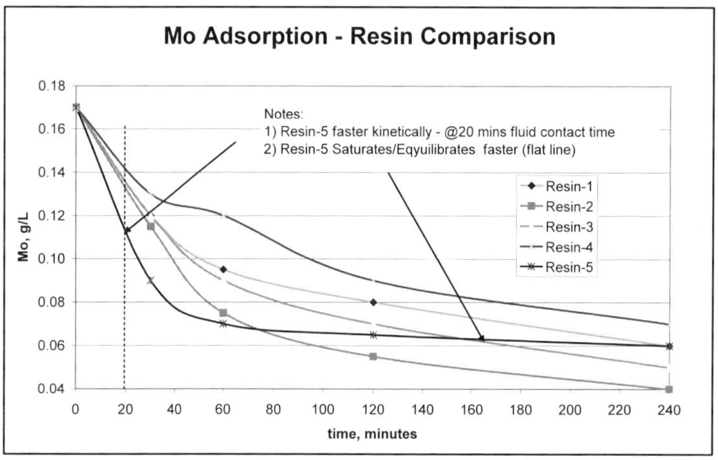


**Figure 1 Resin Evaluation**

From data plot in Figure 1 some conclusions can be drawn. As stated previously, **CIX** typical fluid contact times are of order 2-20 minutes. The vertical dotted line corresponds to a 20 minutes contact time based on a 3BV/hr initial design estimate for the **CIX** adsorption. Consequently, the initial part of the curve is very important. It is clear in this case, that the Resin-5 is superior kinetically up to a contact time > 40 minutes - fluid contact times typical of **CIX.** Choosing this resin translates into faster mass transfer in adsorption (a shorter MTZL - see later). Faster mass transfer results in a smaller resin charge translating into cost savings in terms of plant size.

Be aware also in these preliminary tests that invariably there are competing species and these should also be monitored. For the case above the co-adsorption of arsenic and iron could occur and the analytical results may show potential selectivity differences between resins. Figure 1 only shows Mo adsorption kinetics.

Note also that as the resins equilibrate (resin residence times can be several hours in adsorption) Resin-5 reaches its full capacity earlier than the other resins which continue adsorbing beyond 180 minutes. This lack of full equilibration is potentially an issue with subsequent elution step purity. As contact time increases the curves tend to flatten out. The speed at which a resin saturates with the species of interest can have consequences with respect to product eluate purity. Slow equilibration in adsorption indicates that other species are occupying the resin sites and could adversely impact product purity in elution. This phenomenon causes resins to transfer into elution with some sites still occupied by impurities. The task of purification becomes relatively more difficult and overall this can have an adverse impact on the final eluate purity.

[2] Typically samples are taken at minute intervals of 1, 2, 4, 6, 8, 10, 15, 20, 40, 60, 120, and 240. Total time typically 240 minutes

**Example 2: Mo Selectivity Over As**

Figure 2 shows adsorption behavior of the Mo and As oxy anions in a preliminary contact test. As mentioned above fluid residence times are in the range 2-20 minutes (30-3 BV/hr). Resin contact times are much longer and depend on the rate of saturation of a resin's capacity. Factors that influence this include feed flow rate, feed metal concentration and resin working capacity. One can see that at the start of the contact the resin uptakes both Mo and As; then as contact time increases the Mo begins to displace the As. This is an indication of natural adsorptive selectivity and also points toward a potential pre-elution purification step in the **CIX** scheme. Normally, if a resin adsorbs an element LESS strongly then it will also release that species more easily in the elution step.

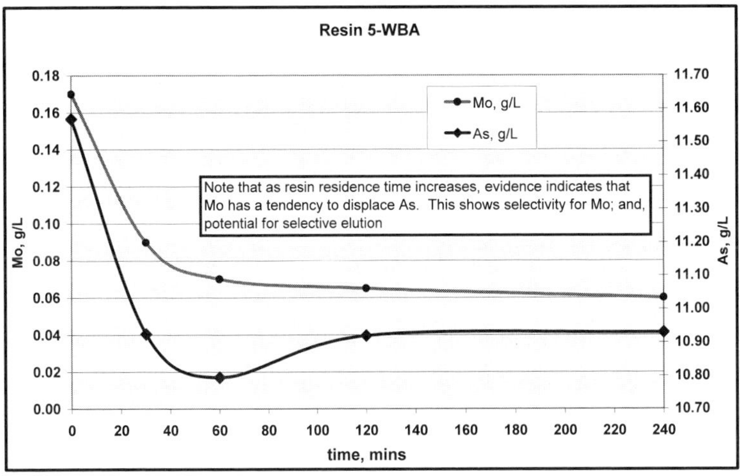


**Figure 2. Contact test showing selectivity for Mo over As**

**Example 3: Silica Behavior in Uranium Extraction**

In treating metallurgical leach solutions it is important to follow the behavior of silica adsorption. Silica fouling of resins can adversely influence the resin performance during its normal life. Figure 3 shows how initial testing can provide information on the performance of resin candidates with respect to silica pick up (or any fouling impurity). In the case of silica adsorption the longer contact time is relevant since this corresponds more closely to the actual process conditions where resin is in contact with fluid for much longer periods than fluid with the resin. In the case shown one can infer that Resin-2 appears susceptible to silica uptake while Resin-3 appears resistant. The observation implies that Resin-3 would only need a silica de-fouling treatment on an intermittent basis whereas the other two resins require more frequent or even continuous treatment for silica fouling.

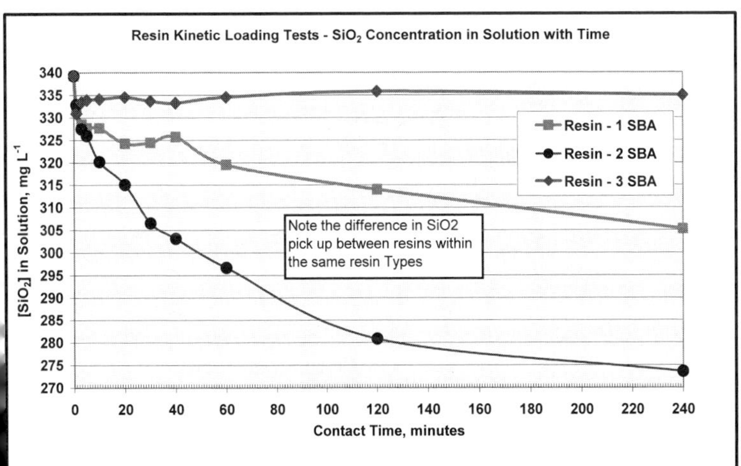


**Figure 3: Silica rejection for various strong base anion resins**

There are many other factors in the selection of resins for a specific application and these include particle size, particle size distribution, resin fouling resistance, relative selectivity for impurities present, etc. It is important therefore to consult with persons of experience in **CIX** system design when selecting resins for service.

For instance, particle size and particle size distribution is an important aspect of resin choice. Smaller particles are kinetically faster in both adsorption and elution. The smaller particles are also more mechanically robust. A consequence of smaller bead sizes is the higher pressure drop characteristic. In **CIX** designs, in contrast to conventional fixed bed designs, higher superficial velocities can be accommodated by using shallower beds arranged in multiple counter-current adsorption stages thereby overcoming the resin pressure drop limitation.

**Stage 2b Isotherm Checks**

When the feed to the process is expected to have a significant variation in feed concentration then an isotherm test is called for. This would be the case for instance when a leach feed is supplied with a variety of materials from various sources with varying levels of the target metal; this is not the case for a typical mine PLS which may vary but usually has been designed to assay within a specified narrow range (+/- 20% around typical value). If a deposit has distinct periods of mining in which ore types lead to changes in PLS grade then each grade should be looked at independently.

The test to do this is simple and uses several beakers. Identical quantities of properly conditioned resin are placed in several beakers and contacted with amounts of fluid spiked to the increasing levels of target metal. The chosen initial concentrations should represent the range of feed concentrations. In this test procedure make sure the resins are well stirred, even fluidized, to remove any bias introduced by variation of diffusion in the liquid phase.

The time of contact for the test is calculated to be not dissimilar from the expected contact time of the resin in adsorption. To estimate time of contact use an operating adsorption rate of 10-15 Bed Volumes per hour (BV/hr) and then the following example provides contact time guidance.

1. Process feed flow = 100 $m^3$/hr
2. Process feed [Metal] = 1 g/L typical
3. Resin loading [Metal] = 30 g/L expected – get advice or use the previous kinetic test as a guide
4. Resin in adsorption at 10 BV/hr = 10 $m^3$
4. Time to saturate resin to 30 g/L = (10 x 30)/ (100 x 1) = 3 hours + 1 (extra time) hour = 4 hours.

Sample the final fluids and analyze. Then by difference calculation one determines the amount adsorbed on the resin. A plot of $[C]_{resin}$ v $[C]_{liquid}$ provides the isotherm over the range of interest.

Run all the tests concurrently for the same time of contact and with the same feed solution; sample at half the contact time and then again at the end of the allotted time. Plotting the two isotherms will provide you information on whether equilibrium has been closely approached at the half-way point.

Knowledge of resins and electrolyte ionic species is useful in understanding the expected kinetic performance. For example, large complex ions like $UO_2(SO4)_3$ are likely to be slow diffusing while a smaller hydrated ion like $Cu^{2+}$ may be faster. On the other hand, the resin functional group may be complex and slow down adsorption on account of difficult stereo-chemistry at the actual adsorption sites. Resin science is still not well defined theoretically and so test work is the only sure way to provide good design data.

The following two plots indicate the likely types of isotherm one could encounter. Figure 4 shows that resin capacity is a strong function of feed concentration across the whole working range of feed solutions. This can be handled by a properly designed **CIX** control system. The two most important parameters to input to a **CIX** control system are the mass flow rate of adsorbing species and the working capacity of the resin under actual feed conditions. Figure 5 shows the typical behavior of a strongly and selectively adsorbed component where the resin capacity is a flat function of feed concentration.

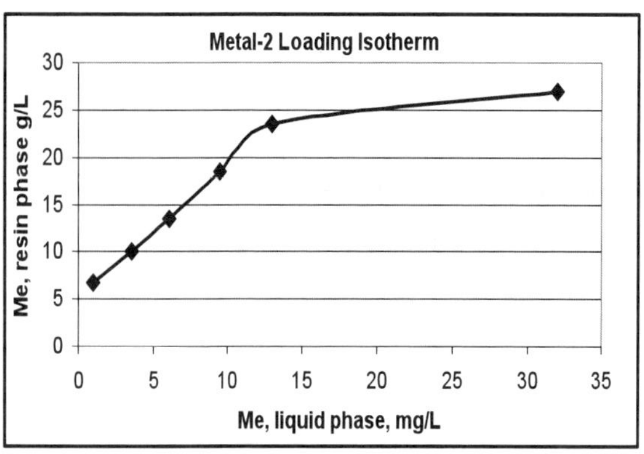


**Figure 4: Metal loading Isotherm – resin working capacity sensitive to feed metal concentration**

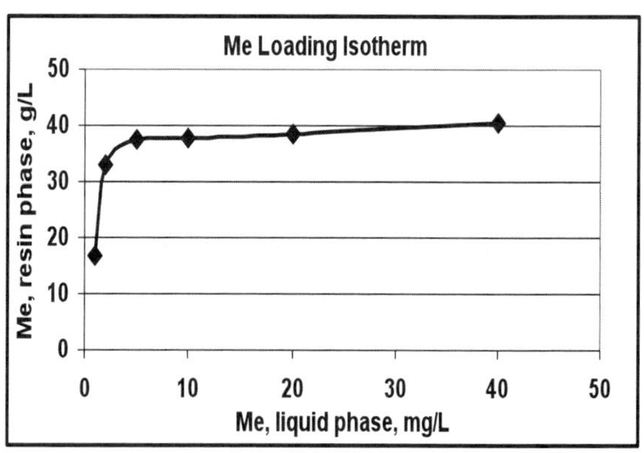


**Figure 5: Metal loading Isotherm – resin working capacity insensitive to feed metal concentration**

Knowledge of the isotherm allows us to construct a feed forward algorithm for the main control of the plant. The actual control of a **CIX** plant also requires feedback information since the isotherm shape is always changing as a function of the feed composition and the feed temperature as well as with the metal concentration. From feed isotherms, the control engineer can calculate the range of resin exhaustion rates or cycle times. This is very useful in determining design parameters for elution treatment and also for pump and line sizing.

**Stage 3 - Single Column work –Development of the Cyclic Process**

From the Stage 2 work usually one or two resins are selected for examination in a single column environment. The column test work will provide design information for the adsorption and elution stages of the cyclic process. In addition this stage will provide a platform to evaluate alternate chemistries in scrubbing and elution.

The target parameters to extract in this stage are:

1. Adsorption **M**ass **T**ransfer **Z**one **L**ength (**MTZL**)
2. Elution eluent strength
3. Elution time required.

First a word on column size for these tests: often for lack of available solution, the tests are run using very small columns sometimes containing as little as 10 milliliters of resin. The recommended minimum size is a 1" diameter clear PVC column containing a minimum of 250 mls of resin. This size column is generally sufficient to provide the initial design information for conceptual plant sizing. Smaller columns are susceptible to severe wall effects that often lead to artificially low loading estimates. Columns should be fully enclosed and fitted with screens at both ends. Both up-flow and down-flow operations will be performed during the test work. Fluids should be pumped through the column(s) at precisely controlled flows during the test work.

**Stage 3a - Adsorption (Breakthrough) and MTZL**

Choose one or more candidates from Stage 2 testing and then run a few cycles of loading and elution in a single column. The adsorption breakthrough (BT) curve will provide sizing information for sizing the adsorption section of the **CIX** flow sheet. The elution profile(s) provide elution time required and eluate peak levels attainable. In addition, the elution profiles are studied to look for potential purification opportunities.

The MTZL is a critical sizing parameter in **CIX** systems since this impacts the resin vessel size. It is always good therefore to run these tests at a minimum of two (2) different velocities to determine the relationship between fluid velocity and mass transfer zone length. Typically in adsorption it has a non-linear relationship as follows:

$$MTZL \; \alpha \; [velocity]^n, \; n = 0.6 - 1.0$$

$$(MTZL_1/MTZL_2) = (v_1/v_2)^{0.6\text{-}1.0}$$

For instance, if the MTZL at an adsorption velocity of 10 m/hr (superficial linear velocity through the bed) is 1 (one) meter and taking $n = 0.6$ then the MTZL for the same operating conditions except for the velocity change to 20 m/hr will be

$$MTZL_{20/hr} = 1 \times (20/10)^{0.6} = 1.51 \; m$$

Effectively this means that by increasing the velocity we have increased the MTZL by 50% while ***doubling the flow***. **This can be quite significant when using expensive resins.** It also reduces the **CIX** vessel size which is a function of resin charge and consequently lowers the capital cost. Further, increasing the rate of mass transfer means the resin spends less time in contact with the PLS or feed solution. Resin life and fouling are often a strong function of this soaking time as was evidenced in the silica adsorption evaluation touched on earlier.

Typically the adsorption column tests are run up flow and samples are taken typically at 5 -10 BV intervals for analysis. The actual breakthrough profile is measured and used to calculate the stable mass transfer zone length under the specific operating conditions; that is, flow rate, and feed conditions and specie concentrations. Figure 6 illustrates two adsorption breakthrough plots originating from column tests using a copper PLS solution.

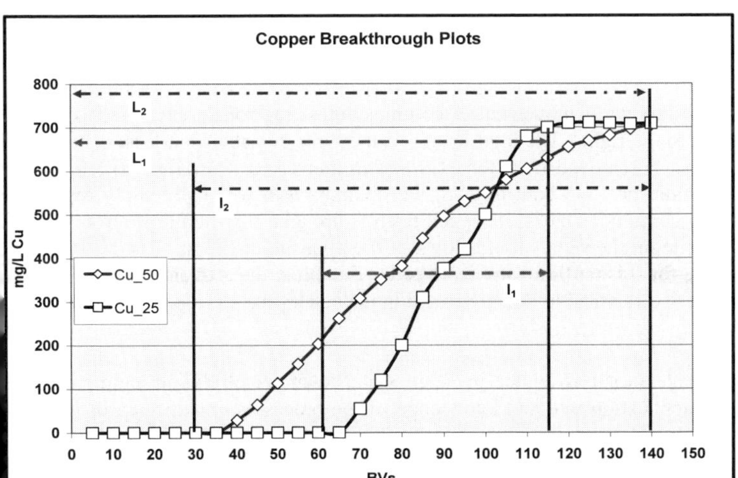


**Figure 6: Typical adsorption breakthrough (BT) to saturation plots**

The breakthrough plot (Figure 6) is typical for recovery of copper & nickel metals generated from sulfuric acid leaching. From summation of the effluent analyses and/or using the area bounded by the BT curve and the horizontal line representing the feed value one can estimate and confirm the resin loading capacity. The ratio of the BVs to breakthrough (where curve begins to rise) and the BVs to saturation (where curve reaches feed level) is used to calculate the MTZL.

$$MTZL = l_i / L_i \times \text{Column Bed Length},\ l_i = \text{BVs to BT} \ \&\ L_i = \text{BVs to Saturation}$$

As a general comment it is the norm to see that in the adsorption step the mass transfer is likely controlled in the main by diffusion in the liquid bulk phase. In such cases, higher fluid velocity will enhance diffusion by reducing the depth of the static film around each of the resin particles. For this reason, the relationship between velocity and MTZL should always be investigated.

Now that we have good data on the adsorption performance one can use the same adsorption breakthrough plot and superimpose on top of the curve a **CIX** column configuration. Figure 7 is a schematic showing this as nine columns arranged in three counter-current passes – each pass consisting of three parallel columns.

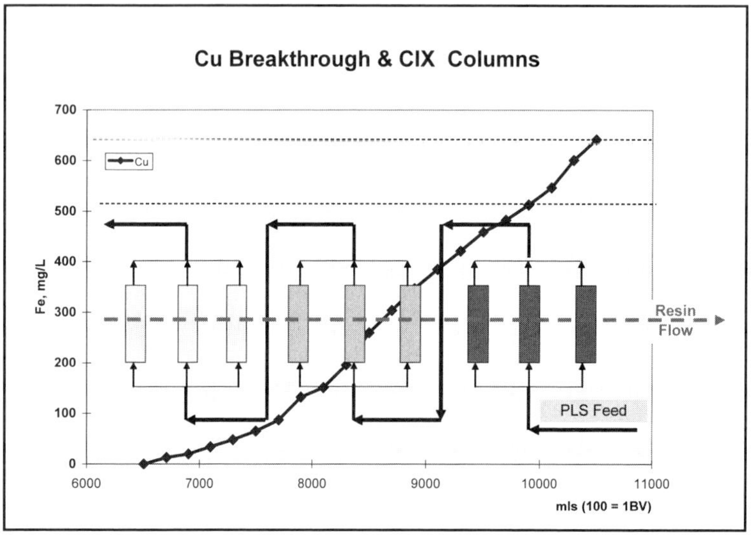


**Figure 7: CIX adsorption concept superimposed on batch column BT curve**

Note that we have only employed that portion of the curve from breakthrough to saturation. This encompasses the zone of mass transfer and, provided we can continuously maintain that ***zone immobilized over a set of columns*** through counter-current flow of both phases, there is no necessity for any extra resin beyond that required to accommodate the MTZL.

In the CIX process resin columns are sequenced to move across the zone from left to right. During the travel from left to right the resin charge gradually loads the metal species until it equilibrates with feed solution just as it exits (fully loaded) to the right.. The PLS initially contacts the almost saturated resin and flows from right to left while transferring its metal content into the resin phase.. This effectively is a counter-current contact between the solid resin and fluid phases.

The rate of resin exhaustion determines the rate of movement of columns across the zone. This rate is timed to fully load the columns by the time they exit to the right. The total length of the three contacts is at least equal to the design MTZL and consequently all the copper is removed from the PLS during its passage from right to left. If the design length of the MTZL is contained within the fluid path through the columns then we shall simply repeat the performance of the static single column on a continuous basis. Stripped resin is brought into the profile from the left and saturated resin is removed from the profile on the right. Fluid flows in the opposite direction. The arrangement **continuously immobilizes the profile inside the adsorption zone** resin beds. Sampling the effluent from each of the columns would produce a [Cu] profile essentially identical to that generated in the batch test.

**Stage 3a - Elution (Resin Stripping)**

A column loaded with the species in Stage 3a is now eluted in Stage 3b. The choice of elution regime requires experience in choosing eluent strength and application rate and knowledge of the processes downstream of the **CIX** plant from which a recycled eluent is often returned.

As a general comment, elution kinetics are mainly a function of internal resin pore diffusion rates and so slower fluid elution rates (BV/hr) which generally produce the desirable higher peak values. Higher temperatures aid in accelerating elution rates.

Some examples of successful **CIX** elution schemes are listed in the Table 2. Examples of downstream eluents and interfaces are shown in Table 3.

**Table 2 Example Elution Schemes**

| Major Element | Typical Eluent | Impurity | Observations |
|---|---|---|---|
| Uranium | $CO_3/HCO_3$ | Mo, V, P, Cl | pH control required |
| Copper | $SO_2/H_2SO_4$ | Fe, Si | Cu:Fe ratio > 1000 desirable |
| Rhenium (ReO4) | NaOH | $SO_4$ | Minimize SO4 |
| Cobalt | $H_2SO_4$ | Ni | Minimize Co Loss |
| Nickel | $SO_2/H_2SO_4$ | Fe | Ni:Fe ratio > 800 target |
| Molybdenum | NH3/NaOH | As | Minimize As co-elution |
| Uranium | $SO_2/H_2SO_4/NaHSO_4$ | Fe, P, Si | |
| PGMs | HCl, | Base metals | |

One can see from Table 2 that the typical regenerants are common acids and alkalis. A word of caution is required when switching from acid to base conditions and back. Metal precipitation can occur easily and so good wash and rinsing techniques are essential to avoid bed fouling.

Often in **CIX** schemes a recycle exists between downstream metal recovery and ion exchange. Examples of this are; recycled electrolyte from a copper EW circuit or, a carbonate or bicarbonate return recycled from a uranium precipitation circuit. Development data for design may only be practical in a pilot operation. The recycle stream composition and properties should be taken into account in this phase of the test work.

**Table 3 Typical Downstream Metal Recovery Operations**

| Element | Eluent | Recovery Method | Comment |
|---|---|---|---|
| Copper | $H_2SO_4$ | EW | Recycle electrolyte |
| Nickel | $H_2SO_4$ | Precipitation | Recycle acid |
| Mo, Re, | NH3/NaOH | Crystallization | Recycle Mother Liquor |
| Uranium | Acid/Alkali | Precipitation | Recycle barren |

Elution testing involves evaluating any pre-elution and resin scrubbing schemes. The experimental set-up is not dissimilar from the adsorption testing. The loaded column from adsorption Stage 3a is first washed free of PLS. Then a scrub of impurities may be used followed by the main elution step. The flow rates for elution are typically much lower than used in adsorption. The elution profile developed at a minimum of two (2) different flow rates which will provide good estimates of resin residence time for the complete elution to occur.

In Figure 8 the flow rate used was 10 BV/hr and from inspection of the graph the elution is complete in about 4 BVs (0.5 to 4.5). This indicates that 24 minutes is sufficient for complete desorption under these conditions. Operating at higher BV/hr will further shorten the elution time but at the cost of a smaller peak value. The peak value indicates closely what the final **CIX** elution product will assay when operated at the same BV/hr rate.

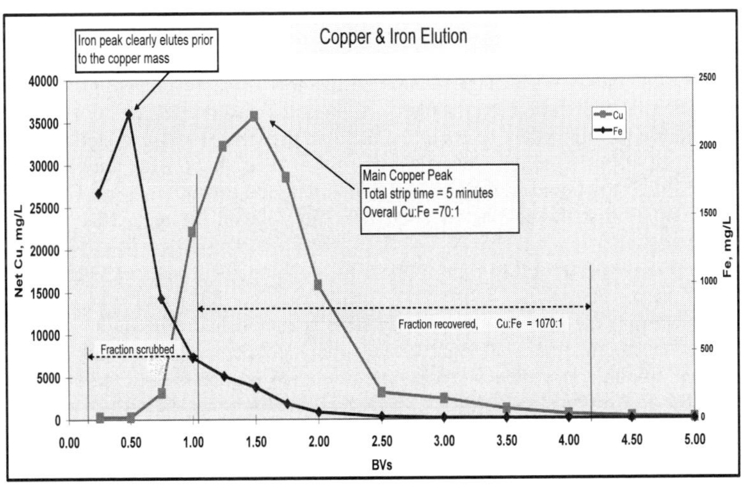


**Figure 8: Copper & Iron Elution showing pre-elution of Iron**

This behavior is typical of metal loaded resins in which the resin selectively extracts a species (Cu in this case) but also co-extracts, albeit less selectively, an impurity (Fe). In the subsequent elution the less selectively held species will usually elute in advance of the more strongly adsorbed element – as is the case here..

Data in Table 4 provides some indication of how important the scrub and pre-elution can be in terms of final metal solution purification. It is this aspect of **CIX** that is little known in the industry and is achievable through state of the art recent (last 15 years) improvements in the equipment technology and process engineering expertise.

The last column in Table 4 shows the ratio of the target metal to impurity; first in the feed; then the same ratio on the resin after adsorption; and, finally, the third number is the achievable ratio in the product eluate after having executed a scrub and/or pre-elution purification of the resin phase. This approach shows the power of using **CIX** to clearly reduce impurity levels far beyond those accomplished solely through natural selectivity of the resin phase.

**Table 4 Scrub, Pre-Elution Purification Effects**

| Metal extracted | Impurity | Purification | Ratio Extracted Metal:Impurity Feed - Resin - Eluate |
|---|---|---|---|
| Copper | Fe | Scrub/Pre-elution | 1:2 → 70:1 → **1000:1** |
| Co[3] | Ni | Pre-elution | 60: 1 → 5:1 → **1:12** |
| Nickel | Fe | Scrub & Pre-elution | 1:4 → 2:1 → **60:1** |
| Uranium_acid | Fe, Mo | Scrub & Pre-elution | 1:5 → 14:1 → **4500:1** |
| Re | SO4 | Pre-elution | 1:10000 → 1:5 → **100:1** |
| Uranium_alkali | V | Scrub | 1:2.5 → 3:1 → **52:1** |
| Mo | As | Pre-elution | 1:30 → 1:2 → **3:1** |

As an example we illustrate the elution of a copper loaded resin (Figure 8). The plot in Figure 8 shows a single column elution of a copper loaded resin using sulfuric acid as eluent. The copper peak is clearly preceded by an iron peak.

Elutions are usually executed at slower flow rates than in adsorption. Some elutions are slow due to large ionic radius and slow diffusion rates. Others are rapid. For the example case (Figure 8) it is quite rapid and was run at 10 BV/hr. The whole elution was complete inside 4 BVs. This equates to about 25 minutes of process contact time. Other elutions can take up to 120 minutes. The elution time required is an important parameter to check. Parameters to optimize in elution are:

1. Elution Time
2. Eluent strength – higher strength better but resin limitations apply
3. Elution temperature – elevated temperature accelerates elution
4. Maximize separation from pre-eluting impurities.

The above parameters are inter-related. Faster flow through elution reduces the peak values but shortens elution time. Elution peak maximum value can be improved by increasing eluent strength and operating at temperatures up to 60 C. Usually the elution strength is maintained in the region of 1 to 2 molar depending on the application. Separation of pre-eluting impurities is often enhanced by using a less strong eluent in advance of the main elution. This is often accomplished by using strong eluate as a pre-elution feed. The possibilities existing in this step require some experience to properly adjust and optimize the process. The experimenter should recognize the natural selectivity as evidenced by a tendency for one species to elute in advance of the others, and then efficiently use this phenomenon to optimize the purity of the harvested peak (Cu peak). Tools to do this include,

1. Scrubbing with dilute eluents
2. Reductive Scrub - switches metal valences and can significantly facilitate resin purification
3. A chromatographic style pre-elution stage
4. Internal recycling of eluate to enrich the resin phase

Taking the example of copper above one can now envisage the immobilization of the developed profile as shown below where a **CIX** counter-current pre-elution and elution arrangement are superimposed on the laboratory profile.

Immobilization of a liquid concentration profile simply refers to the fact that the laboratory generated profile at the bed exit point is also present inside the bed at times earlier. Consequently, if we "feed" the profile with fresh resin at the correct rate and in the opposite direction to the liquid flow then we can maintain the profile stationary in space. The basic principle of a mass balance will dictate that the copper arriving from the right on the resin equals the copper removed as eluate.

Considering the elution step, as resin moves from left to right in the **CIX** stepping sequence (see Figure 9) copper and iron enter the profile from the right on the loaded resin. The rinse water (at right) and eluent (spent EW electrolyte/acid in middle) remove copper from the profile and a steady state is established. The outlet streams from this operation are an iron rich pre-elution stream and a strong relatively pure copper stream suitable for electro-winning. The Fe rich pre-elution effluent amounts to a small recycle load of copper that exits with the iron while the copper is re-adsorbed (continuously) in the adsorption step.

**CIX** technology allows the process to harvest only the peak area. Referring to Figure 9 strong eluate exits from the 3rd and 4th columns counting from the left. The eluate concentration is represented by that fraction of the peak harvested. The profile of the elution tail and the pre-elution peak is immobilized in the columns to either side of the strong eluate peak. By harvesting the elution peak the final product eluate not only has higher concentration than what is achievable in conventional fixed beds but also has less excess eluent. The reduction in excess eluent often has economic benefits downstream in processes that employ precipitation circuits; less acid to neutralize saves cost by reducing the chemical requirements. Higher strength eluate product translates into smaller sized equipment downstream. Smaller more concentrated feeds to precipitation and crystallization produce higher precipitation yields.

[3] Ni removal from cobalt electrolyte. Here the species of interest is really cobalt and the ability to MINIMISE the cobalt in nickel eluate.

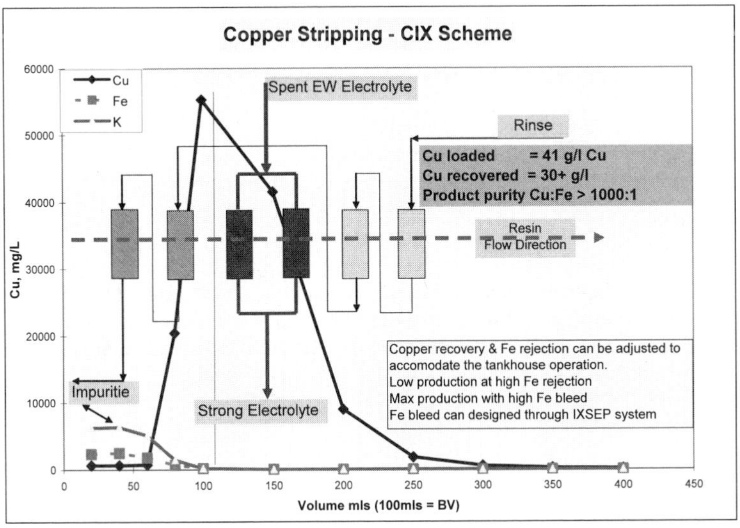


**Figure 9 CIX Elution and Pre-Elution Configuration**

A reductive scrub can also be considered in cases where the Fe levels are high. This can be very effective for removing Fe in copper and nickel processes where the Fe co-loads with the value metal. In alkaline uranium circuits a carbonate ($Na_2CO_3$) post adsorption scrub of the resin effectively removes significant levels of co-loaded Mo, V and other impurities. In acid leaching of uranium the reductive scrub is used to remove Fe. The scrub stages will precede the pre-elution steps

**Stage 4 – Cycle Testing – Process Verification**

In this phase we typically use three columns connected in series and run the adsorption step routing the fluid through all three in series. For the elution steps the lead most saturated column is removed and subjected to the wash-scrub-elution-rinse steps. Then adsorption is re-commenced when the eluted column is returned to the trailing position in the three-column adsorption sequence. Running 6 cycles in this manner allows us to approach a balance where the metal fed during adsorption balances with metal recovered during elution.

This test allows us to monitor the behavior of impurities; it permits changing eluent strength from cycle to cycle; it provides scope for varying scrubbing techniques. A mass balance is conducted for each cycle on all species of importance and as the test progresses the experimenter adjusts the mass of value metal in the feed to balance with mass of metal removed from each of the saturated columns. A pseudo-steady state can be achieved in about 5-6 cycles.

The execution of this phase really needs some input from experienced people but logically follows on from the single column work.

## SUMMARY AND CONCLUSIONS

The various laboratory testing procedures to generate design parameters for **CIX** system design have been outlined. They consist of simple laboratory beaker and small column test work. The cost impact to a developer is relatively minor. Data from tests run in this manner can be interpreted by experienced **CIX** system suppliers to generate a conceptual flow sheet. This level of development will provide budget level capital and operating cost data for project evaluation. If positive then a decision can be made to investigate further either in more extensive laboratory testing or embarking on **CIX** pilot trials.

The testing at the laboratory stage will also provide good process performance information in regard to differences between resin candidates. The test work will also demonstrate the potential for undesirable impurity removal as well as projecting product liquor likely concentration and quality. The projected product liquor quality will influence the costing and design of downstream recovery processes.

In this regard, it should be appreciated that **CIX** is one step in a multi-step process and should be integrated into the whole. Its benefits are not simply those compared to conventional ion exchange but often also serve to reduce cost in downstream metal recovery steps.

The author appreciates the opportunity to present this material and thanks the MPP conference organizers along with several colleagues from the ion exchange community. It is also our intention to further elaborate on the presented material and continue this theme into **CIX** pilot operations and commercial design concepts.

## REFERENCES


1. Rossiter, G., Bio-technology Symposium Antwerp 1997, Continuous Processing using Solid Adsorbents
2. Rossiter G., Carey K., Randol Vancouver 1998. Copper Recovery from Leach Liquors using Ion Exchange
3. Jensen C., Rossiter, G, et alia., Cambridge IEX 2004, The Impact of Small Diameter Resin Beads on Process Kinetics and Economics – A Case Study

# FUNDAMENTAL PROCEDURES TO EVALUATE AND DESIGN INDUSTRIAL WASTE WATER METALS TREATMENT SYSTEMS, CASE STUDY DISCUSSIONS

Jay McCloskey[1], Dr. Larry Twidwell[2], Dr. Hsin-Hsinng Huang[2], and Lucas Goldstien[3]

[1]Center for Advanced Mineral and Metallurgical Processing
[2]Montana Tech of The University of Montana, 1300 West Park Street, Butte, Montana USA
[3]LFR an ARCADES Company, 1900 Powell Street, 12th Floor, Emeryville, California 94608-1827

## ABSTRACT

Methods and techniques to evaluate, design, and implement innovative hydrometallurgical systems for the reduction of dissolved arsenic, barium, selenium, thallium, and other constituents from industrial waste waters are discussed. Case studies utilizing geochemical modelling and matrix design testing to provide efficient methods to identify design parameters for precipitation processes to remove metals from an industrial wastewater are presented.

## INTRODUCTION

The use of geochemical modelling and matrix design testing provides efficient methods to identify design parameters for industrial wastewater treatment approaches. The case study presented in this paper is an example of the standard approach practiced by the Center for Advanced Mineral and Metallurgical Processing (CAMP) to identify appropriate treatment approaches along with optimization of design parameters.

Geochemical modeling for the project waters was performed using the software program STABCAL developed by H.H. Huang at Montana Tech of the University of Montana (Huang, 2008). The program allows the user to evaluate: relative compound solubilities as a function of pH, ion concentration, and solution potential (Eh); adsorption phenomena, potential/pH diagrams (Eh/pH); titration reactions and kinetics of compound formation. The STABCAL program can perform all of the mentioned calculations at any selected temperature from 25 to 200°C.

Design Expert software (StatEase Corporation Version 7.1.1) was utilized to design the test matrix and statistically evaluate the experimental results. Three Design Expert experimental matrices are presented involving the same project water. The design matrix (Minimum Run Resolution IV) allows the user to determine the influence of the variable factors on the measured response. Binary interaction effects are aliased but the main variable factors are cleanly determined. A Discussion of the influence of variable factors on the measured responses is presented along with the statistical validations of the selected models for each selected design space.

## CASE STUDY PROJECT WATER

The water used as a case study for this paper was from groundwater contaminated with solvents from historical dry cleaning operations, and will be referred to as the project water in this paper. The groundwater contains naturally occurring metals, specifically arsenic and barium that are the subject of this study. The flow rate for the proposed pump and treat system is estimated at 150 gallon per minute (gpm).

## TREATMENT OBJECTIVE

The objective for this project was to develop a viable treatment approach to reduce arsenic (As) and barium (Ba) from the project water. The goal was to reduce the As from 15.0 parts per billion (ppb) to less than 5.0 ppb and reduce Ba from 300 ppb to less than 100 ppb. As a secondary goal, the treatment process cannot increase iron (Fe) concentrations from their current levels of 100 ppb. CAMP performed laboratory treatability tests at their facility in Butte, Montana. Laboratory analytical tests were conducted at a state-certified analytical laboratory. Metals were analyzed using inductively coupled plasma mass spectrometry (ICP-MS) methodology.

## GEOCHEMISTRY DISCUSSION

An historical analysis of the project water is provided in Table 1. Prior to a discussion on selected technologies, a short review on aqueous geochemistry of both As and Ba is given below.

**Table 1. Project Water Analysis**

| Parameter | Result | Units |
|---|---|---|
| Total Dissolved Solids | 410 | mg/l |
| pH | 8.1 | SU |
| TEMP | 68 | F |
| Specific Conductance | 760 | umhos |
| Turbidity | 0.1 | NTU |
| Aluminum | 100 | ug/L |
| Ammonia-NH3 | 0.08 | mg/l |
| Antimony | 60 | ug/L |
| Arsenic | 15 | ug/l |
| Barium | 300 | ug/l |
| Beryllium | 2 | ug/L |
| Cadmium | 5 | ug/L |
| Calcium | 100 | mg/l |
| Chloride | 54.2 | mg/l |
| Chromium | 10 | ug/L |
| Chromium VI | 2.88 | ug/l |
| Copper | 2.0 | ug/l |
| Cyanide | 0.01 | mg/L |
| Fluoride | 0.1 | mg/L |
| Hardness (CaCO3) | 390 | mg/L |
| Iron | 100 | ug/l |
| Lead | 1.0 | ug/l |
| Manganese | 50 | ug/l |
| Mercury | 0.5 | ug/l |
| Nickel | 20 | ug/L |
| Nitrate as NO3 | 6.02 | mg/l |
| Nitrite as NO2 | 0.500 | mg/l |
| Nitrogen, Nitrite | 0.05 | mg/L |
| Phosphorous | 2.2 | mg/L |
| Selenium | 5 | ug/L |
| Silver | 5 | ug/L |
| Sulfate | 68 | mg/L |
| Sulfate as S | 26.5 | mg/l |
| Sulfite | 2 | mg/L |
| Surfactants (MBAS) | 0.1 | mg/L |
| Thallium | 5 | ug/L |
| Zinc | 38 | ug/l |

**Arsenic**

Arsenic is a naturally occurring element in the earth's crust, and is found throughout the environment. Arsenic is a redox-sensitive element. This means that arsenic may gain or lose electrons in redox reactions. As a result, arsenic may be present in a variety of redox states. Arsenate (As(V)) and arsenite (As(III)) are the two forms of arsenic commonly found in groundwater (Deutsch 1997). As(V) generally predominates under oxidizing conditions. As(III) predominates when conditions become sufficiently reducing. Under the pH conditions of most ground water, As(V) is present as the negatively charged oxyanions $H_2AsO_4^-$ and $HAsO_4^{2-}$, whereas As(III) is present as the uncharged species $H_3AsO_3^0$ (Deutsch 1997). The strength of adsorption and desorption reactions between these different arsenic species and solid-phase surfaces in aquifers varies, in part, because of these differences in charge. Differences in species charge affect the character of electrostatic interactions between species and surfaces.

The mobility of arsenic under oxidizing conditions is primarily affected by the adsorption of As(V) onto metal oxyhydroxide surfaces. If the applicable adsorption capacity of these surfaces is not surpassed, then arsenic movement will be strongly retarded because of the high affinity of these surfaces for As(V). Under reducing conditions, the dominant arsenic redox species will be As(III), which is not as strongly absorbed. Furthermore, the primary adsorbing solids may not be stable if the redox potential is sufficiently low. As a consequence, arsenic is expected to be much more mobile under moderately low reducing conditions. This mobility may be significantly reduced if arsenic sulfide minerals become saturated and precipitate.

Geochemical modeling of the Project Water was performed using the software program STABCAL and the Eh/pH diagram is shown in Figure 1 for the Lincoln Center Site groundwater iron/arsenic/barium/sulfur/water system. The shaded areas show conditions where iron solid compounds form and illustrate that the adsorption of arsenic onto these solids should occur when conditions are within these regions.

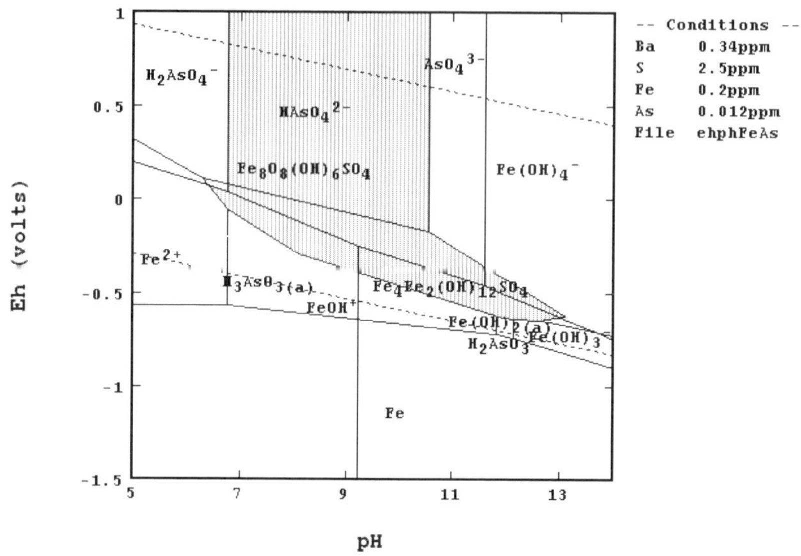


**Figure 1. Eh/pH diagram for the Fe/As/Ba/S water system. The conditions for the formation of iron solid compounds are shown as shaded regions. (Diagram constructed using STABCAL, Huang 2008).**

**Barium**

Barium is a metallic element that is chemically similar to calcium and strontium, but more reactive. The most common naturally occurring minerals are the very insoluble barium sulfate, $BaSO_4$ (barite), and barium carbonate, $BaCO_3$ (Witherite). Barium is not found in nature in its elemental pure form due to its reactivity with air. Figure 2 is a distribution diagram (produced using STABCAL, Huang 2008) that illustrates barium should be removed by adsorption beginning at pH levels above approximately 8 (best at 10). The barium is removed not by compound precipitation but by surface phase adsorption. The removal of barium by iron hydroxide adsorption optimization of the doping done for barium removal has not been previously studied. This is the first study of which we have knowledge where the removal of barium at relatively low concentrations (less than 500 ppb) was studied by adsorption processes.

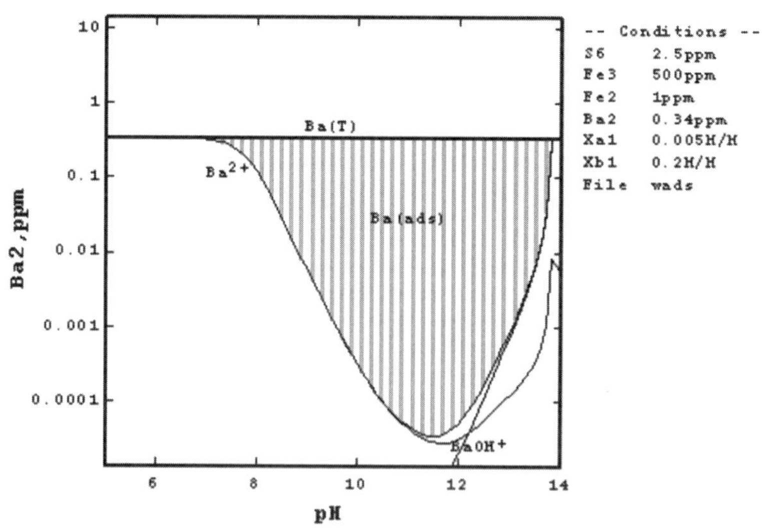


**Figure 2. Distribution Model for the Fe/As/Ba/S water system. (Diagram contructed using STABCAL, Huang 2008).**

## TECHNOLOGY DISCUSSIONS

Three technologies were selected for evaluation. These technologies were selected based on the geochemical modeling information, CAMP's experience working with similar water, and potential capabilities of each of the selected technologies for removing arsenic and barium from the project water. These technologies are:

1. ***Reductive precipitation***, sometimes referred to as iron cementation, followed by ferric or ferrous hydroxide adsorption,
2. ***Ferrous Hydroxide precipitation*** and dissolved specie adsorption process, and
3. ***Ferric Hydroxide precipitation*** and dissolved specie adsorption process.

These technologies fall under the category of Precipitation Processes, specifically, coagulation assisted filtration. As such, waste generated by these technologies includes: backwash water and sludge. This study does not assess the mass or quality of sludge that would be generated. A brief discussion for each technology is provided below. Additional information on each technology is discussed in the Testing and Results section.

### Reductive Precipitation

Arsenic occurs in solution as oxyanions (discussed above). These oxyanions can be reduced by electrochemical reactions. Metals can be used as the reductant. Any metal that has a reduction potential that is less noble than the impurity reduction potential will thermodynamically reduce the impurity specie, e.g., those metals that will reduce arsenic oxyanions include iron, zinc, aluminum, and magnesium. The presence of elemental iron in a slurry reactor or column reactor provides an operating potential sufficiently reducing to remove arsenic oxyanions as elemental arsenic. In the reductive precipitation tests, a stirred reactor with 100 grams of iron powder per liter of process solution is maintained. In a continuous system, the iron powder is replaced as it is used in the iron surface electrochemical reaction and goes into solution. One of the advantages of the reductive cementation technology is that it will remove both arsenate and arsenite species to less than the The Maximum Contaminant Level (MCL) which is the maximum concentration of a chemical that is allowed in public drinking water systems. The MCL is established by the U.S. Environmental Protection Agency (EPA)(Kamrin, 1997). It will also remove other oxyanions, such as antimony and selenium, and associated cationic metal species, e.g., mercury, copper, and nickel. Unlike adsorption processes, the removal efficiency is unaffected by the presence of sulfate. The limitations of this technology is that the water pH must be less than 7, preferably about pH 5.5 and as the electrochemical reactions occur iron is added to the solution phase and must be removed in a subsequent precipitation stage. Either calcium hydroxide (caustic) or sodium hydroxide (lime) can be used to raise the solution pH in the precipitation stage. The reductive cementation approach will not remove barium. However, the barium will be removed via adsorption in the subsequent precipitation stage.

### Ferrous Hydroxide Process

In the Ferrous Hydroxide Process, ferrous sulfate is added to the solution in predetermined amounts. The pH of the solution is then adjusted to approximately 10 or greater using either caustic or lime to form a ferrous hydroxide (green rust) precipitate which is a good adsorber for arsenic and barium. (McCloskey and Twidwell) have demonstrated that Green Rust is an effective adsorbent for arsenic oxyanions (both arsenite and arsenate) over the pH range 9-11 at solution potentials shown in the shaded stability region in Figure 1. Green Rust is a mixed valence ferric/ferrous hydroxysulfate solid $[(Fe^{2+})_4(Fe^{3+})_2(OH)_{12}SO_4]$.

### Ferric Hydroxide Process

In the Ferrous Hydroxide Process, ferric sulfate or ferric chloride is added to the solution in predetermined amounts. The pH of the solution is then adjusted to 6.5 or greater using either caustic or lime to form a ferric hydroxide precipitate that can be a good adsorber for arsenic and barium. Figure 3 is a photo of the ferrous (left/green) and ferric (right/red) hydroxide test solutions.

**Figure 3. Photo of both the ferrous (left) and ferric (right) hydroxide test solutions.**

## EXPERIMENTAL TEST MATRIX DEVELOPMENT

Three Design Expert Test Matrices were followed:

1. The first test matrix was designed to evaluate reductive precipitation in combination with utilizing the ferrous iron in solution resulting from the iron powder and solution chemistry followed by ferrous hydroxide and ferric hydroxide (with additional of peroxide) precipitation tests.
2. The second test matrix was designed to evaluate straight reductive precipitation tests.
3. The third test matrix was designed for ferrous hydroxide and ferric hydroxide tests utilizing the reagents ferrous sulfate and ferric sulfate respectively.

The data collected were evaluated using Design Expert software (StatEase Corporation Version 7.1.1). Each test matrix was set up to evaluate influence of variable factors on the measured responses as a function of As, Ba, and Fe concentrations. When appropriate, a maximum and minimum variable level was determined and in some matrix sets a mean level as used.

## TESTING AND RESULTS

The three Design Expert experimental matrices are presented in the order discussed above in Tables 2-4. Each table outlines the required test conditions and the experimental results. Due to restrictions on paper length requirements, the statistical validation of the selected models for each selected design space is not provided in this paper.

**Table 2. Design Expert Matrix 1 and Measured Responses for Reductive Precipitation Tests Followed by Either Ferrous Hydroxide or Ferric Hydroxide.**

| Standard Run No. | Factors (Variables) | | | | Responses, μg/L | | |
|---|---|---|---|---|---|---|---|
| | A:pH | B:Reduction Time, hrs | C:Type of Adsorbent | D:Adsorption Time, hrs | [As] | [Ba] | [Fe] |
| LFR-SR1-21709-2 | 5 | 1 | Fe2 | 1 | **3.65** | **16.88** | 106.81 |
| LFR-SR2-21809-2 | 7 | 1 | Fe2 | 2 | **3.37** | **11.07** | **80.08** |
| LFR-SR3-21709-2 | 5 | 2 | Fe2 | 2 | **3.69** | **63.00** | 154.41 |
| LFR-SR4-21909-2 | 7 | 2 | Fe2 | 1 | **2.48** | **12.20** | 204.43 |
| LFR-SR5-21709-2 | 5 | 1 | Fe3 | 2 | **3.75** | **20.04** | 255.98 |
| LFR-SR6-21809-2 | 7 | 1 | Fe3 | 1 | **3.37** | **11.99** | **37.52** |
| LFR-SR7-21709-2 | 5 | 2 | Fe3 | 1 | **2.54** | **8.81** | 339.42 |
| LFR-SR8-21909-2 | 7 | 2 | Fe3 | 2 | **2.45** | **15.20** | 204.40 |
| LFR-SR9-21909-2 | 6 | 1.5 | Fe2 | 1.5 | **2.48** | **62.78** | **29.39** |
| LFR-SR10-22009-2 | 6 | 1.5 | Fe3 | 1.5 | **2.46** | **22.29** | **26.34** |
| LFR-SR11-22009-2 | 6 | 1.5 | Fe2 | 1.5 | **1.47** | **70.53** | **62.00** |
| LFR-SR12-22009-2 | 6 | 1.5 | Fe3 | 1.5 | **2.49** | **10.59** | **76.96** |

Note—achieved project goal concentrations; Detection Limit (DL) concentrations taken as data point values for Design Expert evaluations

**Table 3. Design Expert Matrix 2 and Measured Responses for Reductive Precipitation.**

| Standard Run No. | Factors (Variables) | | | | Responses, µg/L | | |
|---|---|---|---|---|---|---|---|
| | A:pH | B:Reduction Time, hrs | C:Type of Adsorbent | D:Adsorption Time, hrs | [As] | [Ba] | [Fe] |
| LFR-SR1-21709-1 | 5 | 1 | Fe2 | 1 | 8.98 | 276.13 | 698290 |
| LFR-SR2-21809-1 | 7 | 1 | Fe2 | 2 | **3.08** | 163.18 | 84108 |
| LFR-SR3-21709-1 | 5 | 2 | Fe2 | 2 | 9.65 | 283.99 | 1305889 |
| LFR-SR4-21909-1 | 7 | 2 | Fe2 | 1 | **1.45** | **45.56** | 3036 |
| LFR-SR5-21709-1 | 5 | 1 | Fe3 | 2 | 5.92 | 286.72 | 708870 |
| LFR-SR6-21809-1 | 7 | 1 | Fe3 | 1 | **2.98** | 153.05 | 81067 |
| LFR-SR7-21709-1 | 5 | 2 | Fe3 | 1 | 8.89 | 305.76 | 1305889 |
| LFR-SR8-21909-1 | 7 | 2 | Fe3 | 2 | **1.46** | **43.54** | 2631.56 |
| LFR-SR9-21909-1 | 6 | 1.5 | Fe2 | 1.5 | **1.45** | 152.81 | 637393 |
| LFR-SR10-22009-1 | 6 | 1.5 | Fe3 | 1.5 | **1.41** | 301.72 | 425160 |
| LFR-SR11-22009-1 | 6 | 1.5 | Fe2 | 1.5 | 5.09 | 270.66 | 414790 |
| LFR-SR12-22009-1 | 6 | 1.5 | Fe3 | 1.5 | 5.91 | 264.10 | 596923 |

Note- achieved project goal concentrations; DL concentrations taken as data point values for Design Expert evaluations

**Table 4. Design Expert Matrix 3 and Masured Responses for Species Removal by Adsorption on Ferrous and Ferric Hydroxide**

| Standard Run No. | Factors (Variables) | | | | Responses, (µg/L) | | |
|---|---|---|---|---|---|---|---|
| | A:pH | B: Time, hrs | C:Type of Adsorbent | D:Amount Fe, mg/L | [As] | [Ba] | [Fe] |
| SR1-21709-1 | 10 | 2 | Fe2 | 800 | 15.23 | **4.09** | 5.19 |
| SR2-21809-1 | 8 | 1 | Fe2 | 800 | **1.46** | **32.70** | 34.43 |
| SR3-21709-1 | 8 | 2 | Fe3 | 800 | **1.45** | **25.28** | 26.57 |
| SR4-21909-1 | 8 | 2 | Fe3 | 400 | **1.45** | **45.62** | **47.26** |
| SR5-21709-1 | 10 | 1 | Fe3 | 800 | **1.45** | **8.00** | 9.11 |
| SR6-21809-1 | 8 | 1 | Fe3 | 400 | **1.46** | **47.65** | 49.32 |
| SR7-21709-1 | 10 | 2 | Fe3 | 400 | **1.56** | **6.47** | **7.54** |
| SR8-21909-1 | 10 | 1 | Fe2 | 400 | 5.58 | **5.27** | **6.36** |
| SR9-21909-1 | 10 | 1 | Fe2 | 800 | 26.22 | **3.77** | **4.84** |
| SR10-22009-1 | 8 | 2 | Fe2 | 400 | **1.45** | **36.22** | **37.45** |

Note- achieved project goal concentrations; DL concentrations taken as data point values for Design Expert evaluations

## INFLUENCE OF VARIABLE FACTORS ON THE MEASURED RESPONSES

### Reductive Precipitation—Ferrous Hydroxide or Ferric Hydroxide Precipitation (Experimental Data in Table 2)

#### *Design Expert Evaluation of Results*

***Arsenic.*** A statistically valid model for arsenic removal is not applicable. The process is so effective that any combination of the variables investigated in this study result in a final arsenic concentration of (<5 µg/L).

***Barium.*** A statistically valid model that describes the design space for the removal of barium shows that the most important variable is pH. Reduction time was shown not to be important as expected because the reductive precipitation (cementation) process is known to be ineffective for removing barium. The influence of pH, reduction time, and adsorption time is presented in the cubic display in Figure 4 (Left for Fe2 treatment and Right for Fe3 treatment). Note the comments presented in the caption. 3-Dimensional plots are presented in Figure 5. Briefly, the conclusion is that all combinations of conditions, regardless of whether cementation is followed by ferrous or ferric hydroxide adsorption, meet the project goal for barium (<100 µg/L).

***Iron.*** A statistically valid model that describes the design space for the removal of iron shows that the most important variables are pH and reduction time. Short reduction times are desired so less iron enters the solution and higher pH is desired so that more ferrous or ferric hydroxides remove iron from solution by precipitation. Cubic displays are presented in Figure 6 (Left for Fe2 treatment and Right for Fe3 treatment). 3-Dimensional plots are presented in Figure 7. Briefly, the conclusion is that only 50% met the <100 μg/L project goal for most combinations of conditions, regardless of whether cementation is followed by ferrous or ferric hydroxide adsorption, meet the project goal for iron (<100 μg/L).

### Cubic Visualizations for Final Barium Concentration

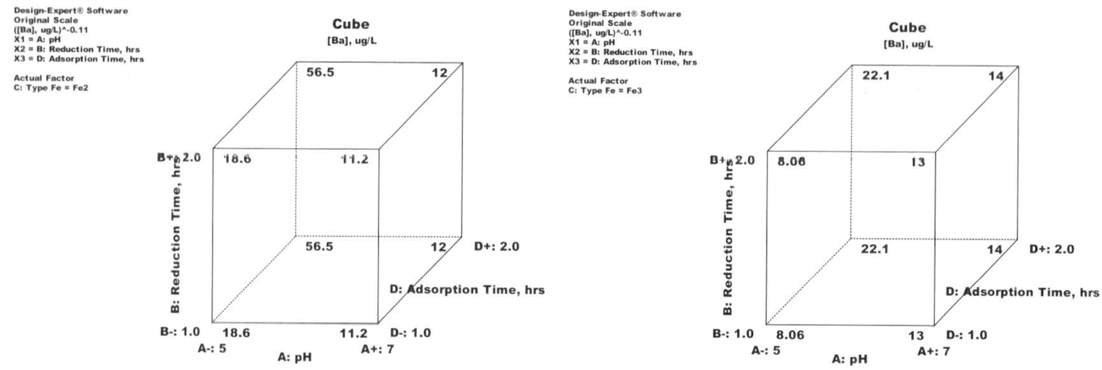


**Figure 4. Cubic representation of the influence of variables on the final barium concentration (Left: ferrous hydroxide (Fe2) adsorbent, Right: ferric hydroxide (Fe3)). Reduction Time and Adsorption Time are unimportant, the pH is important. Most combinations of Fe2 conditions result in barium concentrations <50 μg/L while most combinations of Fe3 conditions result in barium concentrations <25 μg/L. Therefore, the formation of ferric hydroxide (called ferrihydrite and schwertmannite) is more effective for removing barium than ferrous hydroxide adsorption.**

### 3-D Visualizations for Final Barium Concentration

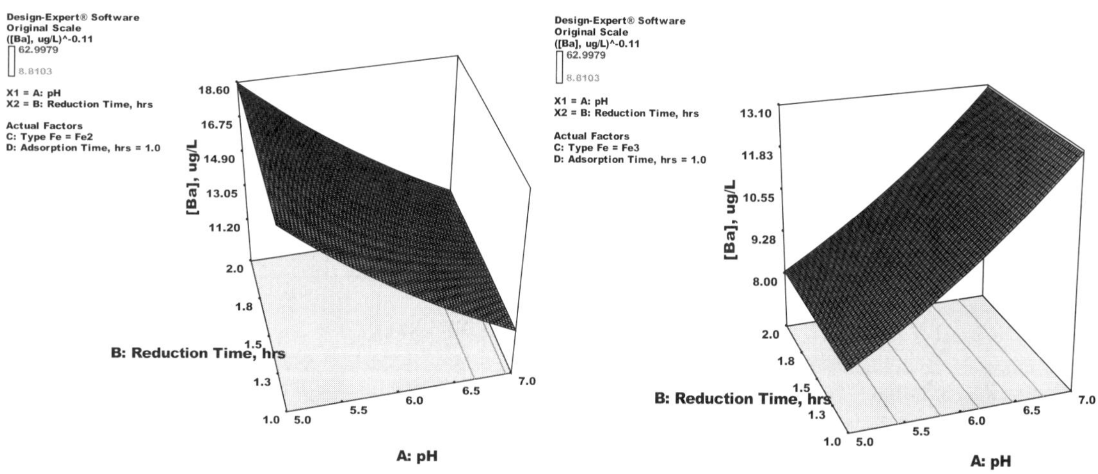


**Figure 5. 3-D representation of the influence of variables on the final barium concentration (Left for the ferrous hydroxide (Fe2) adsorbent, Right for ferric hydroxide (Fe3) adsorbent). For Fe2 the higher pH level is desirable, while for Fe3 the lower pH level is desirable**

**Cubic Visualizations for Final Iron Concentration**

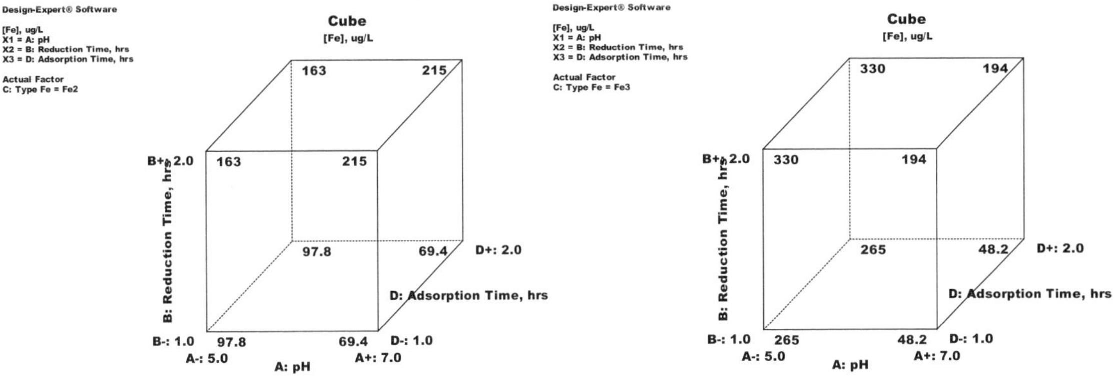


**Figure 6. Cubic representation of the influence of variables on the final iron concentration (Left for ferrous hydroxide (Fe2) precipitation, Right for ferric hydroxide (Fe3) precipitation). For Fe2 precipitation: Reduction Time and pH are important. In general, most combinations of conditions result in iron concentrations less than the project goal of <100 μg/L. For Fe3 precipitation: Reduction Time and pH are important; Adsorption Time is unimportant**

**3-D Visualizations for Final Iron Concentration**

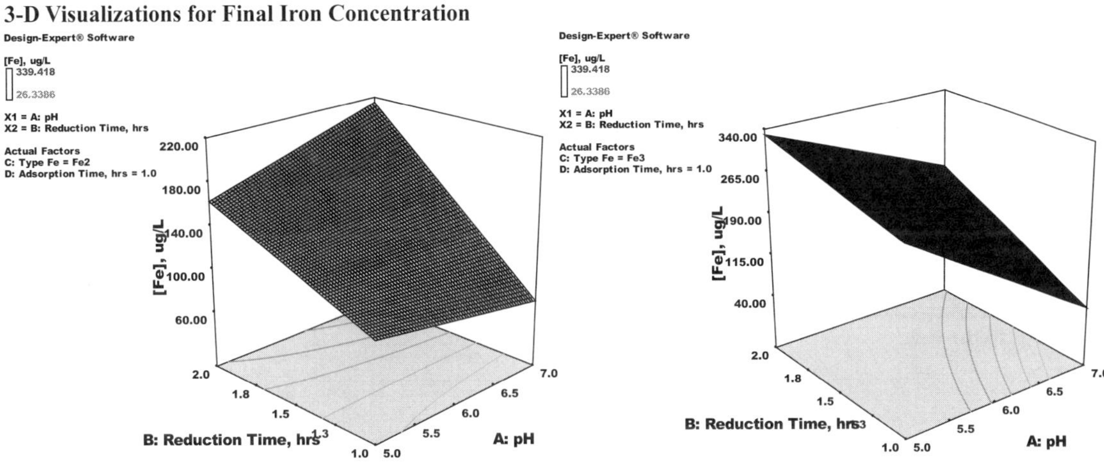


**Figure 7. 3-D representation of the influence of variables on the final iron concentration (Left for ferrous hydroxide (Fe2) precipitation, Right for ferric hydroxide (Fe3) precipitation).**

## Reductive Precipitation (Experimental Data in Table 3)

### *Design Expert Evaluation of Results*

***Arsenic, Barium, Iron.*** The use of reductive precipitation without further subsequent treatment is deemed not appropriate for the present application. The removal of arsenic is excellent; however, this treatment is ineffective for removing barium and the iron content of the solution is greater than the project goal concentrations of 100 μg/L. Therefore, a statistical evaluation will not be further discussed in this report.

## Ferrous Hydroxide and Ferric Hydroxide (Experimental Data in Table 4)

### *Design Expert Evaluation of Results*

***Arsenic.*** A statistically valid model that describes the design space for the removal of arsenic was developed. The adsorption technique is very effective as demonstrated by the data presented in Table 4. The influence of amount of iron added to the system, the time allowed for adsorption to occur and pH on arsenic removal are presented in Figure 8 as cubic representations and in Figure 9 as 3-dimensional representations. Note the comments on the Figure captions. Briefly, the conclusion is that both ferrous hydroxide and ferric hydroxide adsorption is effective for removing arsenic. However, the use of ferrous hydroxide requires lower pH values, whereas the use of ferric hydroxide lowers the arsenic concentration to <2 μg/L for all combinations of conditions in the selected design space.

***Barium.*** A statistically valid model that describes the design space for the removal of barium was developed. The adsorption technique is very effective for removing barium as demonstrated by the data presented in Table 4. The influence of amount of iron added to the system, the time allowed for adsorption to occur and pH are presented in Figure 10 as cubic representations and in Figure 11 as 3-dimensional representations. Note the comments on the Figure captions. The conclusion is that both ferrous hydroxide and ferric hydroxide adsorbents are equally effective for removing barium. Higher pH levels are favored. The project goal for barium is achieved by all combinations of conditions in the selected design space.

***Iron.*** A statistically valid model that describes the design space for the removal of iron was developed. The hydroxide precipitation technique is very effective for removing iron as demonstrated by the data presented in Table 4. The influence of amount of iron added to the system, the time allowed for adsorption to occur and pH are presented in Figure 12 for (Fe2) and (Fe3) as cubic representations and in Figure 13 as 3-dimensional representations. The conclusion is that both ferrous hydroxide and ferric hydroxide precipitation are equally effective for removing iron. Higher pH levels are favored. The project goal for iron (100 μg/L) is achieved by all combination of conditions in the selected design space.

**Cubic Visualizations for Final Arsenic Concentration**

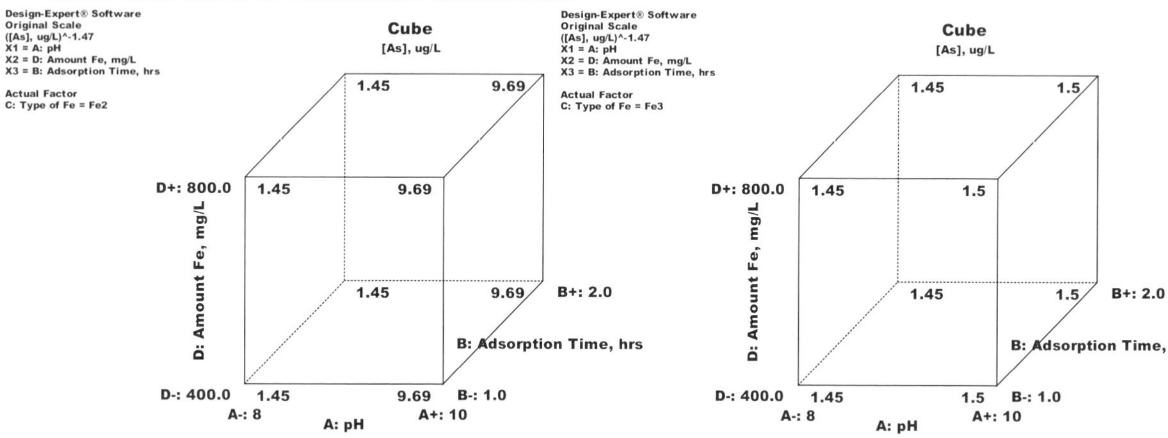


**Figure 8. Cubic representation of the effect of three variables (Left for the Fe2 hydroxide adsorbent, Right for Fe3 hydroxide adsorbent) on the final concentration of arsenic in the solution phase. Note that arsenic adsorption on ferric hydroxide is more effective than on ferrous hydroxide at high pH levels (compare with Figure 6). The adsorption of arsenic via ferric hydroxide is independent of pH, whereas, lower pH levels are better for adsorption on ferrous hydroxide.**

**3-D Visualizations for Final arsenic Concentration**

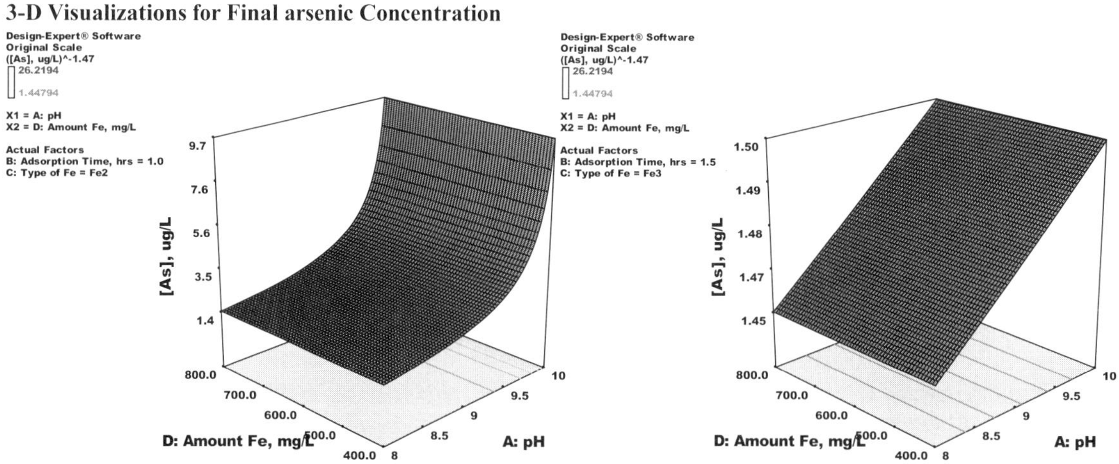


**Figure 9. 3-D representation of the effect of two variables (Left for Fe2 hydroxide adsorbent, Right for Fe3 hydroxide adsorbent) on the final concentration of arsenic in the solution phase. For Fe2 the arsenic adsorption is independent on adsorption time and amount of iron initially present. For Fe3 the adsorption of arsenic is independent of the amount of iron initially present and adsorption time.**

## Cubic Visualizations for Final Barium Concentration

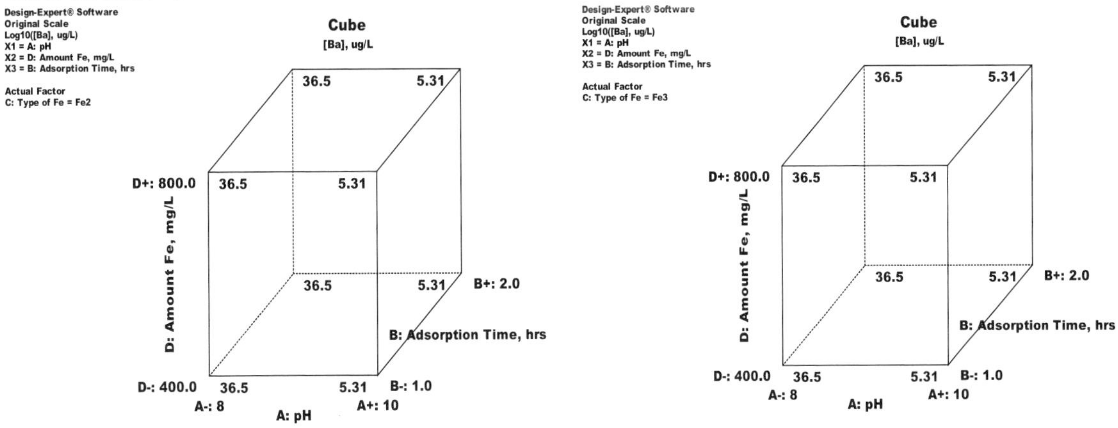


**Figure 10. Cubic representation of the effect of three variables (Left for Fe2 hydroxide adsorbent and Right for Fe3 hydroxide adsorbent) on the final concentration of barium in the solution phase. The only variable of importance is pH; higher pH is favored. There was no difference in the final barium concentration for the ferrous or ferric hydroxide adsorbent solids. The type of adsorbent is unimportant (compare Figure 8).**

## 3-D Visualizations for Final Barium Concentration

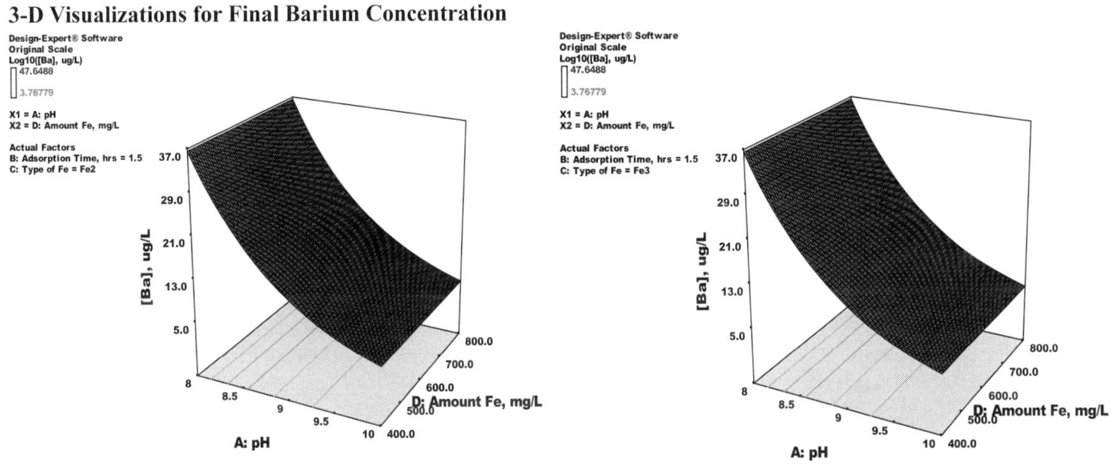


**Figure 11. 3-D representation of the effect of two variables (Left for Fe2 hydroxide adsorbent, Right for Fe3 hydroxide adsorbent) on the final concentration of barium in the solution phase. The removal of barium was independent on adsorption time and amount of iron initially present.**

**Cubic Visualizations for Final Iron Concentration**

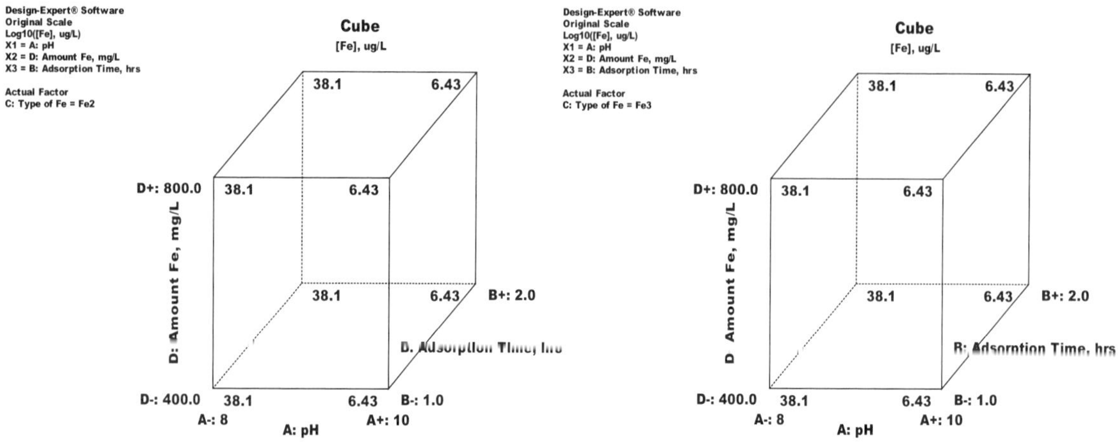


**Figure 12. Cubic representation of the effect of three variables (for Fe2 hydroxide adsorbent) on the final concentration of iron in the solution phase. For Fe2, the only variable of importance is pH; higher pH is favored. The type of adsorbent is unimportant (compare Figure 10). For Fe3, the only variable of importance is higher pH. There was no difference in the final iron concentration for the ferrous or ferric hydroxide adsorbent solids.**

**3-D Visualizations for Final Iron Concentration**

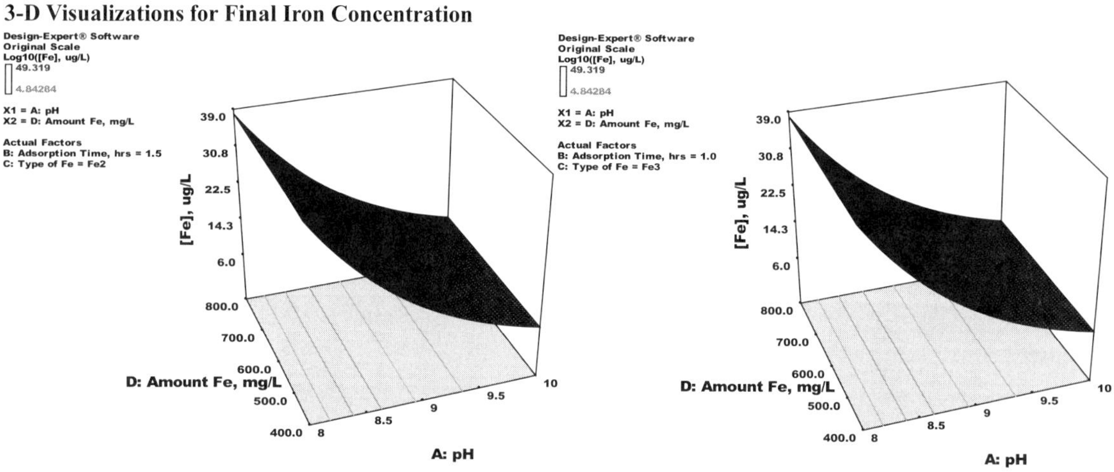


**Figure 13. 3-D representation of the effect of two variables (Left for Fe2 hydroxide adsorbent, Right for Fe3 hydroxide adsorbent) on the final concentration of iron in the solution phase. For both Fe2 and Fe3 hydroxide adsorbent, the removal of iron depends only on the solution pH.**

## SUMMARY AND CONCLUSIONS

The objective of the proposed scope of work was to perform fundamental chemistry evaluations and preliminary bench-scale tests on the Project Water to determine both As and Ba removal capabilities of selected treatment approaches that were capable of reducing the As from 15.0 μg/L to less than 5.0 ppb and reducing Ba from 300 μg/L to less than 100 μg/L. The three technologies that were evaluated were iron cementation (reductive precipitation), ferrous hydroxide, and ferric hydroxide. All three approaches were effective at reducing arsenic to below 5.0 μg/L. The data indicates that barium can be reduced to below 100 μg/L utilizing both ferrous and ferric hydroxide treatment approaches. Both barium and iron removal are a function of solution pH.

### Reductive Precipitation

The iron cementation process resulted in removal of arsenic to less than the ICP-AES detection limit of 0.4μg/L. This was true for most of the conditions investigated in the studied Experimental Design Matrix. Some of the matrix experiments, where the recovered solution after cementation, was treated at higher pH levels (forming green rust from the ferrous in solution) resulted in a significant decrease in the barium concentration (<100 μg/L limit).

Advantages:

- Project Goals achieved ([As]<**5** µg/L, [Ba] <100 µg/L , [Fe] <100 µg/L) using cementation followed by either ferrous hydroxide adsorption or ferric hydroxide adsorption. In fact, conditions can be specified that result in significantly lower concentrations than those required by the Project Goals.
- Only minor amounts of sulfate added to system because elemental iron is source of iron for the hydroxide precipitations.

Limitations:

- Two pH adjustment unit operations required, i.e., cementation followed by hydroxide adsorption as follows:
  - Unit 1: pH must be lowered to 5-6
  - Unit 2: pH must be subsequently raised to about 10
- Hydroxide sludge is produced.
  - Further work needs to be performed to determine amount of sludge that would be produced.

**Ferrous Hydroxide or Ferric Hydroxide Precipitation**

Both the ferrous and ferric hydroxide treatment approaches were successful at reducing the arsenic and barium in solution to below the discharge limits of 5.0 µg/L and 100 µg/L respectively. The solution pH was determined critical for the successful removal of barium.

Advantages:

- Project Goals achieved ([As]< 5 µg/L, [Ba] <100 µg/L, [Fe] <100 µg/L) using cementation followed by either ferrous hydroxide adsorption or ferric hydroxide adsorption. In fact, conditions can be specified that result in significantly lower concentrations than those required by the Project Goals.
- Treatment system is a relatively simple two-step process, i.e., doping with either ferrous or ferric salt followed by solid/liquid separation
- Only one pH adjustment step is required.

Limitations:

- Sulfate is added to system by the addition of iron sulfate salts; sulfate discharge limitations warrant further consideration.
- Hydroxide sludge is produced:
  - Further work needs to be performed to determine amount of sludge that would be produced
  - Further work needs to be performed to determine if recycle of initial hydroxide could significantly reduce amount of sludge, i.e., since the amount of removal of arsenic and barium should not saturate the hydroxide could the hydroxide be recycled to reduce the initial amount of iron sulfate required.

**REFERENCES**

Deutsch, W.J. 1997, Groundwater geochemistry, fundamentals and applications to contaminations, Lewis Publishers, 221 pp.

Huang, H.H., 2000. STABCAL. Montana Tech of the University of Montana, Metallurgical and Materials Engineering Department, Butte, Montana (hhuang@mtech.edu).

Kamrin, M., 1997, ExtorNet FAQ, MCLs and MCLGs, Michigan State University.

Twidwell, L.G., J. McCloskey, M. Gale-Lee, 2002. Removal of Hazardous Constituents from Wastewater, In: *Proceedings Mineral Processing Plant Design, Control and Practice*, Vol. 2, Eds. A.L. Mular, D.N. Halbe, D.J. Baratt, Vancouver, Canada, SME , Littleton, CO., pp 1847-65.

Paul J. Miranda

The Center for Advanced Mineral and Metallurgical Processing, 1300 West Park Street, Butte, Montana, USA

## ABSTRACT

The growing scarcity of worldwide high grade mineral ores has lead to a steady increase in the use of hydrometallurgical techniques. By far, the most popular solution mining technique is solvent extraction, however, in some cases, ion exchange (IX) resins are being utilized for both industrial and environmental purposes. Utilization of IX resins has some distinct advantages along with some potential shortcomings. This paper will discuss several design and operation possibilities using ion exchange technology at bench scale laboratory testing.

## INTRODUCTION

Ion exchange technology has been around for approximately 80 years. Initial ion exchange resins were discovered by English chemists Basil Albert Adams and Eric Leighton Holmes[1]. The resins were created in plastic Bakelite and were made by condensing polyhydric phenols or phelosulfonic acids with formaldehyde. Later, in 1944 Gaetano F. D'Alelio (figure 1) patented todays styrene based polymers[2], with large polymeric molecules in which ionic groups are introduced by chemical treatment. These compounds are usually shown in the following structure where X represents potential ion groups required for ion exchange removal.

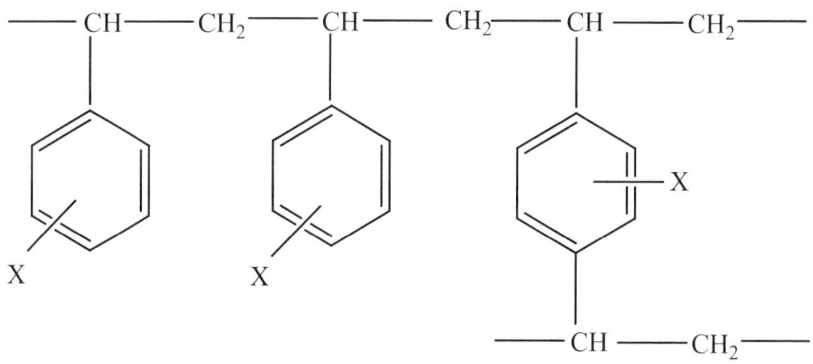


**Figure 1. Chemical Structure of Ion Exchange Resins**

According to the figure, the first two benzene ring structures come from benzene while the third benzene ring is from divinyl benzene which provides cross linking between polystyrene chains thus providing a three dimensional network which can be tight or loose. Overall, the ratio can be adjusted, however, the usual commercial proportion is 8 percent cross-linking.

Potential ion exchange groups include sulfonic acids ($HSO_3^-$) or quaternary ammonium groups ($H_2N^+(CH_3)_3Cl^-$) which account for approximately 90 percent of all ion exchange resins produced[3]. Another useful ion group used for ion exchange resins include iminodiaceteat ($-CH_2N(CH_2COOH)_2$ which forms chelated complexes with all metals except alkali earth metals.

There are four main types differing in their functional groups (figure 2):

- strongly acidic (typically, sulfonic acid groups, eg. sodium polystyrene sulfonate)
- strongly basic, (quaternary amino groups, for example, trimethylammonium groups)
- weakly acidic (mostly, carboxylic acid groups)
- weakly basic (primary, secondary, and/or ternary amino groups, eg. polyethylene amine)

The ion exchange resins are generally round spheres and ranging in size suited for the client.

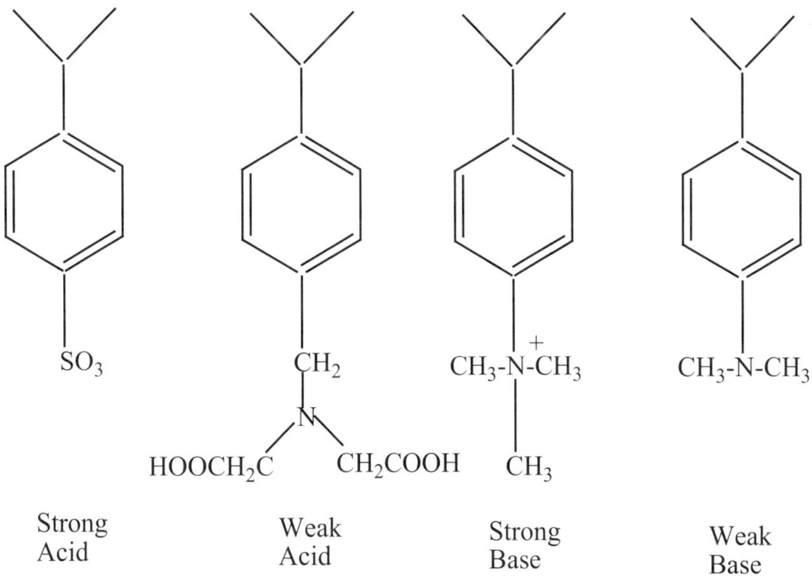


**Figure 2. General Types of Ion Exchange Resins**

**Figure 3. Ion Exchange Beads**

## PROCESS DEVELOPMENT AND IMPLEMENTATION

In order to implement ion exchange resins to large scale facilities, several steps must be taken to ensure a successful application of the styrene based technology. The steps include initial laboratory testing and analysis, followed by pilot plant testing, and lastly a demonstration plant test work.

With each set of test work, overall testing costs will increase. For the laboratory testing, costs are relatively inexpensive, as the test continues pilot plant studies with cost the client more and finally, there is significant cost for the demonstration plant. However, by not doing due diligent starting at the small laboratory scale, major costs (million of dollars) can be avoided a successful project can be implemented.

Costs vary between different types of ion exchange resins. The costs for commercially available resins are shown Table 1.

**Table 1. Ion Exchange Resin Comparable Costs[4]**

| Resin | Cost $/ft |
|---|---|
| Strong acid cation | 70-120 |
| Weak acid cation | 150-200 |
| Strong base anion | 180-250 |
| Weak base anion | 180-200 |
| Chelating cation | 330-600 |

## LABORATORY TESTING

Laboratory testing includes overall resin selection, batch testing, kinetic and selectivity testing, isotherm testing, and cycle testing.

## RESIN SELECTION

For resin selection, the overall hydrometallurgy and metals of interest should be well known for potential process solutions. For example, a potential solution may contain copper and iron, however, both copper can exist in cuprous ($Cu^{+1}$) and cupric ($Cu^{+2}$) ions while iron may be in both ferrous ($Fe^{+2}$) and ferric ($Fe^{+3}$) ions. Understanding the chemistry of the process solution is vital for laboratory testing. Both the pH and Oxidation Reduction Potential (ORP) of the process solutions should be monitored before and after testing. Once the metal chemistry is understood, several types of resins should be evaluated from different resin suppliers due to the fact that each resin will have slightly different characteristics and selectivity when tested on the process solution.

The characteristics of the resins should be evaluated for the following parameters, metal capacity of the targeted element, overall kinetics (loading and stripping), metal selectivity for specific cation, stripping kinetics, and final concentration product for downstream purification steps. These important parameters are required for initial phase of development and testing. An example of potential resin candidates is shown in table 2.

**Table 2. Examples of Resin Candidates**

| Metal of Interest | Solution pH | Impurities | Resin Type |
|---|---|---|---|
| Copper | pH 1 to 3, | Iron | Chelating Cation/Strong Base Cationic |
| Iron | pH < 1 | Copper, Zinc | Chelating Cation |
| Uranium | pH=10-12 | Silicon, Sulfate | Strong Base Anion |
| Molybdenum | pH < 1 | Iron, Sulfate | Weak Base Anion |
| Gold (Cyanide Complex) | pH >12, | Copper, Zinc | Strong Base Anion |

## KINETIC LABORATORY TESTING

Initial kinetic testing employs using laboratory beakers at room temperature (shaker tables) or high temperature testing (hot plates). The potential process solution is analyzed and placed into beakers along with a known amount of ion exchange resin. Samples are taken at 1, 2, 4, 6, 8, 10, 20, 40, 60, 120, 240, 360, and 600 minutes and measured for metal of interest and impurities. Other factors, including pH and ORP, should be identified and tested for kinetic testing. From testing data, graphs are generated to determine optimal resin performance at constant pH and ORP (Figure 3).

According to the data, resin 3 appears to be the best choice for this system. It appears to have fast kinetics and higher capacity compared to the other resin choices. Secondly, resin 4 appears to be the least desirable choice. From the kinetic testing data, equilibrium calculations can be determined from flat curved portions of the graphed data, therefore, determining possible resonance time required for column testing.

## ISOTHERM TESTING

After kinetic testing for identifying the best potential resin for metal recovery from process solutions, isotherm testing should be performed in order to determine the best pH range for both experimental, pilot, and process testing. The simple test uses identical amounts of resin with constant metal concentration. The samples are properly conditioned in several beakers with varied pH ranges targeted to the metals that represent a range of potential feed concentrations.

The experimental contact time for each isotherm test should be kept constant. Initial and final testing solutions should be analyzed, therefore, producing a pH versus concentration plot as shown in figure 2. The sample shown below is an isotherm of ferric iron at pH range 1.0 to 2.0. The bed volume flow rate was measured at 0.2 bed volumes per minute and sample every bed volume for iron using Inductively Coupled Plasma Spectrometer. An example of isotherm testing is shown in figure 4.

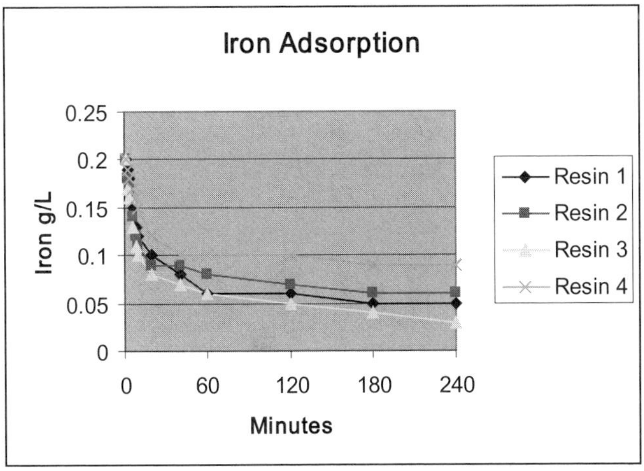


**Figure 3. Kinetic Laboratory Testing Results.**

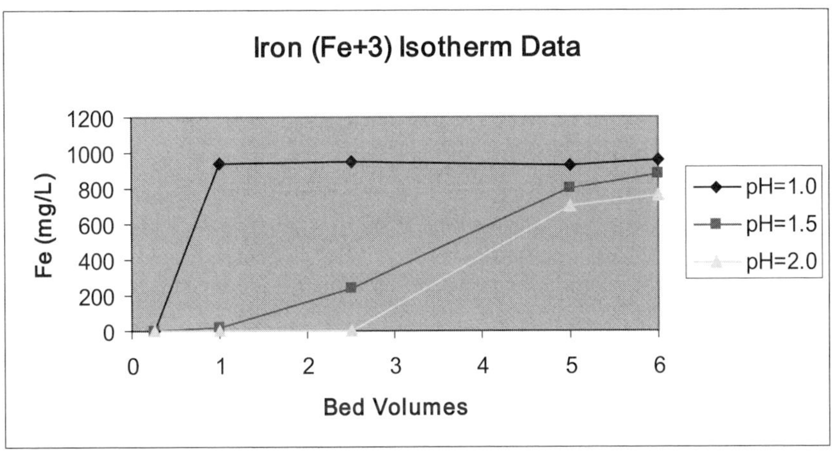


**Figure 4. Ferric Iron Isotherm Testing**

For this example using ferric iron, pH 1.0 had the poorest results while pH 1.5 and 2.0 had better results. In this case, testing above pH 2.0 would have detrimental effects due to the fact that increasing pH would precipitate ferric hydroxide ($Fe(OH)_3$) in solution.

**BREAKTHROUGH TESTING AND MASS TRANSFER ZONE LENGTH TESTING**

Breakthrough testing should be performed to determine overall resin capacity using column testing. Testing parameters should include estimated flow rate for process testing. Generally, contact time used for resin testing in 10 to 15 bed volumes per hour at a metal concentration of potential process solution, for example, a one gram per liter solution. Testing should be done for the estimated contact time of 3 hours or further until overall breakthrough occurs. Previous experimental tests should indicate the whether the metal adsorption is based on resin size, ionic size, or other potential factors. Large complex ions are likely to adsorb and diffuse at a much slower rate than smaller hydrated ions which tend to bond quickly. Secondly, functional groups may be complex which slows down adsorption due to stereo chemical functions of the initial adsorption site.

The breakthrough testing curve will provide potential sizing information for the adsorption section in process columns. The Mass Transfer Zone Length (MTZL) is an important sizing parameter impacts resin vessel size. These tests should be done at several different velocities to determine the relationship between fluid velocity and mass transfer zone length. The equations used for adsorption are the following:

MTZL ά velocity$^n$, $n^3 = 0.6$ to 1.0
$(MTZL_1/MTZL_2) = (v_{1/}v_2)^{0.6\text{-}1.0}$

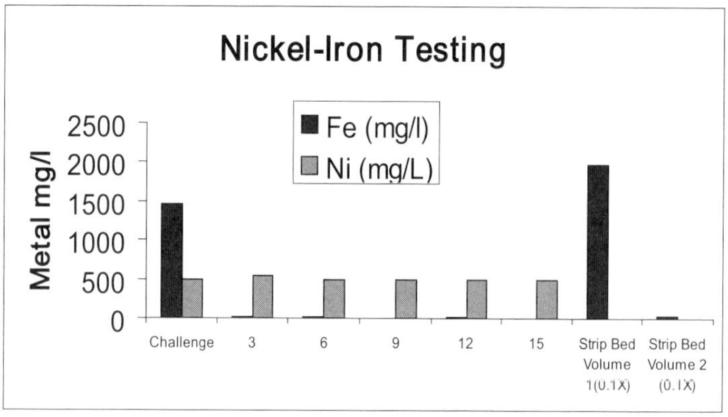


**Figure 5. Nickel Iron Flow Through Testing.**

For example, using an adsorption velocity of 15 m/hr at a 2 meter bed volume and an n value of 0.6, then the MTZL for the same operation conditions for a velocity change of 30 m/hr would be

$$MTZL_{30/hr}=2*(30/15)^{0.6} = 3.01 \text{ m}$$

Observation of the data indicates increasing the velocity by 50% increased the MTZL by 50% and doubled the overall flow rate. This equation indicated lower adsorption by 25%, therefore, lowered the overall capacities.

Further breakthrough testing should be performed in multiple tests to determine the optimal parameter for metal recovery. An example of breakthrough testing is shown in figure 5.

## STRIPPING TESTS

Stripping tests should be done to determine whether the metals are successfully released from solution, therefore, the functional group is able to be utilized for further testing. The test column should be fully loaded with the metal of interest to determine the best parameters for successful removal and concentration followed by potential downstream recovery method. Examples of potential metal recovery, eluent, and method recovery are shown in table 3.

**Table 3. Metal Recovery Options**

| Element | Eluent | Recovery Method |
|---|---|---|
| Copper | Sulfuric Acid | Electrowinning |
| Gallium | Sulfuric Acid | Electrowinning |
| Iron | Sulfuric Acid | Precipitation |
| Molybdenum | Ammonia/Caustic | Crystallization |
| Nickel | Sulfuric Acid | Crystallization |
| Uranium | Acid/Carbonate | Precipitation |

In some cases, ion exchange strip solutions may be placed into a fresh column or recycled for increasing the overall metal concentration required for downstream recovery method. Some examples include recycled copper electrolyte from an electrowinning circuit due to the fact that all the copper is not plated to the electrode and minor amount remains. By recycling and concentrating the remaining copper in solution, the client is able to capture and plate more copper from solution.

A second example of recycling strip solutions for increasing concentrations involves the uranium process. Since uranium carbonate or bicarbonate is contained in solution followed by precipitation processes, the metal must concentrated to higher levels for successful removal as uranium oxide. The pH of the precipitation test must be closely monitored to determine the best parameters for precipitation. The solution may contain slight amounts of uranium which can be recycled and concentrated using the required selective ion exchange resins.

Elution testing should be performed to determine the best parameters for metal stripping and recovery. Initially, the columns are loaded to maximum capacity for the metal of interest. Next, the columns are rinsed with water to ensure removal of impurities. The column is then stripped using a varied of acid/base strengths to determine both removal capability and impurities. For certain processes, such as copper/iron separations (figure 6), the resins will co-load both metals from solution. By selecting the proper eluent and concentration, experimental results will indicate a strong separation with relation to time. From experimental data, the main copper concentration from the eluent can be determined along with overall concentration of the metal of choice.

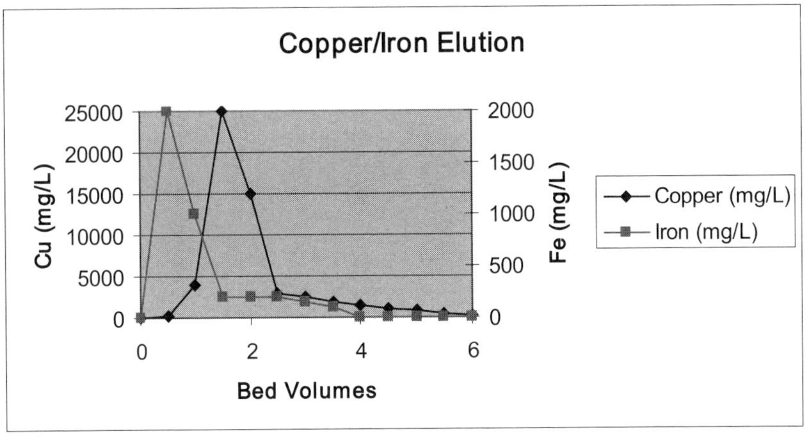


**Figure 6. Experimental Results of Iron/Copper Elution**

From the data, it appears that iron initially is stripped from the loaded resin followed by copper elution. Further experimental work would determine the best parameters for copper iron separation. According to the data, at the highest bed volume peak (1.5), the copper to iron elution ratio is 125.

Elutions are generally slower flow rates than during adsorption due to ion radius and diffusion rates. From figure 5, the entire elution occurred in less than 4 bed volumes with the highest concentration of copper occurring in less than 2.25 bed volumes while iron concentration occurs at less than 1.75 bed volumes. During experimental testing of elutions, important parameters should include total elution time, elution reagents, reagent strength, elevated temperature testing, separation testing, and long term cycle testing.

For elution testing, a STAT-EASE design of experiment could be utilized to determine which factors are important to elution testing. For example, using six parameters at a full matrix, the experimenter would have to use 64 experiments to identify important parameters. The STAT-EASE program allows the experimenter to use either ½ matrix (32 experiments) to determine which parameters are needed for separation and optimization.

From laboratory testing, careful notes should be taken for future reference and pilot scale testing. It is very important to identify potential problems from scale up. By taking careful laboratory notes, the pilot scale testing should identify pilot scale testing problems and provide potential solutions.

An example of a potential flow sheet design is shown in figure 7. The schematic shown in the following figure indicates nine columns with three sets of parallel trains. Solution entering from the left is essentially free of the metal of interest, therefore, the entire system is counter current between process solution and ion exchange resin.

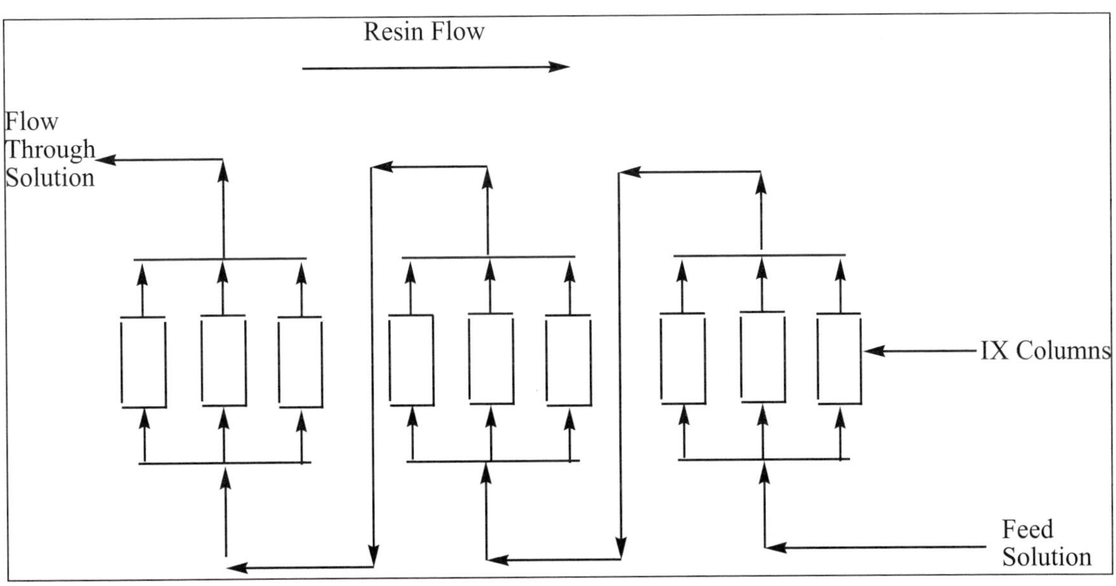


**Figure 7. Potential Process Flow Sheet**

## CONCLUSIONS

When deciding to use ion exchange resin technology, several factors need to be explored and tested for a potential successful project. The baseline experimental laboratory must be thoroughly examined during testing procedures. Since there are 4 general types of ion exchange resins, strong acid, weak acid, strong base, and weak base, careful consideration must be used to identify the proper resin to the process stream. Once a resin type is selected, kinetic testing using several manufactures should be explored to determine the best kinetics. After this step, experimental isotherm tests should to determine the best pH range for selective removal of the metal of interest.

After careful laboratory testing, the next step would be to perform pilot study testing. If the testing is successful, then there is a strong potential for process flow sheet development.

## REFERENCES

Adams, B.A., and Holmes, E.L., *J. Chem. Ind.,* 54, (1935)

F. D'Alelio, U.S. Patent 2366007G, (1944)

C. E. Harland, Ion exchange: Theory and Practice, The Royal Society of Chemistry, Cambridge, 1994.

# AN UPDATE: SELECTION, EQUIPMENT SIZING AND FLOWSHEET APPLICATIONS IN COPPER SOLVENT EXTRACTION AND ELECTROWINNING

Corby G. Anderson[1], Mike A. Giralico[2], Thomas A. Post[3], Tim G. Robinson,[4] and Owen S. Tinkler[5]

[1] Center for Advanced Mineral and Metallurgical Processing, Butte, Montana
[2] LIGHTNIN, Rochester, New York
[3] ConsultDrPost, Rochester, New York
[4] Freeport McMoRan, Phoenix, Arizona
[5] Cytec, Phoenix, Arizona

## ABSTRACT

Since the late 1960s copper solvent extraction coupled with electrowinning has been a growing technical application for production of copper metal. This paper outlined this technology and the fundamentals involved in the selection and sizing of copper solvent extraction and electrowinning circuits. It also illustrates the advancement in flow sheet design for copper leaching scenarios that now include copper pressure oxidation, high acid consuming ores, mixtures of high grade and low grade ores and combined copper/cobalt concentrates.

## INTRODUCTION

The industrial scale copper solvent extraction process is a counter-current multi-stage contacting operation. This usually consists of an extraction and a stripping section involving the use of kerosene based solvents and sulfuric acid to increase the copper concentration and purity of the copper rich electrolyte treated by a process known as electrowinning for copper metal production. The complete solvent extraction chain typically involves two extraction steps (E1 & E2) where copper is extracted from the pregnant leach stream (PLS) into the organic stream, one or two stripping steps (S1 & S2) where copper is stripped from the organic phase into unladen sulfuric acid from the electro-winning house and finally an optional wash step (W1) where further impurities are removed from the PLS prior to being pumped to the electrowinning tankhouse. Feed solutions to SX circuits typically arise from leaching processes such as heap leaching. Typically, copper recovery from the purified copper solutions is done by electrowinning. Electrowinning is precipitation of a metal by electrolytic reduction on a cathode while using an inert insoluble anode. Since the late 1960's copper solvent extraction coupled with electrowinning has been a growing technical application for production of copper metal. This paper will outline this technology and the fundamentals involved in the selection and sizing of copper solvent extraction and electrowinning circuits. It has also been updated to include application flowsheet dvelopments for various copper leaching circuits.

## THE MCCABE THIELE DIAGRAM

The solvent extraction process, as applied to metals extraction, can be defined as:

"The selective transfer and concentration of metal ions from an impure aqueous phase via an organic phase to a second pure aqueous phase from which the desired metal can be recovered in a usable form." Figure 1 illustrates a typical solvent extraction and electrowinning circuit. Oxime based reagents typically form the basis of copper extraction and this is illustrated in Figure 2. The basic reversible chemical reaction is noted as Equation 1.

$$2RH + Cu^{++} + SO_4^{-2} \leftrightarrow R_2Cu + H^+ + SO_4^{-2} \quad (1)$$

To design a solvent extraction circuit and to predict its performance we require the use of McCabe-Thiele stage construction methodology. This allows us to make predictions on extraction and stripping performance under imperfect conditions.

In order to construct a McCabe Thiele diagram, distribution curves or equilibrium isotherms must first be generated in the laboratory. For the extract isotherm (Fig 3), aliquots of feed liquor are mixed with a measured amount of organic reagent, such that a range of say 6-9 samples is available representing various O/A ratios.

The ratios typically cover a range from around 2:1 to 1:10. Mixing should be vigorous and for about15 minutes, to ensure that equilibrium is reached in each case before the phases are allowed to separate for analysis.

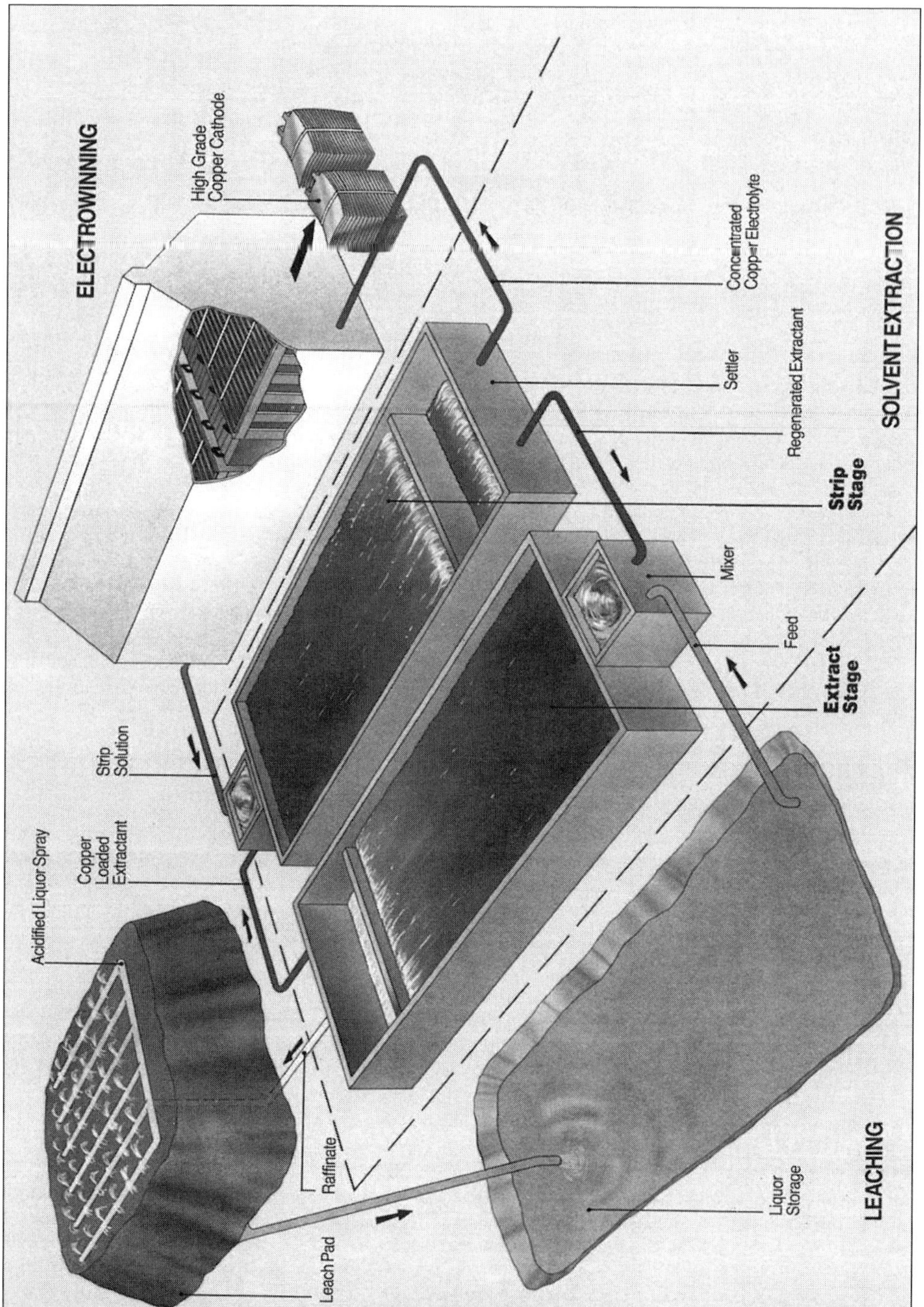


**Figure 1. Typical Industrial Copper Solvent Extraction and Electrowinning Circuit**

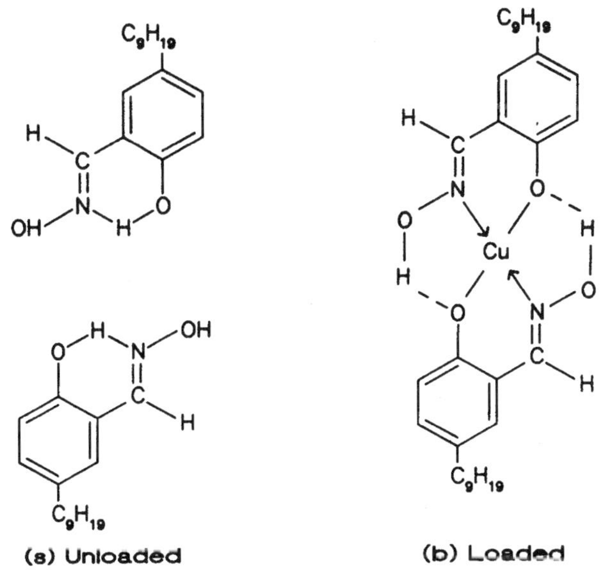


**Figure 2. Nonyl salicylaldoxime molecules and copper complex. Two salicylaldoxime molecules complex with a Cu $^{++}$ cation and release two $H^+$ ions (Biswas and Davenport, 1992)**

10 vol% Acorga M5640 TM, PLS 3 g/l Cu, pH 2

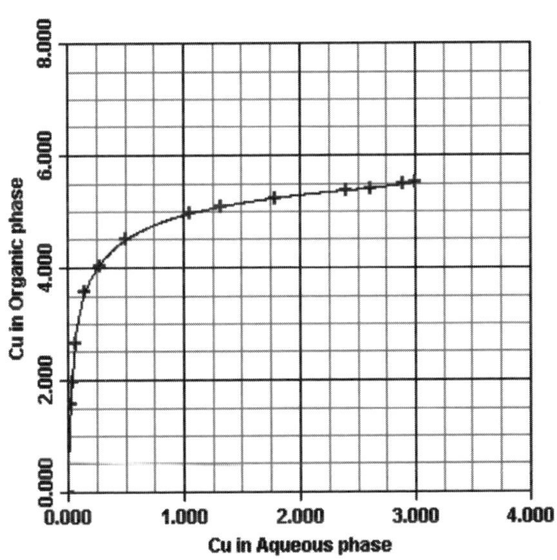


**Figure 3. Typical Extraction Isotherm**

The exact reagent concentration required to match the concentration of copper in the feed solution will depend on factors such as solution pH, circuit configuration (i.e.: the number of extraction and stripping stages) and electrolyte composition. For the purpose of a preliminary evaluation, however, the approximate reagent concentration required (expressed as percentage by volume) can be calculated simply by multiplying the concentration of copper in the feed (expressed in g/l) by three. Current commercial reagents load from 0.52–0.58 g/l copper per volume percent (v/o) reagent and are able to transfer around 60% of this available copper during the stripping process.

A strip isotherm (Fig. 4) is generated in a similar fashion by contacting loaded organic phase with various proportions of the strip liquor (electrolyte). In this case 4-5 points are adequate, typically in the range 5:1 to 1:1. The isotherms, of the types shown in Figures 3 and 4, form the basis of preliminary design for any solvent extraction plant. They are used in conjunction with each other to produce a McCabe Thiele construction, the metallurgical profile of an SX circuit. An example of which is shown in Figure 5.

10 vol% Acorga M5640 TM, Lean Electrolyte 35 g/l Cu 180 g/l Sulfuric Acid

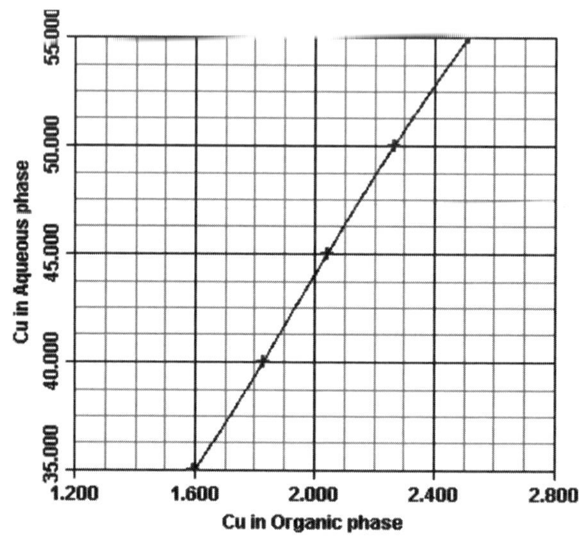


**Figure 4. Typical Strip Isotherm**

The first principle to consider is that of the operating line. The slope of the operating line defines the organic to aqueous flow ratio required to achieve a desired recovery e.g., if the volumes of organic and aqueous were to be equal, then an increase in organic tenor of 1 gpl copper would be matched by a decrease in the aqueous tenor also of 1 gpl; if say we had twice the volume of organic then it would only need an increase of 0.5 gpl in the organic to effect a decrease of 1 gpl in the aqueous tenor. In the first case, the operating line has a slope of 1 and in the other it has a slope of 0.5. Therefore, it is obvious that by choosing the O/A ratio, the concentration in the organic layer will be varied for any aqueous feed tenor. The origin of the operating line is important to locate and this is defined by the stripped organic level and the raffinate level, which latter is obtained by trial and error (if done by hand), or iteration (simulation software will be discussed later).

In the example of Figure 5, an O/A ratio of 1:1 is illustrated so that the slope of the operating line is defined accordingly, passing through the stripped organic value of 2.34 gpl copper and a raffinate level of that required. In order to predict stage performance, a vertical line is drawn from the aqueous feed tenor to meet the operating line, and then horizontally to meet the equilibrium isotherm. This point on the isotherm represents the gpl copper in the aqueous and organic phases leaving the first stage of extraction.

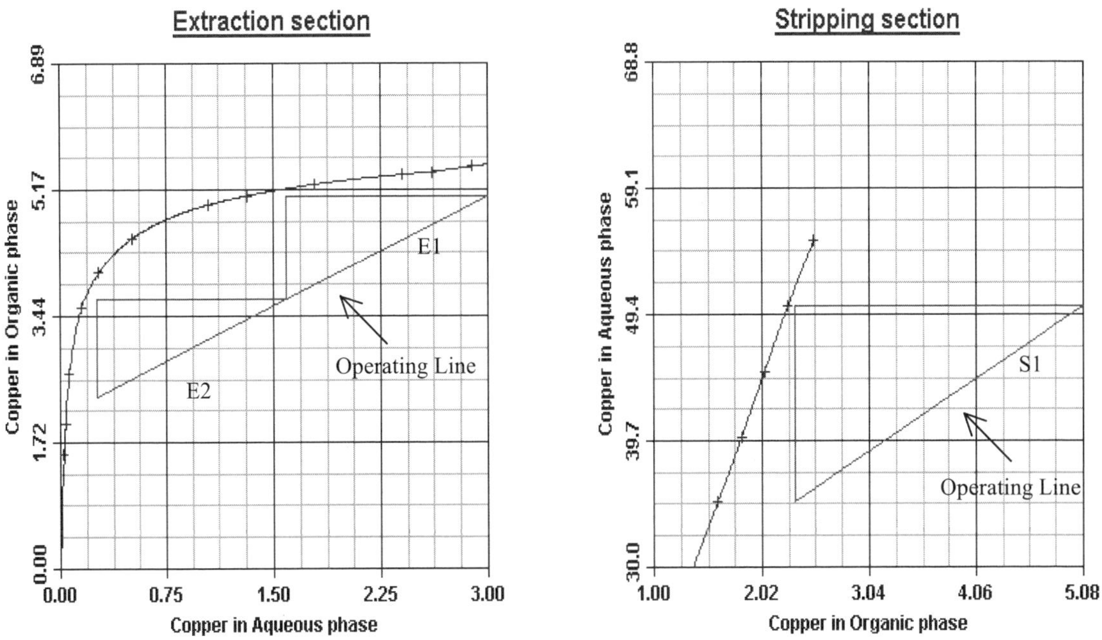


**Figure 5. McCabe Thiele diagrams for extraction and stripping**

In practice, as stated, the reaction will not reach equilibrium exactly and say 90-98 percent efficiency can be assumed for both extract and strip. In Figure 5, a stage efficiency of 95% was used in the extract section and 95% in the strip section. Graphically this is represented by the diagonal of the rectangle completed about the first stage, relative to the extended diagonal touching the isotherm. When done manually, this naturally is also a matter of trial and error. The example shows that the first (E1) stage of extraction would result in a loaded organic of 5.08 gpl copper and an intermediate aqueous raffinate of 1.59 gpl copper. Continuing the stepping-off process to a second extract stage (E2), again allowing for a stage efficiency correction, it can be seen that the aqueous copper tenor drops from 3 gpl in to 0.26 gpl, while the loaded organic rises from 2.34 gpl to 5.08 gpl. Naturally, if this is done by hand, it will be unlikely that the chosen raffinate level will be achieved in an integer number of stages and therefore a trial and error method must be used, changing the raffinate and hence the origin of the operating line. Alternatively, a different O/A ratio could be chosen to change the slope of the operating line.

Clearly this stepping-off procedure, allowance for stage efficiency, and iteration regarding raffinate levels and O/A ratios, can be very tedious to other than a skilled technologist and with this in mind, a computer modeling program have been developed by engineers and suppliers such as Avecia. The benefits and descriptions of a computer model will be described in the following section.

## CYTEC COMPUTER MODELING SOFTWARE

The software can operate in design mode or performance mode. In either case the extraction and strip isotherms are needed as input and then the computer will either decide the O/A ratio, number of stages, etc., to achieve a chosen raffinate, or alternatively will estimate the best achievable raffinate for a given set of plant parameters. The End User Module (MEUMTM) enables users to manipulate data as rapidly as binary file data exchange, via the Internet. The software package includes expert system technology and database and plant control capability. Features include:

1. Isotherm construction from laboratory-generated equilibrium data using the curve-fitting facility integral to the software package. Isotherm prediction.
2. Prediction of copper recovery and generation of corresponding mass balance data for a range of plant configurations.
3. The ability to vary the number of extraction and strip stages employed and arrange them in conventional or "parallel/series" mode as required by the design engineer.
4. Automatic recalculation of isotherm data as the reagent concentration is varied.
5. Automatic analysis of plant performance data and calculation of operating stage efficiencies and O/A ratios from metallurgical data supplied by the user.
6. Display of output data in the form of mass balances, circuit mimic diagrams or constructed McCabe-Thiele plots.

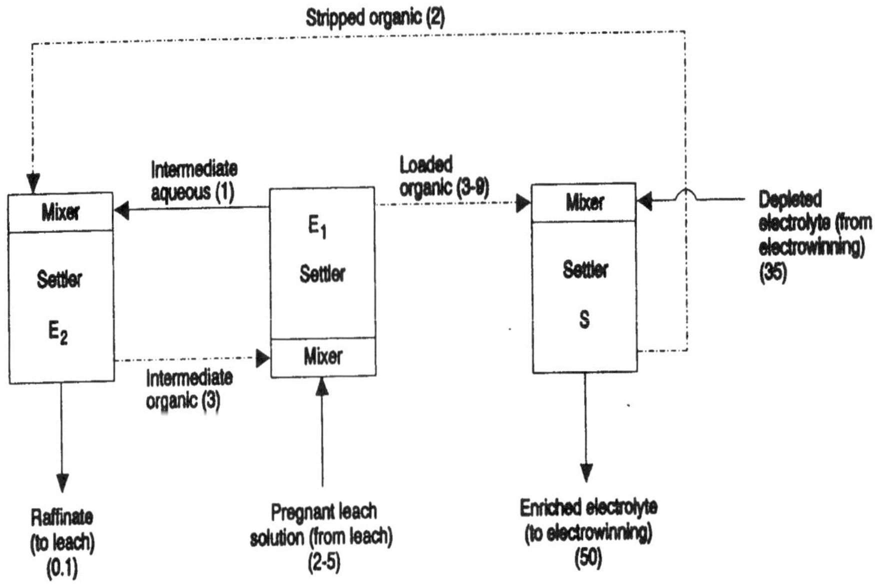


**Figure 6. Industrial Solvent Extraction 2 X 1 Circuit With Typical Copper Tenors Noted In Brackets ( ) ( Biswas and Davenport, 1992)**

1. Local and "Master" database storage and retrieval of isotherms i.e.: flowsheets.
2. Spreadsheet links to allow plant design data to be calculated directly from flowsheet results.

**SX TESTING**

The metallurgical engineer who has been commissioned to develop a metal recovery process flowsheet using SX technology will typically complete a laboratory study to assess the feasibility of the envisaged treatment route as the first step in the process. This study will identify possible problem areas. The scale on which pilot plant testing should be carried out is dependent on many factors. Some of these are:

1. The quantity of aqueous feed solution available.
2. The data required including whether this data will be required for SX plant design purposes.
3. Funding available.
4. Waste disposal facilities available.

It should be noted for a pilot plant trial in which the SX operation is required to be integrated with the leach and EW operations that the scale of these latter operations will determine the size requirements of the SX pilot plant.

For laboratory scale pilot plant testing, suitable equipment is readily available from a variety of sources. One common design is the so-called "Bell rig" which consists of mixer-settler equipment fabricated in clear UPVC or Polycarbonate to allow inspection of dispersion bands and solids activity at the O/W interface. "Bell" units typically have small square mixer boxes of 125–150 ml capacity and a settler design which largely resembles that of commercial plants. However, the dispersion distribution system tends to be fairly basic compared to that on a commercial plant. It has been our experience that this scale of equipment is best used along with good quality metering pumps to better control flow rates and prevent "phase flipping" in the mixers. The purpose of carrying out tests on this scale is to increase confidence generally in SX process technology and more specifically in the effectiveness of the flowsheet under development following the original bench tests. The power input to the mixers on this type of equipment is far in excess of that required. For this reason, testwork on this scale will not provide reliable information for design purposes on either the flow rate per unit of settler area (Specific Flow), the entrainment characteristics or the stage efficiency of the system under test. Useful data, however, may be obtained on performance factors such as metal recovery, selectivity or basic problems e.g.: phase separation under dynamic operating conditions.

For the production of more accurate data for preliminary plant design purposes a pilot plant capable of handling 10-40 liters per minute of combined aqueous and organic flow is considered advisable. Industry experience has shown that pilot plant scale test work can be used effectively. Such equipment has been used to evaluate reagent performance, confirm the suitability of the envisaged flowsheet and define more closely the physical operating characteristics to be expected.

The features of this size of pilot plant are:

1. It allows collection of operating information on a scale which can be used for plant design.
2. The settler walls can be fabricated from transparent plastic to allow operators to assess visually what is happening in the dispersion band and how solids are being controlled. The structure of the plant is sufficiently strong to withstand all but the very worst weather conditions.

3. Standard pumps, valves and fittings are readily available in sizes suitable for this type of operation.
4. It can be integrated easily into a commercial plant circuit to carry out investigational work on problems arising in existing operations.

Where long term effects are being studied, it is preferable that this type of pilot plant operates continuously, although most of the useful data is likely to be gathered during daylight hours.

An important part of the information to be gathered from this scale of operation is entrainment data. Although not strictly representative of the values likely to be obtained on the full-scale plant entrainment values determined will indicate whether the system under evaluation is likely to have major problems in this area. Certainly any improvements observed on this type of pilot plant should translate to the full scale plant.

Samples of exiting phases for entrainment measurement should be collected in equipment specially modified for this purpose. These are designed to fit in a laboratory centrifuge and the amount of entrained phase measured directly on a marked scale calibrated to give results expressed in ppm. The quality of the information which can be obtained on this scale with regard to % metal recovery, selectivity, band depths, entrainment and solids distribution against variables such as staging, phase continuity, phase recycle , temperature, specific settler flow (settler sizing), mixer retention time (mixer sizing) and impeller speed (power input) has been proved to be very reliable. Care should be taken, however, in scaling up from some of this data as solution linear velocities, for instance, on this type of pilot plant tend to be higher than on commercial plants due to differences in settler dimensions.

## SOLVENT EXTRACTION MIXERS AND SETTLERS

While the concept of solvent extraction has existed for centuries, it is the engineering design and application which brought it to prominence in the copper industry. The backbone of this development has been the successful full scale practice of phase mixing and phase disengagement. As illustrated in Figure 7 this is generally accomplished in so called mixer-settlers. The design and operational concepts behind these unit operations is discussed in the following sections of the paper.

## SOLVENT EXTRACTION MIXING CONCEPTS

Since the late 1960s, the Holmes&Narver pumper design has been the standard, serving the SX-market faithfully. Then the Davy BB, with its characteristic draft tube, made some inroads in the late 1970s. For nearly 3 decades nothing changed until the requirement came for reducing organic entrainment and increasing throughputs. In 1986 LIGHTNIN introduced the A310 hydrofoil as an efficient mixer in the auxiliary mix boxes. Then they came up with a new curved bladed pumper called the R320 in 1995. Since then numerous modifications have been made to reduce entrainment, decrease power and increase pumping capacities, while maintaining the head requirement of the system. These new designs greatly improve the hydraulic efficiency of these open pumpers. Most of the mix boxes now consist of both pumper and hydrofoil combinations whereas the hydrofoils in the auxiliaries are now up-pumping. The mix boxes aren't even straight sided anymore. The best designs now call for cylindrical tanks. Figure 8 illustrates a modern mixing design system.

These optimized impeller designs reduce aqueous and organic entrainment losses significantly, reduce or eliminate air entrainment and crud formation, enhance mass transfer and maximize the overall copper yield. When compared to the earlier pumper designs, these new designs require less installed power and better process results. The auxiliary hydrofoil impellers further assist the curved bladed pumpers by maintaining a good dispersion and mass transfer, without increasing the head requirement. The designer should select impellers that require minimal installed power while maintaining good dispersion and mass transfer to minimize entrainment and maximize copper production. Both the pumper and auxiliary impellers can be fabricated in any material including Derakane 470 Vinyl Ester FRP Composite. Currently, these new impeller designs are operating worldwide in over 60 installations with flow rates from 50 to 22,000 GPM and developed heads up to 50 inches.

### Process Requirements of SX Pumpers and Auxiliary Mixers

The pumpers are required to provide the desired head and flow through the plant, while creating the initial liquid-liquid dispersion. The auxiliary mixers are intended to maintain this dispersion and extend the stage residence time to improve mass transfer stage efficiency. Once plant design conditions are met, fine-tuning can make a conventional plant run much more efficiently, in terms of operating cost and lower entrainment losses.

### Pump-Mixer Requirements

SX-pumpers must develop a head high enough to achieve the desired flow rate through the SX plant. If the pumpers fail to provide this key requirement, all other features of the SX-plant are meaningless. This flow rate must also be variable, due to the changing concentrations of dissolved copper in the PLS. The best way to vary the flow rate is with variable speed frequency drive controllers, not by adjusting pipe valves, which increase turbulence. It is also desirable that provisions be made during the design phase that would allow future increases in the flow rates, since most mines will investigate increased production rates.

The pumpers must also create a liquid-liquid dispersion. Ideally, the droplet distribution is uniform; creating droplets small enough to promote mass transfer but large enough to inhibit entrainment losses. Not only must the pumpers create the dispersion, they must maintain a stable dispersion in the primary stage. Too much shear and turbulence creates a wide distribution of droplet sizes. A large percentage are fines; so small that they end up in the other phase and as entrainment.

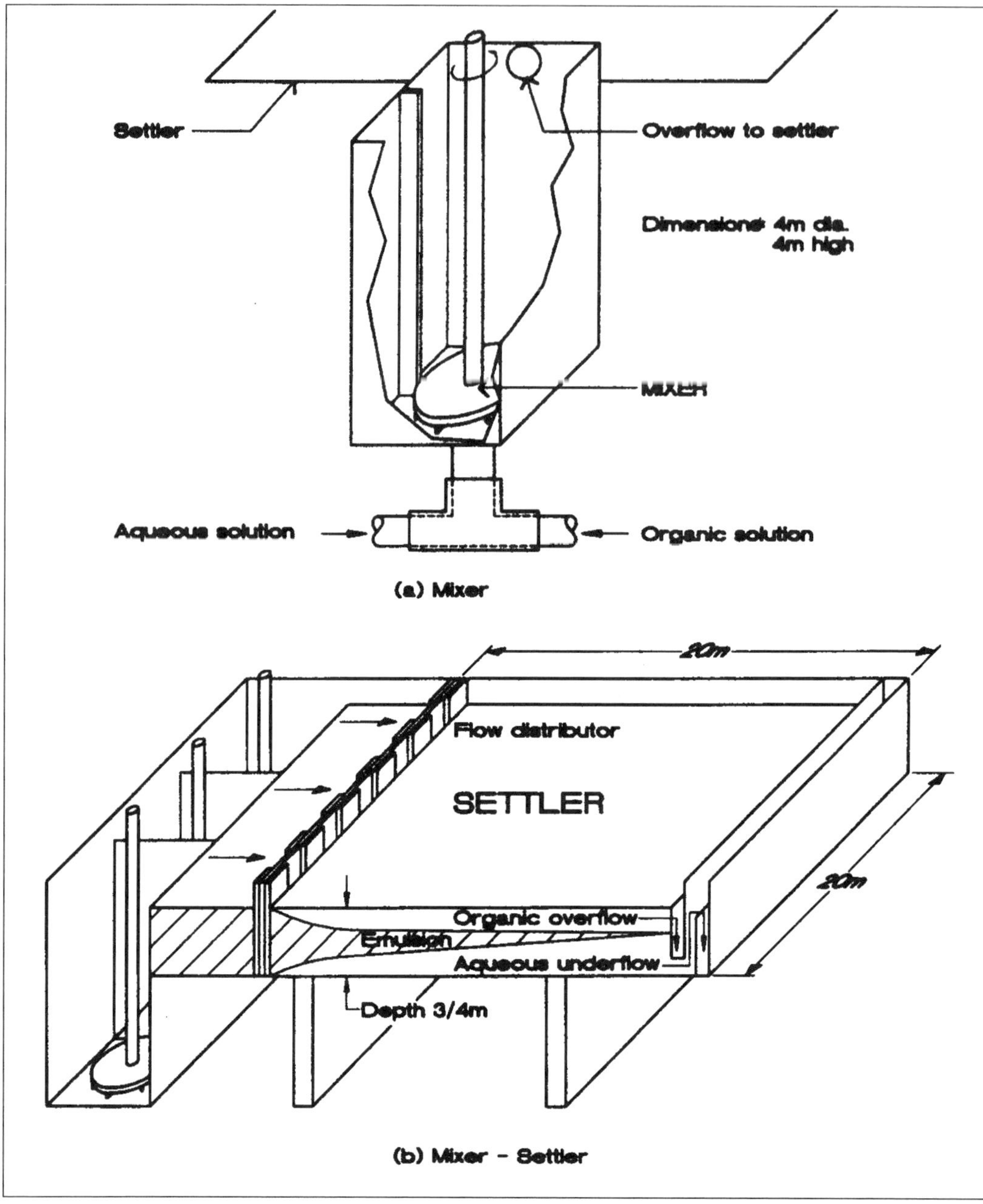


**Figure 7. Typical solvent extraction mixer and mixer—settler (Biswas and Davenport, 1992)**

Too little shear creates unstable large droplets that tend to separate quickly into phases. Phase separation in the mixing stages causes less than optimum copper recovery and limits production capacity.

The droplets are created by the bursting mechanism. This means that large droplets or agglomerates of one phase burst into many smaller spherical droplets in localized areas behind the blades where the pressure drop is the greatest. Elongated (stretched) droplets do not exist in copper SX. Excess power dissipation, as described by power per unit volume or P/V, is responsible for this bursting mechanism. Depending on the local P/V, this bursting of droplets can cascade downward to create very fine and sometimes unrecoverable droplets as entrainment losses. A minimum P/V is needed to maintain a stable dispersion. The local P/V in the shedding vortices is typically an order of magnitude higher than the overall P/V and this inhomogeneity creates the droplet distribution. Inefficient pumpers create very high local P/V and high overall P/V, which causes a wide distribution in droplet sizes, fines, haze, and entrainment losses.

Ideally, the pumper should not operate so violently as to cause air entrainment through vortices and rough air-liquid surface motion. Air that reaches the pumper blades will also disperse into fine bubbles. They attach themselves to aqueous droplets and ruin the function of the settlers and increase entrainment losses. If silicates are present, air-liquid-solid mixtures create crud, which is another leading cause of entrainment loss. Foam has also created problems with organic

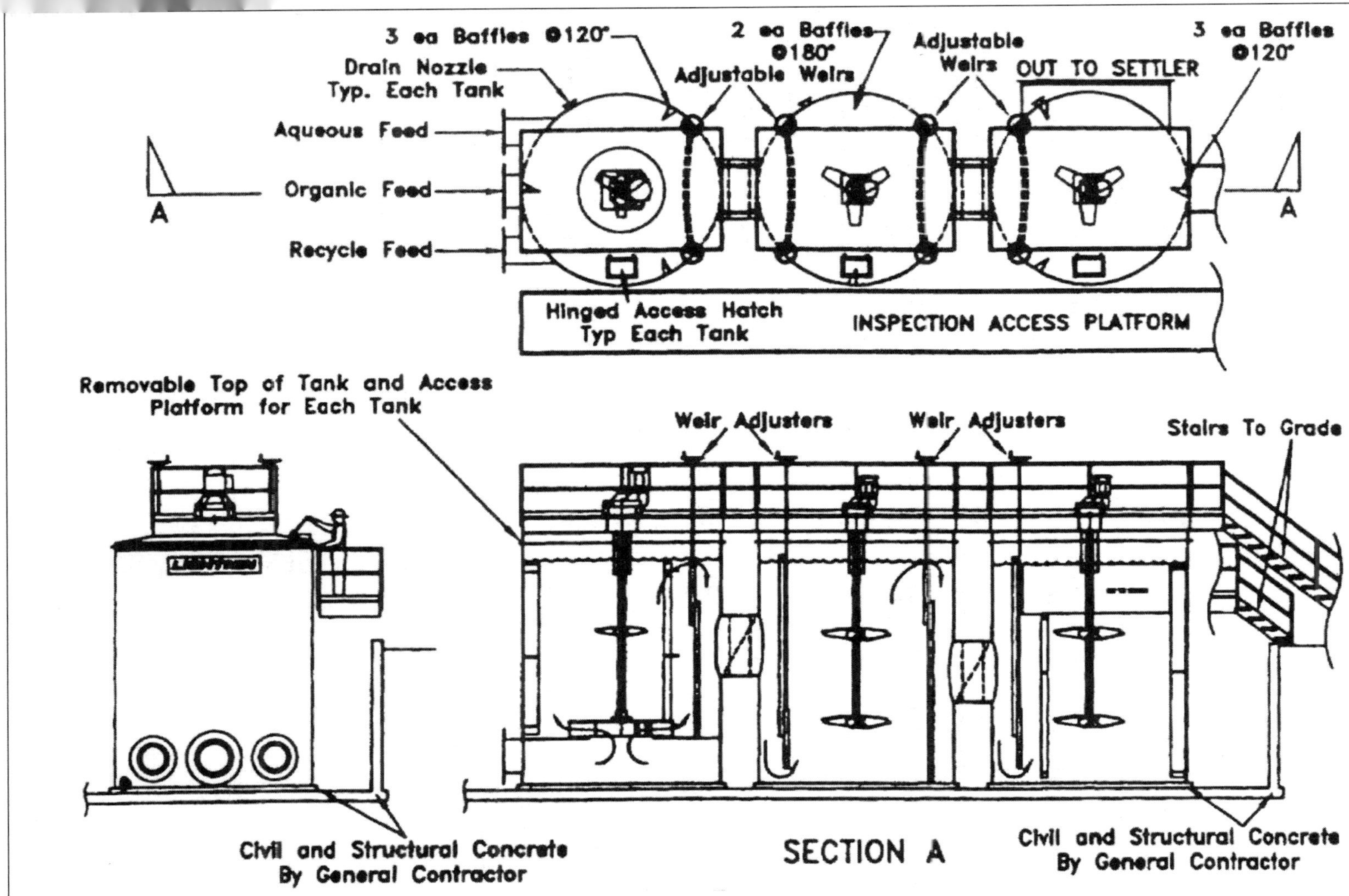


Figure 8. Example Solvent Extraction 3 Mixer Tank Design and Layout

carryover to the auxiliary stages. Air entrainment should be avoided at all costs. Corners in mix boxes are notorious for initiating vortexes. This is one reason to move away from mix boxes and toward stages made of cylindrical tanks. Also, baffles protruding through the liquid surface form vortices behind the baffles. Careful design of the baffles is also important.

**Auxiliary Mixer Requirements**

Auxiliary mixers need to maintain the dispersion created in the primary stage. The auxiliary box provides more residence time and increases the stage efficiency based on the isotherm of the extractant. Proper placement and operation of the auxiliary impeller can reduce the head requirements for the pumper. These mixers create a mixing flow superimposed on the flow through the system created by the pumper. This additional flow creates a better contacting pattern for the two phases, so that maximum concentration gradients can be achieved. This enhances mass transfer and stage efficiency. To achieve a well-mixed dispersion, a CSTR (continuous stirred tank reactor) model can be achieved if the flow generated by the auxiliary mixer is at least three times the flow created by the pumper. To achieve the best results, the auxiliary mixer should also be equipped with a variable speed device. As the flow rate is varied in the pumper, the developed flow of the auxiliary should be changed, too.

A minimum overall P/V must also be obtained in the auxiliary mix stage to avoid phase separation. Radial turbines require similar power levels as the radial pumpers, because they are designed to transform the power into turbulence and shear and not into flow. To get sufficient flow in all areas of an auxiliary mix box, the of P/V of radial turbines must be relatively high. This causes further decreases in the droplet size population and creates emulsions or phase inversions. Axial turbines concentrate their power into developing flow and not turbulence and shear. Thus, axial turbines require less power to activate all areas of the mix box. Properly designed hydrofoil impellers are similar to airplane wings, which reduce the drag (turbulence) and create more lift (flow). Installed in an up-pumping mode in an SX-plant, they can reduce the minimum overall P/V to an order of magnitude less than the pumpers and still achieve 3-13 times the flow rate. Variable speed devices should be installed so the P/V can be adjusted, to allow the droplets to coalesce into larger stable sizes. This is ideal, because it preconditions the dispersion before entering the settler.

Another reason to install variable speed devices on the auxiliaries is to account for changes in temperature and the so-called viscosity effect of these dispersions at cold temperatures. As the temperature gets colder, the viscosity-effect increases the separation times. Thus, the likelihood of phase separation in the mix stages decreases and less P/V is required to maintain that dispersion. The same is valid when discussing the difference between organic continuous and aqueous continuous operations. Typically, a conventional plant is designed for the worst case, i.e., organic continuous operation, when determining settler dimensions and aqueous continuous operation, which is overkill on power demands for the mixers most of the time. With variable speed, the entire unit can be fine-tuned. Like the pumpers, excessive power should be avoided in order to reduce air induction through vortices or surface splashing. The reasons are the same as for the pumpers. Surface splashing can be minimized by careful design of the tanks internals, tank liquid depth to diameter ratio, and proper mixer to tank diameter ratios.

**Advantages of Conventional SX Mixer-Settler Designs**

Conventional SX mixer-settler designs have been around since the late 1960s. As commercial trends push towards larger plants and higher production rates, these designs have shown definite weaknesses in the generation of excessive shear and air entrainment at the higher power levels required to meet production targets. This has left some to believe that the only way to improve performance is to design completely new technologies. Now that the requirements for the mixers have been clearly stated, reassessing the value of the conventional design is important.

The greatest advantage of the conventional design is in its simplicity. Most of the piping is easily accessible as well as the mixing stages and settlers. If a pumper needs maintenance, a complete pumper-shaft-gearbox unit can be exchanged with a replacement in less than 20 minutes. The down time is reduced. Splash walls, covers and tank lids do not need to be removed first. An auxiliary mixer can be replaced without any down time at all, if required.

A change in performance can be easily detected. Where sight glasses in the newer technologies give a very limited view (the dispersion is essentially black) the mixing stages and settlers of the conventional units are open and extend the view over the entire surface. Changes can be detected visually. Detecting surface organic pockets is very easy, indicating the onset of phase disengagement. The effect of changing the mixer speeds of either the pumpers or auxiliaries on the dispersion and air entrainment can be easily monitored. Thus, if a cooler weather period approaches, the impeller speed of the auxiliary mixers can be lowered to deliver the minimum power required to maintain the dispersion. If the pumper speed is lowered to decrease the overall flow rate, the speed of the auxiliary mixers can be increased, since the increased residence times require more power to maintain the dispersion. The open architecture allows immediate visual response of the changes made.

Similarly, the open architecture allows easy access to the fluids being mixed. Samples can be easily taken to detect such things as O:A ratios, fluid continuum, and separation or split times. The open architecture is a great asset for mixers designed to avoid air induction. As flow rates have increased the basic designs of pumpers and auxiliary mixers have required higher tip speeds and excessive power requirements. The additional P/V will cause the fluids to look violent. As a result, the surface begins to splash, and vortices appear behind baffles. Placement of lids on the surface diminishes air induction, but the mixers are still imparting excessive power to the volume. The lids have only solved part of the problem and they have made access to the dispersion and the mixers more difficult. The high P/V is still creating a wide variation in droplet sizes, including those very small droplets (haze or fines) responsible for entrainment losses.

Finally, simple designs are more economic by definition. The plant design in well known, trusted, and has a much lower capital cost due to the simplicity. Using the high efficiency impellers in a standard design produces the same result as if an alternate design is chosen. As such, the conventional design is an overall good investment.

**Optimizing the Design of Pumpers and Auxiliary Mixers**

Conventional pumpers are still in use today. They are called the General Mills or the Holmes&Narver pumper. Attached to the underside of a disk, are six straight blades that extend from the edge of the disk to the center. The inner edge is tapered toward the disk. The height of the blade is always 1/8 of the impeller diameter. The corresponding orifice is without exception between 0.33 and 0.37 times the impeller diameter.

In the late 1970s Davy introduced the Davy BB. This impeller is still in use today and has found wide acceptance. It is placed on top of a draft tube, so that it can be placed in the natural media of which is the desired continuum. It has an upper and lower disk with eight curved blades between them. The lower disk is required to achieve head. The height of the blades is always 1/7 of the impeller diameter. The draft tube opening is always ½ of the impeller diameter.

Since the relative dimensions of these pumpers are fixed, increasing the speed (N), increases flow (Q), tip speed ($TS=\pi ND$), power (P) and power/volume (P/V) proportionately. A scale-up criterion was used for these pumpers that was $N^3D^2$. It can be shown[1] that, for a given impeller type, $N^3D^2$ is proportional to P/V. The critical value of $N^3D^2$ is dependent on pumper type, power number and tank dimensions. Tip speed limits were imposed on pumpers based on operating experience without understanding the underlying mechanism of droplet breakup. Today, we know that tip speed is only part of the whole design criterion and P/V plays a much bigger role in pump-mixer selection than previously thought.

**Characterizing Pumper Design**

The primary requirement of a pumper is to develop enough head to achieve the desired flow rate of an SX plant. Each pumper has a set of characteristic head-flow and power-flow curves (Giralico Post, and Preston, 1995). Since a 50:50 mixture of PLS and Organic has a mixed SG of approximately 1.0, these curves have been developed using only water. Using Bernoulli's law, the hydraulic power ($P_{Hyd}$) required to pump a fluid of average density ($\rho$) at a given flow rate (Q) over a given head (H) is:

$$P_{Hyd} = \rho\, g\, H\, Q \qquad (2)$$

This is the minimum power needed to generate just the head and flow, without any turbulence or shear. Using the power-flow curve, the power (P) needed by the pumper can be calculated. On existing plants, it can be measured. The hydraulic efficiency ($\varepsilon$) of a pumper system can be determined by:

$$\varepsilon = \frac{P_{Hyd}}{P} \qquad (3)$$

The excess power ($P_e$) is defined as the power above $P_{hyd}$.

$$P_e = P - P_{Hyd} \qquad (4)$$

$P_e$ is the power responsible for the droplet breakup mechanism, along with the shear gradients emitting from the blade tips. This power results in turbulence. Most of this power is emitted in close proximity of the pumper's blades, so that once the droplets are formed they usually do not decrease any further in size unless they are recirculated back to within this high shearing zone.

Obviously, pumpers with high values of hydraulic efficiencies are directing their power more toward developing head and flow instead of creating very small droplets. LIGHTNIN's research has shown hydraulic efficiencies achievable up to 67% with a normal operating range of 35-45% with the new R320 family of impellers. Among the conventional pumpers, the Holmes&Narver and other straight bladed turbines are typically in the 15-20% range, while the Davy BB lies between 20-30%. Generally, entrainment losses are indirectly proportional to hydraulic efficiency. Common among all types of pumpers is that hydraulic efficiency decreases as scale decreases. In pilot plants, the range of hydraulic efficiencies is very small, although they compare in the same relative order. Straight bladed pumpers may have efficiencies as low as 5% and the best practical curved bladed pumpers may obtain a maximum of 20%. Much of this is due to poor box geometry design and higher residence times than on full scale. Another reason is that pilot plants do not need to create as much head as their full-scale counterparts. In large enough pilot plants, this can be defeated with novel mixer designs as discussed above.

On the other hand, as scale increases, the range of hydraulic efficiencies increases dramatically as stated above. A mere 6% increase in hydraulic efficiency decreased overall organic-in-aqueous and aqueous-in-organic entrainment by one half over a three-month trial period[1]. Comparison was made at a full-scale copper SX-plant operating two different impeller designs in parallel.

P/V is not the answer to everything. Shear gradients are also important. Determination of shear gradients ($\gamma=\partial v/\partial r$) requires the knowledge of local velocity (v) profiles in the radial (r) direction. A laser operating in a back scattering mode (Giralico, Post, and Preston, 1995). conveniently determines this. Two types of impeller shear gradients are important for scale-up (Weetman and Oldshue 1988) the average shear gradient $\gamma_{avg}$ and the maximum shear gradient $\gamma_{max}$. $\gamma_{max}$ is responsible for the smallest droplets. If a mixer is designed with a uniform exit velocity profile, the droplet distribution will also be uniform. For scale-up, the following rules are typical for radial turbines (Oldshue 1984).

$$\gamma_{max} = \xi_1 \cdot N \cdot D \quad \textit{and} \quad \gamma_{avg} = \xi_2 \cdot N \tag{5}$$

$\xi_1$ and $\xi_2$ are constants, which are impeller dependent. Equation 5 shows that $\gamma_{max}$ is proportional to tip speed. $\xi_1$ has a lower value for long curved bladed impellers than straight bladed impellers[1]. Upon scale-up, the average and maximum shear gradients obviously diverge. Whereas tip speed is often constant, impeller speed always decreases with scale-up. Thus, the drop size distribution is always wider for larger scale equipment unless both gradients are similar, or if P/V is lower in the larger scale.

These concepts show that scale-up is not simple, and that entrainment losses determined in pilot plants do not necessarily predict large-scale entrainment losses. Taking ε, P/V, γ and residence times into account is important. Finding higher entrainment losses upon scale-up is not unusual. A conventional system that develops the desired head and flow with curved bladed pumpers having the highest possible hydraulic efficiency and still maintaining phase stability (minimum P/V) without air induction (maximum P/V) at tip speeds below 1000 ft/min results in the best possible SX-system. Fulfilling all these requirements is possible, provided the mix stage geometries are optimized with the pumper selection.

**Optimized Pumper Designs**

No single pumper design is optimized for all operating conditions. Some generalities can be made, though, to help select the best pumper for each SX-plant. To increase the flow rates of conventional pumpers, only two variables can be changed, either speed or pumper diameter. This is because the geometrical settings of the pumpers are fixed.

$$Q = Nq \cdot N \cdot D^3 \tag{6}$$

Upon scale-up N decreases. Therefore higher flow rates are only possible by increasing D (since Nq (the dimensionless flow number) is fairly constant for geometrically similar impellers). When the impeller to tank diameter ratio, D/T, > 0.7 the radial flow becomes throttled resulting in higher than expected power draws and lower hydraulic efficiencies. At the same time, the power increases dramatically, by the fifth power of D.

$$P = Np \cdot \rho \cdot N^3 \cdot D^5 \tag{7}$$

Since Np, the dimensionless power number, is also relatively constant for a geometrically similar pumper, increasing D to increase flow rates also increases the power.

Other methods are available to increase flow. For example, within certain limits, increasing the blade width can increase the flow rate. This increases both Nq and Np, but because the diameter can be decreased, tip speed and P (and P/V) can be decreased. This invariably increases hydraulic efficiency. Curving the blades does not necessarily increase the developed head over a straight bladed pumper, but Np can be halved. The net effect is higher hydraulic efficiencies at the desired head and flow. The blade length of curved bladed impellers affects the developed head and flow, but has relatively little effect on power. Obviously the curvature of the blades is crucial. Constant radius pumpers are easier to build and can operate over a wider range of flow rates. Pumpers with variable radii are more complicated to build but can achieve the very high hydraulic efficiencies in a narrow range of flow rates.

A big factor in improving the process can be achieved by modifying the other very important part of the pump, the orifice. The conventional straight bladed pumper design is typically accompanied by an orifice to pumper diameter ratio (D0/D) of 0.35. By increasing this ratio, more flow and head can be generated without increasing the power draw (provided the pumper is not throttled by design). This happens up to a ratio of approx. 0.5. Further flow increases require larger D0/D's and higher power but, by increasing the orifice diameter, the systems head-flow performance increases and hydraulic efficiency stays in the 36-45% range.

**Dimensionless Numbers**

Dimensionless numbers are essential for the purpose of scale-up. The two more common dimensionless numbers have already been described. Rearranging equations 6 and 7 give:

$$Nq = \frac{Q}{N \cdot D^3} \quad \text{and} \quad Np = \frac{P}{\rho \cdot N^3 \cdot D^5} \tag{6)(7}$$

During small-scale experiments, where D is constant, the major influence on fluid dynamics is the impeller speed, N. By changing N and measuring the flow rate Q, generated by the system, Nq is determined. If the small-scale unit is attached to a torque cell, the measured torque and the speed yield the impeller power. From this, Np is determined. Plotting Np = f(Nq) is convenient (Giralico, Post, and Preston, 1995). This results in a curve characteristic to the pumper design. If the pumper and its corresponding orifice are built geometrically similar, the small-scale curve can be used to design the full-scale system with equations 6') and 7').

Equation 2 shows how the hydraulic power is related to the head, but does not show how to determine it. LIGHTNIN developed a dimensionless head number for this purpose. It is basically the hydraulic power divided by the tip speed squared.

$$Nh = \frac{2 \cdot g \cdot H}{TS^2} = \frac{2 \cdot g \cdot H}{\pi^2 \cdot N^2 \cdot D^2} \tag{8}$$

During the small scale testing, the liquid will rise directly with the impeller speed. The difference in liquid level from N=0 to N is the head, H. Plotting Nh=f(Nq) is also characteristic of the pumper design.

**Problems with Over-Efficient Pumpers**

If a pumper is designed merely based on optimizing hydraulic efficiency, the SX-plant may run into problems. This is particularly the case when retrofitting an existing unit, because all pump stage dimensions are given, and very difficult to change. Therefore, at a given flow rate, the residence time in a pump stage is also given. The minimum power required to maintain a stable dispersion is dependent on pumper type, location, organic solution properties, tank dimensions and mostly on residence time. Generally, as residence time increases, more power is required to hinder coalescence. Once this minimum power level is determined, the hydraulic efficiency cannot go above a maximum level. Otherwise, the dispersion will already begin to separate in the primary stage, leading to reduced mass transfer. As mentioned earlier, efficiencies of $\varepsilon = 67\%$ are achievable. However, this is often too efficient and can result in phase separation. To maximize hydraulic efficiency, the residence times in the pump stages must be lower than one minute, sometimes as low as 20 seconds. The rest of the residence time must be carried out in the auxiliary mix stages. These design considerations are being carried out on new installations where the SX extractant kinetics is taken full advantage. For existing units, the hydraulic efficiency may need to be artificially reduced so that enough power is available for phase stability. This can be done in several ways; i.e., higher off-bottom clearances, narrower blades, or the addition of radial vanes to the top disk of the impeller.

**Pumper Design for less than Minimum P/V Operations**

If the existing unit has a high liquid depth to tank diameter ratio (Z/T>0.7) the best retrofit can actually exceed the maximum hydraulic efficiency required to maintain a stable dispersion. Tall tanks can be equipped with one or more A510 auxiliary up-pumping mixers in the upper portion of the primary stage. This patented feature (Post, Howk and Preston 1996) has been installed successfully in several full-scale installations, which had suffered from acute phase separation problems. It can maintain the dispersion because the A510 auxiliary mixer requires about 1/10 of the power of a typical pumper. It can keep the surface active, without affecting the head and flow characteristics of the pumper. Thus, the pumper is designed just to deliver the desired head and flow at the lowest possible power consumption. The upper, built-in, auxiliary impeller maintains the dispersion in the otherwise quiet zones away from the pumper's active zone. Retrofits have been so successful, that current design practice is to make the primary stage taller.

**Optimizing Auxiliary Mixer Design**

The optimized auxiliary mixer must achieve a high internal flow rate with the least amount of power. 100% efficient impellers would be ideal, but don't exist. Excess $P_e$ is not required, because the droplet distribution has already been formed in the pump stage. Hydrofoiled impellers are the most efficient flow generating impellers available today. Composite impellers like the A6000 are the most efficient hydrofoils since they have true hydrofoil profiles including variable blade cross-section thickness and optimum twist and camber. They can generate the same flow rate as its metal counterpart, the A510, at about half the power. The composites are also ideal for solutions high in chlorides.

**Table 1. Some Installations With The Dual Impeller Concept**

| Customer | Country | Location | Application | Pumper |
|---|---|---|---|---|
| Sociedad Minera Cerro Verde | PERU | Arequipa | Copper | R321/A310 |
| Mexicana del Cobre | MEXICO | ESDE II | Copper | R320/A310 |
| Soquimich | CHILE | Antofagasta | Iodine | R320/A310 |
| Quiborax | CHILE | Santiago | Borax | R320/A310 |
| IMC | USA | California | Boric Acid | R320/A6000 |
| Phelps Dodge | USA | Morensi | Copper | R320/A310 |
| CS Metals | USA | Louisanna | Vanadium | R320/A310 |
| Cal Energy | USA | Salton Sea | Zinc | R320D/A6000 |

All other types of auxiliary mixers require too much power to achieve high internal flow rates. In fact some radial devices can cause an otherwise aqueous continuous operation to flip over to organic continuous, even at O:A ratios less than one. Too much power will also further decrease droplet size and increase the settling time. This is not necessary, since more than 90% of the equilibrium was achieved in the primary stage.

Because of the low power consumption of hydrofoils, tip vortices are greatly reduced, and instead of further decreasing the droplet size, they can actually achieve some coalescence of the smallest droplets. Pointed in the upward direction, the internal flow generated by the A510 or A6000 goes upward near the shaft and down at the walls. This flow pattern supports the flow coming from the pump stage through the underflow weir or downcomer, reducing the head requirements of the pumper.

The hydrofoil cannot be too small, because the power of the thrust is then focused at the center of the surface. The resulting flow pattern would look like an upward surging spring, which could induce unwanted air. The concentrated local power means that other regions may be underpowered with resulting organic puddles. The hydrofoil cannot be too large either. If the impeller diameter is too large, there will be no room for the downward flow near the walls. The spreading of the upward flow collides with the downward recirculation resulting in a thrashing and violent surface motion with unwanted air induction. This collision of the flows causes an increase in power demand and thus unnecessary turbulence. This chaotic flow pattern is also the result of any sized pitched bladed turbine, because the discharge angle of the flow is not axial like a hydrofoil, but off to an angle. It is not possible to get a top-to-bottom flow pattern with a pitched bladed impeller.

Auxiliary mixer units should have variable speed drives. The added expense is well worth it. Since conditions vary from summer to winter and high grade to low grade PLS, a variable speed mixer is also intuitive. In addition, the simplicity of the conventional mixer settler design makes this addition really worthwhile. Once installed, the optimized operating philosophy should be to adjust the speed on the auxiliary mixers to provide just enough power so that no organic pockets are visible on the surface. This is usually the case when the upward flow in the center of the auxiliary stage is visible but not overpowering. This adjustability allows for the best control of the SX plant and will contribute significantly in reducing not only entrainment losses, but may make it possible to decrease the size of the settlers further. It also eliminates air induction and reduces crud formation.

### Entrainment

With proper consideration to an efficient pumper and auxiliary mixer design, numerous large-scale operators in the copper industry have achieved entrainment levels in the 6 to 20 ppm level. Examples are: Codelco Salvador, Cerro Colorado, Mante Verde, Mantos Blancos Santa Barbara, Aberfoyle/Gunpowder, Girilambone Copper Co, Pasminco Metals/BHAS, El Abra, Compania Minera Zaldivar, Mexicana del Cobre/Nacozari, Southern Peru Copper/Toquepala, BHP Copper/San Manuel, Phelps Dodge/Tyrone, Asarco Silver Bell, El Tesoro, Andacollo, and CS Metals.

### Current Trends in Copper SX-Mixing

Three obvious paths are followed to optimize SX-EW performance with conventional systems (Giralico Post and Preston 1996). One is retrofitting existing plants to the most optimized mixing solutions outlined here. Usually, opening the orifice will be necessary to achieve optimum performance. In any case, auxiliary mixers should be outfitted with hydrofoils like the A510 or A6000 impellers and variable speed drives. The other two paths are reserved for new installations. One is to reduce the size of the primary stage to achieve lower residence times, lower absolute power and higher hydraulic efficiencies. The remainder of the residence time would be mixed in the auxiliary stage or stages. Since the P/V in the auxiliaries is an order of magnitude less than the pumpers, the overall power dissipation to the dispersion is less. This results in less turbulence and less organic entrainment. The other path is to achieve the entire mass transfer in a single tank of larger size, which satisfies the desired residence time. The approach is to size the most hydraulic efficient pumper just to meet the desired head and flow requirements, and install up-pumping hydrofoils on the same shaft to maintain phase stability in the upper regions of the tank. This method reduces the number of mixers required for each mixer-settler while the overall efficiency of the system could well allow either higher specific settler loading (more throughput) or decrease the settler area (capital cost) for a given flow rate.

The conventional mixer-settler design has not yet outlived its usefulness. The proper design of the pumpers and auxiliary mixers, coupled with new settler technology available from consulting engineers, enable conventional plants to meet the requirements of the process without the need for costly and complicated new technologies. The open architecture of the conventional design is ideal for process monitoring, optimization, and control. The documented (Giralico Post and Preston 1995) lower organic entrainment losses from optimized curved bladed pumpers and hydrofoil auxiliaries with their lower overall power consumption, makes the conventional mixer-settler design the most economical solution for copper solvent extraction.

## SOLVENT EXTRACTION SETTLER CONCEPTS

Settlers serve the function of providing quiescent conditions for effective phase separation after mixing. In application this occurs on a continuous basis and occurs in a series of two stages. First there is the so called 'breaking' of the organic phase from the aqueous phase. This is the natural separation of phases due to density differences. Then, the second stage is direction of the separated phases via an adjustable full width weir overflow and underflow system. A conventional settler is illustrated in the following figures. The key elements of design are not complicated. The idea

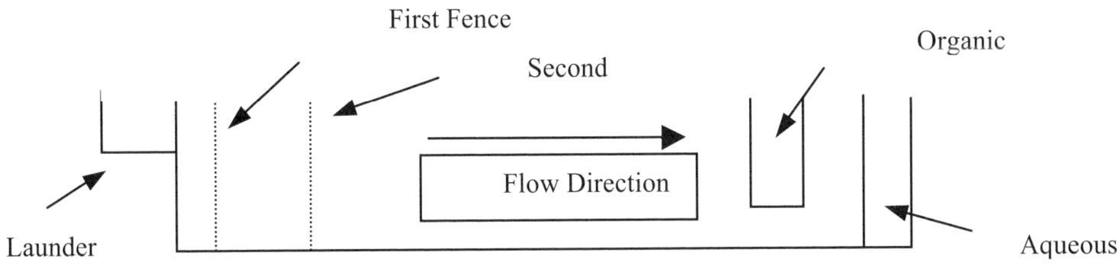


**Figure 9. Conventional Settler Plan View**

is to provide even quiescent flow with enough residence time to allow complete phase separation. Settlers are built on a low profile using stainless steel, concrete lined with HDPE, or reinforced fiberglass. Settlers are almost always square or nearly square. When needed, they are provide with covers. The key component to understand is that the settler width determines the solution or phase velocity. Thus the key concepts are settling rate, or settling specific area and average solution, or space, velocity. These design criteria are normally established for settler designs at about 4 to 5 $M^3/hr/\ M^2$ and 3 to 6 cm/sec respectively. Table 2 (Liguori, A. J. et. al 1997) illustrates a typical design calculation sheet and the typical design criteria for designing and specifying a settler system. It is important to note that the 'total flow' includes both the aqueous and organic settler flows combined together. This calculation produces the general settler size where a maximum settler width is set at about 30 meters. The trade off in any settler design is the higher the settler phase velocity, the narrower the settler will be. This lowers the capital cost and constant flow distribution through the settler is also easier to maintain. But, there is an increased risk of entrainment. As well, it is beneficial to design for a separated organic depth of about 180-300 mm while having the aqueous phase depth at 450 mm to 600 mm. An additional aspect of the design is the inclusion of a settler distribution fence system also know as 'picket fences'. This encourages quiescent and even flow conditions within the settler. Normally two sets of picket fences in series are utilized within the settler.

The organic solution is the key ingredient in the SX plant. It is also a very expensive ingredient. Thus important detail should be taken to avoid an extensively large inventory, include protection against overflows and spillage, include reclaim, clean-up, and spillage measures in the design, and minimize the potential of contamination .Settlers should be sized realistically with reasonable safety factors to base the initial inventory. Include a loaded organic surge tank. Settlers will have adjustable aqueous discharge weirs to enable operators to optimize the organic layer thickness. Install level alarms and high level cutoffs to protect inventory. Lastly, every operation must protect against contamination. Designs must minimize contaminants sources in PLS, water make-up, acid make-up, anti-corrosion chemicals, oils, dust, change in ore types, and insects. When calculating initial organic inventory, be sure to include an allowance for the crest on settlers, 50% of each mix box, and piping, as well as more obvious loaded organic tank and settlers inventory. The operational concept of the settler requires phase separation of the immiscible liquid-liquid dispersion of the organic (kerosene) and aqueous (sulfuric acid) components. Common configurations feed the dispersion through a launder at the center of the settler, followed by an angled picket fence to evenly distribute the inlet stream and followed by a second picket fence located at approximately one third of the total settler length to assure even flow distribution further downstream. The exit section of the settler consists of an overflow weir for the lighter organic phase and an underflow-overflow weir for the heavier aqueous phase. Typical volume ratios of organic to aqueous in the settler are 1:2 to 1:1. Conventional design methodology for rectangular settlers is based on hydraulic theory and empirical results from pilot plants and full scale installations. Generally speaking, the basic settler dimensions are derived from open channel flow theory to develop linear velocity targets, specific flow rate (total flow rate per settler footprint area) and picket fence design. Empirical results have refined the linear velocity, specific flowrate, and total settler depth guidelines derived from theory. The commercial use of solvent extraction technology in copper mining applications supports the claim that the simple rectangular arrangement commonly in use is successful both technically and economically. Nonetheless several issues remain that offer the opportunity for improvement in settler performance. Organic entrainment in the aqueous phase is a cost consideration. Organic lost into the aqueous phase during stripping and wash is sent to the tank house and lost during electro-winning. While there are several ancillary technologies available to recover the entrained organic after the settler, any improvement in the settler itself would be beneficial.

A second concern is the generation of crud or scum. This is a three phase semi-stable emulsion consisting of solid particulate, organic liquid and entrained air. Depending on the continuous phase of the dispersion this emulsion may collect at the organic weir between the organic and aqueous phases and block transport between the two layers. It can thus lead to a shortened effective length of the settler; it also provides a source of organic material that erodes from the crud layer into the aqueous phase. As the degree of crud generation varies between mine sites there are multiple approaches to dealing with this problem ranging from doing nothing to regular removal using specialized suction devices.

A third and less obvious issue is the flow pattern in the settler itself. The flow in a settler is not completely laminar and channeling occurs. Specifically, the flow velocity in the center of the settler can be different than at the edges and reverse flow can often be observed, particularly behind the second picket fence. It is difficult to assess the entrainment impact of this behavior without major modifications to either the shape of the settler or its internals. At minimum however, reverse flow negates the assumption of plug flow used in coalescence review and indicates that for a given total design flow rate existing rectangular settlers exceed the minimum foot print requirement for process.

**Table 2. Typical Settler Sizing and Design Calculation Sheet (Liguori, 1997)**

| | |
|---|---|
| 1. Circuit Configuration | 2 x 1 |
| 2. PLS Flow, Total ($M^3$/hr) | 1,000 |
| 3. No. of SX Trains | 1 |
| 4. O/A Ratio | 1:1 |
| 5. Total Flow Per Settler ($M^3$/hr.) | 2,000 |
| 6. Settling Rate ($M^3$/hr./$M^2$) | 4.4 |
| 7. Organic Depth (mm) | 300 |
| 8. Average Aqueous Depth (mm) | 450 |
| 9. Settler Area (total flow/settler ÷ settling rate) | 454.5 |
| 10. Average Solution Velocity | 3 cm/sec |
| 11. Settler Width (flow/settler ÷ tot. solution depth ÷ solution velocity) | 24.7 (say 25.0) |
| 12. Initial Settler Length (settler area ÷ settler width) | 18.2 |
| 13. Adjust Length for Distribution Fence | 1.0 |
| 14. Adjust Length for Weirs | 2.5 |
| 15. Total Settler length | 21.7 (say 22.0) |

Conversions

$M^3$/hr x 4.4 = GPM

$M^3$/hr./$M^2$ x 0.409 = GPM/$ft^2$

2.54 cm = 1 inch

## SETTLING PHENOMENON

After the organic and aqueous phases have been mixed as a dispersion for a sufficient time to achieve the desired mass transfer, they are pumped into the settler and begin to separate. The dispersion band shrinks until it nearly disappears and there may be left just a thin film between the organic and aqueous layers. This separation is called the primary break time and is a measure of the time, which is required to achieve this primary break, determines the size of the settler. There are different basic settler designs, tank internals, and additives that will reduce the time it takes for the aqueous and organic to separate. It is necessary to understand some of the hydraulics, which determine the depth of the organic layer floating on the top of the aqueous. Any surge in the system can result in one cell being starved of solvent while another cell is overflowing. The level of the aqueous outlet weir governs the depth of the solvent layer, for each settler. To insure smooth operation of the settler, the level of the aqueous outlet weirs should be changed only a little at a time. This is because a change of one inch in the level of the water crest flowing over the weir while a settler is operating will change the depth of the solvent layer in that settler by nearly six inches flooding a great deal of solvent from the system. Conversely, lowering an outlet weir by a large amount will cause solvent to pile up to a much greater depth, limiting the flow of solvent from that stage for a period of time. The best word of advice here is to be patient and make small changes. After each change, let the system reach equilibrium before making further changes.

### Settler Internals

Designs for inlet end presently consist of a launder or a pipe to kill velocities into the settler. Companies like LIGHTNIN are presently evaluating a distribution outlet design that will point the flow to the corners of the settler resulting in smoother plug flow pattern, eliminating reversing flows and channeling. Picket fences usually are supplied in either a straight design or a Chevron type design. Chevron designs develop higher head loss. A coalescence rail further slows down the flow, but more importantly controls the emulsion band between aqueous and organic layer. The band-controlling interface can be adjusted up and down the coalescence rail to locate the optimum setting allowing clear filtering. Some plants even incorporate additional picket fences or coalescence media above the standard two. These should not be needed if the inlet/first picket fence is well designed.

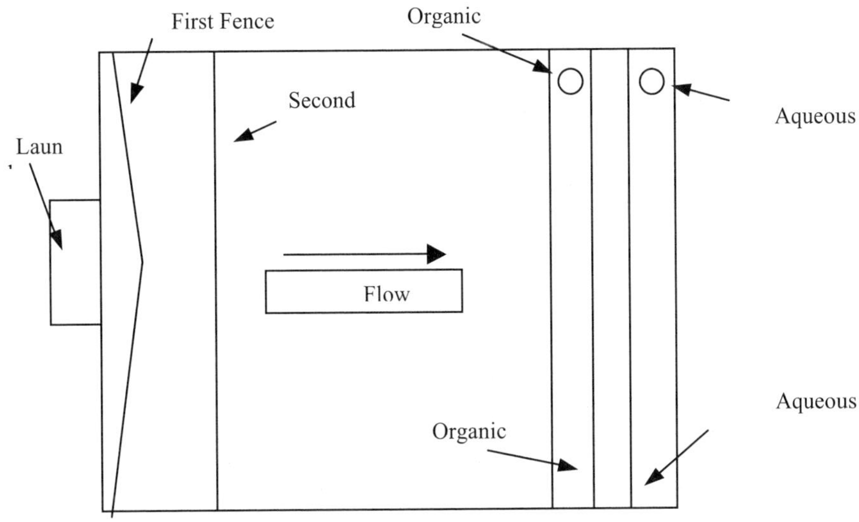


**Figure 10. Conventional Settler Elevation View**

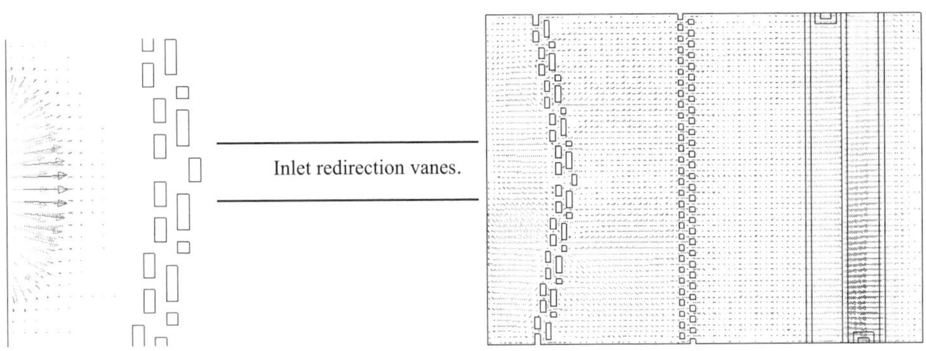


**Figure 11. CFD Study of redirection vanes to the inlet of the settler.**

Designs for the discharge end are affected by the piping and layout considerations. The organic weir sets the static liquid depth. The adjustable aqueous weir sets organic thickness. Fixed advance aqueous weir could split aqueous recycle and advance aqueous flows. Other designs used to achieve the same results. Piping connections located at the bottom fitted with vortex breakers. This will eliminate excess air entrainment. The height of the advance weir usually is set slightly below the normal setting of the adjustable weir to minimize splash and air entrainment.

## COMPUTATIONAL FLUID DYNAMICS (CFD) USE AS A DESIGN TOOL

CFD can be used as an invaluable tool in understanding what actually happens inside a settling tank. This tool is utilized to understand what flow patterns actually take place as it is almost impossible to visually determine what is happening as a result of physical characteristics of the dispersion, then breaking phases. Some examples follow. As seen in Figure 11, study one shows what can happen if install redirection vanes to the inlet of the settler. Without the vanes, there are considerable areas in the corners that see almost no flow at all. If this space can't be utilized, it is just additional holdup volume of chemicals that is expensive and could be eliminated. Adding the redirection vanes let us utilize the maximum volume of the settler by getting better flows in the corners.

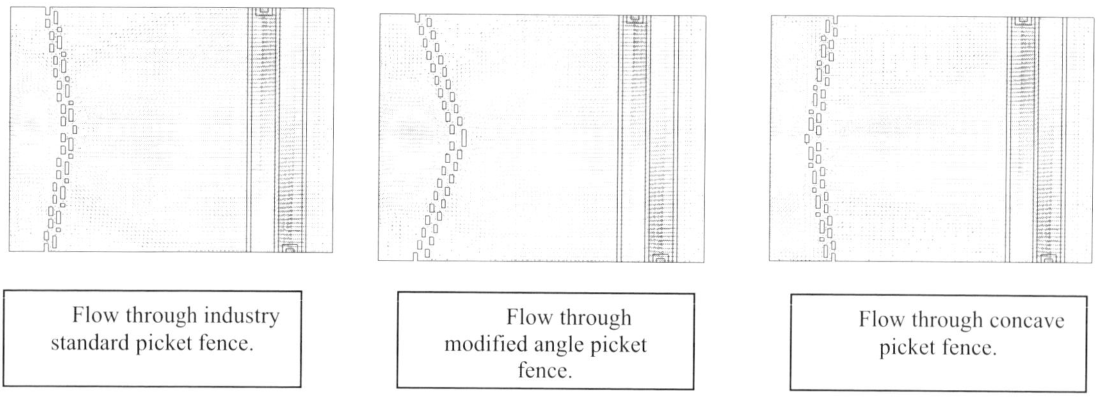


**Figure 12. CFD Study of affect of angle on picket fences**

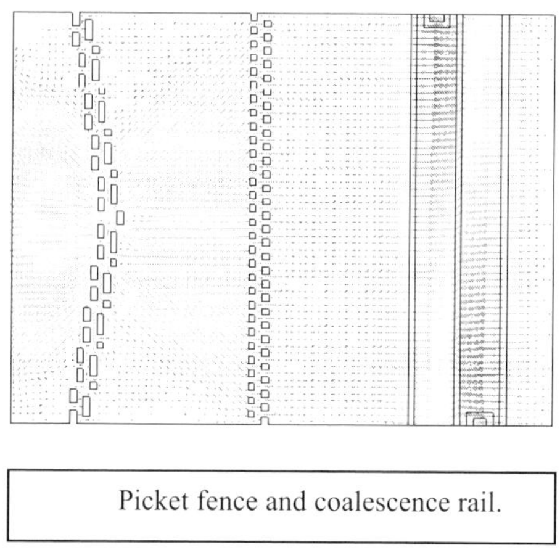


**Figure 13. CFD study of coalescence rail**

As seen in the figures, the study shows the need for a coalescence barrier between the picket fence and the overflow weirs. They straighten the flow past the barrier feeding the weirs. We can also control the emulsion interface between the aqueous and organic phases, benefiting from cleaner split solutions.

## INDIVIDUAL FEATURE DESIGN

The current inability to model the entire flow and process features of a settler requires a reassessment of the usefulness of CFD for application in this process. A number of observations in both small scale experimental modeling and full scale settler analysis indicate that CFD may offer the opportunity to optimize individual features of a settler. Observation of the flow pattern in the inlet region of the settler shows strong channeling toward the center and inlet launder and first picket fence modeling from a fluid dynamic perspective appears useful. Recirculation in the flow field in the section after the second picket fence puts the overall settler shape and second picket fence design into question. Observation of the flow in front of the organic weir points to potential entrainment sources stemming from the weir shape. Finally, CFD allows relatively quick analysis of the flow pattern changes that can be expected from changes in operating flow rate.

### Inlet Launder

Figure 14. models a 9.3 m wide by 0.9 m deep settler from the 2 m wide inlet launder to the front end of first picket fence located 6.9 m into the settler. Inlet flow velocity is 1 m/s corresponding to common full scale application. The model is run 3-D, laminar steady state. Note that only a single phase is modeled. This represents a relatively accurate picture of the flow before the first and second fence as phase separation often has not occurred until after the second fence. Since the inlet launder is assumed to be located in the center of the settler only one half is modeled, the mirror image half is inferred along the centerline symmetry axis.

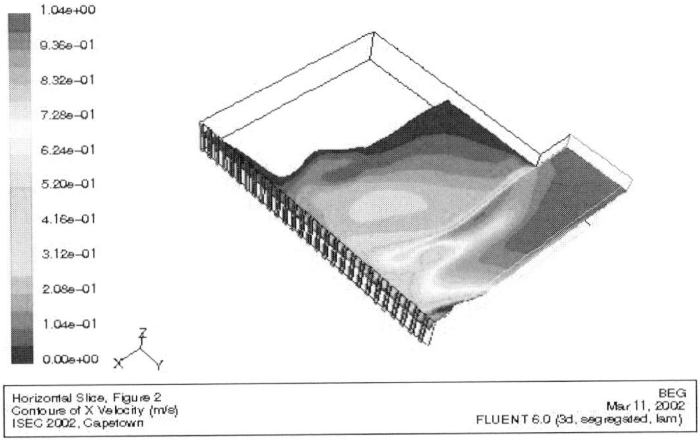


**Figure 14. Horizontal Slice, Velocity Magnitude—Positive X**

The positive x-direction velocity field distribution of this arrangement is shown in Figure 3. Note that clear areas indicate reverse flow in the settler. Here, a plane located 0.35 m off bottom is highlighted showing strong recirculation of the flow within the front section of the settler not only in the horizontal plane but also vertically (plot not shown for brevity).

Plug flow does not hold in the front end of the settler. The axial distribution of the velocity field can be avoided either through the use of an inlet launder spanning the entire height of the settler or by angling the launder from the inlet to the settler floor. The recirculation in the horizontal plane can be avoided through the use of inlet launder directional vanes or through modified picket fence design, or a combination of both.

**Overall Settler Shape**

Figures 15 and 16 show two settler designs with identical overall settler length, launder width, weir width, and liquid depth. The sole difference lies in the trapezoidal shape of the settler in Figure 15 where significant portions of the foot print has been removed. The inlet velocity is set at 1 m/s, overall dimensions of the rectangular settler are appropriate for operating settler design. Figures 14 and 15 show x-direction velocity field distribution in the horizontal plane located 0.35 m off bottom. Note that clear areas reflect flow in the negative x-direction, or flow reversal.

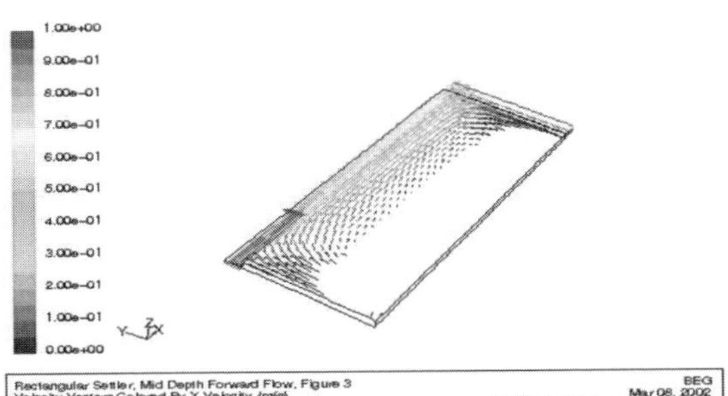


**Figure 15. Positive X-Direction Flow Field in Rectangular Settler**

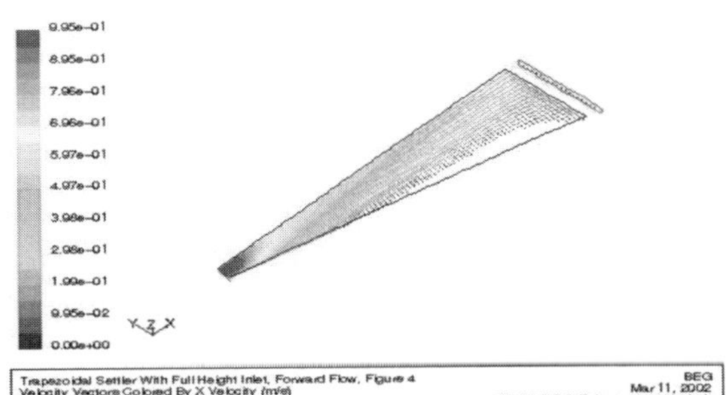


**Figure 16. Positive X-Direction Flow Field in Trapezoidal Settler**

Both designs show recirculation in the outside portion of settler thus necessitating picket fences for flow distribution. The trapezoidal settler shows a lesser degree of recirculation coupled with smaller foot print. While the average residence time of the trapezoidal settler will be shorter due to the reduced volume, the distribution will be closer to plug flow than the rectangular settler of equal weir width.

**Fence Spacing & Design**

In order to facilitate the evaluation of picket fence designs in a settler this section of the work is done in 2D. Due to the relatively small size of the pickets and inter-picket spacing in relation to the overall settler dimensions a 3-D analysis is computationally not reasonable for the scope of this paper.

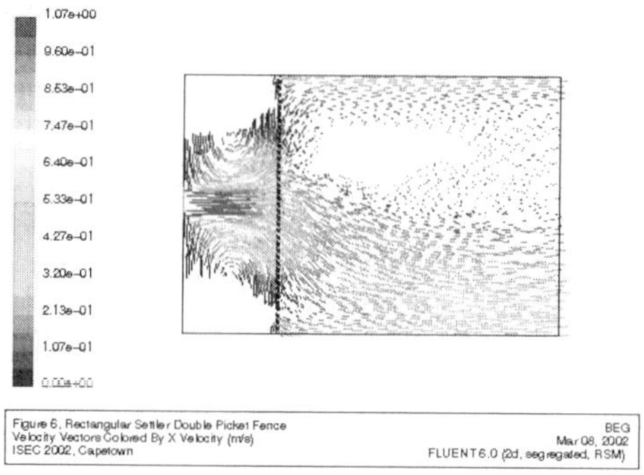


**Figure 17. Rectangular Settler—Single Picket Fence.**

A 2-dimensional plot of a rectangular settler without picket fence is equivalent to Figure 11. Figure 17 adds a single picket fence located roughly 1/3 into the settler. The model pickets are 0.1 m wide with 0.025 m spacing. Overall settler dimensions are 25 m wide by 20 m long; inlet velocity is set at 1 m/s. The resultant flow pattern shows the distribution across the width of the settler. This particular picket arrangement is not optimized as the flow preferentially channels to the outside of the settler. Figure 18 represents the flow field generated in a settler with symmetric double picket fence. For Figure 18 all pickets are centered and completely symmetric. The resultant flow field is uniform and raises the question if single picket fence arrangements are feasible full scale installations.

### Weir Design

As we progress through the settler the final major feature in the design is the weir section. While the basic concept of an overflow weir is simple care must be taken to avoid cross phase entrainment as well as air entrainment. The modeling of air entrainment in the overflow section of the weir requires multiphase modeling that is not addressed here but other features of the weir can be evaluated using single phase modeling. Figure 19 shows a plain rectangular organic weir cross section. By nature of the design the organic-aqueous interface lies on the vertical wall thus forming a stagnation point some distance upstream of the weir itself. Transport of fluid from front to back in the settler locates any accumulated crud in the interfacial region right at this stagnation point. Note the downward flow velocity of the liquid from the interfacial region toward the organic weir underflow. This phenomenon has been validated experimentally and can lead to crud layer erosion and entrainment losses of organic in aqueous. Specific design modifications to alleviate this problem are not addressed here but their evaluation via CFD is a logical starting place.

To date attempts to develop a single comprehensive computational design tool for solvent extraction settlers have met with very limited success. A single phase, multiple species model has shown good agreement with experimental results in small scale open models. Nonetheless, CFD can be used to develop optimized solutions for individual settler components such as inlet launders, picket fence design and location, overall settler shape and overflow weir design. Generalized results are that a trapezoidal shape of the settler offers reductions in overall foot print coupled with improved plug flow performance and reduced organic inventory.

## SIZING AND SELECTION OF COPPER ELECTROWINNING PLANTS

Whereas the combination of leach, solvent extraction (SX) with electrowinning of copper has been used commercially for the past 30 years, direct electrowinning of copper has been implemented for over 85 years. (Davenport, Jenkins, Kennedy, and Robinson, 1999, Robinson, Davenport, and Jenkins, 1997)

Chemically the process can be described as;

Cathode reaction $Cu^{++} + 2\,e^- \rightarrow Cu^O$ (9)

Anode reaction $H_2O \rightarrow \frac{1}{2}\,O_{2\,(g)} + 2H^+$ (10)

The sizing of copper electrowinning plants in this section will take into consideration

- Faraday's law
- electrode active area
- current density
- current efficiency
- time efficiency
- number of cathodes per cell
- rectifier size
- electrode handling machine size

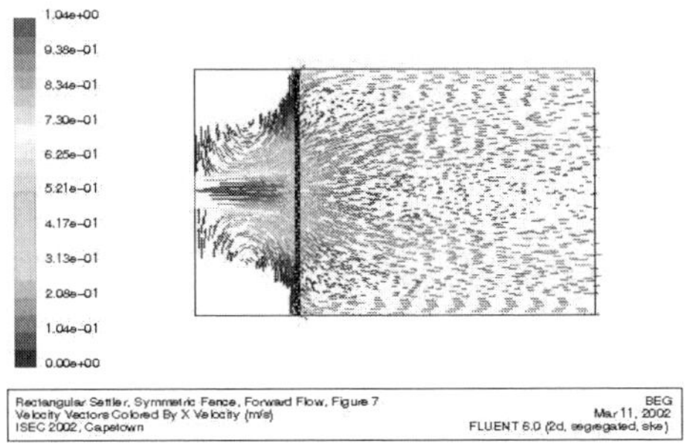


**Figure 18. Rectangular Settler With Double Symmetric Picket Fence**

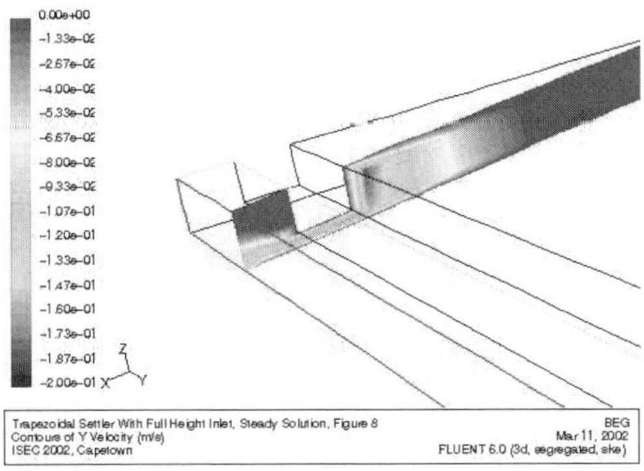


**Figure 19. Organic Weir Cross Section**

**Plant Sizing**

Essentially sizing of the copper electrowinning plant can be easily calculated by using Faraday's Law. The equation is below:

$$\frac{dm_{Cu}}{dt} = \frac{C_D \times A \times TE \times CE \times n}{F} \tag{11}$$

where:

$\frac{dm_{Cu}}{dt}$ = rate of copper produced (capacity), tonnes / hour

$C_D$ = current density, A/m$^2$
A = cathode area, m$^2$
TE = time efficiency, %
CE = current efficiency, %
n = number of cathodes in cellhouse
F= Faraday's constant equal 96, 485 coulombs per gram equivalent mass.

Faraday's Law states that 96 415 coulombs will theoretically plate one gram equivalent mass of copper (or any metal). In example of copper $Cu^{2+} + 2e^-$ ® Cu metal , it is $8.4 \times 10^5$ Amp hr per tonne of Cu

An alternative is to look at electrical equivalent with for $Cu^{2+}$ is 1.18576 gms per amphour

$$M_{Cu} = F \times A \times CE \times TE \times t \tag{12}$$

where

t = time (hrs)
F = 1.18576 gms per amp hours
A, CE, TE is as before.

**Electrode Active Area (A, $m^2$)**

Typical cathode area for copper electrowinning is two times one meter square (on both sides). Due to current 'edge effects', the anode is smaller. Overlaps of 30-50 mm on the cathode edge are typical for a smooth plating finish at the edges.

**Current Density ($C_D$, $A/m^2$)**

Traditional starter sheet operation design current densities were around 200-240 $A/m^2$. With the development of permanent cathode systems with straighter electrodes, design current density can range from 280-300 $A/m^2$ and plants can operate at well over 300 $A/m^2$.

**Current Efficiency (CE, %)**

Current efficiency can be assumed to be 90% for design purposes even if operations can run at 95%. Loss of current efficiency is due to short circuits between anode and cathode and ferric iron concentration in the electrolyte, which competes with the copper electrowinning reaction. Iron can come from the leach solution.

**Time Efficiency (TE, %)**

Time efficiency in the electrowinning cellhouse is close to 100% as cathodes are loaded and unloaded 'live' with current still running through the cell. Capacity loses can occur during the clean out of anode sludge from cells .Two cells per 100 in service are normally cleaned per day. Every third cathode is harvested from the cell.

**Number of Cathodes Per Cell**

Number of cathodes per cell or length of cell is dependent on the amount of copper produced in the plant.
For capacities of

| | |
|---|---|
| 5-15 ktpa | 30-36 cathodes/cell |
| 20-30 ktpa | 45 cathodes/cell |
| 40 and above | 60-66 cathodes/cell |

In colder climates where building space is at a premium low capacity plants can have cathodes of 60 to 80 each in the cell. Standard cathode to cathode spacing is 100 mm.

The number of cells in cell house times the number of cathodes in the cell equals the total number of cathodes.

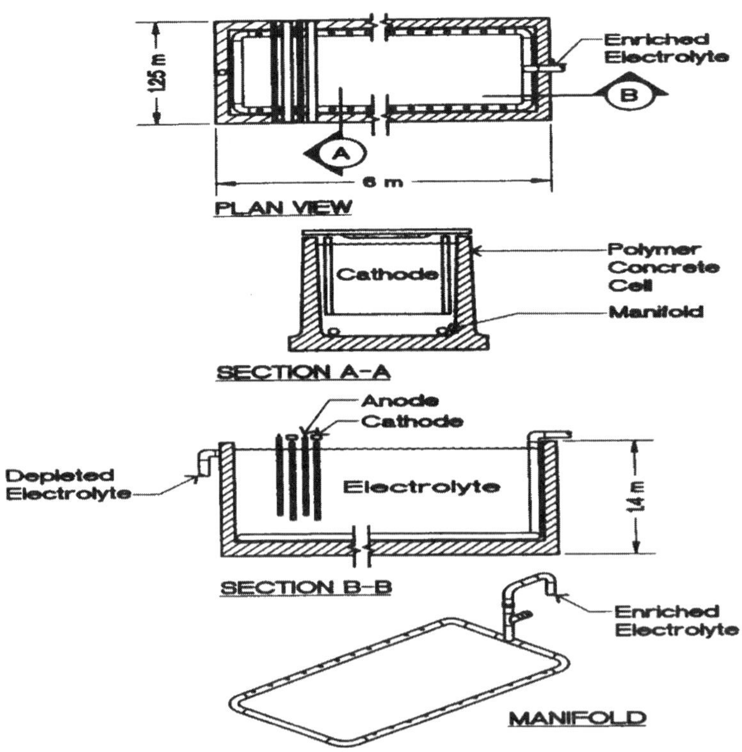


**Figure 20. Typical electrowinning cell components and design (Biswas and Davenport, 1992)**

**Rectifier Size**

Rectifier size is based on total current and total voltage.

Total current is calculated

$$I_T = C_D A \times n \tag{13}$$

$I_T$ = current , kA
$C_D$ = current density , A/m²
A = plating area , m² . (×2)
n = number of cathodes per cell

Total voltage is calculated

$$VT = Vc + VL \tag{14}$$

VT = Total voltage, V
VC = Cell section voltage , V
VL = busbar and losses voltage , V

Cell section voltage is calculated by the total number of cells in the rectifier section by the individual cell voltage (approximately 2.0 V per cell).Voltage losses and busbar voltages can also add 10-15% to design.

Rectifiers may be initially sized for proposed future expansions.

**Electrode Handling Machine Size**

Size of permanent cathode machine can be simply calculated by the following

$$Mc = n/( D \times S \times E) \tag{15}$$

Mc = machine capacity , cathodes/hr.
n = total number of cathodes in cellhouse
D = number of production stripping day/week
S = number shifts per day
E = effective working hours/shift

The target is to strip all of the cellhouse cathodes in one week. Typically in copper electro refineries weekdays from Monday to Friday are the stripping days but in the mine location of a leach SX-EW plant stripping can occur on a virtually continuous basis as suits the labor availability. Total number of cathodes is calculated by multiplying total cells by cathodes per cell.

**Copper Electrowinning Process Selection**

Selection of technology includes

- cellhouse technology
- cells
- ventilation
- contacts

**Cellhouse Technology**

Before the late 1970s copper cellhouses used starter sheets technology these separate starter sheets were plated (1-2 days old), harvested ,stripped, trimmed, straightened and loops attached for commercial production. Disadvantages of this process were high labor demand required to prepare the starter sheets, high machinery maintenance cost and lower productivity

Permanent cathode systems such as the ISA Process (Robinson, O'Kane, and Armstrong, 1995) in the late 70s and Kidd process in late 80s were developed for productivity and cathode quality reasons. The introduction of the ISA Process removed the need to grow starter sheets and so eliminated that whole section from the cell house. Central to the technology were the 316 stainless steel cathode plates and specialized machinery used to strip the cathode. Cathodes would be grown from 6 to 14 days depending on design and operating regime in the commercial cells and harvested from the cells to be stripped in the specially designed machines.

Features of each technology are as follows:

***ISA Process***

- copper plated 304 stainless steel hanger bar
- 3.25 mm thick 316L stainless steel cathode plate.
- "U" shaped rectilinear cathode washing and stripping machine design
- separate cathode sheets from each cathode plate

***KIDD Process***

- solid copper bar hanger bar welded to cathode plate
- 3.0 to 3.125mm thick 316L stainless steel plate
- circular / rotating cathode washing and stripping machine.
- Single enveloped cathode from each cathode plate

ISA Process technology previously required the use of wax to mask the edges of the cathode but now both technologies offer a waxless technology.

Recently other technology derivatives such as an electrometallurgical process developed by Outokumpu are being marketed.

**Anodes**

The first copper electrowinning plants used cast antimonial lead anodes. These anodes would warp, cause short circuits and could contaminate the copper cathode. Cold rolled and alloyed lead anodes have now become the standard anode technology in copper electrowinning.Cold rolling makes the anode stronger, stiffer and straighter than a cast anode. Alloying with calcium and tin also assist with strength and produce a more consistent corrosion product. Life of lead anode now can last six and in some cases ten years. Typical anode blade thickness is 6 mm . The trend to higher current densities has seen the introduction of thicker anode blades.The blade dimensions of an anode are smaller than the cathode due to 'plating overlaps' with overlaps of 30mm on the sides and 30 to 50 mm on the bottom being ideal.

**Electrowinning Cell Technology**

Traditional cell construction material in the past has been monolithic re-enforced concrete and liner systems. Liners were made of lead, FRP and PP, HDPE plastics. Over time liners would wear, break and get damaged and acid would attack the concrete leading to costly maintenance.

Polymer concrete cells were first tested in Zambia in the 1970's but it was in the 1990's where the polymer concrete unicell concept became accepted as the universal standard for copper electrowinning design. (Robinson, Karcas and Weatherseed, 1999)

Unicell's acceptance is due to the following reasons:

- monolithic single casting results in acid resistant homogenous structure
- integrated inlet and overflow piping
- acceptance of the cell to stand as a structural unit where walkways, piping, valves can be attached to inserts embedded in Unicell structure.

Benefits include reduced maintenance and overall cellhouse capital cost.

**Ventilation / Mist Suppression**

Methods of mist suppression available in copper electrowinning today include:

- mist suppression foams
- mist suppression balls and beads
- open tankhouse designs
- cell top hoods
- positive flow ventilation
- combinations of the above

**Busbar Contacts/ Cell Top Insulators**

Busbars that transfer current from one cell to another can vary in design. These include:

- triangular
- 'dogbone'
- double contact

Triangular is the most common and is simple, low cost, easy to replace and requires simple cell top furniture. However it uses an asymmetric anode hanger bar and has low current carrying capacity.

The 'dogbone' contact is a cross section of copper busbar that connects two triangular (most common) contacts and allows for symmetrical anode design and can carry more current. It is a more complicated design and so more expensive and is more difficult to clean.

Recent busbar design developments have lead to the double contact design, which include equalizing bars for anodes or cathodes or both. The equalizing bars equalize current between the electrodes, raising current efficiency and potential current density.

**TYPICAL COPPER SX AND EW PLANT DESIGN CRITERIA LISTING**

As a service to the reader the following pages of the paper list the pertinent design criteria typical for a proposed 13,500 tonne per annum copper cathode production facility (Taylor, 1995).

# Typical Copper SX EW Design Criteria

| | | |
|---|---|---|
| **PLANT DESIGN CRITERIA** | | |
| **GENERAL** | | |
| **PRODUCTION** | | |
| PLANT CAPACITY t/a CATHODE | 13,500 | |
| **OPERATING SCHEDULE** | | |
| | h/d | 24 |
| | d/y | 365 |
| OVERALL UTILISATION/AVAILABILITY | % | 95 |
| **ASSOCIATED PLANT (EXCLUDED)** | | |
| ORE TYPE | | OXIDISED COPPER |
| LEACH SYSTEM | | HEAPS |
| COPPER GRADE | % | 1 |
| NOMINAL CRUSH SIZE | mm | 25 |
| **SOLVENT EXTRACTION** | | |
| **EXTRACTION SECTION** | | |
| FEED SOLUTION | | |
| FLOWRATE | $m^3/h$ | 500 |
| COPPER | g/l | 3.5 |
| TOTAL IRON | g/l | 5.0 |
| FERRIC IRON | g/l | 1.0 |
| pH | | 2.0 |
| CHLORIDE (MAXIMUM) | mg/l | 75 |
| SOLIDS (MAXIMUM) | mg/l | 20 |
| COPPER RECOVERY | % | 93 |
| TEMPERATURE - DESIGN | °C | 20 |
| - MAXIMUM | °C | 30 |
| RAFFINATE COPPER | g/l | 0.245 |
| RAFFINATE FLOWRATE (EXCL. IRON BLEED) | $m^3/h$ | 500 |
| NO. OF STAGES | 2 | |
| MIX BOXES PER STAGE | 2 | |
| MIXER RETENTION PER STAGE | MIN | 2 |
| MIXER O/A RATIO | 1:1 | |
| SETTLER SPECIFIC AREA | $m^3/h/m^2$ | 5.0 |

| | | |
|---|---|---|
| ORGANIC FLOWRATE | $m^3/h$ | 500 |
| STAGE EFFICIENCY | % | 95 |
| **STRIPPING SECTION** | | |
| ELECTROLYTE FLOWRATE | $m^3/h$ | 108.3 |
| SPENT ELECTROLYTE | | |
| - COPPER | g/l | 33 |
| - ACID, MAX. | g/l | 170 |
| - IRON, MAX. | g/l | 2 |
| STRONG ELECTROLYTE | | |
| - COPPER | g/l | 48 |
| NO. OF STAGES | | 1 |
| MIX BOXES PER STAGE | | 2 |
| MIXER RETENTION PER STAGE | MIN | 2 |
| MIXER O/A RATIO | | 1:1 |
| SETTLER SPECIFIC AREA | $m^3/h/m^2$ | 5.0 |
| ORGANIC FLOWRATE | $m^3/h$ | 500 |
| TEMPERATURE - DESIGN | °C | 20 |
| - MAXIMUM | °C | 30 |
| STAGE EFFICIENCY | % | 95 |
| **ORGANIC PHASE** | | |
| REAGENT TYPE | | ACORGA M 5640, LIX 984, OR EQUIVALENT ALLIED CHEMICALS MOC 100 |
| CONCENTRATION (CONFIRM WITH SUPPLIERS) | % | 13 |
| DILUENT TYPE | | HIGH FLASHPOINT KEROSENE SHELLSOL 2046, ESCAID 100 OR EQUIV. |
| FLASHPOINT (ASTM D-93) | °C | 80 |
| AROMATICS, MAX | % | 20 |
| COPPER: IRON TRANSFER RATIO | | 1000 (EXPECTED) 500 (DESIGN) |
| ORGANIC ENTRAINMENT IN RAFFINATE AND STRONG ELECTROLYTE | ppm | 30-75 |
| AQUEOUS ENTRAINMENT IN LOADED ORGANIC PPM | | 250-500 |
| AQUEOUS ENTRAINMENT IN STRIPPED ORGANIC | ppm | 300-600 |

## Typical Copper SX EW Design Criteria - Continued

| | | |
|---|---|---|
| **MIXER – SETTLERS** | | |
| NO. OF STREAMS | | |
| - EXTRACTION | | 1 |
| - STRIPPING | | 1 |
| TYPE | | CONVENTIONAL |
| PRIMARY MIX IMPELLER | | PUMP-MIX |
| SECONDARY MIX IMPELLER | | HIGH EFFICIENCY TURBINE |
| MIXING BOX PHASE CONTINUITY | | |
| - E1 | | AQUEOUS |
| - E2 | | ORGANIC |
| - S1 | | ORGANIC |
| RECYCLES | | AQUEOUS ON ALL STAGES |
| **ELECTROLYTE PREPARATION** | | |
| **ELECTROLYTE CLEAN-UP** | | |
| TYPE | | FLOAT COLUMNS |
| NO. OF STAGES | | 2 |
| DESIGNS AVAILABLE | | JAMESON<br>MAGMA<br>MINFLOAT<br>COMINCO<br>PYRAMID |
| TARGET FINAL ORGANIC CONTENT | ppm | 5 OR LESS |
| **ELECTROLYTE HEAT EXCHANGER** | | |
| TYPE | | PLATE AND FRAME |
| STRONG ELECT. TEMP IN/OUT | °C | 20/40 |
| SPENT ELECT. TEMP IN/OUT | °C | 45/25 |
| **ELECTROWINNING** | | |
| **CIRCUIT DESIGN** | | |
| CURRENT DENSITY | | |
| - NORMAL | $A/m^2$ | 280 |
| - MAX. | $A/m^2$ | 300 |
| CURRENT EFFICIENCY | % | 90 |
| COPPER QUALITY | | LME GRADE A |

| | | |
|---|---|---|
| DESIGN CAPACITYT/D | | 39 |
| OPERATING CYCLE | DAYS | 7 |
| METHOD | | 1/3 CELL LIVE LOAD |
| PRODUCT BUNDLES | t | 2-3 |
| **CELL DESIGN** | | |
| NUMBER OF CELLS: SCAVENGERS | | 11 |
| COMMERCIAL | | 53 |
| TOTAL | | 64 |
| CATHODES PER CELL | | 39 |
| ANODES PER CELL | | 40 |
| CATHODES - TYPE | | 316 STAINLESS STEEL |
| - ACTIVE AREA | | 1 m x 1.1m |
| CATHODE SPACING | mm | 95 |
| ANODES - MATERIAL | | Pb/Ca/Sn ALLOY<br>SOLID ROLLED SHEET |
| - THICKNESS | mm | 6 |
| **ELECTRICAL** | | |
| NO. OF RECTIFIER CIRCUITS | | 1 |
| CELL VOLTAGE: | | |
| - NORMAL | | 2.0 |
| - MAX. | | 2.2 |
| CELL CURRENT: | | |
| - NORMAL | A | 24,000 |
| - MAX. | A | 26,000 |
| BUSBAR - MATERIAL | | COPPER |
| **SOLUTION CIRCULATION** | | |
| FLOWRATE/CELL: | | |
| - SCAVENGERS | $m^3/h$ | 10.3 |
| - COMMERCIAL | $m^3/h$ | 8.5 |
| COPPER DROP PER CELL: | | |
| - SCAVENGERS | g/L | 2.5 |
| - COMMERCIAL | g/L | 3.0 |
| COBALT DOSING | ppm | 100 |
| IRON CONTENT, MAX | g/L | 2 |

## Typical Copper SX EW Design Criteria - Continued

| | | |
|---|---|---|
| CHLORIDE CONCENTRATION, MAX | ppm | 30 |
| MIST SUPPRESSION METHOD | | POLYPROPYLENE BEADS |
| CELL OPERATING TEMP. | °C | 35-45 |
| **TANK AND POND SIZING CRITERIA** | | |
| RENTENTION TIMES: | | |
| SX FEED POND | HR | 24 |
| RAFFINATE POND | HR | 24 |
| LOADED ORGANIC TANK | MIN | 20 |
| STRONG ELECTROLYTE TANK | MIN | 30 |
| SPENT ELECTROLYTE TANK | MIN | 30 |
| SOLUTION HOLDING TANK | CONTENTS OF 1 COMPLETE SX SETTLER | |
| DILUENT STORAGE* | | 1 ½ TANKER LOADS |
| EXTRACTANT STORAGE | | 3 MONTHS SUPPLY |
| COBALTOUS SULPHATE STORAGE | | 3 MONTHS SUPPLY |

(* MAY NEED 2-3 MONTHS FOR REMOTE SITE)

**UTILITIES**

| | | |
|---|---|---|
| EW MAKE-UP WATER | | |
| CHLORIDE CONTENT | ppm | <30 |
| SOLIDS CONTENT | ppm | <10 |
| EW MAKE-UP ACID | | |
| QUALITY | | WHITE ACID |
| HOT WATER | | |
| DUTIES | CATHODE WASHING AND WASH DOWN | |
| TEMP,MAX | °C | 80 |

**COPPER SX FLOWSHEET APPLICATIONS AND DEVELOPMENTS**

As copper solvent extraction has evolved over the past forty years, so to has the design, configuration and layout of the extraction, washing and stripping circuits. This is in response to copper leaching scenarios that now include copper pressure oxidation, high acid consuming ores, mixtures of high grade and low grade ores and treatment of combined copper/cobalt concentrates. The following flowsheet layouts illustrate this.

Figure 21 shows a general solvent extraction circuit. It is configured as a conventional two extraction (2E) one wash (1W) and two strip (2S) circuit. In this case, the PLS is pumped through two extraction stages in series.

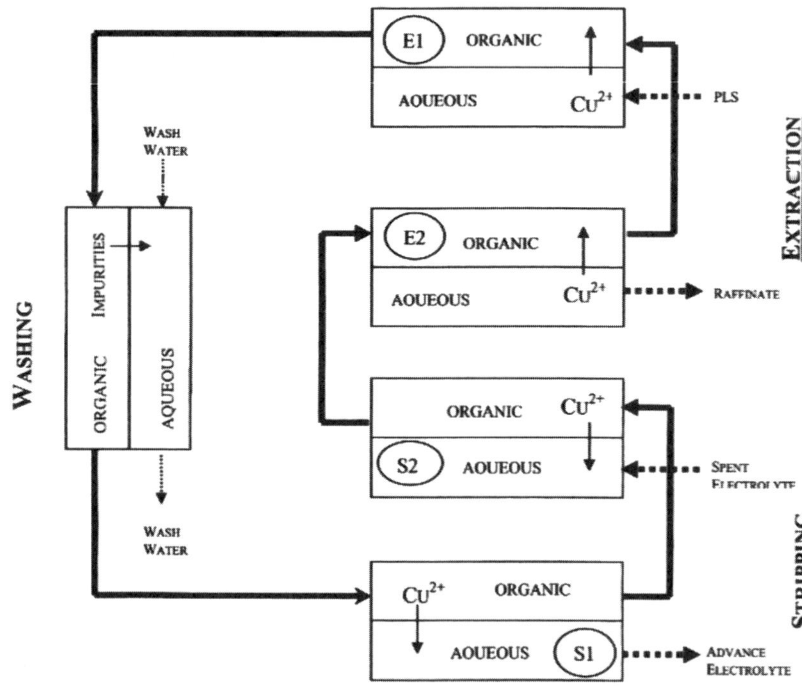


**Figure 21. Typical Industrial Copper Solvent Extraction and Electrowinning Circuit (Dudley, 2008)**

The loaded organic from E1 is washed with acidified water in order to reduce impurity transfer by diluting entrained PLS, in particular the chloride content. The washed organic is then stripped in two stages with spent electrolyte from EW.

Figure 22. also illustrates a traditional copper agitation leach , CCD, SX circuit.

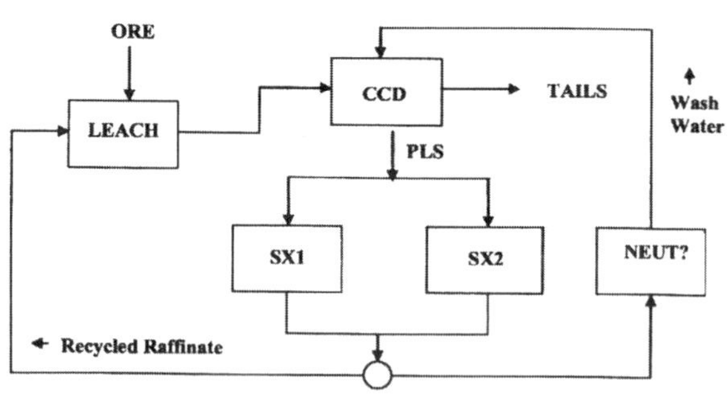


**Figure 22. Traditional Copper Agitation Leach, CCD, Solvent Extraction Flowsheet (Kordosky,2008)**

Figure 23. illustrates the split circuit flowsheet configuration. As seen, the split circuit offer dual SX circuits for simultaneous treatment of both high and low cu tenor PLS streams.

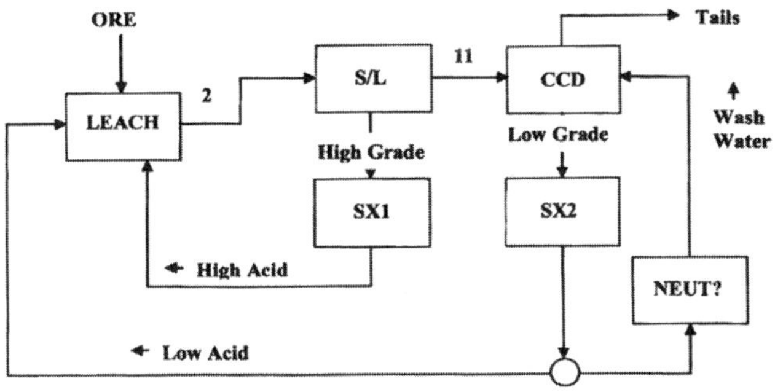


**Figure 23. The "Split Circuit" Copper Agitation Leach/CCD/SX Flowsheet (Kordosky, 2008)**

Figure 24 illustrates the Kanshanshi flow sheet which incorporates copper concentrate pressure leaching and oxide tailings acidic agitation leaching. The SX circuit here incorporates the split circuit concept for treatment of the high and low copper tenor PLS streams.

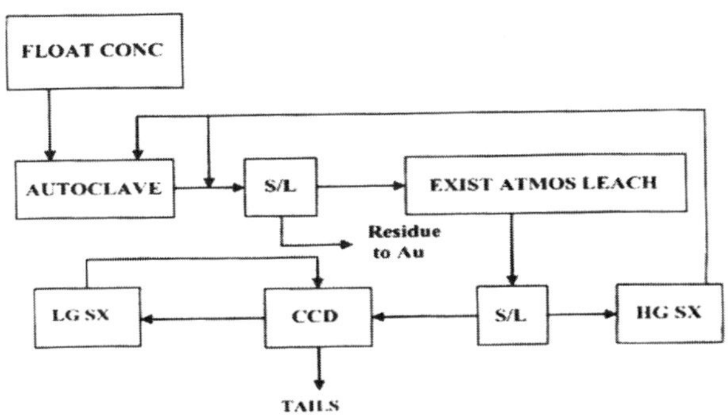


**Figure 24. The Kanshanshi Flowsheet incorporating High Pressure Concentrate Leaching Into Oxide Agitation Leach/SX/EW (Kordosky, 2008)**

Figures 25 through 27 illustrate flowsheets for combined copper and cobalt recovery.

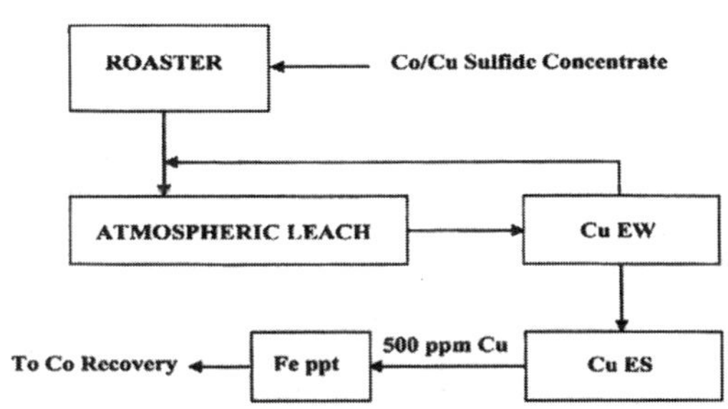


**Figure 25. Traditional Flowsheet For Copper Removal From Cobalt Leach Solutions Derived From Leaching Roasted Copper/Cobalt Concentrates (Kordosky)**

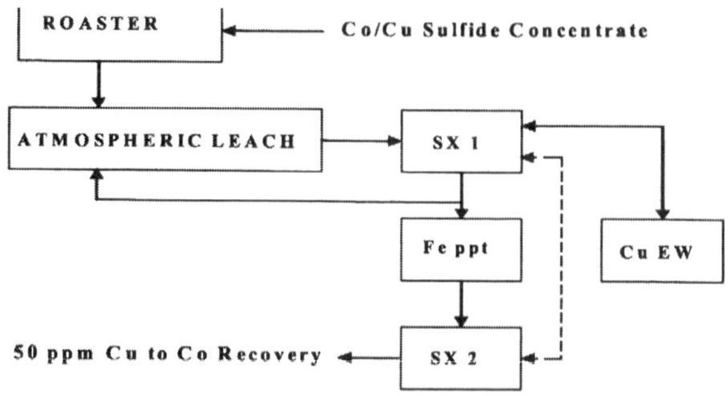


**Figure 26. Copper Solvent Extraction Incorporated Into Cobalt Recovery From A Copper / Cobalt Leach Solution Derived From Leaching Roasted Copper / Cobalt Concentrate**

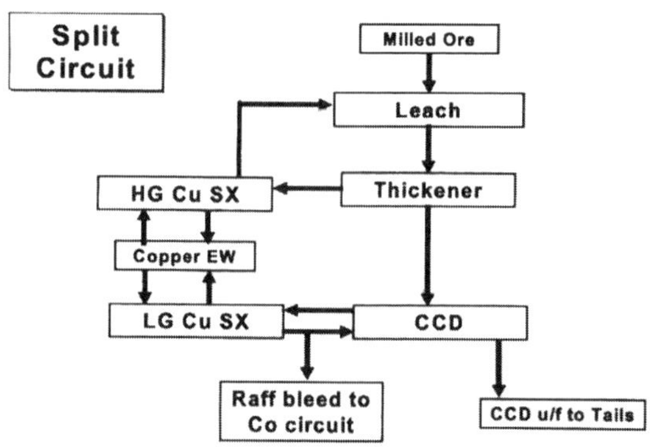


**Figure 27. General Flowsheet For Agitation Leach Plants In The DRC Which Incorporate Cobalt Recovery**

Figures 16, 17 and 18 illustrate comparative METSIM simulations of various flowsheet scenarios applied to high acid consuming copper and cobalt bearing ores (Baxter, 2008).

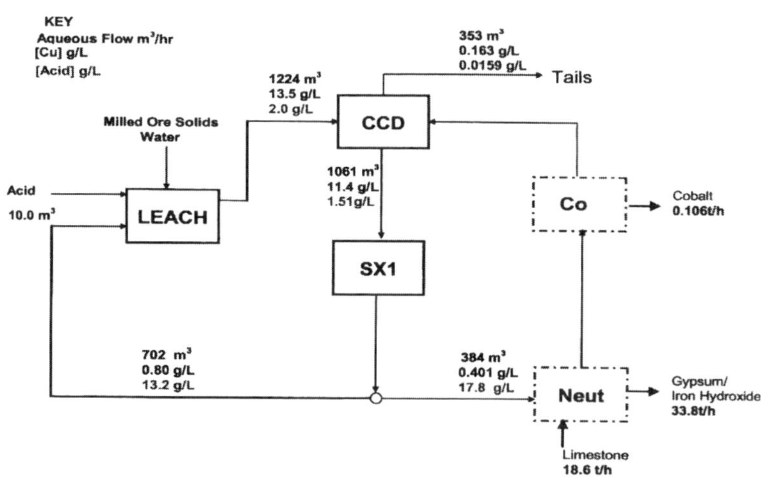


**Figure 28. Conventional Flowsheet As Applied to High Acid Consuming Ores**

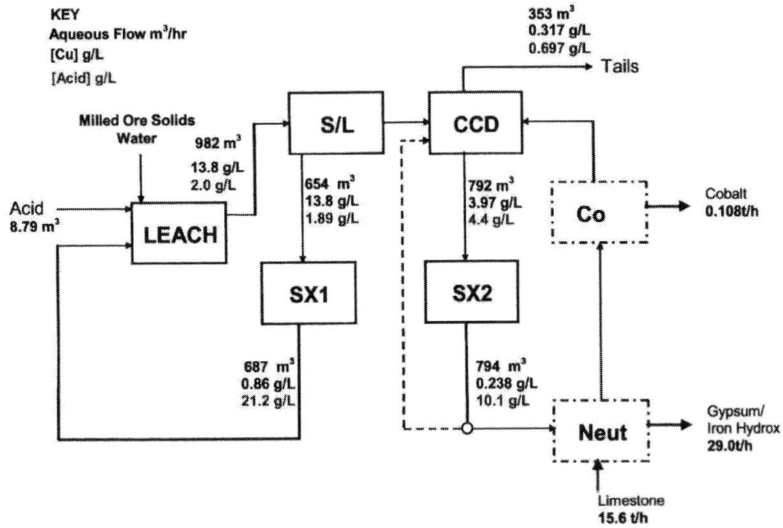


**Figure 29. Split Circuit Flowsheet As Applied to High Acid Consuming Ores**

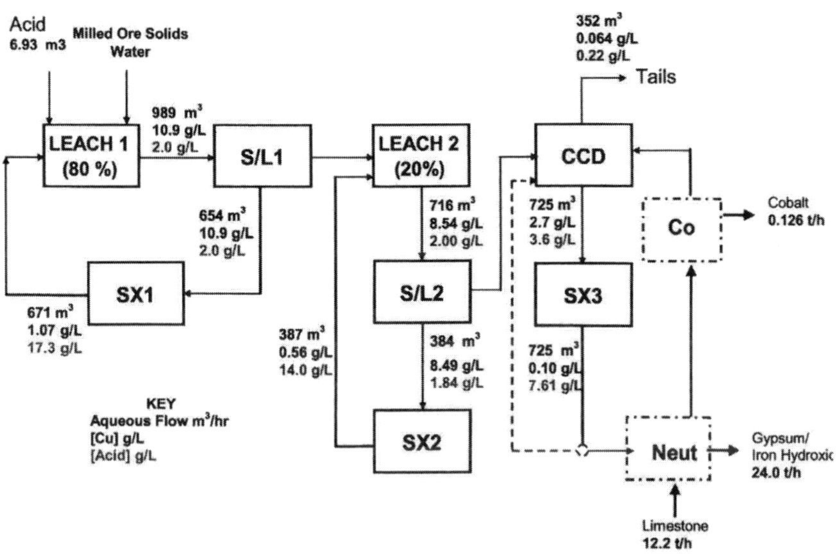


**Figure 30. Sequential Split Leach Flowsheet As Applied to High Acid Consuming Ores**

## REFERENCES


Baxter, K., Nisbett, A., Urbani, M. And Marte, K., 2008, "Flowsheet Alternatives for High Acid Consuming Copper Ores" ALTA 2008 Copper Conference, June 19-20, Perth, Australia

Biswas, A.K., and Davenport, W.G., 1992, *Extractive Metallurgy of Copper*, 3rd Edition, Elsevier Science Inc., New York.

Davenport, W. G., Jenkins, J., Kennedy, B., Robinson, T., 1999, "Leach, Solvent Extraction, Electrowinning of Copper—1999 World Operating Data," Copper 99, Phoenix, Arizona, USA. October.

Davy International, undated, *Qualifications for Copper Leaching Solvent Extraction Elecrowinning Facilities*, Section 4:Technology—SX-EW.

Dudley, K. Horner, A., Virnig, M., Crane, P. and Urbani, M., 2008, "SX-EW Process Optimization at Nyrstar Port Pririe—The Influence of Employing The Correct Regaent Type," ALTA 2008 Copper Conference, June 19-20, Perth, Australia

Gigas, B., and Giralico, M.A., 2000, *SX: CFD Settler Optimisation*, Randol Copper Hydromet Roundtable 2000, September, Tucson, Arizona.

Gigas, B., and Giralico, M. A., 2002, Advanced Methods for Designing Today's Optimum Solvent Extraction Mixer Settler Unit, ISEC 2002, Johannesburg, South Africa, March.

Giralico, M.A., Post, T.A., Preston, M.J. 1995, " Improve the performance of your copper solvent extraction process by optimizing the design and operation of your pumper and auxiliary impellers." SME Annual Meeting, Preprint Number 95-189,,Denver, Colorado, March 6-9,

Giralico, M.A., Post, T.A., Preston, M.J., 1996, "Optimizing the performance of new and existing SX plants through the use of high efficiency impellers for pumper and auxiliary mixers." SME Annual Meeting, Phoenix, Arizona, March 11-14.

Jorgensen, G.V., Editor, 1999, *Copper Leaching, Solvent Extraction, and Electrowinning Technology*, pp 239 ff, Society for Mining, Metallurgy, and Exploration, Inc.; Littleton, Co, USA

Kordosky, G., Feather, A. and Chisakuta, G., 2008, "The Copperbelt of Africa—New Plants and New Technology in Copper Hydrometallurgy," ALTA 2008 Copper Conference, June 19-20, Perth, Australia

Liguori, A. J., 1997, *Copper Heap Leach*, SME Short Course Notes, Denver, Colorado.

Lo, T.C., Baird, M.H.I., and Hanson, C, Editors; 1991, *Handbook of Solvent Extraction;* Section 9, Krieger Publishing Company, Malabar, Florida.

Miller, G., 2001, *Design Tools To Control Transients In Solvent Extraction Plants*, SME Annual Meeting 2001, Denver, March

Oldshue, J.Y., 1984, *Fluid Mixing Technology*, Chapter 2: *Impeller fluid shear rates and pumping capacity*, McGraw-Hill, New York.

Post, T.A., Howk, R.A., Preston, M.J., 1996, "Impeller System and Method for Enhanced-Flow Pumping of Liquids," United States Patent Nr. 5,511,881, Apr. 30.

Robinson, T., O'Kane, J. and Armstrong, R.W., 1995, "Copper Electrowinning and the ISA PROCESS," Copper 95, Santiago, Chile.

Robinson, T., Davenport, W.G., Jenkins. J., 1997, *Electrolytic Copper Electrowinning and Solvent Extraction World Operating Data.,* Randol Copper Hydromet Roundtable Vancouver , November.

Robinson, T., Karcas, G., Weatherseed, M., 1999, *Polymer Concrete Cell Applications in the Electrometallurgical Industry,* ALTA 1999 Copper Hydrometallurgy Forum, Queensland, September.

Taylor, A., 1995, *Copper Leach/SX/EW Project Development*, ALTA METALLURGICAL SERVICES, Brisbane Australia.

Taylor, A., 1995, *Copper SX/EW Seminar: Basic Principles, Alternatives, Plant Design*, ALTA Metallurgical Services, Melbourne, Australia, Section 13

Weetman, R.J., Oldshue, J.Y., 1988, "Power, flow, and shear characteristics of mixing impellers." 6th European Conference on Mixing, Pavia, Italy, 24-26 May.

# Comminution (Crushing and Grinding)

# THE IMPORTANCE OF WATER UTILIZATION FOR MAXIMIZIN GRINDING CIRCUIT CLASSIFICATION SYSTEM EFFICIENCY

Robert E. McIvor

Metcom Technologies, Inc., 1021B W. Baraga Ave., Marquette, MI, USA

## ABSTRACT

The Functional Performance Equation for ball milling circuits allows us to isolate the circuit classification system efficiency (CSE) from the mill's grinding performance. The factors which affect circuit CSE are reviewed, with emphasis on water usage at the classifiers. It is shown that manipulation of such water usage can be carried out to improve plant grinding circuit efficiency, at times by more than 30 percent.

## THE FUNCTIONAL PERFORMANCE EQUATION

Observation of the feed and product size distributions going into and discharging from the mill in grinding circuits with low versus high circulating load ratios initially led to the definition of "circuit classification system efficiency" (or CSE, McIvor, 1988a), and ultimately the discovery of the Functional Performance Equation (McIvor, 1988b, and McIvor et al, 1992). The CSE is the percentage of "coarse" (typically with reference to the plus P80) size material occupying the mill (versus "fines" or product size material, finer than the circuit P80), and can be calculated by taking the average of the percentage of coarse material in the mill feed and mill discharge. As this also represents the percentage of the mill energy expended on targeted, coarse size material, it is directly proportional to overall grinding circuit efficiency and production rate. This is all clearly expressed in the Functional Performance Equation for ball milling circuits, which is derived as follows.

We define "fines" as any product size material, and "coarse" as that targeted to be further ground, the two typically differentiated by the circuit target P80. For any grinding circuit, the production rate of new, product size material (fines) must equal the specific grinding rate of the coarse material (i.e. fine product generated per unit of energy applied to the coarse material) times the power applied to the coarse material.

$$\text{Circuit Production Rate of New Fine Product} = \text{Power Applied to Coarse Material} \times \text{Specific Grinding Rate of Coarse Material}$$

From the above discussion, we know that the power applied to the coarse material is the total mill power times the percentage of coarse material in the mill, the latter also defined as CSE. Therefore:

$$\text{Circuit Production Rate of New Fine Product} = \text{Total Mill Power} \times \text{CSE} \times \text{Specific Grinding Rate of Coarse Material}$$

A measure of the plant ball mill's grinding efficiency is the ratio of the grinding rate of the coarse material in the plant ball mill compared to the grinding rate, or "grindability," of the same coarse material as measured in a standardized test mill set up. That is:

$$\text{Mill Grinding Efficiency} = \text{Specific Grinding Rate of Coarse Material} \; / \; \text{Coarse Material Grindability}$$

Therefore, if we multiply and divide the last term in the previous equation by the material grindability, we have the Functional Performance Equation for ball milling circuits:

$$\text{Circuit Production Rate of New Fine Product} = \text{Total Mill Power} \times \text{CSE} \times \text{Mill Grinding Efficiency} \times \text{Material Grindability}$$

This is a simple yet powerful expression of how a ball mill circuit generates new product size material. Rate of production is a direct function of the material grindability, as well as the total amount of power provided by the mill. It is also a direct function of two separate and distinct efficiencies which are in play. Each of these efficiencies is specifically related to certain physical design and operating variables which we can manipulate. The terms in this equation are readily generated by a circuit survey. Thus, the Functional Performance Equation has provided unprecedented understanding and opportunity for plant ball mill circuit optimization (McIvor, 2006).

It is also noteworthy that the compliment to CSE is the fraction of mill energy being used (i.e. wasted) on further grinding of fines. Such over grinding is often detrimental to downstream process (McIvor and Finch, 1991), and thus an extremely important motivator to achieve of high CSE, even beyond grinding circuit efficiency.

## FACTORS AFFECTING GRINDING CIRCUIT CLASSIFICATION SYSTEM EFFICIENCY

To optimize circuit performance, we clearly want to maximize CSE (within economic constraints, as shown later). This is achieved through the pump and cyclones (or other classifiers). The factors which collectively affect CSE are discussed as follows.

### The "Givens"

For the standard circuit arrangement shown in Figure 1, wherein the new feed reports first to the classification equipment (in this case, hydrocyclones), the "givens" with regard to CSE, over which we essentially have no control, are as follows.

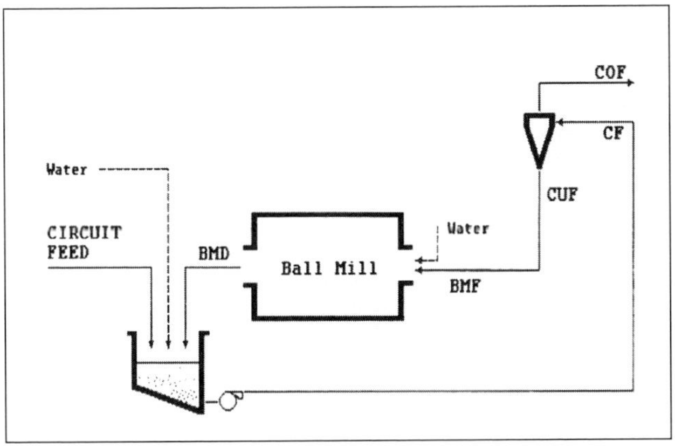


**Figure 1. Typical Ball Mill Circuit**

1. The proportion of "fines" in the circuit feed. For many ball milling circuits, the amount of minus P80 size material in the circuit feed is relatively small, in the range of 10-20%, say, such that once it passes the classifier the amount remaining from the circuit feed contributing to "fines" in the ball mill feed is very small. The exception occurs with very low size reduction circuits, such as many re-grind applications.
2. The product breakage sizing characteristics of the material. Many types of coarse ore particles will produce a very small percentage of minus P80 size material (naturally depending on the starting particle size) during a brief exposure to grinding. Others may shatter or slime, producing substantial fines during a single breakage event. So the natural breakage characteristics of a given ore will affect the amount of fines in the ball mill discharge, whatever the amount of exposure to grinding forces.
3. (Ores consisting of multiple minerals with different characteristics affecting grinding or classification: Different minerals will preferentially grind compared to others, and shape and density may particularly influence behavior in a cyclone.

### Factors That Can be Manipulated

(1) Amount of exposure to grinding forces. Given the ball mill feed sizing, the shorter the grinding time, or, more accurately, the lower the exposure to grinding forces, the lower will be the proportion of fines in the ball mill discharge. This was most dramatically expressed by Davis (1925) and expounded upon by Gaudin (1939) in comparison of ball mill productivity (net production rate of fines) versus the circulating load ratio, as shown in Figure 2. This obviously important circuit performance parameter remains today as the primary way of expressing (the direct inverse of) the relative amount of exposure to grinding forces per pass through the mill. Along with the mill feed and discharge size distributions corresponding to low versus high circulating load (Figure 3), it is also the presentation of experimental data which lead to the writer's discovery of circuit CSE, and the subsequent discovery of the Functional Performance Equation.

(2) Classifier basic (reduced) separation performance. Setting aside for a moment the behavior of water at the classifier, the classifier separation performance is precisely defined by the (reduced or corrected) size by size material split, usually the recovery to oversize (in the case of cyclones, the cyclone underflow, also the ball mill feed stream in Figure 1). The 50%/50% reduced cut size (d50c) is determined by the cyclone physical dimensions and feed slurry specifications (for example, see Heiskanen, 1993), and is the key determinant to select the cyclone and pump needed to achieve the desired circulating load ratio in the Metcom System (McIvor, 1989-2008). The finer the cut size, the greater the proportion of classifier feed reporting to oversize, and thus, by definition, the higher the circulating load ratio. The sharpness of separation is likewise influenced by the cyclone physical design (for example, new models like "Cavex" and "Gmax" provide sharper performance than predecessors) and the feed slurry characteristics, in particular percent solids (again, see Heiskanen, 1993). The sharper the reduced classification (as measured, for example, by the ratio of d25c to d75c), the fewer will be the fines reporting to the coarse product stream.

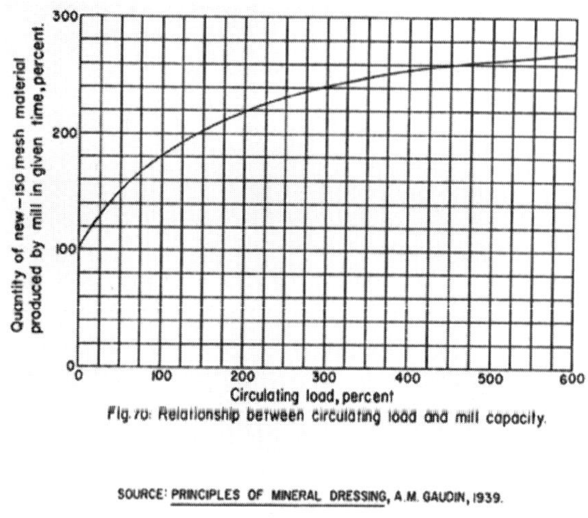


**Figure 2. Circuit Production Rate vs. Circulating Load (Davis, 1925; Gaudin, 1939)**

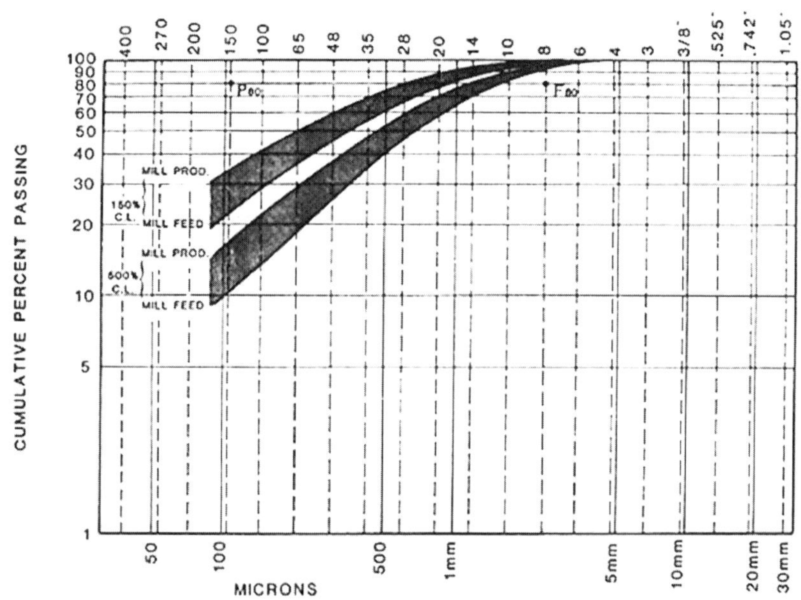


**Figure 3. Mill Feed and Product Size Distributions at High and Low Circ. Loads**

(3) Water balance at the classifier. The portion of the classifier feed water which reports to the oversize stream is normally associated with "bypass" of fine material carried by that water and thusly misplaced and contributing to fines in the ball mill feed. Maximizing cyclone underflow percent solids (without encroaching on roping effects) by suitable apex sizing (for example), will contribute to minimum bypass. However, given fixed and suitably high circulating load ratio and fixed cyclone underflow percent solids, permissible cyclone overflow percent solids (as dictated by downstream processes) will define the permissible amount of cyclone feed water, and thus the fraction of water to oversize and the related bypass.

(4) Alternative circuit arrangements. For a single ball mill circuit, these may include post vs. pre-classification, alternate classification equipment, or multi-stages of classification. The post classification circuit arrangement sends all the circuit feed fines to the ball mill feed, rather than only those that are misplaced by the classifier in the pre-classification arrangement. Therefore, it is only suggested when the circuit feed is so coarse that some grinding is first needed to facilitate its transport in slurry form. The others are special cases which are first listed here for completeness, and further mentioned in the discussions that follow.

## INTERACTION OF CIRCULATING LOAD RATIO, CLASSIFIER BASIC SEPARATION PERFORMANCE AND WATER BALANCE

From the above discussion of factors influencing CSE, the ideal case would consist of a circuit with a very high circulating load ratio, extremely sharp basic classifier separation performance, and minimal water recovery to the classifier coarse product stream. Unfortunately, the first of these is at direct odds with the other two for the ball mill-cyclone circuit depicted in Figure 1. Table 1 shows an example cyclone solids material and water balances for 100% to 400% circulating load ratios. (Note that for the purposes of these discussions, circulating load ratio is defined as the ball mill solids tonnage divided by that of the circuit.) Cyclone overflow solids and water tonnages are assumed equal at 100 each. Cyclone underflow percent solids is assumed constant at 75%. When cyclone overflow percent solids is specified by the downstream process, and cyclone underflow is maintained near the maximum before impinging on roping, cyclone feed percent solids must increase with higher circulating load ratio. This leads to loss in the sharpness of cyclone reduced separation (Heiskanen, 1993). As well, when the amount of water in the overflow is fixed, and percent solids in the underflow is fixed, recovery of water to underflow (and ensuing bypass or entrainment of fine particles) increases with the circulating load ratio.

**Table 1. Hydrocyclone Water Balance at Different Circulating Load Ratios**

| Circ. Load Ratio | COF Water | CUF Water | CF Water | Rec. of Water to Underflow |
|---|---|---|---|---|
| 100% | 100 | 33.3 | 133.3 | 25% |
| 200% | 100 | 66.7 | 166.7 | 40% |
| 300% | 100 | 100 | 200 | 50% |
| 400% | 100 | 133.3 | 233.3 | 57% |

Thus, in this typical circuit arrangement, higher circulating load ratio is necessarily associated with loss of cyclone classifier basic sharpness of separation and increased bypass of fines. This leads to the diminishing returns (in terms of increased CSE and overall circuit efficiency) with increased circulating load ratio, as depicted in Figure 2. When increasing pumping energy and pump and cyclone maintenance costs are balanced against decreasing grinding energy and media costs, all as circulating load increases, the optimum economic circulating load can be identified, as it was at Kidd Creek (Table 2). Note that this was the first of two stages of grinding so downstream processing effects were considered negligible. The circulating load ratio in this case was lowered from an average of approximately 450% to 350% through appropriate, systematic pump and cyclone adjustments (McIvor, 1988b).

**Table 2. Optimum* Economic Circulating Load Ratio Analysis at Kidd Creek**

| C. L. Ratio, % | Relative Circuit Efficiency, % | Relative Operating Costs (c/ton processed) | | |
|---|---|---|---|---|
| | | Grinding | Pumping/Class. | Total |
| 250 | 93.4 | 57.4 | 5.2 | 62.6 |
| 350 | 98.1 | 54.6 | 6.7 | 61.3 (min.)* |
| 450 (Base Survey) | 100.0 | 53.6 | 8.2 | 61.8 |
| 550 | 102.4 | 52.3 | 9.7 | 62.0 |

## EXAMPLES OF PLANT CASE STUDIES INVOLVING WATER USAGE

The case studies that are summarized below demonstrate the effects of water usage and other of the above listed factors on ball or pebble milling circuit classification system efficiency, and subsequently on the overall circuit efficiency.

### Water Usage and Single vs. Two-Stage Cycloning at Selbaie and Dome Mines

Concurrent grinding circuit performance evaluations were carried out at two relatively nearby locations, Les Mines Selbaie and Dome Mines Ltd., in Northern Canada. All the sample analyses were carried out using the same procedures, and the same grindability test apparatus and procedures. Selbaie Survey No.2 ball mill circuit data was found noteworthy, and of ongoing use for general purpose circuit evaluations, as it corresponds to overall the Bond scale up methodology; i.e., the test work index of the ore and the operating work index of the circuit were essentially identical. The ball mill during Dome Survey No.1, on the other hand, was using approximately 30% less than the Bond specified grinding energy.

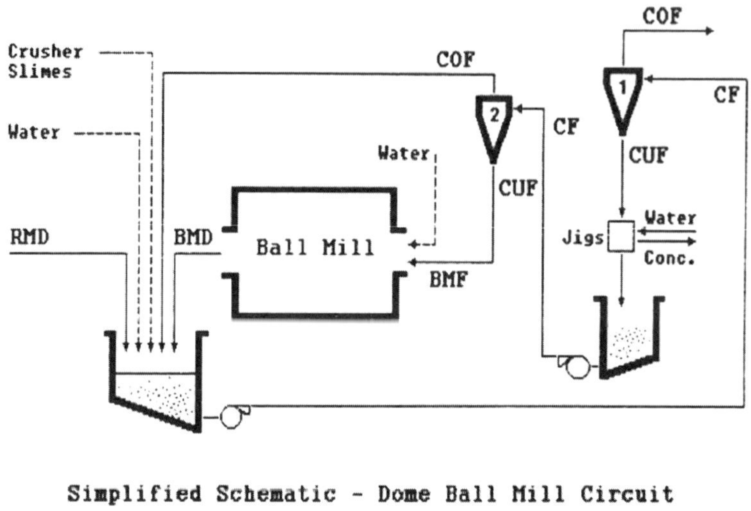


**Figure 4. The Dome Grinding Circuit**

While the Selbaie flow sheet was the same as that in Figure 1, Dome used two stages of cycloning (Figure 4). As the primary cyclone underflow underwent gravity separation for gold, large addition of water in the gravity separation circuit called for a second stage of cycloning (intended for water removal) before entering the ball mill. By functional performance analyses, the two ball mill grinding efficiencies were approximately equal. So there was no doubt that CSE was the main contributor to the excellent Bond work index efficiency of the circuit, as summarized in Table 3. A computer simulation of the Selbaie circuit (see later) showed that its CSE at the P80 could be increased to approximately 80% by water addition to a similar cyclone overflow percent solids as Dome (although this could not be implemented in the Selbaie plant due to required feed percent solids to flotation). This leaves 5 percentage points attributable to other possible ore related and circuit design differences, most notably among them two-stage classification.

**Table 3. Selbaie and Dome Circuit Survey Comparisons**

| Survey Ident. | Bond W.I. Eff. | CLR | COF % Solids (w/w) | CSE @ P80 |
|---|---|---|---|---|
| Selbaie No.2 | 101% | 333% | 42.0%(2.8 SG) | 69% |
| Dome No.1 | 134% | 313% | 21.5%(2.8 SG) | 85% |

**Cyclone Feed Pump Speed Up at Tilden**

Consecutive surveys were carried out on one of the pebble milling circuits at Tilden, interrupted only long enough to make a pump sheave change and for the circuit to stabilize. The control system was set to automatically add water to the cyclone feed sump to maintain level for this testing, and every endeavor was made to hold everything else constant. However, the pump speed increase was associated with a slight increase in tonnage. Note that the P80 was measured to be13 microns for both surveys, while CSE was traditionally evaluated at 25 um for these circuits (Table 4, McIvor et al, 2000).

**Table 4. Tilden Pebble Milling Circuit Performance with Pump Speed Increase**

| Pump RPM | CLR | COF % Solids (w/w) | CSE @ 25 um |
|---|---|---|---|
| 750 | 144 | 6.1% | 52.6% |
| 835 | 173 | 5.8 % | 54.6% |

The CSE increased by 2 percentage points, a relative increase of 4%. While small, this was determined to be several times the experimental error in the measurements, and of major significance in terms of operating cost savings.

**Higher Efficiency Cyclones and Increased Water Addition at Strathcona**

Plant testing was carried out to establish cyclone separation performances for existing and a new generation of "high efficiency" cyclones. These tests showed a significant improvement in the sharpness of the reduced cyclone separation curve. It was also notable fewer very coarse oversize particles (relative to the P80 size) were found in the cyclone overflow (flotation feed). As well, it was felt that the flotation circuit at Strathcona would be tolerant to slightly decreased percent solids, especially given the reduction in coarse particles. This was readily achievable with the existing pump because of reduced pressure drop across the new cyclones at equivalent flow rate. A plant survey and a computer simulation of circuit performance with the higher efficiency cyclones and increased water addition to the grinding circuit were carried out (McIvor and Finch, 2008). Circuit tonnage was held constant. It provided the following results (Table 5).

**Table 5. Circuit Improvements with Added Water and Hi-Efficiency Cyclones***

| | COF % Solids | Cyc. Water Rec. to U.F. | CSE @ 212 um | % - 212 um | % + 600 um |
|---|---|---|---|---|---|
| Reference Survey | 51.5% | 42% | 56.6% | 80% | 2.3% |
| Circuit Perf. Pred.* | 47 % | 38% | 60.9% | 84% | 0.1% |

The cyclones were purchased and installed, and the results have been reported generally as excellent, including unprecedented highs in overall circuit efficiencies (Tuzun, 2008). A follow up circuit survey confirmed that the expected increase in CSE was achieved.

**Evaluation of Addition of a Thickener after Grinding for Barrick Gold**

Discussions around the need to feed CIL circuits at very high (50% w/w) percent solids, and the difficulty this presents for grinding circuit CSE, led to the request for a general evaluation of the effect of water addition rate on grinding performance. The key question was whether the difference in circuit efficiency with substantially more water addition could justify the installation of a thickener between the grinding and CIL circuits. To carry out this evaluation, a very good set of grinding circuit survey data (again, Selbaie Survey No.2) was chosen as the basis of circuit performance simulations using the Finch-McIvor Circuit Modeling System (McIvor and Finch, 2008). The circuit circulating load ratio (approximately 330%), underflow percent solids and overflow product sizing (% minus 106 um) were held constant, while the cyclone water balance and sharpness of separation were changed to match the new water usage as cyclone overflow percent solids was reduced. Since the grind was held constant, the resulting mill power draw (or kWh/t) requirement is then directly proportional to the circuit efficiency. The results are shown in Table 6.

**Table 6. Grinding Circuit Steady State Conditions at Different COF % Solids**

| COF % Sol. w/w | Cyc. Feed % Sol. w/w | Cyc. Water Rec. to U.F. | Mill kW | CSE @ 106 um | Rel. Circ. Eff. |
|---|---|---|---|---|---|
| 50% | 42% | 49% | 580 | 64% | Base Line |
| 42% (B2) | 38% | 41% | 523 | 71% | +11% |
| 40% | 37% | 39% | 515 | 72% | +12.5% |
| 30% | 30.5% | 29% | 478 | 78% | +22% |
| 20% | 23% | 20% | 454 | 82% | +28% |
| 10% | 13% | 10% | 430 | 86% | +34% |

Although not yet fully tuned and optimized, and also combined with a mill feed percent solids adjustment, rearranging the flow sheet to position a thickener ahead of CIL at Tulawaka facilitated lowering the COF percent solids initially from approximately 47 to 35% solids by weight. This has resulted in achievement of unprecedented fineness of grind (6 to11 percentage points finer on 106 um) and approximately a 2 percent increase in gold recovery (Frostiak, 2007).

## NOTES ON OTHER CLASSIFICATION EQUIPMENT AND ARRANGEMENTS

In the first example above, the relative advantage of two vs. a single stage of cyclones was limited to about 5 percentage points of CSE. Although other factors that could be in play mean that this is only an estimate of the benefit attributable solely to two stage vs. single stage cycloning, it is apparent that once a high circulating load ratio and maximum water usage is achieved the remaining potential benefit of two stage cycloning appears rather limited. This complies with the early finding of Kelsall et al (1969) that classifier performance alone does not impact grinding circuit efficiency very significantly. The writer's experiences with multiple stage cycloning with different limitations on circulating load and circuit water addition have met with mixed results, leading to the present conclusion that while the upside of benefits with regard to classifier performance alone are fairly limited (when high circulating load and maximum water usage have been exploited), each case needs to be evaluated on its own merits.

Improved capacity, efficiency and non-blinding performance of new fine screening equipment provide another interesting option for grinding circuits (Valine, 2008). Screens can provide exceptionally low bypass of fines to the screen oversize, generally much higher in percent solids than that of cyclone underflow. Their separation performance can be very sharp, and it is not affected by material particle specific gravity, which can be a prevalent factor in cyclones. Their widespread use in various iron ore processing circuits attests to the importance this attribute, and they are being more and more widely and successfully used in grinding circuits that were once considered the domain of cyclones. When concentration in the ball mill of a heavy (particularly if it is a gangue) mineral can occur, screens may be the ideal grinding circuit classifier. A limitation is the circulating load ratio to be specified. Beyond about 50% of the screen feed reporting to oversize (i.e. a circulating load ratio of 100%), good separation starts to be lost, and screen loading (and therefore needed size of screening area) increases rapidly.

Before computer modeling of grinding circuits was possible, Hukki proposed that the improved classification performance provided by rising current hydraulic classifiers (versus cyclones) would yield an enormous increase in grinding circuit efficiency (1967). That this did not prove out as expected attests to the fact that the benefit of improved classifier performance will be offset by any concurrent lowering the circulating load ratio.

## NOTE ON IMPLEMENTATION PROCEDURES

It should be noted that to improve circuit performance with the factors we can manipulate requires a systematic approach to grinding circuit equipment performance design or adjustment, including the cyclones (or other classifier), the pump, and the grinding mill. Such step-by-step unit equipment process application engineering/design procedures are provided by the Metcom System (McIvor, 1989-2008).

## SUMMARY AND CONCLUSIONS

As defined here, grinding circuit CSE measured at the circuit P80 provides a direct, absolute measurement of the overall efficiency of the classification system, and is directly related to overall energy efficiency of the grinding circuit. It is a simple, valuable measurement that can, and should, be performed by all mill operators.

Closed grinding circuit CSE, and our ability to maximize it, can be regarded as a function of three grinding circuit performance characteristics; 1. the circulating load ratio; 2. the basic (or reduced) classifier separation performance; and 3. the bypass of fines at the classifier to the coarse product stream, as primarily dictated by its water balance.

Circulating load ratio is enormously important to grinding circuit efficiency. Diminishing returns dictate optimum economic operational values of approximately 350% to 400% for most primary ball milling circuits.

Cyclone overflow percent solids is enormously important to grinding circuit efficiency because of the effect it has on the cyclone water balance and the related bypass fraction of fines to the coarse product screen. Addition of a thickener after grinding ahead of CIL in order to allow dilution of the cyclone overflow can provide a 30% increase in CSE and grinding circuit efficiency. However, even when water addition is far more limited, valuable gains are achievable with every extra unit of water added.

Improved sharpness of basic (or reduced) classification equipment performance can be achieved through various means, such as new "hi-efficiency" cyclone designs, two stage cycloning, screening, etc. These will provide higher CSE, but only to a limited extent when a high circulating load ratio and a favorable classifier water balance have already been achieved.

Our available operational data base, supplemented by simulated grinding circuit data, has grown to where the maximum achievable CSE is readily definable for any grinding circuit. Primary among constraints to high CSE is permissible circuit product percent solids. Every additional amount of water added to the grinding circuit will contribute to higher CSE, as well as reduced over grinding. The purpose of this paper is to bring this opportunity to the attention of all plant designers and operators at a time when reducing energy consumption in grinding is so important.

## ACKNOWLEDGEMENTS

Graciously acknowledged are the management and staff of the following operations for their support of process development work and their generosity in sharing plant performance results; Kidd Creek, Selbaie, Dome, Tilden, Clarabelle, Strathcona and Tulawaka. The writer also thanks McGill University for ongoing support and Derrick Corporation for their input on fine screening.

## REFERENCES

Davis, E.W., 1925 "Ball Mill Crushing in Closed Circuit with Screens," *Bulletin of the_University of Minnesota School of Mines Experimental Station*, v.28, no.42, pp.14-18.

Frostiak, J., 2007, internal Barrick summary presentation on "Grinding Improvements at Tulawaka."

Gaudin, A.M., 1939, *Principles of Mineral Dressing*, First Edition, McGraw Hill, New York.

Heiskanen, K., 1993, *Particle Classification*, Chapman & Hall.

Hukki, R.T., 1967, "An Analysis of Mill and Classifier Performance in a Closed Grinding Circuit," *AIME Trans*., Sept., pp. 233-239.

Kelsall, D.F., Reid, K.J. and Stewart, P., 1969, The Study of Grinding Processes by Dynamic Modeling," *Transactions Australia Electrical Engineering*, Mar., pp. 173-186.

McIvor, R.E., 1984, "A Material Balance Calculation Procedure for Grinding Circuit Hydrocyclone Selection," *CIM Bulletin*, pp.50-53, Dec.

McIvor, R.E., 1988a, "Classification Effects in Wet Ball Milling Circuits," *Mining_Engineering*, pp. 815-820, Aug.

McIvor, R.E., 1988b, "Technoeconomic Analysis of Plant Grinding Operations," Ph.D. Thesis, McGill University.

McIvor, R.E., 1989-2008, "The Complete Metcom Training Program on Improving the Performance of Plant Grinding Operations," Metcom Technologies, Inc.

McIvor, R.E. and Finch, J.A., 1991, "A Guide to Interfacing of Plant Grinding and Flotation Operations," *Minerals Engineering*, v.4, no. 1, pp. 9-23.

McIvor, R.E., Lavallee, M.L., Wood, K.R., Blythe, P.M. and Finch, J.A., 1992, "Functional Performance Characteristics of Ball Milling, *Mining Engineering*, pp. 269-276, Mar.

McIvor, R.E., Weldum, T.P., Mahoski, B.J. and Rasmussen, R.S., 2000, "Systems Approach to Grinding Improvements at the Tilden Concentrator," *Mining Engineering*, pp. 41-47, Feb.
McIvor, R.E., 2006a, "Industrial Validation of the Functional Performance Equation," *Mining Engineering*, pp. 47-51, Nov.
McIvor, R.E., 2006b, "Ball Mill Circuit Models for Improving Plant Performance," *Advances in Comminution*, SME, S.K. Kawatra editor, pp. 539-546.
McIvor, R.E. and Finch, J.A., 2008, "The Finch-McIvor Functional Performance Based Grinding Circuit Modeling System," *Comminution '08*, Falmouth.
Tuzun, A., 2008, Xstrata Nickel internal information on grinding improvements at Strathcona.
Valine, S., 2008, Derrick Corporation internal documents and verbal discussion.

# THE BOND STANDARD FOR COMMINUTION EFFICIENCY

C.A. Rowland, Jr.[1] and R.E. McIvor[2]

[1]Deceased
[2]Metcom Technologies, Inc., 1021B W. Baraga Ave., Marquette, MI, USA

## ABSTRACT

Use of the Bond work index to size industrial grinding equipment is widely known. Bond also described its application for measuring grinding circuit efficiency. Refinements to the mill sizing procedure and its use to measure grinding efficiency have also been extensively studied by the primary author of this paper. Work index efficiency has been adopted by industry as a widely used standard. The purpose of this paper is to re-present, clarify and update the application of this tool as a standard for measuring grinding circuit efficiency.

## DEFINITION

"**standard,** ***n.*** something established for use as a rule or basis of comparison in measuring or judging capacity, content, extent, value, quality, etc.; as standards of weight and measure" (Webster, 1958).

## METHOD OF ANALYSIS

In order to quantify the efficiency of size reduction of brittle materials such as ores, three factors must be taken into account:

1. The amount of size reduction that takes place, i.e. the feed and product sizes.
2. The energy expended per unit of material reduced.
3. The resistance to size reduction exhibited by the material.

A standard method for sizing the equipment (or measuring the efficiency) of any size reduction process was created when F.C. Bond successfully established two relationships. The first was the general relationship between energy (per ton) required for size reduction and the process feed and product sizing. He termed this "the third law of comminution" (Bond, 1952), the key discovery being the value of the exponent, 0.5. This first required that a size distribution be represented by a single value. The eighty percent passing size may be considered meaningful because it is close to the modal value of many observed (impact dominated) crushed or ground material size distributions (for example, see Gaudin, 1939). It was first proposed to represent a material's size distribution by Prof. Taggart in a 1941, and then used in his 1945 *Handbook of Mineral Dressing*. It was considered and subsequently adopted by Bond (1952) as an essential enabler of his "third law" equation.

The second needed relationship was the scale-up from a laboratory measurement of grindability to the grinding performance of large-scale equipment. Despite scatter in the data, Bond found there was an approximately fixed factor between the third law equation constants (the work index) for numerous plant versus laboratory data. He made the two equal, on average, by applying what he observed to be the needed "lab to plant" factor for the laboratory value (into the equation used to calculate the work index from the lab test data). This process was no coincidence, as proper sizing of grinding mills was considered to be an important aspect of Allis-Chalmers' business as an equipment provider. A procedure for plant mill sizing from the Allis-Chalmers laboratory grindability test already existed (Michealson, 1944). It was based on lab versus plant circuit product size material production rates. But only with "the third law" could any variation in circuit feed and product sizing now be accounted for in calculating the full scale equipment grinding energy requirement.

It was no small feat that a number of tiered interrelationships were discovered. They were then combined, by Bond to provide his equation for plant comminution specific energy requirement, W (Bond, 1961).

$$\text{Predicted } W = WI_{lab} \times \left( \frac{10}{P80^{0.5}} - \frac{10}{F80^{0.5}} \right)$$

One of the most common (also read 'standard', or 'conventional') circuits of his day consisted of stages of crushing and screening, followed by open circuit rod milling and closed circuit ball milling. Separate laboratory tests provided a WI (work index) value for each of crushing, rod milling and ball milling, each over their respective, normally practiced (or 'standard') size ranges. So via three laboratory WI tests, a fixed (or 'standard') amount of energy per ton of ore to be processed was specified through each equipment stage, and of course the total quantity that would be required for the total conventional plant circuit. That quantity is the *Bond standard specific energy requirement* for comminution processes.

Of course, there is an underlying assumption about efficiency in this scale-up process. Some better designed and operated circuits will use less energy than others, which would account for some of the scatter mentioned above. But for the average (and some specific cases) the full scale equipment will perform as predicted, i.e. the WI measured in the lab will equal the WIo calculated for the plant (Rowland, 1976).

$$\text{Plant } W = WIo \times \left( \frac{10}{P80^{0.5}} - \frac{10}{F80^{0.5}} \right)$$

So in general, according to Bond, WIlab = WIo. This equation expresses the "standard" that plant operating work index equals the laboratory work index.

Also, WIlab / WIo = 1.0 (or 100%). This occurs when the Bond scale-up relationship is exactly "spot on," or when the average efficiency measured in the numerous plants that he observed (to establish his lab scale-up factor) equals the efficiency of the circuit in question. "100% work index efficiency" once again means that the circuit, whatever it happens to be, conforms to the Bond standard in terms of energy usage. This ratio will exceed 100% for more efficient circuits, and of course be less for those that are less efficient. It will also change with process changes, and it is by monitoring gains in this value as plant optimization projects are executed that overall circuit efficiency benefits can be quantified. Importantly, this efficiency closely tracks inversely with grinding costs (typically dominated by energy and media), so cost-benefit analysis of related capital expenditure is therefore facilitated (McIvor, 1988).

In carrying out this efficiency determination, the nature of the industrial circuit does not matter, but only that the range of size reduction is covered by the respective Bond tests. Neither does it matter that the Bond rod mill work index test is performed in a laboratory rod mill, but rather only that the size reduction taking place is from the specified test feed size to the specified test product size. Also, in practice, the coarse (crushing) energy is often such a small part of the total that assuming the rod mill work index applies over the crushing size range does not introduce a significant error in the combined circuit efficiency determination.

There is a fairly common misconception that the "correction factors" used for mill sizing should be taken out of the plant operating work index value before comparison with the laboratory work index in the circuit efficiency calculation. In fact, Bond stated emphatically that they are not to be used for plant efficiency calculations (for example, in 1952 and 1957). Rather, they may be used to either diagnose possible causes of observed poor performance, or to study and make refinements to the basic mill sizing procedure. It is only in the latter context that Rowland has prescribed their use (1976, 1982, 1988 and 2007).

**EXAMPLE**

A SAG-ball milling circuit processes ore from primary crushing, with a SAG circuit F80 of approximately 165,000 um, to flotation feed, with a cyclone overflow P80 of 125 um, as measured during a plant survey. The total specific energy input of the two milling stages was 14.6 kWh/t. The combined circuit operating work index is calculated as follows.

$$14.6\ \text{kWh/t} = \text{WIo} \times \left( \frac{10}{125^{0.5}} - \frac{10}{165000^{0.5}} \right)$$

Solving, WIo = 16.8 kWh/t.

Circuit feed samples from the same survey were sent out for Bond work index testing. The rod mill test done with a closing screen of 1,180 um yielded a value of 14.5 kWh/t, and the ball mill test done with a closing screen of 150 um yielded a value of 13.8 kWh/t. Previous testing showed that the crushing work index was approximately the same as the rod mill work index. If we accept that the rod mill work index applies from the circuit feed size to the rod mill work index test product P80 size (of 1050 um), then the standard Bond energy calculated to reach this size is as follows.

$$W = 14.5 \times \left( \frac{10}{1050^{0.5}} - \frac{10}{165000^{0.5}} \right) = 4.1\ \text{kWh/t}$$

Then from 1050 to 125 um, apply the Bond ball mill work index to calculate the energy to reach the circuit product size.

$$W = 13.8 \times \left( \frac{10}{125^{0.5}} - \frac{10}{1050^{0.5}} \right) = 8.1\ \text{kWh/t}$$

The total Bond standard energy for this amount of size reduction is the sum of these two, 12.2 kWh/t. This can be directly compared to the actual energy usage of 14.6 kWh/t.

WIeff = 12.2 / 14.6 = 84%

Alternatively, we can calculate the combined test work index of this ore as:

$$12.2 = \text{Combined WIlab} \times \left( \frac{10}{125^{0.5}} - \frac{10}{165000^{0.5}} \right)$$

Solving, Combined WIlab = 14.0 kWh/t.

Again, comparing the combined test work index to the overall operating work index, 14.0/16.8 = 83%, the circuit work index efficiency (identical to the above within rounding error). This value is the relative efficiency of this circuit compared to the prescribed Bond standard energy for a 'conventional' comminution circuit. Measured values for industrial circuits have been seen to vary as much as +/- 35% (for example, see McIvor, 2006). Usually the variations are smaller than this, as described by Rowland (1973 and 1976).

## FURTHER OBSERVATIONS AND CONCLUSIONS

A weakness of this analysis is that the typical size distribution represented by the 80 percent passing size was that from closed circuit crushing, rod milling, and closed circuit ball milling. These size distributions can be quite unlike those from autogenous, semi-autogenous or high-pressure grinding circuits. This limitation can sometimes be dealt with by a 'phantom cyclone' (Barratt, 2007), or an estimated 'effective' versus actual 80 percent passing size (MacPherson, 1989). However, this does not normally affect overall circuit efficiency (from primary circuit feed size to final grinding circuit product size) calculations.

It is acknowledged that beyond general observation, this metric does not provide any specific indication of causes of differences in circuit efficiencies, or means of improving them. The exceptions are general correlations that lead to the development of "inefficiency factors," as described by Bond (1961) and Rowland (1982). Tracking work index efficiency changes with process changes can be used to monitor progress in open circuit grinding studies, but the details of grinding process optimization are more the realm of other circuit modeling tools (McIvor, 1988).

Depending on the particular circuit, the reproducibility of the calculated work index efficiency can be very reasonable, approximately plus or minus 5% (relative) under ideal conditions, i.e. using the same plant instruments, well taken samples, and the same laboratory for Bond work index testing (Laplante et al, 1988; McIvor, 1990). The reliability of the third law equation is not usually significantly challenged when it is applied to evaluate the performance of the same plant circuit under different conditions, as the circuit feed and product sizes do not normally change significantly. However, comparisons between different circuits, and involving different laboratories (for example, see Kaya et al, 2003), at least doubles this potential measurement error. This error must be examined and estimated for each application.

Finally, it is sometimes noted that Bond's third theory, as he originally described it as dependent on crack length propagation, does not hold up under detailed theoretical scrutiny. In fact, Bond acknowledged this, granting favor rather to only its empirical correctness in his final (posthumously published) scientific writing (Bond, 1985).

This paper has outlined the application of the "Bond work index," beyond its well known use for industrial equipment selection, to its use as a standard for quantifying industrial size reduction efficiency. These two uses of the same tool, in fact, differ only in purpose, both relying on the same fundamental relationship between plant and laboratory data. That the Bond work index has been so successfully applied for more than half a century attests to its value as a powerful process engineering tool.

## REFERENCES


Barratt, D.J., 2007, presentation at McGill University Mineral Processing Development Seminar, Montreal.

Bond, F.C., 1952, "The Third Theory of Comminution," *Trans. A.I.M.E.*, v. 193, pp. 484-494.

Bond, F.C, 1957, "Non-Cataracting Ball Mill Study," Engineering and Mining Journal, March, pp. 98-105.

Bond, F.C., 1961, "Crushing and Grinding Calculations," *British Chemical Engineering*, June & Aug., pp. 378-385 & 543-548.

Bond, F.C., 1985, "Testing and Calculations," Ch.5, Section 3A, *SME Mineral Processing Handbook*, N.L. Weiss editor, p.3A-18.

Gaudin, A.M., 1939, *Principles of Mineral Dressing*, McGraw-Hill, New York, pp. 125-129.

Kaya, E., Fletcher, P.C. and Thompson, P., "Reproducibility of Bond Grindability Work Index," *Minerals and Metallurgical Processing*, v. 20, no. 3, pp. 140-142.

Laplante, A.R., McIvor, R.E. and Finch, J.A., 1988, "Error Analysis for Bond Work Index Determinations, Part 1: Accuracy and Reproducibility," *Minerals Engineering*, v. 1, no. 2, pp. 113-125.

MacPherson, A.R., 1989, "Autogenous Grinding 1987 – Update," CIM Bulletin, Jan., pp. 75-82.

McIvor, R.E., 1988, "Technoeconomic Analysis of Plant Grinding Operations," Ph.D. thesis, McGill University.

McIvor, R.E., 1990, Skills Application Workshop Training Module on "Measures of Industrial Grinding Efficiency."

McIvor, R.E., 2006, "Industrial Validation of the Functional Performance Equation for Ball Milling and Pebble Milling Circuits," *Mining Engineering*, Nov., pp. 47-51.

Michaelson, S.D., 1944, "Determination of Ball-mill Size from Grindability Data," A.I.M.E. Technical Publication No. 1844.

Rowland, Jr., 1973, "Comparison of Work Indices from Operating Data with those from Laboratory test Data," *Proceedings of 10th International Mineral Processing Congress*, London.

Rowland, Jr., C.A., 1976, "The Tools of Power Power: The Bond Work Index, A Tool to Measure Grinding Efficiency," SME-AIME Fall Meeting, Denver.

Rowland, Jr., C.A., 1982, "Selection of Rod Mills, Ball Mills, Pebble Mills and Regrind Mills," Ch. 23, *Design and Installation of Comminution Circuits*, SME, pp. 393-438.

Rowland, Jr., C.A., 1988, "Using Bond Work Index to Measure Operating Comminution Efficiency," *Minerals and Metallurgical Processing*, Nov., pp. 32-36.
Rowland, Jr., C.A., 2007, personal communication.
Taggart, A.F., 1941, "New Units of Crusher Capacity and Crusher Efficiency," *Trans. A.I.M.E.*, v. 153, pp. 319-323.
Taggart, A.F., 1945, *Handbook of Mineral Dressing*, John Wiley & Sons, New York, p.4-10.
Webster, N., *Webster's New Twentieth Century Dictionary of the English Language*, World Publishing, Cleveland and New York, p.1772.

# IN-PIT CRUSHING—THE MOVE TO CONTINUOUS MINING

Ronald W. Utley

Consultant, FLSmidth Inc. Bethlehem PA, USA

## ABSTRACT

The direct dump, semi-mobile IPCC with a gyratory crusher has been the plant design of choice selected by mine planners for hard ore for the past several years. For soft rock applications including; overburden, coal and oil sands fully-mobile continuous mining systems are becoming increasingly attractive. The crushers most often selected for these applications are low-speed sizers and double roll crushers. Fully-mobile continuous mining systems are investigated along with opportunities to transfer this technology to hard ore crushing and conveying.

## INTRODUCTION

The types of in-pit crushers that are usually being reviewed by mine planners for hard ore are: fixed plants mounted at the rim of the pit and semi-mobile and fully-mobile plants within the pit. The gyratory is the crusher of choice for capacities over 2500 mtph. For soft rock applications including; overburden, coal and oil sands fully-mobile continuous mining systems are becoming increasingly attractive. The crushers most often selected for these applications are low-speed sizers and double roll crushers. The two types of conveyor systems being selected for these systems are the conventional conveyor and high angle conveyor systems. The high angle conveyor system has yet to put in operation in high tonnage mining operation.

## IPCC DEVELOPMENT

A series of events have taken place since 2006 which is changing IPCC again.

- The larger hard rock mines of the world are operating at or approaching 10,000 mtph. Overburden for some of the hard rock mines are exceeding 10,000 mtph and soft rock mines including coal and oil sands are approaching 14,000 mtph.
- Unlike, in the past, just when the IPCC could have been justified from an economy position, truck manufactures would introduce larger haul trucks. It appears that truck sizes have reached their maximum of 400 tons while shovels have continued to get bigger. Truck manufacturers have stated that 500 ton truck is possible, but to date not built. The development of an efficient mine plan can only be done through an analysis of the integration of material removal, loading, hauling, and dumping. With the larger shovels the largest trucks are now the limiting factor in optimizing load and haul production. For large production mines with long haul distances the IPCC has become even more attractive.
- The hard rock crusher of choice, the gyratory crusher, has either reached or is very close to reaching its maximum practical size. Beginning in 1985, the 1.52m x 2.26m (60" x 89") was the crusher of choice for IPCC. The 1.52m x 2.97m (60" x 117") followed in 1990. Manufacturers have been offering the 1.83m x 3.25m (72" x 128") machines for IPCC, but there has not been an order to date.
- The double roll crushers and low speed sizers have been developed for soft rock, coal and oil sands for capacities of 12,000 mtph and designed for 14,000 mtph.
- The world went through another round of oil price increases which in turn caused an increase in mining costs based on fuel for the truck fleets. A reminder that the best way to assure consistent power costs is with electric power as opposed to oil.
- Deliveries of truck tires for the fleet of mining haul trucks extended to over two years, threatening the shutdown of operations.
- The concern over global warming has heightened the awareness of minimizing $CO_2$ emissions. These requirements favor IPCC in general and IPCC continuous mining in particular.
- IPCC plants have been required to add dust suppression and/or dust collection.
- Top Service gyratory crushers became available which provides a less complicated plant design—no need for an eccentric cart and rails. By eliminating the eccentric cart there is an opportunity for lower overall plant height. There is a safety benefit—no need to perform maintenance from below the crusher. The countershaft is higher in the machine resulting in less out-of-balance force.

## CONTINUOUS IPCC SYSTEM DEVELOPMENT

The field of in-pit crushing has come full circle in the past couple of years. Back in 1956 Krupp supplied the first fully mobile crawler mounted crushing plant to a client in Germany. The machine was fed directly by a rope shovel, incorporated a short inclined apron feeder feeding a single shaft hammer crusher and operated at 250 mtph.

Numerous mobile plants would be built during the following years, fed by either a rope shovel or front end loaders. Design capacity was limited to the capacity of the mobile equipment which, for most of the past 50 or so years was in the order of 2,500 mtph.

Geographically it is interesting to see how the countries of operation looked at ore handling. In Europe continuous mining (shovel/crusher/conveyor) has always been the preferred method of operation and few mines incorporated discontinuous methods (truck/shovel). In many cases the mobile plant would be fed direct by the shovel so design capacity was limited to the capacity of the shovel. However, note than many of the operations in Europe are lignite mines where the bucket wheel excavator is a more suitable tool for mining and, thus, hourly capacities far exceed the capacity of shovels and front end loaders.

In North America it is just the opposite, truck / shovel systems are predominant due to many factors, one being that the truck is king in North America, another being that the mines are predominantly hard rock where it is relatively difficult to install a continuous mining system. Granted there are many continuous mining systems installed in North America, the capacity is still limited to about 2,500 mtph.

For many years the truck payload was relatively low by mining standards so a truck / shovel operation was limited by the capacity of the weakest link, this being the logistics of moving trucks from the shovel to the process plant and the time required to spot and dump at the fixed crushing plant.

In the early 80's the trucks had increased in size from 165 ton to the point where a payload of 200 tons was considered routine. However, the cost of fuel was rising and operators were looking for a more efficient means of reducing costs and increasing capacity. The solution was a high capacity semi-mobile crushing plant located in the mine coupled with a series of fixed / relocatable conveyors that would run the ore from the pit to the process plant. Truck haulage was greatly reduced and, accordingly, fewer trucks were required.

Today, the semi-mobile crushing plant can be found in many large mines worldwide and the design capacity remains in the range of 6,000 to 9,000 mtph. Capacity is now limited by the crusher element itself as truck payload has increased to 400 tons and, this, truck cycle time is not a governing factor.

Recently the very high costs of operating trucks (fuel, labor, tires, maintenance costs and replacement costs) have operators looking for alternative solutions. The answer is not an easy one because a truck system tends to have very high availability for the reason that a stand-by truck can easily be called in to replace a truck requiring service. In a continuous mining system each component is depend on the other so if one element is out of service the entire system is down. However, the costs of a continuous system tend to be much less than a discontinuous system so continuous mining schemes are becoming more favorable with many operators.

Considering the above, manufacturers have been developing high capacity shovel/mobile crusher solutions with design capacities approaching 12,000 mtph, limited only by the capacity of the shovel or the capacity of the downstream process plant. The mobile crushing plants need to be as mobile as the shovel so that they do not limit the performance of the shovel. The conveyors need to be shiftable design so that they can be relocated as the bench increases in width. To remain efficient this style of system suits itself to a mine layout where the benches can be long and straight, typical of coal, oil sand and overburden removal operations. This system would be difficult to incorporate in a conventional hard rock mine where the pit design is a deep spiral.

To increase the flexibility of the system one can consider a short (60 m) crawler mounted conveyor downstream of the mobile crushing plants. This would increase the range of operation of the shovel/mobile crushing plants without requiring a relocating of the shiftable face conveyor. To increase the working range one more step one can consider a mobile bridge conveyor typical of the systems RAHCO supply to the large heap leach operations. These machines typically have a lengthy of 300 m which would then greatly increase the operating time before the face conveyor would need to be relocated.

Shovel/mobile crushing plants will gain acceptance with the coal operators (for both ore and overburden) and other mines where they can create long working faces. The hard rock mines will have difficulty with high capacity mobile crushing plants and will continue with semi-mobile crushing plants and, overtime, will rely less and less on fixed crushing plants all in their effort to reduce haulage costs.

## TYPES OF IN-PIT CRUSHING SYSTEMS

### Fully-Mobile Crushing Plant

The fully-mobile crusher is mounted on a steel platform and is self-propelled. The platform houses all auxiliary equipment and sub-systems to operate the crusher. The platform is self-supported and rests on the mine floor. The crusher is located at the working face to minimize truck or front-end loader haulage. Wheels, crawlers or pneumatic pads are integrated into the platform and drive power to move the equipment is included on board. The planned frequency of moves for a fully mobile crusher is between one day and one week.

### Fully Mobile—Continuous Crushing System (IPCCS)

The fully mobile continuous crusher is mounted on a steel platform and is self-propelled. The platform houses the apron feeder, crusher, discharge conveyor and all auxiliary equipment and sub-systems to operate and propel the crushing plant. The feed hopper needs to accept run-of-mine material that is dumped from shovels and/or drag lines. The platform is self-supported and rests on the mine floor. The crusher is located at the working face for direct feed. Crawlers are integrated into the platform and drive power to move the equipment is included on board. The plant moves in tandem with the shovel. The conveying system moves in tandem with the IPCCS.

## OPERATION OF IPCC CRUSHING PLANTS

### Operation: Fully-Mobile Crushers

The fully-mobile crusher is mounted on a steel platform and is self-propelled. Wheels, crawlers or pneumatic pads are integrated into the platform to move the station. The platform houses all auxiliary equipment and sub-systems to operate the crusher. The platform is self-supported and rests on the mine floor. The crusher is located at the working face to minimize truck or front-end loader haulage. The planned frequency of moves for a fully mobile crusher is between one day and one week.

The majority of fully mobile in-pit crushing stations utilize an apron feeder to lift ore into the feed opening of the primary crusher. The use of an apron feeder allows for the crusher station to operate at grade and move without the need for employing self propelled modular transporters (SPMT's). For hard rock, including aggregate and cement, the capacity has been limited to about 2500 mtph with gyratory crushers. The 1.37m x 1.88m (54" x 74") gyratory crusher is the largest unit used for this application.

Mobile crushing plants are favorable installed under the following conditions:

- Clear and undisturbed geological situation (almost even and horizontal coal and waste rock layers).
- Straight benches as long as possible for shiftable face conveyor installation.
- Long term mine planning for the design of face conveyors, collection conveyors at bench end (side slopes) and connecting conveyors to waste dumps or processing plants.

Advantages of fully-mobile crushers include:

- Elimination of truck transport.
- Reduced number of personal.
- Avoidance of high truck maintenance costs,
- Reduction of mine traffic.
- Improvement of the overall safety situation in the mine.

Disadvantages of fully-mobile crushers:

- Increased total capital cost
- Increased maintenance costs associated with adding an apron feeder
- Increased maintenance costs associated with the crusher resulting from using an apron feeder. Due to the nature of the feeder, ore tends to impinge upon small areas within the crushing chamber, causing premature localized wear of the concaves and mantles. The use of alloy steels has mitigated the problem, however, the cost of alloy steel components remains higher than manganese steel and availability is limited.

### Operation: Fully-Mobile Continuous Crushing Systems (IPCCS)

The fully-mobile continuous crushing system is a system with the fully-mobile crushing plant as one component of the system. The components of the system are; the shovel or dragline, the fully-mobile crushing plant and the conveying system that transports the crushed material to the next operation in the flowsheet. For fully-mobile continuous crushing systems handling overburden the conveyors take the material to a dump. For coal and oil sands the material is transported to the processing plant.

The fully-mobile crushing plant takes the material directly from the shovel or dragline. The full bucket load is discharged into the hopper of the fully-mobile crushing plant. An apron feeder elevates the material and discharges it into the crusher. The crusher is a low-speed sizer, a high speed double roll crusher or a hybrid roll/sizer. The product from the crusher is transported to the overland conveyor system. All of the parts of the system are designed to work continuously. This requirement has led to the development of fully-mobile crushing plants that have the ability to continue operation while moving. Older systems in oil sands in Canada and coal in Australia had to stop operations, lift the apron feeder off the ground, move to a new location, set the apron feeder down and resume operation.

Mobile crushing plants are able to work at one, two or three benches with only one shiftable bench conveyor (multi-bench-operation). An important design criterion for this effective technology is the transfer equipment to the bench conveyor which has to bridge not only distances in length but also in height (crusher discharge conveyor, mobile transfer conveyors, mobile transfer bridges). Ramps in the working face or in the side slope are necessary for equipment relocation from on bench to the other.

## FULLY-MOBILE CONTINUOUS MINING SYSTEMS

### FLSmidth

The Fortescue Metals Group installed two FLSmidth Minerals at-the-face fully-mobile crushing systems at their Cloud Break iron-mine in Western Australia. Each complete system has an 8000 mtph capacity, with each incorporating two 4000 mtph Triple Track Mobile Sizers (TTMS). The sizer arrangement with two feed hoppers per TTMS provides great flexibility for the operators. The machines are loaded using Terex O&K 550 ton PH340 hydraulic excavators. The systems are employed for the removal of overburden.

The TTMS design is collaboration among three FLSmidth companies (FLS RAHCO, FLS KOCH/MVT and FLS ABON) with FLS RAHCO providing the crawler tracks, control system and overall system and component integration as part of its design. The units are unique in having dual hoppers and apron feeders to allow greater loading flexibility. The apron feeders carry the material up and into the sizer. Each feeder is driven by two direct hydraulic motors which provide low speed and high torque. The product from the sizer is transferred to twin hoppers on an 80 meter long mobile bridge. The bridge, in turn, transfers the sized overburden to a 45 meter mobile bridge, then onto a 350 meter mobile stacking conveyor. In addition, the TTMS is fully mobile meaning that it can move under load and without having to shut down the sizing operation during relocation.

| **Type of IPCC Plant** | **Fully-Mobile Crushing Plant** |
|---|---|
| Location | Fortesque, Australia |
| Crushing Design Capacity | 4000 mtph per Sizer |
| Material Crushed | Overburden |
| Crusher Type | Low-Speed ABON Sizer |
| Method of Feeding | Apron Feeders |
| Method of Discharge | Belt Conveyors |
| Crawlers | RAHCO w/Berco Undercarriage |

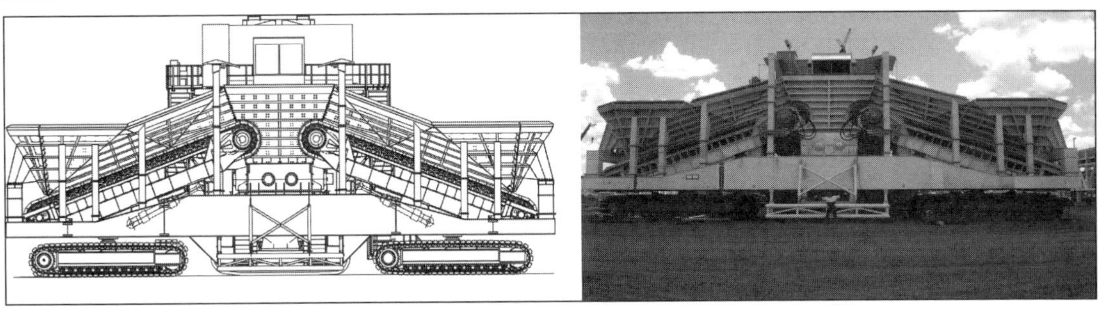

### P&H

P&H mining Equipment is developing a 9,000 to 10,000 mtph capacity IPCC for handling overburden. The heart of the P&H IPCC is a fully-mobile P&H shovel carriage equipped with DC motors and planetary probe transmissions.

**Sandvik Fully-Mobile Crushing Station**

Sandvik has developed a new fully-mobile crushing station, the PF300. The plant can be equipped with a sizer, double roll crusher or hybrid-crusher and is crawler-mounted with a feed hopper and apron feeder. The PF300 can work in combination with a belt wagon, mobile transfer conveyors or load bridges, at the temporary support, keeping the system fully mobile but fully stable under all conditions. The stability concept of the PF300 is based on having two sets of crawlers. One set is located at the hopper area where impacts loads must be expected to result from feeding by the shovel, and the second set is located under the crusher.

The design is aimed at low weight. Due to the low weight approach, the ground pressure of the machine is kept quite low and will be suitable for low ground pressure applications. The two sets of crawlers at the tail end and the front end of the machine reduces the differences between the maximum corner pressure and average ground pressure under the crawler.

The PF300 is designed to be loaded with all types of shovels whether hydraulic excavator or rope shovels. The size and height of the hopper are adapted to the shovel size. The height of the hopper edge takes into account the bottom opening of a rope shovel to avoid interference while slewing. The capacity of the PF300 is aimed at 10,000 mtph but higher and lower capacities will be available. The upper end of the capacity is expected to be 14,000 mtph.

**Takraf**

The Clermont Fully-Mobile Track Mounted In-Pit Crushing System was manufactured by TAKRAF. The Clermont Deposit is located in the Bowen Basin region of North Queensland Australia. The system is handling overburden containing 5% basalt with a compressive strength of 175 MPa. The crusher is an ABON sizer 16/350CCTD which is the reportedly the largest sizer built to date. The crushing plant is fed by dragline at a nominal rate of 10,000 mtph and a design peak of 12,000 mtph. The crushing plant will be located in a hopper slot where it will be fed by rear truck dumps for the first two years of operation. After the initial two year period, the system will then become fully-mobile and will be fed by a single P&H XPC 4100 Electric Face shovel for approximately 15 years.

| Type of IPCC Plant | Fully-Mobile Crushing Plant |
|---|---|
| Location | Clermont, Australia |
| Crushing Design Capacity | 12,000 mtph |
| Material Crushed | Overburden |
| Crusher Type | Low-Speed ABON Sizer 16/350 CCTD |
| Feed Size | 2050 x 2000 x1500 mm |
| Product Size | 350 mm P95 |
| Method of Feeding | Apron Feeder |
| Method of Discharge | Belt Conveyor |

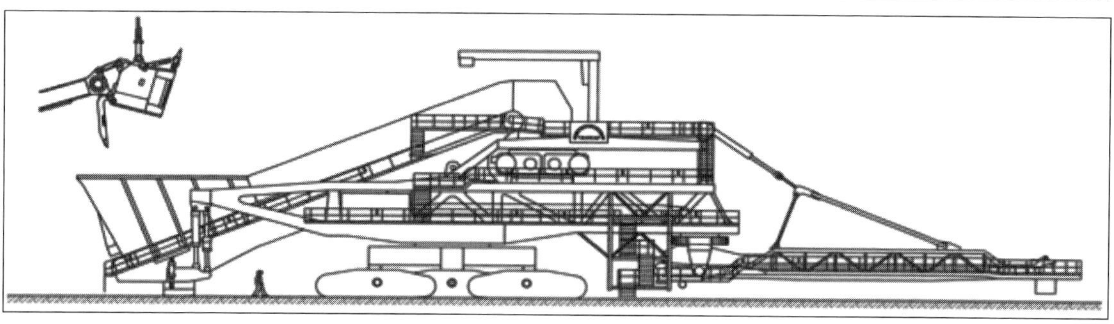

Mobile Crushing Plant 12,000 t/h (TAKRAF)

**THYSSENKRUPP**

The world's first fully-mobile continuous mining IPCC system has been operating at the Yiminhe open-pit coal mine in China since the fall of 2007. The crushing processed ROM coal at the rate of 3,500 mtph. The company is now building four similar systems for The China Power Investment Group for production in the Baiyinhua field in Inner Mongolia. The crushers move in conjunction with the power shovels. The crusher moves on a single two crawler track. The feed hopper does not need to be supported during the loading by the shovel. After crushing, the material is transported away on conveyors, eliminating the need for truck haulage in the pit. This reduces the operating costs and significantly lessens the environmental impact. The systems will each comprise a fully-mobile crusher, a crawler-mounted mobile transfer conveyor with two conveyors, and a rail-mounted hopper car with cable reel car. Three of the four systems will be used to remove overburden at a rate of 6,000 mtph and the fourth system will handle coal at the rate of 3,000 mtph.

**THE FUTURE BEYOND 2009**

The standard 400 ton truck of 2009 and beyond will likely to continue to feed direct dump semi-mobile crushing plants with the standard 1.52m (60") gyratory crushers with 1200 mm (48") feed. There is the ever present possibility that the ore will be fed to the crushing plants utilizing 1.83m (72") primary gyratory crushers or some other type of primary crusher capable of size reduction of ROM material larger than 1500 mm (60"). Efficiency will increase as ore grade decreases. Capacities will increase as a result of the larger crushers while, at the same time, operators will take advantage of the larger crusher to obtain the economy of scales.

The type of in-pit crushing plant has evolved from fully-mobile to semi-mobile to semi-fixed and in many cases to rim mounted stationary structures in the period 1980 to 2001. This progression from fully mobile in the 1970's and early 1980's to less mobility throughout the balance of 1980's and into the 21st Century would seem like a step backwards, considering that the original premise of in-pit crushing was to reduce dependence on truck haulage. The evolution if the in-pit crushing plant can not be considered by itself but must be considered in relation to other changes in the size reduction of ROM ore in the pit and delivering to the secondary stage of comminuation.

- The platform for the development of IPCC in 1980 was the "typical" multi-bench mine.
- The premise was that the design of the plant should be one bench high, i.e., 13.72m (45 feet) for slope stability, economy of dump pocket design and safety.
- Inserting an apron hopper between the dump pocket and the primary crusher could attain the dump height.
- In 1980 the thought was that the crushing plant would be moved every time that the pit became deeper by one or two benches.
- The "typical" mine would be in a position to move the crushing plant in less than 5 years.
- The plant move could be accomplished in as few as 3 days.
- The truck size in 1980 was 165 ton...

The development of the 240-ton, the 300-ton, the 350-ton truck and now the 400-ton haul truck has had a profound effect on the in-pit crushing plant designer and operation of in-pit crushing systems.

- With the larger trucks, the dump pocket was no longer a simple structure.
- The economies of scale and efficiencies of the new larger haul trucks negated the original parameters for moving the plant for every one or two benches of depth, or every few years of operation.
- The operational advantages for the use of apron feeders to elevate the ROM ore was largely gone due to the cost of the dump pocket and the limited number of move required in the life of the mine.
- Maintenance cost showed that the operating costs of the inclined apron feeder were equal to or more than the maintenance cost of the gyrator crusher.
- The capacity of the gyratory crusher was actually less when operating with an apron feeder than when direct dump into the crusher.
- The product from the gyratory crusher is finer when direct dumped than when apron feeder fed.
- Concave and mantle costs are higher when the crusher is fed with the apron feeder.
- Average capacities increased from 3000 to 9000 mtph.

The net effect of these changes was the progression that evolved into semi-mobile direct dump crushing plant designs.

- Plant designs that make use of the slope topography of the pit.
- Plant relocation of 3-5 years of more
- Capacities to 9000 mtph.
- Direct dump formats
- 60" primary gyratory crushers

This mismatch in shovel size, truck size and crusher size will likely manifest itself in fundamental changes to the primary crusher and/or improved and innovative crushing plant designs rather than the installation of the larger and heavier gyratory crushers.

Four of the nine types of primary crushers have achieved operating capacities of 10,000 mtph.

- Gyratory crushers are operating at or near these capacities on copper oxide ores in Chile. A 10% increase in capacity would be achieved with the next size of primary, assuring continuous operation at the 10,000 mtph level. The question for mine planners is if the 10% increase in capacity and the increase in feed size from 48" x 0 to 60" x 0 are economically justifiable.
- Low speed sizers have attained capacities of 14,000 mtph operating in oil sands projects in Canada as well as in the coal fields in Australia.
- Double roll crushers have also attained capacities of 14,000 mtph operating in oil sands projects in Canada.
- Gyratory crushers can crush material with compressive strengths as hard a 600 Mpa (90,000 psi).
- Low speed sizers are crushing material with compressive strengths of 200 Mpa (27,500 psi) on a continuous basis.
- The giant double roll crushers are currently limited to 100 Mpa (14,500 psi).

The designers of the IPCC crushing plants are constantly working toward shorter, more cost effective facilities. The low speed sizer with its inherent low profile design and proven high capacity in crushing materials with characteristics similar to copper oxide, appear to provide the most realistic prospect of transfer of technology for future mining IPCC facilities.

Comparison of the "next larger size" gyratory with a low speed sizer.

- ROM—10,000 mtph
- Top Size—1500 mm (60")
- Product—80% - 9"

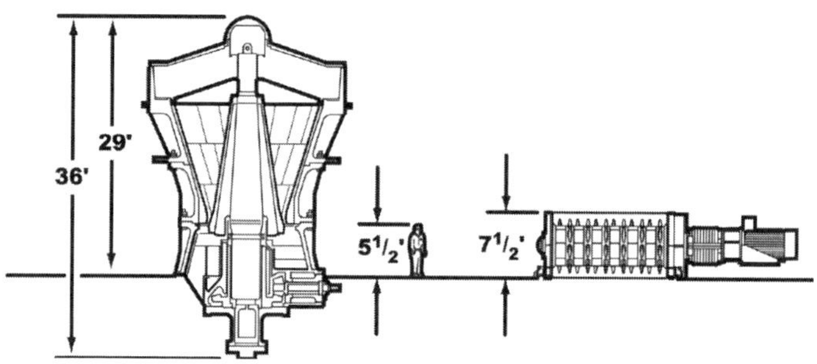


Gyratory Crusher 72 - 124
Weight: 635 tons

Low Speed Sizer 16/350
Weight: 155 tons

There is an emerging market for high production, over 9000 mtph, IPCC systems. This is especially true for the removal of overburden. It no longer makes sense to use large volumes of assets such as trucks, tire, and manpower to move a waste product which has no value. The ability to process high mtph of soft overburden is made possible with the use of high capacity crushers developed for "soft" material, usually less than 125 Mpa. These crushers have a more compact design and significantly lower capital cost than the typical hard rock crusher. These "soft" rock crushers like feeder breaker or sizer make them perfect for the overburden IPCC system.

Sizers in the not to distant future will be crushing harder materials. Sizers are already crushing limestone with compressive strengths of 150-160 MPa. (22,000-23,000 psi). The Lihir Gold Mine ore is between 250-300 MPa and is crushed with a gyratory crusher. In addition to the hard ore, the mine also has wet sticky volcanic ore that is crushed with a sizer. When the gyratory in being maintained the hard ore is run through the sizer. A bauxite mine in Australia crushes bauxite at the rate of 2,500 mtph with 100 MPa ore and 400 MPa dolerite inclusions. The history learned from these and other installations gives confidence that the sizers will be able to handle overburden and ore with 50% 150-200 MPa material at rates of 16,000 mtph.

**ACKNOWLEDGMENTS**

The author is grateful to Sandvik, P & H, Man Takraf and ThyssenKrupp for personal input on continuous in-pit crushing used in this paper.

**REFERENCES**

JOHNSON, R.N., 1980, In-Pit Crushing and Conveying with Mobile Crushers, US Bureau of Mines, Contract No. J0295004, GARD, Inc. Washington, DC, 211 pp.

JOHNSON, R.N., FRIZZELL, E.M., and UTLEY, R.W., Mobile In-Pit Primary Crushers, Mining Congress Journal, Vol. 68, No. 5, pp. 39-30.

**BIBLIOGRAPHY**

UTLEY, R.W., 1997, In-Pit Crushing—A Product of Evolutionary Change—1997, American Mining Congress, Orlando, Florida.

UTLEY, R.W., 2002, Selection and Sizing of Primary Crushers, Mineral Processing Plant Design, Practice, and Control, Society for Mining, Metallurgy, and Exploration, Inc., Vol. 1, 23p. 584-605.

UTLEY, R.W., BOYD, K, 2002, In-Pit Crushing Design and Layout Considerations, Mineral Processing Plant Design, Practice, and Control proceedings, Society for Mining, Metallurgy, and Exploration, Inc., Vol 1, 15 p., 606-620.

UTLEY, R.W., 2003, Alternate Primary Crushing Equipment for Open Pit and Underground Mining, Worland Mining Congress, New Delhi, India.

UTLEY, R.W., 2004, Primary Crusher Optimization for Improved Comminution, The AusIMM, Metallurgical plant design and Operating Strategies, Perth, Australia.

## IMPACT ON GRINDING MILL DESIGN OF RECENT NEW DISCOVERIES

John Starkey

Starkey & Associates Inc., 212-151 Randall St. Oakville, ON, Canada

**ABSTRACT**

At the CIM conference in May 2009, previously unseen differences between macro and micro grindability relationships contained in the SAGDesign database were presented. This database included SAG and ball mill design energy measurements on many samples. These were compared with the extremes of SAG hardness variation from two recent projects. These extremes demand different considerations to be used in the design of the SAG and ball mills needed for these ores. The term 'new discoveries', relates to the wide variation revealed by the new measurements.

Based on those findings it is now evident that historic grindability measurements only reveal the information that the tests were designed to reveal. Bond Rod Mill, Ball Mill, and Crushing Work Index tests have been studied for years to find how to get a good SAG mill design from these (and other) measurements. These efforts have been partially successful.

Other researchers have used breakage parameters to develop SAG mill design data. These are not rotary grinding tests but do give a good estimation of the relative point hardness in an ore body with respect to SAG hardness. These breakage tests do not include ball mill grinding data. Based on the 2009 CIM paper (Starkey and Samuels, 2009), the ratio of SAG to ball mill grindability is an important factor in achieving a reliable SAG mill circuit design because that ratio reveals the best and most practical split for SAG and ball mill energy for the ore being tested.

The reason the SAGDesign test gives an acceptable grinding circuit design result for any ore relates to the 19mm F80 of the ore charge fed to the SAGDesign SAG mill and to the test procedure which includes a Bond Ball Mill Work Index test on SAG ground ore from the first stage of the test. The ratio of SAG pinion energy (to grind to T80 1.7mm) to Bond BM Wi is a key parameter in a SAG mill grinding circuit design, because this ratio controls the power split between SAG and ball mills at the design T80 (transfer size).

### INTRODUCTION

Three recent major projects involving a review and testing by Starkey & Associates, have shown that consulting firms can predict different throughputs in a chosen SAG mill. In one case the same grinding test data was used. In two other cases, different tests were used for mill design, which also provided conflicting results. This paper will show how the new discoveries summarized here, have confirmed that the grinding mill design procedures developed by J. Starkey (JS) in 2004, that were introduced with the commercial use of SAGDesign testing, have worked for every new project. In the author's opinion these design procedures, and the use of results from SAGDesign testing, will lead to client's meeting production schedules for every new project, including those where the SAG hardness is 'off the charts' when compared to existing operations.

The CIM paper (Starkey and Samuels, 2009) was written to analyse the SAGDesign database and show the results from 5 years of SAGDesign™ testing. 232 tests on 35 different projects are included. Grinding mills were designed for each project, so every project in a sense, was a trial of how well the testing/calculation method suited the ore. The results showed that in all cases there were no problems in sizing the grinding mills needed, even for the hardest ores.

There are three key elements in a successful SAG mill grinding circuit design. First is the sampling because proper samples are needed for a proper design. Second, the tests done need to generate meaningful data. And third, the mill sizing procedure should to be standardized to a point where the design calculations relate the empirical measurements made to the proper mill design or prediction of mill performance. Simplicity of sample preparation, total cost of the testwork, and the time required to do the tests are also important. Because SAGDesign technology is patented we are prepared to teach this method to students, engineers and clients.

The SAGDesign test was conceived to accurately measure the required design information on every sample, for SAG milling and ball milling. The calculation method was set up to use these measurements in a mathematically correct way to size the grinding mills required to grind the ore. This method does not require a computer program (other than Excel) or a data base. It is based on results derived from testing the client's ore and the mill design is specific to that ore.

By using these methods, two large areas for mistakes have been avoided. First, the problem of designing for hardest ores has been eliminated because a SAGDesign test identifies SAG hardness up to 34 kWh/t. Second, the use of calculation instead of simulation with other ores has eliminated mistakes caused by comparing ores with different macro/micro hardness ratios. This is the biggest finding from this work because the hardness ratio has not been routinely measured before as a fundamental property that is crucial to achieve a good design. These standard measurements from SAGDesign testing are being seen and reviewed for the first time. It is therefore not surprising that the conclusions have not been noticed before.

### SAGDESIGN DATABASE

One standard way to present SAG hardness variability is to consider the SAG pinion energy (to grind to T80 1.7mm) plotted against the cumulative distribution % of the number of samples tested. Please see Figure 1 below.

These test results demonstrate several important things. There are many samples that are very soft to grind in a SAG mill, and second, there are samples that are harder to grind than any seen to date. The ability of the SAGDesign test to identify extremely hard ores is shown in this graph. No other laboratory SAG test has identified ore that is more than double the 80th percentile hardness. At the low energy end of the curve the hardness increases steadily with no apparent minimum other than the lowest value recorded. Compared to Fig. 2, this is the difference between grinding in a heavy steel chamber and grinding in a semi-autogenous environment. The shapes of the graphs are quite different.

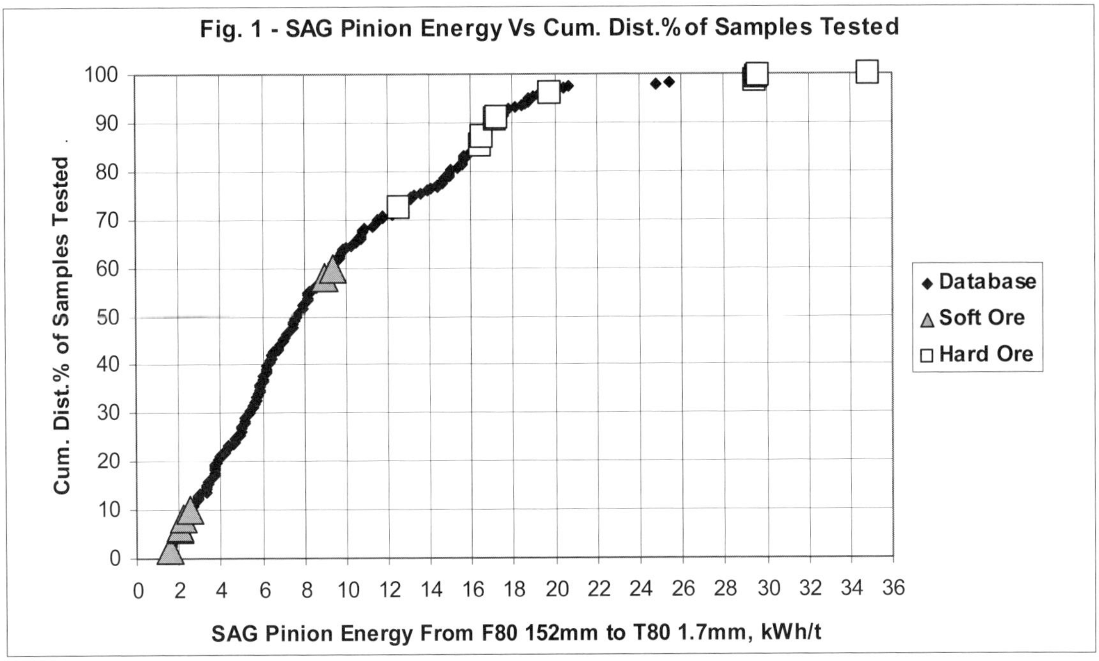


The corresponding Bond Ball Mill Work Index data from the same tests is plotted in Figure 2.

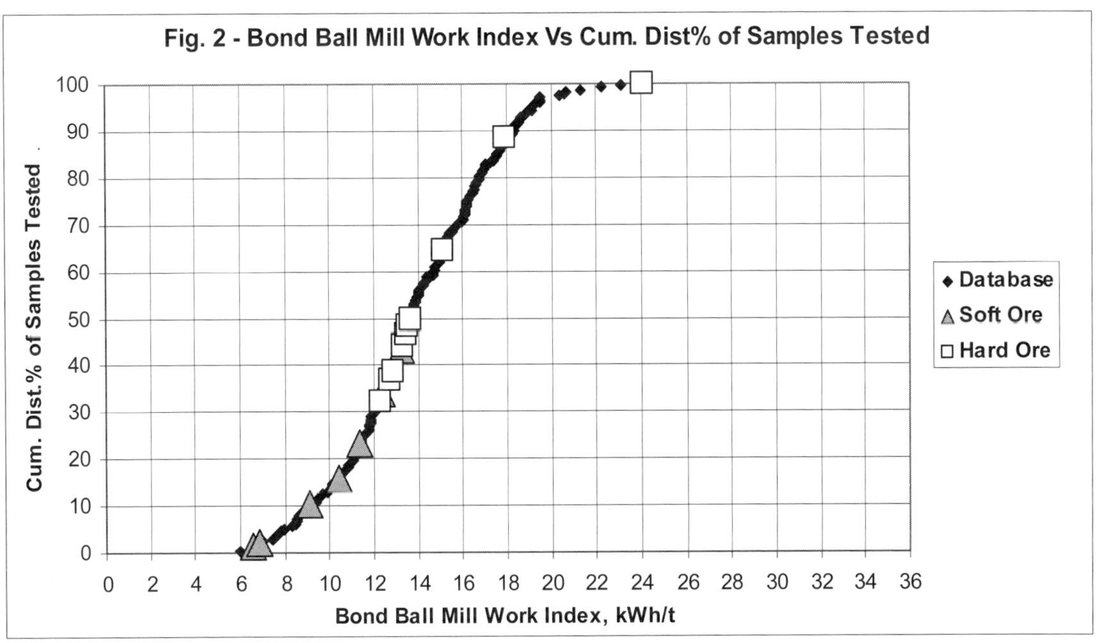


Here both the hard ore and the soft ore (in the SAG mill) overlapped other hardness ranges for the Bond BM Wi database. Because the shape of these curves are so different it proves that the data generated for each sample can only be used in the context of both measurements together. When the Bond Rod Mill Work Index is used instead of the SAGDesign measurement, the Rod Mill Wi variability graph will look more like Figure 2 (McKen et al., 2006) than

Figure 1 because the Rod Mill Wi test is done in an even heavier steel environment than the Ball Mill Wi test. This is why the Rod Mill Wi test should not be used for designing SAG mills. There are other reasons as well that are discussed below.

The graph of SAG pinion energy (to 1.7mm) vs. Bond BM Wi is given in Figure 3. Excluding data from the very soft and very hard projects, shows that the database in general does support a broad correlation between SAG and ball mill hardness but that the function is too ill-defined to be useful in sizing a SAG mill from only a Bond BM Wi measurement. When the hard and soft ores are added to the database, the correlation is invalid. To those who ask if we can design a SAG mill from only a Bond Ball Mill Work Index measurement, we firmly say "no".

It is also seen from Figure 3 that for a constant Bond BM Wi, the SAG hardness can vary from 2 to 34 kWh/t. This 'new discovery' is the reason to recommend that SAG hardness measurements cannot be used alone but only in conjunction with the corresponding Bond BM Wi (preferably done on SAG ground ore as part of a SAGDesign test).

Based on this work we recommend that simulation modeling should consider similar ratios of SAG hardness to ball mill hardness when designing a SAG mill for a specific ore. If the ratios of SAG/ball mill hardness for the new ore are the same as the ores in the database, the design should work. But if the new ore has different ratios (SAG/Bond BM Wi hardness) the design probably will not work well and direct calculation of required mill sizes from the test data should be done instead of using simulation modeling.

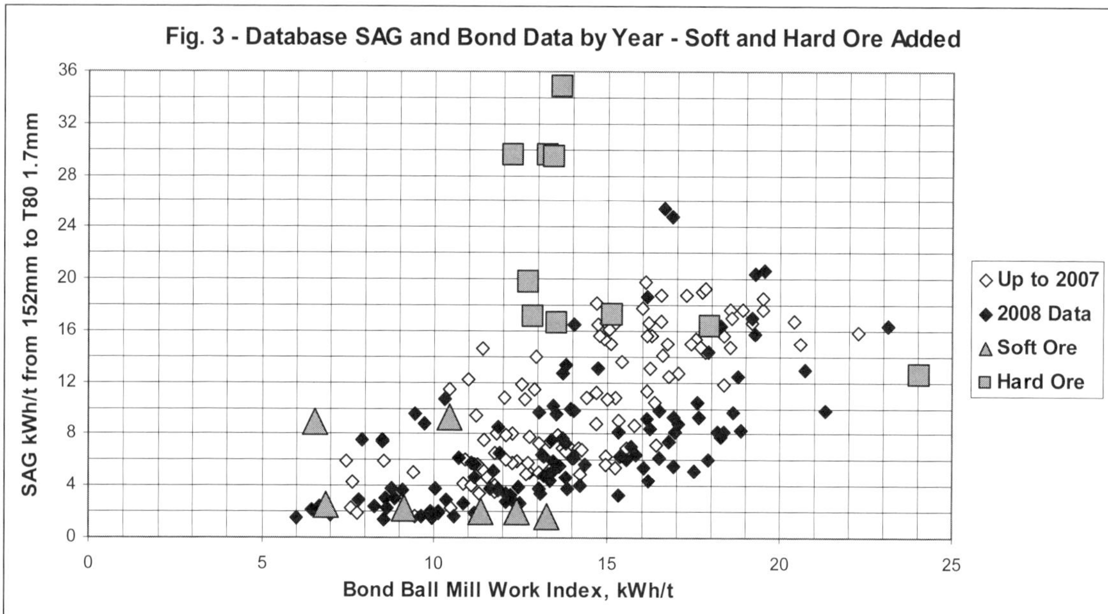


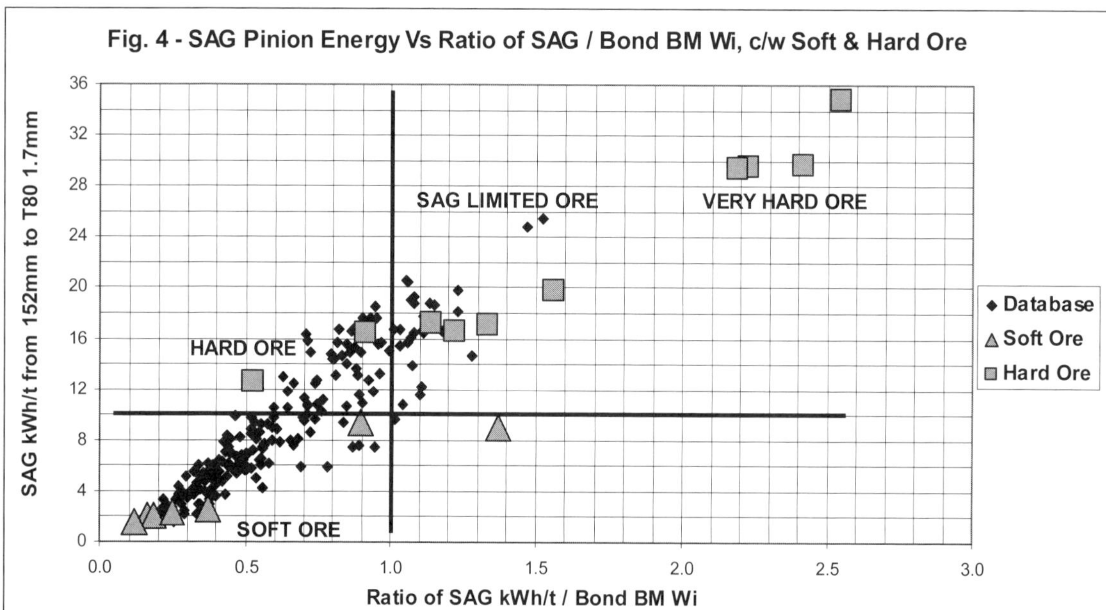

To emphasize this point, Figure 4 has been prepared. Here the SAGDesign database is presented differently with SAG kWh/t being plotted against the ratio of SAG kWh/t to Bond Ball Mill Wi. Looking at the data this way shows the extreme differences between ores at the same ball mill hardness.

The division of this graph into 4 quadrants shows the main areas to be considered in design. 10 kWh/t for SAG grinding to 1.7mm, is considered the dividing line for hard and soft ores. Hardness ratios below 1.0 are also considered to be favorable to basic SAG milling. For ores with SAG kWh/t to Bond BM Wi ratios greater than 1.0, the ore is considered to be hard and SAG limited and pebble crushers should probably be used. Other methods to mitigate the requirement for high SAG energy should also be considered for ores in the top right hand quadrant.

## NEW DISCOVERIES

The new discoveries are highlighted in the following graph (Figure 5) which plots all samples from the database that had Bond BM Wi values between 12 and 14 kWh/t. Samples from both the very hard and very soft projects appear on this plot. In the writer's opinion a calculation method, not simulation is required to properly deal with this situation.

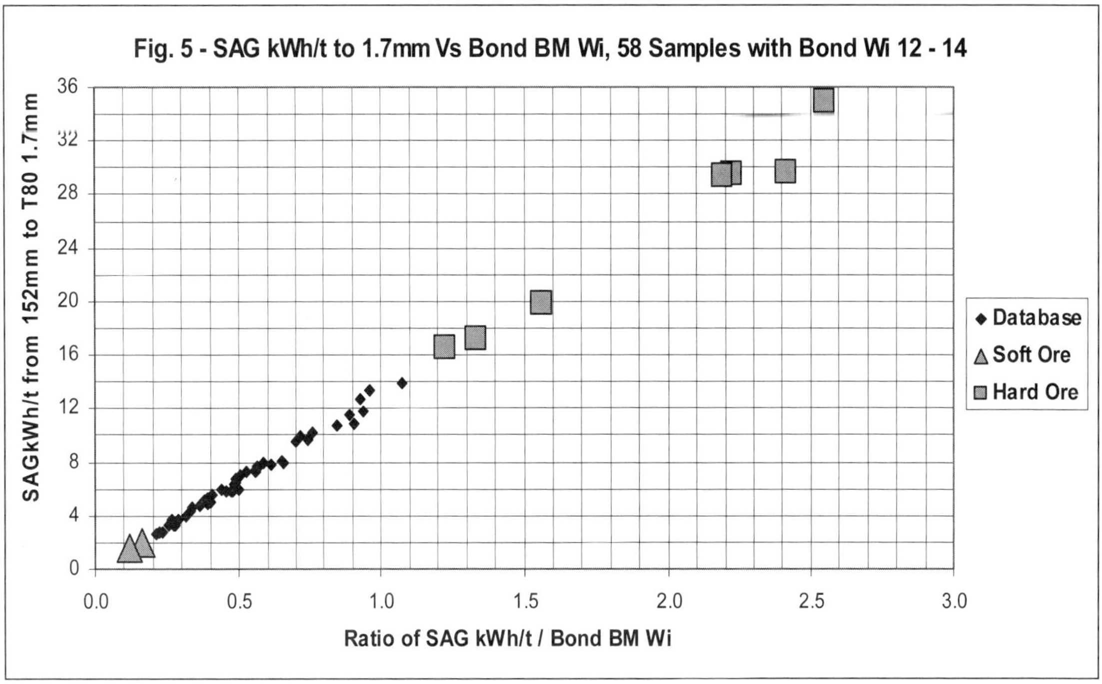


The new discoveries therefore include the wide variance in SAG mill pinion energy from 2 to 34 kWh/t for constant measured Bond Wi values of ~13 kWh/t, and the need to use the ratio of SAG/ball mill hardness as a key factor in designing a SAG mill circuit. To the best of the writer's knowledge of the laboratory methods discussed in this paper, SAGDesign technology is the only laboratory and calculation method that accurately measures and properly considers these factors in the design of new grinding mills.

It has taken 5 years to gather enough data to fully understand SAGDesign test performance on a variety of ores. A clear picture is emerging. Coupled with the success of plant benchmark tests done to date at 6 and 15 kWh/t, we are confident that the SAGDesign test and procedure will work for every ore encountered.

A brief summary of the most common comminution tests are listed below. First listed are the tests that do not use steel in a rotating grinding mill environment. Their function in designing a SAG mill is briefly described. These tests are included to show enough information for readers to judge which tests are likely to give a useful and reliable results for SAG mill design. There are other tests available that are not included here.

- Bond Crushing Work Index Test, CWi, kWh/t. Used by some designers as one of the factors affecting SAG mill power.
- JK Drop Weight Test (DWT), $t_a$, A and b parameters. These are used as inputs to the JK SIMMET program which predicts mill power by simulation and comparison with operating plants. SAG Mill sizing is derived from these simulations. This method is usually good for ores up to ~18 kWh/t for SAG pinion energy but not above that because of a lack of operating data on harder ores. The SAG/ball mill hardness ratio is not considered as a prime input, so errors are possible.
- SMC Test or Drop Weight Index, DWi, A and b parameters. This is a smaller scale test to develop the same parameters as the DWT and the kWh/m$^3$ for each sample tested. Its use allows the sample weight required to be reduced from 100 kg for the DWT to less than 20kg and is excellent for scoping and geometallurgy.

- JKRBT, A and b parameters. This is a highly automated rotary breakage tester that was developed to rapidly generate A and b parameters on a large number of samples by treating sized particles in a controlled environment. This is a new test and the author is not familiar with the results.

The JKRBT is not considered as a rotating grinding mill because it develops breakage information similar to the Drop Weight Test and the SMC Test. It does not grind the ore, it breaks it piece by piece.

The following list includes rotary grinding tests that use steel balls. These mills simulate large scale operating mills, use similar steel loads and dry ore as feed. They are considered by the writer to be more reliable for scale-up.

**Table 1 – Rotating Mill Grinding Tests Using Steel Balls**

| TEST | F80, mm | MILL SIZE | SAMPLE | REMARKS |
|---|---|---|---|---|
| Bond BM Wi, kWh/t | 2 | 12" dia x 12" | 10 kg | Constant volume, BM design |
| Bond RM Wi, kWh/t | 10 | 8" dia x 20" | 20 kg | Constant volume, RM design |
| SPI, Minutes | 12.7 | 12" dia x 4" | 5 kg | Constant weight 2kg, scoping test |
| MacPherson AWi, kWh/t | 19 | 18" dia x 6" | 200kg | Continuous, air swept, pilot scale |
| SAGDesign SAG, kWh/t | 19 | 19.2"dia x 6.4" | 15 kg | Constant vol., SAG/BM design |
| | | Includes SAG Test and Bond BM Wi on SAG ground ore. | | |

The difference in hardness indicated by the Bond BM Wi test and the Bond RM Wi test was the first indication 50 years ago that there is a unique macro/micro grindability ratio for every ore. This has been known since Bond published his methods for mill sizing and relates to hardness above and below a feed size F80 of 2mm. Both of the Bond tests (rod mill and ball mill) are done in a high steel environment. Since the SAGDesign test is done in a much lower percentage steel environment (11 % by volume of steel), it follows that the distribution function of the data will have a different shape. Figures 1 and 2 confirm this fact as noted above.

The other fact of significance here is the F80 for the test concerned. There is a shift in relative hardness for many ores between the macro and micro grinding done in a SAG and ball mill respectively. After noting that the Rod Mill Work Index test on the hardest ore tested did not pick up the SAG hardness of 29 kWh/t (Starkey and Samuels, 2009), it is now realized that the 10 mm F80 for the Rod Mill Work Index tests is probably the reason because the Bond BM Wi was 12.7 kWh/t (at 2mm), the Rod Mill Wi was 19.7 kWh/t (at 10mm) and the SAG pinion energy was 29.6 kWh/t (at 19mm). It is concluded that the macro hardness relative to SAG grinding is not measured properly in tests with feed sizes of 10mm or less.

This raises a new concept in understanding comminution hardness measurements. Comminution tests do a good job determining the hardness in the comminution device for which the test was designed. To use a test for other purposes is an approximation at best and can lead to significant errors as noted above. Because a SAGDesign test includes a Bond Ball Mill Work Index test on SAG ground ore, the SAGDesign SAG test result and its Bond BM Wi result give the required grinding design information.

From Table 1 above one might conclude that the SAGDesign and MacPherson tests should give similar results. In practice however the SAGDesign test uses a 26% load and stops at 80% passing 1.7mm. The MacPherson test produces a much finer product with a P80 of about 0.4mm and the mill is often operated with as much as 45% load as determined from actual test reports. Since the AWi value overlaps Bond in the range of 1.7mm to 0.4mm the method required for interpreting the data from SAGDesign and MacPherson tests is, and needs to be, very different.

## MILL SIZING CALCULATIONS

A typical SAGDesign mill sizing calculation sheet is presented below. Shaded cells in Table 1 are the input numbers, taken from the testing program and the design criteria. Usually the $80^{th}$ percentile is utilized as the design point for a project but the client, given the distribution data, can elect to use hardest ore, median ore or any other point that represents what the mine can deliver to the mill on a consistent basis.

This table is presented to show the precision of the mill sizing calculations once the ore testing data is complete and analysed. By fixing input values, the Case 1 data sheet can be copied. Other cases are then examined to look at balancing the power by changing the transfer size or making other adjustments that are appropriate to the ore body, the hardness distribution, the ratio of SAG to ball mill hardness, and future expansion.

These calculation procedures were chosen to take advantage of the design measurements (for SAG and ball milling) the SAGDesign test program. Bond's diameter correction factor is used by Starkey & Associates because it has been judged to be conservative by determining benchmark operating work indices at Escondida and Candalaria. (Classification efficiencies were not considered). Not all engineering houses use this correction factor. For a 24 ft diameter ball mill, the operating Wio diameter correction factor is 0.8 times the energy calculated from the Bond Wi.

This is design by engineering calculations, not design by simulation. It is understandable and can be examined for accuracy by the client and/or the engineering company responsible for the work. The various factors used are shown and can be adjusted if not appropriate for the duty intended. By avoiding simulation techniques, the quality of the assumptions and algorithms used in the computer program are not a factor in sizing the mills.

Table 2 is an example of how mill sizes are calculated using SAGDesign technology. The SAG pinion energy ($80^{th}$ percentile) required to grind the ore from F80 152mm to T80 1.7mm is used as input. The equation used to calculate the SAG pinion energy can be found in the reference paper below (Starkey et al, 2006). The Bond Ball Mill Work Index on SAG ground ore is also used as input. Note that all 5 inputs to this calculation spreadsheet including T80, P80 and t/h are shown as shaded cells in Table 2. Required SAG and ball mill motor kW are given in the D section of this example and mill dimensions in section E are chosen to provide the exact amount of energy required. The method is precise and only requires selection of suitable mill dimensions to match the grinding to be done.

**Table 2 – Typical Mill Design Calculations**

**CASE 1 - No pebble crusher. Fixed rpm new SAG and ball mills. 80th% ore.**

**A - Calculation of Unit Pinion Energy for SAG Mill and Ball Mill** *Reduction for pebble crush* *0%*

| Item | Units | | 80th% |
|---|---|---|---|
| SAG Mill to 12 mesh US Std. | kWh/t | *design SAG pinion energy* | **29.5** |
| Adjust SAG Energy to "T80"μ | **1700** | *SAGDesign test T80* | **0.0** |
| Total SAG Shell Energy | kWh/t | | **29.5** |
| B Mill Pinion Energy to "P80" μ | **74** | *design P80* | **12.3** |
| *SAGDesign Bond BM Wi* | *kWh/t* | *design Bond Wi* | **13.4** |
| Total Grinding Energy | kWh/t | | **41.8** |
| Design SAG Feed - mt/h | **215** | *design t/h* *Measured Ratio, SAG/BM* | **2.2** |

**B - Calculation of Unit Motor Energy for Fixed speed SAG Mill and Ball Mill (Metered Energy)**

| Item | Factors | | 80th% |
|---|---|---|---|
| SAG Energy kWh/t * @ | 1.06 | * *SAG Motor - corrected for efficiency loss with synchronous motor & clutch* | **31.3** |
| BM Energy kWh/t ** @ | 1.06 | ** *BM Motor - corrected for efficiency loss with synchronous motor & clutch* | **13.0** |
| | 0.88 | *BM diameter correction factor - see Table F* | **11.4** |

**C - Calculation of Unit Motor Sizing for Fixed Speed SAG Mill and Ball Mill**

| Item | Factor*** | | 80th% |
|---|---|---|---|
| SAG Motor kWh/t*@ | 1.10 | *SAG Mill operating allowance factor* | **34.4** |
| BM Motor kWh/t**@ | 1.05 | *Ball Mill operating allowance factor* | **12.0** |
| | | *Installed unit energy* | **46.4** |

*Note *** Includes for SAG operating margin and for ball mill operating margins.*

**D - Motors Required for Fixed Speed SAG Mill and Ball Mill** **215 t/h**

| | | 80th% |
|---|---|---|
| SAG Mill Motor, kW | | **7,394** |
| Ball Mill Motor, kW | | **2,576** |

**E - Mill Dimensions, Motor Specification, Fixed Speed SAG Mill** *Adjust mill D & L to provide req'd kW*

| Mill | Speed | Dia, Ft | A | B | C | EGL | Calculated Motor | | Instal | Mill Dimensions | |
|---|---|---|---|---|---|---|---|---|---|---|---|
| | % Crit | (ID Shell) | ID - 0.5' | 26% Load | 75% Crit | Length | HP | kW | HP | Dia, ft | EGL, ft |
| | *fixed* | | | *10% steel* | | | | | | | |
| SAG Mill | 75 | 32.00 | 986 | 4.17 | 0.1838 | 13.12 | 9,916 | **7,394** | **10,000** | **32** | **13.1** |
| Ball Mill | 75 | 16.00 | 167 | 5.02 | 0.1838 | 22.37 | 3,454 | **2,576** | **3,500** | **16** | **22.4** |

## IMPACT OF NEW DISCOVERIES ON GRINDING MILL DESIGN

The impact of these new discoveries on grinding mill design is clear to the author. But the way forward to use this information may not be easy for others to accept because of the wide array of methods currently used to design SAG mills and SAG grinding circuits. Therefore comparisons of SAGDesign testing with other methods, benchmarking, and field experience using these design techniques, are recommended as an urgent priority.

To date, benchmarking tests at a producing mine in Canada gave a predicted energy requirement of 6.0 kWh/t Vs a measured plant hardness for the ore fed to the mill of the same value. For a hard magnetite ore tested in a 5.5 ft diameter SAG pilot mill in Hibbing, Minnesota, both the observed SAG pinion energy and the measured value to produce 80% passing 1.7mm were 15 kWh/t. It is also known from direct comparison with drop weight test results that the SAGDesign measurements are comparable on a relative basis. (Starkey and Meadows, 2007). On another project, SAGDesign measurements have matched the relative drop weight SAG hardness results on the same hard ore samples.

Given the cost of a single 40ft diameter SAG mill, it is imperative that the grinding mill design procedures used be carefully reviewed for accuracy at the testing and mill sizing stage for every project, and especially the large scale ones. Careful sampling by working with the mine engineers and geologists to generate composite samples that match the early mine production by year, should be done. These samples should be tested using SAGDesign. The advantage of using SAGDesign is that the mill sizing calculations can be checked. This is mill design by calculation from empirical measurements, not simulation based on other ores that may not have the same macro/micro hardness ratio.

## CONCLUSIONS AND RECOMMENDATIONS

1. SAGDesign testing should be used to design the mills for new large tonnage projects where failure to achieve design tonnage and grind will cause severe financial distress to the owner. It is accurate, fast, and affordable.
2. If a client requires confirmation of SAGDesign testing and mill sizing, benchmark testing of a similar grinding circuit should be done by taking a feed sample and comparing the SAGDesign result with plant performance. Other tests can also be done for confirmation if deemed to be necessary.
3. Properly chosen samples are at the heart of any grinding mill design project and are the first requirement for a successful grinding operation, regardless of the tests used. Taking samples is part of the full SAGDesign program.
4. Rotary mill grindability tests produce results that are valid for the purpose intended. Bond tests are good for designing rod mills and ball mills. SAGDesign tests are suitable for designing AG/SAG mills.
5. Conventional impact breakage parameter tests do not measure the SAG pinion energy in kWh/t needed to achieve a stated product size. For ores harder than treated in existing operating plants, this is a problem.
6. The ratio of SAG pinion energy to Bond BM Wi is helpful when selecting the SAG/ball mill power split for a new grinding circuit. We recommend that this ratio be included in the required data for any new SAG/ball mill circuit design and used as a guide in deciding what power split will be chosen.
7. The Bond BM Wi cannot be used alone to design a SAG mill.

## REFERENCES

McKEN, A., VERRET, F-O., and WILLIAMS, S. 2006. Digging Into a Large Testing Database. International AG and SAG Grinding Technology, 2006, Vol. IV. 16 p. 331-346.

STARKEY, J.H., HINDSTROM, S., and NADASDY, G.N., 2006. SAGDesign Testing – What It Is and Why It Works. International AG and SAG Grinding Technology, 2006, Vol. IV. 15 p. 240-254.

STARKEY, J.H., MEADOWS, D., 2007. Comparison of Ore Hardness Measurements for Grinding Mill Design for the Tenke Project. CMP Conference Proceedings, January 2007. 13 p. 19-31.

STARKEY, J.H., SAMUELS, M. 2009. New Discoveries in the Relationship Between Macro and Micro Grindability. CIM Conference in Toronto, May 13, 2009. 8 p.

# RECENT TRENDS IN THE DESIGN OF COMMINUTION CIRCUITS FOR HIGH TONNAGE HARD ROCK MINING

Persio P. Rosario, Ken Boyd, and Mike Grundy

AMEC Mining & Metals, 111 Dunsmuir St, Vancouver, BC, Canada

## ABSTRACT

The high capacities and simple circuits of autogenous (AG) and semiautogenous grinding (SAG) mills opened the door to the high-tonnage, low-grade operations that predominate in today's mining industry. From the early 1980s to the early 2000s, the use of SAG mills prevailed in high tonnage hard rock operations throughout the world; this paper begins by tracing the evolution of SAG circuits through to the preferred configurations of the present day. More recently, crushing circuits using high pressure grinding rolls (HPGR) as their tertiary stage have been also applied to the treatment of hard and very hard ores. Recent SAG and HPGR circuits, and their main features and limitations, are described.

A flowsheet with an AG mill/scrubber and parallel trains of crusher-grizzly-HPGR is proposed to treat hard weathered ores with a high proportion of clays and moisture. Trends in the use of stirred mills in coarser applications, various trends in circuit operation, and the contribution of recent increases in crusher capacity are discussed.

## INTRODUCTION

Some facts are self-evident: commodities prices fluctuate; high-grade, large deposits with easy-to-process ore exist only in dreams; and energy efficiency is a public matter. However, mineral processors adapt as well as most to the changing economic environment, especially in the field of comminution. Comminution is the largest energy consumer in mineral processing, and, if the ore is hard, requires the largest capital and operating cost. In modern low-grade mining operations, the scale of the use of energy and other consumables is unprecedented.

Proper design of the comminution circuit is a critical task in a large-scale hard-rock project. Today, several options are analyzed when designing such a circuit. Some are based on long-established technologies, and others are based on more recently developed technologies, or technologies that have been adapted from other types of projects. Selecting the most appropriate circuit is of paramount importance, not only in deciding the equipment, but also how it is configured. The design task can be quite different in greenfields projects than in expansions, or in modifications of existing circuits.

## RECENT HISTORY OF COMMINUTION

From the 1920s to 1950s, most comminution circuits were designed with several stages of crushing, followed by rod and ball mills. During the 1960s, the use of rod mills declined, as larger diameter ball mills, accepting coarser feeds, became available. The 1960s also saw the advent of autogenous, and, later, semiautogenous mills, and by the early 1970s, large-diameter autogenous grinding mills (AG), and semiautogenous grinding mills (SAG), often together with ball mills, became the accepted norm. Although the power consumption was generally higher, the simpler circuits with fewer components and smaller footprints made the overall economics of SAG mills superior to three-stage crushing in most cases (Bond 1985), and opened the door to the high-tonnage, low-grade operations that have characterized the base metal industry for the past 40 years. The application of these large tumbling mills increased in such a way that from the early 1980s to the early 2000s most new or expansion mining projects have selected some circuit configuration that includes either an AG or a SAG mill (Barratt and Sherman 2002).

More recently, two factors have driven a change in this trend, especially in hard ore operations. Firstly, the wish to reduce energy consumption intensified, driven not only by economics, but also by public interest in climate change, greenhouse gas emissions and carbon footprint.

Secondly, high-pressure grinding rolls (HPGR) became more attractive as their manufacturers developed roll-wear protection systems to better deal with hard and abrasive ores. As HPGRs are more energy-efficient than conventional grinding mills, and because large HPGRs can deliver higher unit throughput at higher reduction ratios than tertiary cone crushers, some projects are now using HPGRs in combination with secondary cone crushers instead of SAG mills.

Stirred milling technology was developed in the 1950s but has only been applied for mineral processing during the last couple of decades. There are a few different models of stirred mill machines in the market and they have been mostly used for regrind applications. The stirred mill presents better energy efficiency than ball mills for fine grinding and during the last few years there has been an increasing interest in applying this technology in coarser ranges (Valery and Jankovic 2002).

## SAG MILL CIRCUITS

Before SAG milling entered the scene, large grinding plants consisted of many trains of two or three stages of crushing, rod milling, ball milling, and all the associated conveyors, screens and surge bins (Figure 1). The SAG mill gained its preeminent status in large mill operations because of its ability, in a single unit, to receive coarse primary crusher product and deliver adequate ball mill feed at high operational availability (approximately 93%) (Figure 2). Development since the early days has centered on increasing the amount of ball mill feed that a single unit produces.

Since their appearance in the 1970s, SAG mills have increased in size and power, their drive systems are more advanced, they are equipped with better control systems, and their benefits and shortcomings are better understood. Such development resulted in new circuit configurations and programs to improve the quality of feed; many of the more significant advances were made by operators determined to wrest more from what they were given.

Large diameter SAG mills have been selected for new hard rock projects and expansions (Los Bronces Development Project, Phoenix Project, San Cristobal) which indicates that, depending on the ore type and project specifics, the SAG mill may still be the preferred circuit for some time.

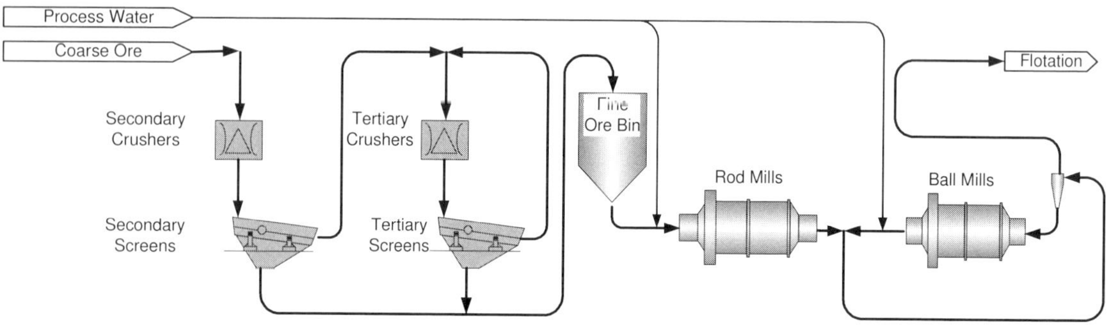


**Figure 1. Three Stages of Crushing, Rod mill, Ball mill**

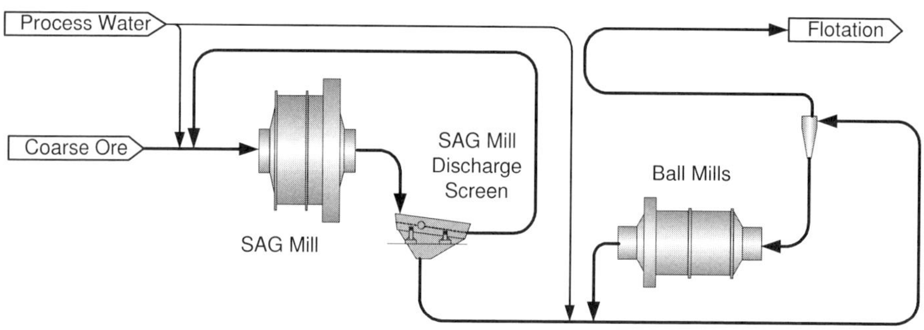


**Figure 2. SAG-Ball Mill Circuit**

### Operational Parameters

Steel ball charges range up to 20% by volume, and a typical value for SAG is 12%. The total charge (balls plus slurry) is usually between 20% and 35%, and the slurry is usually between 65% to 75% solids. The most frequent ball size for large mills is 127 mm diameter, but it can vary from something around 90 mm to a maximum of 152 mm.

The trend has been to operate at increasingly high ball loads, and at increasingly low total loading—it has been observed that a lower total charge improves capacity. Today some operations operate with ball charges up to 20%; total mill volumetric loading has decreased from around 35% in the early days to as low as 24% or below (Sepulveda 2008).

High ball charges have only been made possible by the advent of variable-speed drive, one of the most significant advances in SAG milling. Variable-speed drive was first installed on a SAG mill at Afton (1977) (Thomas 1989) and is now almost universally used. An early graphic example of the advantage of variable-speed drives was at Lornex (now Highland Valley Copper), where a variable-speed mill installed in 1981 was operated at up to 19% ball load, compared to 12% for fixed-speed mills in parallel circuits. The operators could drive the new mill harder, confident that if the ore suddenly became softer, they could slow the mill down to protect the shell.

### Pebble Crushing

Competent rocks in the 12 mm to 75 mm range (critical size) present reduced breakage rates in autogenous (AG) mills. A significant contribution of grinding media in a SAG mill is to accelerate the breakage of critical size material to reduce its tendency to accumulate in the mill. Another, nowadays less common, method of preventing the build-up is the Autogenous Mill-Ball Mill-Crusher (ABC) circuit, where the critical size material is extracted from the mill, crushed, and returned to the mill. These two techniques were combined during the 1980s, when there were several successful attempts by operating mines, to improve their SAG mill performance by using pebble crushers—the Semiautogenous-Ball Mill-Crusher (SABC) circuit. Examples are Los Bronces, Similkameen (Major and Wells 2001) and Chino (Vanderbeek 1989). Inclusion of a pebble circuit has become almost standard in the design of grinding circuits (Figure 3). Even if it is not thought appropriate to install pebble crushers at the outset, it is usually considered prudent to leave space should circumstances require pebble crushing later in the operation.

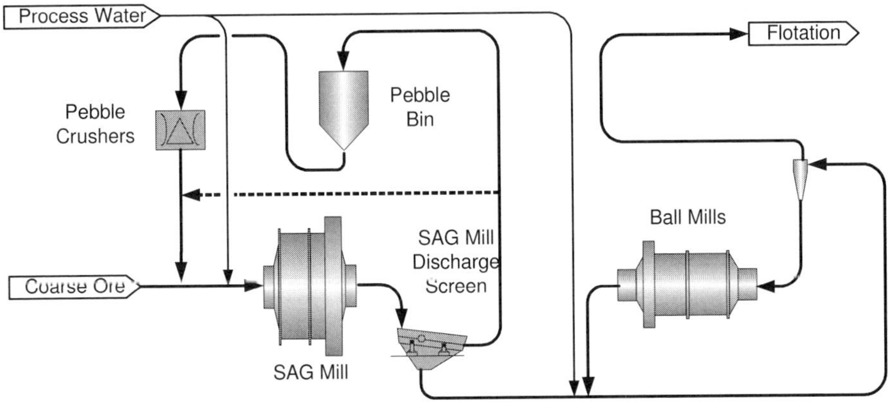


**Figure 3. SABC circuit**

For hard and very hard ores (JK Axb values below 40 and Bond Work Indexes above 16 kWh/t), correct forecasting of the production of critical-size material, and of its extraction rate through the mill grates, is still difficult. There have been reports of operations that spent great effort to achieve the designed pebble extraction, and therefore the design throughput, for quite some time after startup—for example Cadia Hill (Hart et al 2001) and Sossego (Delboni et al 2006).

Until recently, AG mill and SAG mill circuits were invariably closed by screens, oversize being recycled to the mill feed. Recently, however, some SABC installations have been operated in open circuit, in that screened crusher product reports to the ball mill circuit (Figure 4). The effect of opening the circuit is to pass more tonnage at coarser size to the ball mill circuit; consequently in most cases it has been used to increase throughput of an existing operation which had extra ball mill capacity or could tolerate a coarser grinding-circuit product size. There are also new installations (most at the planning stage) where the largest available SAG mill could not reach desired capacity with the pebble crusher in closed circuit, and open-circuit SABC was chosen. An example is the El Teniente Colon Concentrator with an 11.6m diameter SAG and four parallel pebble crushers in SAG open circuit configuration (Spann and Ottergren 2004).

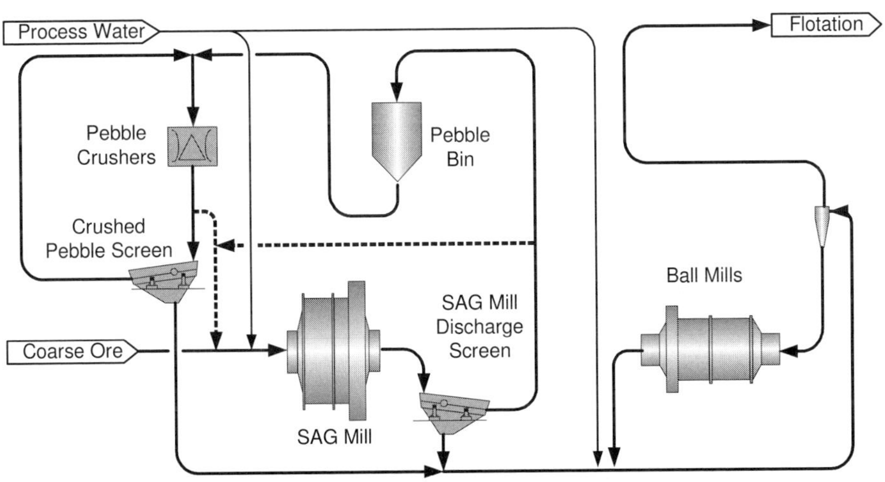


**Figure 4. Open-circuit SABC**

A design where the pebble crusher can either be used or bypassed provides the operator with some external operating control of the SABC circuit. The ability to open or close the circuit during operation provides additional flexibility. The authors recently completed a study for a property where the run-of-mine ore had zones of greatly fractured ore, and zones of very competent ore, both with high ball-mill work indices. It was proposed to lay the plant out so that the operator could bypass the crushers and operate the SAG mill in closed circuit when receiving fractured ore (to maximize SAG mill power) and, when the ore was competent, use the pebble crushers and even open the circuit, to pass more of the work to the ball mills.

Another recent development in pebble crushing is the addition of HPGRs to treat the pebble crusher product (Figure 5). The pebbles are reduced to a much finer product, decreasing ball mill power requirements. Depending on the original circuit, opening the SABC circuit and adding an HPGR stage may achieve a significant capacity increase, without increasing the ball mill duty requirement. This concept can also be applied to a circuit that will ramp-up after startup; for example the Peñasquito project will start up with a single SAG line in mid 2009, a second SAG line will be added in mid 2010, and later one HPGR will be added (Goldcorp 2009).

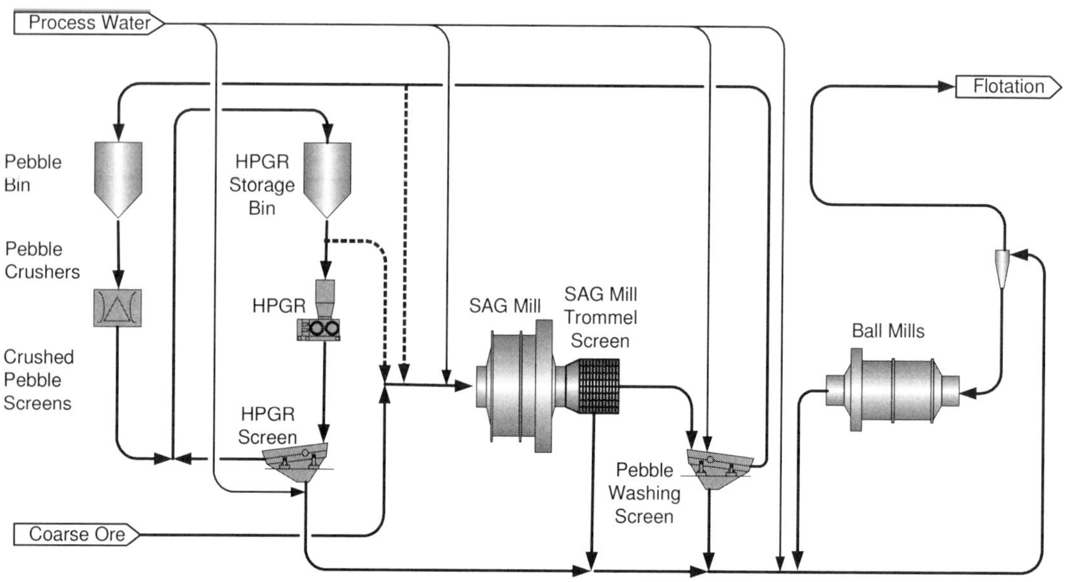


**Figure 5. SABC with HPGR**

***Screens***

Early designs for the screens closing the AG/SAG mill circuit were either trommel screens with water jets returning pebbles through the discharge end of the mill, or vibrating screens with a series of conveyors returning the oversize to the feed end of the mill. Some (such as Lornex and Copperton), used a combination of trommel screens, pumps and vibrating screens. Since pebble crushing circuits have become common, the trommel screen/water jet has become less so, although some (Alumbrera, Antamina) added recycle conveyors and pebble crushers to their trommel screen system, turning the water off when the pebble crushers are in use. More recent large operations employ trommel screens to remove most of the slurry, followed by vibrating screens to wash the pebbles before discharging them onto the recycle conveyors.

***Pebble Surge Capacity***

Early SABC circuits incorporated crushers retrofit into SAG mill recycle conveyor systems, and often had no surge capacity. Surge capacity is highly desirable, enabling the crushers to be choke fed by controlling the feed rate, and pebble bins are now included in circuits as a matter of course. The scale of many recycle operations is now at the point where a pebble stockpile is more economic than a pebble bin.

**Feed Preparation**

In early days, typical SAG ball charges were in the range of 3% to 7% of mill volume, and the general consensus was that large rocks in the feed were always necessary to assist in breakage. Under present operating conditions of high ball load and low total loads, the contribution of large rocks as grinding media is insignificant. It is now realized that tuning up the blasting and primary crushing phases to deliver a steady and finer feed to the mill are cost-effective contributions to the overall comminution system. Several cases have demonstrated substantial improvement in SAG production by feed preparation programs (mine-to-mill), improving production by a factor of up to 15% (Lam et al 2001).

SAG throughput is very susceptible to changes in the hardness of the ore and this should be assessed at early stages of design. In cases where the ore body presents a high variability of friability, provision for blending may be an option to minimize high fluctuation in production (Dance 2004). If a well elaborated plan to maintain reasonably uniform ore hardness is not possible, the operation should be prepared to sustain fluctuations in tonnage.

In some cases, pre-crushing—scalping off and crushing coarse material in the SAG mill feed—has been applied to manipulate the SAG feed size distribution (Figure 6). Pre-crushing can be used where there is limited scope to optimize blasting and primary crushing, as in block-caving underground mines. In addition, pre-crushing has been applied to maintain designed production levels at mines where the ore hardness has increased over time or to expand production (Sylvestre et al 2001).

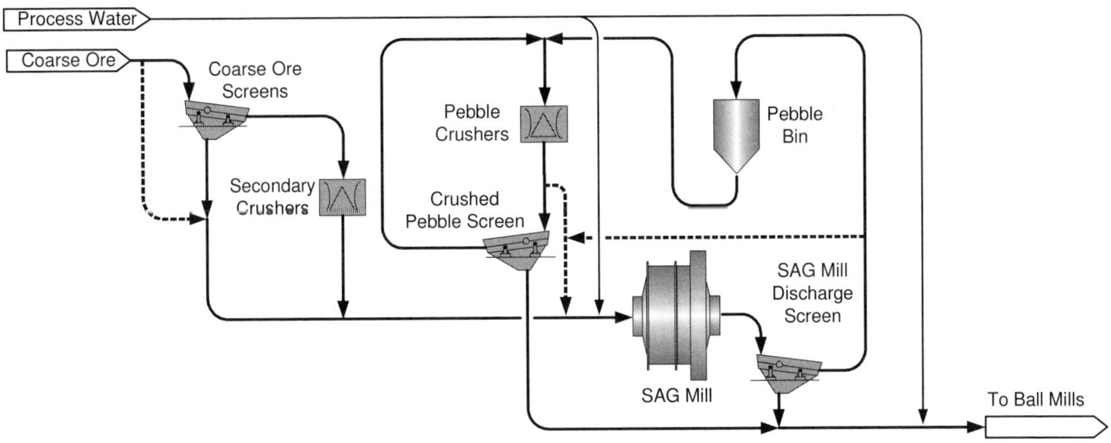


**Figure 6. Pre-crushing in a SABC circuit**

## HPGR CIRCUITS

Fundamental and applied research on fracture phenomena demonstrated that interparticle comminution at high pressure consumes substantially less energy than conventional grinding processes, and served as the origin for the development of the HPGR (Schonert 1988 and 1991).

The application of HPGRs in comminution circuits increased over the past two decades and is well established in processing cement, diamonds and iron ore (Broeckmann and Gardula 2005). The major benefits supporting this trend are better energy efficiency, improved grinding capacity, and higher metal recovery in downstream processes such as heap leaching and flotation. In the last few years HPGR-based plants processing hard ore in high-tonnage precious and base metals operations have been designed. SM Cerro Verde has already started production and Boddington is expected to start in 2009.

### Flowsheets

Of the many possible flowsheets that have been proposed for HPGR, those using HPGR as tertiary crushers, in closed circuit with fine screens, are expected to provide maximum energy efficiency. In addition, the secondary crushing product is screened before feeding the HPGR to avoid oversized material damaging the rolls (Morley 2006a). This configuration is illustrated by the Boddington and Cerro Verde flowsheets (Figures 7 and 8).

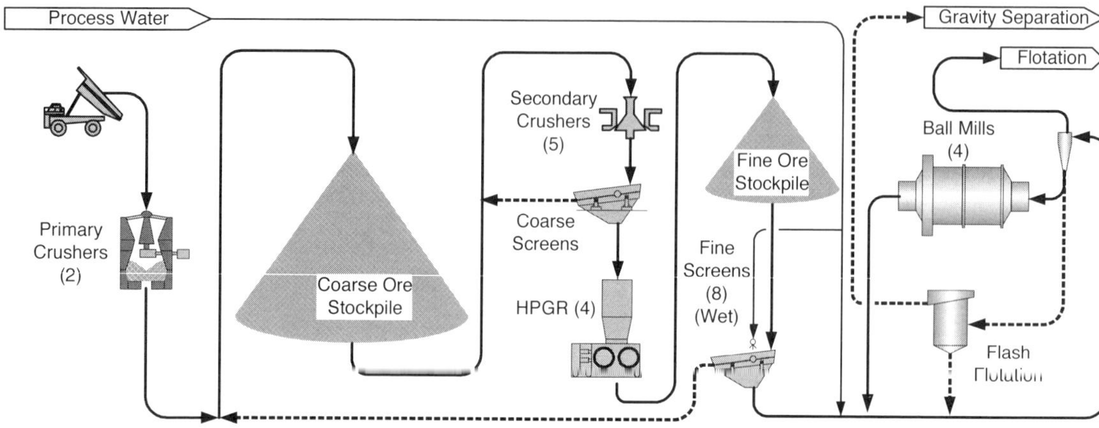


**Figure 7. Boddington HPGR (Dunne et al 2007)**

Boddington has a design capacity of 35Mtpa (approximately 96,000 t/d) and processes two very hard gold ores with average Bond ball mill work indices (BWi) of 15.1 and 16.6 kWh/t, Bond rod mill work indices (RWi) of 22.8 and 24.2 kWh/t, and JK Axb values of 27.9 and 25.5. The circuit is comprised of: five 746 kW cone crushers, four 2.4 m diameter(D) x 1.65m length(L) 5.5 MW HPGRs, and four 7.9m D x 11.9m L ( 26 x 39 ft ) 15.6 MW ball mills (Dunne et al 2007). The projected roll surface wear life was estimated in 4,250 hours. A 2006 tradeoff study showed that a preliminary SABC circuit would have 7% lower capital costs than the HPGR circuit, and that the HPGR circuit provided 12% savings in comminution operational costs. Furthermore, the study concluded that the lower operational costs of the HPGR circuit offset its higher capital costs (Seidel et al 2006).

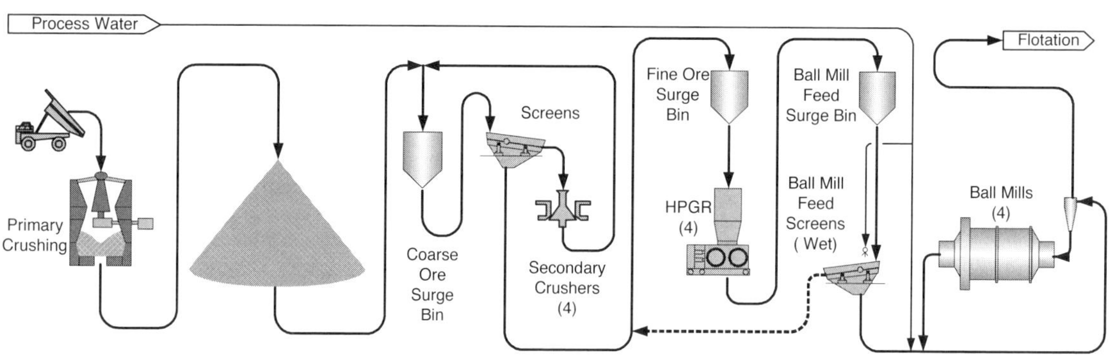


**Figure 8. Cerro Verde (Vanderbeek 2006)**

Cerro Verde has a design capacity of 108,000 t/d of hard copper-molybdenum ore (average BWi 15.3 kWh/t). The circuit is comprised of: four 746 kW cone crushers; four 2.4m D x 1.65m L 5.0 MW HPGRs; and four 7.3m D x 10.7m L ( 24 x 35 ft ) 12 MW ball mills. The projected roll surface wear life is 6,000 hours. In 2006, just prior to startup, it was reported that although estimated capital costs were higher for the HPGR circuit than an equivalent SAG circuit, the estimated total comminution operational costs were US$ 1.33 and US$ 1.70 for the HPGR and SAG circuits respectively, being the main contributors for this difference the costs of power and grinding media. The estimated total comminution circuit specific energy for the SAG circuit was 20.1 kWh/t and for the HPGR circuit 15.9 kWh/t. In addition, risk analysis results and internal rate of return factors were responsible for the decision of building an HPGR circuit instead of SAG (Vanderbeek 2006).

AG circuits are notoriously sensitive to changes not only in ore hardness, as previously noted, but also in feed size; SAG mill circuits are more stable, and SABC circuits more stable still, but SAG-based circuits are still very sensitive to feed variations (Morrell and Valery 2001). Anglo Platinum, at the Mogalakwena North concentrator, selected crushing technology in large part because it gave stability in feed rate and product size, and HPGR in particular because of its economic advantages over tertiary and quaternary crushing (Rule 2006).

In a variation of this flowsheet, the HPGR is equipped with a dividing chute; the product from the centre of the rolls is directed to the ball mill, and the material produced at the edges of the rolls, which is coarser, is returned to the HPGR feed. This form of HPGR product recirculation has already been applied in iron ore projects and has been recently developed for some base metals projects (Gruendken et al 2008).

**Ore Properties**

Like other crushers, the HPGR is not ideal for ores with a high proportion of very fine material (clayish material), ores with elastic properties, or significantly soft ores. These substances tend to cushion the crushing action and make the process inefficient. High moisture in the HPGR feed is also problematic as it may cause slippage of the material on the rolls surface, accelerating wear. Tramp metal may be catastrophic as it can damage the roll surface (Morley 2006b).

Both Boddington and Cerro Verde ores produce a moderately friable HPGR cake product that is de-agglomerated in wet screening by the action of water sprays and vibration. However, if an ore has the tendency to create competent product cakes it may be necessary to have a more powerful process to break it down prior to sending to downstream processes. Scrubbers are standard in the diamond industry, not only to wash out the clay prior to the HPGR feed, but also after each HPGR stage. Flowsheets incorporating low-speed tooth roll sizers (MMD sizers) used as de-agglomerators downstream of the HPGRs have been developed (Valery and Jankovic 2002).

Some base metal ore bodies contain a mixture of very hard rocks, softer material, and clays, in proportions that can vary greatly throughout the deposit. The authors have been involved in such a project and have been developing a flowsheet to take advantage of potential operating cost savings of the HPGR circuit for hard rocks whilst mitigating issues caused by the soft and clayish materials. The proposed flowsheet is shown in Figure 9. The scalping screens (coarse vibrating grizzlies) protect the HPGRs when the cone crushers release coarse material to relieve jams. The autogenous mill acts as a scrubber, and is either an overflow mill or open-ended grate discharge. The mill scrubs the hard rocks and breaks soft rocks. If necessary, crusher-grizzly-HPGR clusters can be put on- or off-line as the proportion of different materials fluctuates. Potential advantages: it is essentially a crushing circuit with its inherent stability in terms of throughput and product size; much of the softer ore feed bypasses the crushing circuit; the scrubber should cope with any clays and do away with possible issues with high capacity wet fine-screening; dust generation is reduced; no steel ball scats; any competent HPGR cakes are de-agglomerated in the mill/scrubber before screening.

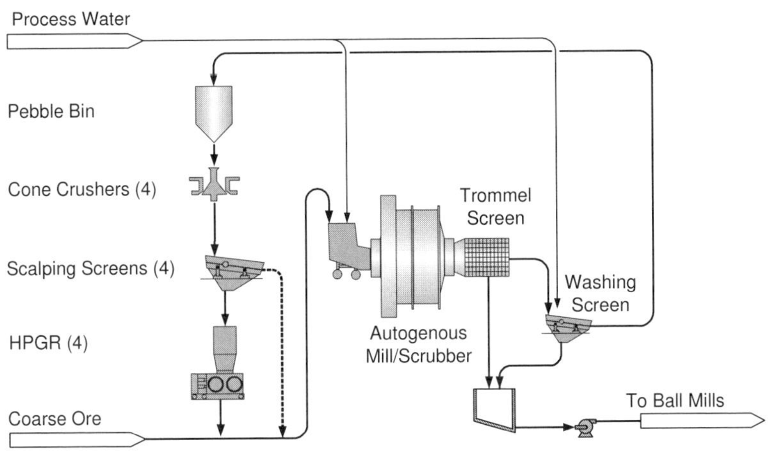


**Figure 9. Proposed Flowsheet for Hard Ore with Clay**

**Machine Sizes**

Over the past four decades, tumbling mill sizes increased considerably; the largest mills were 6,750 kW in the early 1970s, and now there are mills operating at 23,000 kW. The increase appears to have stalled somewhat, not because of a lack of need for larger units, or that the mills have reached a manufacturing limit, but seemingly because of a lack of confidence in the industry that motor manufacturers can produce reliable drives in larger sizes. On the other hand, crusher sizes are increasing; recently crusher manufacturers launched larger machines than were previously available. FLSmidth XL crushers are available up to 895 kW and Sandvik have two new heavy duty cone crushers, one available to 1044 kW. Both Polysius and KHD produce HPGR up to 6.6 MW. The increasing sizes of crushing units mean fewer units and a simpler, less costly, plant, and in doing so reduce the chief advantage of the SAG mill.

## STIRRED MILLS

The stirred mill presents better energy efficiency than ball mills for fine grind, one of the reasons being that it can operate with smaller media and provide a better match between the particles and the media. It has been demonstrated that media sizing is a key factor for ball mill comminution efficiency (McIvor 1997). The Metso Vertimill® has been used in tertiary grinding for many years, for example, at Red Dog and Chino (Vanderbeek 1997) and has now been successfully used in secondary milling (Valery and Jankovic 2002; Jankovic and Valery 2004). A large scale secondary grinding circuit is now being built and the manufacturer has recently launched a higher capacity machine equipped with a 2,240 kW motor, claiming that this mill can handle 6mm top size in the feed (Metso 2009).

The horizontal stirred mill IsaMill™, manufactured by Xstrata Technology is also now available in a larger scale unit equipped with a 3,000 kW motor (Isamill 2009). This type of mill has been tested in the platinum industry as a primary mill receiving the product of two HPGRs in series; as well as high energy efficiency, this experimental flowsheet is also aiming at metallurgical efficiencies by having a comminution circuit free of metal media, as the HPGR rolls are built to hold a layer of ore and the IsaMill™ utilizes ceramic grinding media (Ayers et al 2008).

## FULLY AUTOGENOUS GRINDING

The same economic and social factors that have driven the incursion of HPGR into the realm once almost exclusive to SAG mills, have driven the reversion of some operations to two-stage fully autogenous milling (Koivistoinen and Levanaho 2006). It may be that the current high cost of steel will persist or strengthen, in which case the replacement of ball milling with pebble mills may grow into a trend. Provided that the ore is reasonably competent, pebbles may be obtained from the feed to either ABC or HPGR circuits. Figure 10 shows one of the circuits in the now defunct Bethlehem Copper mill; this use of pebble mills was commonly practiced in the Canadian iron ore and uranium business.

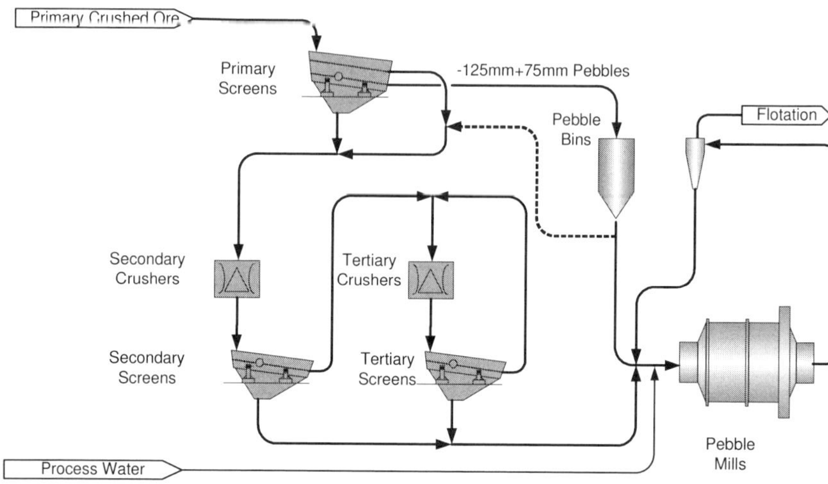


**Figure 10. Pebble Extraction and Milling**

## ACKNOWLEDGEMENT

This paper was prepared with the support and permission of AMEC Americas, for which it is thanked.

## REFERENCES

Ayres, C., Knopjes, L., and Rule, C.M. 2008. Coarser feed applications of MIG IsaMilling. *MEI Comminution'08 Conference*, Falmouth, Cornwall, UK, June 17-20 2008.

Barratt D. and Sherman M. 2002. Factors which influence the selection of comminution circuits. *Mineral Processing Plant Design, Practice, and Control Proceedings*, eds. Mular A.L., Halbe D.N., and Barratt D.J., Society for Mining, Metallurgy, and Exploration, Inc. (SME), 1: 539-565.

Bond F.C. 1985. General aspects of comminution – History. *SME Mineral Processing Handbook*, ed. Weiss N.L., vol.1, section 3A, chapter 1: 1-4.

Broeckmann C. and Gardula A. 2005. Developments in high-pressure grinding technology for base and precious metal minerals processing. *Proceedings of the 37th Annual Meeting of the Canadian Mineral Processors*, Jan. 18-20 2005, Ottawa, Canada.

Dance, A. 2004. Remove the disturbance: stabilising SAG mill tonnage through proper feed preparation. *SME Annual Meeting February 2004, Denver, USA*. Preprint 04-110:1-5

Delboni H., Rosa M.A.N., Bergerman M.G., and Nardi R.P. 2006. Optimisation of the Sossego SAG mill. *SAG2006-SAG Mill Circuit. Int. Conf. on Autogenous and Semiautogenous Grinding Technology* 1: 39-50.

Dunne, R., Hart, S., Parker, B., and Veillette, G. 2007. Boddington gold mine – an example of sustaining gold production for 30 years. *Proceedings of World Gold 2007 Conference*, Cairns, Australia, 22-24 October 2007.

Goldcorp Inc. 2009. Peñasquito Project Technical Report March 10, 2009. *Via the internet, April 2009* http://www.goldcorp.com/_resources/project_pdfs/penasquito2009.pdf

Gruendken, A., Matthies, E., and van der Meer, F.P. 2008. Flowsheet considerations for optimal use of high pressure grinding rolls. *MEI Comminution'08 Conference*, Falmouth, Cornwall, UK, June 17-20 2008.

Hart, S., Valery Jr., W., Clements, B., Reed, M., Song, M., and Dunne, R. 2001. Optimisation of the Cadia Hill SAG mill circuit. *SAG2001-SAG Mill Circuit. Int. Conf. on Autogenous and Semiautogenous Grinding Technology* 1:11-30.

Isamill 2009. Isamill overview. *Via the internet*, April 2009: http://www.isamill.com/index.cfm?action=dsp_content&contentID=24

Jankovic A. and Valery Jr. W. 2004. Design and operation of Vertimill® for secondary Grinding. Presented at the 36th International Conference on Mining and Metallurgy, 29 Sepetember-2 October 2004, Bor Lake, Serbia and Montenegro.

Koivistoinen, P. and Levanaho, J. 2006. The role of critical-sized material in AG and SAG grinding. *SAG2006-SAG Mill Circuit. Int. Conf. on Autogenous and Semiautogenous Grinding Technology* 2: 246-257.

Lam, M., Jankovic, A., Valery, Jr. W., and Kanchibotla, S. 2001. Maximizing SAG mill circuit throughput at Porgera Gold Mine by optimising blast fragmentation. *SAG2001-SAG Mill Circuit. Int. Conf. on Autogenous and Semiautogenous Grinding Technology* 1:271-288.

Major, K. and Wells P. 2001. SABC cone crusher retrofits: where does the crusher go? *SAG2001-SAG Mill Circuit. Int. Conf. on Autogenous and Semiautogenous Grinding Technology* 1:215-233.

McIvor, R.E. 1997. The effect of media sizing on ball milling efficiency. *Comminution Practices*, ed. Kawatra, K.S., Society for Mining, Metallurgy, and Exploration, Inc. (SME), Chapter 35: 279-292.

Metso 2009. Vertimill® for fine grinding. *Via the internet*, April 2009: http://www.metsominerals.com/inetMinerals/mm_home.nsf/FR?ReadForm&ATL=/inetMinerals/mm_grin.nsf/WebWID/WTB-041122-2256F-74BB9

Morley C. 2006a. Flowsheets for HPGR. *SAG2006-SAG Mill Circuit. Int. Conf. on Autogenous and Semiautogenous Grinding Technology* 4: 172-189.

Morley, C. 2006b. High-pressure grinding rolls- a technology review. *Advances in Comminution* , ed. Kawatra, K.S., Society for Mining, Metallurgy, and Exploration, Inc. (SME). 15-40.

Morrell, S. and Valery Jr. W. 2001. Influence of feed size on AG/SAG mill performance. *SAG2001-SAG Mill Circuit. Int. Conf. on Autogenous and Semiautogenous Grinding Technology* 1:203-214.

Rule, C. 2006. The development of a process flowsheet for the new Anglo Platinum, PPRust North concentrator, incorporating HPGR technology. *SAG2006-SAG Mill Circuit. Int. Conf. on Autogenous and Semiautogenous Grinding Technology* 4: 94-109.

Schonert K. 1988. A first survey of grinding with high-compression roller mills. *International Journal of Mineral Processing* 22: 401-412.

Schonert K. 1991. Advances in comminution fundamental, and impacts on technology. *XVII International Mineral Processing Congress*, Dresden, Sept. 23-28 1991, 1: 1-21.

Seidel, J., Logan, T.C., LeVier, K.M., and Veillette, G. 2006. Case study – investigation of HPGR suitability for two gold/copper projects. *SAG2006-SAG Mill Circuit. Int. Conf. on Autogenous and Semiautogenous Grinding Technology* 4: 140-153.

Sepulveda, J.E. 2008. The theoretical rationale behind current trends in operating practice of semi-autogenous grinding operations. *Presentation at V International Mineral Processing Seminar -Procemin 2008*, Santiago, Chile

Spann, C., Ottergren, C. 2004. Is tertiary crushing replacing SAG milling as the optimal grinding technique? - The great debate – *5th Annual Crushing & Grinding. Jupiters, Townsville March 2004.*

Sylvestre, Y., Abols, J., and Barratt, D. 2001. The benefits of pre-crushing at Inmet Troilus Mine. *SAG2001-SAG Mill Circuit. Int. Conf. on Autogenous and Semiautogenous Grinding Technology* 3: 43-62.

Thomas, P.D.F. 1989. Development of S.A.G. mill motor drives. *Advances in Autogenous and Semiautogenous Grinding Technology*, Mular A.R. and Agar G.E., Eds, Department of Mining and Mineral Process Engineering, University of British Columbia 2:621-638.

Valery Jr. W. and Jankovic A. 2002. The future of comminution. *Proceedings of 34th IOC on Mining and Metallurgy Conference*, 30 Sept.-3 Oct. 2002, Bor Lake, Yugoslavia.

Vanderbeek, J.L. 1989. SAG mill recycle crushing plant performance at Chino Mines Company. *Advances in Autogenous and Semiautogenous Grinding Technology*, Mular A.R. and Agar G.E., Eds, Department of Mining and Mineral Process Engineering, University of British Columbia 1:379-404.

Vanderbeek, J.L. 1997. Tertiary grinding circuit installation at Chino Mines Company. *Comminution Practices*, ed. Kawatra, K.S., Society for Mining, Metallurgy, and Exploration, Inc. (SME), Chapter 31: 241-248.

Vanderbeek, J.L. 2006. HPGR implementation at Cerro Verde. *SAG2006-SAG Mill Circuit. Int. Conf. on Autogenous and Semiautogenous Grinding Technology* 4:45-61.

# FACTORS INFLUENCING THE SELECTION AND SIZING OF CRUSHERS

Ken Major

KWM Consulting Ltd., Maple Ridge, BC, Canada

## ABSTRACT

There are a number of factors that can influence the selection of crushers for specific applications. These can include operational considerations, location of the mine and ore characteristics. This paper identifies a number of these factors and discusses how they influence plant design and operations.

## INTRODUCTION

Crushing is an integral portion of the comminution flowsheet for mineral processing operations and is critical for the preparation of ore for downstream processing. The selection of the right crushing equipment for a specific application is influenced by many factors some of which are upstream of the crushing plant (blasting pattern and mining method) and others which are downstream of the crushing plant (heap leach or mill, grinding circuit selection). For most applications there is a flowsheet previously designed that will match the requirements.

The basic design of the various types of crushers uses gravity to clear the crushing chamber. The larger the crusher and/or crusher product the easier it is to clear the chamber for a specific ore. As the ore conditions vary; for example the amount of fines, the moisture content, particle size distribution, etc; the efficiency of the crusher to clear the cavity can have a significant influence on the crusher performance. The crusher or crushing circuit successful for one project does not necessarily translate to success on another project.

There have been a number of hybrid applications that have been developed to improve unit operating costs and/or increase capacity. The development of the SAG mill was primarily to reduce the costs related to the multi-stage crushing circuit. This trend has been reversed with the addition of a cone crusher in a SAG circuit (SABC) to crush the recirculating pebbles. These pebbles tend to be more resistant to impact breakage in the SAG mill and the crusher creates ore surfaces more conducive to breaking and grinding in the SAG mill. HPGR has been introduced after two or three stages of crushing to generate a finer feed to the ball mill circuits allowing an increase in throughput for the same grinding horsepower.

A crusher has a more efficient transfer of applied power to the breakage of rock than a grinding mill. This has been seen with the development and acceptance of the SABC circuit and also led to Nordberg's (Metso Minerals) development of the "Waterflush" cone crusher. The application of Waterflush technology has resulted in the production of a finer product than normally achievable with an equivalent shorthead cone crusher. Part of this was the ability to quickly relieve the fine ore using water so that it did not result in packing in the crushing chamber.

The high pressure grinding rolls (HPGR) have been included in most diamond processing flowsheets as they have been effective in crushing the Kimberlite ore to liberate the diamonds while minimizing diamond breakage. The HPGR has had recent successes in the iron ore industry as a substitute for the AG or SAG mill in preparing ball mill feed. The equipment suppliers are working very hard to adopt this technology for base metal and precious metals operations. The primary application proposed for the HPGR will be to provide one more stage of crushing prior to the rod mill or single stage ball mill reducing the unit energy consumption.

## FACTORS AFFECTING CRUSHER SELECTION

The factors that affect the selection of size and type of crusher for a specific application include:

- Plant throughput, ore delivery schedule
- Size of feed
- Desired product size
- Ore characteristics; hard rock, clay, gravel, variability, etc.
- Climatic conditions
- Downstream processes
- Capital Costs
- Operating Costs
- Experiencea

**Plant Throughput and Ore Delivery Schedule**

Plant throughput and ore delivery schedule will form the baseline for flowsheet design and equipment selection. From this information the size, type, number of stages and number of crushing units per stage required for an application can be identified. For example, a primary jaw crusher will be better suited for a conventional underground mining operation because:

- Tonnages are typically lower
- Feed material size is smaller, more fines are generated in blasting and mucking
- Less headroom and a smaller excavation is required

Higher throughput, open pit operations will typically use one or more gyratory crushers for the primary crusher application and product will be delivered to the mill feed stockpile for delivery to a SAG mill grinding circuit. Throughput limits for the design of secondary and tertiary crushers have a significant impact on the capital costs and operating costs for higher tonnage operations because of the number of crusher units per stage required to address capacity issues. As a result as plant throughputs increased the fewer the 3-stage crushing plants that were designed.

Process operations are normally designed for nominal throughputs. Grinding circuits generally achieve availabilities of 93% (SAG) to 98% (Rod Mill / Ball Mill). Crushing circuit availability will typically be 75% to 85% depending on the complexity of the circuit and application of surge piles and/or bins. Availability is a measure of time that the crushing circuit is in operation. The two components that impact the availability are scheduled downtime and unscheduled downtime. Scheduled downtime is typically attributed to routine plant maintenance including preventative maintenance, crusher rebuilds and clean-up and maintenance on ancillary equipment such as conveyors, chutes and screens. Unscheduled downtime is an interruption in the continuous operation that might result from equipment failures or plugged chutes.

The installation of surge capacity between unit operations permits the continuous independent operation of the equipment upstream or downstream of the surge pile or bin. The time period available for continuous operation will be dependent on the capacity of the surge. Without the surge capability unscheduled downtime on one stage of the crushing circuit will impact the other stages. For a three stage crusher circuit with no surge capacity between stages the overall plant availability will be lower than what is capable from any one unit operation. If one assumes that each unit operation has an 80% design availability based on 10% each for scheduled and unscheduled maintenance and assuming scheduled maintenance is conducted on all units at the same time, the overall plant availability could be 65%. To achieve the design throughput in these conditions larger crushing and ancillary equipment will need to be selected. This will result in higher equipment and installation costs. The inclusion of surge capacity in the circuit design will also impact the size of the crews required to maintain the equipment. If all of the equipment has to be maintained at the same time larger crews will be required. Whether or not to include surge piles or bins in the circuit design will be a capital and operating cost trade-off of increased costs for equipment versus cost of installing surge capacity for each application.

There are also occasions where the scheduled delivery of the ore will have a significant impact on crusher sizing and selection. Examples will be mines that are developed on the basis of delivering ore 2 shifts per day or 5 days per week. With the expectation of maintaining continuous 7 day feed to a downstream processing plant there will be a need to install larger crushers with surge capacity.

It is also important to consider upstream risks when selecting the crusher equipment. If one key piece of loading equipment in the mine or a conveyor in the ore delivery system has an unexpected failure that will result in extended downtime the surge capacity that has been provided in the crusher circuit or ahead of the grinding circuit will allow continuous operation for a period based on the status of the surge capacity at the time of the failure. Assuming the unscheduled failure results in using all of the surge capacity the risk to the operation will be in how long it takes to rebuild the live and dead load capacities.

**Feed Size**

For different mining methods and different ores, feed sizes to the crusher can vary significantly. The crusher selected must be sized for throughput and also the largest piece of rock that will be expected from the mine. The smaller the crusher the smaller the dimensions of the feed ore that can enter the crushing chamber. A balance is often required to ensure that the plant capacity and size of crusher are matched. For example, the installation of a 1220mm jaw crusher underground for an operation that the throughput dictates a 760 mm crusher would be sufficient, because of expected blast size. The ore generated from blasting should be evaluated to determine if the blasting or drilling concepts should be changed or a method of scalping oversize prior to the crusher should be installed.

In multi-stage crushing circuits the product of the preceding stage will be a determining factor in the selection of the type of crusher and the configuration of the crusher liners.

The duty of a cone crusher as a secondary crusher or as a tertiary crusher will need to consider both the feed size and the desired product size. The liner configuration in a cone crusher (extra coarse, coarse, medium or fine) will determine the largest piece that can enter the crushing chamber and the minimum closed side setting.

**Product Size**

The target product size required from the crushing circuit will determine the number of crushing stages and the types of crushers to be used for a specific application. For a SAG mill grinding circuit application requiring a coarse 150 mm feed a single stage primary crusher can be used.

For a low tonnage rod mill / ball mill or single stage ball mill application requiring a 12 mm feed size or finer a two stage crushing circuit using a primary jaw crusher and a secondary (standard) cone crusher in closed circuit with a vibrating screen may be appropriate. The vibrating screen ensures that the oversize generated in a single pass through a standard cone crusher does not create inefficiencies in downstream processes. A standard cone crusher operates with a larger throw and a smaller crushing chamber that when added to operator awareness and liner wear will result in a coarser crusher product.

A similar, higher tonnage operation will require a three stage crushing circuit with the third stage using shorthead cone crushers to produce the 9 to 12 mm product.

The ability to crush finer has been required for specific applications. The "WaterFlush" cone crusher concept was developed by Nordberg (Metso Minerals) using shorthead cone crusher designs to produce finer feed for grinding circuits allowing increased throughput at lower capital costs than costs typical of increasing throughput by adding grinding capacity. Water is added with the feed to the cone crusher. The water acts as a medium to flush fine particles from the crushing chamber and prevent packing enabling the crusher to be set with a tighter closed side setting.

For fine product sizes in dry process applications flowsheets have incorporated vertical impact crushers operated in closed circuit with vibrating screens. At a project in Uzbekistan vertical impactors were used as a quaternary stage of crushing to produce a 3 mm product to optimize recovery from the heap leach process.

**Ore Characteristics**

There are many ore characteristics that must be considered when sizing and selecting the equipment for inclusion in the crushing flowsheet including hardness, toughness, abrasiveness, moisture content and mineralization. The work done by the geologist in defining the ore body and the types of mineralization will provide information that can help define the approach to circuit design. The geologist should be able to provide some indication as to the different rock types and relative amounts of the different rock types that will be processed during the life of the mine. Together with the mining engineer a mine plan should be developed to indicate the short and long term schedule for delivering the ore. It is not unusual to find a situation that the best crushing circuit design for the early years of mine operation will not be the best circuit for later years requiring significant changes in how the circuit is operated and often requiring additional equipment. At one project a drill core rock study completed by the geologist, mining engineer and rock mechanics engineer determined that with a specific blasting procedure less than 6% of the ore would be coarser than 400mm and the decision was made to design a materials handling system that would permit feeding of -400mm direct to the SAG mill.

In evaluating the different ore characteristics an ore can be hard but can be relatively easy to crush because the rock will break along mineral boundaries. Similarly, a soft ore can be tough to break because there are no defined mineral boundaries. The minerals in the ore will provide an indication as to how abrasive the ore will be, the higher the silica content the higher the wear rate on crusher liners, transfer chutes, etc. High moisture content in the ore may lead to higher equipment downtime because of equipment plugging from fine material. Similar situations will also result if the ore contains significant quantities of clay or clay type minerals.

For hard ores with minimal fines conventional types of crushers that rely on gravity to draw the ore through the crusher, such as jaw, gyratory and cone crushers, can be used successfully. For softer ores with higher fines or moisture contents the use of crushers that propel the ore through the crushing chamber, such as rotary breakers (MMD), rolls crushers, horizontal impactors, may be a better selection using a rotating feed system for delivering ore through the crushing chamber. Rotary breakers have been used very successfully for fine soft ores with high clay contents such as nickel laterite operations or oil sands where the volume of rock in the feed is low. The application of traditional crushing equipment with high clay ores typically requires the use of water to help move the fines through the crusher or the installation of a wet screening plant before the crusher circuit to remove the clay.

One of the advantages of the SAG mill circuit was to improve ore materials handling problems that were a result of soft, clay like ore conditions that were not amenable to multi-stage crushing because of screen, crusher and/or chute blockages.

For some operations the ability to use specific methods to classify the feed has provided the opportunity to use conventional crushing equipment. For a clay ore gold heap leach operation it was determined that if the ore was delivered to the jaw crusher using a wobbler feeder the rock component of the ore could be fed to a single stage jaw crusher with the undersize of the feeder and the jaw crusher product recombined prior to feeding an agglomeration drum.

There will also be operations where the ore conditions may vary significantly based on the mining zone in the bit or seasonal conditions. The crusher selected in this situation may not be ideal but should allow processing of the range of ores.

**Climatic Conditions**

Climatic conditions have a different type of impact on selection of a crushing plant flowsheet. If the plant is located in a dry warm climate a crushing plant can be designed for an outside installation. In a wet climate location the crushing plant will need to be enclosed for operator accessibility and equipment protection. Depending on the ore type the wet conditions can also result in materials handling and crushing problems where the use of prescreening equipment and/or rotary breakers should be considered.

In very cold climate conditions the plant will need to be enclosed for operator accessibility and to contain heat. There are significant capital and operating cost implications when the crushing circuit has to be enclosed. A large building enclosure will be required to be heated and ventilated. This in turn will impact the design of the dust collection system. Larger operations and maintenance groups will be required as the number of unit operations increases.

Poor climatic conditions will favor the installation of a SAG mill grinding circuit requiring only a single stage of primary crushing to prepare the ore feed size. Operating costs for a SAG grinding circuit may be similar to a three stage crushing, ball mill grinding circuit but the capital costs will reduce significantly with the elimination of the crusher building and ancillary equipment favoring the selection of the SAG mill.

**Downstream Processes**

The downstream recovery processes will also influence the selection of crushing equipment. If there is a heap leach, the crusher product size will be specified for optimum recovery. The crusher flowsheet will be determined by the optimum feed size to leach and the throughput. Feed to heap leach pads can range from run-of-mine ore (blast control) to very fine product requiring four stages of crushing.

Most of the recent process plant grinding circuit flowsheets have focused on SAG mill installations. There are still occasions where rod mill / ball mill or single stage ball mill circuits are preferred. The number of stages and complexity of the crushing circuit will be determined by the design feed rate for the grinding circuit and the targeted grinding circuit feed size.

In either case the number of stages of crushing and the size, type and number of crushers required will be selected as discussed previously.

**Capital Costs**

Preliminary sizing and selection of the crusher flowsheet typically occurs during various stages of feasibility studies. Plant design capacity and product requirements are defined. From the design criteria the complexity of the crushing plant flowsheet is defined. After developing the preliminary flowsheets Meetings are held with vendors and knowledgeable people with experience in operations and/or maintenance of the equipment. Based on the various experiences that are presented the crusher is selected, budget equipment costs are provided, preliminary layouts are completed and a capital cost is compiled.

After the project is approved Vendors are requested to submit proposals based on Detailed Equipment Specifications developed by the Owner's engineer and approved by the Owner. For most projects the start of engineering also becomes the start of identifying cost savings opportunities. The crusher is the first process in the flowsheet, generally requires large expensive equipment and is the one process where everyone has expertise becoming an easy target for capital optimization.

From the Vendor's perspective the ability to have a unit cost less may favor them during the evaluation. From the Engineer's perspective they are tasked to control costs for the project and there are always some things that are going to cost more. From an Owner's representative perspective there an opportunity to save money in one area will provide money that can be used in another.

In general, although comparative cost estimates for various size crushers in the same application may be significant, 10% to 15% for gyratory crushers this estimated difference does not have a significant impact on the overall project costs. Before deciding to reduce the size of a crusher that was recommended and selected during the project feasibility study a complete and thorough assessment of the associated risks considering the knowns (ore conditions, surge capacity) and unknowns (equipment failures) should be completed.

**Operating Costs**

As crusher flowsheets become more complex with the addition of stages of crushing, screening and materials handling the equipment the maintenance requirements and the number of maintenance people increase. Each additional drive is something else that can go wrong and cause plant downtime. Each new transfer point is another spot requiring regular clean-up effort. The number of operators required to keep the crushing plant operating smoothly will increase with complexity (not necessarily size)

**Experience**

During the development of the crushing flowsheet design concepts and continuing through detailed design there is generally significant prior experiences that can be drawn on to ensure a successful operation. This experience can often be found on the Owner's team, the plant design engineer and most significantly the Vendors. Most Vendors have developed models that provide the opportunity to test different crusher flowsheet configurations. Using these models the ore feed conditions can be varied to be able to complete a preliminary evaluation showing how the variables affect the circuit performance.

Most experienced plant operators have seen crushers and so compared to most unit operations in a project the crushing plant will generally have the most people willing to provide input based on their prior experience. Based on the availability of all of this experience it can be assumed that the crusher flowsheet will always work – a poor assumption. Recent projects have seen:

- The installation of a rotary breaker in what typically would be a gyratory / jaw crusher application and subsequent addition of other hard rock crushing capability after plant commissioning.
- The reduction in size of the crusher identified in a feasibility study to reduce capital and subsequent installation of a larger crusher because of equipment failure and high operating costs. The replacement crusher was larger than the one identified in the feasibility study due to observed rock conditions after blasting.
- The insertion of an open circuit secondary standard cone crusher to feed a rod mill / ball mill circuit. The potential impact on production from poor maintenance and operating practices has been speculated but until the plant is commissioned the magnitude of the problem can't be defined.

A complex multi-stage crushing plant includes a large number of unit operations. It is not possible to account for all of the things that can go wrong with every piece of equipment but there can be reasonable assumptions made with respect to the overall availability of the plant. The amount of surge capacity provided in stockpiles and bins will influence the design capacity however it must be remembered that the plant needs to be able to refill the stockpile in a reasonable time period after it has been depleted. After the equipment has been selected a risk assessment should be completed to determine the economics of installing additional capacity by selecting a larger crusher to minimize the potential impact on the downstream recovery processes.

# Financial and Feasibility Studies

# FINANCING YOUR MINE

Deepak Malhotra[1] and Nick Michael[2]

[1] Resource Development Inc., 11475 West I-70 Frontage Road North, Wheat Ridge, Colorado 80033, USA
[2] SRK Consulting (U.S.), Inc., 7175 W. Jefferson Ave., Suite 3000, Lakewood, Colorado 80235, USA

## ABSTRACT

This paper presents a brief overview of the process required to progress a mining project from exploration through to financing, commissioning and startup.

## INTRODUCTION

Mining has been described as the art of turning rocks into dollars, though it is sometimes the other way around. The trick of course is to be able to distinguish the good projects from the bad projects. Generally, the lender or the investor would like to invest only in successful projects.

What makes a good project good? A good project in the eyes of some could be a bad project in the eyes of another. However, we could generally agree that the attributes of a good project could include high tons and good grades, favorable processing characteristics, minimal environmental impacts, etc. But it is how these attributes are determined by the owner and then presented to the potential lender which will ultimately win his confidence. A CEO of a mining company once said: "you can take a perfectly good, practical experienced mine superintendent out of the mine, and put him in a project evaluation team and he becomes a raging optimist.

Here are some typical comments made by project owners:

- "Excellent potential; we may be sitting on a district."
- "We have been very conservative in our estimates."
- "The permits will be in by Tuesday."
- "Joe Doe is the best geologist/miner/metallurgist/operator/manager… on the planet."
- "We are looking for a banker to be our long-term partner; someone who understands the business and recognizes opportunity."

But the lenders perspective will generally be different. In the back of his mind, he may be thinking what was quoted from a London Merchant Banker: "All I know is that if it begins in "Z" or ends in "stan"—I don't want to be there."

The lender, however, is interested in investing in the mining sector and if he is impressed with the project, he will want to learn more. So, what the lender will say is:

- "We are impressed with you and your project and can arrange a financing that will address all your needs… very competitively."
- "Now may we see your feasibility study?" and "We will tell you whether or not your feasibility study is bankable."

What are the steps required to prepare a bankable feasibility study for a project?

## PROJECT DEVELOPMENT STAGES

In order to win the confidence of the investor, it is important to quantify the attributes of a project using sound, industry-accepted engineering and scientific principles and to present them in a clear and documented manner which can be independently verified by either the lender or an independent third-party consultant of his choosing. The results of these efforts are of course, presented to the lender in the form of the feasibility study.

However, the most cost-effective path to a potentially bankable, or saleable, feasibility study is through the preparation of a series of studies and reports, each building from the prior as more information is obtained, and more is learned about the project, and more engineering work is completed. The systematic path to the ultimate goal of producing a bankable feasibility study is provided in Figure 1. A typical decision process is provided in Figure 2.

Early stage reports would include a scoping study followed by a conceptual study (or initial pre-feasibility study). Only when more is learned about how best to advance the project should the pre-feasibility and the feasibility studies be completed.

The data obtained for the project and what is assumed for the various levels of study are given in Appendix 1. These tables have been circulating in the mining industry for a long time (and hence the credit goes to whoever says they originally developed the information).

Scoping Study: The scoping study is the initial study where minimal technical information is available and the study is based on assumptions for the mining method, process flowsheets and production throughputs for determining the project economics. Some surface geology and sampling has been completed. Typically, no drilling has been done; hence no resource or reserve estimates have been determined and no metallurgical test work has been performed.

Capital and operating costs estimates are factored from similar operations. No formal quotes are obtained or design specifications are prepared. High contingency factors on the order of 30% to 50% are added to each estimate. The accuracy of the study is on the order of minus 30 plus 50%.

Information requirements are typically:

- Geographic location of the proposed operation.
- Type of ore to be processed.
- Magnitude production capacity of the project.
- Basic mining method and processing flowsheet.
- Products from the processing plant.
- Factored cost estimates.

The results of the scoping study are typically used to reject a project, but are seldom adequate for a positive, go decision for the project. In essence, the scoping study gives the operator an indication whether to proceed to the next project level with additional studies.

At this stage in the project development, minimal dollars and time have been expanded and the financial risk to the company has been minimal. The major risk associated with the scoping study is that the input assumptions into the study were so erroneous that they resulted in abandonment of the property.

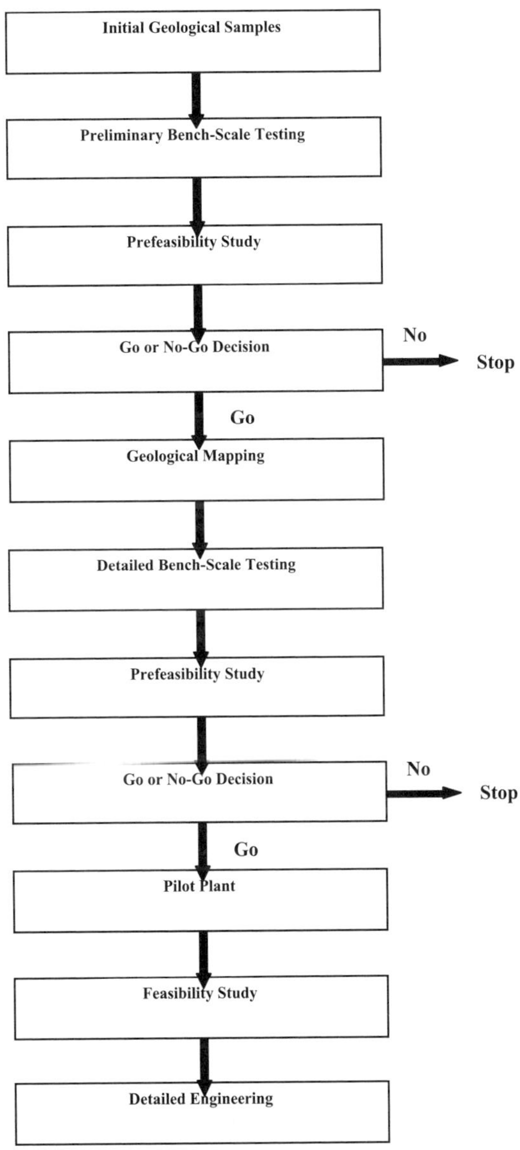


**Figure 1. Systematic Program for Process Development**

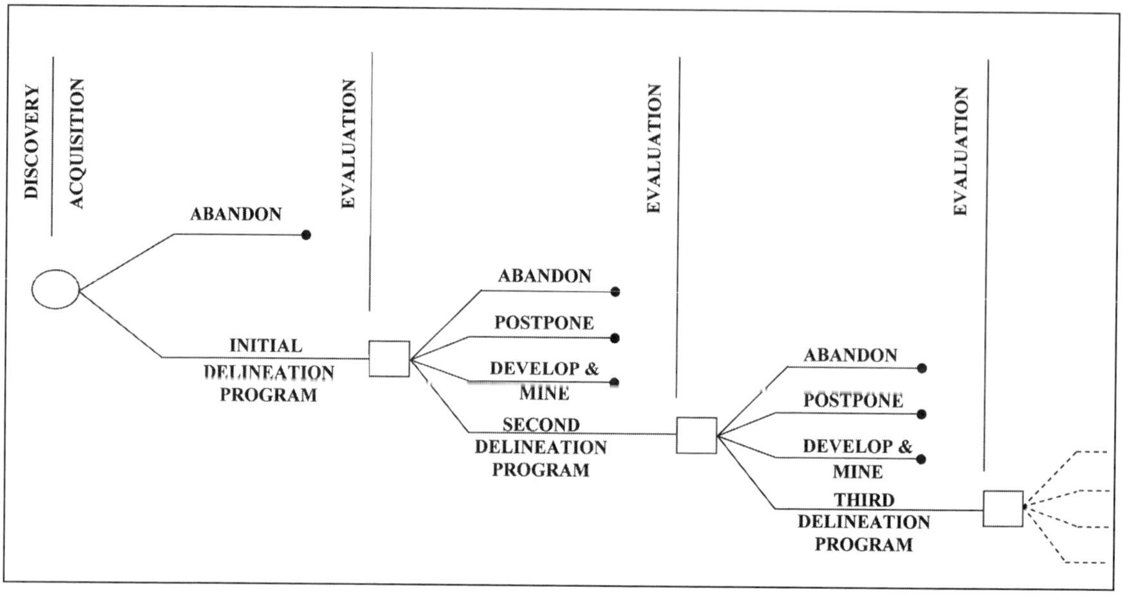


**Figure 2. Systematic Pathway to Evaluation Process**

Conceptual or Initial Pre-Feasibility Study: The conceptual study follows the scoping study and is performed when there is sufficient engineering work to prepare a conceptual mine plan based on an industry mining method, process flowsheet and production throughput for determining the project economics.

Surface geology and sampling has been completed; typically, some initial drilling has been completed. Sufficient geological information is available to make an initial resource estimate but no reserve estimates have been prepared.

Major equipment items have been sized and cost based on historic budget quotes or file data. Capital and operating cost estimates are factored as percentages.

Infrastructure requirements have been identified. However, no formal quotes are obtained. An initial assessment of the environmental factors has also been addressed.

Contingency factors on the order of 25% to 30% are added to each estimate. The accuracy of the conceptual study is of the order of minus 25 plus 40%.

In addition to the scoping study, the additional information requirements will typically include:

- A resource estimate.
- Initial bench-scale metallurgical test work.
- Determination of the type or types of ore to be processed.
- Conceptual production capacity of the project.
- Conceptual mining method and processing flowsheet.
- Products from the processing plant.
- Infrastructure identification.
- Environmental factors.
- Historic budget or file cost estimates.

Similar to the scoping study, the conceptual results are not of sufficient detailed accuracy to be used for positive, go decision for the project.

An initial environmental fatal flaw analysis for the project should also be conducted in this phase. In essence, the conceptual study gives the operator an indication whether to proceed to the next project level with additional studies.

At this stage in the project development, the project expenditures and engineering time have been moderate and the financial risks to the company have been moderate. Again, the major risk associated with the conceptual study is that the input assumptions into the study were so erroneous that they resulted in the abandonment of the property. No financing considerations have been discussed so far because they are irrelevent at this project stage.

Pre-Feasibility Study: Given positive results in the early stages and once trade-off studies have been completed and the path to advance the project becomes clear, it is time to prepare the Pre-Feasibility Study.

At this stage, sufficient engineering work to prepare preliminary engineering design and general arrangement drawings has been completed. Sufficient geologic work and drilling has been completed to prepare initial resource and reserve estimates for the deposit. In addition, bench-scale metallurgical test work has been completed.

Major equipment items have been sized with design specifications and costs based on vendor quotes and first principles. However, some equipment may not be sized in sufficient detail since definitive data does not exist as yet.

Preliminary environmental studies have been initiated and infrastructure requirements have been defined. Contingency factors on the order of 15 to 25% are added to each estimate. The accuracy of the pre-feasibility study is on the order of minus 15 plus 25%.

In addition to the conceptual study, additional information requirements will typically include:

- Resource and reserve estimate.
- Advanced bench-scale metallurgical test work.
- Preliminary construction schedule.
- Preliminary design drawings.
- Production capacity of the project
- Mining method and processing flowsheet
- Products from the processing plant.
- Preliminary environmental studies.
- Local labor rates and schedules.
- Local rates for utilities.
- Historic off-take agreements for products.
- Vendor quotes for consumables.
- Preliminary assessment of tax regime.
- Unit consumption rates based on metallurgical test work or historic data.
- Vendor quotes for major equipment items.

The Pre-feasibility study provides an increased level of accuracy because of reduced contingency levels. The results should be of sufficient detail and accuracy to be used for positive decision to proceed with a feasibility level study for the project where major commitments of capital and labor will occur.

At this stage in the project development, the project expenditures and engineering time are becoming sizeable and the financial risks to the company are increasing.

At this time, preliminary project financing discussions and options can and should be explored with financial community.

Feasibility Study: Given the discussion to move ahead with the project, the Feasibility Study will be performed. The generic outline for the Bankable Feasibility Study is given in Appendix 2. Most of the studies generally use this generic outline with minor changes for both Pre-Feasibility and Feasibility Studies.

By the feasibility stage, there is sufficient engineering work to prepare engineering design and general arrangement drawings supported by piping and instrumentation drawings, mechanical arrangements and single line electrical drawings.

Sufficient geologic work and drilling will have been completed to prepare definitive resource and reserve estimates for the deposit.

All geotechnical work for the mine design and processing requirements such as the tailings dam, heap leach pad, etc have been completed.

Bench-scale metallurgical and vendor tests work have been completed. Pilot plant studies if warranted, on the bulk sample have also been done.

All relevant environmental studies have been completed and the permitting process is well underway. Major equipment items have been sized with design specifications and costed based on firm vendor quotes. Capital cost estimates are based on material take-off calculations.

Infrastructure requirements have been defined and detailed manpower schedules are developed. Contingency factors on the order of 10 to 15% are added to each estimate. The accuracy of the feasibility study is on the order of minus 10 plus 15%.

In addition to the Pre-feasibility study, additional information requirements will typically include:

- Signed-off resources and reserve estimates.
- Completed geotechnical studies.
- Bench-scale and vendor metallurgical test work and pilot plant of a bulk sample if warranted.
- Detailed construction schedule with CPM.
- General arrangement design drawings.
- Material and energy balances.
- Defined mining method and processing flowsheet.
- Products from the processing plant.
- Defined infrastructure requirements.
- Local labor rates and schedules.
- Local rates for utilities.
- Indicative off-take agreements for the products.
- Vendor quotes for consumables.
- Detailed assessment of tax regime or regimes.
- Unit consumption rates based on metallurgical test work or historic data.
- Firm vendor quotes for equipment items.

The results of the Feasibility Study should be of sufficient detail and accuracy to be used for a go-no go decision for proceeding with detailed engineering for construction and development for the project.

A detailed engineering level study for the project is a major commitment of capital and labor for the company. Prior to this commitment, the feasibility study is generally presented to interested financial institutions.

## DUE DILIGENCE

The financial institution will undertake due diligence of the project. A due diligence is a thorough appraisal of all material aspects of an existing or proposed mining project by a team of appropriately qualified and experienced persons.

By the due diligence stage, advanced project financing discussions and options have been completed with potential financing institutions and financing is contingent to a positive due diligence outcome. The purpose of the due diligence is to:

- Examine the fundamentals of the project, to check for fatal flaws and to confirm that the project is technically feasible and economically viable
- Confirm that the project has been designed and planned in accordance with the sound engineering judgment on adequate data and realistic assumptions.
- Prove that the project schedule is achievable.
- Assure that project risks can be practically mitigated.

The due diligence will confirm the:

- Resources and Reserves.
- Mining planning and design.
- Access and infrastructure.
- Processing.
- Tailings design.
- Geotechnical and hydro-geological designs.
- Environmental and permitting status.
- Project status and schedule.
- Management and organizational structure.
- Technical and financial project economics.

However, the due diligence will also consider the influence of issues which may arise on the project such as:

- Authorship/ownership of changes if they occur.
- Compliance issues such as Equator Principles.
- Disclosure requirements for corporate reporting.
- Professional/corporate responsibility code of ethics.
- Integrated team/interaction during the feasibility study and for construction.
- Consultant support during the feasibility process.
- Guarantees/warranties for completion and performance.
- Last minute crisis and their resolution.
- Independent sign-off of reserves, designs and engineering.
- Ongoing monitoring as required.

Other factors, considered by the lender during the due diligence, process include:

- Inflation/Currency Devaluation.
- Exchange Rate.
- Discount Rate.
- The calculation of NSR.
- Hedge Book.
- Compliance with Term Sheet.
- Deferred Capital/Operating Costs.
- NPV vs. Market Capitalization.
- Exploration Prospects and.
- Political Risk.

## POST DUE DILIGENCE

Finally, once financing is in place, the lender may appoint an independent engineer to oversee the project through the detailed engineering, selection of the EPCM contractor, construction, start-up and completion testing phases as prescribed in the lender's term sheet.

## CONCLUSION

The authors have attempted to present an overview of the process, which is required to take a mining project from early exploration stage to building a profitable operation. In conclusion, despite careful planning and due diligence at various levels, projects do fail and often do not meet expectations. However, through systematic and detailed engineering, design, analysis and economic planning, project risk is minimized, and the probability of successful and timely financing is increased.

## APPENDIX 1: MINIMUM REPORT REQUIREMENTS FOR ENGINEERING STUDIES

**Minimum Report Contents for Engineering Studies**

| DESCRIPTION | Conceptual (Scoping) Study | Pre-feasibility Study | Feasibility Study |
|---|---|---|---|
| ***INTRODUCTION*** | | | |
| ***Location, Topography and Climate*** | | | |
| Site Location Map | Basic map | Preliminary map showing claims and boundaries | Detailed map showing all claims and boundaries |
| Topography Map | Basic map showing site topography | Preliminary map showing site topographic features | Detailed topographic map; aerial surveys verified with ground controls and surveys |
| ***Property Ownership*** | Review of property lease | Review of property lease; claims list provided; mineral rights known | Property lease and rights secured and controlled; claims list and map provided; mineral rights secured |
| ***Current Status and History*** | | | |
| Historical Chronology | Basic presentation | Full presentation | Detailed presentation |
| Past Production (if any) | Basic presentation | Full presentation | Detailed presentation |
| ***EXPLORATION AND GEOLOGY*** | | | |
| ***Geologic Description*** | | | |
| Review | Preliminary review | Preliminary site-specific analysis | Detailed site-specific analysis |
| Data Posting | Review of available existing maps | Detailed geologic mapping with cross-sections | Deposit well-defined with three dimensional mapping, geologic maps, long sections, level plans |
| Geologic Assessment | Preliminary | Basic assessment and review | Detailed assessment of structures/rock contacts, alteration, mineralization, deposit trends |
| Mineralogical Sampling & Analysis | Limited sampling; preliminary assessment | Preliminary mineralogical sampling and analysis; preliminary mineralogical study | Detailed mineralogical sampling and mapping; detailed mineralogical study |
| ***Drilling, Sampling and Assaying*** | | | |
| Drill Hole Parameters | Wide spaced drilling as appropriate | Initial in-fills of wide spaced drilling; preliminary grid patterns | Close spaced drilling on a detailed grid pattern to support calculated reserve categories |
| Underground Drilling | Review of existing data | Drilling if accessible | Detailed drilling if accessible |
| Samples | Preliminary; some outcrop samples | Geophysical and geotechnical sampling; test pits | All sampling programs complete |
| Drilling/Assay Data | Preliminary check of existing drill hole data | Check of drill holes (coordinates, elevations, angles, etc.), check assays, angled hole vs vertical hole comparison; assay flow diagram, dependable database | Check of drill holes (coordinates, elevations, angles, etc.), check assays, angled hole vs vertical hole etc.), check assays, angled hole vs vertical hole comparison, twin hole drilling; assay flow diagram; validated database |
| Condemnation Drilling | None | None | Areas under waste dumps, tailings and plant drilled |
| ***RESOURCES AND RESERVES (Internationally Recognized Standards [see note 1])*** | | | |
| ***Resources*** | Indicated and Inferred | Measured, Indicated, and Inferred | Indicated and Measured |
| Geologic Controls | Assumed | Established from geologic data and/or variograms | Well established from geologic data |
| Tonnage Factors | Preliminary assessment if available | Preliminary analysis and determinations | Detailed analysis and determinations |
| Statistical Analysis | Not performed | Preliminary analysis and determinations | Detailed analysis and determinations |
| Geostatistical Analysis | Not performed | Preliminary analysis and determinations | Detailed analysis and determinations |
| ***Reserves*** | Only resources estimated | Proven and Probable | Proven and Probable |
| Calculation Parameters | Usually no reserves are estimated | Known or estimated | Detailed analysis and determinations |
| Cutoff Grade (COG) Equations | Usually no reserves are estimated | Calculated from floating cone parameters | Optimized using mining/processing parameters |
| ***MINING*** | | | |
| Mining Method | Assumed between open pit and underground | Specific method identified | Method and mine plan finalized |
| Open Pit Mine Plan | | | |
| Pit Slopes | Assumed | Preliminary estimates by rock type and basic geotechnical data | Defined by geotechnical data from structural mapping and oriented core holes |
| Pit Design | Simple outline of final pit | Preliminary pit design from optimized analysis; preliminary haulroad incorporated | Detailed pit designs with phases and access for equipment operation |
| Waste Dumps | Simple outline of final dumps | Preliminary design for total waste tonnage; incremental and final outline of dumps | Dump sites identified from geotechnical data; final waste tonnages determined with incremental phases, yearly and final dump outlined |

**Minimum Report Contents (continued)**

| DESCRIPTION | Conceptual (Scoping) Study | Pre-feasibility Study | Feasibility Study |
|---|---|---|---|
| Underground Mine Plan | Assumed mining system; general outline of mine plan and development | Preliminary mining system identified from geologic and geotechnical data preliminary outline of mine plan and development including mine access | Specific mining system identified from geologic and geotechnical data; detailed outline of mine plan and development including mine access |
| Production Schedule | Basic schedule based on assumed mine life | Yearly and mine life ore and waste tonnages and grade | Detailed annual schedules showing ore / product quality and waste tonnages and grades |
| Capital Cost Estimate | Order-of-magnitude, factored or from similar operations | Preliminary equipment list; budget or historical price quotes; some factoring | Detailed equipment list; firm price quotes for all major equipment items; all capital items identified |
| Operating Cost Estimate | Order-of-magnitude; factored or from similar operations | Quantified estimates for labour, power and consumables; budget or historical price quotes; some factoring | Detailed engineering estimate by project area based on quotes and studies |
| ***PROCESSING*** | | | |
| ***Ore Sampling and Test Work*** | Minimal sampling; conduct lab bench scale process characterization tests on collected samples (if available) | Sampling of core; preliminary bench scale testing to determine preliminary recoveries, ore characterization and processing parameters for flow sheet development | Sampling of core for different ore body zones; confirm flow sheet; comprehensive beneficiation test program to determine recoveries, ore/product characterization and finalize processing parameters |
| ***Process Engineering and Design*** | | | |
| Production Rate and Product(s) | First estimate of production rate and product(s) | Preliminary mining and processing rates and plant product(s) | Fixed mining and processing rates and plant product(s) |
| Design Basis | Preliminary using factored estimates | General design basis; preliminary engineering drawings; trade-off studies optional | Complete design basis; basic engineering drawings essentially complete; trade-off studies performed |
| Design Concept | Outline of design criteria and specifications incorporating area/regional climatic conditions | Design criteria established for construction site incorporating known site climatic conditions | Design specifications defined incorporating known site climatic conditions |
| Process Description | General | Narrative; 1 to 2% of detail engineering complete | Detailed; 5 to 15% of detail engineering complete |
| Layout | Approximate geographic locations and site map; no general arrangement drawings | Optimization of facility locations on site map showing topography; simple general arrangement drawings of major equipment items | Exact geographic locations on site map with topography; detailed general arrangement drawings; detailed layout of all facilities |
| Flow Sheets | Assumed flow sheet from known processes; simple block diagram | Establishment of probable flow sheet from preliminary test work data; major process flow diagrams; initial determinations of material and heat balances. | Detailed flow sheet based on comprehensive beneficiation test program, detailed equipment list; diagrams for all process flows; material and heat balances finalized |
| Civil Work | Rough topographic maps; no soil conditions considered or quantities estimated | Rough topographic maps; soil conditions report for foundation determinations; basic preliminary quantities | Detailed topographic maps with soil conditions identified for foundation design, loadings and quantities |
| Equipment Specifications | Major equipment items listed | Preliminary listing of major equipment items with initial sizing and specifications | Complete listing of major equipment items with detailed sizing and specifications |
| Architectural | None | Sketches | Exterior elevations only |
| Piping/HVAC | None | Preliminary P&ID | Major P&ID |
| Electrical Distribution | None | Basic one-line diagram | All design one-line diagram |
| Motors | None | General description | Detailed list of major items with horsepower |
| Instrumentation | None | General description | Detailed list of components |
| ***INFRASTRUCTURE*** | | | |
| Facilities | General overview with types of support facilities described | All required support facilities identified, sizes and quantities estimated | All necessary support facilities identified, sized and costed |
| Communications | Communications requirements identified | Communications systems study | Communications licensing and standards known |
| Power | Overview of power availability and regional unit power costs | Power sources and requirements identified; unit costs obtained from power source | Power requirements and unit costs derived from detailed engineering study; unit costs from quotes |
| ***HYDROLOGY*** | | | |
| Water Sources | Estimated using regional data | Preliminary hydrology study | Specific water source identified |
| Water Usage | Factored plant volume and unit costs | Required plant water volumes and unit costs estimated | Requisite plant volumes and unit costs derived from detailed engineering/geotechnical studies |
| Dewatering | Dewatering parameters identified | Dewatering parameters estimated | Dewatering parameters confirmed and plan defined |
| ***ENVIRONMENTAL*** | | | |
| Setting | Preliminary evaluation of project setting for potentially significant environmental or permitting constraints for site data | Preliminary evaluation of the project's impact on the environment; schedule of environmental and/or other permitting requirements; evaluate project setting for potentially significant environmental and/or permitting constraints from site data | Characterization of all the project's potential impacts on the environment; finalize schedule of environmental and/or other permitting requirements; evaluate project setting for potentially significant environmental and/or permitting constraints |

**Minimum Report Contents (continued)**

| DESCRIPTION | Conceptual (Scoping) Study | Pre-feasibility Study | Feasibility Study |
|---|---|---|---|
| Data | Collect and review all available, existing data for environmental studies assessments or audits; regulatory inspections, waste handling practices, management plans, and all applicable environmental laws and regulations; no social, training or safety programs identified | Collect and review available environmental data from existing databases for environmental studies, assessments or audits; regulatory inspections, waste handling practices; management plans; and all applicable environmental laws and regulations; plans; initiate baseline data gathering social, training, and health /safety programs identified | All requisite environmental data for project are identified; site sampling and analyses are complete; detailed review of the type, scope and schedule for producing environmental and/or government reports; comprehensive gathering and evaluation of baseline environmental conditions; social, training, and health/safety programs confirmed |
| EIS/EA | None | Draft EIS/EA initiated | Draft EIS/EA submitted to regulatory authorities |
| Reporting and Plans | Conceptual plans for managing any identified environmental issues | Preparation of environmental plans and monitoring programs; preliminary sediment and erosion control plan; conceptual reclamation plan; evaluation of acid rock drainage; geotechnical stability review of waste dumps and tailings dam; preliminary impact mitigation plan; preliminary spill and emergency response plan | Environmental characteristics used in project design; environmental plans and monitoring programs are finalized; sediment and erosion control plan; management plan finalized for solid and hazardous wastes; finalize impact mitigation plan; geotechnical stability analysis of all major facilities; finalize reclamation plan; final analysis of acid rock drainage; finalize spill and emergency response plan |
| Monitoring | Not considered | Outline of a site environmental monitoring plan | Complete environmental monitoring plan |
| ***PERMIT REQUIREMENTS*** | General overview | Comprehensive overview and listing of required permits | Detailed evaluation of all pertinent environmental and permitting requirements and schedule for obtaining operating license |
| ***PROJECT DEVELOPMENT SCHEDULE*** | | | |
| Development Plan | Development period and mine life estimated | Development period and overall schedule estimated; mine life determined development schedule set | Detailed development schedule; mine life known; development schedule finalized |
| Project Master Schedule | Estimated showing start and end of construction; Gantt bar chart of major work elements | Gantt bar chart with overall time frames; schedule outline for detailed engineering; QA/QC program outlined; preliminary construction schedule; preliminary project execution plan | Gantt bar chart with overall time frames and project flow planning; detailed project level schedule showing project deliverables and detailed engineering; CP schedule; major milestones identified; project control system outlined; QA/QC and safety program finalized; preliminary project procedures manual; project design basis finalized |
| ***CAPITAL COST ESTIMATE*** | | | |
| Basis<br>Civil Structural<br>Piping/HVAC<br>Architectural Electrical<br>Instrumentation<br>Construction Labor<br>Construction Labor Productivity<br>Material Volumes/Amounts<br>Material/Equipment Pricing<br>Infrastructure | Order-of-magnitude based historic data or factoring | Estimates from historical factors, percentages and vendor quotes based on materials volumes | Detailed from estimates; engineering 15 to 25% complete; multiple vendor quotes |
| Contractors | Included in unit cost or as a percentage of total cost | Percentage of direct cost by cost area for contractor; historic for subcontractors | Written quotes from contractor and subcontractors |
| EPCM | Percentage of estimated construction cost | Percentage of detailed construction cost | Calculated estimate from EPC(M) firm |
| Pricing | FOB mine site including all taxes and duties | FOB mine site including all taxes and duties | FOB mine site including all taxes and duties |
| Owner's | Historic estimate | Estimate from experience factored for similar project | Estimate prepared from detail zero based budget |
| Environmental Compliance | Factored from historic experience | Estimate from experience factored for similar project | Estimate prepared from detail zero based budget for design engineering and specific permit requirements |
| Escalation | Typically not considered | Based on company's current budget percentage | Based by cost area with risk |
| Working Capital | Factored from historic experience | Estimate from experience factored for similar project | Estimate prepared from detail zero based budget |
| Accuracy | +/- 50% | +/- 25% | +/- 15% |
| Contingency | 25% | 15% | 10% |
| ***OPERATING COST ESTIMATE*** | | | |
| Basis | Order-of-Magnitude estimate | Estimates for unit rates and quantified estimates with some factoring | Detailed from zero-based budget; minimal factoring |
| Operating Quantities | General | Quantified by estimates with some factoring | Detailed estimates |
| Unit Costs | Historic unit costs and factoring | Estimates for labor, power and consumables; some factoring | Letter quotes from vendors; minimal factoring |
| Accuracy | +/- 35% | +/- 25% | +/- 15% |

**Minimum Report Contents (continued)**

| DESCRIPTION | Conceptual (Scoping) Study | Pre-feasibility Study | Feasibility Study |
|---|---|---|---|
| ***ECONOMIC EVALUATION*** | | | |
| Financial Analysis | Preliminary assessment of principal economic parameters | Assessment of the principal economic parameters | Full assessment of all principal economic parameters |
| Commodity Price(s) | Estimated based on 3yr average minimum | Estimated based on 3yr average minimum | Estimated based on 3yr average minimum or detailed market studies |
| Royalties and Taxes | Preliminary assessment | Preliminary analysis | Detailed analysis with tax authority opinion |
| Smelting, Refining and Freight | Historic data | Budgetary quotes | Firm quotes |
| Cash Flow Analysis | Simple cash flow | Preliminary cash flow | Formal, detailed cash flow |
| Economic Criteria | Simple IRR and NPV (pre- and after-tax) | Preliminary IRR and NPV (pre- and after-tax) | Fully defined IRR, NPV, ROI, and payback period (pre- and after-tax) |
| Sensitivity Analysis | Basic analysis to minimal amount of project variables | Preliminary to selected key project variables | Numerous analysis to all key project variables |
| ***RISK EVALUATION*** | | | |
| Risk Assessment | General overview | Fatal flaw analysis | Formal Monte Carlo analysis and fatal flaw analysis |
| Project | Preliminary overview of geology, engineering, and environmental | Preliminary environmental, country, permitting, technology, and business; detailed geology and engineering | Detailed geology, engineering, environmental, legal, permitting, country, technology, business, and financial |

Note 1: Internationally Recognized Standards include:

1. Canadian National Instrument 43-101 and 43-101 CP.
2. Australasian Code for Reporting of Mineral Resources and Ore Reserves - prepared by the Joint Ore Reserve Committed (JORC)
3. U.S. Securities & Exchange Commission Industry Guide 7
4. SME Guide for Reporting Exploration Information, Mineral Resources and Mineral Reserves

## APPENDIX 2: TYPICAL PRE-FEASIBILITY AND FEASIBILITY STUDY REPORT OUTLINE

- INTRODUCTION
  - Project Overview
  - Company Profile
  - Scope of Work and Terms of Reference
  - Basis of Study
  - Qualifications of the Consultant
    - Project Team and Responsibilities
  - Reliance on Information
  - Disclaimer
  - Units of Measure and Abbreviations
- PROPERTY DESCRIPTION, LOCATION, HISTORY AND OWNERSHIP
  - Property Description and Location
  - Access
  - Physiography and Climate
  - History
  - Ownership
    - Land, Water and Mineral Rights
- SUSTAINABILITY
  - Infrastructure
    - Transportation
    - Water
    - Services
  - Social Issues
    - Skilled/Unskilled Labor Forces
    - Hospitals
    - Schools
- GEOLOGY
  - Regional Geology
    - Lithology
    - Structure
    - Mineralization
  - Local Geology
    - Lithology
    - Structure
    - Mineralization

- Exploration
  - Historic Exploration
  - Current Exploration
  - Proposed Exploration
- MINERAL RESOURCES
  - Geologic Database
    - Mapping
    - Drilling
    - Logging
  - Sample Analyses
    - Sample Security
    - Sampling, Sample Preparation and Assaying
    - Quality Assurance/Quality Control
    - Specific Gravity Determinations
  - Exploration Database
    - Database Verification
    - Data Statistics
    - Assay Compositing
  - Block Model
    - Geologic Model
    - Variography
    - Grade Interpolation
    - Block Model Validation
    - Adequacy of Resource Estimation Methods
  - Mineral Resource Statement and Classifications
    - Resource Estimate
    - Resource Classification
- MINERAL RESERVES
  - Reserve Estimation
  - Conversion of Resources to Reserves
- MINE PLAN
  - Selection of Mining Method
    - Mine Design Parameters
  - Optimization
  - Pre-Production Mine Development
  - Mine Production Schedule
  - Mine Equipment
    - Operating Parameters
    - Mine Fleet
  - Mine Labor
  - Mine Support and Utilities
  - Capital Costs
  - Operating Costs
- METALLURGY
  - Metallurgical Test Programs
- PROCESS
  - Process Design Criteria
    - Process Assumptions
  - Alternative Processes
  - Process Description
  - Site Layout Considerations
  - Materials Handling
  - Recoveries
- GEOTECHNICAL
  - Tailings and Waste Rock Storage
    - Tailings and Waste Rock Facility Design
    - Phasing
    - Volumetrics
    - Slope Stability
    - Hydrogeological
    - Closure

INFRASTRUCTURE
- Access Roads
  - Mine Access Roads
  - Plantsite Access Roads
- Minesite Facilities
- Plantsite Facilities
- Tailings
- Power
- Water
- Product Transportation
- Communications
- Fuel Storage
- Security

ENVIRONMENTAL
- Environmental Regulations
- Environmental Permitting
- Health and Safety
- Environmental Monitoring and Management
- Environmental Baseline Data
- Environmental Management Plan
- Closure
- Reclamation

MARKETING
- Commodity Specification
- Probable Uses and End Products
- Market Demand and Supply
- Price
- Prospective Users

CAPITAL COSTS
- Exploration
- Pre-Production Development
- Working Capital
- Mining
- Process
- Infrastructure
- Owner's Costs
- Contingency
- Exclusions

OPERATING COSTS
- Mining
- SProcess
- G&A
- Contingency
- Exclusions

ECONOMIC EVALUATION
- Project Funding
- Assumptions
- Operating Costs
- Capital Costs
- Taxes and Royalties
- Dividends
- Economic Indicators
- Economic Results
- Sensitivity Analysis

PROJECT PLANNING
- Implementation Strategy
- Project Organization
  - Project Management
- Project Schedule
- Responsibilities and Scope
- Construction
- Start-up Sequence
- Process Commissioning

- Acceptance Testing
- Insurance
- Policy
- Contractor Insurances
- Project Controls
- Operations
- Training

PROJECT OPPORTUNITIES, RISKS, CONCLUSIONS AND RECOMMENDATIONS

- Opportunities
- Risks
- Conclusions
- Recommendations

REFERENCES

# EARLY-STAGE RISK MINIMIZATION IN PROCESS FLOWSHEET DESIGN

Will R. Goodall[1] and Mike D. Adams[2]

[1]MinAssist Pty Ltd, Australia
[2]Mutis Liber Pty Ltd, Australia

Of increasing importance in the design of extractive metallurgical process flowsheets is the early identification of techno-economic, occupational safety and environmental risks potentially associated with the project. This is particularly the case for semi-refractory low-grade, high-volume mineral resources, where it is essential to constrain both capital and operating expenses to the minimum.

An approach is presented that serves to minimize the upfront costs of project assessment, allowing the stakeholders either an early exit or decision to proceed to the next level. This is achieved via the integration of several modern tools available to the process engineer, including historical data mining and analysis, quantitative mineralogy techniques such as QEMSCAN® or MLA, 'sighter' level physico-chemical testwork and short-form process and economic modelling. The current global economic position is limiting access to mineral process development funding; hence, the ability to produce meaningful process development decisions on a limited budget is of great value.

## INTRODUCTION

The rapid and economic evaluation of new projects is an integral part of business for mining companies looking to expand their resource and project base. The global economic climate has meant that many opportunities have become available for discerning operators. However, it has also become necessary to rapidly evaluate these targets at minimal cost, allowing best use of limited capital reserves The paper presented examines the concept of utilizing desktop analysis, along with targeted mineralogical analysis to provide rapid evaluation of technical risk associated with a project, the impacts on process design or optimization and economic viability.

Due diligence on potential projects is often simply a desktop exercise that relies upon existing plant or testwork data. The proposed method adds cost effective and rapid mineralogy-based physico-chemical analysis to the desktop process in due diligence. This allows a total evaluation of both historical and present technical and economic risks. Armed with this information rapid go/no-go decisions can be made early in the evaluation cycle, thereby reducing costs and streamlining the process. While it is imperative to consider all aspects of risk when evaluating new projects, this paper chiefly examines rapid techniques for evaluation of technical risk specific to extractive metallurgical issues.

A strong approach to early-stage project evaluation should be efficient, providing maximum useable data for minimal capital outlay. By focusing on mineralogy-based analysis early in the project life cycle it can be possible to generate a good understanding of the expected behavior of material through existing or proposed process flowsheet options. This focus allows operators and investors to effectively identify potential impacts on value recovery, which becomes especially important in low-grade refractory materials. These materials are often characterized by low margins and requirements for economies of scale to produce a viable resource. A good understanding of the impact that refractory effects may have on recovery can therefore mean the difference between a viable project and an uneconomic one.

The testwork required to prove up a potential process is directly related to the novelty of that process, the complexity of the ore and nature and quantity of design criteria (Cambell, McConaghy and Vardill 2004). This means that it can be important to define these parameters early in the project life-cycle so effective planning and budgeting for feasibility. The proposed process utilizes rapid quantitative mineralogical analysis, generated by systems such as QEMSCAN®, in conjunction with traditional metallurgical and analytical processes. The outcomes of this analysis provide a qualitative understanding of key material parameters. When combined with desktop due diligence studies this can provide a first-pass overview of the technical risk associated with a project. The information generated can then be readily input into a simple economic model of the proposed process to generate economic parameters suitable for making the decision to proceed with or terminate a project.

An example undertaken by the authors is the evaluation of a low-grade gold calcine project located in Western Australia.

## RISK MINIMIZATION IN PROJECT EVALUATION

The definition and minimization of risk in evaluation of any project is an important consideration. This is especially true in difficult economic climates where only a small number of potential projects may be evaluated due to capital constraints. It is therefore important to quickly identify the viability of proceeding with projects at an early stage to ensure the most efficient use of capital reserves. In these situations an efficient process for project evaluation and risk minimization means that additional projects can be evaluated under the same capital budget, increasing the probability of positive financial return for the investors.

The process proposed by the authors utilizes a decision-tree based flowsheet. The process is presented in Figure 1 and follows a logic path defined by a series of go/no-go decisions open to the review team. Following this process has allowed the authors to clearly define to both management teams and investors the logic behind decisions to proceed with or terminate projects at an early stage. The goal of using this approach is to provide an avenue for early termination of unviable projects before unnecessary expenditure is made.

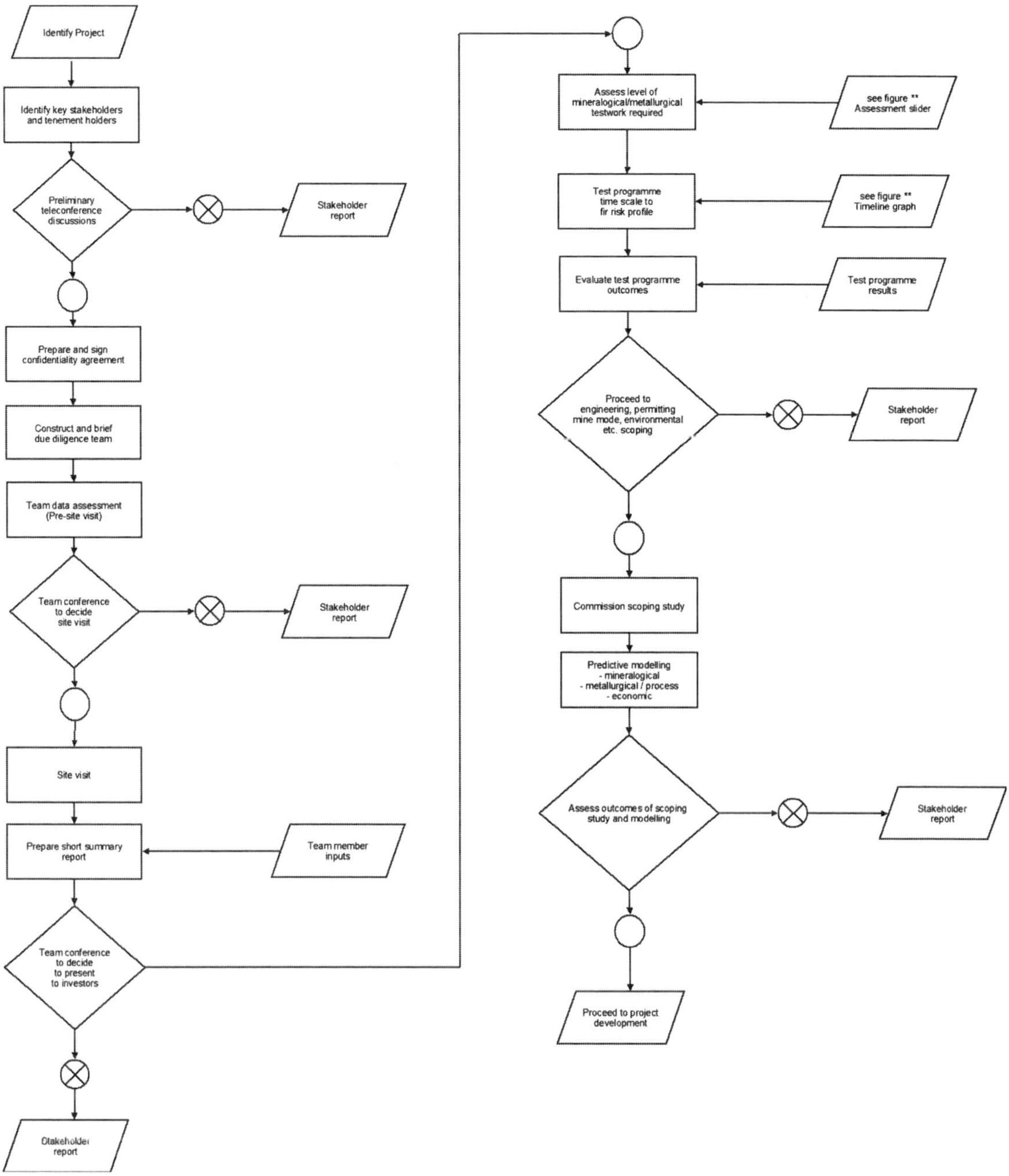


**Figure 1. Decision tree flowsheet for risk minimization in project evaluation**

As an example, application of the process presented in Figure 1 to evaluation of a gold calcine tailings project allowed an efficient evaluation pathway to be established. Early contact and collaboration with the tenement holders, who were themselves developing a hard-rock project at the site, was initiated. Before proceeding it was important to establish the basis for implementation of development outcomes with the existing tenement holder and sign confidentiality agreements to protect both parties. A core team was put together including a mineralogist, metallurgist and financial analyst. Historical feasibility and process reports were provided by the tenement holder and the review process was undertaken. The initial focus of due diligence evaluation was on definition of the potential project size and value. This was supported by assessment of where more in-depth data was required, as a measure for development of the subsequent testwork programme. The outcomes of initial analysis identified sufficient reserve and value to proceed with the evaluation and undertake a site visit to establish site layout and existing amenities.

The process for developing a targeted and efficient testwork programme was based around key parameters to define the most appropriate focus of analysis. This was achieved using the testwork slider process presented in Figure 2. Using this process, key variables were analyzed and combined to establish the outcomes required from the testwork programme and consequently the form that the programme should take. This provided a structured approach to testwork development that could reduce the risk associated with undertaking testwork that was not directly relevant to the final decision-making process. The evaluation process addressed four key areas in the evaluation of new projects and assigned them a value in a sliding scale based upon the amount of data already existing. This has been presented in Figure 2 and provided a benchmark for which areas need to be targeted to ensure that sufficiently robust data was available for the decision process in pursuing a project.

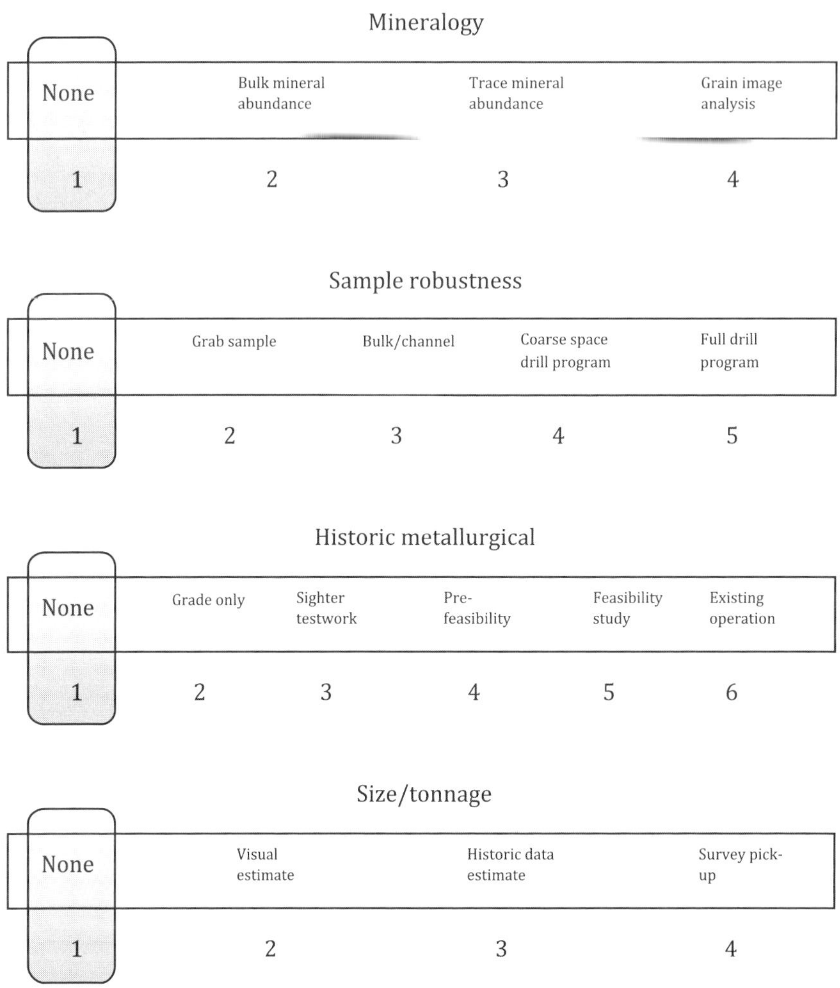


**Figure 2. Slider selection process for testwork programme development**

A key consideration in the evaluation of potential metallurgical processes and risks for a potential ore or tailings material should be the process mineralogy. The proposed process provides a strong focus on the investigation of process mineralogy as a tool to define the key parameters that will define the most appropriate metallurgical process for recovery of contained value. To achieve this, rapid and economic evaluation of mineralogy is required. This has become feasible in the last decade with the growth of automated mineralogical systems, such as QEMSCAN® or MLA, which can be used to provide statistically relevant mineralogical, locking/liberation and grain size information. These tools have been used effectively in ore characterization and process evaluation. A number of examples have been presented where they have been used to rapidly identify problems by Miranda and Anderson (2007), Latti et al (2001) and Goodall (2008). All these studies, along with numerous others, show use of automated mineralogy techniques in evaluation of existing process issues but provide the benchmark for use of the systems in project assessment.

The outcomes of analysis for the calcine tailings example under examination can be seen in Figure 3. From this simple visual analysis it was clearly identified that although a reasonable amount of quality data was available for metallurgical analysis on robust samples, there was a gap in any mineralogical investigation. For a complex material, such as calcine tailings, an understanding of the mineralogy and key mineral deportment can have a significant impact on the most appropriate processing route to maximize recovery. The key findings from this analysis were that the testwork programme was warranted and should focus on measurement of mineralogy. This could be used in association with some specific metallurgical testwork targeted at preliminary investigation of innovative ways to win gold value from this material.

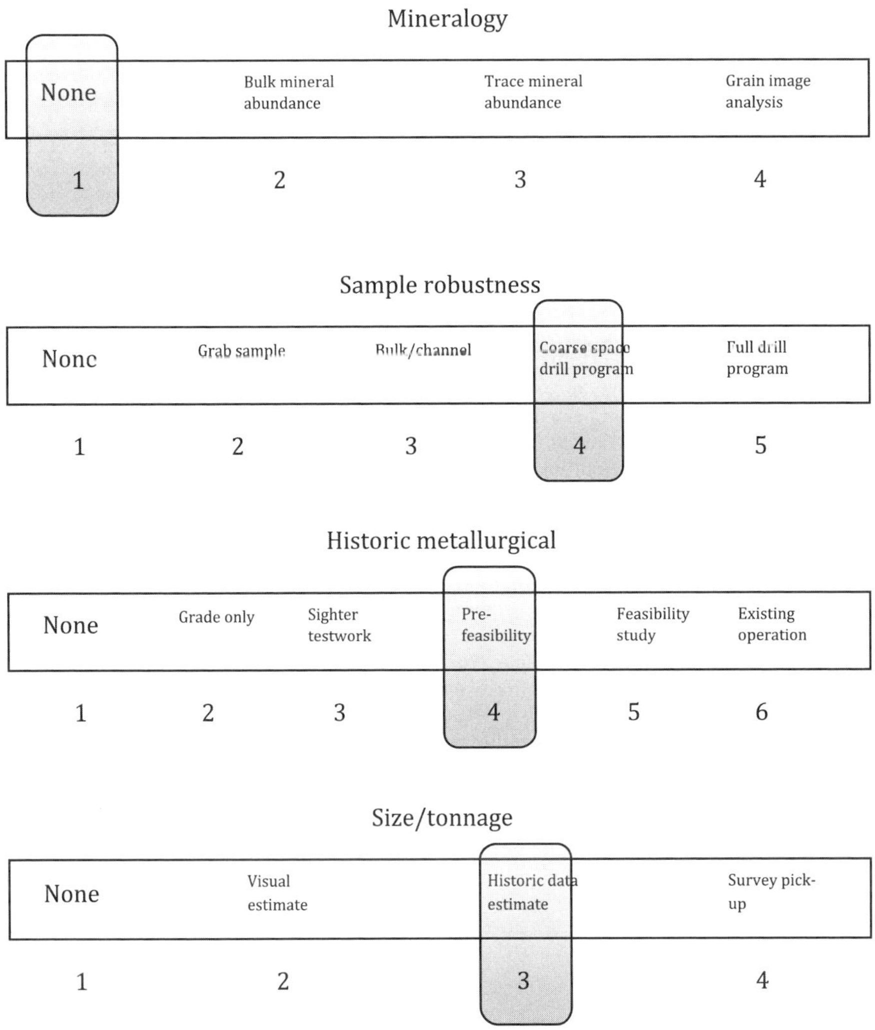


**Figure 3. Slider selection process for testwork programme development for a calcine tailings project**

The implementation of the mineralogically focused testwork programme gave sufficiently robust data for a scoping study to be commissioned, examining the viability of a number of innovative process flowsheets. The scoping study was designed to provide cost estimates to ±30% accuracy and give a basis for preliminary economic analysis. This would allow an informed decision to be made on whether to proceed to full feasibility analysis or terminate the project before significant outlay had been made.

## ECONOMIC MODELING

Once sufficient historic, mineralogical and metallurgical information has been gathered, economic evaluation of options can commence. The workflow proceeds as follows:

1. Base case input data is typically formulated into a table for input into the developed spreadsheet model.
2. A mass-balance simulation is constructed for each of the process options under consideration.
3. Capital and operating cost estimates are developed for each option, based on available site or region-specific data.
4. Comparative analysis is now used, inclusive of specific scenario testing (e.g., varying capital amortization periods, upset conditions).
5. Cashflow modeling over life of mine is used to determine potential issues using the selected minimum risk option.

This approach was used by Adams (2006) in consideration of sequential counter-current decantation (CCD) stage elimination and replacement with resin-in-pulp stages. Adams and Lloyd (2008) used a similar approach in assessment of a variety of cyanide destruction and recovery options against the novel base case of tailings washing and pond stripping.

In the case of the calcine tailings example referred to earlier, a process flowsheet was selected and the economic model tested various scenarios of gold recovery and earnings split between the tenement holder and the process operator. In this case it was concluded that while an acceptable gross margin could be realized over the anticipated three-year project life, the risk profile was unacceptable in the following aspects:

- Gold recovery variability
- Relatively low unit recoveries using conventional technologies.
- Poor cashflow in the first half of project life.
- Earnings split precluded a satisfactory earnings scenario for both tenement holder and process operator.

## BRINGING IT TOGETHER

To make an informed decision on the viability of a project under evaluation the review team should draw on as much technical and economic information as possible. The process described allows the operator to systematically evaluate important parameters for the project related to overall risk minimization early in the project life-cycle. The example presented in this paper showed the process for evaluation of a gold calcine tailings project. This project showed sufficient up-side to warrant progressing through to the scoping study and economic analysis stage. The outcomes of this deeper analysis showed that return on investment (ROI) was borderline for the project when all capital and operating costs were evaluated. Using the evaluation philosophy proposed this defined that expenditure should be halted on the project.

It can be seen from the procedures implemented that a structured approach to new project evaluation can allow small mining companies to get the best use of limited funds available. By using a disciplined approach decisions can be made quickly and justified to investors with supporting information.

## REFERENCES


Adams, M.D. Towards a virtual metallurgical plant. 1. CCD vs RIP case study. In: *ALTA Ni/Co 06, Proceedings*, ALTA Metallurgical Services, 2006. 17 pp.

Adams, M.D. and Lloyd, V.B. Cyanide recovery by tailings washing and pond stripping. *Proceedings, Precious Metals 07, Brisbane, August 30-31 2007; Minerals Engineering*, vol. 21, pp. 501-508 (2008).

Campbell, F,, McGonaghy, E. and Vardill, W, 2004. Startup and reliability of nickel laterite plants, in *International Laterite Nickel Symposium 2004* (Eds: W P Imrie and D M Lane), pp25-42 (The Minerals, Metals and Materials Society: Warrendale).

Goodall W.R. (2008) "Characterisation of mineralogy and gold deportment for complex tailings deposits using QEMSCAN" *Minerals Engineering* 21 518-523

Latti D., Doyle J and Adair B.J.I (2001) "A QEM*SEM study of a suite of pressure leach products from a gold circuit" *Minerals Engineering* 14 (12) 1671-1678

Miranda P. and Anderson C. (2007) "MLA: the new Sherlock Holmes of the mining and metallurgical engineering industry" In: *Proceedings of the 39th Annual Canadian Mineral Processors Conference 2007* Ottawa, Canada 23-25 January, 2007 pp 553-566

# GOING GREEN IN MINING PROJECT FINANCE—WHAT DOES IT MEAN IN TERMS OF THE ENGINEERING STUDY?

Terry Braun

SRK Consulting (U.S.), Inc., 7175 West Jefferson Avenue, Suite 3000, Lakewood, CO USA

## ABSTRACT

Since 2003, a growing number of international lending institutions have endorsed a set of environmental and social policies and guidelines termed the Equator Principles. As of December 2008, sixty-four financial institutions from twenty-six countries adopted the Equator Principles. Funding via external project financing is increasingly likely to be subject to an overall project assessment with respect to environmental and social impacts. In this case, part of the "bankability" status of an engineering study will be contingent on satisfying the Equator Principles. This paper will present an overview of the Equator Principle program and identify opportunities for integration of the program objectives during development of an engineering study.

## OVERVIEW

Project financing in the mining industry has always relied on sound engineering and economic modeling in the "bankable" Feasibility Study to define the terms and conditions of a lending agreement. Within the last 10 years, this scope expanded into consideration of social and environmental issues associated with a given project. Such consideration includes concepts of sustainable development, and "green building" design as well as environmental, health and safety guidelines. Some argue that it is the failure of mining companies to include the social and environmental costs of their activities in their feasibility studies that has upset so many environmental and sustainability advocates (Botin, 2009).

In 2003, a consortium of international financial institutions developed the Equator Principles (EP) as an initial framework for lenders to evaluate the social and environmental issues of a proposed project. A financial institution that officially subscribes to the Equator Principles is known as an "Equator Principles Financial Institution" or EPFI. The Equator Principles define a process for the EPFI to appraise the relative social and environental aspects of a given project based on established international guidelines. This appraisal can translate into an Action Plan and Management System which present compliance requirements that become loan convenants. Similar to conventional loan covenants, the EPFI reserves the right to implement remedies as it deems necessary to correct non-compliance with these loan covenants.

Projects that will require external funding will benefit from careful review of the EP guidance at each stage of the engineering study. Many of the design concepts associated with Equator Principles represent the standard of care currently applied in engineering studies for mining projects around the world. However, much like the innovation of the mining industry, the EP guidelines are reviewed and updated to reflect lessons learned and emerging good practice. This paper highlights a number of EP guidelines and observed trends that are relevant to the mining industry and of particular importance during the engineering study.

### Engineering Study Defined

The composition of an engineering study varies based on the phase of the project. Specific definitions for each level of study may differ slightly in the mining industry; however, for the purposes of this paper, the levels of engineering study can be defined as scoping, prefeasibility or feasibility level.

A scoping study (e.g., conceptual) typically takes place during the early phase of the project and relies on preliminary resource estimate data. Capital and operating costs are normally factored are assumed based on similar operations. Processing and metallurgy aspects are based on preliminary test work. The project proponent may consider a series of smaller "trade-off" studies to compare and contrast alternative mining or process methods. There is no formal environmental permitting activity; however, the regulatory context of the project is defined and some level of stakeholder engagement may take place. The common level of cost accuracy for a scoping study is approximately ±35% although this range may vary.

The prefeasibility study represents an increased level of analysis of a specific mine plan. At this phase of the project, the preferred mining and process method is identified. Capital and operations costs are based on preliminary supplier quotes for major items and project-specific estimates of productivity and equipment requirements. The resource estimate is based on additional drilling and the anticipated mining method. Processing and metallurgy aspects are based on additional bench scale test of representative drill core. Based on the preferred mining and process method, the proponent may initiate early lead time activities associated with major environental permit requirements. The prefeasibility defines major equipment requirements. The common level of cost accuracy for the pre-feasibility study is approximately ±15%.

The feasibility study is the final phase of detailed engineering design prior to development of a bid package for construction. The mine plan and process plant are defined in detail. At this phase of the project, the draft EIS or EA for the project is under review by the lead regulatory agency and other environmental permits are in progress or complete. Capital and operating cost estimates are based on specific supplier quotations. The target level of cost accuracy for the feasibility study is approximately ±5%.

In general, a project proponent relies on a "bankable" feasibility study to obtain project financing. At this phase of the project, a proper engineering study will allow the EPFI to efficiently assess compliance with the Equator Principles and

**Table 1. High-Income OECD Members (June 2009)**

| | | |
|---|---|---|
| Australia | Greece | New Zealand |
| Austria | Hungary | Norway |
| Belgium | Iceland | Portugal |
| Canada | Ireland | Slovak Republic |
| Czech Republic | Italy | Spain |
| Denmark | Japan | Sweden |
| Finland | Korea, Republic of | Switzerland |
| France | Luxembourg | United Kingdom |
| Germany | Netherlands | United States |

The World Bank Data and Statistics, Country Classification, 2008.

other performance criteria related to social and environmental issues. With a working knowledge of the key elements of the Equator Principles, the design team can make the feasibility study more "bankable" and gain other project efficiencies (e.g., favorable loan covenants). This approach during the early project phases (e.g., scoping and pre-feasibility) will also help the team define a mine plan that anticipates schedule and cost risks associated with social and environmental issues for the project.

### Equator Principles Defined

As an international guideline, the Equator Principles provide a benchmark of performance criteria for project finance opportunities around the world. These criteria may be superceded in countries which have regulatory, permitting and public comment process requirements that generally meet or exceed the requirements of the International Finance Corporation (IFC) Performance Standards and Environmental Health and Safety Guidelines (Equator Principles, 2006). In general, these countries meet the World Bank criteria for High-Income and are members of the Organization for Economic Co-operation and Development (OECD).

Table 1 presents the list of High-Income OECD countries as defined by the World Bank Development Indicators Database. Proponents for projects located in countries listed in Table 1 can demonstrate compliance with relevant host country laws, regulations and permits in lieu of the IFC standards. This demonstration must verify that the benchmark criteria of the Equator Principles are met.

Per the overall statement of principles (Equator Principles, 2006), EPFIs will only provide loans to projects that conform to Principles 1 through 9 listed in Table 2 (Principle 10 is the reporting obligation assumed by the EPFI). Each principle represents a logical progression within the Equator Principle framework; however, this paper will focus on Principle 3: Applicable Social and Environmental Standards. Specifically, Principle 3 incorporates IFC performance standards and Industry Specific EHS Guidelines which are particularly relevant to the engineering study.

## IFC PERFORMANCE STANDARDS EXPLAINED

The IFC maintains and updates the inventory of performance standards on social and environmental sustainability which are broadly applicable to private sector projects throughout the world. There are 8 performance standards published by the IFC; however, in terms of the engineering study, Performance Standard 3: Pollution Prevention and Abatement presents most applicable design criteria for mine planning and process design.

In general, this IFC standard promotes control technologies and practices that minimize or reduce adverse impacts on human health and the environment while remaining technically and financially feasible and cost-effective. The standard tends to rely on narrative-style descriptions rather than quantitative criteria. The independent reviewer retained by the EPFI must be in a position to identify and assess the degree of compliance of the engineering study with respect to this IFC standard.

Performance Standard 3 includes provisions that are consistent with a typical Pollution Prevention Plan for hard rock mining operations in the United States. Such plans are typically affiliated with a site waste management or storm water protection program. Key design criteria in a pollution prevention plan include:

- Efficient use of energy;
- Efficient use of water; and
- Waste avoidance and utilization.

Leading, global mining trade groups recognize the value of materials stewardship or the responsible production, use, re-use, recycling and disposal of materials (ICMM, 2007). In terms of pollution prevention and materials stewardship, such groups seek sustainable consumption and production. Under a policy of continuous improvement, good practices associated with these concepts will change over time and the Equator Principles and IFC will likely revise the performance standard accordingly.

### Mining Industry Specific EHS Guidelines Explained

The IFC publishes EHS Guidelines that are applicable to underground and open pit mining. The guideline is broad enough to address conventional as well as alluvial and solution mining applications. The mining industry specific guideline borrows elements of the general IFC Performance Standard 3 described above.

**Table 2. Equator Principles Summary**

| Principle | Title | Description | Comment |
|---|---|---|---|
| 1 | Review and Categorization | Assess magnitude of potential impacts and risks of the proposed project | Based on generic environmental and social screening criteria of the International Finance Corporation |
| 2 | Social and Environmental Assessment | Conduct an assessment to address impacts and risks of the proposed project | Generally consistent with international Environment Imact Assessment methodologies |
| 3 | Applicable Social and Environmental Standards | Demonstrate the project's overall compliance with, or justified deviation from, applicable standards | Incorporates by reference the IFC Performance Standards and applicable Industry Specific EHS Guidelines if project is located in a non-OECD country or not designated as High-Income |
| 4 | Action Plan and Management System | Plan to address relevant findings of the Assessment | Applicable to projects located in a non-OECD country or not designated as High-Income |
| 5 | Consultation and Disclosure | Confirmation of consultation with project affected communities in a structured and culturally appropriate manner | Projects with potentially significant adverse social or environmental impacts that are diverse, irreversible or unprecedented |
| 6 | Grievance Mechanism | Commitment to continue consultation, disclosure and community engagement throughout the project and to establish a grievance mechanism as part of the management system | Projects with potentially significant adverse social or environmental impacts that are diverse, irreversible or unprecedented |
| 7 | Independent Review | Retain an independent social or environmental expert to review the Assessment, Action Plan and consultation process documentation assist in lender due diligence | Audit function |
| 8 | Covenants | Incorporate covenants in the financing documentation that are linked to compliance with EP 7 | In the event of non-compliance with the covenants, the EPFI reserves the right to exercise remedies as they consider appropriate |
| 9 | Independent Monitoring and Reporting | EPFI will appoint an independent environmental and/or social expert to verify monitoring information | Applies during the life of the loan |
| 10 | EPFI Reporting | Annual reporting by the EPFI of implementation processes and experience | Public disclosure |

Equator Principles, July 2006

In addition to industry-specific impacts and management, this guideline describes performance indicator and monitoring protocols to assess compliance with environmental and safety standards. Under impacts and management strategies, the guideline provides narrative and quantitative standards for Environmental, occupational health and safety and community safety. Table 3 presents a summary of environmental impact categories and example management strategies. The IFC standard provides significant narrative and quantitative guidance based on what is considered "Good International Industry Practice."

**Engineering Study Relevance**

Figure 1 illustrates the relationship between the Equator Principles described above (or equivalent Environmental Impact Assessment) and the phases of the engineering study.

During the scoping level phase of a mining project, opportunities to minimize or reduce adverse impacts on human health and the environment should be identified by the project proponent. These opportunities can be factored into trade-off studies that compare and contrast engineering options related to mine development. This decision-making process is documented prior to proceeding to the prefeasibility study of a project.

**Table 3. Mining Industry Environmental Impacts and Management Strategies (December 2007)**

| Environmental Category | Sample Requirements |
|---|---|
| Water Use and Quality | • Efficient oil and grease traps or sumps should be installed and maintained at refueling facilities, workshops, fuel storage depots<br>• Permanent stormwater drainage systems should be designed for a 100 year/24 hour recurrence period<br>• Sediment control facilities should be designed and operated for a final Total Suspended Solids discharge of 50 mg/L<br>• Pipeline systems carrying pregnant solutions should be designed with leak detection systems and secondary bunded containment |
| Wastes | • Dumps should be designed based on the nature of the material and local geotechnical considerations to minimize erosion and reduce safety risks<br>• Potential change of geotechnical properties in dumps due to chemical or biologically catalyzed weaterhing should be considered<br>• Consider use of natural or synthetic liners for tailing systems to minimize risks<br>• Non-hazardous waste should be collected for recycling or disposal at an approved sanitary landfill<br>• Non-hazardous solid waste should not be disposed of together with waste rock or overburden except under exceptional circumstances |
| Hazardous Materials | • To minimize the risk assocated with accidental spills from storage tanks and pipelines:<br>• Construct pipelines with double-walled or thick-walled sections at critical locations<br>• Installing shutoff valves to minimize spill volumes and isolate flow in critical areas |
| Land use and Biodiversity | • Habitat-specific impacts will be defined by the Social and Environmental Assessment (or equivalent). |
| Air Quality | • Dust suppression techniques for roads and work areas<br>• Loading, transfer and discharge of materials should take place with a minimum height of fall, and be shielded against the wind and consider use of dust suppression spray systems<br>• Conveyor systems for dusty materials should be covered and equipped with measures for cleaning return belts |
| Noise and Vibrations | • Implementation of enclosure and cladding of processing plants<br>• Installation of proper sound barriers and/or noise containments with enclosures and curtains at or near the source equipment<br>• Use of specific blasting plans, correct charging procedures and blasting ratios, delayed/ microdelayed or electronic detonators |
| Energy Use | • Use of non-invasive technologies such as remote sensing and ground-based technologies to minimize exploratory digging and drilling<br>• Correctly sizing motors and pumps used in the excavation, ore moving, ore crushing and ore handling process as well as using adjustable speed drives in applications with highly varying load requirements |

Environmental, Health, and Safety Guidelines, Mining, December 10, 2007

At the prefeasibility level, potential environmental mitigation measures are subject to a more detailed technical and financial evaluation during the design effort. This analysis will form part of the environmental impact assessment (or equivalent) which is conducted independent and during the latter stage of the prefeasibility study.

During the feasibility level design, the project proponent proceeds with detailed planning of proper pollution prevention and abatement measures for the selected mine and process design. At this point, the draft environmental impact assessment (or equivalent) for the project is filed with the lead regulatory authorities. Environmental permits that involve significant lead times (e.g., baseline data collection, discharge or disposal approvals, stakeholder consultation) are pursued.

The feasibility study forms the basis for lending institutions to review and appraise a project prior to the funding of major capital required for project development. A key part of the "bankability" of the feasibility study is the degree of compliance with IFC performance standards referenced in the Equator Principles and assessed by the EPFI.

## RELATED TRENDS

Mining today is typically an international endeavor with many mining companies operating multiple projects in developed and undeveloped countries. Starting with World Bank guidelines, followed by the IFC performance standards

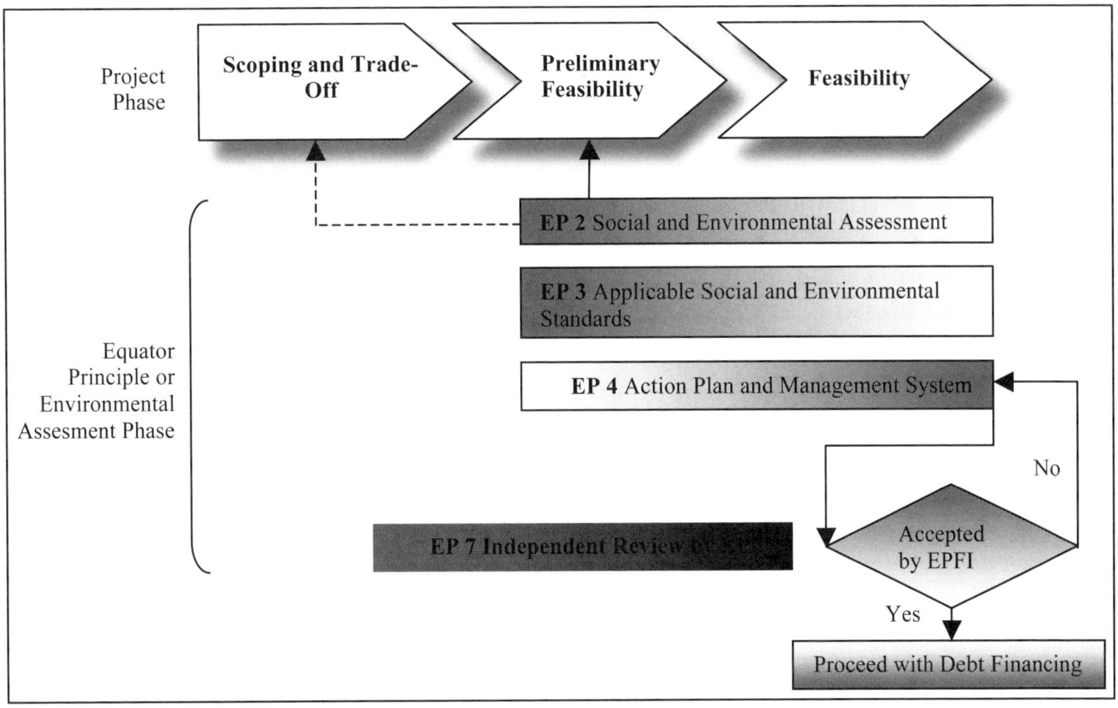


**Figure 1. Relationship of Equator Principles to the Engineering Study**

and the Equator Principles, mining companies recognize international benchmarks for social and environmental performance. In addition, mining trade organizations (e.g., International Mining and Minerals Council) promote innovation and continuous improvement in terms of "Good International Industry Practice" and sustainable development concepts. Given that the IFC guidance and Equator Principles update what is considered good practice on a routine basis, an overview of another international standard related to green building construction may offer insight into future trends.

The U.S. Green Building Council (USGBC) is one of several organizations focused on Green Building design in commercial and residential spaces. Organizations like the USGBC endorse the Leadership in Energy and Environmental Design (LEED) as an international standard for green building practices. The LEED standards address various aspects of environmentally sustainable construction. Version 3 of the LEED 2009 rating system includes the following evaluation criteria with respect to engineering design and construction practices:

- Sustainable Site—how will the construction phase address pollution prevention, will habitat within the project site be protected or restored, how does the stormwater design promote quantity and quality control?
- Water Efficiency—how does the design achieve water use reduction, are innovative wastewater technologies applied, is there water efficient landscaping?
- Energy and Atmosphere—does the design optimize energy performance, is there on-site renewable energy, to what extent is green power used?
- Materials and Resources—how are materials re-used during the contruction and operating phase of the project, what is the recycled content of the building materials, is there an opportunity to use regionally available materials?
- Indoor Environmental Quality—are low-emitting building materials used, is there an air quality management plan during contruction and before occupancy?

As indicated under the IFC performance standards, water quality protection and energy conservation are key factors in mining and process design today. The concepts of water use reduction, recycled material use and "green power" are of growing importance in commercial engineering design. Such concepts will advance and ultimately contribute to a new "Good International Industry Practice" which will be used to benchmark future engineering studies.

## CONCLUSION

Environmental design practices will play an increasing role in the success of future engineering studies conducted for the mining industry. The Equator Principles incorporate a combination of narrative and quantitative design criteria which will be assessed in the event of debt financing for a project. EPFI's recognize the significance of environmental and social assessments, particularly in the context of the mining industry. The assessment conducted by the EPFI is a critical part of a successful project financing. For projects subject to EPFI financing, the project proponent should addressing the Equator Principles from the scoping phase through the feasibility study of a project.

## REFERENCES

International Council on Mining and Minerals (ICMM), 2007. *Materials Stewardship, Eco-Efficiency and Product Policy*.

IFC, 2006. *Environmental, Health and Safety Guidelines for Mining*. December 10.

Botin, J.A. ed., 2009. *Sustainable Management of Mining Operations*. Published by the Society for Mining, Metallurgy, and Exploration, Inc.

## SUGGESTED READING

For the complete (and most up to date) version of the Equator Principles, visit: http://www.equator-principles.com/documents/Equator_Principles.pdf

For the specific IFC EHS guidelines, including the industry specific standards for mining, visit: http://www.ifc.org/ifcext/sustainability.nsf/Content/EnvironmentalGuidelines

For more information on the USGBC and the LEED program, visit: www.usgbc.org

# CONCENTRATED SOLAR POWERED MINERAL PROCESSING

Jeff Claflin, Rod Leonard, Doug Rogers, and Fernando Urrutia

Lycopodium Minerals Pty Ltd, 1 Adelaide Terrace, Perth, Western Australia, Australia

## ABSTRACT

Mineral processing facilities are all strongly dependent on power. Remote facilities often depend on diesel for their power requirements. Given the high variability in the price of oil and subsequently the price of diesel, the future viability of projects may be uncertain. Concentrated solar power can be strategically applied to decouple the cost of power from the cost of oil thereby reducing the project's exposure to oil's highly volatile price. This paper explains the methodology used to select the solar technology and its application to a remote mineral processing facility with a 30 MWe load. Preliminary capital and operating costs are presented and used to determine a levelized cost of electricity and ultimately an equivalent price of diesel fuel for a breakeven point for concentrated solar power.

## INTRODUCTION

Lycopodium has investigated a range of power supply options for a proposed mineral processing plant in a remote region of Western Australia. Concentrated solar power generation was identified as being potentially viable for this remote project. Various solar technologies were considered prior to selection of a preferred configuration for this particular application.

### Importance of Power

Mine profitability is directly related to plant operating cost. For most mineral processing facilities the largest single operating cost is power (Figure 1) and by far the greatest consumer of power is comminution (Figure 2). Power use may be optimized especially in the comminution area with up to 16% reduction in the power or increased plant capacity being realized, by the judicious use of energy efficient crushing and grinding equipment. However, the overall power demand must still be satisfied.

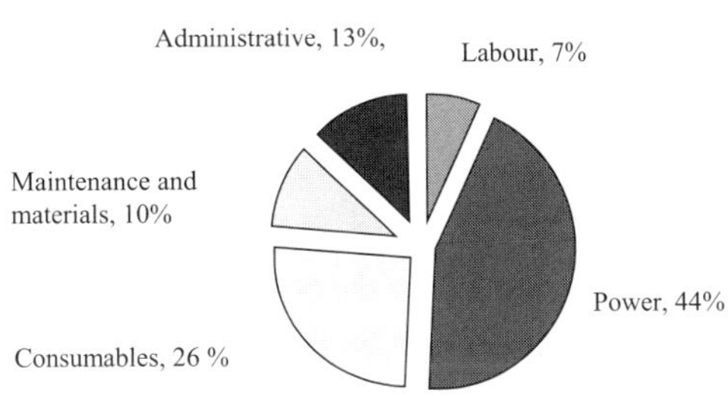


**Figure 1. Typical Mineral Processing Operating Costs**

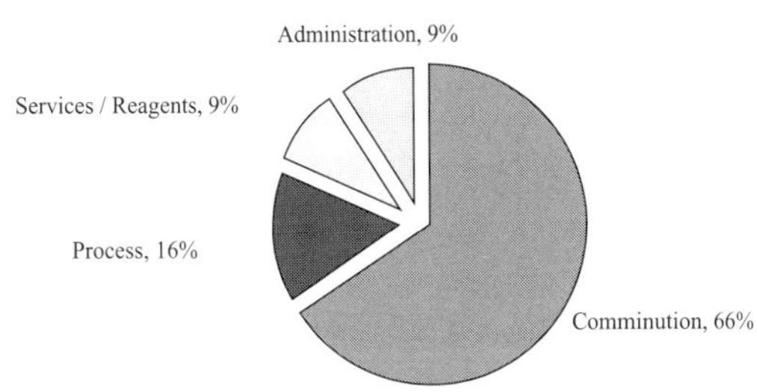


**Figure 2. Typical Mineral Processing Power Usage**

### Power Alternatives

Many countries including the United States of America and Australia, are heavily dependent on fossil fuels as shown in Table 1 due to the low cost, availability and convenience of fossil fuels. However, unless a grid network is well established, inexpensive power is often not available where it is needed.

Many remote operations rely on diesel as their primary fuel as it is easily transported and may be used for both mining and power generation. However, diesel price is directly linked to the price of oil and in recent years the price of oil has varied from less than 20 US$/bbl to over 135 US$/bbl (see Figure 3)—more than a 600% variation within the life of a typical mineral processing project. Notably, the ability to predict the price of oil is not good as evidenced by Figure 4 where in 2001 the price of oil was forecast by Gaffney, Cline & Assoc. to drop rather than rise. Perhaps conservatively it may be said that since 1999 the trend (inserted arrow of Figure 3) is upwards at a rate of US$4 per barrel per year. At this rate the price of oil will easily double over the next 20 years and the price of diesel will double with it.

As power represents the largest portion of the operating cost, the high variability in the cost of energy can put the viability of projects in doubt even when the capital cost of the project and the current cost for energy suggest the project is sound.

Power may be provided from many other sources of energy including hydro, biomass, wind, geothermal, tidal, wave and solar. These alternative forms of energy are not strongly supported in Australia and the USA (Table 1) and generally are only used to support the grid.

Remote mineral processing facilities are often not connected to the power grid and must be treated as power 'islands'. Suitable alternative forms of energy are usually location specific, e.g., tidal energy can only be used adjacent to coastlines

**Table 1. Energy Sources**

| | Australia(Ref 1,2), % | USA(Ref 3), % |
|---|---|---|
| Coal | 75.6% | 49.1% |
| Natural Gas | 15.0% | 16.3% |
| Oil and Diesel | 1.8% | 0.7% |
| Petroleum Coke | 0.0% | 1.1% |
| Nuclear | 0.0% | 19.4% |
| Combustion Turbine | 0.0% | 3.6% |
| Hydro | 6.4% | 6.9% |
| Biogas | 0.1% | 0.0% |
| Biomass | 0.4% | 1.3% |
| Incinerators | 0.0% | 0.3% |
| Geothermal | 0.0% | 0.3% |
| Wind | 0.7% | 0.7% |
| Solar | 0.0% | 0.1% |
| Renewables | 7.7% | 9.7% |

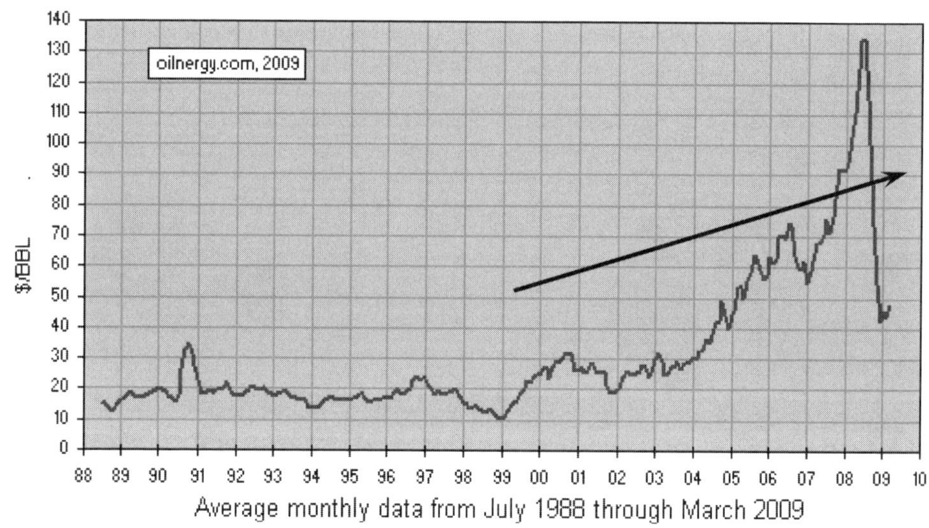


**Figure 3. Historical Brent Crude Oil Price, Ref 4**

with significant tides. By far the most prevalent energy resource is the sun and solar radiation is available at sufficient levels to be technically feasible over vast areas across the world (lighter regions in Figure 5).

Concentrated Solar Power (CSP) is the group of technologies that can capture solar radiation reaching the surface of the earth and convert it to thermal energy suitable for power production. Photovoltaic (PV) power also uses the sun's radiation directly, but, for daytime only operation, PV generally requires of the order of 50% more capital than CSP and, if storage is considered, as much as 500% more capital. Concentrated Photovoltaic (CPV) is competitive with CSP on an up front capital cost for a daytime only basis, (refer to time equals zero of Figure 6,) but is uncompetitive on an operating cost basis, (compare slopes of CSP and CPV in Figure 6—long dash and dotted lines respectively).

Figure 6 also compares a number of power station alternatives to CPV and CSP for our base case 30 MWe facility located 350 km away from an existing power grid or natural gas line. The cost and time values were preliminary and were used only to indicate where to apply further engineering effort. Nonetheless Figure 6 clearly demonstrates that, for a typical mine life of ten years or more, CSP is worthy of consideration over any diesel fuel scenario and would likely be equal to the cost of a grid connected plant, should the grid be able to support it, or a gas turbine, should a pipeline have spare capacity. For the project being evaluated, the grid was expensive to extend and required additional generation capacity and there was insufficient spare natural gas available in the pipeline located 350 km away, unless new capacity was brought on stream. Hence CSP was investigated more fully.

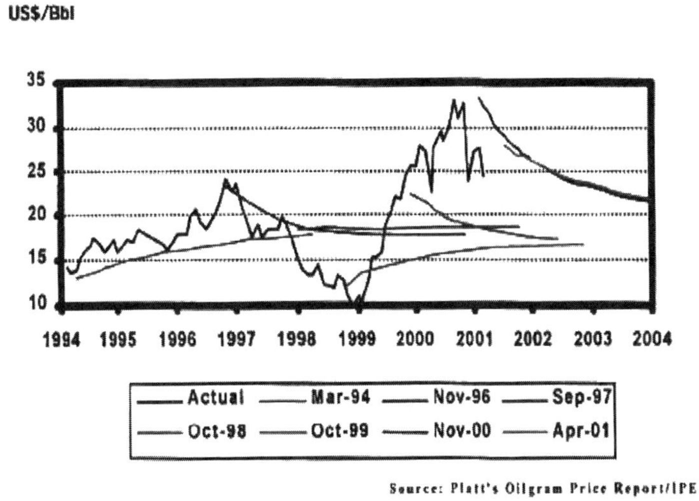


**Figure 4. Forecast for Brent Crude Oil to 2004, Ref 5**

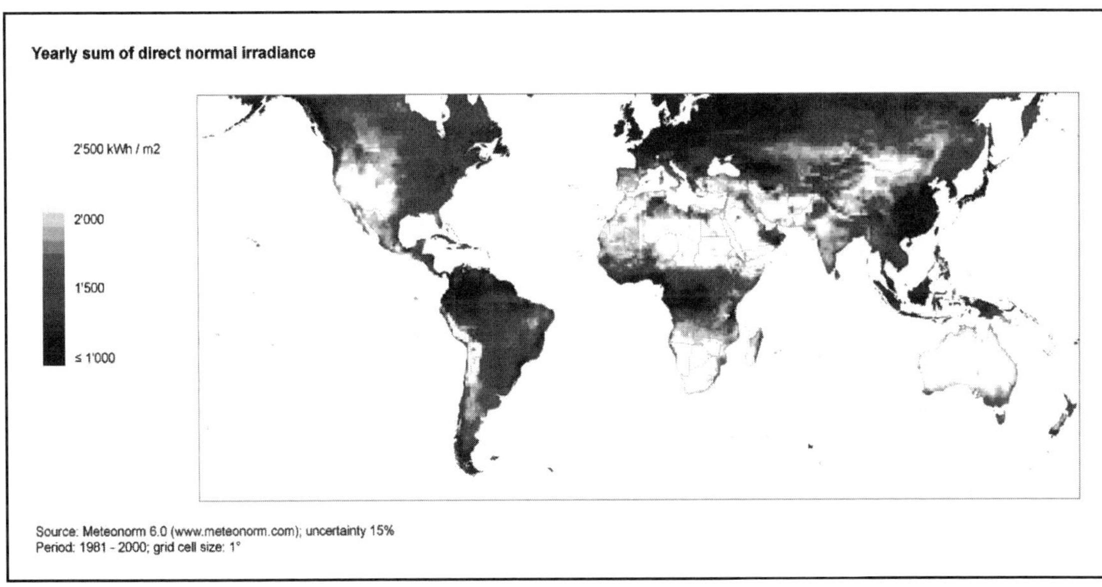


**Figure 5. Annual Direct Normal Irradiation**

## ELEMENTS OF CSP

CSP involves three fundamental unit operations: radiation collectors (the solar field), energy storage, and the power block, schematically represented in Figure 7.

### Solar Field Alternatives

The purpose of the solar field is to collect radiation in a form that is useful. Radiation can be collected in flat plates, e.g., domestic solar hot water systems, but the temperature that can be achieved is relatively low and the potential for converting this heat into work is limited (see Figure 8). The higher the temperature that is achieved, the higher the Carnot efficiency, which represents the best conversion of thermal energy to mechanical work that can be achieved. Quite high temperatures can indeed be generated using different technologies, however, as the temperature rises, so too do the cost of materials of construction and rate of heat loss. Temperatures far in excess of 550 °C become technically challenging and expensive.

The solar field must also contend with the fact that collectable radiation is seasonal due to weather and the height of the sun throughout the day and at different times of the year (see Figure 9). The data shown in Figure 9 is for a 64 MWe (nameplate) Parabolic Trough CSP plant. In the summer months the plant exceeds the nameplate capacity for 10h of the day, however, during the winter actual production averages just 25 MWe for 6 h of the day. The more north or south of

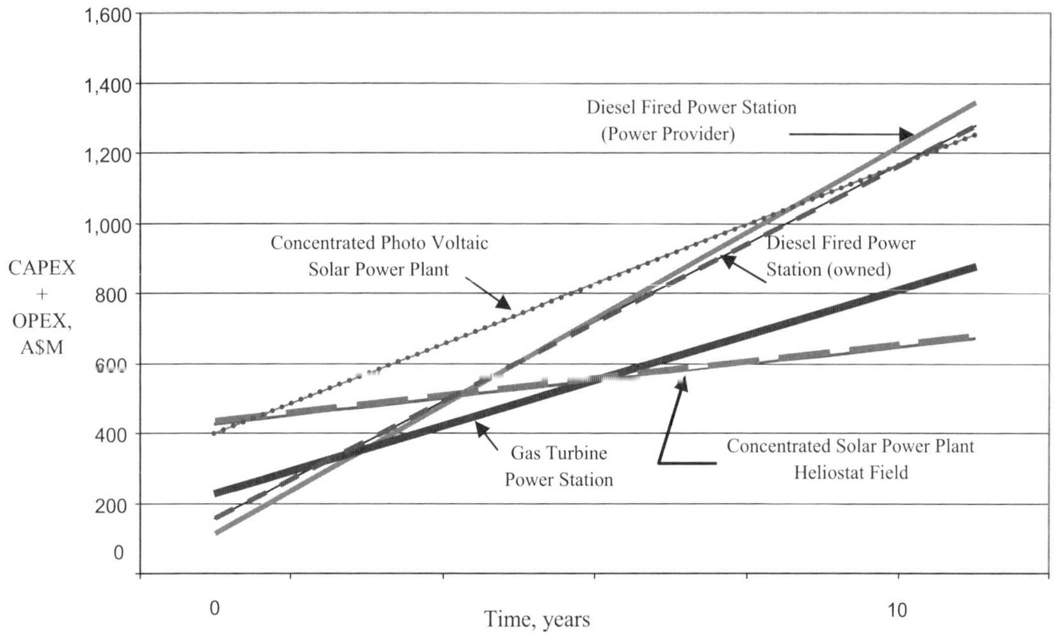


**Figure 6. Relative Power Plant Options, 30 MWe Capacity**

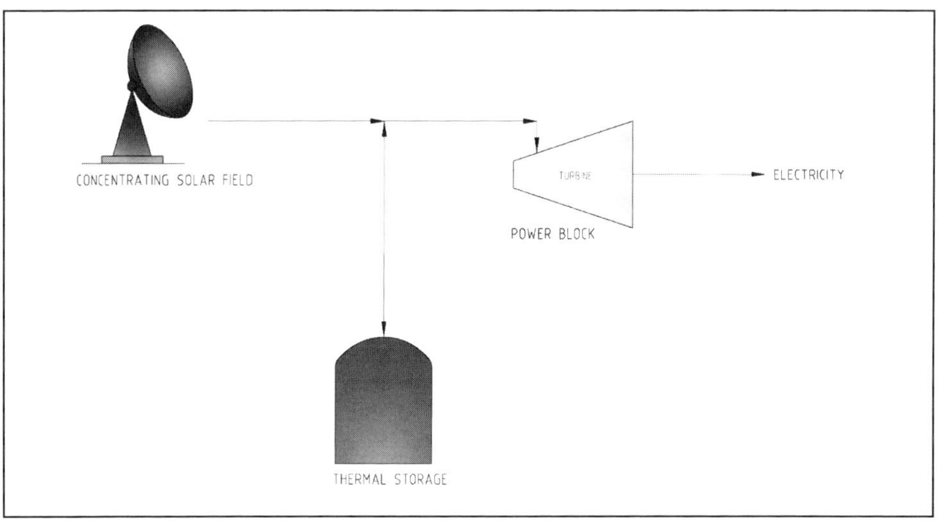


**Figure 7. Schematic of CSP**

the equator, the more pronounced the variation from winter to summer. For the engineer, this highly variable feed into the solar facility requires a back-up power plant for island power applications to consistently meet the load demands.

CSP Solar fields may be categorized as linear collectors, e.g., the parabolic trough and Fresnel collectors, both with a tubular receiver, or point collectors, e.g., a central receiver in a field of heliostats or a dish field each with its own receiver, see Figure 10. The linear collectors are suitable for lower temperature applications and typically use a thermal oil to transfer the heat from the tubular receiver to the power block or storage device. Linear collectors are therefore limited to less than 400 °C. Point collectors can achieve much higher temperatures, which lend themselves to higher engine efficiencies and smaller sized thermal storages and chemical storage systems. Materials of construction tend to limit the point collectors to less than 700 °C. A central tower system was adopted as satisfying the basic criteria for the West Australian project.

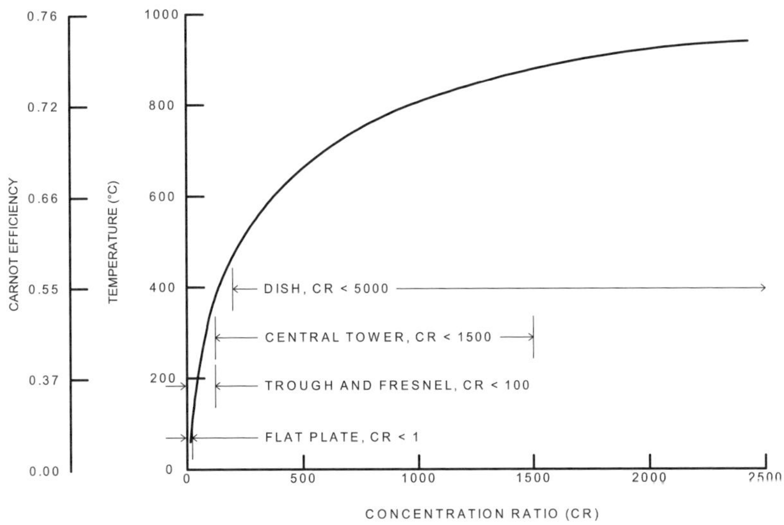


**Figure 8. Solar Field Technology Carnot Efficiency**

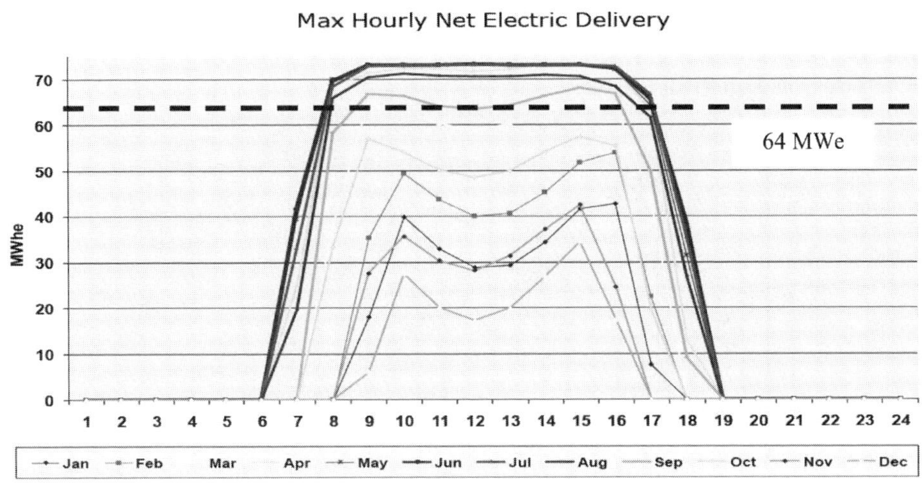


**Figure 9. Nevada Solar One Parabolic Trough Electricity Production, Latitude 35.5 N, Ref 6**

**Energy Storage Alternatives**

The purpose of energy storage is to provide for electricity generation at night for the mineral processing plant, plus load shifting from one day to the next or to provide only sufficient time to get the back-up power plant spinning, or anything in between. The decision is very much an economic one where the additional cost of the storage device plus the additional solar field to supply the heat must be weighed against the cost of operating the back-up power plant to supplement the solar power. To minimize the storage volume and ultimately optimize the engine efficiency, the temperature is preferably high, see Figure 11.

Therefore storage alternatives include:

1. No storage—daylight operation only
2. Store heat in steam or hot gases
3. Store heat in solids
4. Store heat in liquids
5. Chemical storage
6. Phase change materials

Thermal storage in gases or vapors is excellent in that high temperatures are reasonably easily achieved and this technique may be suitable for transient storage, e.g., as an accumulator. However, for storage of more than a few minutes, this technique is often not suitable. Storing heat in solids is available commercially and can achieve temperatures up to 800 °C, however, the cost is currently high and the technology is still in development. Using the CSP to provide heat for a reversible endothermic chemical reaction is being pursued by a number of technology providers. This technique, when developed, will provide a source of energy when reaction is reversed, e.g., combining nitrogen and hydrogen to re-form

**Figure 10. CSP Solar Field Alternatives, Ref 7**

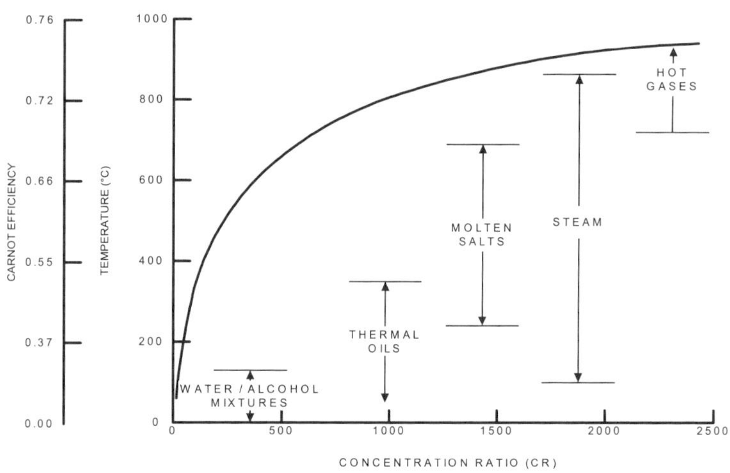


**Figure 11. Working Temperatures for Fluids**

ammonia. However, this technology is still in a developmental stage. Similarly the use of phase change materials is being developed to release heat at a given temperature. The phase change technology is also not available for commercial use at this time.

Thermal storage in liquids is common. However, because most of these heat transfer fluids cannot be used at high temperatures, a very large volume is required to store the quantities of heat required. Most of these liquids are relatively expensive and require pressurizing to keep them from boiling at temperatures over nominally 300 °C, hence they are not ideal. Molten mixtures of potassium and sodium nitrate, however, have been used successfully at 565 °C and atmospheric pressure and are not expensive relative to other heat transfer fluids. Figure 12 shows the molten salt storage tanks for Andasol 1 during construction.

Molten salt technology including tank design, salt pumps, heat exchangers, etc is well established in the fertilizer and nuclear power industries, though it would be new technology for most mineral processing facilities.

## Power Block Alternatives

The heart of the power block is the engine that converts thermal energy to mechanical work. Different engines lend themselves to different temperature regimes as per Figure 13. The Brayton engine is used for natural gas turbines and for jet airplanes. The use of this engine with CSP is still under development. The Stirling engine is an excellent concept, but

**Figure 12. Molten Salt Storage Tanks, Andasol 1, Spain, Ref 7**

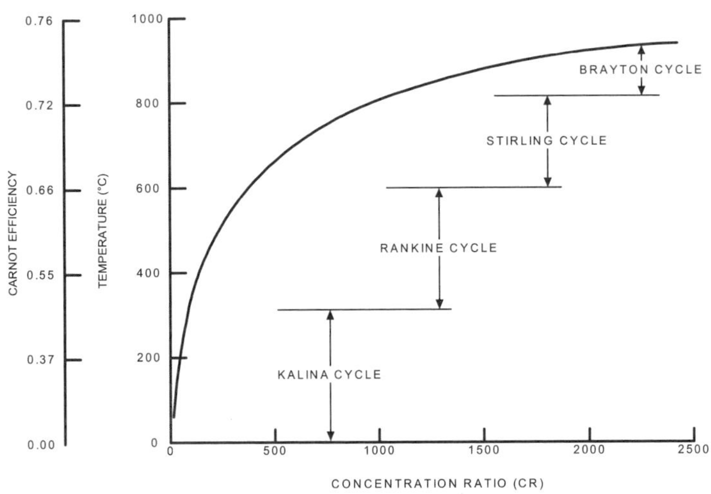


**Figure 13. Engine Alternatives**

has not yet been proven at a large scale and tends not to have a long service life. The Kalina engine is also known as an organic Rankine cycle engine. The Kalina engine can have excellent efficiency, but is not usually used for large engines due to the amount of organic fluid required that can be both hazardous and expensive.

The steam Rankine engine is the work horse of the power industry. The efficiency of the power block is driven on one side by the inlet steam conditions and pulled on the other side by the condensate conditions. For many remote applications only dry-cooling is available for condensation and on hot days, the design condensate temperature can be as high as 65°C. This high condensate temperature drives down the overall Rankine engine's efficiency to below 35%.

## CSP FOR MINERAL PROCESSING

The final selection of CSP technology must consider the operation of the mineral processing facility. In particular start-up load profiles for the mills and emergency shutdown load profiles for the solar field must be accounted for. The marriage of a heliostat field, molten salt central receiving tower, molten salt thermal storage, air-cooled condenser and a conventional high temperature steam turbine with no re-heat circuit has been selected. Full load back-up power will be provided using a conventional diesel power plant in the first few CSP applications, until the reliability of the technology gains acceptance.

Thermodynamic simulations have been used to evaluate the facility as a whole and provide sizing information to feed back into the financial model and permit an economic optimum to be determined.

A 30 MWe solar facility would nominally involve 35 to 45 Ha of mirrors spread over 250 Ha of land reflecting heat onto a receiver located on top of a concrete tower. Provision of 24 h thermal storage permits heat to be collected during one day and used that night and the following day, thereby greatly increasing the solar share of power generation for relatively little additional capital cost. The amount of storage required is dependent on the specific site's weather and the amount of morning and evening cloud that can be expected.

The initial field sizing provides for a facility capable of a 77% solar share of the power (the result of designing on summer conditions with 24 h thermal storage) with only two thirds of the solar field installed. However, the plant has been designed to be capable of having the solar field expanded to 100 % resulting in a solar share of 90%.

The operating cost for a solar facility at 77% solar share is estimated at 0.035 A$/kWh, see Table 2, with labor making up 43% of the total cost. Accounting for the diesel generators providing 23% of the power results in the estimated operating cost for power to the mineral processing facility of 0.076 A$/kWh.

Installing the remainder of the field to achieve a solar share of 90% reduces the solar facility's operating cost to 0.030 A$/kWh and the cost of power to the mineral processing facility to 0.048 A$/kWh.

**Table 2. CSP Operating Cost Estimate, 77% Solar Factor**

| Cost Area | Proportion, % | Cost, A$/kWh |
|---|---|---|
| Plant Labour | 12.6 | 0.015 |
| Contract Labour | 2.9 | 0.001 |
| Maintenance Materials | 20.4 | 0.007 |
| Consumables | 7.0 | 0.002 |
| G&A | 27.1 | 0.010 |
| Total | 100.0 | 0.035 |

The capital cost of the 30 MWe power plant operating with a 77% solar share is estimated to be A$M472 including all site works, water treatment (from hyper-saline bores) and a 30 MWe back-up diesel power station. The heliostat field is the single largest capital item costing just under 50% of the total, see Table 3. The solar field has thousands of individual heliostats all provided with power and communications.

Of particular noteworthiness is the cost for thermal storage. It was identified that, for this particular site, the solar share is strongly affected by the thermal storage capacity. An increase of 8 h storage capacity provides sufficient energy reserve to produce electricity during typically cloudy periods and results in an increase in solar capacity of 8% for a relatively small increase in capital of 3%.

**Table 3. Capital Cost Estimate (± 30 %, 1Q09)**

| Cost Area | Capital Cost, A$M | Proportion of Total, % |
|---|---|---|
| Heliostat Field and Central Receiver | 234 | 49.6 |
| Thermal Storage & Steam Generation | 131 | 27.7 |
| Power Block & Water Treatment | 59 | 12.5 |
| Back-up Diesel Power Generation | 48 | 10.2 |
| Total | 472 | 100.0 |

Optimizing the heliostat field and thermal storage in terms of capital cost and solar share is critical during the design stage of the solar facility design. Levelized cost of electricity (LCoE) is a means of estimating the price for which power may be sold over the duration of the project to ensure all costs and benefits are adequately taken into account, including taxes, carbon credits, depreciation, profits, etc. LCoE may be used to establish energy prices by power providers 'over the fence' and is an objective way to compare different energy options.

For a 30 MWe plant the LCoE is detrimentally affected by the capital cost, all else being equal, see Figure 14. The LCoE of a diesel power generation plant operating under the same conditions has also been plotted on Figure 14, however, rather than vary the installed capital cost as a function of LCoE, the LCoE has been used to show the equivalent price of diesel fuel.

Figure 14 may be used to estimate a levelized cost of electricity given a capital cost of a solar facility or the forecast price of diesel for the duration of the project. Alternatively given a capital cost for a solar facility, Figure 14 may be used to determine the price of diesel fuel above which the solar facility is a more attractive option. Or a forecast price of diesel may be used to estimate how much capital may be beneficially spent on the solar facility .

For reference, the current price of diesel for non-transport use is plotted on Figure 14. For a 30 MWe solar facility to provide electricity at a lower cost than a 30 MWe diesel generator over a 20 year life, the solar facility would need to cost less than A$M440, if the price of diesel stayed constant over the next 20 years. However, it was previously noted that the price of diesel will likely at least double over the next 20 years with the likely result being that solar power will be more attractive relative to diesel power in remote locations in the very near future.

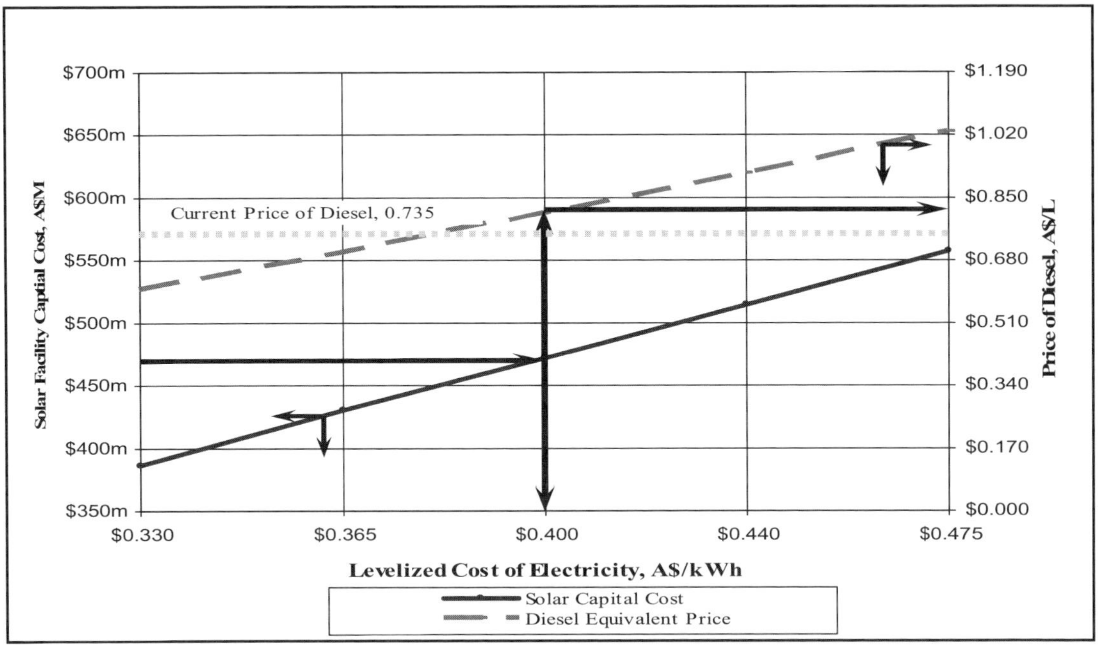


(20 Year project life, 10% Discount rate, Inflation 2.5%, 100% Equity Funding)

**Figure 14. Effect of Capital Cost on Levelized Cost of Electricity**

## CONCLUSIONS

1. Concentrated Solar Power can be used to mitigate the effect of high future diesel prices on the viability of a mineral processing facility.
2. A viable technology to be applied to a mineral processing facility is a central molten salt receiver, heliostat field, molten salt thermal storage and steam generation, and a conventional steam turbine power block.
3. The operating cost for the solar power portion is forecast to be less than 0.035A$/kWh based on a 77% solar share including all labor and on-costs. The total operating cost of the power generation facility including diesel power generation costs is forecast to be 0.076A$/kWh.
4. The capital cost for a 30 MWe CSP plant capable of supplying round the clock power, but on a 77% solar share basis is estimated to be A$M472.
5. The levelized cost of electricity is estimated to be A$0.40/kWh, which is the equivalent to paying A$0.83/L for diesel fuel.
6. Given conservative trends in the price of oil, the price of diesel will likely rise through A$0.83/L within two years rendering a 30 MWe solar facility the preferred option relative to diesel power generation For a long life mining operation.

## REFERENCES

Key Facts—Australian Energy Industry. Australian Government, Department of Resources, Energy and Tourism. 2005-06

ABARE Research Report 07.24. Australian Energy—National and state projections to 2029-30. December 2007

Annual Energy Review 2007. Department of Energy. www.eia.doe.gov

www.oilnergy.com/1obrent.htm

www.ret.gov.au/resources/fuels/alternative_transport_fuels/Documents/GTL_Part9.pdf

Cohen, G.E., Acciona Solar Power, "Solar Steam at Nevada Solar One," SolarPaces 2008, Las Vegas, NV

Neberera, J.A., ACS SCE, "Solar Thermal Power Generation," Genera08, Madrid, Spain, 26 Feb 2008

# Particle Separation

# IMPACT OF FLASH FLOTATION ON GRINDING AND MAIN FLOTATION CIRCUITS—DESIGN BY SIMULATION AND CASE STUDIES

Pertti Lamberg[1], Peter Bourke[2], and Christian Kujawa[3]

[1] Senior Technology Adviser - Mineral Processing, Outotec Research Oy Pori, Finland
[2] Principal Metallurgist – Flotation, Outotec Pty Ltd., Level 2, 1 Walker Avenue, West Perth, 6005, Australia
[3] Global Flotation Support – North America, Outotec (USA) Inc., 10771 E. Easter Ave., Suite 120, Centennial, CO, USA

## ABSTRACT

Flash flotation has proven to be highly successful with yielding improved overall recoveries for gold or minerals that liberate preferentially, are fast floating, prefer a deslimed slurry environment and tend to be classified inefficiently by grinding circuit cycloning due to their high specific density.

Whilst the suitability and impact of flash flotation technology on overall recovery and circuit operation was determined empirically in the past, suitable models and design methodology have been developed now for the more accurate simulation of grinding circuit flash flotation, showing impact on the grinding and main flotation functions. Reliable simulations require that flotation kinetics of each mineral are determined size-by-size using special SkimAir® laboratory tests. In addition, size-by-mineral data is required on the mill discharge and split of the cyclone. By simulation, it can be demonstrated where and how improved recoveries are achieved, how grinding throughput and classification efficiency change with flash flotation and how the main flotation capacity requirement is reduced. Case studies are given.

## INTRODUCTION

Flash flotation has proven to be highly successful with yielding improved overall recoveries for gold or minerals that liberate preferentially, are fast floating, prefer a deslimed slurry environment and tend to be classified inefficiently by grinding circuit cycloning due to their high specific density.

The SkimAir®, Outotec's flash flotation machine, is placed in the closed circuit grinding circuit, typically treating the underflow of the separation cyclones. The SkimAir unit floats the valuable liberated particles from the circulating load of the grinding process whilst the particles are still relatively coarse. These liberated particles, recovered early, are not ground further and do not enter the main flotation process as very fine particles with resultant poor recovery. Plants fitted with the Outotec SkimAir flotation machines therefore show additional pay-mineral recoveries.

Further, the use of flash flotation in the grinding circuit removes spikes in head grade, resulting in a more stable feed to the conventional main flotation circuit. This creates extra capacity and improved control in the main flotation process. Alternatively in new flotation plants, the main flotation capacity can be reduced as a result.

The main flotation concentrate is blended with the higher grade and coarser SkimAir concentrate to produce an overall final concentrate, which dewaters more easily, giving lower moisture values and therefore lower transportation costs. Due to the different size distributions joined cleaning of the concentrates is not recommended. Should cleaning of the flash flotation concentrate be envisaged this is done effectively in dedicated cleaning flotation, the FlashCleaner tailings typically being fed back to the mill discharge sump.

Along with flotation improvements, the grinding mill throughput can be increased due to the controlled high density of the SkimAir underflow.

SkimAir flash flotation is mainly used in gold processes but has been successfully installed also in sulfide copper and nickel applications. It should be noted that while the benefits of flash flotation are substantial, that not all mineral suites are suitable candidates necessarily. A thorough understanding of the behavior of valuable minerals in the grinding-classification-flash flotation arrangement is required.

Similarly it takes thorough understanding of the proper design of the SkimAir flotation machine itself and implementation to cater for the extreme coarse cyclone underflow, the SkimAir feed and tailing handling and the flotation and handling of the coarse concentrate requirements, as well as the general automation and control of the flash flotation machine.

## TESTING PROCEDURE

Outotec has developed a special laboratory test procedure to collect detailed information for modeling and simulation in order to forecast the metallurgical performance in full-scale operation and giving required data for flowsheet configuration and sizing. The problem in testing and designing the flowsheet is that the feed of the SkimAir, i.e. cyclone underflow, cannot be produced without piloting. Fresh plant feed will show much lower recoveries than in the full-scale process whereas cyclone underflow from a circuit without SkimAir will give too high recoveries. Therefore test results have to be studied size-by-size and models have to be created for the mill and the cyclone.

**Background Information**

For the background information following data is needed:

**Table 1. Required Background Information**

| Data | Description | Units | Entries |
|---|---|---|---|
| Plant Feed | Dry Mass | t/h | |
| Head Grade | Au | g/t | |
| | Cu | % | |
| | S | % | |
| | [..] | [...] | |
| Mineralogy | Ore Minerals | % | |
| | Sulphide Gangue | % | |
| | Non-Sulphide Gangue | % | |
| Mill Discharge | P80 | mm | |
| | %Solids | % | |
| Cyclone Underflow | P80 | mm | |
| | %Solids | % | |
| Cyclone Overflow | P80 | mm | |
| | %Solids | % | |
| Circulating Load | Cyclone Underflow/Fresh Feed | % | |
| Recipe for Standard Flotation Test (Rougher Feed) | pH | - | |
| | Reagents and Dosages | -, g/t | |
| | Flotation Times Used | min | |

**Sample**

The test feed sample should preferably be taken from freshly crushed ore. A sample size of at least 20 kg, all less than 5 mm particle size, is required. The sample should be representative of the ore and the element grades should match the plant feed. In addition, for higher confidence, the cyclone underflow should be sampled as well. The cyclone underflow samples should be of 5 kg slurry mass, packed hermetically.

**Mineralogical Study**

The mineralogical study of the sample includes chemical assaying, x-ray diffraction analysis, optical microscopy and quantitative estimation of the mineral grades. One objective of the mineralogical study is to determine the mineral liberation size. Sometimes, liberation data by size class is required. For gold, for example, the mode of occurrence is determined by diagnostic leaching of relevant size classes. The method then requires quantification of the mass proportion of gold as: 1) exposed, 2) locked in sulphide minerals, 3) locked in non-sulphide gangue minerals.

**Laboratory Flash Flotation Test**

To start off, a number of laboratory flash flotation tests are generally carried out to determine the kinetics of the minerals with flash flotation for certain reagent regimes. Gold is analyzed by fire assay technique to ensure necessary accuracy. The concentrate and tailing of one test are classified into 3-5 size fractions to allow determination of performance by size. For higher confidence, two additional tests with the cyclone underflow sample can be carried out too.

The following procedure should be used as a guide:

- A standard laboratory flotation machine with a cell of 2.5 liters is used. If using a different sized cell, the weight of the feed sample used in the tests is scaled to match with 45% solids in the slurry.
- The 20 kg sample is crushed stepwise to –1 mm and split into 2.2 kg sample bags and stored in a freezer. One sample is taken for mineralogical study, sizing and chemical assay analysis. Grinding tests are carried out to determine the grinding time for a matching P80 of the cyclone underflow.
- For each flash flotation test a 2.0 kg sample is ground to the desired P80. The grinding should be done just before the flotation test and special care should be taken to prevent the oxidation of sulphides during the sample preparation.
- The reagents for flotation are made up. This will vary depending on what minerals are involved. The reagents must be added separately and as quickly as possible. Conditioning time should not exceed 10 seconds in total. Flotation is done at 45% solids in the slurry.
- The flotation machine is run at 1200-1600 rpm to keep coarse particles in suspension. Water is added to maintain pulp level during the test as required. The concentrate is removed from the cell by continuously raking froth using a paddle in a consistent and repeatable matter.
- At the beginning, one should determine to what degree materials are free floating, e.g. by adding frother only and determining the percentage material free floating. A second test should involve only frother and one collector. Reagent mixes can then be scaled up once these basic scans are done.

- A series of tests are made using a normal reagent mix. It is advised not to try to optimize reagent additions until at least 3 test results are known. Results can be scattered particularly on gold at same reagent addition.
- After finding of the proper reagent mix the kinetics are determined by collecting concentrate samples at 30, 60, 90 and 120 seconds. Water is added to maintain pulp level in the laboratory flotation cell during the test as required. The concentrate is removed from the cell by continuously raking of the froth using a paddle in a consistent and repeatable matter.
- The kinetic test should be duplicated.
- After the kinetic testing, a test is designed to collect enough material to size the 60 seconds concentrate. At least 100 g of concentrate should be collected. This might mean, that the test may have to be repeated several times. All concentrates are combined and all tailings are combined. The concentrate and tailings are sized using the following screens: 37 mm, 74 mm and 105 mm. For gold assays, the whole size fractions are submitted for Fire Assay analysis and another series of tests is done for the base metal assays.
- Head sample, concentrate samples, flotation tail samples are assayed. All dry weights are recorded. For gold, analysis by fire assay should be done. In addition, at least the base metals (Cu, Zn, Ni, Co) and sulphur should be assayed for.
- Recoveries and size-by-size recoveries for elements are calculated. Required also are the mineral conversions and the calculations of the grades and recoveries for the ore minerals, gangue sulphides and non-sulphide gangues. If talc or some other floatable non-sulphide gangue minerals are present, the grade and recovery of the mineral should be estimated. This may require additional chemical analyses or other quantitative mineralogical techniques, such as for example quantitative x-ray diffraction.

**Cyclone Performance**

As the SkimAir is placed on the cyclone underflow, it is important to know the separation of minerals, especially gold, between cyclone overflow and underflow. The key issue in flash flotation is that gold concentrates preferentially in the cyclone underflow and therefore a moderate stage recovery in the flash flotation can mean significant overall recovery.

In existing operations cyclone feed, underflow and overflow should be sampled, samples should be sized and analyzed size-by-size. Finally, a mass balance by mineral of the cyclone will be used in modeling.

In the case of Greenfield projects, the estimate on the cyclone split by mineral can be made by mineralogical study. This requires liberation analysis.

**Modeling and Simulation**

Based on background information, laboratory test results and cyclone performance, models need to be created for the mill, cyclone and flash flotation unit operations.

The models, which are used, are:

- Fixed particle size distribution for the mill
- Efficiency curve by mineral for the cyclone
- Floatability component of minerals by size for flash flotation

Finally, the following scenarios are simulated:

- Feed rate vs. metallurgical performance
- By-pass ratio vs. performance
- Water balance with and without dual outlet

The simulation provides a metallurgical balance of the circuit, recommendations for the by-pass and how the water balance can be controlled in the circuit without jeopardizing the mill throughput.

## CASE STUDY ESPERANZA

**Background**

Esperanza copper-gold deposit is located near the town of Sierra Gorda, 2nd Administrative Region in Chile. Deposit hosts 535million tonnes of ore grading at 0.57 % copper and 0.27 g/t gold. The main copper minerals are chalcopyrite and bornite. Gold is carried as native gold and in solid solution with copper.

Antofagasta Minerals S.A. conducted a series of flotation studies at SGS Minerals Services. The studies included both laboratory and pilot scale work. In the pilot scale study by SGS Mineral Services, the applicability of the flash flotation technology was studied with high-grade andesite and high-grade porphyry samples. The flash flotation pilot scale work gave significantly higher recovery for gold than the conventional flotation circuit. The recoveries for copper were slightly higher using only conventional flotation.

**Modeling**

The test results were used to determine the best layout, size the units properly and estimate the metallurgical performance. For that purpose, in the representative laboratory flotation tests, elemental grades were converted to mineral mass proportions and mineral recoveries were calculated for copper sulphide mineral, pyrite, gold and gangue.

From the flotation time vs. cumulative recoveries, kinetic constants were fitted according to the following equation:

$$R = M_{Fast} * (1-e^{-kFast * t}) + M_{Slow} * (1-e^{-kFast * t}) \quad (1)$$

In the equation, R is recovery of a mineral in a size fraction, $M_{Fast}$ is the mass proportion of fast floating mineral, $k_{Fast}$ is the kinetic constant of the fast floating mineral, $M_{Slow}$ is the mass proportion of slow floating mineral, $k_{Slow}$ is the kinetic constant of slow floating mineral. Flotation kinetics of copper and gold in flash flotation are depicted in Figures 1-3.

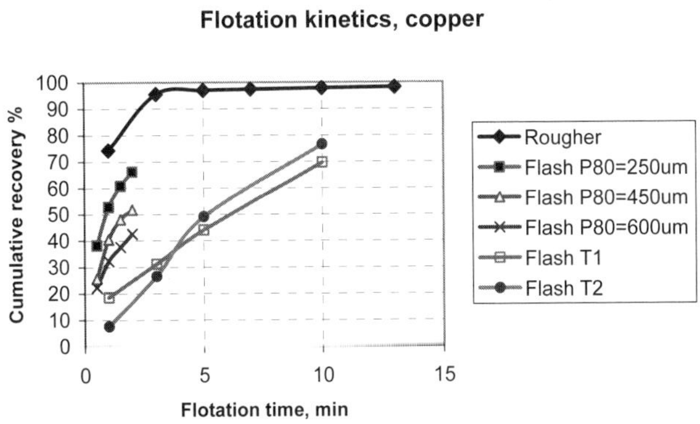


**Figure 1. Flotation kinetics of copper in the flash flotation (Flash) and rougher flotation (Rougher).**

In the rougher flotation, gold is slower than copper but in the flash flotation it is the opposite, i.e. gold is faster than copper, especially when the feed gets coarser (Figure 3).

The kinetic constant fitted with the laboratory flotation tests were fine-tuned using the pilot test results. The laboratory test results were available only for porphyry ore; therefore kinetic constants for the andesite were solved using the pilot test results.

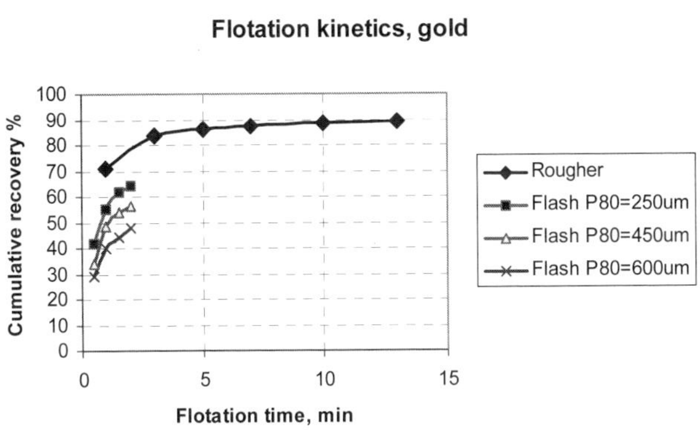


**Figure 2. Flotation kinetics of gold in the flash flotation (Flash) and rougher flotation (Rougher).**

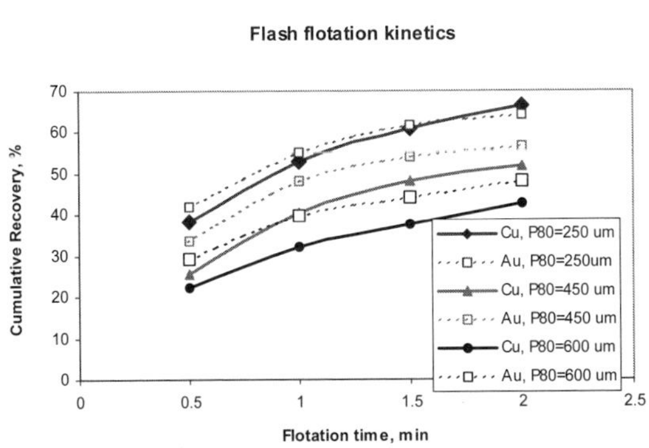


**Figure 3. Flotation kinetics of gold and copper in flash flotation according to laboratory tests.**

## Simulation

Simulations were done using the HSC Sim software of the HSC Chemistry program family developed by Outotec. For the cyclone, a three-parameter efficiency curve by mineral was used. The enrichment of gold in the cyclone underflow in the conventional flotation circuit without the SkimAir is demonstrated in Figure 4. For the gold, the circulating load is 402%.

The model parameters of the flotation units defined based on laboratory tests were scaled and fine-tuned using the pilot plant data. The residence time in each flotation unit was calculated based on volumetric flow rates and flotation cell sizes.

## Results and Discussion

The basic case simulation with the Porphyry ore reproduced the pilot test results and the gold recovery in the circuits without and with SkimAir is shown in Figures 4 and 5. The circuit including SkimAir produces 7.6% higher overall recovery for gold and 2.2% for copper.

However, if the SkimAir concentrate is not cleaned the final concentrate grade will be significantly lower than in the conventional case, 17.6% Cu (Figure 6). Two different cases were simulated. In the first simulation, the SkimAir concentrate was cleaned in a FlashCleaner (Figure 6) and in the second simulation the SkimAir concentrate was directed into the cleaner circuit (not shown). In both cases high final copper concentrate grade about 28% can be achieved with 87.6% recovery.

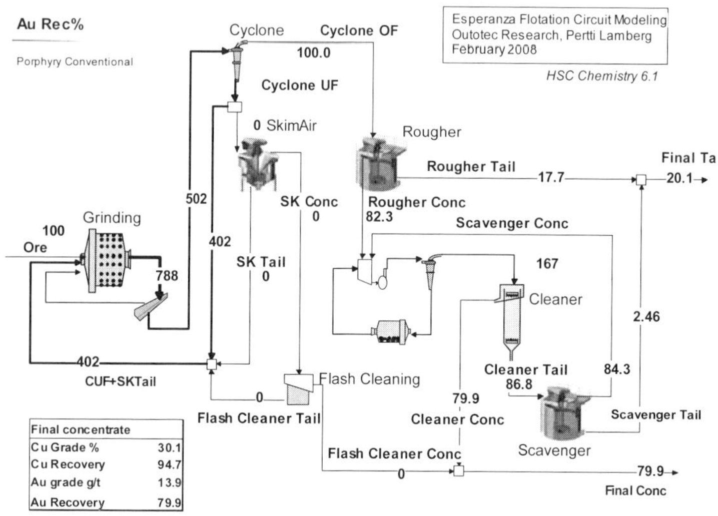


**Figure 4. Gold recovery (distribution) in the Esperanza flotation circuit in the case where the SkimAir is by-passed, i.e. conventional flotation circuit without flash flotation.**

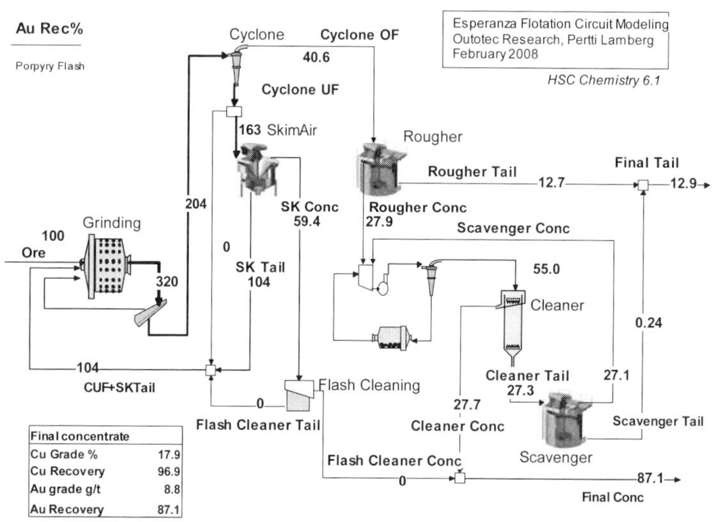


**Figure 5. Gold recovery (distribution) in the Esperanza flotation circuit in the case where cyclone underflow is directed to the SkimAir flotation machine. The FlashCleaning unit is set off in the simulation.**

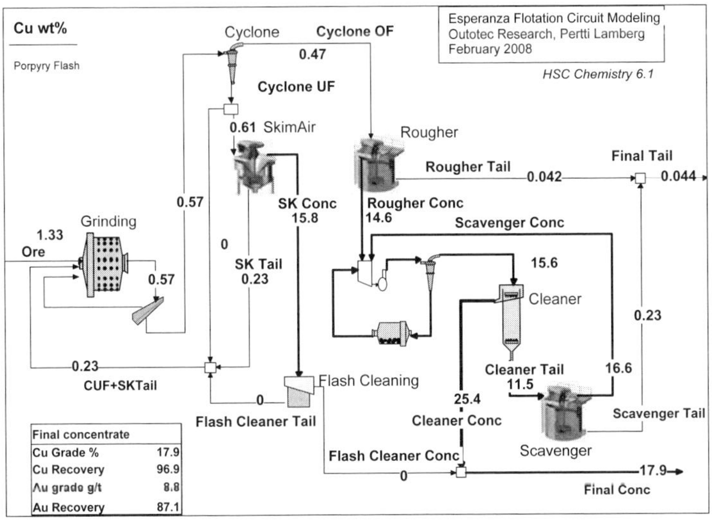


**Figure 6. Copper grade in the Esperanza flotation circuit in the circuit with SkimAir and without FlashCleaning.**

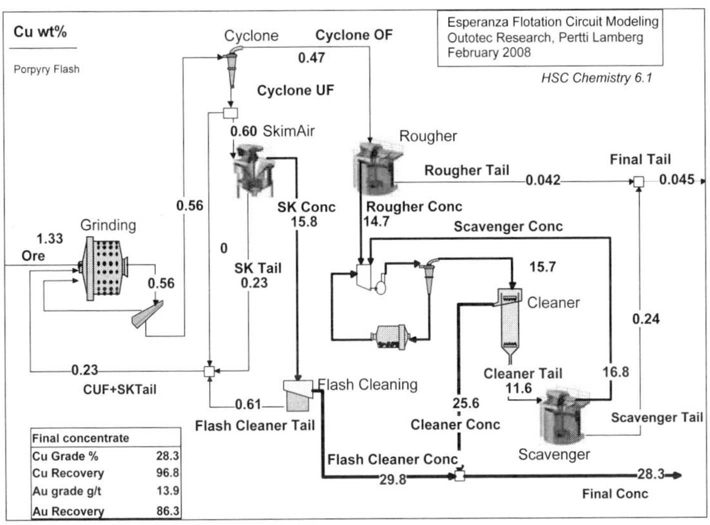


**Figure 7. Copper grade in the Esperanza flotation circuit in the circuit with SkimAir and with FlashCleaning.**

In the case of the andesite ore, the recoveries were significantly lower, but the recoveries increases were higher in the circuit configuration with the SkimAir. As the SkimAir gives additional better recoveries, especially for gold, a series of simulations investigated scenarios that minimize capital expenditure, were run. Using flash flotation, how much shorter could the main rougher flotation section be compared to a conventional circuit, but still reach similar overall gold recovery. In the configuration with SkimAir, the rougher bank of the main flotation circuit was shortened and the reduction in the recoveries were simulated. For gold, equal recoveries compared to the circuit without the SkimAir could be achieved with a 50% shorter rougher section (Figure 8). For copper, the circuit with the SkimAir would require a 15% shorter rougher bank and would give equal recovery compared to the conventional only flotation circuit configuration without SkimAir.

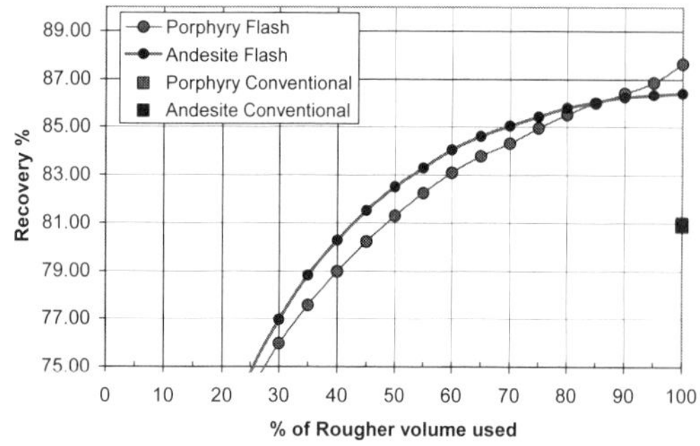


**Figure 8. Gold recovery in the Esperanza flotation circuit with porphyry and andesite ore processed in conventional flotation circuit (squares, almost on top of each other) and in the circuit with SkimAir flash flotation. X-axis shows how the shortening of the rougher bank in the flotation circuit will effect on recoveries in the circuit with SkimAir.**

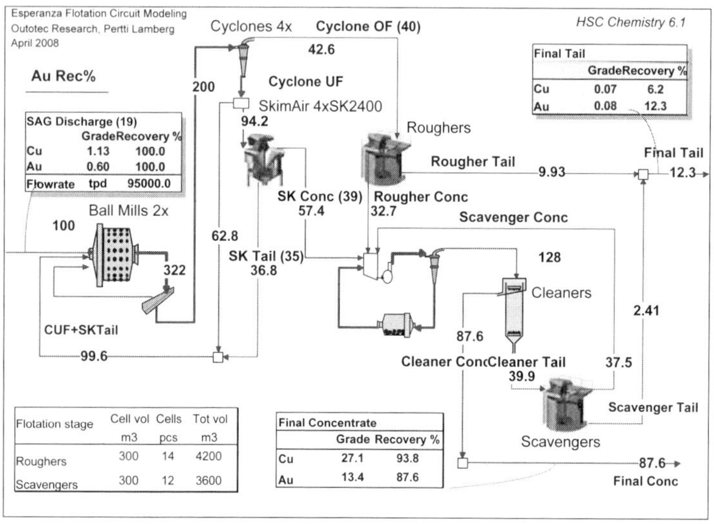


**Figure 9. Simulated metallurgical performance (gold distribution) in the flotation circuit with Outotec's SkimAir flash flotation unit.**

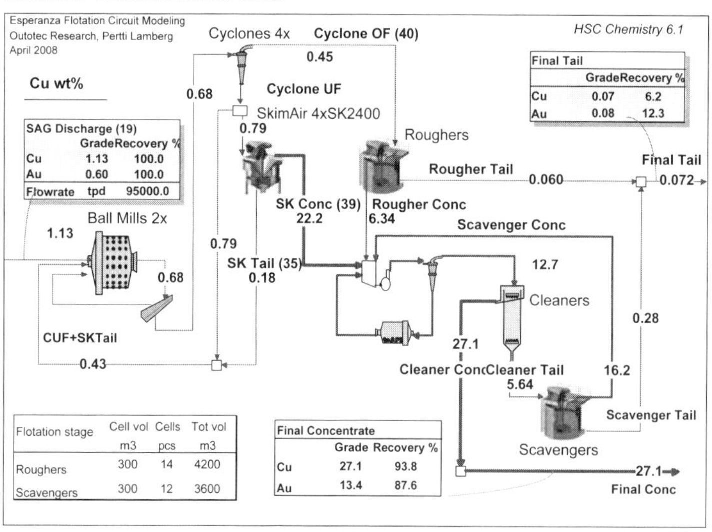


**Figure 10. Simulated metallurgical performance (copper grade) in the flotation circuit with Outotec's SkimAir flash flotation unit.**

The final simulations were made to find the metallurgical result in a full-scale process with the average ore; i.e. with 70% andesite and 30% porphyry mixture. In addition, the purpose was to define the required flotation volumes for desired production levels. Figure 9 shows required flotation volumes for rougher and scavenger flotation. A total of four SkimAir® 2400 are placed to treat 60% of the cyclone underflow. Projected concentrate grades were 27.1% Cu and 13.4 g/t Au with the recoveries of 93.8% for Au and 87.6% for copper (Figures 9 and 10).

**Conclusions**

Simulation verifies that flash flotation will provide higher recovery for gold and it indicates that there is potential to reach higher recoveries also for copper. The increase in recovery compared to conventional flotation is 5% for gold and 2% for copper. The results of the pilot test are in accordance with the simulation for gold but for copper the results in the pilot tests were about equal with and without SkimAir.

In the flotation circuit with flash flotation, the simulation shows that a separate final grade product can be produced by using FlashCleaning to upgrade the SkimAir concentrate.

Simulation indicates that in the flotation circuit with SkimAir flotation there is potential to decrease the installation volume of the main rougher section by 40-50% and still reach equal gold recovery with corresponding copper grade than in the conventional only circuit.

The pilot tests serve as the back up for the installation of four Outotec SkimAir® 2400 flash flotation machines in the Esperanza circuit for the process of 95000 tpd of copper-gold ore.

The study shows that HSC Chemistry is a powerful simulation tool in comparing different processing scenarios and to define required unit sizes. When flash flotation is considered, the model level has to be mineral-by-size since the flotation kinetics of minerals are strongly size dependent. Also, a reliable distribution model of minerals-by-size is needed for the hydrocyclone classification. The applicability of flash flotation for a given ore can be tested with a laboratory flotation test and, if properly done, the result can be used in simulating the metallurgical performance of SkimAir for the full-scale process.

## CASE STUDY GRAVITY SEPARATION VERSUS FLASHFLOTATON

**Background**

In a Cu-Au ore sample grading 0.89 g/t gold and 0.55% copper gold occurs as native gold grains varying from 0.7 um to 70 um. Heavy liquid separation shows that about one third of gold is attached with gangue rich particles with SG less than 3.3 g/cm3. Chalcopyrite is the dominant copper mineral with some bornite, chalcocite, cigenite and covellite. Pyrite is the main iron sulphide and its grade is low at 0.27% compared to the chalcopyrite grade of 1.6%.

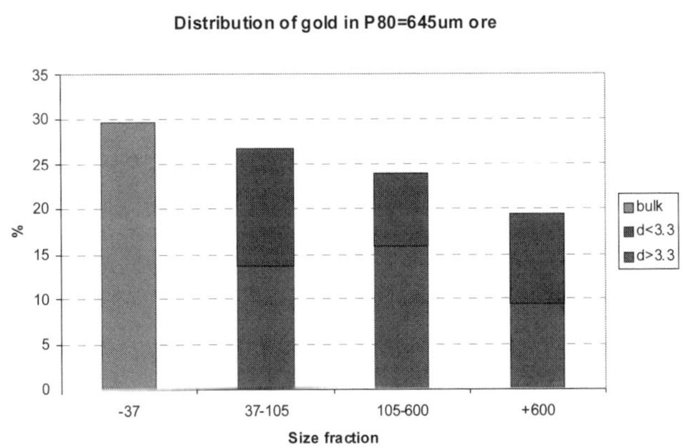


**Figure 11. Distribution of gold in P80=645um ore.**

**Testwork**

Laboratory gravity separation tests and flash flotation tests are compared in Figure 12. It demonstrates that flash flotation gives higher recovery especially in the finest, -37 um, fraction than gravity separation.

Batch laboratory tests were made and test feed and process products were studied carefully to quantitatively characterize the mode of occurrence of gold and recovery of different type of gold, chalcopyrite, pyrite, magnetite and gangue silicate particles.

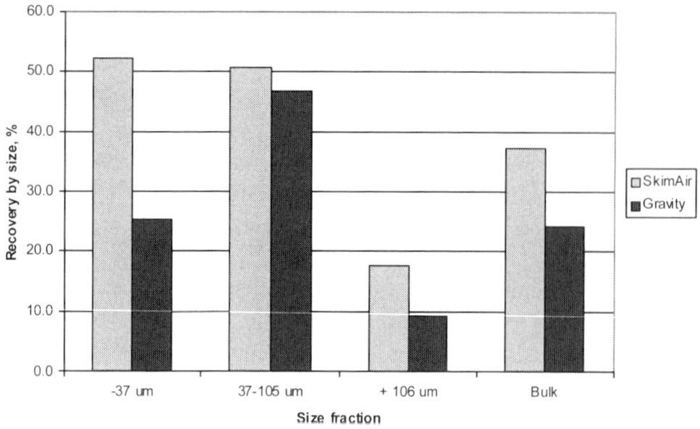


**Figure 12. Recovery results of flash flotation versus gravity separation.**

### Modeling

Mineral by size models were created for the mill, hydrocyclone, gravity separator and SkimAir for the gold, chalcopyrite, pyrite, magnetite and gangue silicate particles.

### Simulation

Steady-state simulation with HSC Sim shows that a SkimAir placed on the cyclone underflow gives a higher recovery for gold (21.7%, Figure 14.) than by gravity separation (11.6%, Figure 13.) because of higher capacity and better ability to recover the fine gold.

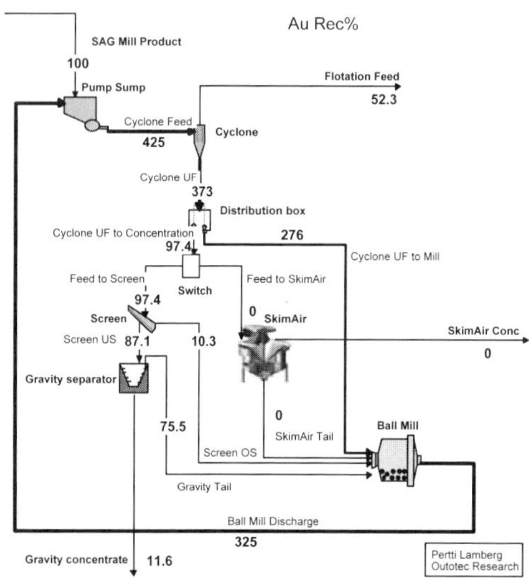


**Figure 13. Gold recovery with gravity separation, SkimAir flash flotation not used.**

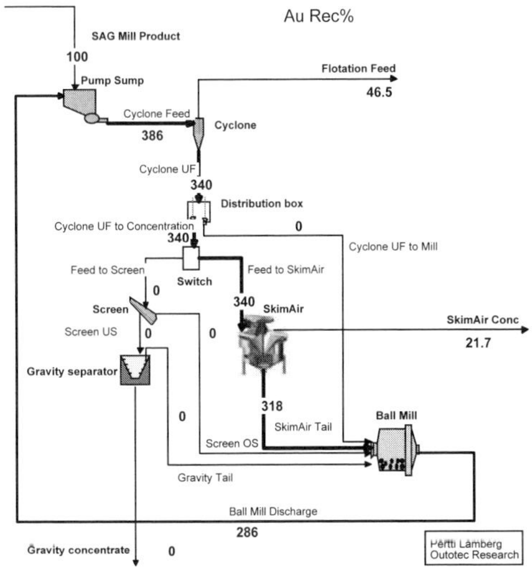


**Figure 14. Gold recovery with SkimAir flotation, gravity separation not used.**

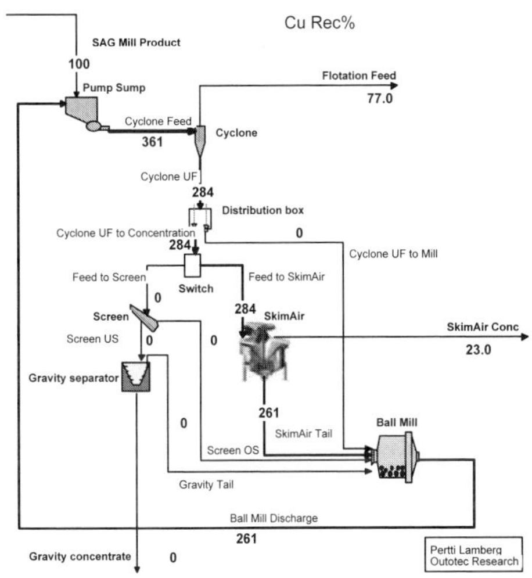


**Figure 15. Copper recovery with SkimAir flotation, gravity separation not used.**

## Conclusion

The second case study demonstrates the advantages of flash flotation compared to gravity separation.

## REFERENCES

SGS Mineral Services, October2006. Experimental Pilot Scale Flotation Campaign to Evaluate Flash Flotation Technology. Project 3563 – Report #2.

# RHEOLOGY IN MINERAL PROCESSING

Bern Klein[1] and Don Hallbom[2]

[1] UBC Norman B. Keevil Institute of Mining Engineering, 517 - 6350 Stores Road, Vancouver, BC, Canada
[2] Pipeline Systems Incorporated, 200 - 1140 W. Pender St., Vancouver, BC, Canada

## ABSTRACT

Concentrated suspensions of fine minerals often exhibit behavior that is markedly different from that of common Newtonian fluids, such as water. Failure to allow for this non-Newtonian behavior may cause problems in the design and operation of process units. Failure to recognize the causes and effects of non-Newtonian behavior may make it difficult to troubleshoot problems. Therefore, it is important that mineral process engineers understand the behavior of non-Newtonian fluids, which is to say "rheology." This paper reviews the fundamentals of rheology, its application in mineral processing and the chemistry of additives that can be used to modify the rheology of mineral suspensions. Rheological measurement methods are described; without proper measurement the effects of rheology and opportunities to modify the rheology for improved processing cannot be determined.

## INTRODUCTION

Rheology is the science of the flow and deformation of matter. In practice, the flow of true viscous (Newtonian) and elastic (Hookean) flow are considered in the fields of fluid mechanics and mechanics of materials respectively. The rheologist is, thus, relegated to studying a subset of materials that behave "oddly." It is natural for a mineral process engineer to ask "why learn something as abstruse as rheology if it only deals with odd (strangely behaving) materials?" There are, in fact, several reasons:

- There are a lot of oddly behaving materials including: modeling clay, peanut butter, polymer melts, gelatin, blood, printing ink, river mud and, of course, most mineral slurries. These are all suspensions of fine solids in a carrier liquid.
- There is a general trend to process finer (smaller sized) particles. As easy to process ores become depleted, new processing plants are being designed for poly-metallic sulfides and refractory gold ores that require ultra-fine grinding or fine grained materials like nickel laterite. As the particle size in slurry gets finer, the non-Newtonian effects become more pronounced.
- There are various forces driving the industry to processing ore at higher solids concentrations: equipment size (higher concentrations mean lower volumes to process and store), limited water availability (thickened tailings in arid regions), additive reduction (binder in paste backfill), and equipment availability (large capacity high compression thickeners). As the concentration of slurry is increased, the non-Newtonian effects become more pronounced.

Much of mineral processing involves moving non-Newtonian slurries in (or between) operating units. Table 1 summarizes the rheological properties that affect various mineral processing unit operations and their significance.

**Table 1. Importance of rheology to mineral processing applications**

| Application | Rheological Property | Significance |
|---|---|---|
| Pumping and Transport | Apparent viscosity, flow-curve type, yield stress, time dependence | Pressure heads, pipe diameters, saltation, turbulent transition velocity (Hanks, 1981, Wasp 1977, Wilson, 1997, Slatter, 1999, Usui et al, 1997, Klein and Hallbom, 2002) |
| Grinding | Flow curve type, yield stress | Power, capacity, cataracting, breakage rate, progeny size distribution, media consumption (Klimpel,1982, Shi et al, 1999, Hintikka et al, 1999, Yue and Klein, 2004) |
| Classifying Cyclones | Apparent viscosity, yield stress | Cut size, % to underflow, separation efficiency (Shi et al, 1999, Horsley and Allen, 1987) |
| Gravity Concentration | Yield stress, apparent viscosity | Separation efficiency, cut point shift, medium stability (Klein, 1992, He et al, 2004) |
| Flotation | Apparent viscosity | (Prestidge, 1999, Laskowski, 2001) |
| Mixing / Leaching | Apparent viscosity, time dependence | Oxygen and mass transfer, leach kinetics, pulp density (Klein and Laskowski, 1999) |
| Thickening | Yield stress, apparent viscosity, time dependence | Pumping pressure head to initiate flow (Nguyen and Boger, 1998, Boger, 1999) |

Despite this knowledge, little attention is paid to rheology by practicing mineral process engineers. Rheology has gained a reputation for being difficult, primarily due to complexity of mathematical rheology, the multitude of models available, the use of technical jargon, and a tendency for researchers to focus on more difficult or unusual materials.

The objective of this paper is to review the fundamentals of rheology and to show how it can be applied in mineral processing. The focus will be on bridging between the complexity of quantitative mathematical rheology and the simplicity of qualitative applied rheology. Rheological measurement methods are reviewed since this is required to determine the role of rheology when designing or selecting equipment and when attempting to modify the rheology for improved processing.

## DEFINITIONS OF KEY TERMS

As with most disciplines, rheology uses technical jargon to concisely convey information. The following is a glossary of some of the more important terms.

**Aggregate**: A general term to describe a cohesive mass of particles, which includes agglomerates, coagulates and flocs. The formation of aggregates is called aggregation.

**Apparent viscosity**: The ratio of the shear stress to the shear rate. The apparent viscosity changes with shear rate for non-Newtonian fluids and with time for time-dependent fluids. Newtonian fluids have a constant apparent viscosity (the "viscosity")

**Bond**: A general term for the net forces that hold aggregated particles together. Strongly bonded particles require a high shear stress to cause mechanical degradation. Bonds that cannot be broken and reformed indefinitely are referred to as links.

**Clay**: Hard particles less than 2 μm in size.

**Coarse**: Particles that are sufficiently large that they settle rapidly and surface chemistry related bonding is not significant. Nominally this includes sand sized and larger particles.

**Consistency**: A general term for the flow properties (i.e., the shear rate vs. shear stress relationship) of a fluid. For a Newtonian fluid, consistency is equivalent to viscosity.

**Constitutive equation**: An algebraic equation to describe the consistency of a fluid.

**Fine**: Particles that are sufficiently small that they settle very slowly and surface chemistry related bonding may dominate behavior. Nominally this includes clay sized and smaller particles.

**Newtonian fluid**: Any fluid (liquid or gas) with a linear relationship between shear rate and shear stress. It describes a "normal" or "true" fluid, such as water or alcohol.

**Non-Newtonian fluid**: Any fluid (usually a liquid) that does not have a linear relationship between shear rate and shear stress.

**Sand**: Hard particles in the 63 to 2000 μm size range.

**Shear rate**: The rate of change of shear strain (i.e., the relative in-plain displacement of two parallel layers in a material divided by their separation distance).

**Shear stress**: The component of stress that causes parallel layers of a material body to move parallel to each other in the plane of shear.

**Silt**: Hard particles in the 2 to 63 μm size range. For practical purposes, silt particles are often divided between "coarse" and "fine" groups at 37 μm (i.e., a 400# sieve).

**Suspension**: Solid particles suspended in a liquid medium ("carrier fluid").

**Viscoplastic**: A property in which a material behaves like a solid (or an extremely viscous fluid) when the stress is below some critical value, the yield stress, but flows like a viscous fluid when this value is exceeded.

**Yield stress**: The critical shear stress, below which an ideal viscoplastic will not flow. Once the yield stress is exceeded, the material flows like a fluid. In most real viscoplastic fluids, the fluid flows slowly (i.e., is extremely viscous) below the apparent yield stress.

## FLOW CURVES, CONSTITUTIVE EQUATIONS

Consider a viscous fluid between two parallel planes spaced $\Delta x$ apart (Figure 1). If the top plane is moving relative to the bottom plane, a force must be applied to the top plane to balance the viscous drag force. The shear stress ($\tau$) is the force (F) applied per unit area (A); $\tau = F/A$. The shear rate ($\dot{\gamma}$) is the change in velocity ($\Delta v$) divided by the distance between the two planes ($\Delta x$), $\dot{\gamma} = \Delta v/\Delta x$.

The relationship between the shear stress ($\tau$) and shear rate ($\dot{\gamma}$) describes the consistency or rheological flow properties of the fluid between the plates. This relationship can be described with a constitutive equation with the general form: $\tau = f(\dot{\gamma})$. The most common relationships between shear stress and shear rate are presented graphically as flow-curves in Figure 2.

Most concentrated mineral slurries containing fine particles are viscoplastic. This essentially means that the apparent viscosity is low at high shear rates and high at low shear rates. For this reason, the characteristic shear rate of a process should be known so that an appropriate apparent viscosity value can measured and used for engineering purposes. At very low ("zero") shear rates concentrated slurries behave like a soft solid until the applied stress exceeds the yield stress. The constitutive equations for viscoplastic fluids that have come into common engineering usage (i.e., Bingham plastic, power law and Casson) can be grouped into two overlapping families described by the following three-parameter models, yield power law (Herschel and Bulkley equation) and yield plastic (Hallbom 2008, Hallbom and Klein, 2009):

$$\tau = \tau_o + K\gamma^n \qquad \text{Yield power law} \qquad (1)$$

$$\tau^k = \tau_o^k + \left(\mu_\infty \gamma\right)^k \qquad \text{Yield plastic} \qquad (2)$$

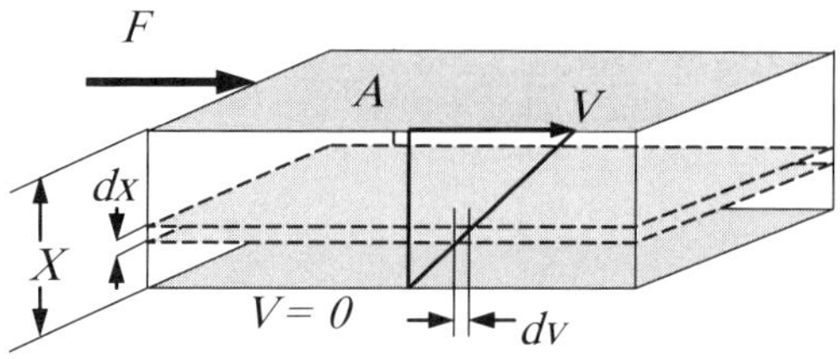


**Figure 1. Application of force *F* to a plate of area *A* and distance *x* from a parallel plate causes it to move at velocity *v*.**

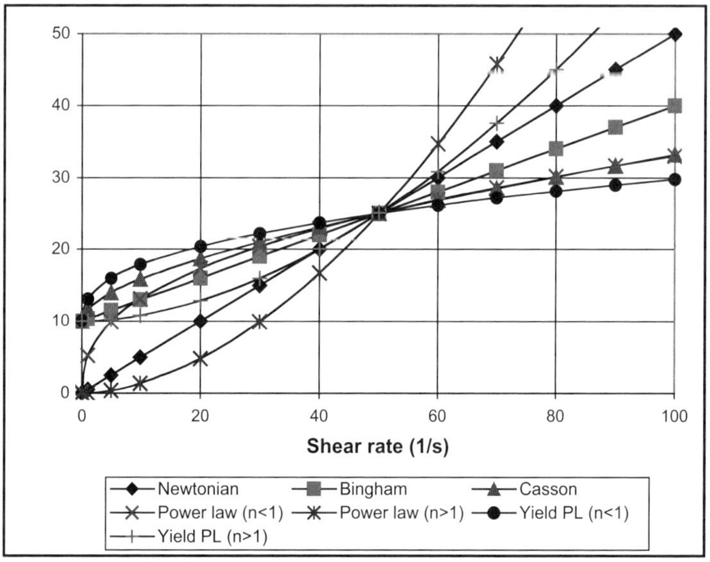


**Figure 2. Typical flow-cures showing shear stress – shear rate relationships**

Where, $\tau$ is shear stress (Pa), $\tau_o$ is yield stress (Pa), $K$ is consistency coefficient (Pa-s$^n$), $n$ is flow behavior index (-), $\mu_\infty$ is plastic viscosity (Pa-s), $k$ is scaling factor (-).

Both models have the Newtonian and Bingham plastic models as special cases. The yield power law model has the power law model as a special case. The yield plastic model has the Casson model as a special case. The two families are compared in Table 2.

If one is merely doing mathematics, the choice of models is a matter of personal preference. Either model will fit a range of rheogram data within the accuracy limits of measuring devices. The yield power law model is older and has a larger body of research (i.e., test data and engineering equations) available. It also has the advantage that it can be used to represent dilatant fluids over a limited range of shear rates. However, if one is trying to describe how homogeneous mineral slurry actually behaves, the yield plastic family has several advantages.

The first advantage is that, for $n < 1$, the yield power model predicts erroneously that the apparent viscosity will drop below viscosity of the carrier fluid at high shear rates and vanishes at the limit. Such predictions are never observed and are physically inadmissible. Real slurries tend towards a constant apparent viscosity at high shear rates (the "upper Newtonian plateau"). The yield plastic family models all tend towards a limiting viscosity at high shear rates (i.e. $\mu_\infty$).

The yield power law model will fit the upper Newtonian plateau if $n = 1$. If the flow behavior index is one for the entire shear rate range, then the model reduces to either the Newtonian or Bingham plastic model, which are also in the yield plastic family. If the flow behavior index changes with shear rate it becomes a shear rated dependent parameter. Furthermore, the consistency coefficient ($K$) not only changes in magnitude with the shear rate, it changes its units (Pa-s$^n$). This is problematic when trying to describe the behavior of a yield power law liquid based on its parameters and when trying to form the dimensionless groups that are ubiquitous in fluid mechanics. The yield plastic parameters are constant throughout the shear range and the units are coherent (Pa and Pa-s).

Therefore, advantages of the yield plastic model over the yield power law model are:

1. It provides a more reasonable estimate of the high shear apparent viscosity,
2. The coefficients have physical significance and,
3. The equation is a general case of the well-known Bingham and Casson equations, which are commonly, used rheological equations.

**Table 2. Comparison of Yield Power Law and Yield Plastic Models**

| | Yield Power Law | Yield Plastic |
|---|---|---|
| Model basis | Rheogram curve fitting | Aggregate breakdown |
| Parameters | 3 | 3 |
| Newtonian | $\tau = K\dot{\gamma}$ | $\tau = \mu_\infty \dot{\gamma}$ |
| Bingham plastic | $\tau = \tau_o + K\dot{\gamma}$ | $\tau = \tau_o + \mu_\infty \dot{\gamma}$ |
| Power law | $\tau = K\dot{\gamma}^n$ | No |
| Casson | No | $\tau^{1/2} = \tau_o^{1/2} + (\mu_\infty \dot{\gamma})^{1/2}$ |
| Limiting high shear rate viscosity | No | Yes |
| Shear thickening | Yes | No |

## A PHYSICAL MODEL NON-NEWTONIAN BEHAVIOUR

An engineer needs to be able to predict how the behavior of homogeneous slurry will change when a physical parameter changes. Specifically, they are interested in physical parameters that they can practically measure or predict (e.g., temperature, pH, solids concentration, etc.). A constitutive equation describes how a slurry behaves, but it does not explain why it behaves the way it does. This is explained with a physical model that relates the behavior at the micro-scale to the observed behavior at the macro-scale. These physical models are invariably simplifications, but they can accurately describe the major features of the behavior.

The rheological properties depend on the physical composition of the suspension and the degree of aggregation and/or dispersion and are the basis for the yield plastic constitutive model (Hallbom 2008). Particle crowding increases with solids content resulting in more particle interactions and therefore viscous energy dissipation. Three types of particle interactions contribute to the rheological properties of the suspensions: physical interactions, hydrodynamic interactions and or colloidal interactions. With decreasing particle size, the colloidal interactions that determine the degree of aggregation and dispersion become more significant. Most mineral suspensions contain a portion of particles that are fine (sub 10 μm) and these particles often contribute most to the rheology of the suspension through colloidal interactions. Particle aggregation results when inter-particle forces of attraction (van der Waals, electrostatic opposite charged surface) dominate over repulsive forces (electrostatic same charged surface).

Particle aggregation results in the formation of a "network" structure and is responsible for the yield stress. The strength of the attractive forces relates to the strength of the aggregates and network structure and is indicated by the scaling factor (k) in the yield plastic equation. Shearing breaks the attachments such that at a high shear rate particles will be dispersed. The rate of structure build-up and breakdown relates to time-dependent (thixotropic) properties of the suspension. At a given shear rate, an equilibrium structure exists that determines the rheological properties. In general, with increasing aggregate strength, the yield stress and time dependent properties will become more significant. As will be described in the subsequent section, rheology modifiers affect the state of aggregation-dispersion and thereby the rheological properties of the suspension.

The effect of volume concentration ($\phi$) on apparent viscosity has been known since the early work on dilute suspensions by Einstein (1905). Various models for moderately concentrated suspensions have been proposed, with the empirical model by Thomas (1965) being commonly used:

$$\frac{\mu_\infty}{\mu} = 1 + 2.5\varphi + 10.05 + 0.00273\exp(16.6\varphi) \approx \left(1 - \frac{\varphi}{0.64}\right)^{-1.8} \quad \text{Thomas equation} \qquad (3)$$

The approximation to the Thomas equation above is preferable in that it reveals the effect of approaching the maximum packing density (in this case 64%), where the apparent viscosity rapidly increases towards "infinity." The Thomas equation can be used to estimate the high shear rate viscosity ($\mu_\infty$) in the yield plastic model, where shearing induces particle dispersion. The contributions of particle behavior to the observed rheological flow curve are presented graphically in Figure 3.

## RHEOLOGY MODIFIERS

Chemical additives can be used to modify the rheological properties of concentrated mineral suspensions (Klein and Pawlik, 2005). These additives interact with the particle surfaces to affect the degree of particle aggregation or dispersion. As implied in Figure 3, rheology modifiers do not significantly effect the contributions from the carrier fluid or the dispersed particles, but influence state of structure due to aggregation. At high solids contents, decreasing aggregation results in a lower yield stress and apparent viscosity. Improving the degree of dispersion in this manner also reduces the magnitude of the time dependent responses. For some applications, the objective is to increase the yield stress by inducing aggregation in order to prevent particle settling. In this case, coagulants or flocculants may be added. As shown in Table 3, the chemical additives may be either inorganic or organic.

Inorganic additives can be used to either increase or decrease the magnitude of electrostatic repulsion forces. Dispersants increasing electrostatic repulsion and coagulants decrease the repulsion. Common dispersants are poly-

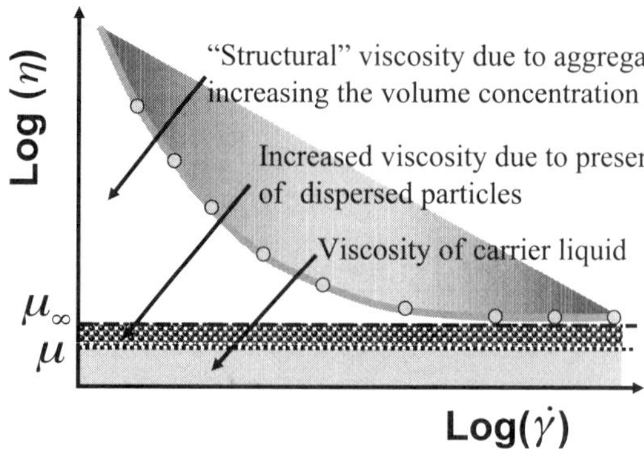


**Figure 3. Contributions particles and aggregation to rheological flow curve.**

**Table 3. Viscosity modifiers for mineral suspensions.**

| Reagent | Function |
|---|---|
| Inorganic | |
| pH modifiers | Dispersant, Coagulant |
| Lime | Coagulant |
| Poly-phosphates | Dispersant |
| Poly-silicates | Dispersant |
| Electrolytes | Dispersant, Coagulant |
| Organic | |
| Non and An-Ionic | Dispersant |
| Low Molecular Weight | Flocculant |
| High Molecular Weight | Coagulant |
| Cationic Flocculant | |

**Table 4. Classification of polymeric rheology modifiers**

| **Low MW<$10^5$ Dispersants** | **Low MW<$10^5$ Dispersants** | **High MW>$10^6$ Flocculants** | **High MW>$10^6$ Flocculants** | **Medium MW $10^5$-$10^6$ Coagulants** |
|---|---|---|---|---|
| Organic | Inorganic | Natural | Synthetic | Synthetic |
| Dextrins<br>Polyacrylates | Poly-phosphates<br>Poly-silicates (Water glass) | Starches<br>Gums | Polyacrylamides<br>Polyalkylene Oxides | Polyamines and cationic derivatives of various polymers |

silicate and poly-phosphates. Some additives can act as either a dispersant or a coagulant depending on their dosage. For example, at a pH near the iso-electric point (pH at which zeta potential is zero), coagulation will result. Increasing or decreasing the pH from this point will increase electrostatic repulsion resulting in particle dispersion.

Lime acts as a coagulant, which is believed to be due to the adsorption of calcium ions onto mineral surfaces that reduces or reverses the surface charge of particles. The use of lime, as a viscosity modifier, is complicated due to dual effects resulting from the presence of calcium ions and increasing the pH. It has been used to improve the flow properties of tails (Hoffert and Poling, 1985). Conversely, for lateritic gold ores, the addition of lime can cause a fluid suspension (e.g., 45% solids) to become excessively viscous likely due to reducing the surface charge where coagulation can occur (Vickel, 1999, Klein and Laskowski, 1999).

As shown in Table 4, there is a range of polymeric rheology modifiers. Poly-phosphates and poly-silicates represent inorganic polymeric additives, both of which are low molecular weight and act as dispersing agents. Organic viscosity modifiers can be divided into three groups. The first group is low molecular weight non-ionic and anionic additives, such as polysaccharides and polyacrylates. These additives behave as dispersants through electrostatic and steric effects. Steric effects refer to the adsorption of the polymer to create a physical barrier that prevents particles from approaching to a distance where van der Waals attraction forces can act. The result is a reduction in the yield stress and apparent viscosity. While these types of dispersing agents are thermodynamically more stable than inorganic dispersants, there are issues related to effective mixing of the reagents with concentrated mineral suspensions and high dosage requirements.

The second group of organic viscosity modifiers includes high molecular polymers such as starches, polyacrylamides and polyethylene-oxides. These flocculants form bridges between particles. At low solid contents, they increase settling rates, but at high solid concentrations, they can prevent settling.

The third group is cationic flocculants, which are of intermediate molecular weight. Examples include cationic dextrins, polyamines and cationic derivatives of various polymers. Although referred to as flocculants, the function may more accurately be described as coagulation.

## RHEOLOGICAL MEASUREMENTS

Concentric cylinder rheometers and tube viscometers are the most commonly used instruments to obtain rheological flow curves. Other devices, such as the vane impeller and cylinder slump test apparatus, are used to estimate the yield stress. It is important to understand what is being measured and possible measurement errors that can result from improper use of these instruments. If not considered or properly accounted for, measurement errors can produce difficult to interpret and misleading results. It is the authors' belief that improper measurements, leading confusing and conflicting interpretation of the results, is the main reason that mineral process engineers do not apply concepts of rheology to practice.

### Concentric Cylinder Rheometers

The most common rotational viscometer geometry used for mineral suspensions is the concentric cylinders. They can be used to measure most rheological properties including viscosity, yield stress and time dependent properties. Other geometries such as cone-plate and plate-plate are not suitable for most mineral suspensions due to settling and wall slip as the particles are displaced away from the shear surfaces.

With concentric cylinders (Figure 4), the shear stress and shear rate depend on the geometry of the fixture and the rotational speed. The shear stress is proportional to the torque exerted by the suspensions on the cylindrical shearing surface (Equation 4). The shear rate will depend on the type of rheological flow-curve but for Newtonian liquids can be represented obtained from Equation 5.

$$\tau = T/2\pi r^2 h \tag{4}$$

$$\dot{\gamma}_N = f(\tau) = 2\omega R_2^{\,2}/R_2^{\,2} - R_1^{\,2} \tag{5}$$

Where, T is torque, r is radius, h is cylinder height, $\dot{\gamma}_N$ is shear rate assuming Newtonian flow curve, $f(\tau)$ is the flow-curve relationship between shear stress and shear rate, $\omega$ is the angular velocity, R is the cup radius, $R_1$ is the bob radius

Suspensions are mixtures of finely ground mineral particles in water and behave like a liquid, enabling their transport and the processing of particles. One important advantage of rotational viscometers is the small sample sizes that is required ranging from less than 50 ml up to one liter. These samples can be generated by bench scale testing to simulate the actual samples for an undeveloped full-scale operation.

The presence of the particles, however, complicates rheological measurements with rotational viscometers. Several possible errors can contribute to the overall errors including end effects, temperature effects, non-Newtonian shear rate effects, wall slip and particle settling effects. These errors can be corrected or minimized by the adjustment of the data and the proper design of the fixture. The following describes the main types of measurement errors and methods to correct these errors.

End effects error result from the additional torque exerted by the fluid on the bottom surface of the rotating bob (Whorlow, 1980). As a consequence, the measured shear stress and therefore apparent viscosity are erroneously high. To reduce the magnitude of these end effects, the bottom surface of the bob is usually recessed creating a trapped air pocket, which has negligible resistance to shear.

Since the viscosity depends on temperature, it must be maintained constant during measurement. For high viscosity slurries, viscous energy dissipation due to shearing produces heat. The cup and bob are often surrounded with a temperature-controlled water jacket to help maintain constant temperature during measurement.

Since many suspensions exhibit non-Newtonian flow properties, the shear rate across the annular gap does not decay in a linear manner. Despite this, shear rates are often incorrectly obtained assuming Newtonian flow using Equation 5, resulting in a measurement error referred to as the "non-Newtonian shear rate effect." The use of Equation 5 may be acceptable if the gap size is very small (<1 mm) and if the suspension is not strongly non-Newtonian (Klein et al, 1995). However, the particle size also determines the gap size below which bridging across the annular gap can produce erratic high torque measurements. The recommended gap size to particle size ratio is > 10.

Prior knowledge of the flow-curve relationship $\dot{\gamma} = f(\tau)$, will allow the shear rate to be calculated by substituting the relationship into Equation 5 (Whorlow, 1980). If this relationship is unknown, an alternative method (Krieger, 1968) can be used to estimate the shear rate from the shear stress and rotational velocity data that involves the use of the shear stress and rotational speed.

For some instruments, the rotating bob is immersed in the vessel containing the suspension, such as a beaker. For these instruments the geometry is not defined and therefore, although they may provide a useful index, they cannot produce quantitative rheological information.

Measurement errors resulting from large or undefined gap sizes are most pronounced for suspensions exhibiting a yield stress. Under these conditions, the shear stress will decay with distance from the rotating bob to a position where it equals the yield stress. If the magnitude of the yield stress is unknown and therefore the location of this position is unknown, the shear rate cannot be accurately determined.

Wall slip is caused by the displacement of particles away from the cup and bob surface since they cannot overlap with the surface. Particle displacement causes the formation of dilute layer adjacent to the surfaces. The effect may be enhanced by the centrifugal acceleration causing particles to migrate away from the surface of the bob. The particle-depleted layer has a lower resistance to shear than the bulk of the suspension and results in a low shear stress and therefore apparent viscosity measurement. Several methods have been developed to mathematically correct for wall slip that rely on assumptions about the flow (Yoshimura, A. and Prud'homme, 1988). Alternatively, roughening of the shearing surfaces has been shown to reduce the magnitude of this error (Klein et al, 1995).

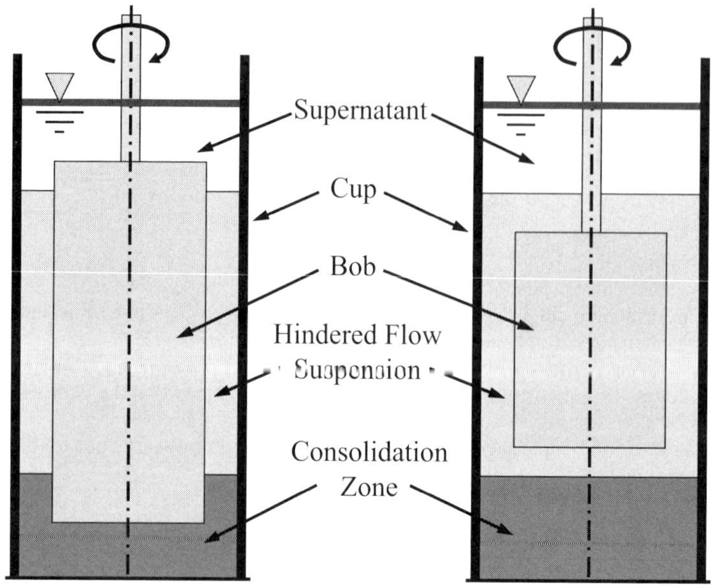


**Figure 4. Section view of conventional concentric cylinder cup and bob viscometer with settling suspension and elongated cup with bob positioned within the constant density zone during measurements.**

For mineral suspensions, perhaps the most significant measurement error is due to particle settling. In a conventional cup and bob, particle-settling results in the formation of a particle depleted zone in the upper annulus between the cylinders and a sediment at the bottom of the cup (Figure 4). The particle-depleted zone has a lower resistance to flow than the original suspension causing erroneously low shear stress measurements. Conversely, the build-up of sediment can contact the bob impeding its rotation causing a falsely high stress measurement. It cannot be assumed that these errors are offsetting. The magnitude of the errors will depend on the composition of the suspension and the geometry of the fixture.

Several devices have been developed to measure the rheological properties of settling suspensions that maintain the composition by introducing an upward flow in the annular gap to counteract settling. Overend et al (1984) used a cup with slots in its walls and an open bottom, which is placed in a mixing bowl containing the suspension. The mixed suspension flows into the annular gap through the slots in the cup wall and thereby maintains the composition of the suspension in the annular gap. They showed, using Newtonian fluids, that at low agitation rates in the mixing bowl, the results compared well to those obtained without mixing.

Clarke (1967) and Purohit and Roy (1968) positioned a cup with an opened bottom in a vessel. An impeller at the bottom of the vessel forced the suspension to circulate up through baffles in the walls of the vessel into the top of the cup and down through the annular gap between the concentric cylinders. Similarly, Haake developed an arrangement with a pump, which forces the suspension up through the annular gap at a rate greater than the settling rate of the particles. Valentyik (1971), Ferrini et al (1979), Klembowski (1988) and Reeves (1990) supplied the suspension to the top of the cup, allowing it to flow through the annular gap and out through the bottom during measurement.

Klein et al (1995) developed a device for suspensions exhibiting zone-settling properties (Figure 4). The device is an elongated double gap cup and bob. The bob is positioned in an elongated cup so that it is completely within the constant density zone for the duration of a measurement. The fixture uses a double gap geometry with the bob supported by spokes attached to the shaft thereby allowing particles to settle through the top of the bob into both the inner and outer gap. The design including vertical grooves in the shearing surfaces to reduce wall slip errors and a narrow gap to lessen non-Newtonian shear rate effects. The advantage of this device is that it does not require the introduction of mixing or flow during measurement.

### Vane

The vane test has become the standard for measuring the yield stress of thickened slurry and paste. The method is an adaptation of a standard soil mechanics test that can is able to measure the much lower yield stresses of interest in mineral slurries. The vane method uses a paddle shaped impeller or "vane" (Figure 5) and a rotational rheometer capable of rotating at very low speeds. The open design of the vane allows it to be inserted into the slurry without breaking down the inter-particle structure. When the vane is rotated, it shears the slurry along the surface of a cylinder with the same length and diameter as the vane. The torque required to turn the vane typically increases with the angle of rotation until it reaches a peak and then drops off (Figure 5).

The yield stress is obtained from the maximum torque value and the area of the shear plane using the following relationship (after Nguyen and Boger, 1983).

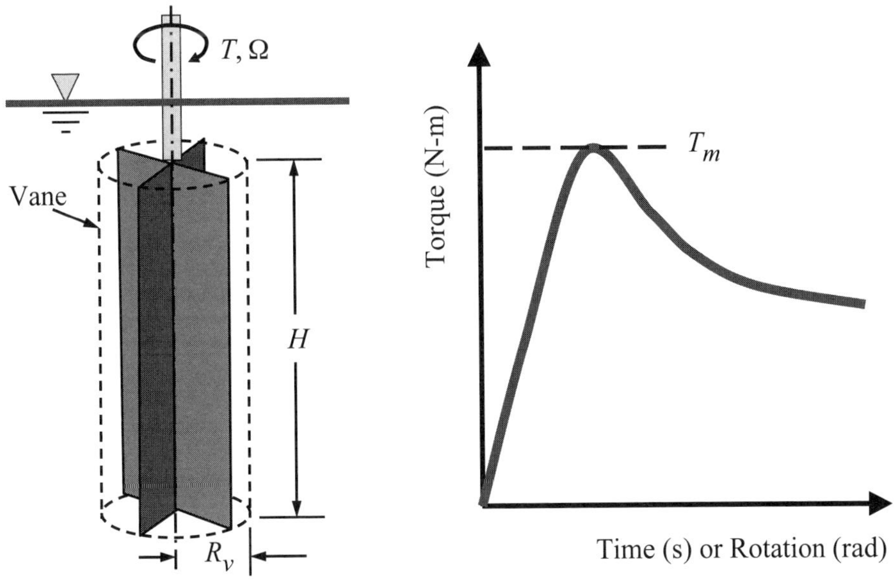


**Figure 5. Torque vs. time (rotation angle) for vane rheometer**

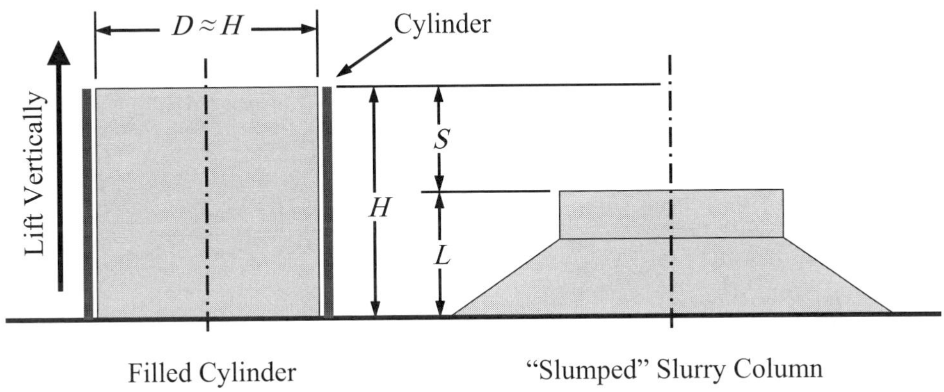


**Figure 6. Cylinder slump test**

$$\tau_o = T_m \Big/ \left(2\pi R_v^2 \left(H + \frac{2}{3} R_v\right)\right) \quad \text{Vane test} \tag{6}$$

Where $\tau_o$ is the yield stress (Pa), $T_m$: the maximum measured torque (N-m), $R_v$ is the vane half-width or cylinder radius (m), $H_v$: the actual height of the vane (m)

It should be noted that getting repeatable yield stress results is problematic and some researchers have even questioned whether it exists at all. The vane test has become something of a standard for measuring the yield stress of high concentration mineral slurries (i.e., the yield stress is what the vane test says it is).

**Cylinder Slump Test**

The ASTM conical slump test is a standard test for the "workability" of concrete. The slump test has been modified to make it appropriate for testing the yield stress of paste and thickened slurries. The main changes are the use of a cylinder (often a short piece of pipe) instead of a cone and a reduction in the size to make it suitable for laboratory work. The height to inside diameter ratio of the cylinder should be approximately one.

To perform the test, the cylinder is placed on a flat surface and filled with slurry. Any air bubbles in the slurry can be removed by stirring with a rod. The cylinder is then lifted and the slurry column allowed to collapse ("slump") under its own weight. The distance of the slump ($S$) is measured with a ruler (Figure 6).

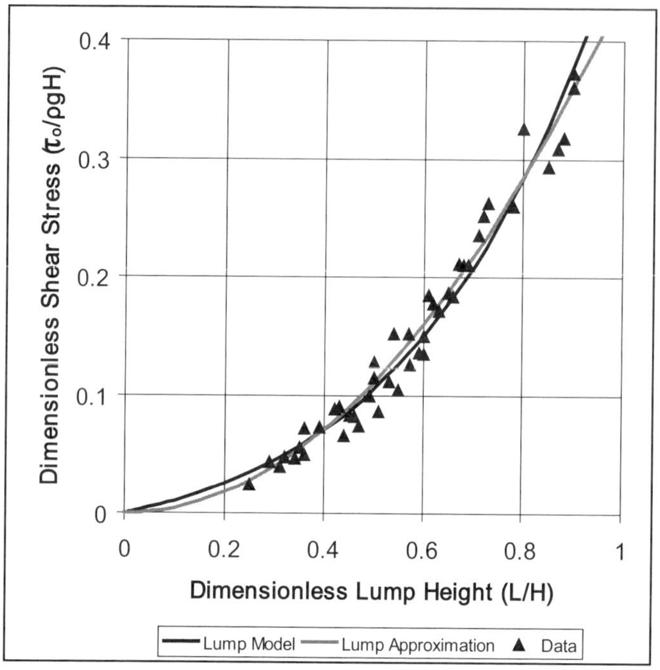


**Figure 7. Comparison of cylinder slump test results to vane data**

The yield stress of the slurry is found from the following non-dimensional relationship base on compression forging theory (Hallbom 2005):

$$\tau' = 0.5(1-S')\exp\left(-\sqrt{3}S'\right) \quad \text{Slump test} \tag{7}$$

$$\tau' = \frac{\tau_o}{\rho g H} \tag{8}$$

$$S' = \frac{S}{H} \tag{9}$$

Where, $\tau_o$ is the yield stress (Pa), ρ is the slurry density (kg/m$^3$), $g$ is the local gravitational acceleration (m/s$^2$), $S$ is the measured slump (m), $H$ is the initial height of the cylinder (m). When the slump increases, the yield stress decreases. The effect of yield stress is made conceptually easier if it is related to the final height of the slumped slurry column ($L$):

$$L' = \frac{L}{H} = \frac{1-S}{H} = 1 - S' \tag{10}$$

The yield stress relationships based on this remaining "lump" of slurry become:

$$\tau' = 0.5L'\exp\left(-\sqrt{3}(1-L')\right) = \frac{L'}{11.3}\exp\left(\sqrt{3}L'\right) \approx \left(\frac{2}{3}L'\right)^2 \quad \text{Lump test} \tag{11}$$

The lump test relationships are compared with vane test data from, Pashias *et al.* 1996, in Figure 7.

If the height of the cylinder and the final column height are measure in millimeters ($h$ and $l$ respectively) and the specific gravity of the slurry ($SG$) is used, then the lump approximation gives the yield stress directly in Pa:

$$\tau_o \approx \frac{4.36(SG)}{h} l^2 \tag{12}$$

This implies that there will be a maximum yield stress that can be measured in any given sized slump cylinder, although a 75mm or 100 mm cylinder is usually adequate for most slurries of interest. While the slump test results are approximate, it is quick, cheap, and the accuracy is sufficient to characterize the slurry behavior.

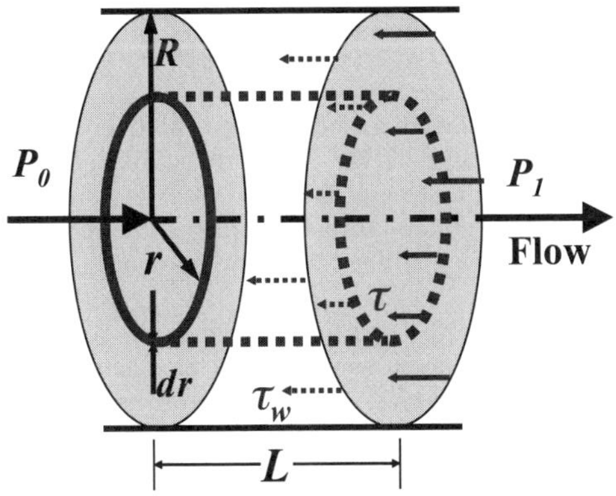


**Figure 8. Principle of tube viscometer.**

The vane test and the cylinder slump test only measure the yield stress. This may be adequate for thick slurries (paste backfill), where the yield stress dominates the behavior but for most unit operations it is necessary to get the flow parameters.

**Tub Viscometers and Loop Tests**

Tube viscometers are the simplest and least expensive devices used to measure viscosity. They can provide very precise measurements for low viscosity fluids and on a large scale can simulate actual slurry transport conditions. The viscosity can be obtained from the pressure drop and flow rate through a tube of known diameter and length (Figure 8).

The volumetric flow rate (V) through a tube of radius r is related to the shear stress ($\tau$) according to equation 13.

$$V/pr^3 = 1/\tau_w^{\,3}\ \tau^2 f(\tau)d\tau \qquad (13)$$

Where, V is volume flow through cross section, r is-radius, p is pressure drop per unit length, $\tau$ is shear stress, $\tau w$ is shear stress at the wall, $f(\tau)$ is the function relating shear rate to shear stress.

For a Newtonian fluid $f(\tau) \cong \eta_N\,\dot{\gamma}$ and the expression simplifies to the Poiseuille Equation:

$$\dot{\gamma}_{WN} = 4V/\pi r^3 \qquad (14)$$

$$\tau_W = rp/2 \qquad (15)$$

$$\eta_N = r4pp/8V \qquad (16)$$

Where, $\dot{\gamma}_W$ is shear rate at tube wall, $\tau_W$ is the shear stress at the tube wall and $\eta_N$ is the Newtonian viscosity.

Slurry sample sizes can range from less than one liter for measurements in a capillary viscometer to several thousand liters for large scale loop tests. The appropriate scale of testing will depend on the physical characteristics of the suspension, such as particle size, and the application to which the data will be applied.

Tube viscometers are considered to be similar to and therefore more suitable for obtaining information for applications involving slurry transport through pipes. They can be used to obtain shear stress – shear rate data, but cannot provide information about time dependent properties. Results from these devices must be adjusted to account for non-Newtonian flow properties, end effects and wall effects. Particle migration and settling can also influence the results and must be considered in the interpretation of the data.

Since most mineral suspensions are non-Newtonian, Equations 14 to 16 cannot be used to determine the rheological responses. If the shear stress-shear rate relationship is known, expressions can be derived in a similar manner. For example, Buckingham derived the expressions for a Bingham plastic fluid according to equations 17 and 18.

$$\tau_W = \tau_Y + h_{Pl}\,\dot{\gamma}_W \qquad (17)$$

$$V/\pi r^3 = 1/\,h_{Pl}\,[4/\tau_W - \tau_Y/3\{1–¼\,(\tau_Y\,/\,\tau_W\,)^3\}] \qquad (18)$$

Where, $\tau_Y$ is the Bingham yield stress, $h_{Pl}$ is the plastic viscosity

In most cases, the shear stress-shear rate relationship is unknown. In this case, the shear rate can be approximated from the flow rate and pressure drop using the following expression (Whorlow, 1980).

$$\dot{\gamma}_W = \dot{\gamma}_{WN}\ \{3/4 + b/4) \qquad (19)$$

where $b = d[\log(\gamma_{WN})]/d[\log(\tau_W)]$

In the above expression, $\dot{\gamma}_{WN}$ and $\tau_W$ are obtained from equations 14 and 15. The derivation of expressions for non-Newtonian flow through tubes can be found in several references and will not be described further in this paper (Whorlow, 1980, Van Wazer et al, 1963).

Pressure drop resulting from flow into and out of a tube are responsible for end effect errors and may be significant if the L/D ratio for the tube is small. Tests using tubes of varying length but fixed diameter will demonstrate the magnitude of this error. A plot of pressure drop versus tube length should produce a straight line passing through the origin. The line will be displaced along the pressure axis by the magnitude of the end effect. This pressure can be subtracted from recorded pressure measurements. Alternatively, the tube length can be selected such that the relative magnitude of the error is below and acceptable limit.

Since particles cannot overlap the tube wall, a particle depleted "lubricating" layer forms adjacent to the wall. Conducting measurements using tubes of fixed length while varying the diameter will demonstrate the magnitude of the slip error. If results are affected, the response will change with the diameter, the effect becomes less significant at large diameters. In this way an appropriate diameter can be selected. Alternatively, an estimate of the slip velocity can be included in the shear rate expression (Oldroyd, 1956. Whorlow, 1980).

In addition to particle displacement causing wall slip, particles will migrate toward the center of the tube creating a solids concentration profile that ranges from near zero at the wall to a maximum at the center of the tube. This "tubular pinch" effect complicates the determination of shear rates and may not be accurately accounted for, which may limit the applicability of results obtained using tube viscometers.

For the design of slurry pipelines, flow loop tests provide a greater degree of confidence than tests with laboratory tube or rotational viscometers. With proper design of instruments and adjustment of the data, these instruments should produce the same rheological information as large-scale tests. Loop tests allow direct determination of parameters such as the deposition velocity, transition velocity and hydraulic gradient (McKibben al, 2000, Gillies et al, 2000, Gillies and Shook, 2000, Shook et al, 1984). They also allow demonstration of the homogenous / heterogeneous character of the suspensions which cannot be confirmed in smaller scale tests.

For loop tests, large quantities of representative sample must be obtained. In the case of a new project, this would involve pilot scale processing to generate concentrates or tailings with sample size up to several tons. For suspensions with coarse particles, the particles must be removed by screening prior to measurement in rotational viscometers. Due to the large sample requirements, it is impractical to run loop tests for a wide range of material types as they may be costly or impossible to collect. For this reason, the trend has been on designing pipelines using rheological data from properly conducted concentric rheometer tests (Hallbom, 2008).

**Online Rheological Measurements**

Several devices have been developed for online rheological measurements to allow these properties to be used for control purposes (Reeves, 1990, Cowper, 1994, Kawatra and Bakshi, Nguyen et al 1999). While these devices are not widely used, their use has been demonstrated in industrial operations.

Two tube type instruments developed are the U-tube viscometer (Cowper, 1994) and the pressure vessel rheometer (Kawatra and Bakshi, 1998). The U-tube viscometer feeds slurry at a constant flow rate and measures the pressure drop to obtain apparent viscosity data; by varying the flow rate, complete flow curve data can be generated. The instrument is used in slurry transport applications. As implied, the pressure vessel viscometer varies the pressure to obtain a range of measured shear rates and thereby a flow curve. The instrument has been used to monitor viscosity of concentrate thickener underflow.

Reeves (1985, 1990) developed the Debex on line concentric cylinder viscometer in which a sample of the slurry is pumped upward through the annular gap between rotating concentric cylinders. The device has been used for online measurements on dense media (Napier-Munn et al, 1989, Shi et al, 1996) and in grinding circuits (Shi et al, 1999). Similarly, Nguyen et al (1999) developed a helical concentric cylinder system that was designed for settling suspensions. Both instruments give full rheological flow curves.

**SUMMARY**

The flow properties of mineral suspensions are fundamental to the handling and processing of mineral suspensions. They are becoming increasingly important for design and operation of pipelines, which is being used increasingly for applications such as the transport of paste backfills and concentrates. This holds true for mineral processing such as in the cases of processing nickel laterites and ultrafine grinding. The role and control of rheology will become more important as the industry moves to processing more complex fine grained ores.

Most mineral suspensions exhibit non-Newtonian rheological properties, which can be characterized using flow curves and simple empirical flow curve equations. The apparent viscosity may be useful, but it does not provide complete information and can be misleading. Therefore, the coefficients from simple empirical flow curve equations are recommended for characterizing rheological properties. The yield stress has been identified as a rheological parameter that is important to design and operation. Consequently, indirect and direct methods have been developed to more accurately determine its value.

The physical and chemical interactions between the particulate and aqueous phases are responsible for the rheological properties. Knowledge of these interactions allow us modify the rheology by changing such parameters as solid content and state of aggregation/dispersion, by adding rheology modifiers.

Rheological measurements are performed using two main types of instruments: tube viscometers and rotational viscometers. Tube viscometers range from small-scale capillary apparatus to large pilot scale loop test facilities. Various geometries exist for rotational viscometers and the most appropriate for minerals suspensions is the cup and bob. Several potential errors are inherent to these measurement techniques. Rheological test results that have not been adjusted to account for these measurement errors must be treated cautiously. Instruments for online measurement allow the use of rheological properties for control purposes.

## REFERENCES


Boger, D.V., 1999, Rheology and the minerals industry, Min. Pro. Ext. Met. Rev, Vol. 20, pp 1-26.

Clarke, B. 1967, Rheology of coarse settling suspensions, Trans. Inst. Chem. Eng., Vol 45, T251-T256.

Cowper, N.T., 1994, Continuous on-line measurement of fluid or slurry rheology, US Patent 5,315,863.

Ferrini, F., D. Ercolani, B. de Cindio, L. Nicodema, L, Nicolais, S. Ranauda, 1979, Shear viscosity of settling suspensions, Rheol. Acta. Vol 18, pp 289-296.

Gillies, R.G., Shook, C.A., 2000, Modeling high concentration settling slurry flows, Can. J. Chem. Eng. Vol. 78. pp. 709-716.

Gillies, R.G., Schaan, J., Summer, R.J., McKibben, M.J., Shook, C.A., 2000, Deposition velocities for Newtonian slurries in turbulent flow, Can. J. Chem. Eng. Vol. 78. pp. 704-708.

Hallbom, D.J., 2008, Pipe flow of homogeneous slurry, PhD Thesis, University of British Columbia, Vancouver, Canada.

Hallbom, D.J., and Klein, B, 2009, A physical model for yield plastic fluids, Particle Science and Technology—An International Journal, Vol. 27 Issue 1, 1, 2009

Hanks, R.W., 1981, Hydraulic design for flow of complex fluids, RW Hanks Associates Inc., Orem, Utah.

He., Y.B., Laskowski, J.S., Klein, B., , 2004, Particle movement in non-Newtonian slurries: the effect of yield stress on dense media separation, Chem. Eng. Sci. Vol. 59, Issue 3, February 2004, Pages 731-732

Hintikkka, V.V., Kalapudas, R.P., Viitanen, P..I., 1999, Effect of rheology in grinding efficiency in the laboratory scale continuous classifying mill, Min. Pro. Ext. Met. Rev, Vol. 20, pp 133-154.

Hoffert, J.R., Poling, G.W., 1985, The action of lime in promoting pipeline flow in tailing slurries, CIM Bulletin, August.

Horsley, R.R., Allen, D.W.1987, The effect of yield stress on hydrocyclone performance in the mining industry, Proc. 3rd Int Conf on Hydrocyclones, Oxford, P.Wood ed, Elsevier, NY, pp269-275.

Horie, M., Pinder, K.L., 1979, Time-dependent shear flow of artificial slurries in coaxial cylinder with a wide gap, Can. J. Chem. Eng. Vol 57. pp 125-134.

Kawatra, S.K., Bakshi, A.K., On-line measurement of slurry rheology in a thickener at a copper concentrator, Minerals and Metallurgical Processing, Vol. 15, No. 4, pp37-40.

Klein, B., 1992, Rheology and stability of magnetite dense media, PhD Thesis, University of British Columbia, Vancouver, Canada.

Klein, B., J.S. Laskowski, S.J. Partridge, 1995, A new viscometer for rheological measurements on settling suspensions, J Rheol. Vol 35., No 5, pp 827-840.

Klein, B., Laskowksi, J.S., 1999, Rheological measurements on settling suspensions: characterization of a cyanide leach pulp, Min. Pro. Ext. Met. Rev, Vol. 20, pp 41-56.

Klein, B., Hallbom, D.J., 2002, Modifying the rheology of nickel laterite suspensions, Minerals Engineering, 15, pp. 745-749.

Klein, B., Pawlik, M., 2005, Rheology Modifiers for Mineral Suspensions, Minerals and Metallurgical Processing, Vol 22, No. 2, pp 83-88.

Klimpel, R.R., 1982, Laboratory studies of the grinding and rheology of coal water slurries. Powder Technology, 32 pp267-277.

Klembowski, V., 1988, Continuous measurement of viscous properties of suspensions with settling particles, Progress and Trends in Rheology II, Prague, ed. H. Giesekus , Springer, NY, pp 177-179.

Krieger, I.M., 1968, Shear rate in the Couette viscometer, Trans. Soc. Rheol., Vol 12, pp. 5-11.

Laskowski, J.S., 2001, Rheological measurements in mineral processing related research, New Developments in Mineral Processing Proc. 9th Balkan Min. Proc. Congr. Istanbul, Turkey, 2001.

McKibben, M.J., Gilies, R.G., Shook, C.A., 2000, Predicting pressure gradients in heavy oil-water pipelines, Can. J. Chem. Eng. Vol. 78. pp. 752-756.

Napier-Munn, T.J., Reeves, T.J., Hansen, J.Y., 1989, The monitoring of medium rheology in dense medium cyclone plants, AusIMM Bulletin, Vol. 294, No. 3, pp85-93.

Nguyen Q.D., D.V. Boger, 1983, Yield stress measurement in concentrated suspensions, J. of Rheology., Vol, 27, pp. 321-349.

Nguyen Q.D., D.V. Boger, 1998, Application of rheology to solving tailings disposal problems, Int. J. Min. Proc.Vol. , 54, pp 217-233.

Nguyen, Q.D., C. Devasagayam, D.J. Brown, 1999, Development of an on-line flow rheometer for mineral slurries, Min. Pro. Ext. Met. Rev, Vol. 20, pp 75-91.

Oldroyd, J.G., 1956, Non=Newtonian flow of liquids and solids, Rheology: Theory and Applications, Vol.1, F.R. Eirich. Academic Press, New York, pp. 653-682.

Overend, I.J., R.R. Horsley, R.L. Jones and R.K. Vinycomb, 1984, A new method for measurement of rheological properties of settling suspensions, Proc. 9th Int. Congr. On Rheol., ed. Mena, Mexico City, Vol. 2 pp. 583-590.

Pashias, N.; Boger, D.V.; Summers, J.; Glenister, D.J. (1996) A fifty cent rheometer for yield stress measurement Journal of Rheology, 40 (6) pp. 1179-1189

Prestidge, C.A., 1999, Rheological investigations of DLVO and non-DLVO particle interactions in concentrated mineral suspensions, Min. Pro. Ext. Met. Rev, Vol. 20, pp 57-74.

Purohit, N.K., A.N. Roy, 1968, Studies on the rheological properties of rapidly settling suspensions including minerals, Trans IMM, C201-C208.

Reeves, T.J., 1985, On-line viscometer for mineral slurries, Trans IMM, Vol 94, C201-208.

Reeves, T.J., 1990, On-line viscosity measurement under industrial conditions, Coal Prep. Vol. 8, pp 1-9.

Shi, F.N., Napier-Munn, T.J., 1996, Measuring the rheology of slurries using an on-line viscometer, Int. J. Miner. Process, Vol. 47, pp. 153-176.

Shi, F.N., T.J. Napier-Munn, I.K. Asomah, 1999, Rheological effects in grinding and classification, Min. Pro. Ext. Met. Rev, Vol. 20, pp 123-131.

Slatter, P. 1999, The role of rheology in the pipelining of mineral slurries, Min. Pro. Ext. Met. Rev, Vol. 20, pp 281-300.

Shook, C.A., Gilies, R.G., Husband, W.H.W., Small, M., 1984, Flow of a coal slurry with a yield stress, J. of Pipelines, Vol. 4. pp. 289-297.

Thomas, D.G., 1965, Transport Characteristics of Suspension: VII. A Note on the Viscosity of Newtonian of Uniform Spherical Particles, J. Coll. Sci, Vol20, pp 267-277.

Usui, H., T. Saeki, K. Hayashi, T. Tamura, 1997, Sedimentation stability and rheology of coal water slurries, Coal Prep. Vol. 18, No. 3-4, pp 201-214.Valentyik, L., 1971, Instrumentation and control of the specific gravity and the rheology of heavy-media suspensions, Min. Sci. Eng. Pp 38-44

Valentyik, L., 1971, Instrumentation and Control of the Specific Gravity and the Rheology of Heavy Media Suspensions, Min. Sci. Eng., 38-44.

Van Wazer, J.R., Lyons, J.W., Kim, K.Y., Colwell, R.E., 1963, Viscosity and flow measurement, John Wiley & Sons, New York.

Vickel, G., 1999, Challenges & improvements in milling at Omai gold mines, Proc. 31st Annual Meeting of the Canadian Mineral Processors, CIM, pp19-32.

Wasp, E.J., Kenney, J.P., Gandhi, L.R., 1977, Solid-liquid flow slurry pipeline transport, Trans Tech Publications, Clausthal, Germany.

Whorlow, R.W., 1980, Rheological Techniques, Ellis Horwood Ltd, Sussex, England.

Wilson. K.C., Addie, G.R., Sellgren, A., Clift, R., 1997, Slurry transport using centrifugal pumps, 2nd Edition, Blackie Academic and Professional, London.

Yue, J., Klein, B., 2004, Influence of rheology on the performance of horizontal stirred mills, Particle breakage kinetics in horizontal stirred mills, Minerals Engineering. Vol 17, pp. 1169-1177.

Yoshimura, A., R.K. Prud'homme, 1988 Wall slip corrections for Couette and parallel disk viscometers, J. Rheol., Vol 32, No.1 pp. 53-67.

# ONLINE SLURRY PARTICLE SIZE ANALYZERS

Matti Kongas and Kari Saloheimo

Outotec - Minerals Processing, Riihitontuntie 7 D, FI-02200 Espoo, Finland

## ABSTRACT

New technology has opened up new applications for on-line particle size analysis of mineral slurries. The control of grinding circuits based on product particle size is essential for downstream process performance. In some cases the particle size distribution of the final concentrate is an important quality issue. Cost-effective use of energy in grinding also needs more attention. This paper discusses established on-line particle size analysis technologies and presents experiences on the use of laser diffraction in several minerals processing operations. The on-line applications discussed are; Blaine number measurement of Fe concentrate before pelletizing, grinding of variable batches of ore and paste thickener control.

## INTRODUCTION

In most cases of ore grinding processes, the primary control variable for the particle size distribution is the fresh ore feed rate to the grinding circuit, and there is a tradeoff between the tonnage throughput and the fineness of the grinding product. With variable grindability of the ore, this gives a possibility for optimization – the softer the ore, the more tons can be pushed through the mill to achieve adequate liberation. Figure 1 shows calculated capacity values of a single-stage ball mill circuit producing a constant product particle size. The practical implementation of such control is often operator based, but there are a growing number of closed-loop expert system applications doing the job for the operators. In both cases an online particle size instrument is needed to give timely feedback to optimize the mill feed, with a potential average throughput increase of about 10% in typical a case.

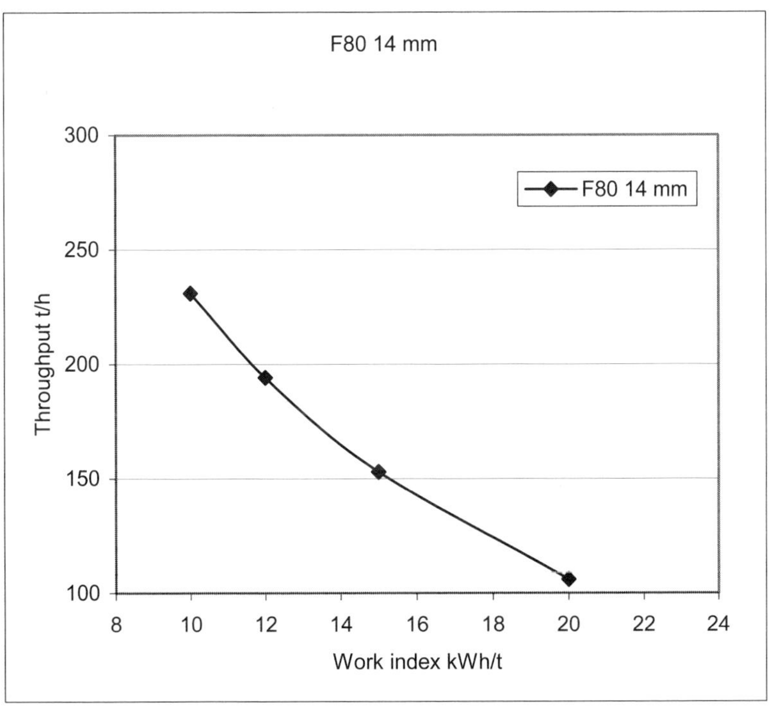


**Figure 1. Effect of varying ore work index on 5m (D) by 6.8m (L) closed circuit ball mill throughput with a fixed product particle size, P80=75 μm**

A grinding circuit that is not producing consistent grind may cause significant metallurgical losses downstream. A gold plant for quartz based ore with a typical closed circuit ball mill followed by thickener and CIP requires a relatively fine grind of 60 – 80% under 74 μm. Too coarse grind increases gold losses exponentially as leach residence time is not long enough to dissolve all gold contained in coarse particles, see Figure 2. Too fine grind is waste of energy/ton processed, reduced plant capacity and problems with thickening, filtering and tailings disposal.

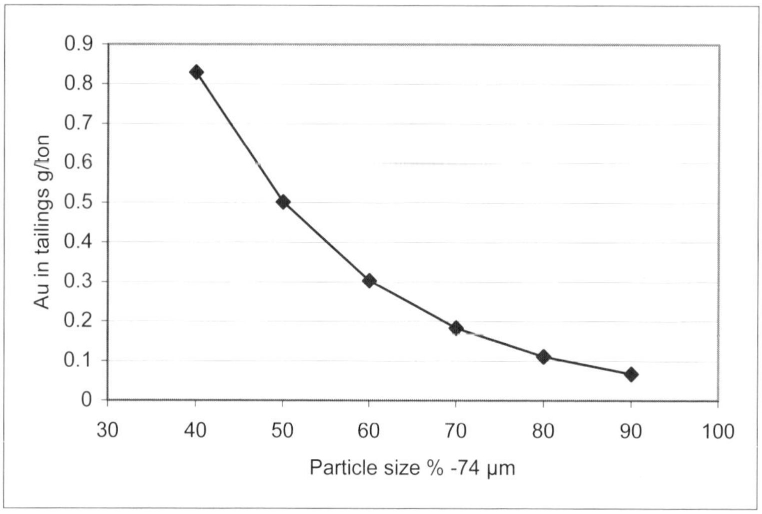


**Figure 2. CIP leach plant tailings as function of grind particle size**

The particle size distribution of the feed to a flotation process has an influence on the flotation recovery through limited liberation of valuable minerals and reduced flotation rates of both too coarse and too fine particles. In addition, the particle size distribution has an effect to the flotation reagent consumption and to the dewatering of the slurry. That is why the control of the preceding grinding process usually includes constraints or target values set to the particle size distribution of the grinding product. Problems in grinding circuit operation typically result in a significant change in product size distribution. An example of a Zn/Pb flotation shown in Figure 3 shows the effect of particle size on lead recovery. The recovery drops quickly with finer grind by the tendency of heavy minerals to produce fines.

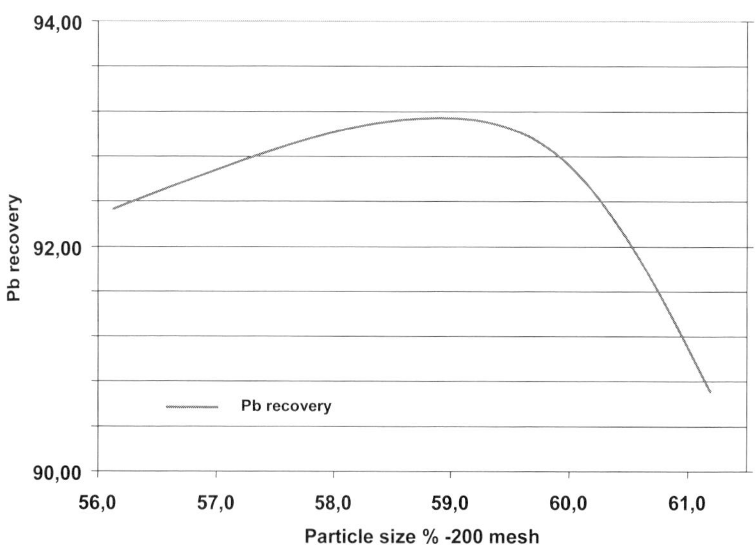


**Figure 3. Effect of -74 μm size on recovery of Pb**

The above examples illustrate the need for online particle size analysis as part of grinding control. General requirements for the particle size measurement instruments can be summarized as follows,

- Reliability – this is the foremost requirement of online analysis from the operator's point of view. If false control decisions are done based on incorrect readings or if the analyzer needs continuous maintenance to keep working, the operators soon lose their confidence on the system altogether. Reliability is closely tied to equipment availability; the whole system needs to be designed for 24-hour, 7-day operation in the process environment.
- Representativeness – the measurement results need to reflect the changes of the overall particle size distribution in the bulk process flow. Thus careful design is required in the sampling of the slurry and the presentation of the sample to the measurement technique to get a reliable measurement of especially the coarser (>100 μm) fractions.
- Precision – the analysis should be sensitive enough to indicate the process changes that are relevant for the purposes of control.
- Speed – the analyzer needs to deliver new results frequently enough that the process control can react to the feed and process changes. The typical average residence time of the material in the whole grinding circuit is about 10 minutes, and the analysis delay and cycle time should be less than or about half of that for effective control.
- Ease of use – With the increasing demand of effectiveness in the industry, easy operation and maintenance are among major requirements. The personnel specialized in other tasks than the production is more and more rare at the plant, and there is not much room for extra maintenance items.

One fact of particle size analysis in general is that each particle size measuring technique describes a 3-dimensional particle differently, as illustrated in Figure 4. Sample preparation procedures can change the level of agglomeration of particles in the sample actually analyzed. The absolute result might be the same for perfect, homogeneous glass beads but the effect of particle shape, specific gravity variation or other factors on the size measurement is dependent on the method used. In on-line analysis full size distribution is not always required and the results can be correlated to a relevant laboratory standard method. It is more the precision of the analysis that counts - the method must give a consistent response to process changes, and the repeatability should be in proportion to the needs of the control.

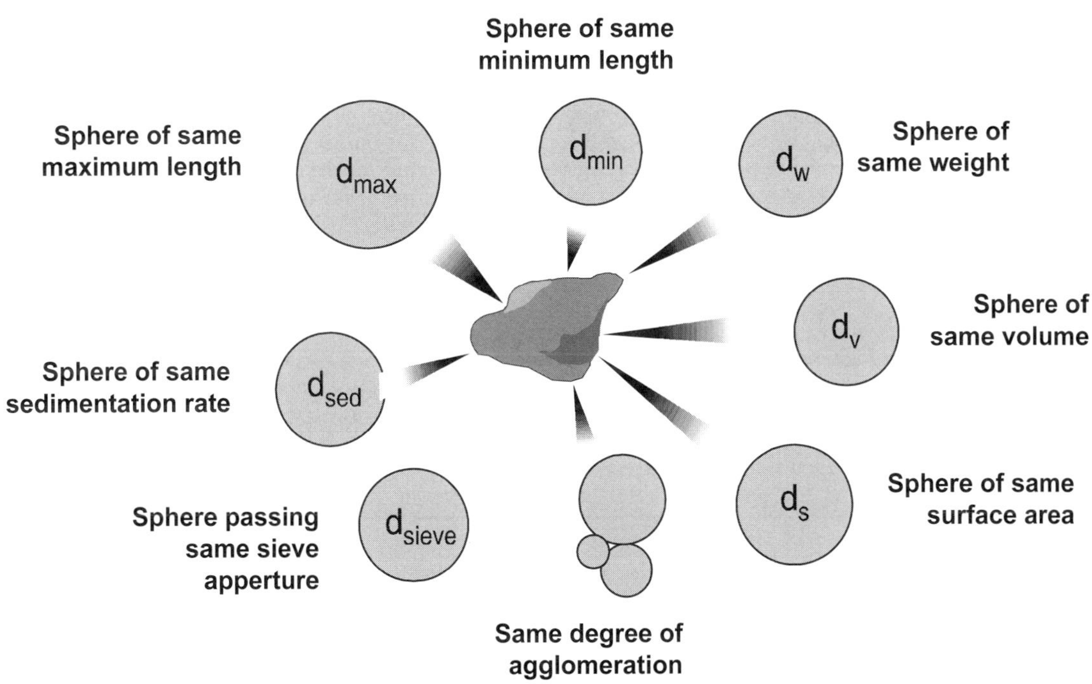


**Figure 4. The result of measuring the size of three dimensional particles depends on the method**

## PARTICLE SIZE MEASUREMENT TECHNOLOGIES

### Online Sampling

Online particle size analysis requires reliable and representative sampling. One of the challenges of the on-line analysis system is to place a representative sample of the solids into the active measurement volume of the measurement technology. This can be achieved by successive sampling and sample reduction steps (Kongas et al, 2003). Static samplers are used, taking advantage of the mixing caused by process pumps and/or pipe geometry changes, and isokinetic flow. The subject is discussed in more detail elsewhere in this publication.

### Laboratory

Particle size has to be measured, so that it can be controlled. Manual sampling, sample preparation and analysis using sieves or more modern instruments in a laboratory can be done. However, manual methods are labor intensive and take time, so most operations can only afford to do them once per shift. Process control requires more frequent measurements, e.g. the reaction time of a grinding circuit is typically 5-15 minutes. On-line particle size measurement has therefore become common practice in industry because the return on investment for applying process monitoring and control based on such instruments is very high.

### On-line Ultrasonics

Ultrasonic attenuation has been applied in wet mineral processes since the 1970s. Because ultrasonic particle size measurement does not tolerate air bubbles in the sample, sample has to be de-aerated before measurement. Recently ultrasonics has been combined with gamma-ray transmission and sound velocity measurement for more accurate solids content compensation (Coghill, Millen and Sowerby, 2002). The sample volume measured can be relatively large, which is an advantage of this technology. The major limitations of ultrasonic particle size analysis in mineral processing operations are its sensitivity to entrained air bubbles, flaky particles, solids content variation and viscosity changes. These interferences necessitate frequent calibrations.

### On-line Image Analysis

Many optical image analysis methods based on particle counting have been proposed and applied to specific needs. With the wide size distributions present in mineral processes they are quite inefficient in achieving the desired volume or weight distribution because the number distribution is so heavily weighted towards fine particles. Significant dilution is required in order to overcome the opaqueness of mineral slurry and to separate fine particles from each other.

### On-line Direct Mechanical Measurement

The quantitative particle size analyzer based direct mechanical measurement principle employed in the Outotec PSI 200 particle size instrument, has become the most common method for on-line particle size analysis in mineral processing plant slurries. A representative sample is flowing between ceramic tipped moving part and fixed ceramic tip (Saloheimo and Anttila, 1994). An electronic sensor measures the position of the moving part. A randomly selected particle is trapped between the ceramic tips and its diameter is measured twice a second. The simplicity of its operation means it has high availability and minimal maintenance. New updated model of this analyzer has just been released,

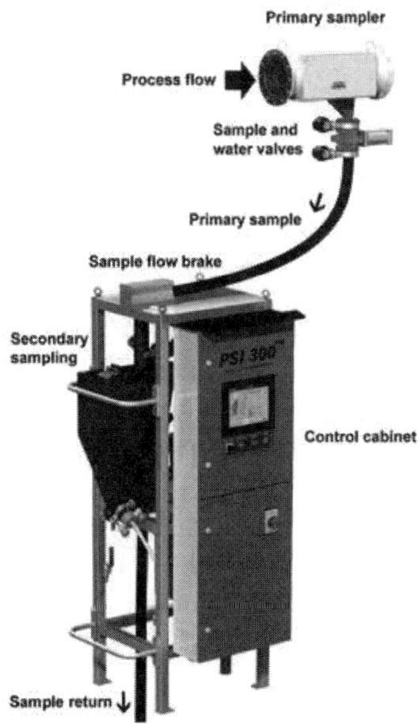


**Figure 5. PSI 300 on-line particle size analyzer based on mechanical particle size measurement**

Most applications are in primary grinding, where the control strategy is based on tonnage maximization against a constraint of the coarsest particle size fractions to prevent cyclone blockage, ensure adequate liberation or fast enough leach rate in gold CIP circuits.

**On-line Laser Scattering Technology**

Low-angle laser light scattering, or laser diffraction, has been known as a laboratory technique since the 1960s. It is routinely used in the laboratory, especially when wide distributions need to be analyzed. The size analysis is based on an intensity distribution measurement of coherent laser light scattered by the particles. The form of the scattering pattern is described by the Mie theory and the width of the pattern is dependent on the size distribution. When laser light meets a population of particles, volumetric size distribution can be calculated back from the scattered light distribution. The principle is illustrated in Figure 6.

A significant advantage of laser diffraction is that it gives a consistent volumetric particle size analysis result without any external calibration. There is a difference in the particle size analysis results measured by other methods based on different size criteria. For example, sieve analysis measures particles by passing them through square holes and it is sensitive to specific weight differences in particle size fractions, size accuracy of holes and wear. As discussed above, repeatability and precision are the most important features in process control applications, both of which are very good over a wide particle size range for laser scatter method. For example, one can reach 1-2% relative precision of the distribution median over the 1-500 μm size range with the on-line system described below. Also the speed, non-contact nature and robustness of the method to ambient conditions make laser diffraction ideal for on-line mineral slurry applications.

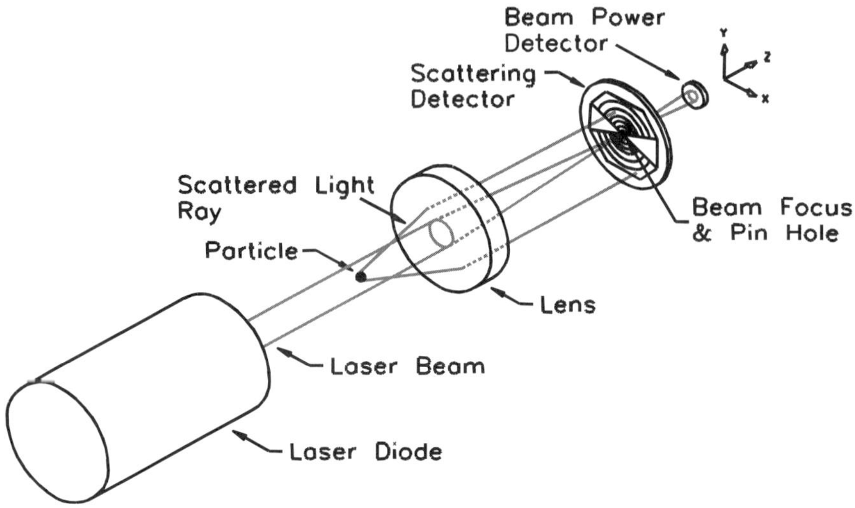


**Figure 6. Principle of laser diffraction particle size distribution measurement (Figure courtesy of Malvern Instruments Ltd)**

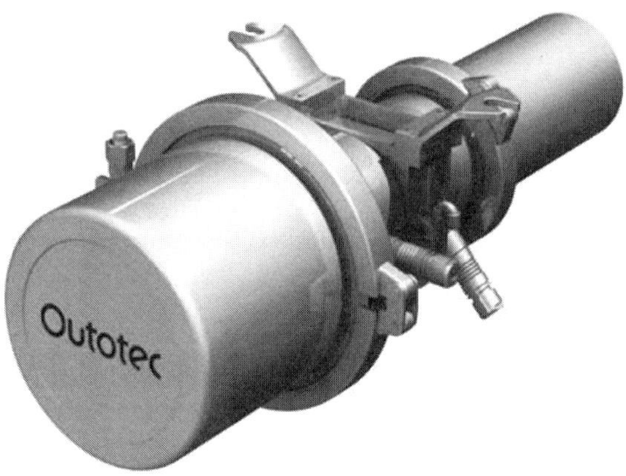


**Figure 7. Measurement sensor of on-line laser diffraction analyzer**

Mineral slurry in industrial applications is opaque so that it must either flow in a very narrow channel and needs dilution in order to allow the scattered light through it. At the analyzer the primary sample is further divided into smaller batches by moving the sample line across a cutter. The cuts are then mixed and diluted with water in a diluter tank. Thus, the flow cell below the diluter receives the sample in a turbulent condition to ensure that the solids have an even distribution and the speed of particles of various sizes is the same in the active measurement region. The sensor shown in Figure 7 is connected to an industrial PC that performs all controls of the measurement cycle and calculations for the particle size analysis.

## Interfacing

The measurement and status data is sent to the plant automation system through Ethernet or serial ports by various protocols. Besides the DCS system link, online particle size analyzers usually include other ways of accessing its data and operation. For instance, a local user interface implemented via a touch screen display, Figure 8. A web server provides access to diagnostic and primary measurement data logged to the system through a remote link.

**Figure 8. A laser scatter particle size analyser installation at a concentrator producing Fe concentrate for pellet production**

## LASER SCATTER PARTICLE SIZE ANALYZER FOR FE PELLETIZING FEED GRINDING

### Background

The iron pelletizing industry use the specific surface area for indication of fine particle size of slurry feed to balling process. This measurement result is generally expressed as $cm^2/g$. As a criterion, a value about 1800 $cm^2/g$ is considered ideal for pelletizing.

The available equipment in market for on-line measurement of particle size of ground pellet feed are based on two principles: attenuation of an ultrasound in a pulp or diffraction of a laser beam by diluted suspension of particles. These analyzers are good trend indicators and are extensively used for control of milling and classify circuits. The main reasons to choose the laser diffraction particle size analyzer are mainly the possibility to analyze smaller particle size and measure the specific surface area. As it is a new technology to be adapted for slurry on-line analysis, it was necessary to put the equipment through an approval process, (Dias et al, 2006).

### Laboratory Reference Methods

The on-line analyzer results were compared with several laboratory instruments. The size fraction determinations were carried out by the following methods:

**Dry Screening (*dry sieve*)**, utilizing around 5 grams. The screening run 3 minutes using an Alpine automatic vibrator and sieve 325# (45μm) from Tyler series, after weighting the passing material and comparing it with the initial weight (%).

**Wet Screening (*wet sieves*)**. 100 grams of dry sample was used for wet screening. Two sieves of Tyler series were used: 325# (45 μm) and 200# (75 μm). The screening run 30 minutes, using an automatic vibrator. The oversize (retained) of sieves is dried in an oven 20 minutes, weighted and compared with the initial weight (%).

**Microtrac (laser).** The sample was homogenized and split until it reached around 1,3 g. Distilled water with Hexadecyltrimethyamoniumbromid was used as a dispersant. The sample was measured using the laboratory laser diffraction analyzer. The measuring process was repeated and the averages of the results were calculated for ranges < 8 μm, < 13 μm, < 38 μm, < 45 μm and < 75 μm.

**The specific surface** determination was carried out using a Helium picnometer and a Fisher permeabilimeter:

**Density measurement** was done using 6 grams of sample. The picnometer was initially calibrated by 5 measurements of two steel spheres used as standard. Then the sample was measured by 5 times and the average of results was calculated. This value was used later for measurement of specific surface by Fisher determination.

**Specific surface area measurement**, the quantity of sample used is indicated through a density measurement, in grams. The equipment gives as output the average diameter of particle. The measurement is repeated and the average was calculated. The specific surface area was calculated using the following formula:

$$S.E. = \frac{60.000}{specificgravity * averageparticlesize}$$

For minimizing the laboratory analysis errors, all measurements were carried out by the same technician on the same equipment.

### Data Analysis, Results and Discussion

Analyzing Figure 9, we can conclude that there is a good correlation between results from PSI500 (DAC38) and from dry screening on mesh 0,045mm. Comparing with the wet screening, the correlation is not as good in Figure 10.

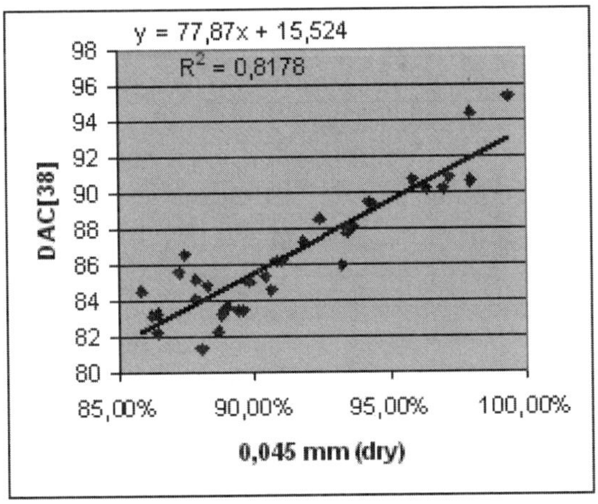


Figure 9. PSI500 vs Alpine dry sieve

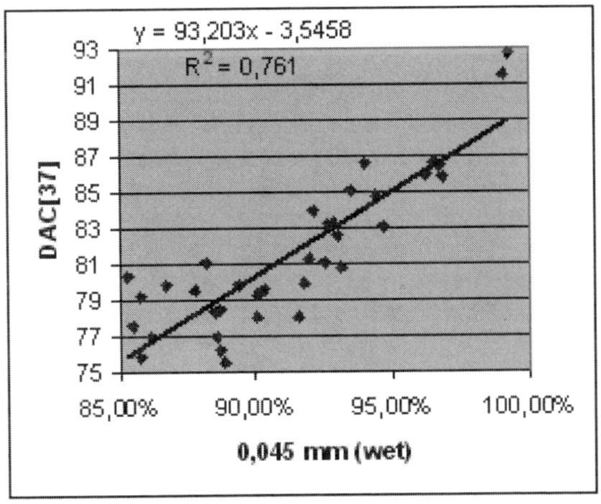


Figure 10. PSI500 vs Wet Sieve <0,045μm

Table 1 shows the regression analysis for specific surface area. Of all analyzed models, the one that presented the best value adjusted of $R^2$ was the model that has as independent variables outputs numbers 05, 23, 25 and 40 from PSI500. All these variables are significant, that is demonstrated by p-value less than 0,05. The comparison of the analyzed results via Fisher laboratory analysis and that ones calculated by model is show on Figure 11.

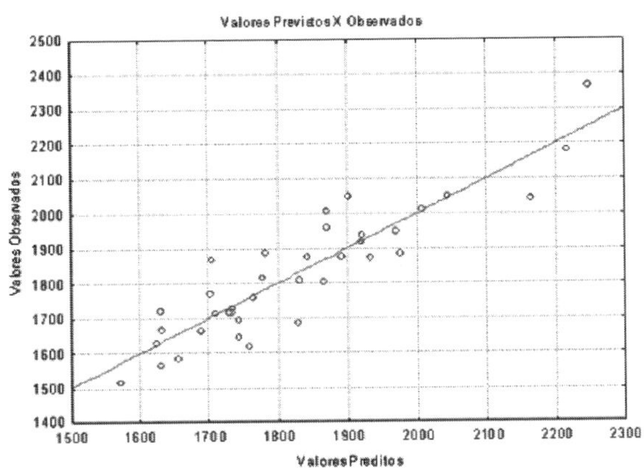


**Figure 11. Predicted vs observed surface area values.**

**Table 1. Specific surface area performance PSI 500 as compared with Fisher permeabilimeter**

| Summary Statistics; DV: Fisher cm²/g (PSI - Dados.sta) | |
|---|---|
| | **Value** |
| **Multiple R** | 0,90580 |
| **Multiple R²** | 0,82048 |
| **Adjusted R²** | 0,78458 |
| **F(6,30)** | 22,85240 |
| **p** | 0,00000 |
| **Std.Err. of Estimate** | 84,41629 |

A performance test for the on-line analyzer was also done by using manually introduced external glass granule samples. These reference samples were made by Malvern Instruments and were checked using a lab laser analyzer Mastersizer 2000. The results on Table 4 can be considered as comparison of on-line analyzer against a reference.

**Table 2. Glass bead reference sample measurement**

| Particle size | Mastersizer 2000 (lab.) | PSI 500 |
|---|---|---|
| Dv(10) | 35.6 um | 32.2 um |
| Dv(50) | 46.7 um | 47.2 um |
| Dv(90) | 61.2 um | 68.2 um |

The effect of a filter in the on-line analyzer PSI500 causes some difference for Dv(10) and DV(90) and has to be taken into consideration. However, the filter does not affect the DV(50) reading, as this is located in the middle of distribution range and the filter effect is cancelled. Dv(50) result indicates that this range represents a good indication of absolute accuracy of PSI500. From these tests we can conclude that the measurement Dv(50) of PSI 500 has a good correspondence to the laboratory measurements.

In order to evaluate the repeatability of PSI 500 with real process samples, a dry sample was prepared. The sample was homogenized and divided into 6 parts, following the same procedures as used in the laboratory, including using distilled water and Hexadecyltrimethyamoniumbromid for de-aggregation. The results of these repeatability tests

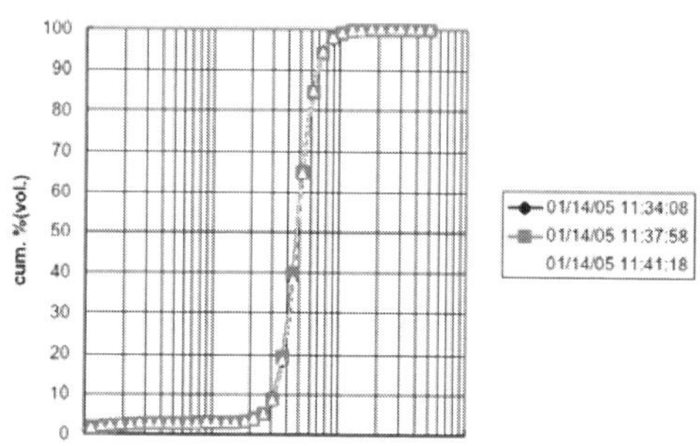


**Figure 12. Repeatability of PSI500 with glass granules, cumulative distribution**

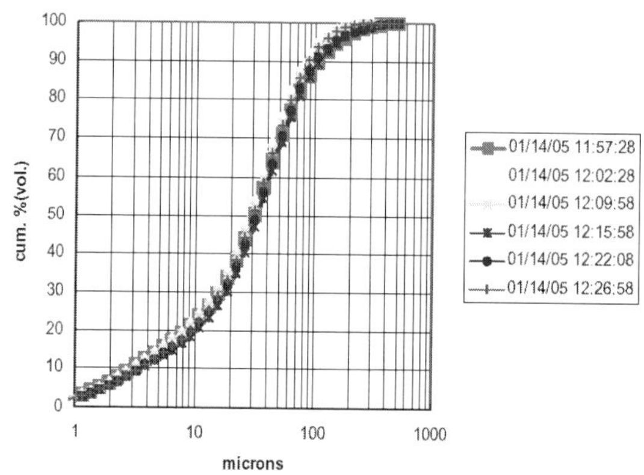


**Figure 13. PSI 500 repeatability with plant samples, cumulative distribution**

are shown in Figure 13. We can conclude that the repeatability of PSI500 is good and the small deviations are due the changes in sample preparation and the rate of addition of sample in the dilution tank.

A process test carried out by the operational team on 27/01/2005 as a "step test", varying the feed rate of the mill, and observing the changes in on-line analyzer reported particle sizes. When the feed rate was changed from 168 t/h (average of the last 30 minutes) to 203 t/h, we noted a change in the range of passing particles less than 45μm reducing his initial rate from 74% to 66%. The same behavior also can be verified for the other ranges. Doing the opposite operation, that is, a new step reducing the feed rate to 153 ton/h, we can see the positive variation of the particles size range less than 45μm and other ranges, what was expected. With this test we can conclude that the equipment presents a good sensitivity for the process variations.

## CONCLUSIONS

The size measurements carried out by PSI500 present satisfactory correlations with the laboratory data. PSI500 can be used as a trend indicator for improved control of the milling stage at CVRD. Similar results have been reported by Samarco, another Fe operation in Brazil, (Resende and Vieira, 2005).

## MONITORING CHANGES IN GRINDING AFTER HOMOGENIZATION BED CHANGE

### Motivation for Online Particle Size Analysis and Control

The Yara phosphate mine and phosphoric acid plant (former Kemira GrowHow operations) are located in Siilinjärvi in Eastern Finland. The capacity of the Siilinjärvi mining operation is around 10 Mt/a of ore. The main products are 0.8 Mt/a apatite concentrate, 0.1 Mt/a calcite concentrate and 10 kt/a mica. The apatite concentrate production is used totally in Siilinjärvi as a raw material in the phosphoric acid and fertilizer production. The main minerals of the igneous carbonatite deposit are 10% apatite, 20% carbonate minerals (calcite + dolomite) and 65% phologopite mica. Especially the carbonates/mica ratio varies throughout the deposit; the average is about 0.3 but it can range between 0.15 and 0.7. This ratio has a major influence to the grindability of the ore and the capacity of the concentrator. The mine has an ore homogenization plant, now with short batch change times as the production has been increased from the original design criteria (Aaltonen et al, 2004).

The concentrator has been undergoing changes to increase apatite production while improving the apatite recovery from the existing level of 83%. The apatite recovery losses are mainly in the coarse fraction – two thirds of the losses lay in the +210 µm fraction and one-third in –40 µm fines. The apatite recovery can be increased especially by reducing the amount of oversize apatite in the flotation feed without increasing the amount of fines excessively. On the other hand there seems to be possibilities to improve the coarse apatite recovery to a certain extent by modifying the flotation cells and the circuit configuration. A project is underway to split the flotation feed to fine and coarse streams and to process them separately.

The particle size distribution of the flotation feed is 25% -74 µm as an average, and at the same time roughly 30% of apatite is in oversize fractions. The particle size distribution of apatite and carbonates is much finer than that of phologopite. The energy consumption in phologopite grinding is high and on the other hand flaky phologopite particles will deteriorate the sharpness of the separation in hydrocyclones. High flaky phologopite content is a challenge to particle size analysis. On-line particle size analyzers based on ultrasound and image analysis were not successful at this site.

The recovery of apatite has been above 83% even if the fineness of the feed has varied between 22 to 30% -74 µm, depending on the ore type and conditions in the flotation stage. With the history described above, it has been difficult to specify for each ore batch the optimal particle size distribution and feed level that would provide the best economical and metallurgical results. This is especially challenging in conditions where the quality of ore feed may change quite much every second day without possibilities to characterize the new ore adequately before it enters the grinding section.

### On-stream Analyzer Installation

The targets for the PSI 500 particle size analyzer installation were set as follows:

1. Get reliable and real-time information about the changes in particle size distribution, especially when new ore batch enters the grinding process; there are no means to characterize the ore and its grindability in advance only by chemical analysis, for example.
2. Analyze and control the particle size distribution of both grinding circuits separately especially because of the different size of the grinding lines and differences in the particle size distributions of the feed to the rod mills.
3. Analyze the whole distribution, not only the percent passing 74 µm screen but also for example the amount of solids between 20 and 250µm, or d95 and d50 sizes. With this information it should be easier than today, together with the response from the flotation circuit, to adjust the feed level fast so that maximum amount of apatite would be in the size range from 20 µm to 250 µm.
4. Fine-tune the grinding process by optimizing for example the pulp densities in the mills and the circulating loads by changing the pulp density in cyclone feed and number of cyclones in use.

The PSI 500 analyzer was installed at Siilinjärvi in the grinding circuit in March 2004. Two Launder Box Samplers were installed in the cyclone overflows of each grinding circuit. Figure 6 shows the analyzer located on a service platform about 5 meters below the cyclone battery. The primary sample lines flow by gravity, and the sample return flows to a conditioner tank directly below the analyzer. The samplers are automatically backflushed with water every 4 hours to make sure that the cutters keep open. At the same time, the analyzer makes a background check in the measurement system by running only dilution water through the flow cell. The analysis cycle with the two sample lines is 5.5 minutes.

The plant automation system reads two fractions of each sample line, %-74 μm and %+210 μm. The Modbus serial link also carries status information and remote commands for the control room operators to switch the samplers and the analyzer to standby at shutdowns.

### Verification Tests Results

The precision of the analysis was estimated by recording the analysis results over a 30 min period of stable process conditions. Standard deviations of successive analyses are given in Table 3. Since all analyses are independent and no filtering is performed on the signals, the numbers present the overall short-term random error component from sampling to the actual analysis.

**Table 3. Statistics of Successive Sample Measurements (over 30 minutes)**

| Sample line | Analysis | Average, % | Std. Deviation, %(abs) |
|---|---|---|---|
| J1 | %+210 μm | 57.1 | 0.86 |
| | %-74 μm | 33.0 | 0.63 |
| J2 | %+210 μm | 53.5 | 0.81 |
| | %-74 μm | 30.3 | 0.55 |

Changes in the particle size distribution occur when the ore batch from the homogenization stacks changes. Figure 14 shows one of these instances. In this case, the grinding product gets finer, and the operator has increased the ore feed to the mills. The new ore allowed 15% more tons into the grinding circuit and it still produced a finer distribution. The particle size distributions before and after the ore change are also of different shapes.

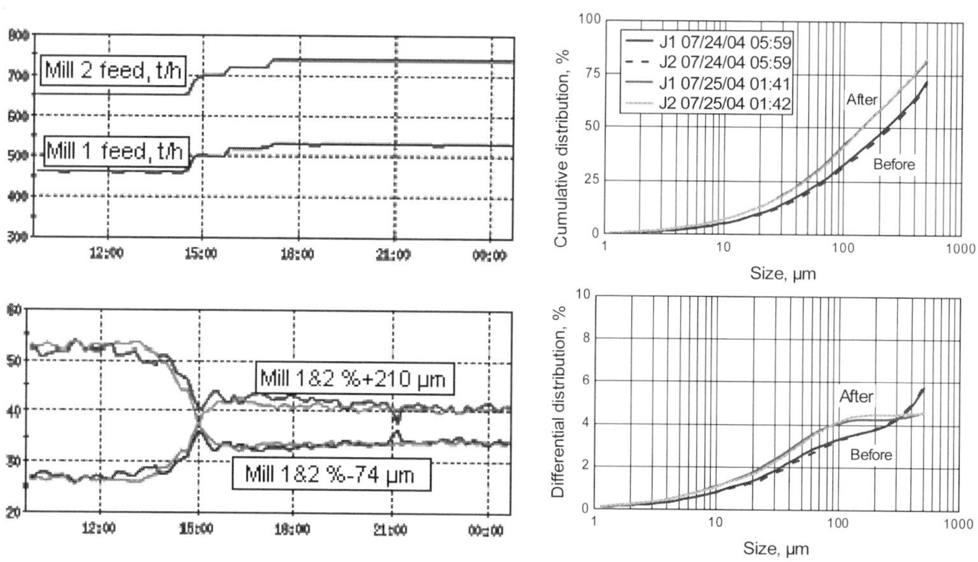


**Figure 14. Ore Change in July 24; Process Trends on the Plant Automation System and Particle Size Distributions**

The on-stream particle size analyzer proved to be a valuable tool for reacting to large changes in the mill feed when homogenization bed changes.

Similar experiences have been reported from a mineral processing plant switching frequently between a complex sulphide ore supply from 5 mines all with different mineralogy (Viklund et al, 2006).

## PASTE THICKENER CONTROL BASED ON PARTICLE SIZE

### Background

Water is a strategic resource for many mining operations located in dry areas. As much as possible water has to be re-circulated from plant tailings for reuse. Paste thickening is the most effective way to accomplish this. The particle size distribution of tailings material is variable with varying amount of fines.

The availability of online particle size analyzers has made it possible to implement the control system described here (Patterson and Rule, 2006). An expert control system has been configured to integrate all of the control loops, its the detailed description is beyond the scope of this paper.

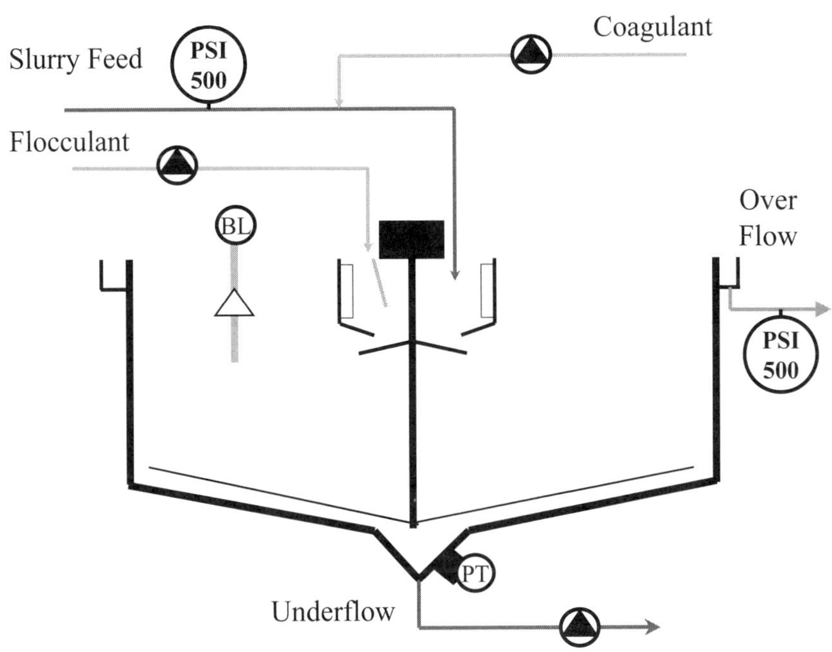


**Figure 15. Bed Level and Flocculant Addition Control Loop**

### Bed Level and Flocculant Addition Control loop

The purpose of this control loop, shown in Figure 15 is to provide the ideal amount of flocculant for the desired bed density.

The required flocculant dosage is calculated from the particle size distribution in the paste thickener feed. The turbidity of the overflow is used as a feedback signal for the control loop. The particle size is expressed in terms of the specific surface area. As the surface area (measured in m2/cm3) increases, the average particle size is getting finer.

The PSI 500 has been used, as it is ideal for the ultra-fines measurement as well as being able to output the specific surface area signal directly. The PSI generates a new measurement every two minutes, which is ideal for online control.

When the average particle size is less then ideal for direct flocculation the coagulant control loop is activated.

### The Coagulant Control Loop

The coagulant control loop forms part of the flocculant control. The purpose of this loop is to coagulate the finer particles, and increase the average size of the solids in the feed to within the ideal operating range of the flocculant used.

When dealing with reactive ores, for example, kimberlite, the particle size within the feed varies continually due to many factors including the chemical nature of the ore and the water quality. In the case that the measured ultra-fine or clay content is too high for effective flocculation, the coagulant dosing rate is increased to "grow" the average particle size to be greater than the minimum for flocculant to be effective.

In the event that the measured ultra-fine content is low for effective flocculation, this coagulant control loop is disabled.

This controller will result in the flocculant consumption being optimized. As this is an open control loop, the turbidity signal in the thickener overflow is used to indicate if this control loop is sub-optimal.

The PSI 500 can be configured to measure the presence of particles in the overflow, or turbidity, as well as point to the cause of the problem, i.e., the coagulant control loop or the flocculation control loop, as in Figure 15.

## REFERENCES

Aaltonen, J, Turunen, J, Kongas, M and Saloheimo, K, 2004. New on-stream slurry particle sizer, Proceedings *Beneficiation of Phosphates 2004*, Miami, USA

Coghill, P J, Millen, M J and Sowerby, B D, 2002. On-line measurement of particle size in mineral slurries, Minerals Engineering, Vol. 15, No.1-2, p. 83.

Dias, F B, Ribeiro, M R, Avila, H A M, Filho, J X S, Pessanha, S, Lipper, G Kongas, M and Timperi, J, 2006. Evaluation of PSI500 laser difraction particle size indicator iron ore grinding process of plant 1 at CVRD Vitoria (Avaliação do uso do analisador de tamanho de partículas por difração a laser - PSI500 – no processo de moagem de minério de ferro da usina 1 da CVRD, in Portuguese). Proceedings X Seminário de Automação de Processos 4-6.10.2006, Belo Horizonte, MG, Brasil.

Kongas, M, Saloheimo, K, Pekkarinen, H and Turunen, J, 2003. New Particle size Analysis System for Mineral Slurries, Preprints *IFAC Workshop on New Technologies for Automation of the Metallurgical Industry*, Shanghai, China.

Paterson, A and Rule, R, 2006. Proceedings Paste Thickener Control, The South African Institute of Mining and Metallurgy Solid Liquid Colloquium 17 May 2006, Saxonwold, South Africa

Resende, A and Vieira, J M, 2005. Validation of a particle size analyzer for process control in the Samarco's remilling (Validação de um analisador de tamanho de partículas para controle do processo de remoagem da Samarco, in Portuguese), *Proceedings VI Simpósio Brasileiro de Minério de Ferro – 30.8 – 02.9.2005 ABM, 2005*, Florianópolis, SC, Brasil

Saloheimo, K and Anttila, K, 1994. Utilization of particle size measurement in flotation processes, Expomin 1994, Santiago, Chile.

Viklund, T, Albertsson, J, Burstedt, J, Isaksson, M and Söderlund, J, 2006. Evolution of ag mill control system at Boliden Mineral Ab, Proceedings *SAG 2006* Conference [CD ROM], Department of mining engineering University of British Columbia, Vancouver, B. C., Canada

# FINE SIZING WITH THE DERRICK® STACK SIZER™ SCREEN

Steven B. Valine, Jobe E. Wheeler, and Laercio G. Albuquerque

Derrick Corporation, 590 Duke Road, Buffalo, NY 14225, USA

## ABSTRACT

The high capacity Stack Sizer screening machine consists of up to five decks positioned one above the other and all operating in parallel. Its use together with urethane screen surfaces as fine as 75 microns (200 mesh) has made fine wet screening a practical reality in mineral processing operations worldwide. The application of this technology in closed circuit grinding is demonstrated with specific application examples.

## INTRODUCTION

Screening is the process of separating particles by size and fine screening typically refers to separations in the range of 10 mm (3/8 inch) to 38 microns (400 mesh). Fine screening, wet or dry, is normally accomplished with high frequency, low amplitude, vibrating screening machines with either elliptical or straight-line motion. Various types of wet and dry fine screening machines and the factors affecting their operation have been discussed previously (Valine and Wennen 2002).

In fine particle wet screening, the undersize particles are transported through the screen openings by the fluid and the fraction of fluid in the slurry will therefore affect the efficiency of the separation. From a practical standpoint, the feed slurry to a fine screen should be around 20% solids by volume to achieve reasonable separation efficiency. As most of the fluid passes through the screen openings rather quickly, the fine screening process can be completed in a short screen length. Therefore screen width, rather than screen area, is an important design consideration for fine wet screening.

Recognition of this concept led to the development of multiple feed point fine wet screening machines. For example, the Multifeed screen consists of three screen panels mounted within a rectangular vibrating frame and is actually three short screens operating in parallel. Each screen panel has its own feed box and the oversize from each panel flows into a common launder and then to the oversize chute. Similarly, the undersize from each of the three panels flows into the undersize hopper. The popular 1.2 m (4 ft) wide by 2.4 m (8 ft) long version has a total effective width of 3.0 m (10 ft). In general, multiple feed point machines have been shown to have 1.5 to 2 times more capacity than a single feed point machine of equivalent size and screen area.

## ADVANCES IN TECHNOLOGY

### Stack Sizer Screening Machine

Expanding further on this concept, the Stack Sizer screening machine was introduced in 2001. With a capacity considerably greater than any other type of fine wet screening machine previously available, the Stack Sizer has up to five vibrating screen decks operating in parallel for a total effective width of 5.1 m (17 ft). As shown in Figures 1 and 2, the decks are positioned one above the other and each deck has its own feed box. A custom-engineered single or multiple-stage flow distribution system is normally included in the scope of supply to representatively split the feed slurry to each Stack Sizer screen and then to the decks on each machine. Ample space is provided between each of the screen decks for clear observation during operation and easy access for maintenance and replacement of screen surfaces. Each screen deck, consisting of two screen panels in series, is equipped with an undersize collection pan which discharges into a common launder with a single outlet. Similarly, the oversize from each of the screen decks collects in a single hopper with a common outlet. Two Super G® vibrating motors rated at 1.9 kW (2.5 HP) each and rotating in opposite directions produce a uniform high frequency linear motion throughout the entire length and width of all screen decks for superior oversize conveyance.

As mentioned above, the fluid passing through the openings carries the undersize particles through the screen openings. The screening process is essentially complete when most of the fluid has passed through the openings. Any remaining undersize particles adhere to the coarse particles and are misdirected to the oversize product. An optional "repulping" system is available for the Stack Sizer in which spray water is directed into a rubber-lined trough located between the two panels on each deck. With this feature, oversize from the first panel is reslurried and screened again on the second panel. This repulping action maximizes the correct placement of undersize particles and its use will depend upon the particular objective of the screening machine.

To date, over 600 Stack Sizer screening machines are in operation at mineral processing plants worldwide. Dry mass flow capacity typically ranges from 100 to 350 t/h. This is roughly equivalent to 3 or 4 of the older style Multifeed screens discussed above. Like all screening machines, capacity depends upon many factors such as screen panel opening, weight recovery to oversize, the amount of nearsize particles, particle shape, and slurry viscosity.

Figure 1. One of 38 Stack Sizer screening machines operating at an iron ore mine in Brazil

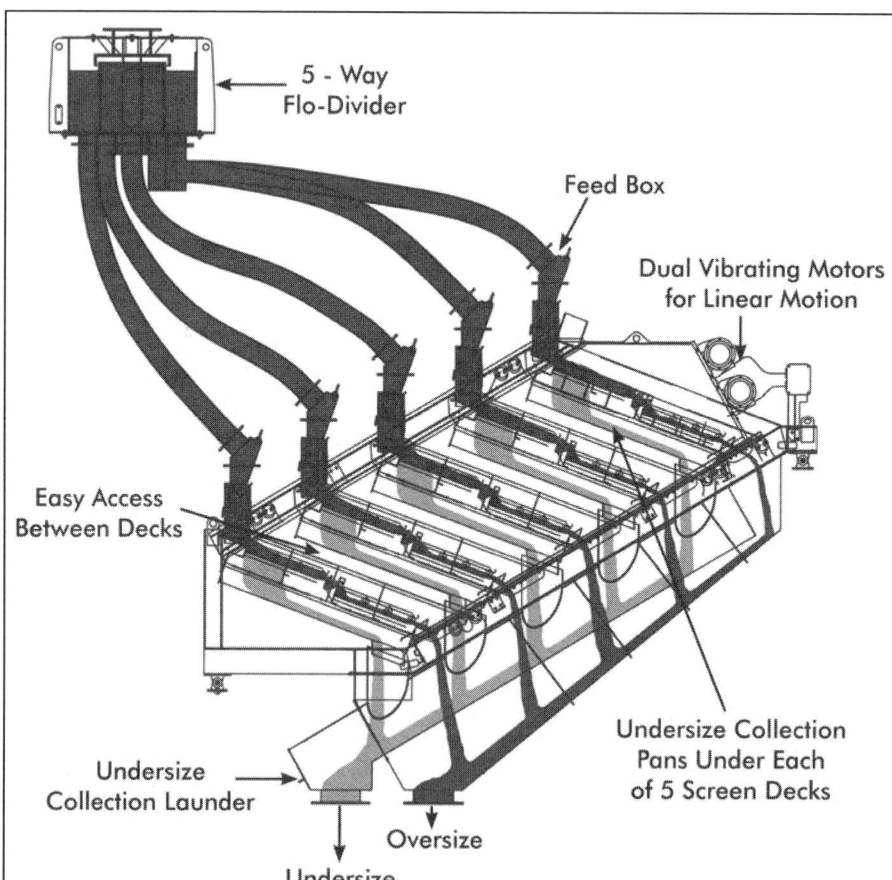


Figure 2. Principle of operation

**Polyurethane Screen Surfaces**

The availability of high open area, long wearing polyurethane screen surfaces with openings as fine as 75 microns (200 mesh) has made fine wet screening more feasible than previously thought possible. In the past, plant designers typically avoided the use of fine screens due their relatively low capacity, the high consumption rate of woven steel wire screen panels, and the tendency of these types of panels to blind with nearsize particles. This view has now changed as demonstrated by the following case studies describing process improvements through the use of Stack Sizer screening machines and fine urethane screen surfaces in place of conventional particle classification equipment.

## CASE STUDIES

As an introduction to these four case studies which all describe the use of fine screens in closed circuit grinding, it is interesting to note that research work over many decades has supported the concept of closing grinding circuits with screens. In 1925, E. W. Davis at the University of Minnesota Mines Experiment Station conducted closed circuit grinding tests with a rake classifier, a screw classifier, and a screen. He concluded that the use of screens would result in higher mill capacities with less overgrinding (Davis 1925). Other studies concluded with statements such as "lower grinding costs are possible with screen circuits" (Albert 1945) and "the master key for great improvements in capacity and in energy consumption in closed grinding circuits in improved sharpness of classification" (Hukki and Eland 1965). A recent presentation on flowsheet development for the planned Essar Steel iron ore concentrator project in Minnesota which included results from pilot scale tests and simulation work demonstrated a significant coarsening of the grind/grade relationship through the use of screens in the grinding circuit instead of hydrocyclones (Murr, Wennen, and Nordstrom 2009). This study predicted that a high grade concentrate could be produced at a reasonable but significantly coarser grind than would be required if the mills were closed by hydrocyclones. In fact, this study concluded that the desired product grade could not be obtained with hydrocyclones, even with extremely fine grinding. This coarsening of the grind/grade relationship with improved classification in grinding was also demonstrated on a commercial scale with the addition of screens to all secondary mill lines at the nearby National Steel Pellet Company, now U. S. Steel Keetac, resulting in the production of the same product grade at a significantly coarser grind, specifically, 72% passing 45 microns with screens compared to 79% passing 45 microns in the previous circuit (Wennen, Nordstrom, and Murr 1997).

**Mina Colquijirca - Sociedad Minera El Brocal S.A.A.** (Aquino and Vizcarra 2007)

Sociedad Minera El Brocal, located at Tinyahuarco, Pasco Province, Peru, produces lead, zinc, copper and gold. The concentrator flowsheet begins with three rod mills operating in parallel and in open circuit followed by three ball mills operating in parallel and in closed circuit with a bank of ten 10-inch hydrocyclones. As illustrated in Figure 3, rod mill discharge and ball mill discharge are combined in the same sump and pumped to the cyclones. The circulating load is about 350% and overgrinding of the high specific gravity minerals such as galena results in significant slime losses before flotation. Cyclone feed and product particle size distributions are provided in Table 1.

**Table 1. Brocal hydrocyclone feed and product particle size distributions**

| Size | Cumulative %Passing | | |
|---|---|---|---|
| (microns) | Cyclone Feed | Cyclone Underflow | Cyclone Overflow |
| 500 | 78.0 | 64.0 | 96.0 |
| 300 | 55.2 | 39.3 | 94.4 |
| 212 | 42.7 | 27.0 | 81.9 |
| 150 | 35.6 | 21.4 | 77.0 |
| 75 | 25.9 | 15.1 | 53.3 |

Brocal desired to increase capacity and was considering the addition of a fourth ball mill. As an alternative, Brocal considered ways to improve grinding efficiency and conducted full-scale Stack Sizer tests. Encouraging test results led to the installation of two 5-deck Stack Sizer screens fitted with 0.50 mm urethane panels in place of the 10 hydrocyclones. The effect was immediate as production increased 11% and lead recovery increased 9% even though Brocal shut down two of the three operating ball mills.

To take advantage of the increased grinding capacity, Brocal added additional crushing, flotation, and filtration capacity, as well as three additional Stack Sizer screens. One of the two ball mills originally shut down was put back into operation and rod mill rotational speed was increased (Aquino and Vizcarra 2007). As shown on Figure 4, production has increased from 138 t/h with the cyclone circuit to 245 t/h with the screen circuit, an increase of over 75% with less total grinding energy. The circulating load is about 60% and overgrinding has been minimized; the slime content (particles finer than 10 microns) in the flotation feed has been reduced from 18% to 10%, resulting in signification increases in metal recovery. Screen feed and product particle size distributions are provided in Table 2.

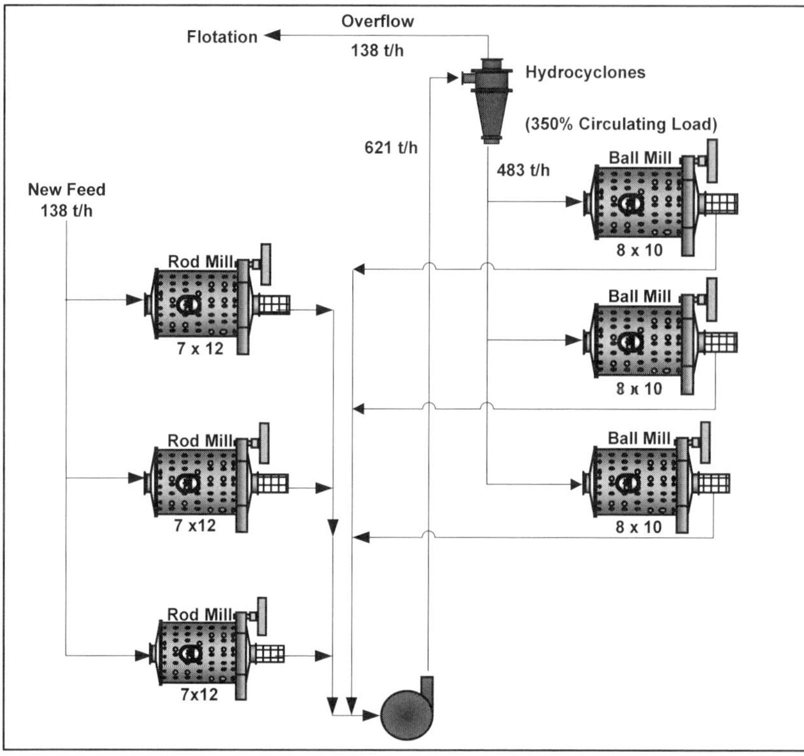


**Figure 3. Brocal grinding circuit with hydrocyclone classification**

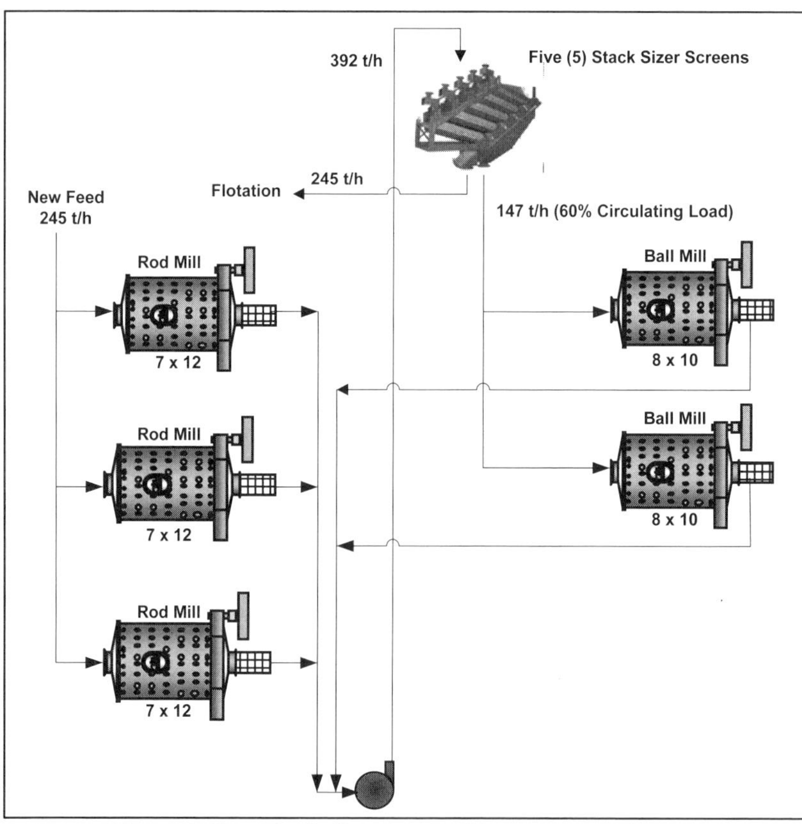


**Figure 4. Brocal grinding circuit with screen classification**

**Table 2. Brocal screen feed and product particle size distributions**

| Size | Cumulative %Passing | | |
|---|---|---|---|
| (microns) | Screen Feed | Screen Oversize | Screen Undersize |
| 500 | 77.0 | 63.0 | 98.0 |
| 300 | 59.1 | 31.8 | 95.1 |
| 212 | 48.4 | 19.6 | 85.2 |
| 150 | 39.6 | 13.4 | 72.8 |
| 75 | 27.9 | 8.7 | 52.0 |

A comparison of the particle size distributions of the grinding circuit product (flotation feed) from the hydrocyclone circuit and the screen circuit is shown in Figure 5. Note the coarsening of the grind/grade relationship as the concentrate grade in both cases was similar.

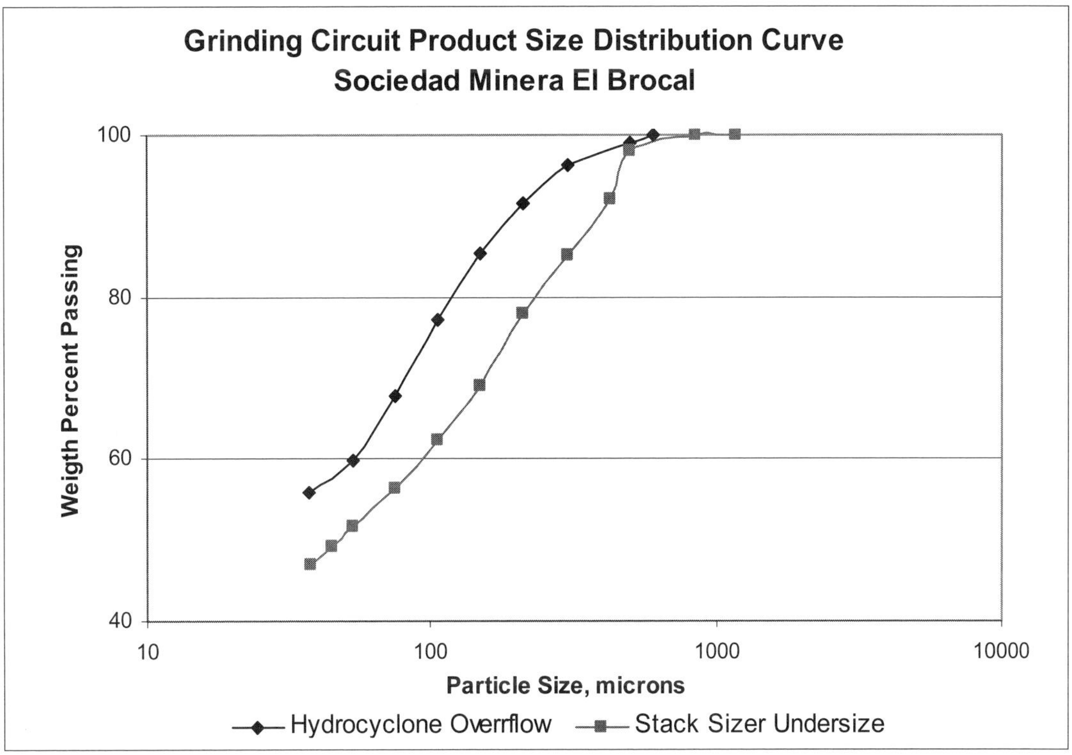


**Figure 5. Comparison of circuit products from hydrocyclone and screen circuits at Brocal**

**PhosAgro - OJSC Apatit (OJSC Apatit 2009)**

OJSC Apatit (Apatit), a division of PhosAgro, is a phosphate producer located in the Murmansk Region of the Kola Peninsula in Russia. At Apatit, multiple parallel lines of single stage ball mills in closed circuit with either screw classifiers or single and double stage hydrocyclones grind crushed ore. The product from multiple grinding lines is combined and pumped to the flotation plant. Over time, Apatit has experimented with various flowsheet options to improve classification efficiency and increase mill production rates. In 1996 and 1997, tests were conducted with a full-scale Multifeed screen and various urethane screen panel openings. Although the test results were encouraging, screens were eliminated as a possible alternative due to the high number of machines required and the limited space available in the vicinity of the grinding mills.

With the development of the high capacity Stack Sizer, Apatit desired to reevaluate screens as a possible method to improve grinding efficiency and increase production rates. They began in 2005 with continuous pilot-scale grinding tests with a ball mill in closed circuit with a hydrocyclone. This was to establish a correlation between the pilot plant and commercial plant operation. Next, a single-deck Stack Sizer screen was installed in place of the hydrocyclone in the pilot plant grinding circuit. The pilot tests demonstrated that the use of screens in place of hydrocyclones would increase production rates and reduce overgrinding and justified the next phase of evaluation, the modification of an existing production line to screen classification.

Mill 24 was selected for the modification because it was a smaller line and this would minimize the cost of the industrial-scale test. Typical performance of Mill 24 with hydrocyclones is illustrated in Figure 6. The normal production rate was 95 t/h with a circulating load of 430%. Hydrocyclone feed and product particle size distributions are provided in Table 3.

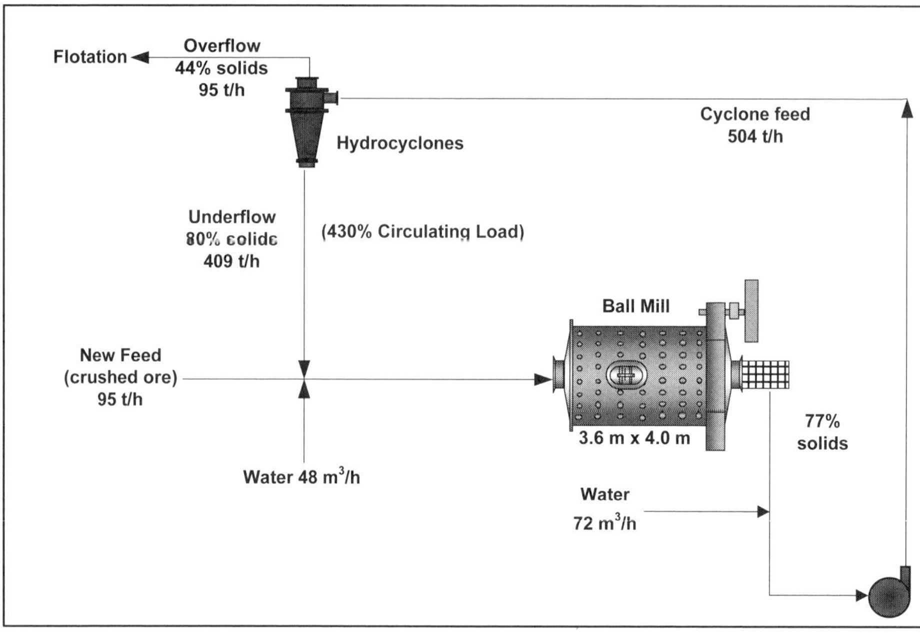


**Figure 6. OJSC Apatit Mill 24 grinding circuit with hydrocyclone classification**

**Table 3. OJSC Apatit Mill 24 hydrocyclone feed and product particle size distributions**

| Size | Cumulative %Passing | | |
|---|---|---|---|
| (microns) | Cyclone Feed | Cyclone Underflow | Cyclone Overflow |
| 450 | 68.8 | 61.5 | 100.0 |
| 320 | 57.3 | 49.0 | 93.0 |
| 200 | 41.1 | 31.7 | 81.3 |
| 160 | 32.3 | 23.0 | 72.1 |
| 100 | 21.2 | 13.3 | 55.3 |
| 71 | 16.8 | 9.9 | 46.4 |

In 2006, four 5-deck Stack Size screens fitted with 0.39 mm urethane panels were installed with Mill 24 in place of the hydrocyclones. Upon the completion of industrial-scale testing, Apatit concluded that grinding mills closed with Stack Sizer screens will see production rate increases of at least 30% compared to mill lines using conventional classification equipment. The modified circuit is shown in Figure 7. The production rate of Mill 24 increased from 95 t/h when closed with hydrocyclones to 124 t/h when the mill circuit was closed with screens. The circulating load decreased from well over 400% to about 135%. Power consumption per ton dropped about 20% to 25% as total mill power was about the same for both circuits. Urethane screen panel life was in the range of 10 to 12 months. Particle size data for the screen feed and products is provided in Table 4.

**Table 4. OJSC Apatit Mill 24 screen feed and product particle size distributions**

| Size | Cumulative %Passing | | |
|---|---|---|---|
| (microns) | Screen Feed | Screen Oversize | Screen Undersize |
| 450 | 61.5 | 32.8 | 100.0 |
| 320 | 49.3 | 18.0 | 91.4 |
| 200 | 38.2 | 9.9 | 76.2 |
| 160 | 31.8 | 6.9 | 65.2 |
| 100 | 24.4 | 4.7 | 50.9 |
| 71 | 19.3 | 3.6 | 40.5 |

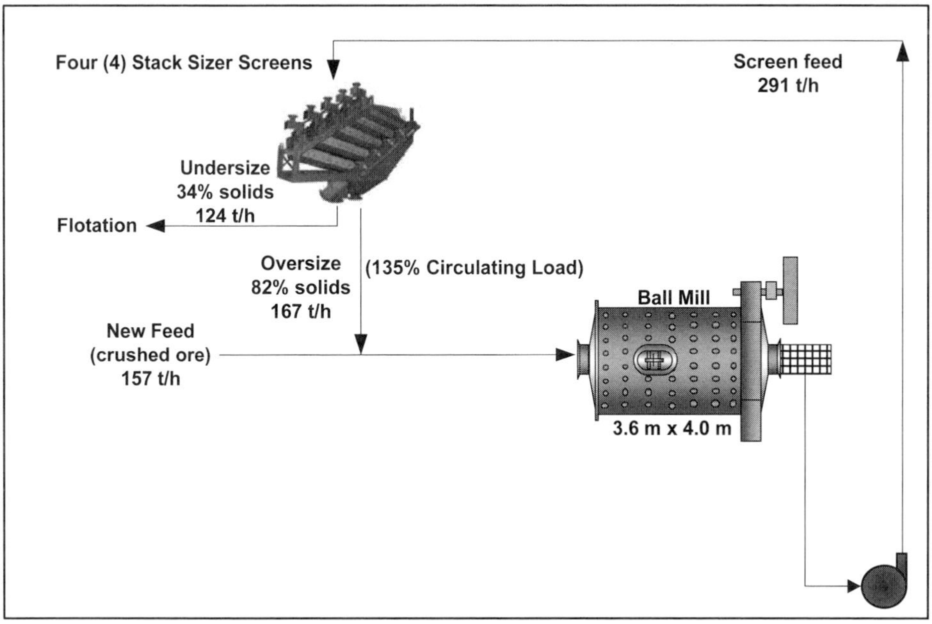


**Figure 7. OJSC Apatit Mill 24 grinding circuit with screen classification**

The coarsening of the grind/grade relationship is illustrated in Figure 8 with a comparison of the original circuit product (hydrocyclone overflow) and that produced in the screen circuit (screen undersize). Final product grade from the flotation circuit was similar even though the flotation feed is measurably coarser.

As the space required for the screening equipment is now much less with the Stack Sizer, Apatit is moving forward with plans to upgrade two larger mill lines, each requiring 10 Stack Sizers per line. With the anticipated 30% increase in mill capacity, Apatit is expecting a corresponding decrease in grinding hours to produce the same quantity of flotation feed.

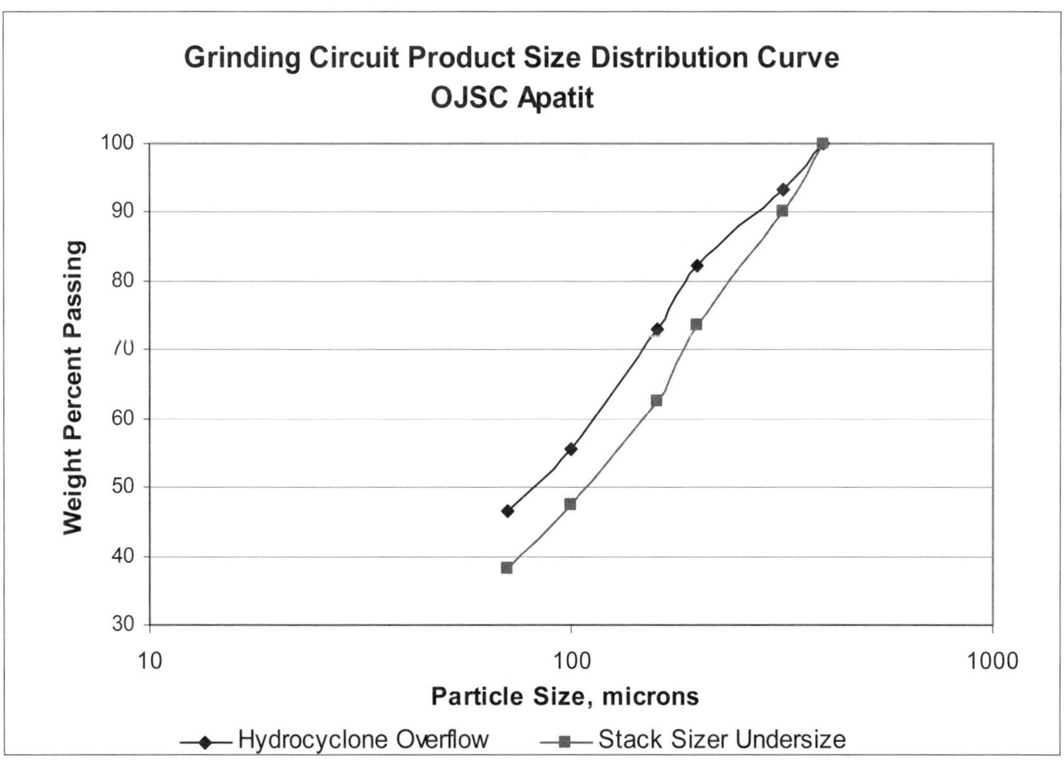


**Figure 8. Comparison of circuit products from hydrocyclone and screen circuits at OJSC Apatit**

**OJSC KMAruda** (Pelevin and Lazebnaya 2009)

KMAruda operates an iron ore mine in Russia and run of mine ore contains about 34% total iron. The concentrator flowsheet includes two stages of grinding and three stages of magnetic separation. The primary grinding circuit on each line consists of two ball mills in parallel operating in closed circuit with spiral classifiers. Similarly, the secondary circuit has two ball mills in parallel in closed circuit with hydrocyclones. All four mills are identical in size. The secondary circuit is illustrated in Figure 9. The first stage magnetic concentrate feeds the cyclones and the cyclone underflow is distributed to the two secondary mills. The mill discharge is fed to the second stage magnetic separators and the magnetic concentrate circulated back to the cyclones. The circuit product goes on to the third stage of magnetic separation.

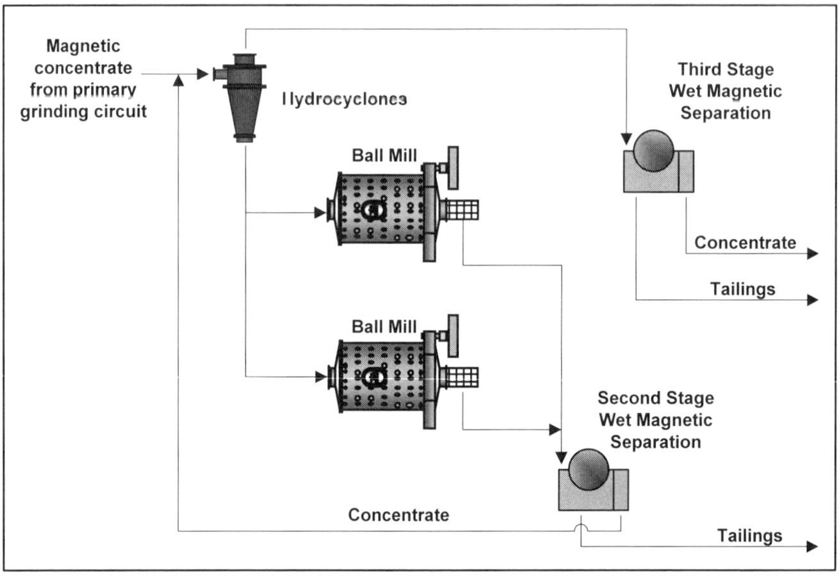


**Figure 9. KMAruda second stage grinding circuit with hydrocyclone classification**

Two Stack Sizer screens fitted with 100 micron urethane panels were installed in place of the hydrocyclones on one concentrator production line. Subsequent testing and optimization of the screen circuit resulted in the shut down of one of the two second stage ball mills. The screen circuit is illustrated in Figure 10.

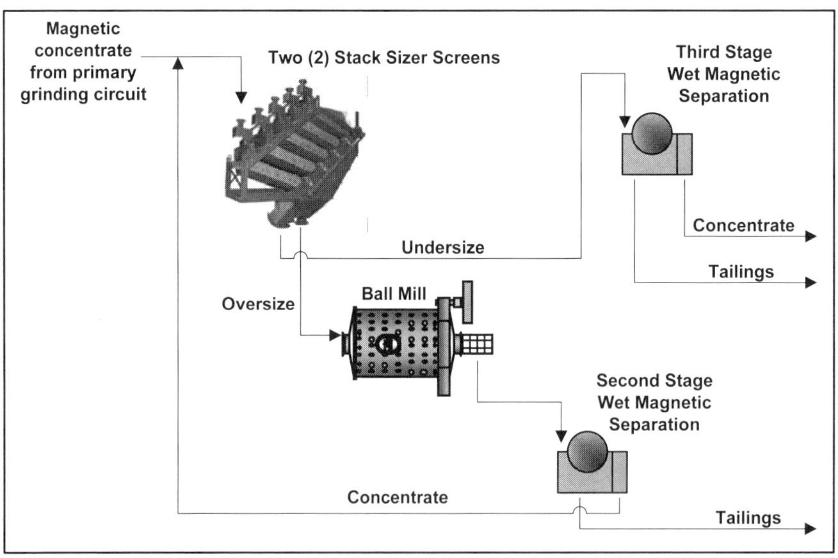


**Figure 10. KMAruda second stage grinding circuit with screen classification**

As summarized in Table 5, improved classification efficiency in the second stage circuit resulted in greater line capacity and a significant reduction in power consumption per ton with the elimination of one mill. The circulating load in the screen circuit was lower and considerably coarser at 22.9% minus 71 microns compared to 51% minus

71 microns in the cyclone circuit. While the goal was to maintain the same final product grade, the final concentrate from the screen circuit was also coarser at 82% minus 71 microns compared to 92.5% minus 71 microns on a typical cyclone line. KMAruda also noted that while cyclone underflow was normally around 65% iron, in the screen circuit the screen oversize ranged from 36 to 45% iron, leading to lower iron losses in the second stage of magnetic separation.

**Table 5. KMAruda concentrator performance with hydrocyclones and screens in the secondary mill circuit**

| Parameter | With Hydrocyclones | With Screens |
|---|---|---|
| Capacity (t/h) | 120 | 140 to 150 |
| Secondary mill power draw (kW) | 1260 | 630 |
| Secondary mill circulating load (%) | 210 to 220 | 130 to 160 |
| Secondary mill circulating load size (% minus 71μm) | 51.0 | 22.9 |
| Final concentrate size (% minus 71μm) | 92.5 | 82.0 |
| Final concentrate grade (%Fe) | 65.85 | 66.00 |
| Final concentrate iron recovery (%) | 82.2 | 84.6 |
| Final tails grade (%Fe) | 10.9 | 9.09 |

## Minera Cerro Lindo

Minera Cerro Lindo is located southwest of Lima, Peru and produces copper, lead, and zinc concentrates. As shown in Figure 11, the grinding circuit at Cerro Lindo consists of a 14.5 ft x 23.5 ft ball mill operating in closed circuit with a bank of 26-inch hydrocyclones. The circulating load is about 260%. Cyclone feed and product particle size distributions are provided in Table 6.

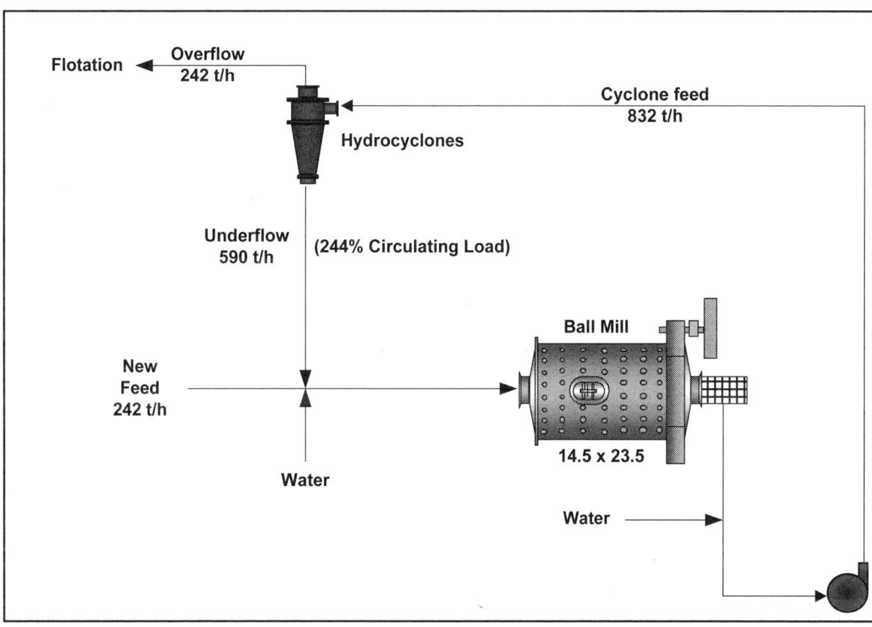


**Figure 11. Cerro Lindo grinding circuit with hydrocyclone classification**

**Table 6. Cerro Lindo hydrocyclone feed and product particle size distributions**

| Size (microns) | Cumulative %Passing | | |
|---|---|---|---|
| | Cyclone Feed | Cyclone Underflow | Cyclone Overflow |
| 500 | 100.0 | 100.0 | 100.0 |
| 300 | 100.0 | 100.0 | 100.0 |
| 212 | 63.9 | 52.1 | 94.4 |
| 150 | 46.1 | 31.0 | 85.4 |
| 75 | 22.5 | 8.4 | 58.9 |

Following the success of screen classification in grinding circuits at Brocal and other polymetallic mining operations in Peru, Cerro Lindo initiated a program to improve concentrator performance. Following full-scale testing, four Stack Sizer screens were installed in place of the 26-inch diameter hydrocyclones. Three screens are fitted with 0.23 mm urethane panels and one with 0.18 mm urethane panels. As shown in Figure 12, the circulating load has decreased to 108% and line capacity has increased 13.6% to 275 t/h. Screen feed and product particle size distributions are provided in Table 7.

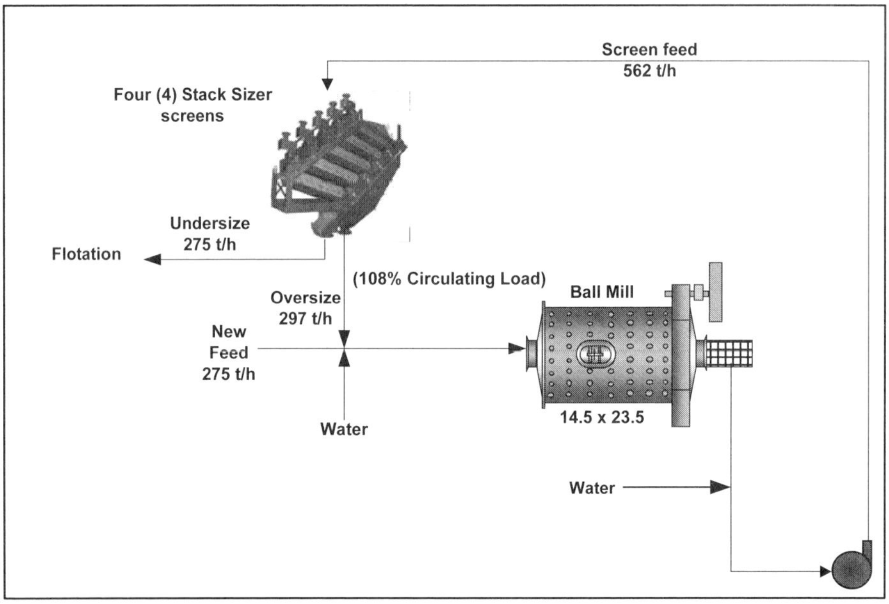


**Figure 12. Cerro Lindo grinding circuit with screen classification**

**Table 7. Cerro Lindo screen feed and product particle size distributions**

| Size | Cumulative %Passing | | |
|---|---|---|---|
| (microns) | Screen Feed | Screen Oversize | Screen Undersize |
| 500 | 91.0 | 88.0 | 100 |
| 300 | 75.3 | 54.9 | 99.3 |
| 212 | 59.4 | 31.1 | 90.6 |
| 150 | 47.8 | 20.5 | 75.0 |
| 75 | 30.4 | 9.9 | 47.0 |

Table 8 highlights some of the benefits of the change to screen classification. In the screen circuit, while the mill discharge product has more minus 75 micron material, flotation feed has less minus 75 micron. The $P_{80}$ is coarser in the screen circuit at 160 microns compared to 141 microns with the cyclone circuit. Steel consumption per kwh has decreased. Tailings filtration capacity also increased as the tailings are now coarser and more homogenous.

**Table 8. Comparison of Cerro Lindo concentrator performance with hydrocyclones and screens**

| Parameter | With Hydrocyclones | With Screens |
|---|---|---|
| Capacity (t/h) | 242 | 275 |
| Circulating load (%) | 244 | 108 |
| Steel consumption (g/kwh) | 778 | 696 |
| Mill discharge particle size (% minus 75μm) | 24 | 30 |
| Mill solids (%) | 83 | 79 |
| Flotation feed particle size (% minus 75μm) | 55 | 47 |
| Flotation feed $P_{80}$ (μm) | 141 | 160 |

## REFERENCES


Valine, S. B. and J. E. Wennen. 2002. Fine Screening in Mineral Processing Operations. *Mineral Processing Plant Design, Practice, and Control.* A. L. Mular, D. N. Halbe, and D. J. Barratt, editors. SME.

Davis, E. W. 1925. Ball Mill Crushed in Closed Circuits with Screens. *Bulletin of University of Minnesota, Mines Experiment Station Bulletin No. 10.* 28:42.

Albert, E. 1945. Characteristics of Screen-Circuit Products. *Min. Tech.* T.P. 1820.

Hukke, R. T. and H. Eland. 1965. The Relationship Between Sharpness of Classification and Circulating Load in Closed Grinding Circuits. *Transactions.* Society of Mining Engineers.

Murr, D. L., J. E. Wennen, and W. J. Nordstrom. 2009. Essar Steel Minnesota, Concentrator Flowsheet Development. *70th Annual University of Minnesota Mining Symposium.* Duluth, Minnesota.

Wennen, J.E., W.J. Nordstrom, and D.L. Murr. 1997. National Steel Pellet Company's Secondary Grinding Circuit Modifications. *Comminution Practices.* 19-25.

Aquino, B. and J. Vizcarra. 2007. Re-engineering the Metallurgical Processes at Colquirjirca Mine, Socidad Mineral El Brocal S.A.A. *Mineria*, 30-38.

OJSC Apatit. 2009. Written communication.

Pelevin, A. E. and M. V. Lazebnaya. 2009. Application of Derrick Screens in Locked Grinding Circuit at the KMAruda Mining Complex Concentrating Plant. *Mineral Processing Journal* (Russia). No. 2.

Ticse, F. 2009. Milpo-Cerro Lindo. Information provided during plant visit.

# MAXIMUM CLASSIFICATION EFFICIENCY THROUGH DOUBLE CLASSIFICATION USING THE RECYCLONE

E. Castro[1], J. Lopez[1], and D. Switzer[2]

[1] Weir Minerals Sud-America, San Jose 0815, San Bernardo, Santiago, Chile
[2] Weir Minerals North America, 2701 S. Stoughton Road, Madison, Wisconsin, USA

## ABSTRACT

Double classification has been shown to be beneficial to many mineral process applications in reducing the amount of misplaced fines. However, these systems can be costly due to the additional equipment required to support them. The ReCyclone hydrocyclone is a double classification unit, working in a single stage, without intermediate pumping. The ReCyclone has proven to be extremely efficient and the classification efficiency is reflected in the capability for generating very low by-pass of fines in the underflow while reducing the misplacement of coarse particles to the overflow. In this work, practical experiences in mining plants have been included. In tailings applications, the ReCyclone has been shown to consistently achieve a desirable 15% or lower passing 200 Mesh in the underflow. These benefits can be applied to various applications include comminution, sand generation for tailings walls, filtering, etc.

## INTRODUCTION

Whether it is on grinding circuit applications, deslime applications or sand production, classification plays an important role in the quality of the final product as well as the profitability of the process. Utilizing more efficient equipment will result in making the entire process in which it is inserted more efficient.

Since the 1990s with the introduction of the involute head of the CAVEX® hydrocyclone, which increased classification efficiency and reduced the amount of misplaced coarse particles to the overflow, there has been no innovation with proven practical application like the ReCyclone. The ReCyclone has taken an important step with respect to conventional hydrocyclones in terms of all the parameters that model the classification of slurry particles, as presented further in this paper. Introducing more typical variables with which the operation of a hydrocyclone is controlled has allowed the grinding classification processes to be more flexible, increasing the throughput.

### An Overview of the ReCyclone

The ReCyclone stands out for being double classification equipment—internally a control strategy has been implemented in order to impact the cleansing of the viscous layer and the mass partition between the first and second stages of cyclones. This concept, which has been the key to success, distinguishes the ReCyclone from the old double classification models. Previously developed double classification models could never find the solution to wear problems in the transition zone and mass partition.

Operating cyclones in series, as Figure 1 shows, primarily has an impact on the efficiency, classification and the by-pass of fines. The by-pass of fines defined as a fraction of the fine particles that produces a short-cut in the classification not being classified and these fines report to the underflow. As proposed by Kelsall and afterwards by Bradley, it is a connection between what is classified minus the mass of by-pass of fines (Bpf), which has produced a short-cut within the system.

Likewise, overall efficiency is the link between the efficiency of each of the classifiers, according to what is shown in equation (1).

ReCyclone Overall Efficiency can be defined as follows:

$$E_{Overall} = \frac{Uu_i}{Ff_i}$$

$$E_1 = \frac{Tt_i}{Ff_i} \rightarrow \quad Tt_i = E_1 Ff_i$$

$$E_2 = \frac{Uu_i}{Tt_i} \rightarrow \quad E_2 = \frac{Uu_i}{E_1 Ff_i}$$

$$\rightarrow \quad E_1 E_2 = \frac{Uu_i}{Ff_i}$$

$$E_{Overall} = E_1 E_2 \qquad (1)$$

where:

$E_{Overall}$ = Overall efficiency of the ReCyclone
$U$ = Total mass in the underflow.
$u_i$ = Percentage of retained mass in size "i" in the underflow.

$F$ = Total mass in the feed.
$f_i$ = Percentage of retained mass in size "i" in the feed.
$T$ = Total mass in the transfer zone.
$t_i$ = Percentage of retained mass in size "i" in the transfer zone.
$E_1$ = Efficiency of the first stage.
$E_2$ = Efficiency of the second stage.

Global analysis of the by-pass of fines within the ReCyclone is identified as follows:

$$Bpf_{Overall} = \frac{W_U}{W_F}$$

$$Bpf_1 = \frac{W_T}{W_F} \rightarrow \quad W_T = B\,pf_1 W_F$$

$$Bpf_2 = \frac{W_U}{W_T} \rightarrow \quad Bpf_2 = \frac{W_U}{B\,pf_1 W_F}$$

$$\rightarrow \quad Bpf_1\,Bpf_2 = \frac{W_U}{W_F}$$

$$Bpf_{Overall} = Bpf_1\,Bpf_2 \tag{2}$$

where:
$Bpf_{Overall}$ = Overall by-pass of the ReCyclone.
$W_U$ = Water in the underflow.
$W_F$ = Water in the feed.
$W_T$ = Water in the transfer zone.
$Bpf_1$ = First stage by-pass.
$Bpf_2$ = Second stage by-pass.

As an example of these equations and correlations, data from a copper tailings application carried out with a 500/400CVX DE ReCyclone hydrocyclone can be found below in Tables 1 and 2. Table 1 provides the grading profiles, solids concentrations and mass for a copper tailings application. Table 2 provides the efficiencies for this same application. As shown below, the relationships explained in equations (1) and (2) are clearly confirmed with respect to the link between the by-pass of fines and the efficiency of overall classification.

As shown in the efficiency curve of the global classification, as shown in Figure 1, rates of clarity and the by-pass of fines are exceptionally good with respect to conventional classification equipment. Figure 1 plots the efficiency of both cyclone stages as well as the overall efficiency with the percent of solids going to the underflow on the y-axis and particle size in microns on the x-axis.

**A Smart Way of Reclassifying Fines**

One of the most important features that differentiate the ReCyclone from double classification equipment designs that existed in the past is the cleansing chamber. This component, which is located in the lower part of the primary hydrocyclone, consists of a water injection and a moveable inverted dart valve. The dart valve adjusts the annular space for the pulp to go through, between the valve and the hydrocyclone wall, as shown in Figure 2. Figure 2 illustrates the components of the cleansing chamber found just below the cylinder section on the 1st stage of classification.

The purpose of the cleansing chamber is to modify the viscous layer as a consequence of the upstream of the pulp in the water-cleansing, which descends close to the hydrocyclone wall. As described by Bloor and Ingham (1976) and afterwards by Concha (1994), the limit layer attached to the hydrocyclone wall contains not only coarse particles, but also a fraction of fines which are trapped in this highly viscous layer. These same studies assign curious reasons to the particle behavior in the mix zone in the inlet head. The viscous head is the one that breaks or changes, allowing fine particles to be released and, therefore, giving an opportunity to create a new classification. The particles are dragged into the zone of the ascendant vortex to be discharged to the overflow. At the same time, bigger and heavier particles have more energy and are able to overcome the attractive force of the water-cleansing, following its path towards the transition zone.

The variable annular space of the ReCyclone increases or decreases the inlet area between the walls, generating a modification in the rate of acceleration of liquid suspension and upstream particles, as if it were a variable apex. This variation adjusts the cut size between the two stages and therefore the mass partition between them. This phenomenon is illustrated in Figure 3. Figure 3 shows the rejection of some of the fine particles from the wash water in the transition zone of the ReCyclone.

Table 1. Feeding profiles and discharge of ReCyclone Industrial Trial of Copper Tailings

| | Wt% Solids | Standard Deviation |
|---|---|---|
| Feed | 47.00 | 5.00 |
| Common Overflow | 34.00 | 5.00 |
| Underflow | 75.00 | 5.00 |
| Primary Overflow | 30.00 | 5.00 |
| Secondary Overflow | 18.00 | 5.00 |
| Wash Water ($m^3$/hr) | 80.0 | 5.00 |

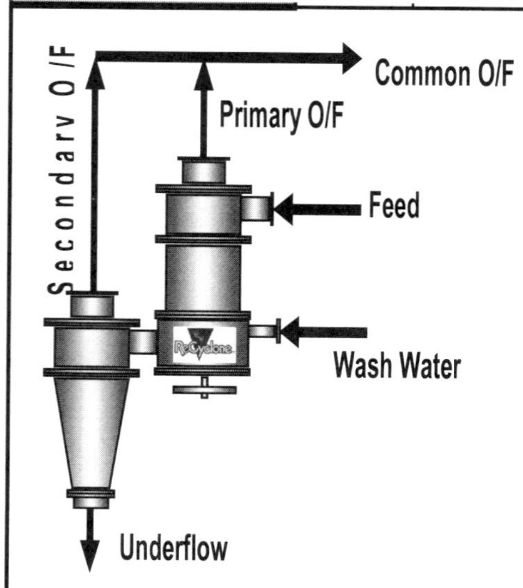


| Mesh | Microns | Feed Cumulative % Passing | Common O/F Cumulative % Passing | Primary O/F Cumulative % Passing | Secondary O/F Cumulative % Passing | Underflow Cumulative % Passing |
|---|---|---|---|---|---|---|
| 0.53 | 13500 | 100.00 | 100.00 | 100.00 | 100.00 | 100.00 |
| 3/8 | 9510 | 100.00 | 100.00 | 100.00 | 100.00 | 100.00 |
| ¼ | 6350 | 100.00 | 100.00 | 100.00 | 100.00 | 100.00 |
| 4 | 4760 | 100.00 | 100.00 | 100.00 | 100.00 | 100.00 |
| 6 | 3360 | 100.00 | 100.00 | 100.00 | 100.00 | 100.00 |
| 8 | 2380 | 100.00 | 100.00 | 100.00 | 100.00 | 100.00 |
| 10 | 1680 | 100.00 | 100.00 | 100.00 | 100.00 | 100.00 |
| 14 | 1190 | 100.00 | 100.00 | 100.00 | 100.00 | 99.40 |
| 20 | 841 | 99.66 | 100.00 | 100.00 | 100.00 | 99.23 |
| 28 | 595 | 97.24 | 100.00 | 100.00 | 100.00 | 99.06 |
| 35 | 420 | 91.38 | 99.49 | 99.27 | 99.59 | 85.86 |
| 48 | 297 | 83.08 | 97.70 | 96.24 | 99.40 | 64.62 |
| 65 | 210 | 72.94 | 92.76 | 89.34 | 99.20 | 41.36 |
| 100 | 149 | 63.10 | 83.68 | 80.02 | 95.94 | 25.36 |
| 150 | 105 | 56.14 | 76.76 | 71.54 | 88.74 | 19.12 |
| 200 | 74 | 48.16 | 66.58 | 61.92 | 76.84 | 14.76 |
| 270 | 53 | 42.06 | 58.60 | 54.14 | 68.32 | 12.70 |
| 325 | 44 | 40.36 | 55.98 | 51.66 | 65.52 | 11.86 |
| 400 | 37 | 39.36 | 53.84 | 49.86 | 63.06 | 10.78 |

**Table 2. Efficiency of primary cyclones, secondary cyclones and global classification**

| Actual Efficiency | | | |
|---|---|---|---|
| **Size (microns)** | **Efficiency of Primary Cyclone** | **Efficiency of Secondary Cyclone** | **Global Efficiency** |
| 13500 | 100 | 100 | 100 |
| 9510 | 100 | 100 | 100 |
| 6350 | 100 | 100 | 100 |
| 4760 | 100 | 100 | 100 |
| 3360 | 100 | 100 | 100 |
| 2380 | 100 | 100 | 100 |
| 1680 | 100 | 100 | 100 |
| 1190 | 100 | 100 | 100 |
| 841 | 100 | 100 | 100 |
| 595 | 93 | 100 | 91.6 |
| 420 | 83 | 100 | 83.0 |
| 297 | 69 | 100 | 68.7 |
| 210 | 54 | 89 | 47.9 |
| 149 | 46 | 61 | 27.8 |
| 105 | 43 | 39 | 16.6 |
| 74 | 42 | 29 | 11.9 |
| 53 | 40 | 25 | 9.9 |
| 44 | 39 | 24 | 9.2 |
| 37 | 38 | 23 | 8.7 |

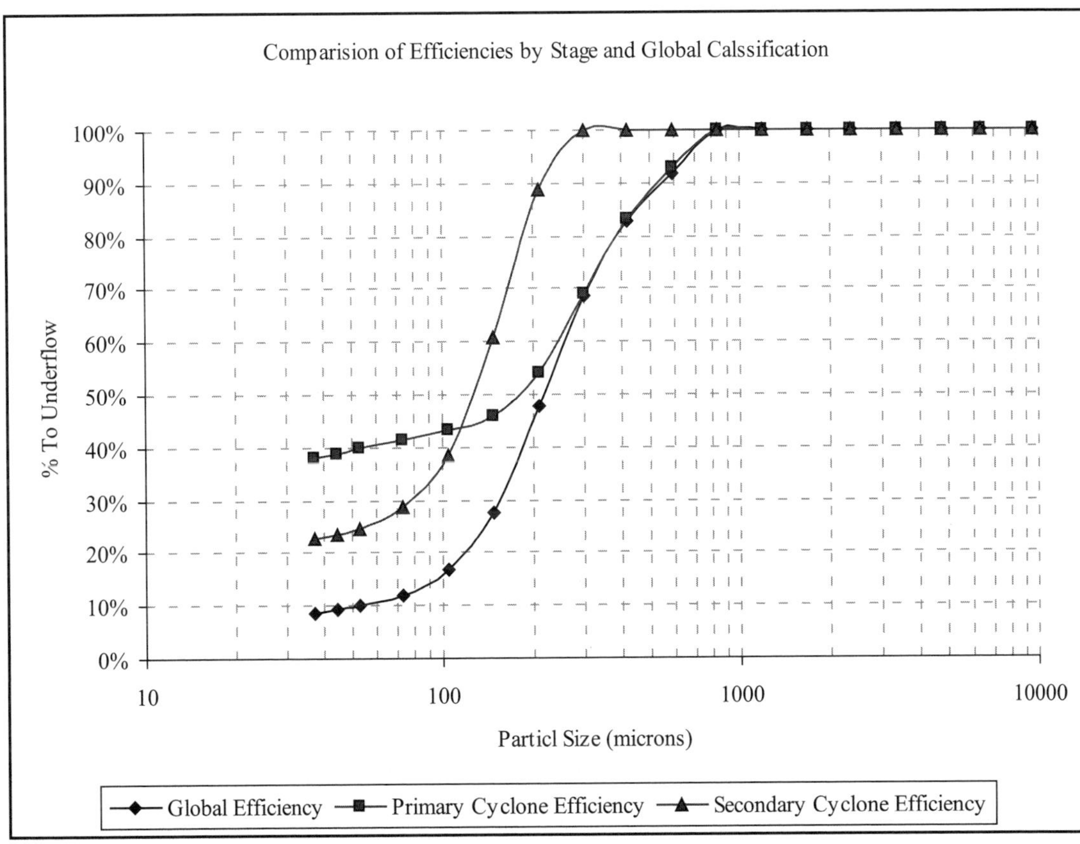


**Figure 1. Efficiency curves for primary cyclones, secondary cyclones and global classification**

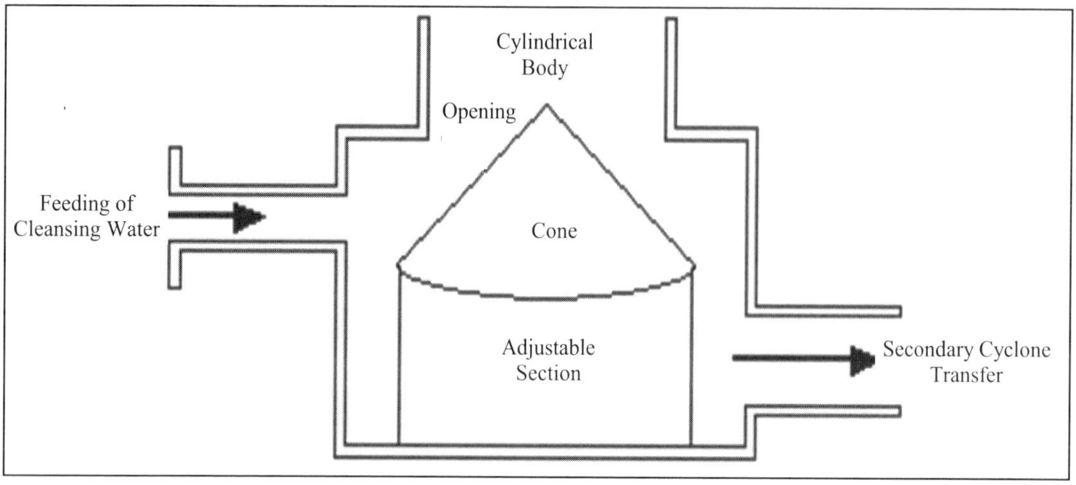


**Figure 2. Cleansing chamber**

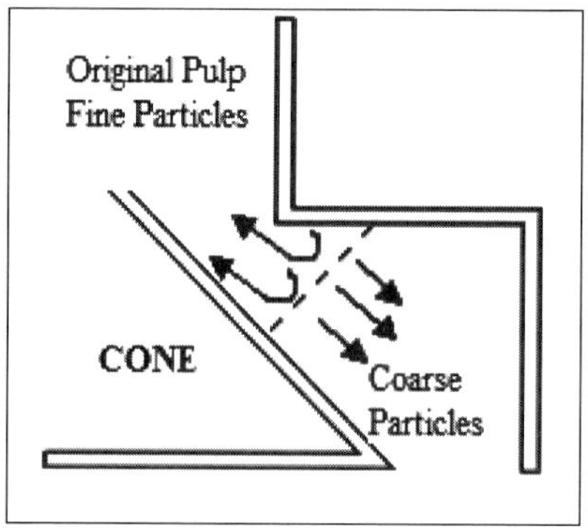


**Figure 3. Effect of wash water on the particle separation**

Synergy produced between the in-series classification plus the effect of the release of the viscous layers is what physically supports why the ReCyclone can reach lower by-pass of fine values in comparison to a conventional hydrocyclone. Example of industrial applications later in this paper will further support this statement.

**ReCyclone Design**

The ReCyclone was developed to take advantage of two-stage classification as well as those benefits realized using the CAVEX® hydrocyclone geometry. These hydrocyclones utilize an involute head without sharp corners or abrupt edges to allow for a more laminar flow within the hydrocyclone. In addition, these hydrocyclones operate with decreased turbulence inside the hydrocyclone resulting in increased wear life, increased hydraulic capacity and improved separation efficiency. The reduction in turbulence within these hydrocyclones as compared to a conventional cyclone is shown in Figure 4.

The ReCyclone design follows many of the same principles of that of a conventional hydrocyclone. The first classification stage is generated by a cylindrical hydrocyclone with conical bottom. Water injection is then used to control the fine material fed to the second stage. The second stage is carried out in a cylindrical hydrocyclone, which is fed through a transfer zone. This transfer zone, along with the other components, is defined in Figure 5 below.

In order to understand the basic functions of the ReCyclone, it is important to focus on the design of the primary hydrocyclone. This first classification stage is carried out in a cylindrical hydrocyclone finished in an expansion zone, in which the material fed is located with an inverted conical body. In this expansion zone, fine particles are trapped by the dragging of the coarse particles on the walls of the hydrocyclone to be released and, therefore, to generate the possibility of dragging them through the ascendant vortex reporting to the overflow of the primary hydrocyclone. This possibility increases when water is injected into the chamber or expansion zone. The water allows carrying out the cleansing of the ascendant coarse material and, at the same time, allows obtaining an appropriate dissolution for the second classification stage. An illustration of the cleansing chamber, specifically detailing the location of cleansing zone and water injection, is provided in Figure 6.

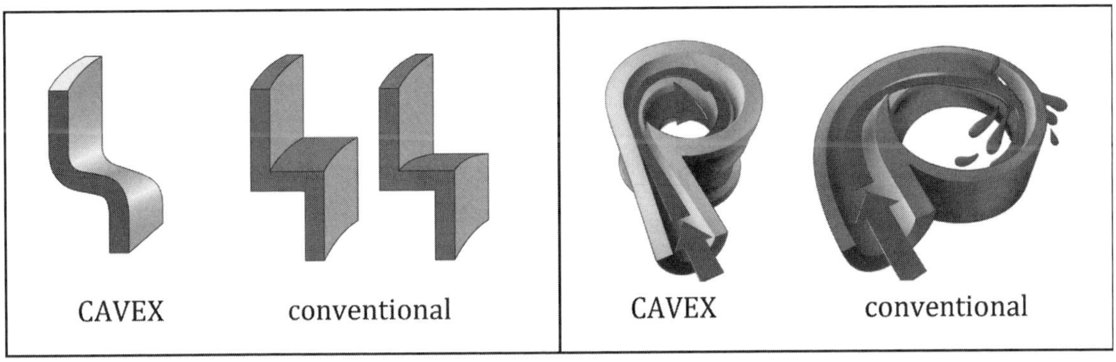


Figure 4. Illustration of the edges and corners within a CAVEX hydrocyclone compared to that of a conventional cyclone. The impact of these differences is further illustrated showing the introduction of slurry.

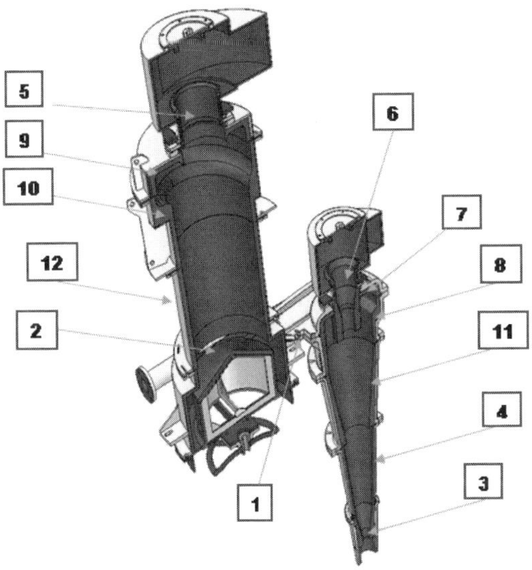


| Ref# | Description |
|---|---|
| 1 | Transition Zone |
| 2 | Adjustable Cone |
| 3 | Secondary Cyclone Apex |
| 4 | Lower Cone of Secondary Cyclone |
| 5 | Primary Cyclone Vortex Finder |
| 6 | Secondary Cyclone Vortex Finder |
| 7 | Secondary Cyclone Cover Liner |
| 8 | Secondary Cyclone Feed Chamber Liner |
| 9 | Primary Cyclone Cover Liner |
| 10 | Primary Cyclone Feed Chamber Liner |
| 11 | Upper Cone of Secondary Cyclone |
| 12 | Primary Cyclone Cylinder Section |

Figure 5. Illustration of the ReCyclone Components

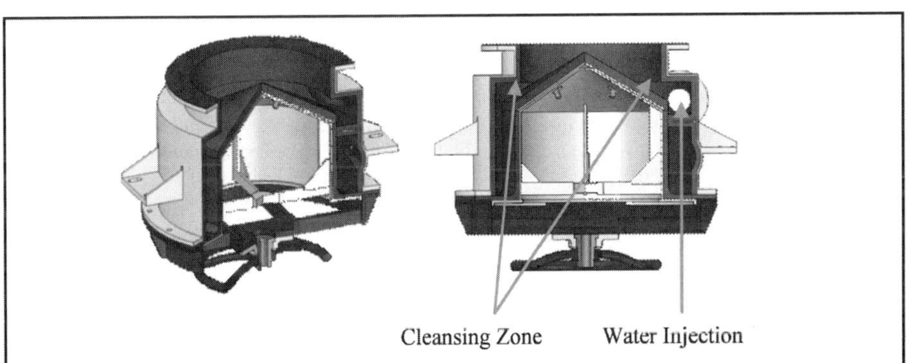


Figure 6. Cleansing chamber, with details of expansion zone and inverted cone

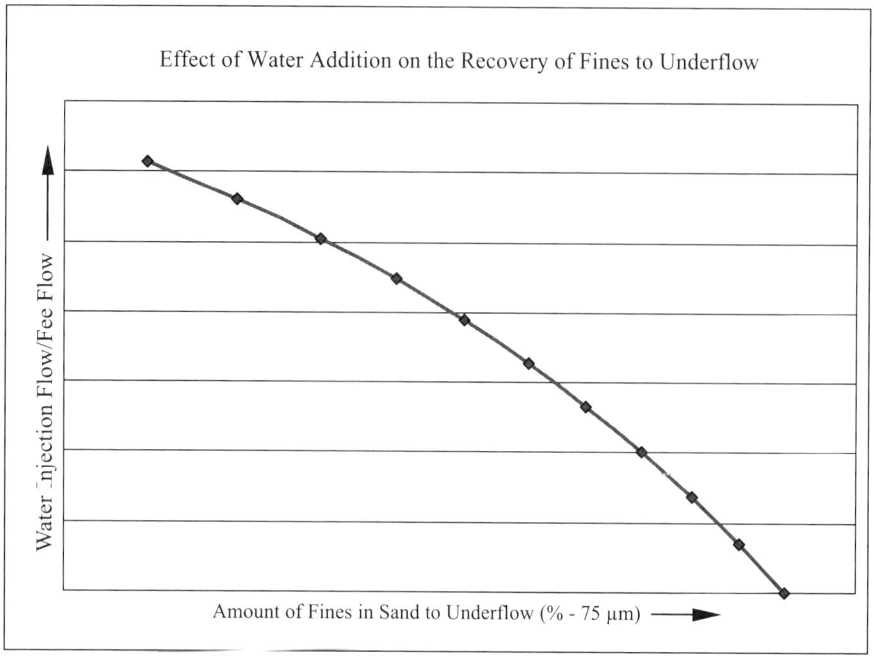


**Figure 7. The impact of water injection on the recovery of fines to underflow**

Another feature of the primary hydrocyclone is the ability to modify the distribution of the burden between both hydrocyclone units through a hand wheel, attached to the bottom exterior of the primary hydrocyclone. By using this wheel, a change of the elevation between the vortex finder and the bottom of the hydrocyclone is generated.

The adjustment of the opening in the cleansing zone, caused by the increase or decrease of elevation of the inverted cone, generates a modification in the area where the particle cleansing occurs. This, along with the amount of water injected, allows for controlling fast and efficient cut sizes without changing the apex or vortex finder under different operation circumstances. This feature provides high flexibility to the process. Figure 7 provides the general performance trend illustrating the increase in water injection along the y-axis and the increase in amount of fines in the sand to the underflow along the x-axis. As the amount of water injected increases, the amount of fines going to the underflow decreases.

The impact of water addition in the cleansing chamber allows controlling the amount of fines in the discharge. This is an extra variable for the ReCyclone operating in applications where a specific particle size in the underflow is required, which gives more operational flexibility in comparison with conventional equipments. As the amount of water injected to the cleansing chamber increases, the amount of fines dragged by the coarse material to the underflow decreases,

Due to the innovative design of the ReCyclone and the variables introduced to the operation, it is possible to obtain better results by handling the operation with an operational strategy which will accomplish the purposes of the plant. With the current operational experience and industrial tests carried out, an operational model, possible to adapt to the needs of each application, has been developed. Figure 8 illustrates this model with feed density plotted against the cumulative % passing 200 mesh in the feed. The percentage of wash water required for both the ReCyclone and a conventional cyclone is given.

The operational model compares the operation of conventional equipment with ReCyclone, under the same feed conditions. It is possible to observe that for the same size distribution, the water requirement decreases between the two teams and discharges have a higher percentage of solids. The results are higher underflow densities which are favorable for some applications such as tailing walls.

## INDUSTRIAL APPLICATION EXAMPLES

### Las Tortolas—Anglo American

#### Application: Tailing Classification

The west wall of the tailing reservoir "Las Tortolas" has a station with a cluster of six ReCyclones for the supply of sand in the wall. Sand produced at this site for the project reaches 15% passing 200 mesh.

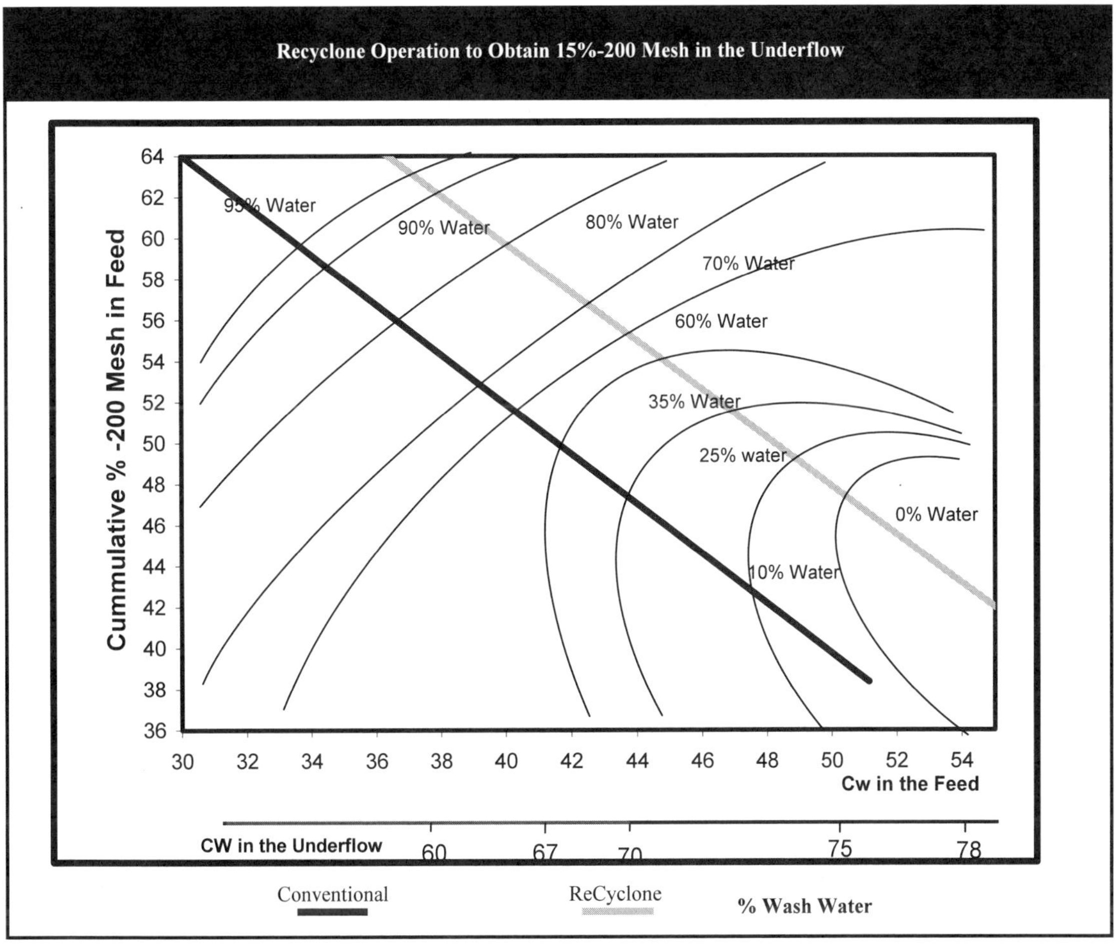


**Figure 8. Operational model illustrating the wash water required to obtain 15% passing 200 mesh in the underflow of the ReCyclone.**

**Photo 1. Photo of the installed six—place cluster of 500/400CVXDE ReCyclone units at Las Tortolas.**

**Photo 2. Photo of the installed twelve—place cluster of 500/400CVXDE ReCyclone units at Codelco.**

## Codelco—Andina Division

### Application: Tailing Classification

The expansion of Codelco's Andina division installed ReCyclone equipment after successful experience and results obtained after installing two of these equipments in their old cyclone tailings plant. The ReCyclone proved it's merit by increasing the amount of sand produced and decreasing the amount of fines present in the sand placed on the wall. These factors have influenced the final decision of acquiring new classification equipment.

**Los Pelambres Mine**

**Application: Tailing Classification**

The ReCyclone was installed in the old tailing reservoir of the Los Pelambres mine, where it was tested under several operational conditions. Results proved that the ReCyclone responded to all the operational necessities, producing the required quantity and quality of sand. After several tests, it was further concluded that the ReCyclone provided an important savings of water due to the possibility of operating it at a greater percentage of solids than conventional equipment. Under normal operating circumstances at this site, the ReCyclone also does not require water injection to its cleansing chamber.

**Photo 3. Photo of the installed ReCyclone units at Los Pelambres**

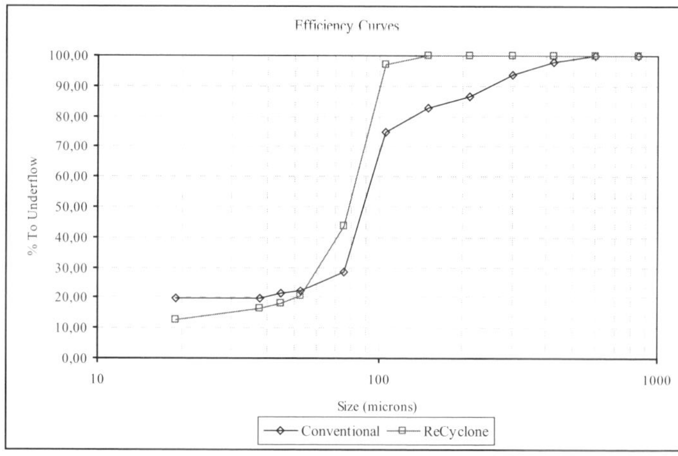


**Figure 9. Efficiency curve for both the conventional cyclone and ReCyclone at Los Pelambres**

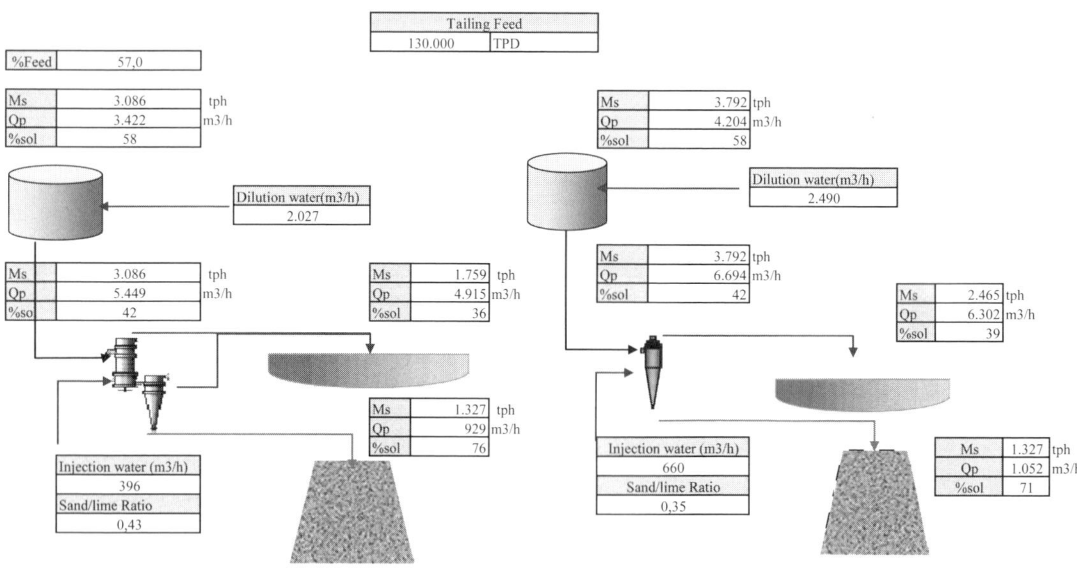


**Figure 10. Flowsheet from Los Pelambres illustrating the tailing dam operation with both a ReCyclone on the left and a conventional cyclone with a cyclowash on the right.**

Using the Recyclone in this application, the projected annual water savings is 10,973,920 ($m^3$ per year). This savings in water can also be translated into additional reduction in energy and equipment consumption due to the lower requirement of water pumped to make the classification.

**Table 3. Performance results of the ReCyclone at Quebrada Honda**

| ReCyclone Data | | | | | |
|---|---|---|---|---|---|
| Date | Wt% Solids Feed | Pressure (psi) | Wt% Solids Underflow | % -200 Mesh to Underflow | Injection Water (l/s) |
| 27-05-2006 | 55 | 19 | 74.2 | 15.32 | 10 |
| 31-05-2006 | 56.5 | 18 | 75.5 | 16.56 | 8.38 |
| 31-05-2006 | 56.5 | 20 | 74.1 | 16.06 | 7.6 |
| 01-06-2006 | 55.6 | 19.5 | 74.8 | 16.72 | 9.5 |
| 03-06-2006 | 55.1 | 15 | 75.4 | 14.76 | 9 |
| 06-06-2006 | 54 | 14.5 | 75.5 | 11.98 | 12.6 |
| 07-06-2006 | 54 | 19 | 76 | 14.18 | 6.3 |
| 08-06-2006 | 53 | 14.5 | 75.5 | 14.3 | 6.6 |
| 09-06-2006 | 54.5 | 18.5 | 76 | 13.8 | 6.3 |
| 09-06-2006 | 54 | 20.5 | 76 | 13 | 6.3 |

**Photo 4. Photo of the installed 500/400CVXDE ReCyclone units at Quebrada Honda. A total of twenty-five ReCyclones have been installed at this site.**

**Photo 5. Photo of the final installation of 650/500CVXDE ReCyclone units at Tintaya Xstrata**

**Southern Peru Copper Corporation—Quebrada Honda**

**Application: Tailing Classification**

After testing the ReCyclone technology, the project for the tailing reservoir at Quebrada Honda of Southern Peru was able to increase the percentage of solids in the feed from 38 wt% in the conventional equipment to 54 wt% with the ReCyclone. The recovery of sand (RAL) to the underflow also increased from 32% to 42%.

**Tintaya Xstrata**

**Application: Tailing Classification**

The stable operation of the new plant of ReCyclone equipment, verifies the outcomes of the pilot tests. Operational results confirms the ability to produce sand for the project at 15% passing 200 mesh, facilitating operation and allowing the tailing reservoir to continue its operation. The ReCyclone will continue to be tested at this facility to verify equipment response due to the variable grading of the feed tailing.

## CONCLUSIONS

ReCyclone is the only double classification equipment proven in the market. The double unit also incorporates other features in its design to maximize performance. High efficiency classification illustrated by the ReCyclone is related to the involute head, increasing the in time of performance and handling of the by-pass of fines which is produced in the cleansing chamber.

As explained in the design section, the ReCyclone incorporates a couple of more typical variables that govern the classification in conventional equipments. These principles provide several possibilities to make comminution processes more flexible. The classification efficiency and handling of the by-pass of fines allows the comminution circuit to handle overloading conditions, making the overall circuit more efficient.

In tailing walls and deslime applications, this equipment has proven to be extraordinarily flexible and yields a high production capacity. Moreover, in comparison with the current equipment used for this process, the ReCyclone has proven to drastically reduce the water consumption as well as a significant savings of energy and equipment.

## REFERENCES

Concha, F., Barrientos, L., Muñoz, O., Bustamante and O. Castro. 1994. A phenomenological model of a hydrocyclone , *Int. Jl. Mineral Process.*, submitted 1994.

Kelsall D. F .Trans. Inst . Chem Engrs 30, 87 (1952)

Bloor, M.I.G. and Ingham, D.B. 1973. Theoretical investigation of the flow in a conical hydrocyclone, *Trans. Inst. Chem. Eng.* 51(1973a)173.

Bloor, M.I.G. and Ingham, D.B. 1973. On the efficiency of the industrial cyclone, *Trans. Inst. Chem. Eng.* 51(1973b)173.

Bloor, M.I.G. and Ingham, D.B. 1973. Turbulent spin in a hydrocyclone, *Trans. Inst. Chem. Eng.* 53(1973a)1.

Bloor, M.I.G. and Ingham, D.B. 1973. The leakage effect in the industrial cyclones, *Trans. Inst. Chem. Eng.* 53(1973b)7.

Bradley, D. and Pulling, D.J. 1959. Flow patterns in the hydraulic cyclone and their interpretation in terms of performance. *Trans Inst. Chem. Eng.* 51(1959)263.

Heiskanen, K. and Oveskainen, P. 1988. *New developments in multistage cycloning.* Frankfurt: ACHEMA 1988, int Meeting on Chem Eng. and Biotechn.

Klimpel, R.R. 1982. Slurry theology influence on the performance of mineral/coal grinding circuits. Part 1. *Min. Eng.* 34(12): 1665-1668.

Shi, F.N. and Napier-Munn, T.J. 1996. Measuring the theology of slurries using an on-line viscometer. *Int. J. Miner. Process.,* in press.

Shi, F.N. and Napier-Munn, T.J. 1995. A model for slurry theology, *Int. J. Miner. Process.,* in press.

# PASTE THICKENING OF TAILINGS: PROCESS AND EQUIPMENT DESIGN FUNDAMENTALS RELATIVE TO DEPOSITION GOALS

Fred Schoenbrunn, Mark Niederhauser, and Frank Baczek

FLSmidth Minerals, 2850 S. Decker Lake Drive, Salt Lake City, UT, USA

## ABSTRACT

Deposition of tailings as a Paste or Thickened Tailings has numerous advantages for recovery of more process water, lowering of the cost of containment structures, decreasing the time for drying and reclamation, and making mine backfill more economical. Each site has its particular mineral type, particle size distribution, and preferred tailings disposal objectives. However, the production of a "manageable" Paste or Thickened Tailings to meet the disposal objectives begins with the disposal requirements, site conditions and tailings characteristics, and works upstream towards the equipment design to achieve the ultimate goal for an economic disposal plan. Many sites and process conditions are amenable to "manageable" Paste or Thickened Tailings disposal. Thickener technology has advanced greatly as far as the capability to produce higher underflow densities. The thickener design includes a number of variables that can be optimized depending on the process requirements.

The paper discusses the process fundamentals required for producing a manageable Paste or Thickened Tailings and also reviews thickening and filtration equipment designs used in Paste production operations.

## INTRODUCTION- THE BENEFITS OF PASTE AND HIGH DENSITY THICKENING

Conventional tailings disposal is a major cost in mining operations from the standpoint of providing safe, stable containment, loss of re-usable water, and eventual reclamation of the tailings site.

If additional water is removed from the tailings prior to deposition, a number of benefits become apparent. The first is the additional process water that is immediately recovered at the plant and not lost to evaporation or seepage from the tailings pond. The second is the reduction in volume of the tailings flow and this leads to additional tailings pond capacity. The third is that the tailings slurry will begin to develop a yield stress as water is removed, and if a substantial yield stress is produced, the tailings pond dam requirements become less. In some cases, the dam structure has been eliminated and a runoff water containment dike was sufficient in managing the disposal area. The thickened slurry that exhibits a substantial yield stress is called a "paste" and also has the properties of being a non-segregating suspension. The fourth is that with a non-segregating suspension and with proper layout management, the drying and consolidation of the tailings deposit occurs at a much more rapid rate, thus allowing for a shorter time to access and reclaim the area.

### Water Recovery Benefit

Figure 1 shows the additional process water recovery for typical copper tailings that are thickened above the normal value of 55wt% solids up to about 65-70wt% solids. At a tailings production rate of 100,000 mtpd, an additional 1200-1600 m$^3$/hr of process water is recovered.

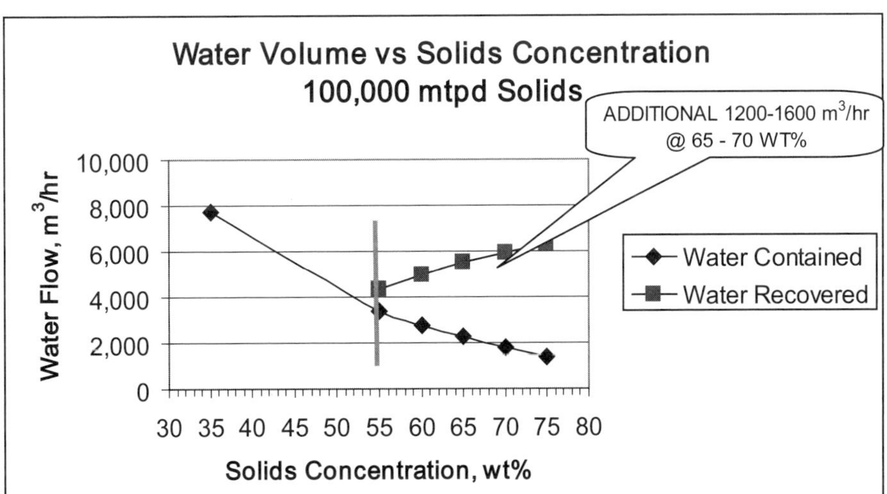


**Figure 1. Water Recovery with Additional Thickening**

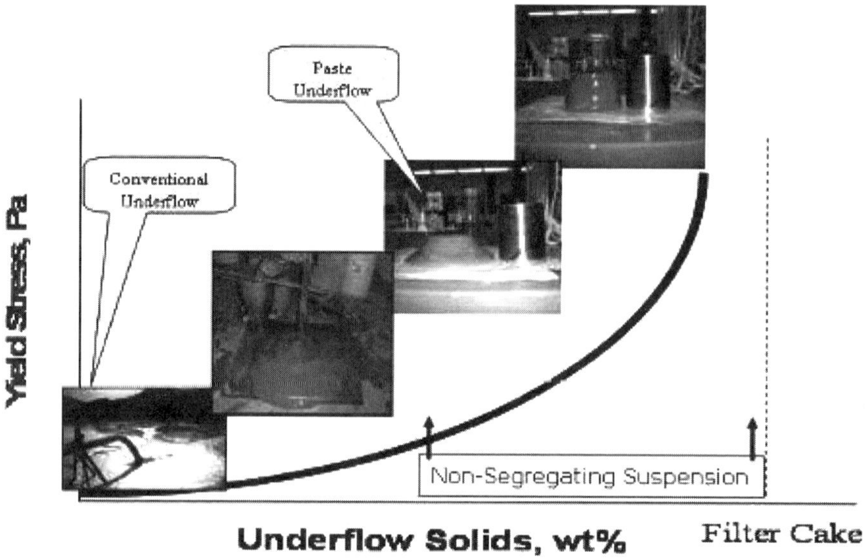


**Figure 2. Typical Yield Stress Behavior as Solids Concentration is Increased**

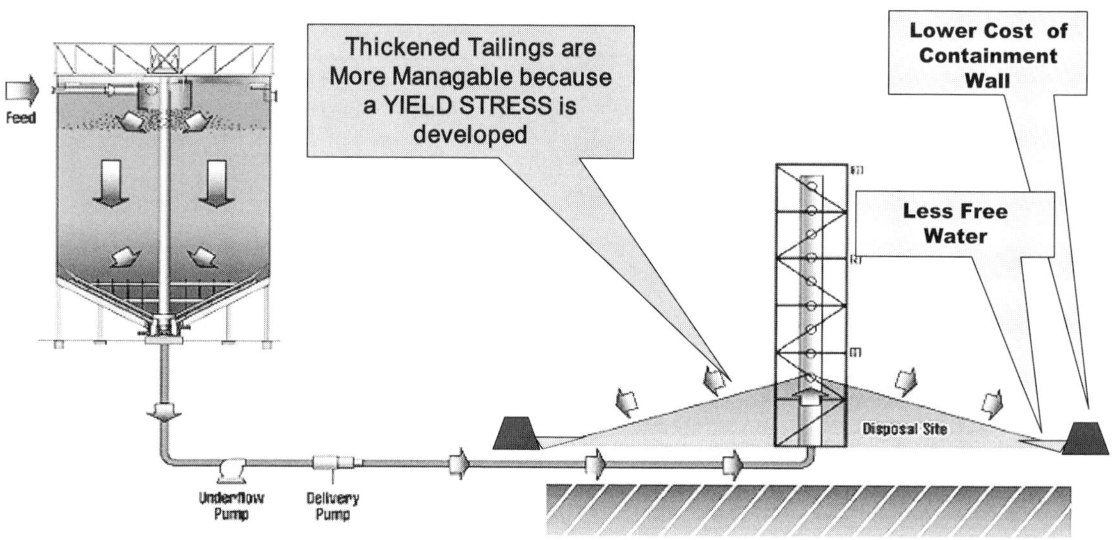


**Figure 3. Schematic of Above Ground Tailings Stacking**

**The Nature of Paste and Its Strength Benefit**

Figure 2 shows the typical difference in strength and flow behavior between "conventional" tailings thickener underflow slurry and paste thickener underflow. The example photos are overlaid on a yield stress curve that shows an increasing underflow yield stress as the solids concentration is increased.

The caption of conventional underflow in Figure 2 shows the almost Newtonian behavior of the tailings slurry due to the large volume of contained water. This characteristic, in addition to the high daily volumes of discharge, leads to the requirement for costly containment dam structures. Comparing the caption of paste underflow, one can see the strength of the mud due to the yield stress and the noticeable absence of free water. These characteristics can then lead to a tailings deposit as shown in Figure 3.

Figure 3 shows a schematic example of an above ground stacking operation using a distribution tower. If the thickened mud has a substantial yield stress, it will begin to form a cone and flow into thin layers over the cone. The thin layer deposit, with a low water content, will also dry more quickly. If the deposition area is designed with multiple discharge points, then management of the area to allow one point to dry while fresh flow is added to another will lead to quicker drying and consolidation of the tailings. If the flows are managed sequentially among multiple points, allowing

**Figure 4. Alumina Red Mud Tailings Deposit. Courtesy Rio Tinto Alcan**

**Figure 5. Copper Tailings Deposit after 4 months. Courtesy Cobriza Doe Run**

drying and consolidation, then a large containment dam structure can be eliminated and a run-off water containment dyke may be sufficient.

**Quicker Reclamation Benefit**

Figures 4 and 5 show consolidated tailings areas that have used above ground stacking as a result of thickening to a paste consistency.

The strength of the deposit to carry heavy equipment loads is shown in the above figures.

Overall, paste and high density thickened tailings disposal is gaining momentum as a viable technology in the minerals processing industry. Many sites around the world have adapted the practice and noted the operating and cost benefits, in addition to achieving a more environmentally acceptable reclamation plan.

**The Design and Decision Process**

Each site has its particular mineral type, particle size distribution, and preferred tailings disposal objectives. However, the production of a "manageable" paste or high density thickened tailings to meet the deposition objectives begins with the disposal requirements, site conditions and tailings characteristics, and works upstream towards the equipment design to achieve the ultimate goal for an economic disposal plan. Many sites and process conditions are amenable to "manageable" paste or high density thickened tailings disposal.

The following, and the remainder of this paper, assumes the decision has been made to use paste/high density thickened tailings disposal.

**Figure 6. Multiple point deposition from a hill side. Doe Run Corp., Cobriza**

**Disposal Requirements.** The tailings disposal objectives need to be defined first: above ground deposition or mine backfill.

Mine backfill typically uses cement addition to a paste slurry to create an engineered backfill suitable for placement in stopes. The technique is very site-specific but can have a number of advantages over conventional backfill methods including reduced water drainage and ore dilution.

For above ground deposition, the natural topography of the tailings site will play a role in determining the methods of distribution and the mud rheology that is required. For example, if the tailings site is in a valley or depression, the tailings feed could come from multiple points along the rim of the valley area and flow generally towards one direction forming a beach angle for management of run-off water into a diked area. If the site is on flat terrain, perhaps multiple distribution towers in a center area are more appropriate. Numerous geotechnical firms, who are experienced in paste and high density thickened tailings deposition by various means and under different topographical conditions, can advise on the alternatives available and give guidance during the decision making process. Typical weather conditions will also need to be considered in the approach to designing the tailings deposition system.

The disposal plan selected will then define the mud slurry rheological characteristics and the solids concentration. Figure 6 shows deposition from the side of a hill into a flat area with a containment dike at the far edges of the site.

**Rheology.** The required rheological properties of the thickened tailings will be dependent on the mineral type, particle size distribution, and solids concentration. Great care should be taken in understanding how potential variability in the ore type and changes in particle size distribution during the operating life of the mine will affect the ability to produce a manageable paste or high density thickened tailings, and how the yield stress will vary under those changing conditions. Once data is collected, an "operating rheology window" can be defined and compared to the process requirements for liberation and recovery of values in upstream operations.

Figure 7 gives an example of an "operating rheology window." The figure shows that the fines fraction (less that 37μ) affects the paste characteristics region of manageable operation. At high solids concentrations, a greater amount of minus 37μ material is required to produce a slurry that can be pumped using conventional centrifugal pumps.

Figure 8 shows an example of a typical yield stress curve with the yield stress increasing as the slurry solids concentration increases. Each mineral type and particle size distribution will have its own characteristic curve. The amount of flocculant added in upstream thickening operation, as well as the shear history of the slurry being handled, will also have an effect on the yield stress. In most cases, as the mud is sheared the yield stress will decrease as is depicted in the figure. The study of shear effects on rheology is also necessary in understanding how pumping and handling of the thickened tailings ahead of the deposition site will affect the ability to distribute the flow in the manner desired for the site design.

The pumping system can be designed once the rheology characteristics are determined. The shearing within the pumping system will then affect the yield stress for transportation and at the deposition site. Once all the above information is gathered, the requirements for the thickening system, or combined thickening and filtration system, can be set. If management of particle size distribution of the slurry being thickened or filtered is also required, a cyclone classification system may precede the dewatering operations.

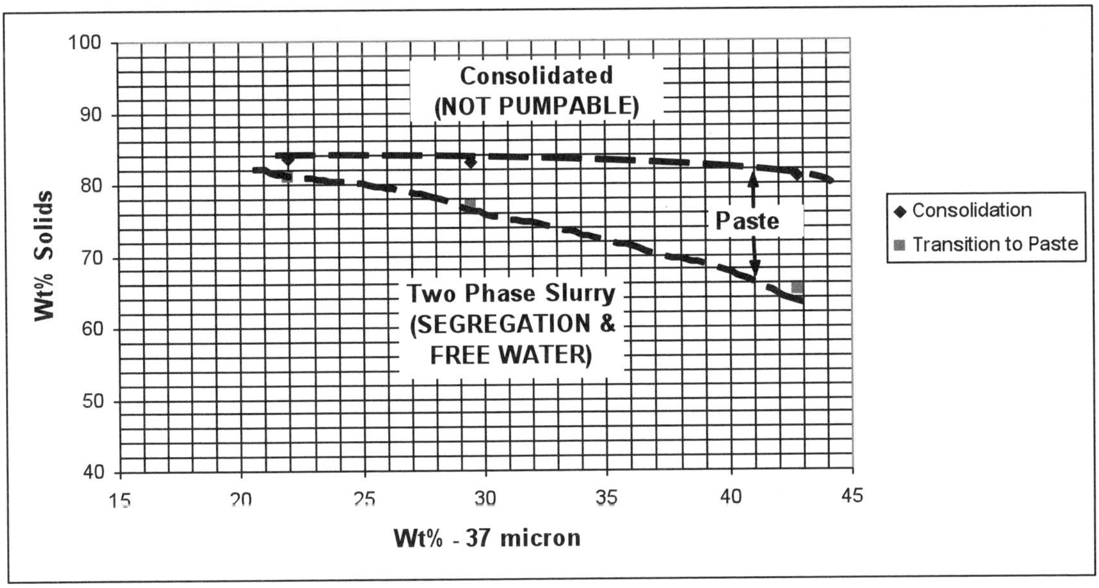


**Figure 7. "Paste Widow" for a particular mineralogy and size distribution**

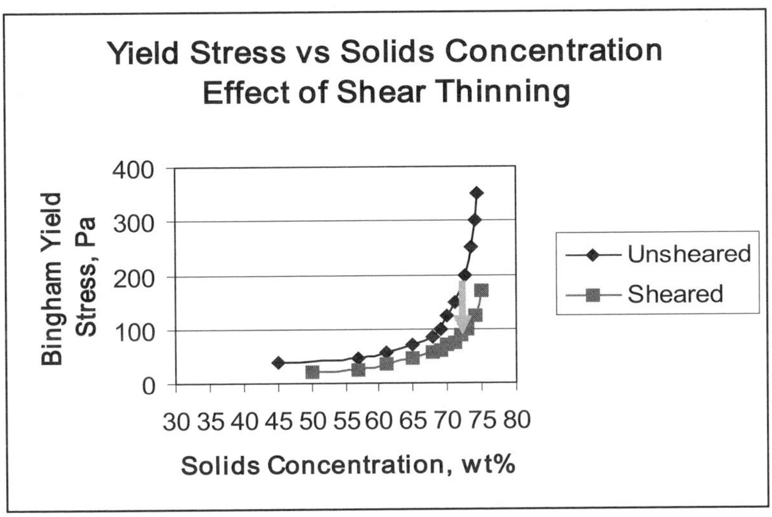


**Figure 8. Example of Yield Stress Behavior vs Solids Concentration**

**Equipment Design for Paste or High Density Thickening**

The design of thickening equipment has evolved over the past 20 years to produce and manage higher underflow densities. Paste and High Density thickeners are now designed up to the range of 50m to 60m diameter, however the application of these larger diameter units depends on the underflow solids concentration, the associated rheology and the required rake mechanism drive torque.

**Important Elements of Paste and High Density Thickener Design**

There are six items in the design of thickeners that require careful consideration:

1. The feedwell and flocculation system
2. The retention time necessary to thicken to the target density
3. The rake mechanism design to rake high yield stress slurry
4. The tank floor slope to assist in mud transport to discharge
5. The rake mechanism drive torque
6. The underflow discharge design to remove a high yield stress underflow

**Figure 9. View inside a High Density Thickener**

**Figure 10. View inside a Deep Cone® Paste Thickener**

Figure 9 shows a High Density thickener and Figure 10 shows a Deep Cone® Paste thickener for reference in the design discussions that follow.

**Flocculation and Feedwell System.** Proper flocculation of the thickener feed is critical in achieving the design unit loading and underflow density. The feedwell and flocculation system is the heart of thickener performance.

Flocculation is not a simple design item or operating function. Figure 11 presents the concept and the important aspects of flocculation and feedwell design.

Design of the flocculant preparation system is often guided by the flocculant vendors and many flocculant supply companies also provide packaged systems for flocculant preparation. Typically, flocculant make-up solutions are at a concentration that has substantial viscosity and will not disperse readily into the slurry at the feedwell system. Further

**Thickener High Performance begins with Efficient Flocculation........**

Think of it as a chemical reaction.......

Reagent + Reagent = Product of Reaction
Flocculant + Solids = Flocculated Particles

Flocculant Solution: Correct Preparation and Activation Time
The Proper Concentration for Dispersion
The Proper Pump Delivery System
Control to the Optimum Dosage

Feed Solids: The Optimum Concentration

Reaction Conditions: Mixing Intensity for Rapid Dispersion
Mixing Time and Intensity for Floc Production

Goal: Maximize Settling Flux and Underflow Density
(Product of Reaction)

**Figure 11. Flocculation Concept and Design Items to be considered**

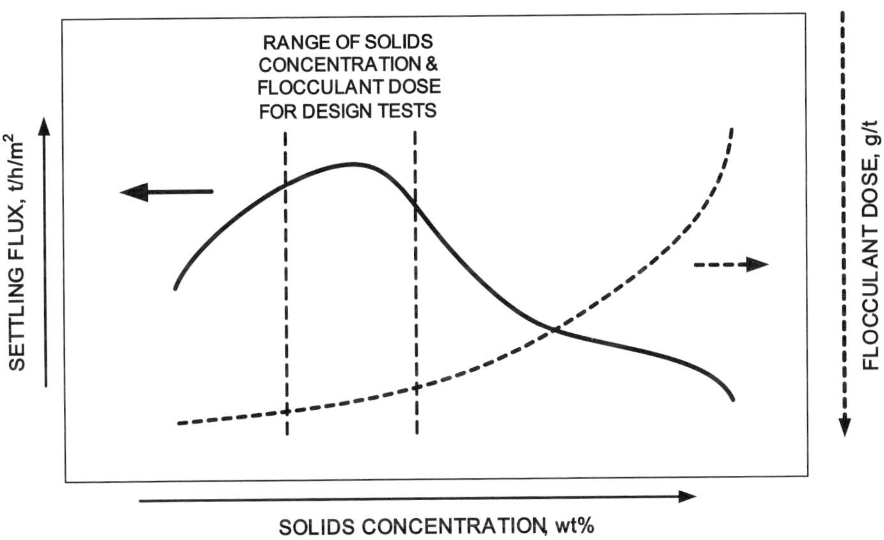


**Figure 12. Effect of Feed Dilution on Flocculant Dose and Settling Flux**

dilution is typically required down to the range of 0.03% to 0.02% to lower the solution viscosity to assist in good dispersion into the feedwell system.

The feed slurry will also have an optimum solids concentration for flocculation that produces the highest settling flux. Determining the optimum conditions for flocculation that produce the highest settling flux will allow the thickener designer to know the minimum diameter at which to safely design the unit for a given design tonnage. Figure 12 shows how dilution of the feed will result in achieving a high initial settling flux in the thickener due to improved flocculation, but will also reduce the required flocculant dose for flocculation.

The feedwell system needs to provide the proper feed dilution, flocculant solution delivery to the diluted slurry, mixing intensity and mixing time. An example of this type of diluting feedwell system is the E-Duc® Feedwell System which is diagrammed in Figure 13.

The design of the E-Duc® Feedwell System has the concentrated thickener feed introduced into a diluting mix tube or open channel through an educting nozzle. The educting nozzle draws dilution water from the body of the thickener so there is no need for bringing in external water. Flocculant solution can be added at the entrance to the mix tube along with the diluting water, and can also be added in stages along the mix tube and at the feedwell. The mix tube is designed to give

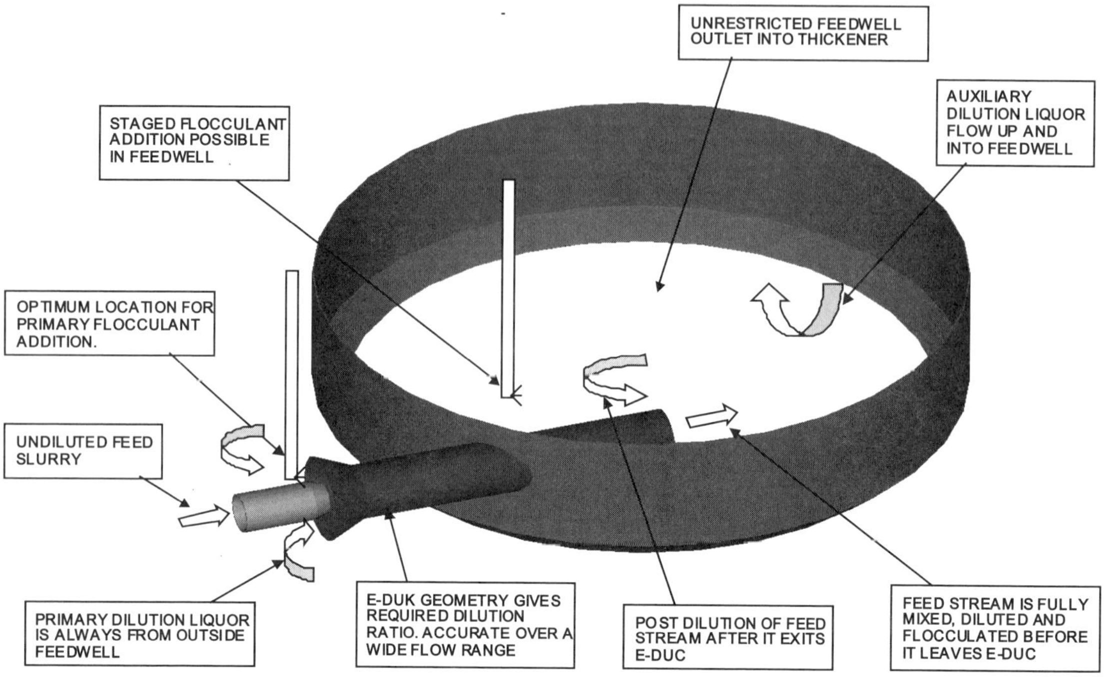


**Figure 13. E-Duc® Self-Diluting and Flocculating Feedwell System**

the mixing intensity and time required for flocculation. Once the flocculated slurry enters the feedwell body, the majority of the flocculation is complete and the flocculated slurry begins the initial settling. Testing of representative samples in the laboratory will determine the flocculation conditions and give the design engineer the criteria for the feedwell system design. Figure 14 is a photo of an E-Duc® feedwell system.

**Retention Time for Thickening.** Conventional High Rate thickeners are designed for lower density, lower yield stress underflow. Typical retention time for thickening in these units is on the order of 2 hours or less. In paste and high density thickening, retention times are substantially longer to achieve the higher underflow densities. Once the thickener diameter is established from the thickening tests, the sidewall height is then adjusted to provide the mud holding volume to achieve the required retention time. A typical paste/high density thickener will have a higher sidewall than a conventional or high rate thickener.

Figure 15 shows a consolidation rate curve for a gold tailings sample tested for design of a paste thickener. Note that reaching the "conventional" 50-55 wt% underflow solids requires only about one hour retention time, but increasing the design underflow density to, say, 65 wt% requires about 7 hours mud residence time. The thickener tank sidewall is then increased to allow for the additional residence time.

**Thickener Rake Mechanism.** The rake mechanism is designed with rakes blades to transport the thickened mud to the center discharge. It is mechanically designed to meet the maximum torque capable of being delivered by the rake drive assembly. The actual operating torque is a function of the thickener mud yield stress, the speed of rotation of the mechanism, and the amount of mechanism steel exposed to the high yield stress mud. For this reason, paste and high density thickener rake mechanisms are designed to minimize the amount of steel exposed to the high yield stress mud.

Figure 16 shows a low profile paste thickener rake mechanism in a 15m diameter Deep Cone® Paste thickener. The photo is taken from the top of the tank looking down into the thickener. The thickener has a 45 degree floor slope.

**Tank Floor Slope.** For paste thickeners up to the range of 24 to 30m diameter, floor slopes of 25 to 45 degrees are used to assist in the transport of the underflow to the discharge point. As the diameter of paste thickeners increases, the steep floor slope can limit the diameter as the elevated steel tank becomes very costly.

For high density thickeners, the floor slopes range from 10 to 15 degrees since the thickened mud yield stress is less.

**Rake Mechanism Drive Torque.** The drive torque capability is selected based on the underflow yield stress. The historical method of selecting a drive torque is to use the equation:

$$\text{Torque, ft-lbs} = K \times (\text{diameter, ft})^2$$

The "K" factor in the above equation is an empirical number based on experience in the design and operation of conventional and high rate thickeners. In Paste thickening, the same equation is used, but the K factor is selected based on the underflow yield stress; and, a rough rule of thumb is that the K factor is equal to the un-sheared mud yield stress in Pascals.

Figure 14. E-Duc® Feedwell System

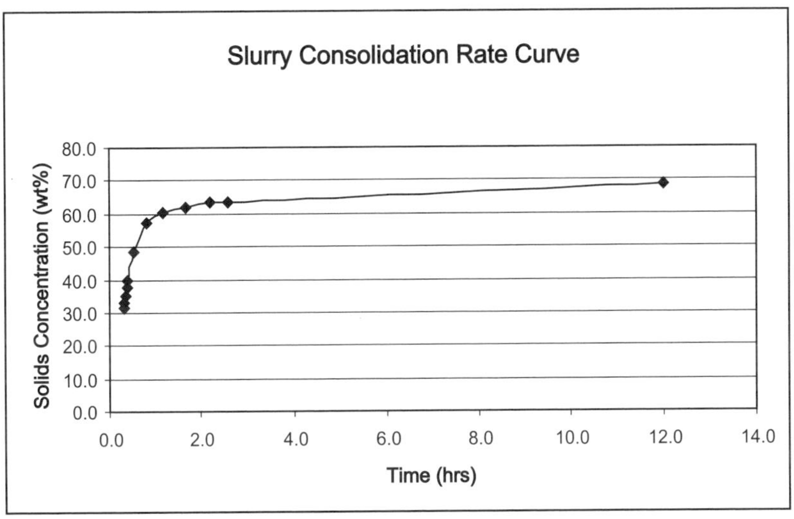


**Figure 15. Slurry solids rate of consolidation curve**

Center drive mechanisms are currently available up to a duty rated torque of 7.2 million ft-lbs. Current design practice then considers a peripheral drive for torque requirements over 7.2 million ft-lbs.

**Design for Underflow Discharge.** A true Paste thickener will have an underflow discharge arrangement that includes a shear-thinning recirculation loop. An example of the yield stress reduction from shear-thinning was shown in Figure 9. The design of this system considers the underflow discharge rate, un-sheared yield stress, and maximum probable recirculation flow needed for yield stress reduction. All of the design criteria are established during the testing of samples.

The paste thickener underflow discharge system will include the recirculation shear-thinning centrifugal pump with a variable speed control, necessary isolation valves, a flow measuring device, and perhaps a density meter, if one is not included on the underflow transfer line.

**The Total Paste Thickener System.** A typical Paste thickener system schematic is shown in Figure 17. The control instrumentation shown in Figure 17 is recommended for reliable control of the thickener system. Mass flow into and out of the thickener is required to maintain the mass balance around the thickener while also controlling and knowing the solids residence time. The feed mass flow is also important for controlling the flocculant dose to the optimum for thickening operation. The combined signals from a bed level instrument and bed pressure sensor are used to determine the inventory of solids and average bed density in the thickener. The flow meter on the recirculation shear-thinning loop allows control of the pump speed for shear thinning.

**Figure 16. Low Profile Rake Arm**

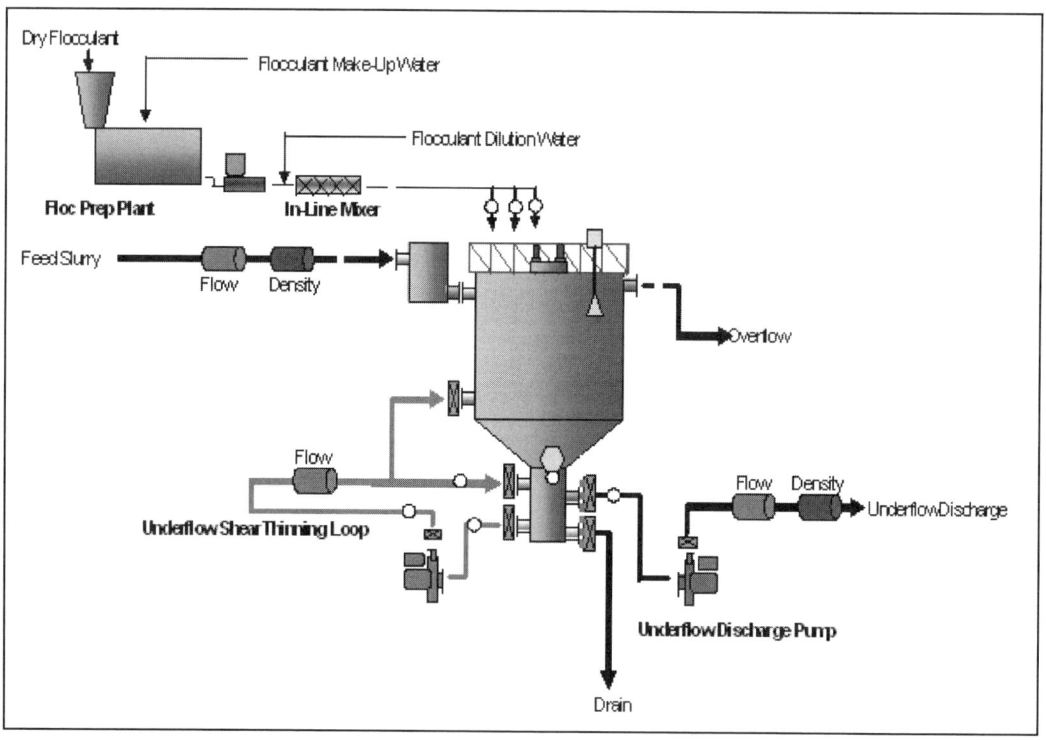


**Figure 17. Paste Thickening System**

## CONCLUSION

Thickened tailings disposal has developed as a reliable means of improving tailings disposal over the past 20 years. With a growing numbers of plants incorporating the technology around the world in a variety of minerals processing applications, the engineering know-how and equipment design has also advanced to meet the demand and minimize the risks and improve the economics.

# Other Separation Processes

# INSTALLATION AND OPERATION OF THE ERIEZ CRYOGEN-FREE SUPERCONDUCTING MAGNET AT SIBELCO UK

Daniel Norrgran[1], Daniel Ashton[2], and Paul Fears[3]

[1] Eriez, 2200 Asbury Road, Erie, Pennsylvania, USA
[2] Sibelco UK, Lovering Lodge, East Gold Works, Kingsteignton Road, Newton Abbot, United Kingdom
[3] Eriez Europe, Bedwas House Industrial Estate, Bedwas, Caerphilly United Kingdom

## ABSTRACT

Sibelco UK (formerly WBB Minerals) serves a wide range of end markets with products drawn from its reserves of clays, silica sands, and dolomite. In 2003, Sibelco UK evaluated an expansion of the economic reserves of kaolin clay in order to sustain the current product specifications for the ceramics portfolio. The assessment indicated that current reserves of kaolin would have to be upgraded through a high gradient magnetic separator to maintain the product quality.

With the development of a new industrial cryogen free superconducting magnet by Eriez, Sibelco UK conducted extensive test work on a range of kaolin clays to characterize the separation parameters. After proving the separation capabilities, Sibelco UK subsequently purchased and installed an Eriez production scale superconducting magnet. Engineers from Sibelco UK and Eriez collaborated on the project from the design and development through the installation and commissioning of the separator.

The paper details the following:

- Assessment of the magnetic separation capabilities of the kaolin clay deposits
- Development and design of the control systems required for the HGMS plant
- Design of the ancillary plant required to run the magnet
- Commissioning the magnetic separator and validation testing

The installation and operation of the magnet has opened up new market opportunities for the sale of the kaolin and extended the life of the deposits.

## INTRODUCTION

Sibelco UK, formerly known as WBB Minerals, provides advanced industrial minerals for a broad range of industries including ceramics, glass, foundry, paper, and construction. Sibelco UK operates sites throughout the United Kingdom mining and processing minerals from reserves of silica sand, ball clay, kaolin and dolomite.

A project was undertaken in order to assess the expansion of the kaolin clay reserves. Several factors were considered in the expansion.

- Substantially increase the fired reflectance (brightness) of selected feedstocks within the reserves.
- Sustain current product specification for Sibelco UK ceramics portfolio.
- Expand the mine to the full extent of the current concession.
- Expansion of the reserves must be economic. This would require the development of a beneficiation process for the lower grade reserves.

## MAGNETIC SEPARATION DEVELOPMENT – LABORATORY

There are several process variables that need to be taken into consideration when specifying a magnet separator for clay processing. Parameters such as magnetic field strength, velocity through the magnet or retention time in the magnetic field, matrix type, and canister arrangement need to be determined. These process parameters are made more complex when the kaolin feed properties are then taken into account. Properties such as mineralogy, morphology, rheology, and particle size distribution will all affect the separation to some degree. Laboratory trials are essential to quantify the separation parameters.

Magnetic separation is applied to the clay slurry in order to capture the "dark" colored iron bearing constituents. Removing these "dark" minerals improves the brightness and consequently the quality of the clay. The collection of the iron bearing constituents is not always straightforward. Typically the iron occurs in different forms within the mineralogical make up of a kaolin feedstock. These iron contaminants can be classified into three main groups:

1. Associated iron-bearing minerals such as mica, tourmaline, hematite, goethite, rutile, and magnetite. These are the easiest to separate out as they are either ferro or strongly para-magnetics.
2. Iron bearing layered minerals such as smectites. These minerals are often associated with kaolinite and may occur in-between the lattice planes of the kaolinite stacks; these contaminants can be removed with a magnet, however care must be taken to avoid removing too much kaolinite along with the contaminants.
3. Iron hydroxyls or iron staining on the surface of minerals such as kaolinite and feldspar. Where this type of iron occurs, the iron is very weakly magnetic and other process methods can be used.

Layered minerals such as smectites can cause processing problems and affect the susceptibility of a kaolin to magnetic separation. Kaolins with problematic mineralogy can be difficult to disperse and the magnetic constituents may be hard to liberate from the feedstock.

**High Gradient Magnetic Separation**

Bench scale and pilot scale testing was carried out on selected samples of kaolin clay using High Gradient Magnetic Separation (HGMS) in order to quantify the separation characteristics. HGMS refers to the canister–type magnetic filters that are batch operated. HGMS is also a common term for high capacity magnetic filters used in the beneficiation of clay and other associated fine grained earthy minerals. These separators were developed with specific features to provide peak separation effectiveness while operating at relatively high feed capacities. A schematic of a High Gradient Magnetic Separator is illustrated in Figure 1.

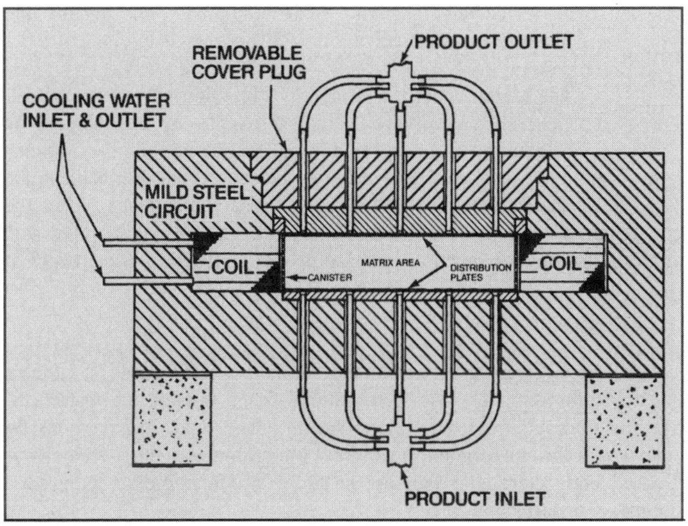


**Figure 1. Schematic of a High Gradient Magnetic Separator detailing electromagnetic coil, steel housing, and feed and product distribution.**

A High Gradient Magnetic Separator consists of a solenoid coil encased in steel. The steel housing is necessary to shape and contain the magnetic field. The coil generates a uniform magnetic field throughout the bore. A matrix is housed in the bore of the solenoid coil and provides the collection of weakly magnetic minerals. The matrix is a ferromagnetic material that provides a high level of magnetic induction. It also has a high degree of open area to provide slurry flow without physical entrapment. The function of the matrix is actually three fold:

1) Amplification of the applied magnetic field.
2) Produces localized regions of extremely high gradient.
3) Provides the collection sites for paramagnetic particle capture

Typical matrix material consists of screen cloth, expanded metal, wedge wire, steel balls, and steel wool. The matrix is selected based on the size distribution of the ore. The optimum matrix for a given application should provide a high magnetic loading capacity, high magnetic gradient, and have a relatively open structure for flow through capacity. The matrix should be coarse enough to allow the largest particle to pass through unimpeded yet fine enough to generate high magnetic field gradients and provide sufficient collection sites. An engineered matrix resembling steel wool is typically used in the beneficiation of clay minerals.

HGMS is operated with a controlled slurry flowrate through the matrix. The slurry is pumped into the bottom of the matrix canister and discharges out of the top. In this manner, the retention time in the matrix is controlled. The slurry must be well dispersed and the particles held in suspension to operate in this manner. As feed material filters through the matrix, the paramagnetic particles are captured and consequently removed from the particle stream. The fluid drag provides the separating force between the magnetic contaminants and the nonmagnetic medium. When the magnetic contaminants eventually build-up on the matrix, the separator is de-energized and the matrix is flushed clean.

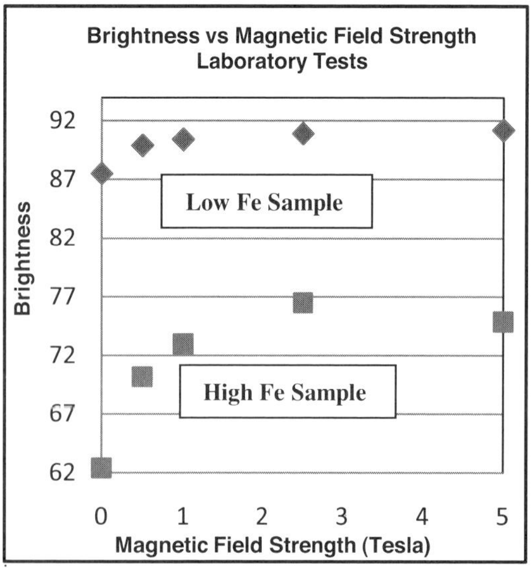


**Figure 2. Brightness as a function of Magnetic Field Strength. Magnetic separation tests on two different samples with varying iron content.**

The magnetic separation test work was conducted to evaluate the following variables.

- Magnetic field strength
- Slurry feed velocity (retention time)
- Total processing time
- Type of separation matrix

The test work is summarized below.

**Magnetic Field Strength**

The background magnetic field strength or the magnetic field strength in the bore of the solenoid is a first order variable in the collection of iron bearing minerals. Magnetic separation tests were conducted on two different samples representing the extremes of the deposit. One sample represented the low iron deposit while the other sample represented the high iron deposit. A series of magnetic separation tests were conducted at increasing magnetic field strength. The results are illustrated in Figure 2.

The brightness of the product is a function of the iron content. Generally lower iron content results in an increase in brightness. As expected, the brightness of the product increased with increasing magnetic field strength. The higher magnetic field strengths decreased the iron content and subsequently increased the brightness. The most significant increases in brightness occurred at the lower end of the magnetic field strength spectrum ranging up to 1 Tesla. There was little increase in brightness beyond this point with the low iron sample. The brightness of the high iron sample continued to increase up to a magnetic field strength of 2Tesla.

**Slurry Velocity – Retention Time**

The velocity of the slurry through the magnetic field or conversely the retention time in the magnetic field is a variable that relates to both the separation efficiency and the throughput. As feed material filters through the matrix, the paramagnetic particles are captured and consequently removed from the particle stream. The fluid drag provides the separating force between the magnetic contaminants and the nonmagnetic medium. Increasing the slurry velocity is equivalent to decreasing the retention time. Subsequently there is an increase in the hydrodynamic drag acting on a particle and a decrease in the effectiveness of the magnetic attractive force.

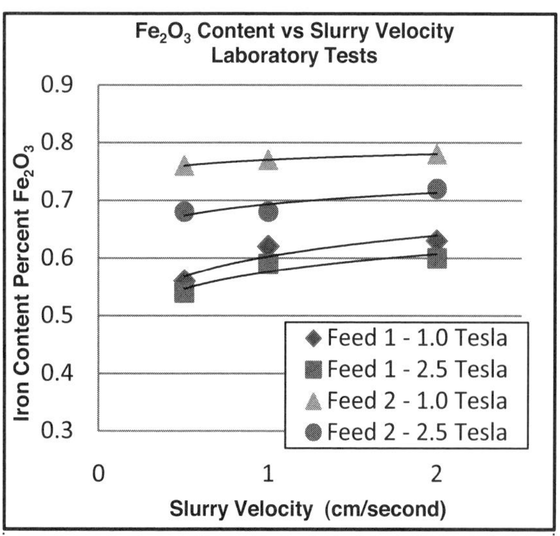


**Figure 3. Iron Content as a function of Slurry Velocity. Magnetic separation tests on two different samples at increasing slurry velocity and magnetic field strength.**

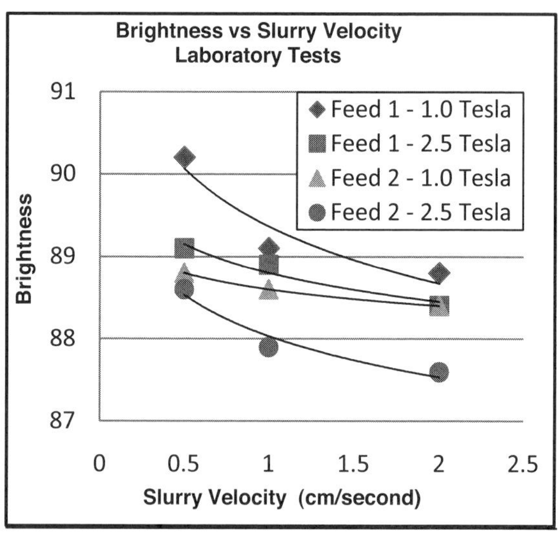


**Figure 4. Product Brightness as a function of Slurry Velocity. Magnetic separation tests on two different samples at increasing slurry velocity and magnetic field strength.**

The process velocity can form part of a flexible approach to dealing with a deposit varying in quality. By altering the velocity of the feed through the magnet, the upgrade can be decreased or increased depending on the quality of the starting material allowing for a constant final product quality to be achieved. If the velocity through the canister is slowed, the feedstock being processed is exposed to the field for a longer period of time and as a result there is more opportunity for separation of magnetic particles.

Magnetic separation tests were conducted on two different samples at increasing slurry velocity. Magnetic field strengths of 1.0 Tesla and 2.5 Tesla were tested. The results of this test work is illustrated in Figures 3 and 4.

Figure 3 is details iron content in the product as a function of slurry velocity. The iron content increases as the slurry velocity through the separator increases. This again is to be expected. As the slurry velocity increases, the retention time decreases proportionately and the probability of capturing magnetic particles decreases.

Figure 4 is details product brightness as a function of slurry velocity. The iron content increases as the slurry velocity through the separator increases adversely affecting the brightness.

The two feedstock samples are apparently similar in quality but with different mineralogical composition. Feed stock 1 has a high feldspar content, whereas feedstock 2 has a high kaolinite content along with a higher layered mineral content. Feedstock 1 provides a more definitive upgrading than feedstock 2.

**Processing Time – Duty Cycle**

Another parameter that can alter the final product quality and the capacity of a HGMS system is the processing time. As the kaolin slurry passes through the separating canister the magnetic particles are collected on the matrix. After a certain period of time, the matrix becomes saturated, and the separation efficiency declines. Since the HGMS is a batch process, the matrix needs to be backflushed periodically to maintain throughput. The processing time or duty cycle is dependent on the iron content and consequently the amount of magnetics that will be collected on the matrix. The processing time can be optimized for any given feedstock type and velocity, to give a required product quality.

Figure 5 details the cumulative iron content in the product as a function of processing time. Various slurry velocities and magnetic field strengths were tested to allow the process to be optimized. The iron content in the product increases with increasing processing time. This is simply a matter of the iron bearing particulates collecting in the matrix reducing the available collection sites. The collection of iron bearing minerals then diminishes.

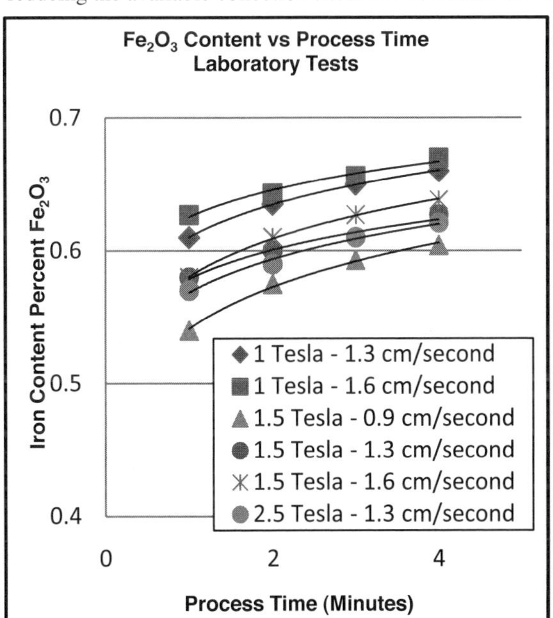


**Figure 5. Iron Content as a function of Process Time. Magnetic separation tests at increasing slurry velocity and magnetic field strength. Feed is 0.77 percent $Fe_2O_3$.**

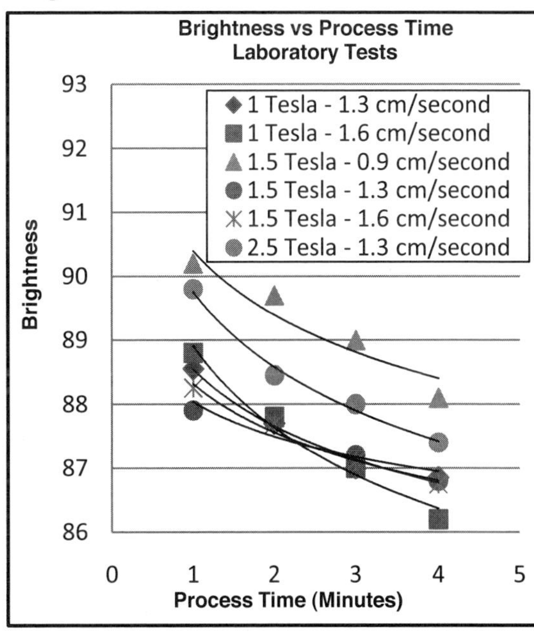


**Figure 6. Brightness as a function of Process Time. Magnetic separation tests at increasing slurry velocity and magnetic field strength. Feed is 79.1 Brightness.**

Figure 6 details the cumulative product brightness as a function of processing time. The graph shows how the product quality progressively decreases over time. The iron content increases as the processing time is extended adversely affecting the brightness.

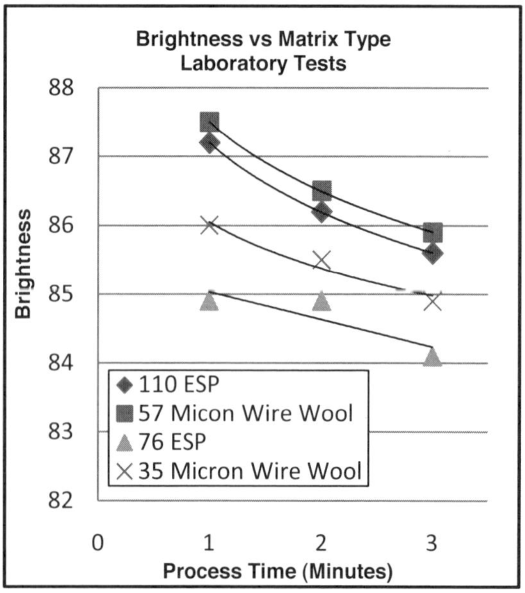


**Figure 7. Brightness as a function of Process Time for various matrices. Feed is 77.7 Brightness.**

In the case of this particular feedstock the optimum set of conditions appears to be using a magnet at a field strength of 1.5 Tesla and a velocity of 0.9 cm/sec.

### Matrix

The type of matrix used as the separation media is a very important parameter from both a process and an operational point of view. There are two main types of matrix commonly used in an HGMS processing clay. The traditional matrix is a wire wool matrix, packed into a canister to a certain packing density. There is also an engineered matrix or an "ordered" matrix. This ordered matrix has more open area and an ordered structure.

Two different sizes of steel wool and two different sizes of the ordered matrix were tested. The ordered matrix is termed "ESP".

As can be seen in Figure 7, the cumulative brightness yielded the best results with the 57-micron wire wool matrix. The 110 ESP matrix followed the same pattern only slightly lower. The 76 ESP matrix and the 35-micron wire wool matrix were shown not to have as high a separation efficiency. It is anticipated that these matrices have a much finer opening and were quickly loaded with magnetics.

## PRODUCTION SCALE MAGNETIC SEPARATOR

Based on the results of the laboratory and pilot scale magnetic separation test work, a production scale separator was specified. The project objectives are listed below.

- Design and build a HGMS plant to process a selected proportion of the reserves annual output.
- Fully integrate the HGMS plant inline into the current process.

## ERIEZ POWERFLUX SUPERCONDUCTING HGMS

An Eriez PowerFlux HGMS was specified for the application. The PowerFlux is a relatively new concept in HGMS. The PowerFlux operates essentially energy free at superconducting temperatures without the use of liquid cryogens.

### Emphasis on Simplicity

The PowerFlux magnetic separator provides the simplest and least expensive superconducting HGMS production plant presently available. The separator utilizes a niobium-titanium conductor to generate the magnetic field. The conductor becomes superconducting at a temperature below 11 degrees K. In the superconducting state the conductor operates without resistance producing the magnetic field without energy consumption.

A schematic representation of the PowerFlux magnetic separator is illustrated in Figure 8.Conventional superconducting technology requires that the coil be submerged in liquid helium to establish superconducting temperatures. The design of the PowerFlux superconducting magnet is unique in that it requires no liquid cryogen for cooling. The coil is housed in a cryostat which uses a vacuum insulation to thermally shield the coil. Cooling is accomplished by conduction with the use of a cryocooler. The cryocooler leads are cooled to superconducting temperatures with helium gas in a closed loop system with a helium refrigeration unit. The cryocooler has a copper thermal pathway that subsequently cools the coil by conduction to superconducting temperatures. The cryocooler reaches a temperature of 5 degrees K and the coil reaches a temperature of 7 degrees K through conduction. Figure 9 is a photograph of the cryostat containing the superconducting coil. Also shown is the cryocooler mounted on the side of the cryostat.

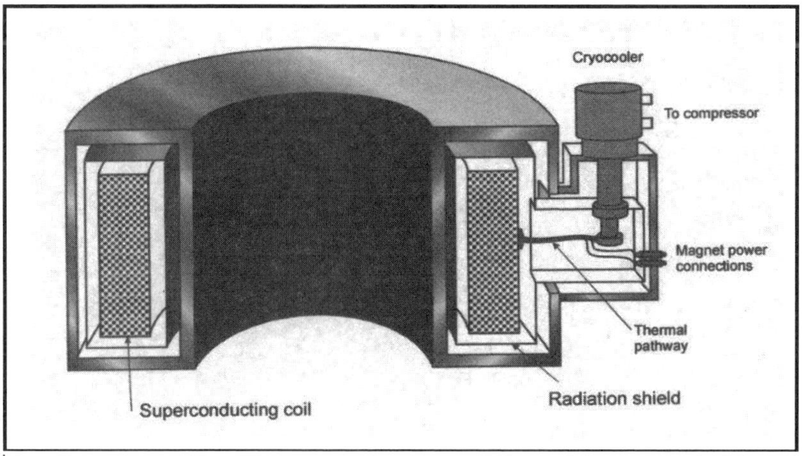


**Figure 8. Schematic of superconducting system. Shown is the coil enclosed in the cruostat. The superconducting coil is cooled by conduction with a cryocooler.**

The superconducting magnet itself was designed to provide straightforward operation and limited maintenance. As shown in Figure 10, the PowerFlux is very compact and relatively easy to install. The whole separation system is totally enclosed in a steel housing and there is no material spillage keeping the working environment clean. The top section can be easily removed for matrix removal or matrix canister exchange, reducing downtime and thus costs. A view of the top slurry distributor that collects the product is shown in Figure 11.

**Figure 9. Cryostat containing the superconducting coil. Cryocooler in place in foreground.**

**Figure 10. Complete PowerFlux superconducting magnetic separator with the cryostat, cryocooler, canister, matrix, and slurry distribution system enclosed in a steel housing**

**PowerFlux Magnetic Separator Specifications**

The specifications for the production scale separator are provided in Table 1. A few notable items regarding the specifications follow.

A magnetic field strength of 1.5 Tesla was selected. Based on the test work, it was felt that there was little advantage in going to a higher magnetic field strength. Ferromagnetic matrices saturate at a nominal 1.2 Tesla background field. An increase in background field beyond saturation does not lead to an increase in the magnetic field gradient.

The bore or canister diameter of the PowerFlux magnetic separator is 1.0 meter to provide production rates of 5 MTPH to 15 MTPH depending on the area of production.

The 110 ESP ordered matrix was selected. The ordered matrix, or woven mesh matrix, is very easy to handle, can easily be manually cleaned and inspected if necessary, and will last longer than wire wool.

Flexiblity is required in the design of the separator in order to process clays from different sites. Variability in the slurry velocity (feed rate) and the duty cycle were considered in the process scheme.

**Figure 11. PowerFlux superconducting magnetic separator slurry distribution system. Top of the separator where the product discharges.**

**PowerFlux Specifications**

- Magnetic field strength of 1.5 Tesla.
- Operating bore of 1.0 meter.
- A monobloc canister with 110 ESP woven mesh matrix.
- "Quick" ramp-up and ramp-down times while cycling.
- The magnet operates at 6 watts and the cryocooler operates a 6500 watts.
- The PowerFlux processes between 5 and 15 MTPH.
- Information on the velocity and processing time was obtained allowing flexibility to be designed into the plant.

**Table 1. PowerFlux superconducting magnetic separator specifications.**

**Operating Parameters**

When designing a HGMS plant it is very important to assess the processing and preparation of the kaolin slurry to be fed into the magnet. The density of the slurry should be constant to avoid variations in product quality. Prior to feeding the magnetic separator, contingencies need to be put into place to ensure that there are no oversized material in the feed, as this could cause problems with blockage or reduce capacity. Adequate buffering capacity is advantageous, as this insures the supply of a fully dispersed, conditioned stable feed.

Depending on the system type, a moderate amount of water (typically $3m^3$ per flush) are required to rinse the matrix and canister free of magnetics deposited during processing. The quality of water fed to the magnet is a very important factor to consider. Many kaolin clay processing plants do not have large quantities of potable water available in the process. Pond water is often times the only water supply. Most kaolin plants recycle the plant water and consequently water quality can be a major issue and design consideration. If the quality of the water used to rinse and clean the matrix is poor, the plant can be plagued with performance problems affecting the throughput and product quality. The pH of the water may cause problems with de-flocculation, essential for liberation and hence separation of paramagnetic magnetic particles present in the slurry. Any solids within the water can also prove detrimental to the efficiency of the system with the matrix facing the possibility of blockage or blinding. These factors reduce separation efficiency and hence quality as well as throughput. The frequency of cleaning and maintenance may also be increased.

The pH along with the presence of salts in the water, from either the deposit or other processes such as bleaching, can damage the matrix used in the canister. This may significantly increase downtime and reduce the separation efficiency of the magnet system.

The magnetic separator receives a prepared feed in the process. At the outset, the clay slurry is drawn into the process directly as thickener underflow at 25 percent solids. This slurry is mechanically "de-flocculated" and scereen at 106 microns to remove any oversize material. The slurry then reports to a buffer tank where it is monitored to the magnetic separator. The slurry is drawn from the buffer tank to the magnetic separator at a controlled rate with a variable speed drive pumping system. During the magnetic separation process cycle, the non-magnetic product reports to the product tank where it is homogenized, re-floccuolated, and pumped to a product storage tank. During the magnetic separation flush cycle, the magnetic fraction from the separation reports as a final tailing. Sample points were installed in the magnetic separator feed and product lines

**Magnetic Separator Processing Cycle**

The magnetic separator operates in a batch cycle. A schematic detailing the magnetic separator operation on a time line is illustrate in Figure 12. The processing cycle is as follows.

1. The magnet starts to ramp up to the operating magnetic field strength. This will vary depending on the magnetic field strength required.
2. The canister is prepared with slurry before the magnet fully reaches magnetic field strength.
3. When the magnet reaches the operating magnetic field strength, the processing time starts. Slurry is pumped through the matrix at a predetermined velocity and a time.
4. At the end of the designated process time, the canister is rinsed with water to displace any kaolin slurry still in the canister. This helps to recover as much processed kaolin as possible.

5. At the end of the canister rinse the magnet starts to ramp down.
6. When the field is fully dissipated the cleaning sequence starts. A sequence of air and water blasts rinses the matrix in the canister, removing any magnetics present.
7. During the cleaning sequence the coil cools prior to ramping up to field again and the sequence restarts.

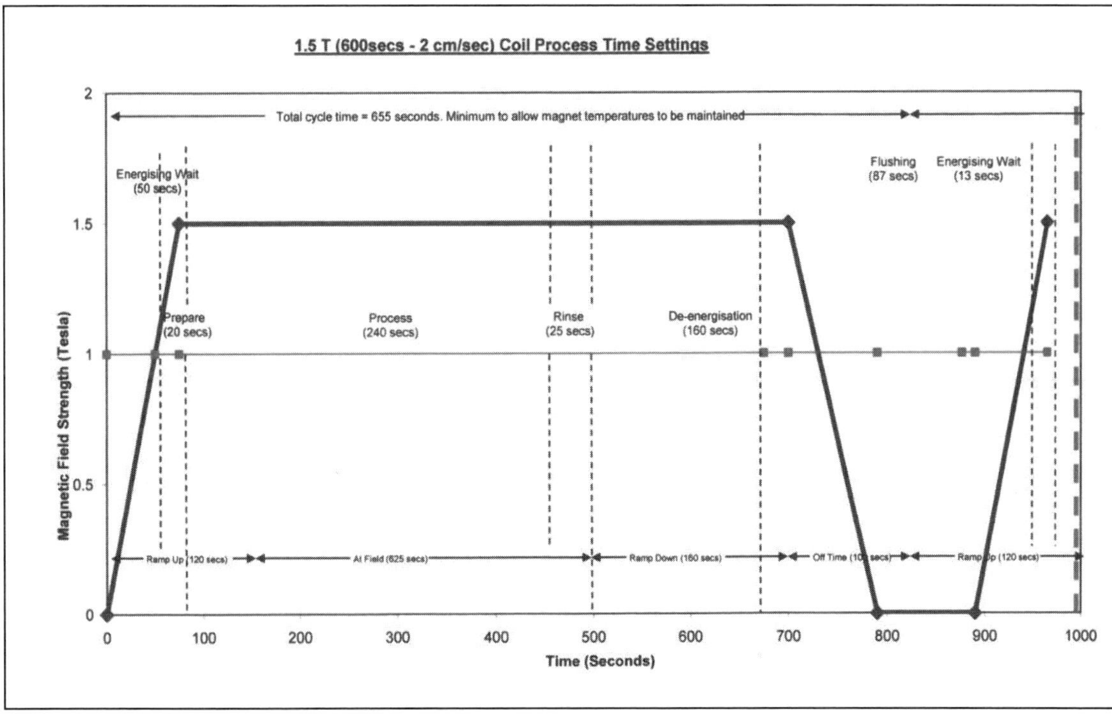


**Figure 12. PowerFlux processing cycle schematic detailing magnetic separator operation on a timeline.**

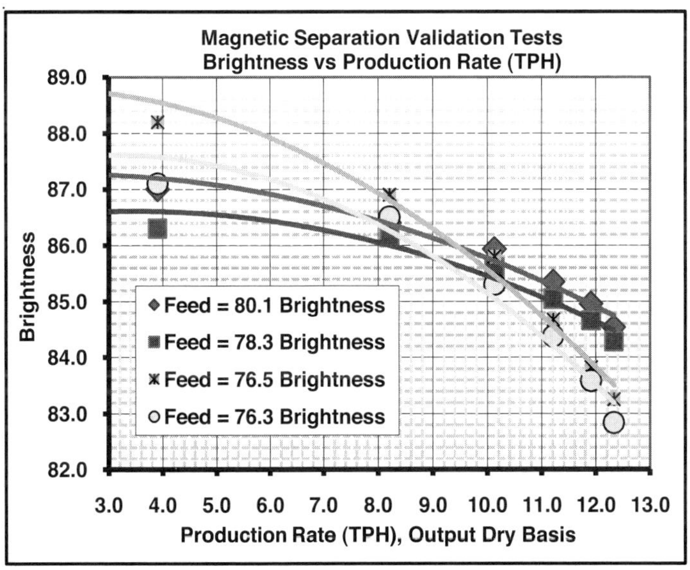


**Figure 13. Validation tests on the PowerFlux. Tests were conducted at a magnetic field strength of 1.5 Tesla at increasing slurry velocity or production rates.**

Once the PowerFlux magnetic separator was integrated into the process and operational, a series of validation tests were conducted. The validation tests were conducted on four different materials representing different areas of the reserves. The production runs were conducted at a magnetic field strength of 1.5 Tesla with a 110 ESP matrix at increasing production rates or slurry velocity. The results of the test work is illustrated in Figure 13. The graph illustrates the product brightness as a function of production rate or feed rate.

As expected, there was a significant increase in the brightness of the clay product. The brightness decreased as the slurry velocity or production rate increased. Further, the results achieved with the magnetic separation process are as predicted through the testwork.

Several production campaigns were conducted testing the operating parameters and flexibility and of the magnetic separation system. The tests were conducted on various areas of the reserves in order to facilitate production planning. The results of the production runs are shown in figures 14 through 19.

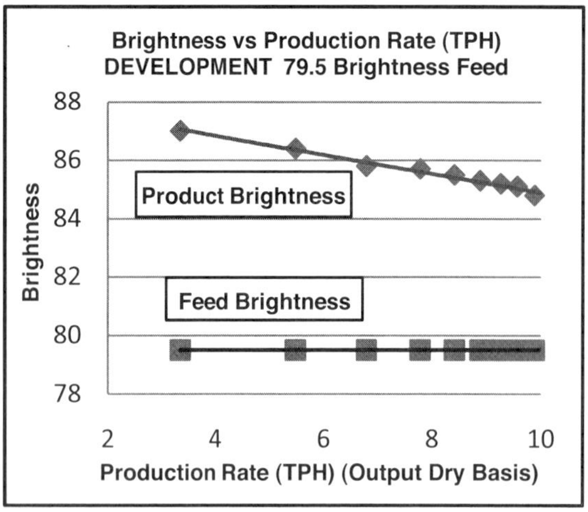


**Figure 14. Production results. Product Brightness as a function of Production Rate. Production Development Area.**

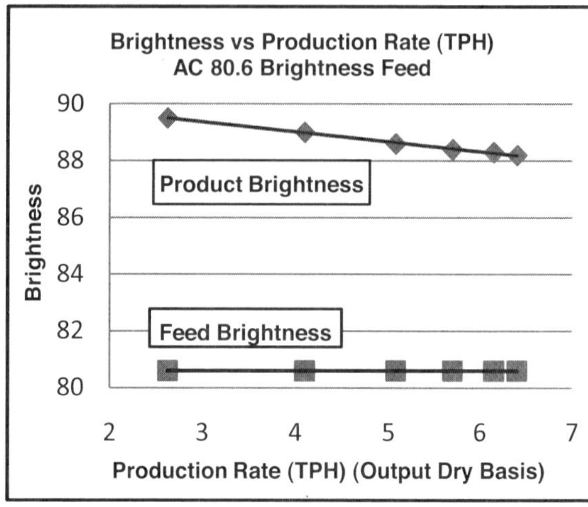


**Figure 15. Production results. Product Brightness as a function of Production Rate. Production AC Area.**

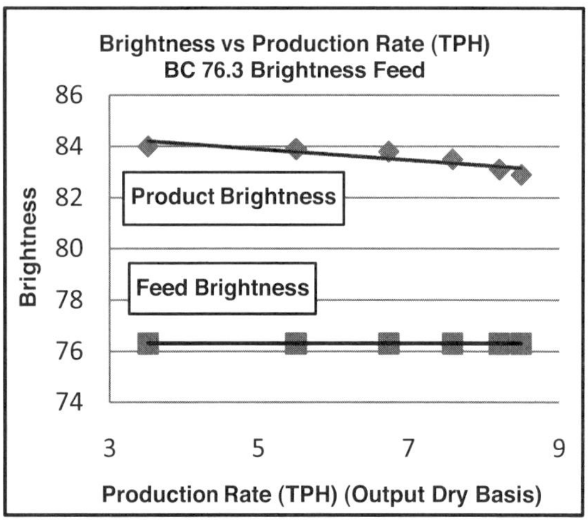


**Figure 16. Production results. Product Brightness as a function of Production Rate. Production BC Area.**

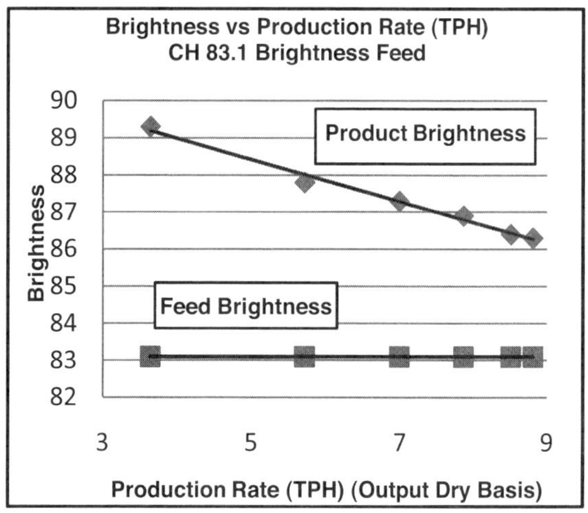


**Figure 17. Production results. Product Brightness as a function of Production Rate. Production CH Area.**

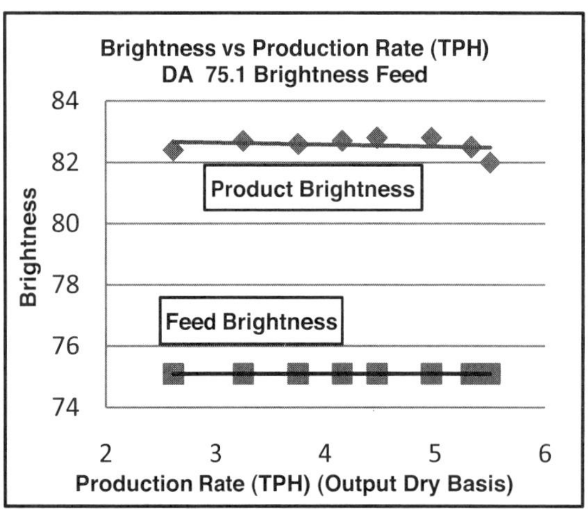


**Figure 18. Production results. Product Brightness as a function of Production Rate. Production DA Area.**

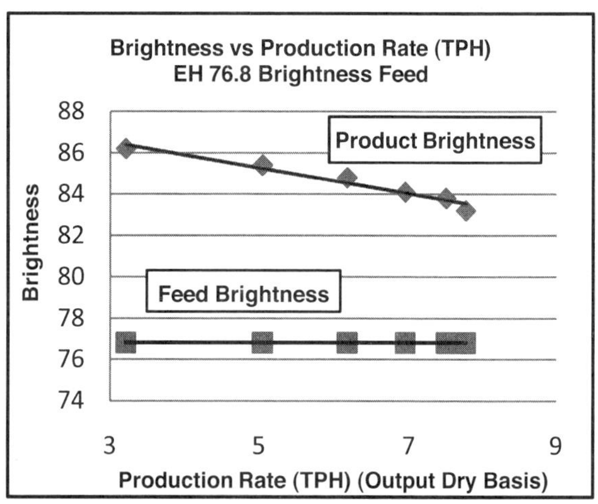


**Figure 19. Production results. Product Brightness as a function of Production Rate. Production EH Area.**

These production runs were again conducted at a magnetic field strength of 1.5 Tesla and at increasing slurry velocity or production rates. The results of the test work is illustrated in Figure 13. The graph illustrates the product brightness as a function of production rate. For comparison, the feed brightness is also shown.

In each case there was a significant improvement in the product brightness. The product brightness decreased at varying rates when the production rate was increased. It is interesting to note that the upgrading of the clay and the sensitivity to the feed rate is very site specific.

**Summary**

The plant is very efficient and flexible and is proving effective at upgrading clay feedstocks that were previously unable to be included in product portfolio. Due to the importance placed on flexibility in the design stages of the project, the plant is being used to develop new products and investigate new opportunities. The usable reserves have been increased with the introduction of the magnet.

The results of the PowerFlux magnetic separation project are highlighted in Table 2.

**PowerFlux Project Results**

- The PowerFlux magnetic separator has been in operation for seven years.
- It is now fully integrated into the process and a straightforward operating scheme.
- It is an integral tool in the development of the reserves and is allowing the processing of clays that would have been stripped in the past.
- It has significantly increased the mine life.
- It has achieved the goal of substantial increases in fired reflectance (brightness).

**Table 2. PowerFlux magnetic separation project results**

# NEW DEVELOPMENTS IN SENSOR-BASED SORTING

Lütke von Ketelhodt and Kai Bartram

Commodas Inc., 47B Harbour Square, P.O. Box 81097, Toronto, Ontario, Canada M5E 1W7

## ABSTRACT

Optical Sorters using color line-scan cameras have been around for over 20 years. These sorting machines are successfully used in various mineral applications as a pre-concentration method. Particularly in industrial mineral applications color sorting machines are an established technology. In recent years other mineral applications such as platinum, gold, coal etc started using sensor-based sorting as a beneficiation process. This has caused a surge in research and development in finding new sensors and to improve existing machines and designs. Today sensor based sorting is considered a significant innovation in mineral processing technology. New sensor systems have been developed, which exploit material properties such as electrical conductivity, magnetization, molecular structure, thermal conductivity etc. Now exciting new pre-concentration and upgrading possibilities have emerged for applications in nickel, gold, copper and other metal ores as well as industrial minerals, coal,diamonds and other gem stones.

## INTRODUCTION

Although hand-sorting ore has been used since the start of mining, electronic sorting, as we know it today, was first adapted for use in mining in the 1970s. Early electronic ore sorters include the RTZ Model 19 Conductivity/Magnetic Response sorter and the combined photometric and conductivity sorter from Ore Sorters (Canada) Ltd. Since then a variety of sensors including, but not limited to, scintillometers, cameras, conductivity/magnetic susceptibility sensors, photomultiplier tubes, microwave attenuation sensors, xrf and infrared spectrometers have been used for sorting.

Commodas and UltraSort have been active in sensor-based sorting since 1993. Their first sorting machines processing industrial minerals such as calcium carbonates and magnesite were based on camera technology and digital image processing. Despite these initial applications in the mineral industry, it was the recycling market which drove the rapid growth of sorting technology. Real advancements in mineral applications of Commodas' technology really began in the year 2000. Of the over 500 sorting machines operating in various fields today, approximately 150 are used in the mining and mineral processing industry.

## MECHANICAL DESIGN

Two types mechanical design of the sorting machines are available; the chute and the belt system. It is most important that the feed material moves/flows past the sensors adequately spread out in separate particles and at a uniform speed to achieve the most efficient sorting results.

- The chute or free-fall type machines are used for larger particle size ranges -300mm to +20mm where the flow properties of the material are suitable for this approach and where such bulk materials are easy to convey.
- The conveyor-belt type is used for materials which have difficult characteristics such as texture, particle shape or moisture content which would have an uneven flow on a chute.

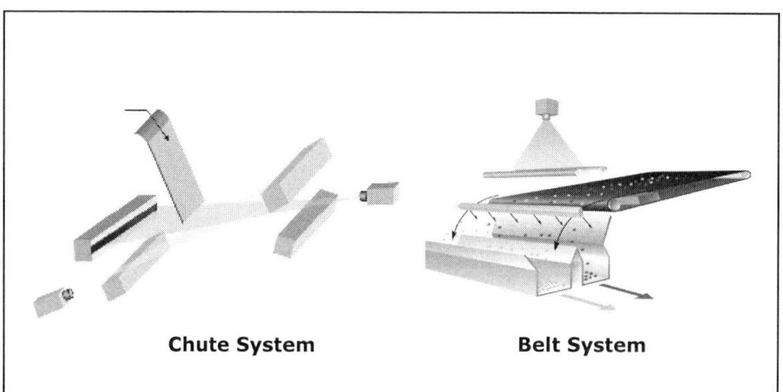


**Figure 1. Two types of sorting systems – Chute vs. Belt**

## SENSORS

Apart from the diamond industry which uses photo multiplier tubes to detect x-ray fluorescence induced diamonds, the most common sorters used in mineral processing applications are either optical or conductivity sorters. Modern color sorters have a large installation base in industrial minerals, specifically in the filler industry where ore is often graded by color. Similarly, conductivity sorters have proven successful in nickel ores with a high enough metal content to produce a strong conductivity signal.

In optical sorting applications there needs to be sufficient correlation between color differences and mineral composition to be able to attain a decent separation. Sensor sorting is based on single particle evaluation of the information the sensor detects.

Experience has shown that in many potential sorting applications there is either an insufficient correlation between color differences and mineral composition, or the old conductivity sensors are not suitable to separate low-grade ores.

New sensor systems have been developed in recent years which exploit material properties such as electrical conductivity, magnetization, molecular structure, thermal conductivity etc. These sensor systems must work in non-contact conditions, sufficiently fast and with a good particle resolution in order to be suitable for single-item sorting. All these sensors are working within the electromagnetic spectrum where the characteristics of various wavelengths and hence variable energy levels are utilized.

The schedule below presents the range of sensors currently used as well as the new methods which show promise for sorting in the mineral industry.

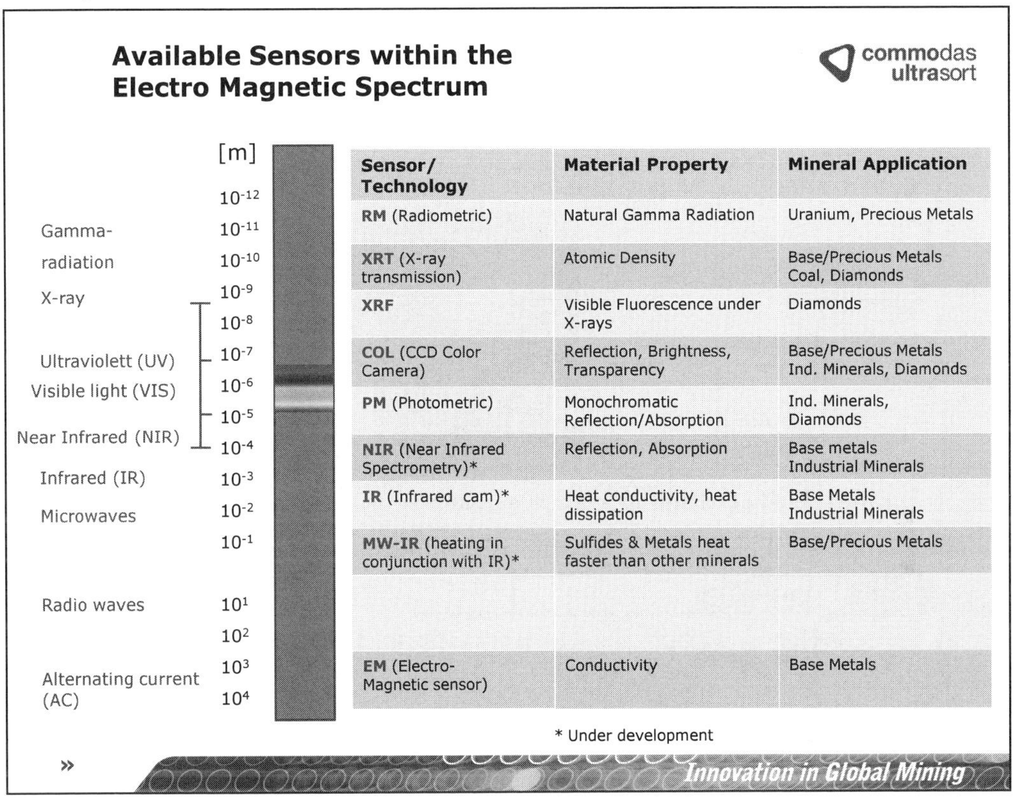


| Sensor/ Technology | Material Property | Mineral Application |
|---|---|---|
| **RM** (Radiometric) | Natural Gamma Radiation | Uranium, Precious Metals |
| **XRT** (X-ray transmission) | Atomic Density | Base/Precious Metals Coal, Diamonds |
| **XRF** | Visible Fluorescence under X-rays | Diamonds |
| **COL** (CCD Color Camera) | Reflection, Brightness, Transparency | Base/Precious Metals Ind. Minerals, Diamonds |
| **PM** (Photometric) | Monochromatic Reflection/Absorption | Ind. Minerals, Diamonds |
| **NIR** (Near Infrared Spectrometry)* | Reflection, Absorption | Base metals Industrial Minerals |
| **IR** (Infrared cam)* | Heat conductivity, heat dissipation | Base Metals Industrial Minerals |
| **MW-IR** (heating in conjunction with IR)* | Sulfides & Metals heat faster than other minerals | Base/Precious Metals |
| **EM** (Electro-Magnetic sensor) | Conductivity | Base Metals |

**Figure 2. Sensors within the Electro Magnetic Spectrum**

The sensors/methods displayed in Figure 2 that can be categorized as generally accepted and 'proven' technology includes the following:

- photometric sensors (PM)
- color line scan cameras (COL)
- visible fluorescence under x-ray (XRF).
- radiometric (RM)
- electromagnetic (EM)

The newest sensor types applied to minerals include:

- XRT sorters which have been introduced fairly recently to a handful of mineral processing applications such as coal, diamonds and base metals

- NIR sorters are widely applied in recycling. Recent test work has shown some promising results for mineral sorting.

Sensor technologies under research and development

- Microwave-Infrared (MW-IR) sorting shows promise, but it is still in the experimental stage of development.

X-ray fluorescence spectrometry (XRF-S), not shown on this chart is the least developed for mining and mineral processing. However, it has potential in qualitative and quantitative material identification, as is known from laboratory analytical work and a handful of sorters in the recycling industry.

Development in recent years has not only focused different sensors but also equipment design to make these machines robust and reliable to operate efficiently in a tough mining environment.

## SENSORS AND MEASUREMENT PRINCIPLES USED IN THE MINERAL INDUSTRY FOR SORTING

Over the past 20 years developments of the sensors have gone hand in hand with the advancements achieved in different mineral sorting applications. The need for new mineral sorting applications has fueled the drive to search and improve sensors to fit into sorting machines. The following describes the different types of sensors and the related latest applications achieved in the mineral processing industry.

## RADIOMETRIC SORTERS (RM)

The primary use of these sorters is to upgrade ROM uranium type ores.

### Radiometric Sensing

- Detectors are used to measure the natural gamma radiation of the uranium ore
- Detectors consist of a NaI scintillator and a photomultiplier tube (PMT)
- There is a direct correlation between number of counts (gamma rays) and grade of uranium ore

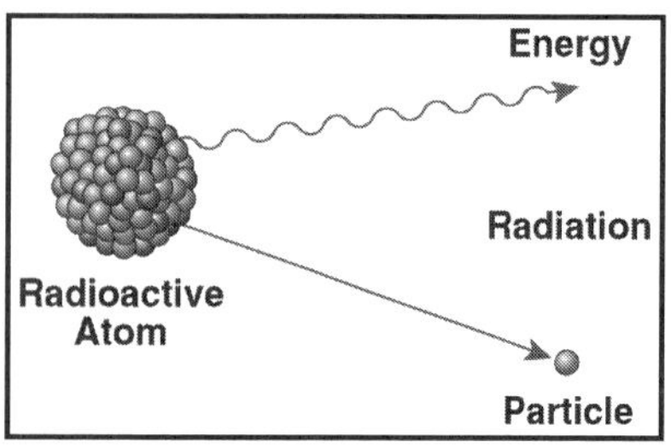

**Figure 3. Radioactive particle and NaI scintillator and a photomultiplier tube (PMT)**

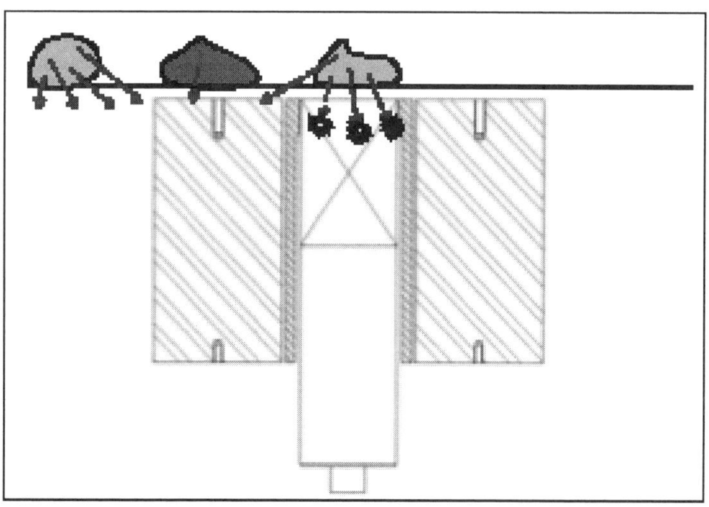

**Figure 4. Gamma counts per NaI scintilator**

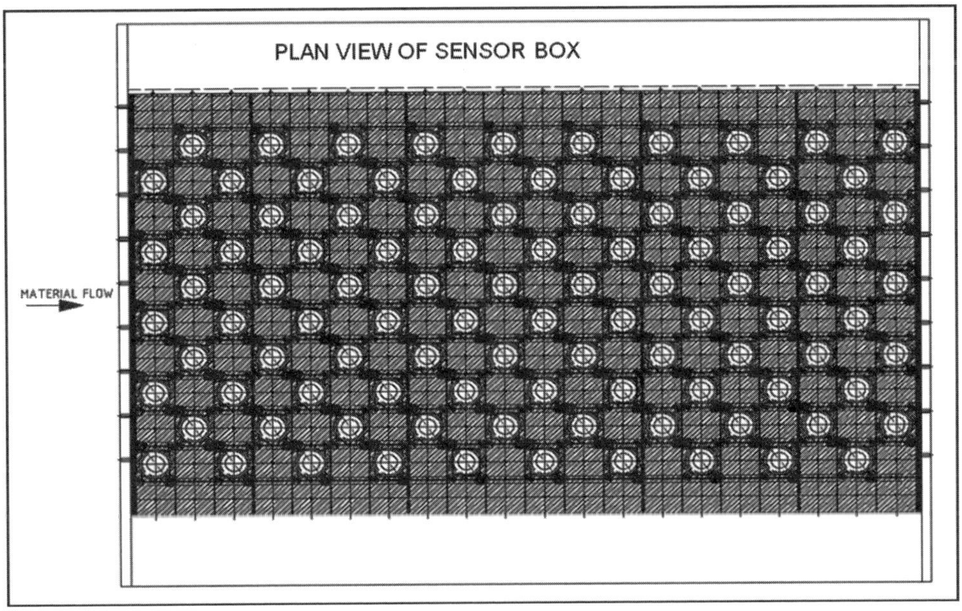


**Figure 5. Bank of NaI scintilators detect and accumulate gamma count**

The radiometric sensor is a battery of NaI scintilators that are sensitive to gamma radiation. The accumulated gamma counts of each particle travelling over these sensors results in accurate differentiation between waste rock and uranium bearing rock.

**Radiometric Sorting Plant at ERA Ranger – North West Territory, Australia**

- Feed rate: 70 t/h
- Size range: -80mm +25mm
- Rejection rate: 30% to 50%
- Efficiency: 98%

**Figure 6. Radiometric sorting plant at ERA Ranger**

## X-RAY TRANSMISSION (XRT) SORTERS

The x-ray transmission method works according to the DR-XRT principle (Dual Energy X-Ray Transmission) as it is known from airport baggage inspection scanners. Commodas/UltraSort has drawn upon this basic principle and has developed a sensor system suitably adapted to sorting techniques.

### XRT Sensor

The broad-band x-ray radiation of an electrical x-ray source is applied to the material to be sorted while it is moved along the scanning area at a rate of 3m/s (see image below). The x-ray sensor system, which works like a line scan camera, picks up the x-rays penetrating the material to be sorted and converts them into digital image data. The sensor system consists of two channels, each capturing the image of material in different x-ray energy spectra. The material to be sorted attenuates the x-ray radiation received, thus decreasing the modulation amplitude of the sensor, so that these image areas appear dark.

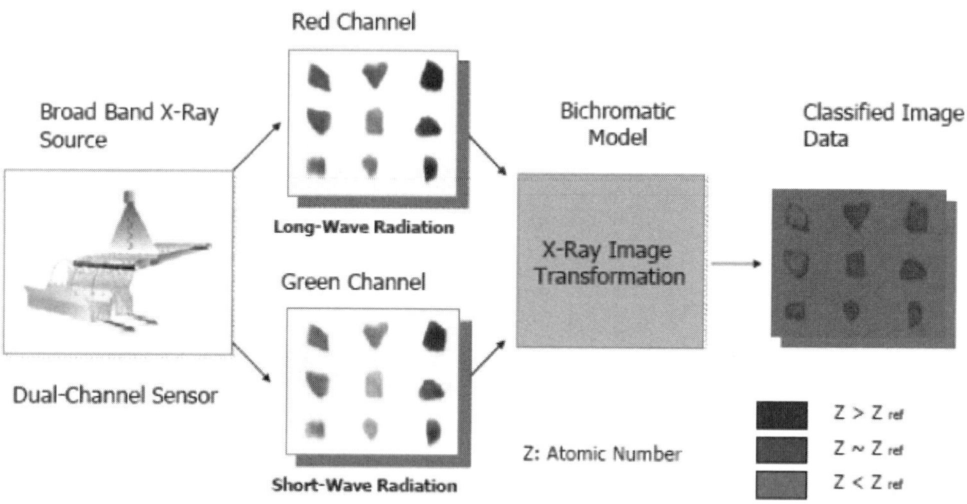


**Figure 7. XRT Sorting Principle**

Attenuation depends on both the thickness and atomic density of the material. An image transformation of the density images of dissimilar spectral ranges makes it possible to classify each pixel according to atomic density. This is accomplishable almost regardless of the thickness of the material. The example above shows pieces of iron ore with variable iron content and an either low, moderate, or high ore content (from left to right). The image on the right displays a color-encoded classification of the pixels based on densities. As the picture shows classification of the ore pieces on the basis of their iron content is feasible.

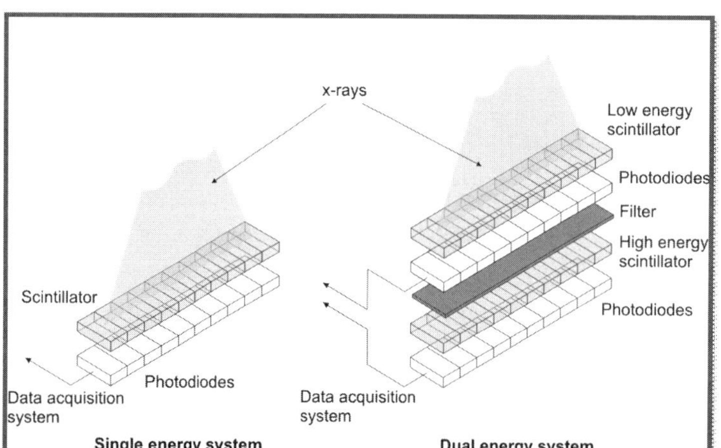


**Figure 8. XRT – the Dual-Energy X-Ray sensing principle**

The sensor's high local resolution of 0.8mm or 1.6mm (depending on the model) also permits the evaluation of particle shape, particle size, material thickness and texture of the gray-scale image (this represents inclusions of material of various densities). X-ray transmission image processing provides a highly efficient sensor system for classifying different materials/minerals.

**XRT Sorting Applications**

The beauty of XRT sorting is that it is a dry sorting process. No washing/cleaning of the feed material is required which saves water and water-treatment costs. A number of different mineral applications have proven to be very effective:

- Sulfide type base metal ores such as nickel, copper, lead, zinc
- Vein-type sulfide gold/silver ores
- Diamond sorting
  - accurate detection and separation of +4mm size diamonds
  - independent of fluorescence
  - clean concentrate
- De-shaling of coal

**XRT Application 1: Coal**

The ideal size range DE-XRT Sorters is between >1" <4" where throughputs of approximately 50 t/h will be achieved (latest tests have achieved 80 t/h at a coarse <6"). Traditionally analysis of heavy liquid separation at cut points increasing in 0.1 increments from 1.3 to 1.8 will give indications of shale/ash content vs. coal purity. Results achieved show a direct correlation between the SG of the individual particles of a typical Run-of-Mine coal feed and the x-ray image of these particles. The atomic density of coal (consisting mainly of carbon) is clearly distinguishable from Shale (consisting of Silica and other rock bearing minerals).

Line scan x-ray images were taken of the samples reporting to the -1.3 and +1.7 SG classes as these provided the best examples of pure coal and pure shale.

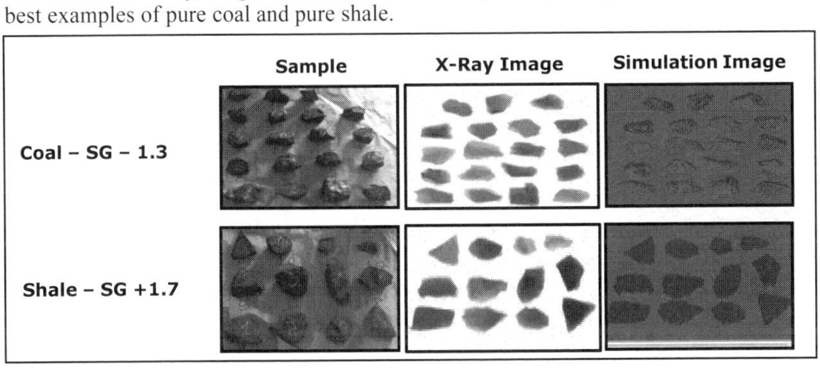


**Figure 9. XRT line scan images of coal and shale with simulation analysis**

The partition curves for the separation of coal from shale are presented below:

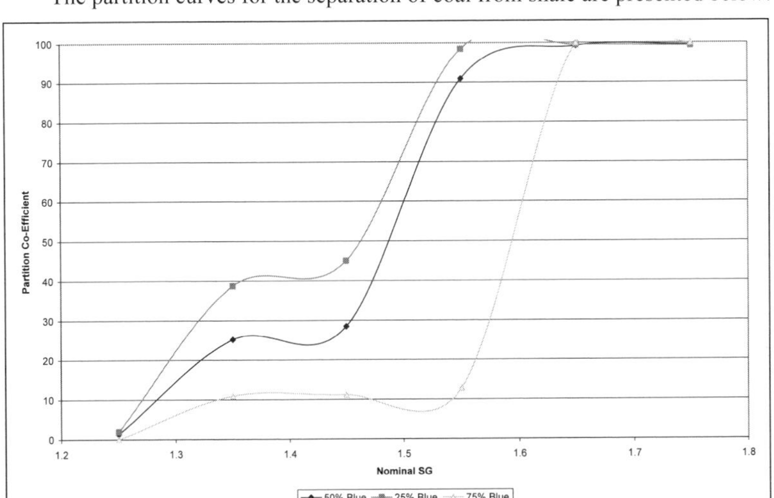


**Figure 10. Partition curves for three x-ray sorting tests**

The cut point of the separation is controlled by the sensitivity of the sorting program. For example, at a sensitivity of 75% blue, a cut point of approximately1.6 is attained. This cut is reduced to approximately 1.5 and 1.45 for sensitivities of 50% and 25% respectively. As the sensitivity is increased, so the efficiency of the separation is improved.

| | | Product Distribution | | | Ash Content | | |
|---|---|---|---|---|---|---|---|
| S.G Class | Nominal SG | Accept Stream (Coal) | Reject Stream (Shale) | Reconsituted Feed | Accept Stream (Coal) | Reject Stream (Shale) | Feed (Calc) |
| -1.3+1.2 | 1.25 | 0.63 | 0.00 | 0.63 | 8.20 | 0.00 | 8.20 |
| -1.4+1.3 | 1.35 | 19.57 | 0.08 | 19.65 | 10.2 | 36.20 | 10.30 |
| -1.5+1.4 | 1.45 | 21.18 | 5.07 | 26.25 | 12.6 | 52.20 | 20.25 |
| -1.6+1.5 | 1.55 | 40.63 | 7.10 | 47.73 | 22.1 | 50.20 | 26.28 |
| -1.7+1.6 | 1.65 | 0.01 | 5.08 | 5.09 | 8.5 | 41.20 | 41.13 |
| -1.8+1.7 | 1.75 | 0.00 | 0.65 | 0.65 | 0.0 | 54.60 | 54.53 |
| | | 82.02 | 17.97 | 100.00 | 16.70 | 48.32 | 22.38 |

**Figure 11. XRT sorting results**

The results show that the coal was upgraded from 22.38% ash to 16.70% ash. This can be further improved upon by reducing sensitivity further.

**XRT Application 2: Diamonds**

Just as coal, the diamonds are made up of carbon atoms. Thus the x-ray image of a piece of coal and a diamond is very similar. The atomic density differences between a diamond (carbon) and non-diamond waste rock (consisting of a mix of Si, Fe, Ca etc) are vast. The XRT images shown below clearly show these differences.

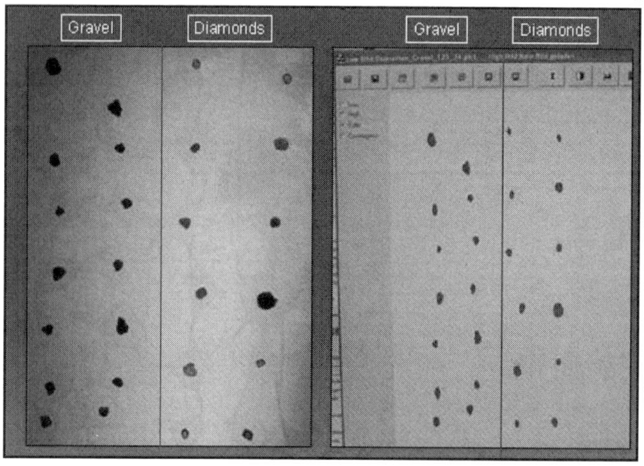


**Figure 12. XRT sorting results**

The ease with which this sensor technology identifies and differentiates diamonds from gravel is outstanding. The low yield of concentrate with little misplaced waste pieces will improve the final hand sorting of diamonds where less waste material needs to be handled. With the throughput of a feed size -8mm +4mm is 6 to 8 t/h on a standard 600mm wide sorter makes this an attractive piece of equipment for diamond recovery. The only drawback at this stage is that this sensor is not yet able to identify and treat the fine size range -4mm +1.5mm.

**X-RAY FLOURESCENCE (XRF) PHOTOMETRIC PRINCIPLE**

X-ray fluorescence sorting is well entrenched throughout the diamond industry for many years. These sorters range from high throughput sorters to single-particle sorters treating the final diamond concentrate. There are large diamond sorters, dry and wet sorting arrangements. All of these make use of the fluorescing properties of diamonds when they are exposed to x-rays.

| Type | Material Handling | Sensor | Resolut./ channels | Width [mm] | Size range [mm] | Capacity [t/h] |
|---|---|---|---|---|---|---|
| GEM Fine XRF | Drum | XRF/PMT | 2 X 6 | 2 x 300 | 0,8..12 | Up to 12 |
| GEM Mid XRF | Belt | XRF/PMT | 12 | 600 | 1..32 | Up to 15 |
| GEM Large XRF | Belt | XRF/PMT | 20 | 800 | 25..65 | Up to 80 |
| GEM SPS XRF | Particle Singulation | XRF/PMT | 4 | N/A | 1..30 | N/A |
| GEM Basic XRF | Wet/Dry Chute | XRF/PMT | 3 | 150 | 1..30 | 4 |

**Figure 13. Commodas/UltraSort range of XRF Sorters**

**XRF – The Photometric Sensing Principle**

The PMT is a glass vacuum cathode tube fitted with dynodes (anodes) arranged in cascade positions. Inident photons strike the cathode layer and cuse free electrons (photoelectric effect). The electrons are accelerated in the electric field, hit the dynode and cause 3 to 5 secondary electrons. A PMT has typically 10 dynodes. The signal produced by the incident light is typically multiplied by 108.

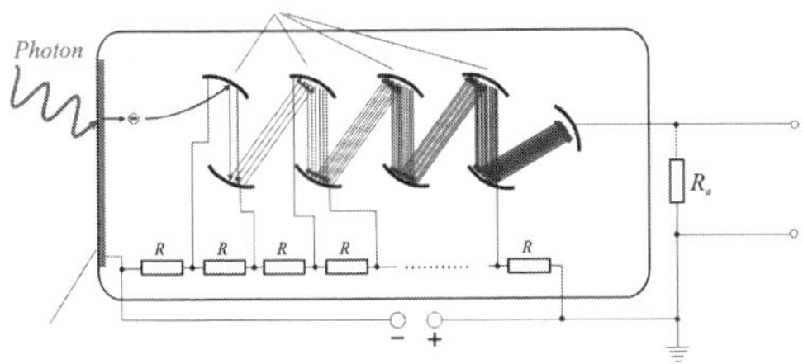


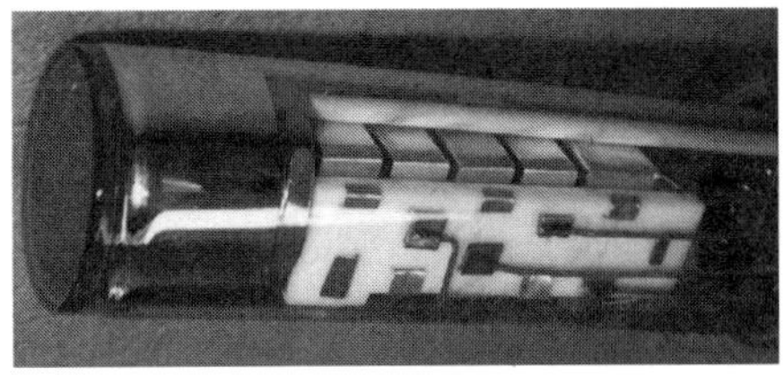

**Figure 14. Photo Multiplier Tube (PMT)**

Various models and makes of XRF diamond sorters are supplied to the market. One example shown below is the UltraSort GEM Mid XRF sorter.

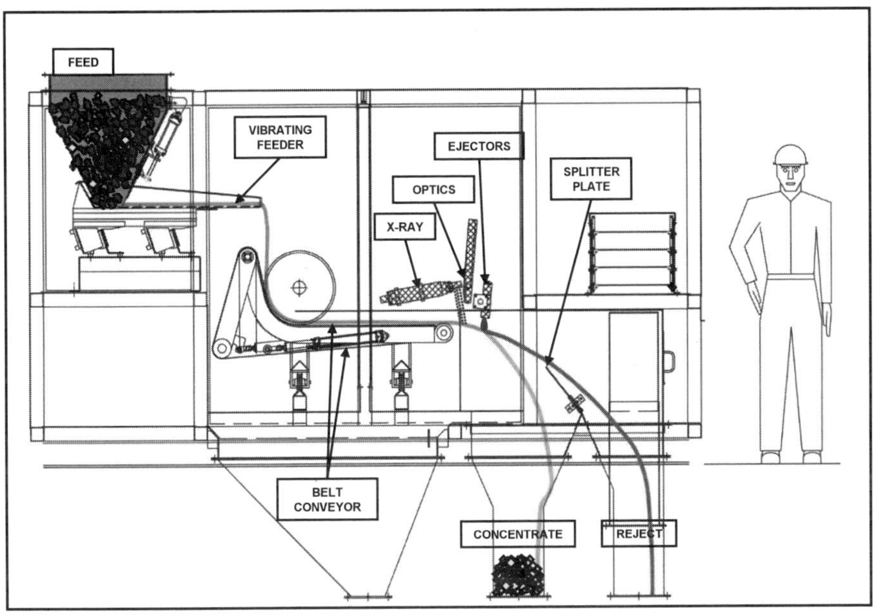


**Figure 15. The UltraSort GEM Mid XRF Sorter**

## COLOR SORTERS (COL)

Optical sorters using line-scan cameras are widely used in industrial mineral applications such and calcium carbonates, quartzite or magnesium. These are mainly used in the -80mm +20mm size range. In recent years more robust sorters have been developed to handle large size rocks of -250mm.

The Commodas/UltraSort sorting machine "PRO Primary COLOR" is designed for the particle-size range of 80 to 250mm. The machine is equipped with 2 color line scan cameras to enable a double-sided screening of the material which covers approximately 80 percent of the particle surface. The material is fed directly from a vibrofeeder to the chute causing the particle to accelerate and spread apart. At the end of the chute the material is in free-fall. This is the line-scan detection zone where bright fluorescence lamps are installed to illuminate this area.

A few centimeters below the scanning area a series of aligned of compressed-air valves are triggered to release a pulse of compressed air at the right time and at the right position to the deflect single particles over the splitter bar in the separation chamber. The reject and accept conveyors below the separation take away the sorted material.

The processing height of this machine, from the top of the chute to the discharge outlets is 4.5 meters. The machine achieves a throughput rates between 200 and 250 tons/h.

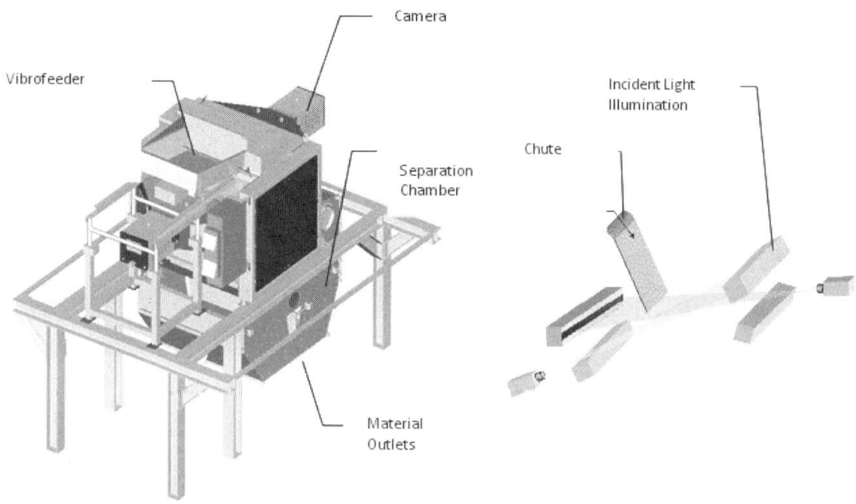


**Figure 16. Commodas/UltraSort sorting machine "PRO Primary COLOR"**

**Color Sorting Application 1. Platinum**
At Anglo Platinum's Waterval section a sorting plant was installed to treat ROM UG2 ore.

- Size range – 250mm + 60mm
- Feed rate 200 to 300 t/h
- Rejection ratio 20 to 60% waste

**Figure 15. ROM sorting at Anglo Platinum**

**Color Sorting Application 2. Gold Waste Rock Dumps**
During 2003/4 a pilot plant was operated at Kloof Gold Mine recovering all the gold bearing pieces of ore.

**Figure 16. Waste dump sorting at Kloof Gold Mine using PRO Secondary Color sorter**

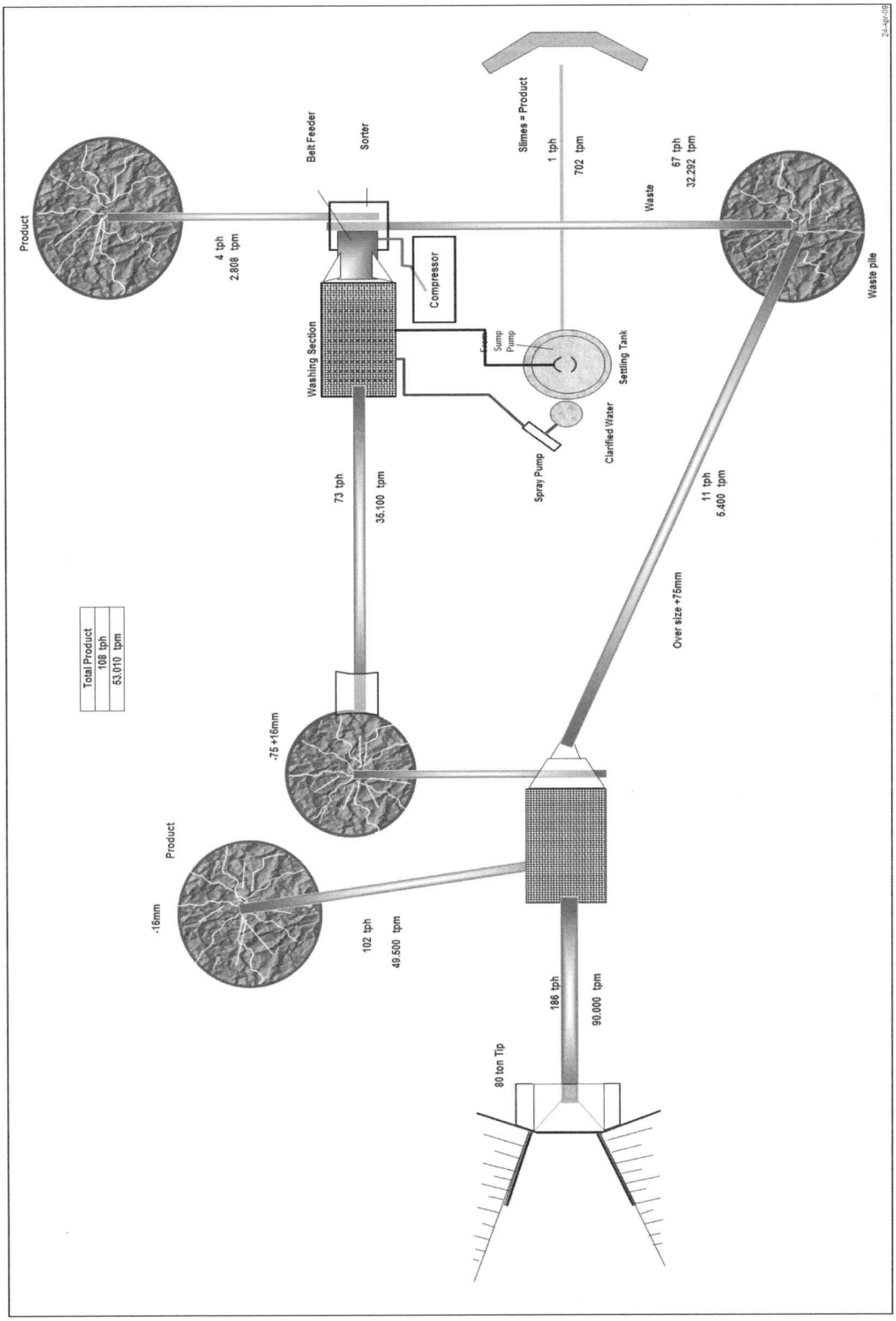


**Figure 17. Block diagram of screening and sorting plant at Kloof Gold Mine**

**Color Sorting Application 3. Diamonds**

Color sorting is a fairly new approach to sorting diamonds. The sorter is set to detect any transparent material i.e. free diamonds. This has the distinct advantage that the detection is not reliant on the fluorescence of the diamond like in the XRF approach. Not all diamonds fluoresce, particularly the purer diamonds. With the optical sorting approach one can generally say that the purer the diamond, the more transparent it is, the better the detection and recovery.

A number of color sorters are in operation in the South African diamond industry. Commodas' GEM Mid COLOR sorter achieves high through-put rates of 10 to 14 t/h in the fine -8mm +2mm size range.

**Figure 18. Containerized diamond sorter treating recovery tailings dump material – Trisano Mine, South Africa**

**Figure 19. Commodas GEM Mid COLOR sorter sorting at 12 t/h - size range -8mm +2mm**

## NEAR-INFRARED (NIR) SPECTROSCOPY

NIR spectroscopy has gained increasing interest, especially in process control applications. NIR sorting has its roots in the recycling industry with more than 1500 NIRS sorters installed. Portable near-infrared spectrometers have been used in the mining industry to evaluate deposits and to log drill cores. The material to be analyzed is illuminated with an infrared light source. The spectrum of the reflected radiation displays strong decreases in intensity at characteristic wavelengths which map particular minerals. Commodas has adapted this technology for mineral sorting which requires a higher processing rate, a higher spatial resolution and extended frequency range. Some novel arrangements have been implemented to extend the measurement range to include wavelengths from 1μm and 2.5μm with a resolution of 10nm.

The spectrometer works, as shown in the schematic figure below. It has two measurement chambers, each of which record 16 channels in two different wavelengths ranges. An optical splitter ensures that the incident radiation is distributed to the two measurement chambers. A rotating polygonal mirror scans the passing sample and an image is created for analysis. The existing measurement rate of 40,000 spectra per second thus produces a symmetrical pixel resolution of 4.6mm x4.6mm at an interval velocity of 2.8m/s and a scan width of 500 pixels. Additionally, lining up several scanners in a row allows an improved signal quality over a feed width of 2m. These innovations are based on the many years of experience in applying the technology in the recycling industry. The reader is referred to the paper "Near-Infrared Sensor-Based Sorting in the Minerals Industry" by H. Wotruba,M. Robben, D. Balthasar, 2009.

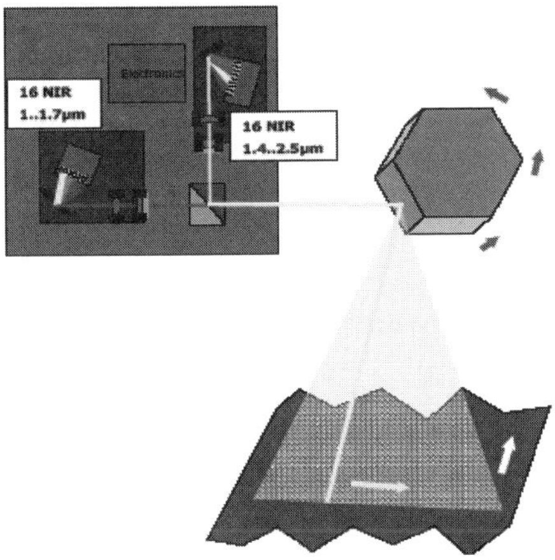


**Figure 20. Schematic of a NIR sensor**

**NIR Example 1: Talcum, Quartz and Carbonate**

Separation of talcum from waste material, namely carbonate and quartz cannot be done with optical sensors since there is no discernible color difference. However, the NIR spectra have strong absorption features that are differentiable and repeatable for these rock types. Representative spectra of the different rock types can be seen below. Talcum has strong diagnostic features and therefore it is possible to sort based on those results, especially in higher wavelengths of the spectrum (from 2200 till 2600 nm).

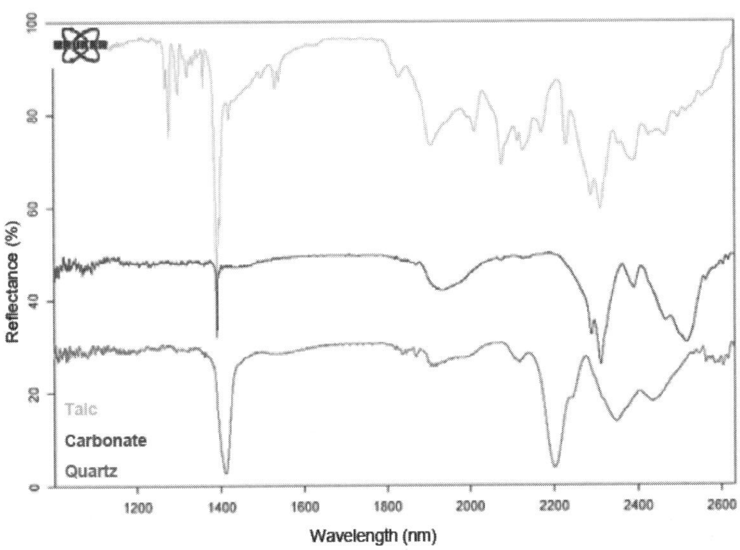


**Figure 21. Reflectance spectra of talcum, carbonate and quartz**

## MICROWAVE-INFRARED (MW/IR)

Most sulfidic base metal ores contain only a small percentage of base metals. The sulfide ores are not readily recognizable on the rock surfaces, and the surface material may not be indicative of grade as ore particles may lie in the interior of an individual rock. This suggests that surface sensing technologies are not likely to be useful for sorting sulfide ores, especially low grade sulfide ores.

The principle behind microwave-infrared sorting is straightforward. The samples are warmed with a microwave source, which heats up the conductive component of the rocks. An IR-camera records the hotspots on each particle and measures the rock's temperature. The hot spots identify where sulfide minerals are embedded in the rocks. The amount of heating energy required can be small since the current infrared sensors can detect very small differences in temperature, and the measurement does not require contact with the rocks. Sulfides, carbon and graphite have been reported as responding to microwave radiation. The reader is referred to the paper "Upgrading Molybdenite Ores Between Mine and Mill" by Van Weert and Kondos presented at the CMP 2009.

**ELECTRO-MAGNETIC (EM)**

The EM Sorters has been successfully operated on material from major Nickel producers and uses field proven technology with some innovative adaptations to accurately sort Nickel Sulfide and other Ores. The new system works on a basis of measuring changes in the fields generated by a matrix of very high frequency coils. This new patented technology means that these new machines can now sort Ores that are amenable to EM detection in a Random Stream, thus significantly increasing feed rates through each machine. These machines can now sort material at down to 6mm. The EM sorting machine has proved to be successful on ROM, Dumps and on Smelter slag. Valuable metals can now be recovered from the many dumps, slag heaps and low yield deposits around the world. This series of high speed machines is designed for simple operation and for use in remote areas. The electronics, optical and EM systems are reliable, fast and easy to maintain.

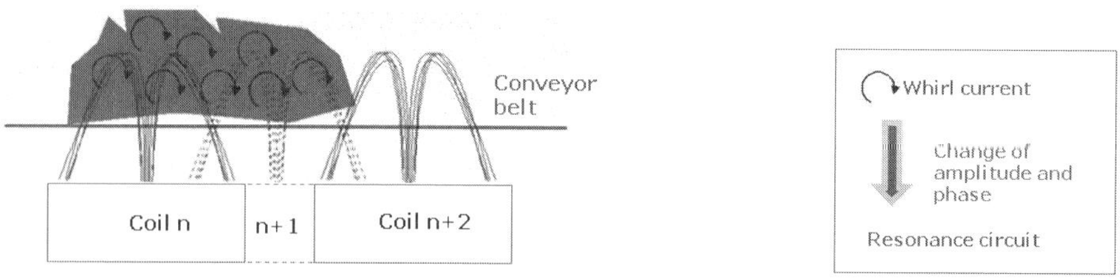


**Figure 22. High frequency coils measuring changes magnetic fields**

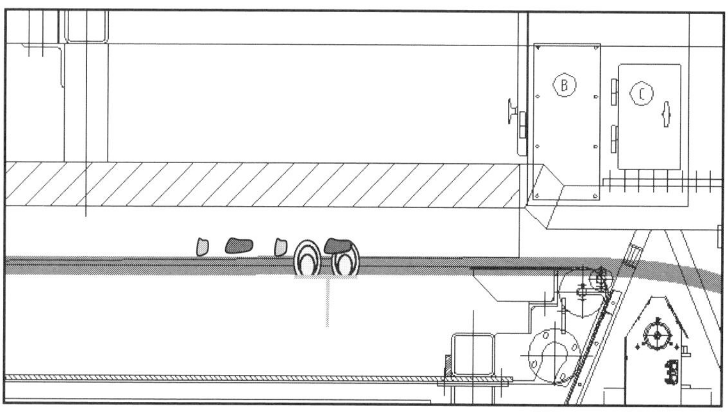

**Figure 22. EM Sensors detect variations in both frequency and amplitude of signal**

**EM Sorting Application - Nickel**

At Consolidated Minerals Kambalda (WA) the nickel ore sorting plant features the Commodas/UltraSort ROM Secondary EM Sorter:

- Size range – 80mm + 25mm
- Feed rate 60 t/h
- Rejection ratio 40% waste

**Figure 23. Nickel Ore Sorting at Consolidated Minerals Kambalda (WA)**

## CONCLUSION

In today's and in the future's mining and mineral processing industry we face increasing challenges such as lower grade ore reserves, high energy cost, water shortages and increasing environmental legislature. Sensor based sorting is today a very significant innovation in mineral processing technology and up-front beneficiation. However, there are new technologies that are only at the beginning of their market introduction. These novel sensor technologies are necessary to extend the application of sensor-based sorting and develop applications that are not yet economically viable. To reap the rewards possible from these new technologies, mining and processing companies must be aware of the range of possible sorting methods that they might use and should be doing testing at an early stage in their mineral projects to assess their potential. The uptake of existing sorter technology can improve the economics of mines today.

By placing a sorter between the mine and the milled will provide numerous benefits when not having to spend effort and energy on crushing, grinding and treating barren material. Mill the ounces, not the waste!

## REFERENCES

Harbeck H., Kroog H. New Developments in Sensor Based Sorting, *Montan University Loeben, Austria, January 18, 2008*

Strydom H., von Ketelhodt L. Dual-Energy X-Ray Sorting for: Coal/Torbanite Separation and Coal De-Shaling, *Fossil Fuel Foundation – Coal Indaba 2009, Johannesburg March 11, 2009*

Wotruba H., Robben M., Balthasar D. Near-Infrared Sensor-Based Sorting in the Minerals Industry, *RWTH 2009 Aachen, Germany,*

von Ketelhodt L. Sensor Based Sorting Beneficiation Examples in Gold Mining Applications, *SME Mineral Processing Division Meeting 2009 – Colorado Spring, May 1, 2009*

van Weert G. and Kondos P., Upgrading Molybdenite Ores Between Mine and Mill, *41st Annual Meeting of the Canadian Mineral Processors 2009, Ottawa, Ontario, Canada.*

Bartram K. and Kowalczyk M. New Developments in Sensor Based Sorting in Mining, *CIM Conference and Exhibition May 200, Toronto, Canada.*

## SELECTED READINGS

de Jong T.P.R., DE-XRT Sorting of Coal, Sensor Based Sorting *2006, Aachen, Germany.*

van Weert G. and Kowalczyk M. Mill feed: by 8 m ore block assay, or by 8 cm rock sensing? Sorting Technology can fill the gap between mine and mill, *SME Annual Meeting 2008, Denver, Colorado, USA*

Wotruba H., Sensor Sorting Technology – is the minerals industry missing a chance, *IMPC 2006, Istanbul, Turkey.*

Bergmann C. Optical Sorting of Gold Waste Dumps, *Mine Metallurgical Managers Association of South Africa, April 2004*

# REAL TIME INFORMATION MANAGEMENT INFRASTRUCTURE—COLLABORATION AT MINE-MILL FOR ASSET OPTIMIZATION

Osvaldo A. Bascur and Ales Soudek

OSIsoft, Inc., Houston, TX and San Leandro, CA

## ABSTRACT

Recent advances in industrial automation require a new approach to get plants reducing their start up times and adapting to the varying ore types. Integration of many subsystems is a requirement for improved operational management in metallurgical complexes operations.

For example, drastic changes in ore type at the mine require a proactive evaluation for the optimal crushing requirements for improved leaching of the ore. Access to the internet and faster networks are assisting in providing high visibility of the operations. Gathering the information for real time analysis and presentations on the web is changing the way that we operate and collaborate. The implementation of competences centers is now feasible to collaborate at the enterprise level.

This paper will detail some innovative ways of organizing the information for improved overall performance. The integration of the information enables the calculation of indices to track in details the operations and equipment availability. Examples in the mineral processing will be presented. An example of a metallurgical mass balance is proposed. The major benefits are increased overall process equipment effectiveness, reduction of organic losses and improved energy and quality management of the cathodes.

The use of the portal for collaboration in the large metallurgical complexes will be discussed.

## WHERE ARE THE OPPORTUNITIES

Increasing competition in the global market place has forced companies to seek new ways to achieve cost-effective production. Previous projects emphasized production rather than asset availability and reliability. Cutting maintenance budgets was often seen as a fast easy way to reduce production costs, but it is a short-term strategy.

Preventive maintenance combined with reliability analysis provides large opportunities for simultaneous cost reduction and productivity improvements. Until now, predictive maintenance has been a good concept that could not really be implemented because of low investment and elementary tools. Future systems must emphasize repairing root causes of failures rather than just the results. This provides greater equipment availability while simultaneously improving product quality.

The relationship between losses and equipment effectiveness parallels production quality and equipment availability. The Overall Production Effectiveness (OPE) is a measurement of equipment and process productivity. OPE is part of a total productive manufacturing (TPM) methodology to improve equipment productivity. The elements of TPM strategy include cross functional teams to improve equipment uptime, rigorous preventive maintenance programs, improved maintenance-operations management efficiency, better process and equipment training, and providing information to help develop better next generation equipment.

There are developments in both extending both process capabilities as well as equipment availability. Equipment problems need to be identified quickly; only then can they be solved. Most companies already have reams of data about their equipment coming from:

- Maintenance Management systems
- Cost Systems
- Predictive Systems
- Production Systems
- Manufacturer specifications and reliability data

What is needed is not necessarily more data (the problem is there is too much of it), but an environment that simplifies integration of the data with tools available to understand and analyze it.

Asset optimization seeks improved operating practices through the use of process analysis and diagnostic monitoring to notify operations and maintenance systems of quality deviations and to permit further improvements. Asset optimization involves the manipulation of real time process and equipment status to improve performance, equipment availability and overall process effectiveness.

The strategies behind predictive maintenance and asset optimization have been around for more than a decade, but the major drawbacks have been traditional management of the collected information and the limits of the typical plant organization. Usually, independent functions and islands of automation have precluded their implementation.

The next section reviews the advances in technology that enable a simplified environment to close the loop. The closing of the loop is at the industrial desktop within an analysis framework. This graphical, adaptable user environment promotes continuous improvement, provides tools and facilities to help the user analyze and make discoveries about the plant and business processes, and most importantly, helps the user to implement the findings. In a nutshell, it promotes both continuous improvement and innovation. (Bascur and Kennedy, 1995,1998).

## EQUIPMENT DEGRADATION AND FAILURE PATTERNS

As the performance of the belt conveyors, crushers, mills, pumps, compressors, thickeners, and other mineral processing equipment deteriorates over time, equipment efficiency decreases, power consumption goes up, throughput is reduced and operating costs rise. Process and equipment performance event monitoring using desktop computers and remote access via the Internet offers an effective way to offset that toll.

Condition based monitoring has been decoupled in six different patterns (Moubray, 1997). Studies revealed that 68% of all failures were attributed to pattern F, while only 5% of the failures were attributed to pattern B. Pattern F is defined as: at the start, high rate of failure, and then consistent operation with very slow increase of conditional based failure. Pattern B is defined as: at the start, stable, followed by continued stability, then a wear-out period. Moubray feels this study is substantial enough to translate across industries.

Djuric, 2002, has shown that a new failure paradigm can be used for conditioned based equipment monitoring. Based on the asset condition curve (or equipment degradation in time), he defines several stages. A failure condition starts early in the curve. Between this early start and the old definition (equipment broken) a time interval between **Potential Failure and Functionally failed** (P-F interval) is defined. The new definition of failure is based on equipment not performing the intended function.

This change in paradigm calls on a more general approach to use the process variables and equipment condition based data. There are many ways to transform the same data into information.

The ability to use the process plant topology to store equipment connectivity information and the use of equipment templates to store the equipment attributes, and the links to detailed asset information from the manufacturer can connect to real time process information to continuously monitor the asset under process conditions. Figure 1 shows the software infrastructure to enable people in the organization to access the necessary data based on their roles to help them to perform their jobs more effectively.

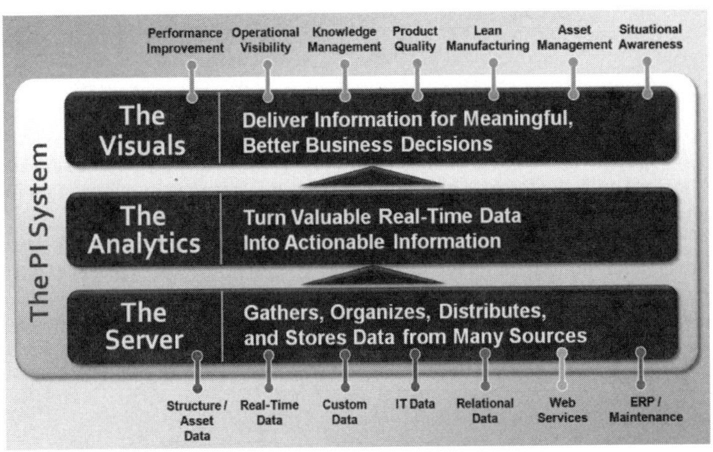


**Figure 1. Empowering people to access their information according to their roles and responsibilities**

The major requirements for such a change in culture are:

- An enterprise data server to collect and archive the information from laboratory systems, maintenance systems and control systems.
- A common analysis framework environment to access the real time data according to context.
- A common view to access the integrated information by operation, maintenance, engineering and management.

## AN ANALYSIS FRAMEWORK AND THE INDUSTRIAL DESKTOP

The three-tier architecture is a logical, not physical, computer architecture that was a key software development, without which large client/server systems would not be possible. The client server architecture may access data and methods from dissimilar sources in order to provide these applications, but it is the three-tier architecture that addresses how one can maintain client/server systems and permit them to operate reliably in a distributed environment.

PI Analysis Framework (PI AF) allows users to create templates of their assets, embed calculation logic and methodology into the templates, and create the asset structure. This provides a standardized data directory of the assets allowing for building models of the processes, maintaining the assets from a centralized library of templates, and combining PI and non-PI data in a common toolset. PI AF isolates the client applications from the data sources, allowing the continuous improvement of the underlying technologies and embedded know how, allows for creating templates of assets for easy maintenance of the assets. It also integrates the process data, the "Application Models" (plant, physical, organizational, etc.), and the scheduling of calculations and procedures, providing a scalable and evolving infrastructure for the implementation of the client applications. Plant models are time referenced, including specific business rules as defined from the "Cases."

Figure 2 shows an example of a pump created from a template. The pump template has the definitions of the attributes of a pump, like suction pressure, the logic for data access and calculation methods for items like efficiency. This ensures that every pump created from this template has the exact same attributes and calculation methods. Any changes to the template get propagated to all the existing pumps. This way the maintenance of the assets is controlled by the template to ensure standardization.

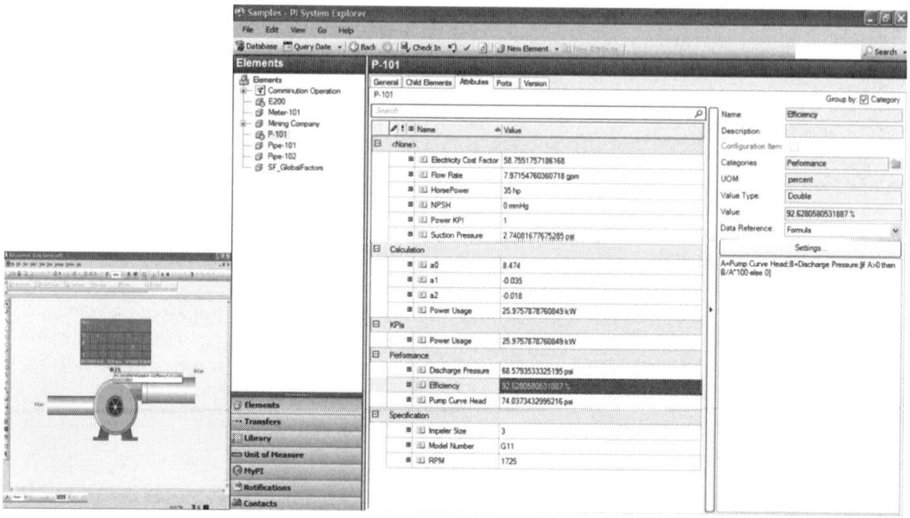


**Figure 2. Pump created from a PI AF template**

This infrastructure for the operational data management must go from a strategic decision a "Package that solves a set of identified needs" toward an infrastructure where the needs are implemented, while allowing the addition of new applications, plug-ins and modules.

Figure 3 shows an overview of the real-time information management infrastructure. A strategic decision involves the identified set of needs in addition to the framework to allow the system evolution, as new technologies arise, for the support of business and operational needs. The increasing availability of "Web-Services" (and the increasing availability of communication bandwidth) will completely change the type and scope of applications, allowing reuse between sites, support from third parties, even those that are remotely located.

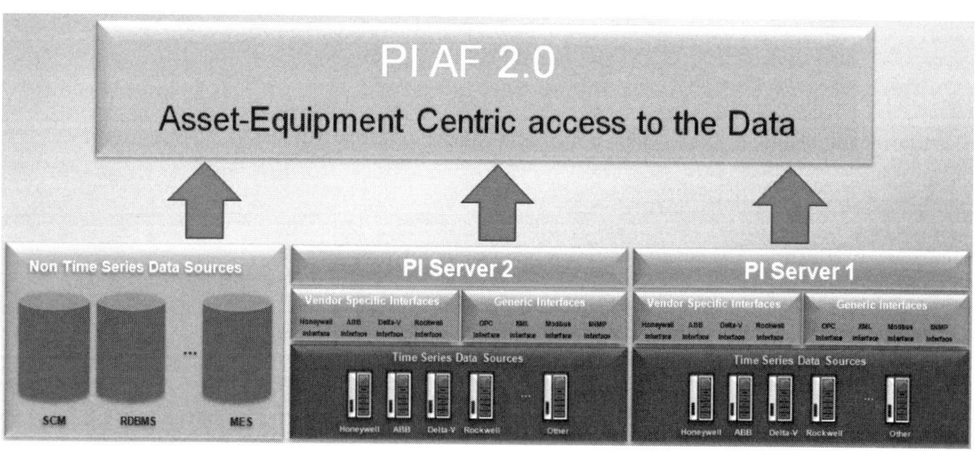


**Figure 3. Real time information management infrastructure**

From a strategic perspective, the deployment of an infrastructure for managing the operational data is a completely different scenario than the one based on applications based on a core data management system (i.e. a RDB). This infrastructure allows for new implementations as the needs arise and for the implementation of "Organization Models," which can have access to Plug-Ins, Modules and Applications.

To build the plant connectivity the basic concepts are the typical block and the process flow diagrams guidelines used in economic process evaluations, process design and process analysis. The PI AF captures the plant topology and the links to the plant data according to the objectives of the application.

With the large number of data it is not realistic to expect the users to be continually watching computer screens for changes in data or abnormal trends. The only way to manage this is to create triggers on key attributes and notify the users of the problems. PI Notifications is an add-on to the PI AF, It uses the defined asset structure and the attributes and allows the users to create triggers. The Notification infrastructure is depicted in Figure 4.

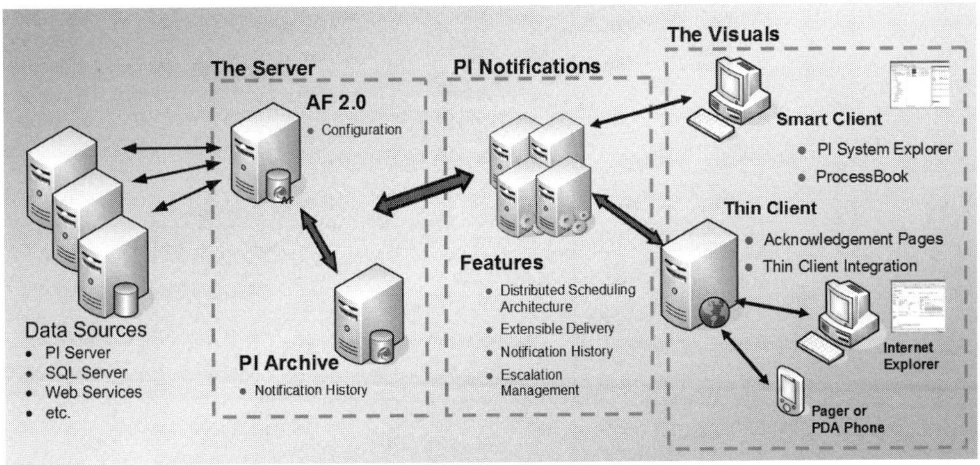


**Figure 4. Notification information management infrastructure**

## BOTTOM LINE, YIELDS AND INVENTORY MANAGEMENT

One of the biggest challenges to process plant management is the accumulation of accurate information on process operations. This information is necessary for any analysis and decision making within the plant and enterprise. Therefore, there is a requirement for meaningful, accurate and consistent data. Material balances calculated from data measured at various locations around process units, tanks inventories, stockpiles, silos, bins, and assays are useful for many purposes, such as yield accounting, on-line control, and process optimization (catalyst selections, reagent schemes, liner replacements, water management, utilities management). To achieve material balances, gross errors or anomalies in the production data must first classified, detected, and the source of the data examined.

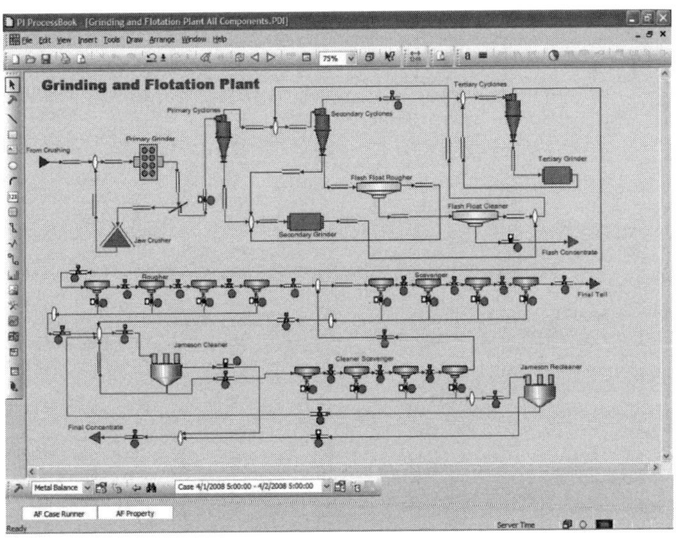


**Figure 5. Metallurgical balance on grinding and flotation plant**

Often, it is possible to calculate material balances by several independent procedures when excess measurement information, i.e., redundant data, is available. Clearly, if the data were collected without measurement errors (a theoretical condition never found in practice) all material balances calculated from redundant data would be in agreement. The real situation is that errors exist in practical measurements, so that the results of material balances determined from available optional procedures differ. Consequently, best-fit computational procedures to adjust the material balances by taking measurement errors into account can be implemented (Bascur and Soudek, 2001a, 2001b). Figure 5 shows an example of a metallurgical balance on a grinding and flotation area. The measured data is validated, gross errors eliminated; instrument faults and biases identified, water consumption calculated, and possible losses quantified.

Once the process topology is defined, the PI AF is used to define the assets and create a connectivity model. An example of the asset structure and the connectivity is shown in Figure 6.

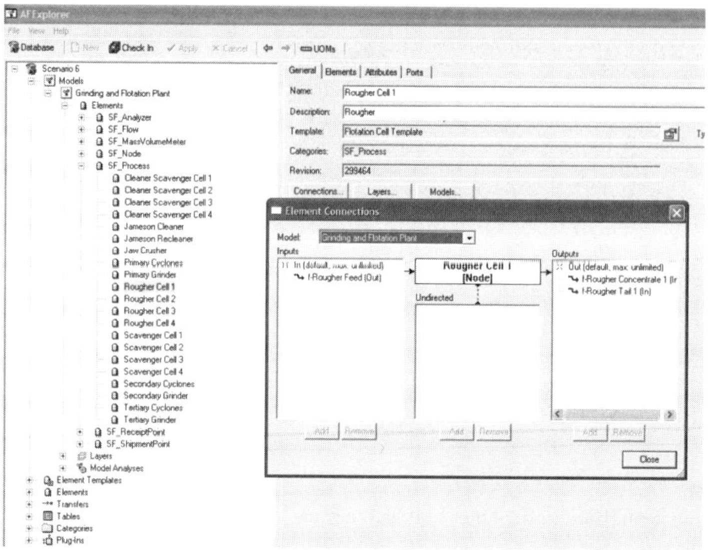


**Figure 6. Asset structure and connectivity**

The Sigmafine plug in can be used to reconcile the data from inventories, flows, and compositions by performing a mass balance. Figure 7 shows the results in a WEB environment for access by management, personnel and external resources. PI WebParts can then be used for allowing of viewing of the information in the Web.

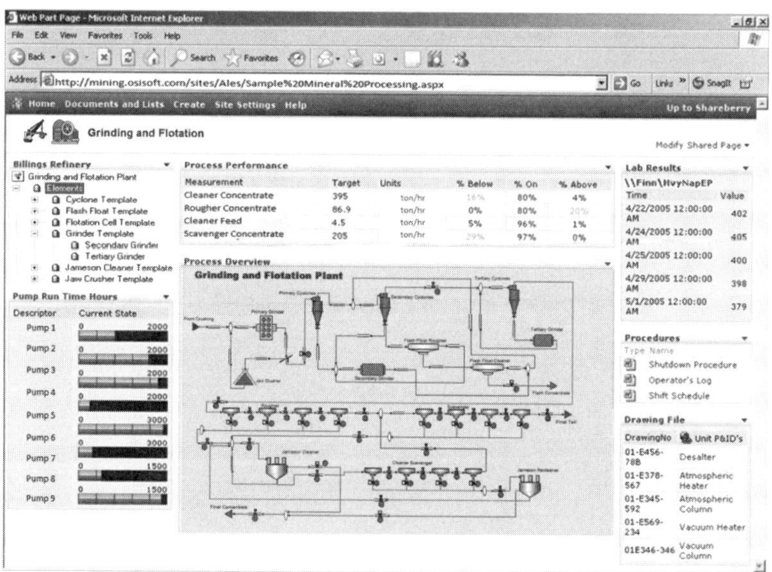


**Figure 7. Web browser showing real time performance indicators**

This integrated approach enables collaboration between operations, engineering, accounting and management to drive the organization's bottom line according to their business strategy. At the same time personnel can look for opportunities using alternative processing strategies (grinding efficiency, reagents, and blasting methods) to adapt to the changes in ore type to produce the least cost concentrates.

## PROCESS AND EQUIPMENT EVENT MONITORING

The PI AF extends the use of the traditional event management subsystem. The most common structure to build the process and equipment event monitoring includes a main module with the key performance indicators such as ore production, overall equipment availability, and specific power consumption. Sub-modules are defined for all equipment in the area (the grinding section). For each grinding mill, triggered events are defined (calculated digital sets) for operating conditions (running good, in trouble, overloaded, not running) and equipment conditions (operating, unavailable), operator comments, manual data entry, alarm status and process values. These data and events are captured based on a defined time interval or on a main event such the production of a lot or the movement of certain feed material. As such, all consumables and production records can be reported together with the process operations, laboratory data and any additional information included in the asset modules.

Transformations of data such as these add valuable information for asset management decisions. With both events and historical data available, sophisticated joined queries can be resolved rapidly for analysis of critical equipment. Human response is seldom based solely on the current value; we need to classify and look for trends and changes in the patterns. Many of the new events are simple alerts, but adding the batch event and transfer record into a structured environment and combining them with history affords unique views.

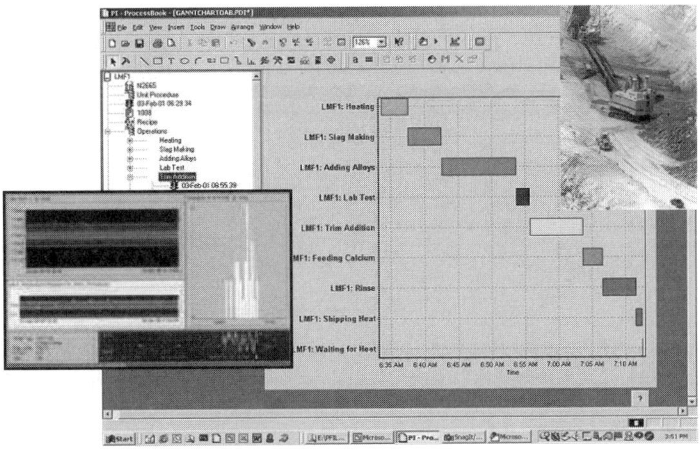


**Figure 8. An Event based query for an asset showing operating states in a Gantt Chart**

The batch event defines a set of coordinated data over a period of time, rather than data at an instance of time. Batch events are a result of rules that are used to define triggers marking the start and stop times (e.g. an outage or production run) and then cause the resulting events to be stored. For example equipment startup can be considered a batch event. Another kind event is the transfer record that includes a source and sink as well as the times. This event is used to define a high level view of material movements that are essential for reconciliation of the flow data. Examples of the data added to the historian are operations within a batch process (charging, cooling, heating, flooding), detected conditions for equipment (cavitations, fouling, high vibration, high amperage, down, running, maintenance, troubleshooting), product quality (on spec, out spec, cycling), movements (receipts, shipments, transfers), exception (process/quality alarms, collections of operator comments, collection of manually entered unmeasured data) and calculations (performance indicators). Modern systems must have much wider scope (e.g. financial and structural information) to provide these views. They also need a sophisticated computational layer to prepare the presentation.

Figure 8 shows the results of a query for the unit displayed in the tree view on the left. The tree view contains all the events such as operating conditions, manually entered data, operator comments and the associated sensor data connected to the unit. The operating events for the time interval are shown in a Gantt chart. Availability, downtime and cycle time analysis is performed to check that the standard procedures are consistent with the economic targets. A statistical control chart (SQC) to plot the quality variable to check that the process is under the statistical process control limits or within the normal bounds of the process is shown on the left hand side of the figure. SQC events can be automated to generate process or equipment alerts to notify maintenance or metallurgical operations.

Events add information needed for diagnostic actions, but special tools are required to analyze the combination of these events and real time data. Forms or query systems based on relational databases are simply not up to the task. Tools must allow the non-specialists to get all events and data by product, order numbers, from any unit in the plant without programming or writing SQL queries. A good example of combining financial and plant data is marginal economics where operations can see the margin on each order in real time before it is shipped.

The actions, rules and calculations are themselves important data and must be versioned, agreed to and shared by the entire organization.

## CONDITION BASED ANALYSIS AND NOTIFICATION

Downtime incidents are often not correctly documented by operators. Many times the most painful downtimes are those that are of short duration but occur frequently. Many of these incidents are caused by problems in the operation that could have been detected and corrective action taken to prevent the unscheduled downtime. There is a need to create rules and triggers that monitor the process in real time and send alerts to the appropriate people before the downtime incident.

This example uses the PI AF and PI Notifications for condition based analysis to monitor the engine component conditions on a dump truck. The conditions are based on rules to indicate if there has been a violation in the current design practice. When the alert is triggered it stays active until it is acknowledged by the recipient, and can be escalated if the recipient does not respond in a preset time frame.

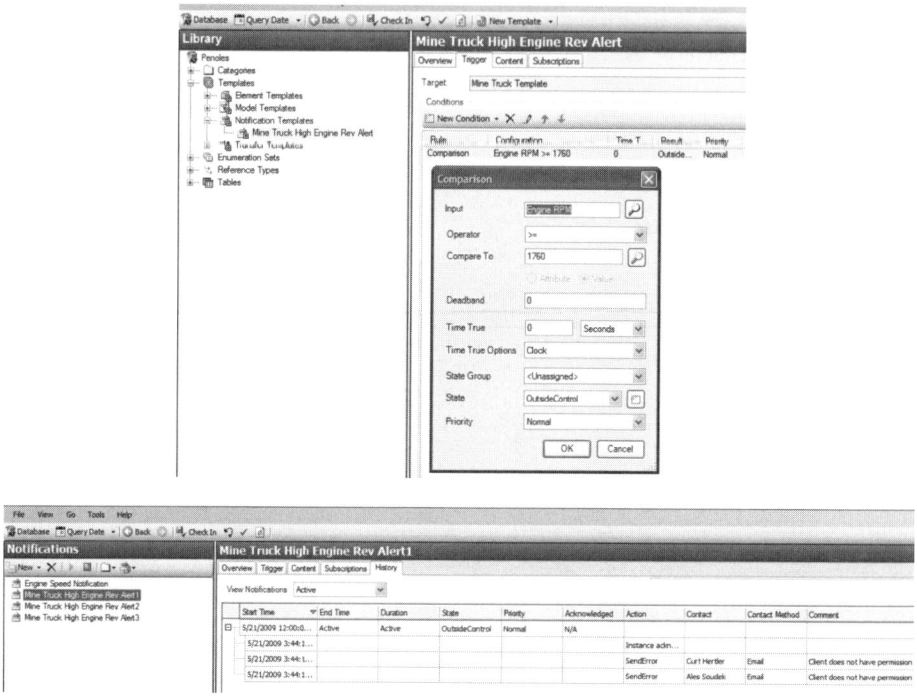


**Figure 9. Real time graphical interface and spreadsheet reporting**

Figure 9 shows an example of a trigger set for the engine speed, together with a history of the notification. The information is available to look through the history of the notification, how many times this incident happened, and the duration of the incident. This way the analysis leads to corrective action to prevent the engine speed to run at too high rpms, and prevent having to possibly do maintenance earlier than at the scheduled interval.

All the basic templates are available as objects to be reused. The applications support changes to the elements without programming.

## CONCLUSIONS

There is a critical need to integrate legacy systems into real time information management infrastructures. This environment should enable users to transform process data into actionable information. A methodology based on adding the process structure (plant topology) and knowledge of the measurement system and its strategic locations will minimize the global error based on satisfying the material balance constraints. Process topology is the key to defining the operational management database for implementation of variable cost management; yield accounting, dynamic process and equipment performance monitoring, downtime analysis and asset monitoring.

Figure 8 shows the integration of real-time information with metallurgical laboratory, maintenance systems, mining system and financial systems (Bascur, 1990, 1991). The tools have drastically evolved facilitating the integration of systems, data collection, and easy access by users and applications.

The technologies are available today to rethink plant operations and to increase the performance of current production systems. Data unification simplifies the integration of information from the process, laboratory, receipts and manual data entry. It generates high quality performance information from process data. The synergy of combining process data with transactional data provides a deeper understanding of the data for continuous improvement and innovation. It simplifies and speeds up the accounting process and detection of gross errors. It simplifies the identification of process, equipment, quality problems and opportunities. These data can be used for plant optimization, environmental reporting, costing, gross error detection, accounting, instrument management and assessment of the global measurement strategy.

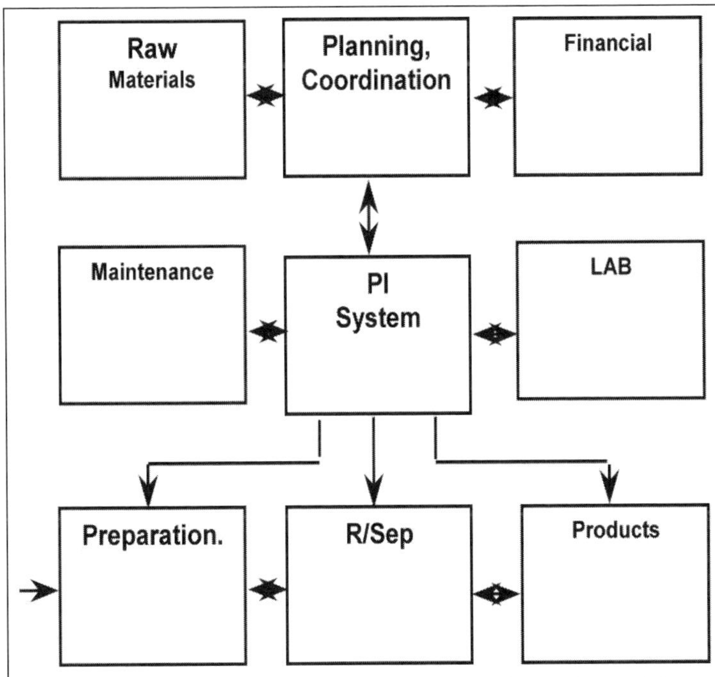


**Production**
Track Key Performance Indicators based on real time data
Aggregate production, material consumption
Generate Operational Reports
Inventory and Composition Tracking of tanks and stockpiles
Yields, Recoveries
Energy Consumptions

**Maintenance**
Delay Logs, Runtimes
Equipment Event Management
Maintenance Planning based on Equipment Condition
Equipment Alerts based on history for work orders

**Quality**
Certificates of Analysis
Batch and Composition Tracking
Statistical Process Control, Non-Linear SPC, Multivariate Statistics.

**Figure 8. Integration of metallurgical laboratory, maintenance system, mining systems, and financial systems with real time information infrastructure**

Linking process and asset information is the key to extending equipment availability and reducing operating costs. This application environment enables users to identify the best application and to continuously improve the application without disruption of the real time information infrastructure.

The unified information can be exposed into web centric environments. The Web is quickly becoming a key driver of data cleanliness (or dirtiness) as it gains ground as a way for knowledge personnel, managers, research groups, service providers and customers to input and access business information.

The key to re-engineering is linking people, business processes, strategies and the best enabling technologies. It is important to recognize that the cleaning data is a process. As such, several groups (instrumentation, maintenance, process engineering, accounting and managers) collaborate in the data unification process. This team effort should be rewarded.

The successful application should be judged on how it provides added value to the overall information system, such as new ways of storing data and events, classifying, aggregating, and combining and visualizing existing data and information.

## REFERENCES

Bascur, O.A., 1990, " Human factors and aspects in process control use," *Plant Operators Symposium Proceedings*, SME, Littleton, CO.

Bascur, O.A., 1991, Integrated Grinding/Flotation Controls and Management, *Proceedings of Copper 91*, Volume II, Ottawa, Canada, pp.411-427.

Bascur, O. A., 1993, "Bridging the Gap Between Plant Management and Process Control," *Emerging Computer Techniques for the Minerals Industry*, B.J. Scheiner et.al. Eds. SME, Littleton, CO, pp. 73-81.

Bascur, O.A., (1999). "The Industrial Desktop – Real Time Business and Process Analysis to Increase Productivity in Industrial Plants,"*IPMM99 Proceedings*, Prof. John Meech, Editor, UBC, IEEE Conference, BC, Canada.

Bascur, O.A., and Herbst, J.A, 1985, Dynamic Simulators for Training Personnel in the Control of Grinding/Flotation Systems, Automation for Mineral Resource Development, *IFAC Proceedings*, July.

Bascur, O.A. and J.P. Kennedy, 1995, "Measuring, Managing and Maximizing Performance in Industrial Plants," *XIX IMPC Proceedings*, SME, Littleton, CO.

Bascur, O.A. and J. P. Kennedy, 1996, "Industrial desktop - information technologies in metallurgical plants," *Mining Engineering*, September.

Bascur, O.A. and J.P. Kennedy, 1998, "Overall Process Effectiveness in Industrial Complexes," *Latin American Perspectives: Exploration, Mining and Processing*, Bascur, O.A., Ed., SME, Littleton, CO.

Bascur, O.A. and J.P Kennedy, 2002," Web Enabled Industrial Desktop: Increasing the Overall Process Effectiveness in Metallurgical Plants," Preprint 02-134, 2002 SME Annual Meeting, Phoenix, AZ.

Bascur, O.A, Soudek, A., 2001a, "Improving Metallurgical Performance: Data Unification and Measurement Management," *Proceedings VI SHMT*, Rio de Janeiro, pp. 748-755.

Bascur, O.A, Soudek, A., 2001b,"Integration of Mineral Processing Operations via Data Unification and Gross Error Detection," *International Autogenous and Semiautogenous Grinding Technology 2001*, Editors Barratt, D., Allan, M. and Mular, A., IV-34-47.

Bialkowski, W., 1996, "Auditing and reducing process variability in your plant," Fisher Rosemount System Users Group '96, November, Houston.

Djuric, V, 2002, "How PI Played a Key Role in Achieving Maximum Equipment Reliability at Dofasco," OSIsoft Users Conference, Monterrey, CA.

Fuenzalida, R.E., 1998," Economic Operations Management in Copper Concentrators," *Latin American Perspectives: Exploration, Mining and Processing*, Bascur, O.A., Ed., SME, Littleton, CO, pp311.

Kennedy, J.P., 1996,"Building the industrial desktop," *Chemical Engineering*, January.

Llao, M, 2001, " PI Uso y Expectativas en la Industria Minera," OSIsoft Technical Users Seminar, Buenos Aires, Argentina.

Luyt, C., 2002, "BHP-Billiton Escondida RLINK Plant Maintenance Implementation," OSIsoft Users Conference, Monterrey, CA.

Mah, R.S.H., 1990, Chemical Process Structures and Information Flows, Boston: Butterworks.

Martin, J., 1995, The Great Transition, Using the Seven Disciplines of Enterprise Engineering to Align People, Technology, and Strategy, AMACON, New York.

Moubray, J., 1997, Reliability Centered Maintenance, 2nd Edition, New York, New York. Industrial Press Inc.

Potts, D.R, 1996,"Reliability and Availability Analysis," OSI Software PlantSuite Seminar, Microsoft Advanced Technical Center, Houston, TX, February.

Rojas, J.H., and H.M. Valenzuela, 1998, " Strategic Plan for Automation and Process Management at the Chuquicamata Mine," *Latin American Perspectives: Exploration, Mining and Processing*, Bascur, O.A., Ed., SME, Littleton, CO, pp281-292.

Senge, P.M., 1990, The Fifth Discipline, The Art and Practice of the Learning Organization, Currency Doubleday, New York.

Stengl, B and R. Ematinger, 2001, SAP R/3 Plant Maintenance, Addison Wesley, New York.

Turton, R., Bailie, R., Whiting, W., Shaeiwitz, J., 1998, Analysis, Synthesis, and Design of Chemical Processes, Prentice Hall, New Jersey, 7, 34.

# GRAVITY RECOVERY OF GOLD FROM WITHIN GRINDING CIRCUITS

Ish Grewal[1], Mark Van Kleek[2], and Steven McAlister[2]

[1] Met-Solve Laboratories Inc., 8515 Eastlake Drive, Burnaby, BC, Canada
[2] Falcon Concentrators, 9663 199A Street, Langley, BC, Canada

## ABSTRACT

Gravity recovery of gold from within grinding circuits has become common in many operations around the world. Effective gravity concentration effort is a function of numerous factors including the quantity and size distribution of the Gravity-Recoverable-Gold (GRG) present, downstream processes, water balance and gravity section operating costs. This paper describes test work protocols, mathematical modeling techniques and the practical considerations that lead to the 'right' gravity circuit being installed in a given grinding circuit. The right gravity concentration effort is that which produces the best economic outcome for the operator.

## INTRODUCTION

Gold recovery using batch-type centrifugal concentrators (BCC's) from within grinding circuits is a well established practice in many operations where the presence of gravity-recoverable-gold has been identified to be present in the ore. The benefits of including gravity concentration to supplement cyanidation and flotation are well understood and accepted by the industry. Considerable research and fundamental work (Banisi, Laplante and Marois, 1991) in understanding the behaviour of gold within grinding circuits has contributed to the development of mathematical models and testing requirements for predicting gravity gold recovery. Installation practices, on the other hand, vary widely. In addition to describing effective test and mathematical modeling procedures, economic benefit calculation parameters and best installation practices are presented based on the authors' extensive operating experience.

## MODELING GRAVITY GOLD RECOVERY

Batch centrifugal gravity concentrators, which produce very high concentration ratios, have become the devices of choice for recovering gold from within grinding circuits. These are usually installed in the mill discharge, cyclone feed or the cyclone underflow streams where a fraction, typically 10-40%, of the stream is treated by the gravity circuit. The key factors affecting gravity recovery of gold from within grinding circuits are:

- the gravity-recoverable-gold (GRG) content of the ore
- the size distribution of the GRG
- the recovery efficiency of the gravity concentration circuit
- cyclone efficiency (as defined by the partition curve of GRG)
- the fraction of the circulating load treated by the gravity concentrator
- the amount of GRG converted to non-GRG in the grinding mill (grinding behaviour of the gold)
- Gravity Section availability
- gold room efficiency

A population balance model can incorporate each of these key factors. The power of the model not only helps to predict gravity gold recovery but can also assist in trouble shooting existing circuits.

### The GRG Test (Ore Characterization)

The gravity-recoverable-gold content of an ore, as obtained via a GRG test, provides a quantitative theoretical limit of gold that can be recovered using a BCC. The test itself consists of a sequential liberation via grinding followed by gravity concentration using a lab scale BCC (see Figure 1). The concentrates and the final tails products are screened and analyzed for gold by particle size class. The progressive grind approach limits the smearing of gold particles and allows for the recovery of GRG as it is liberated. The results from the test are presented as a cumulative GRG distribution as ell as GRG distribution by particle size class as shown In Figure 2.

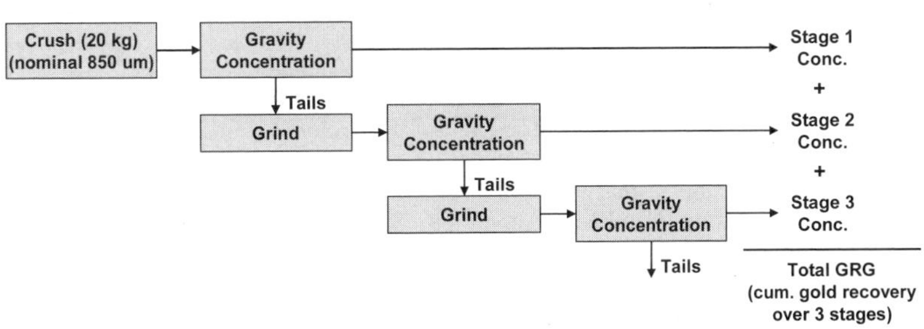


**Figure 1. Standard procedure for a GRG test using a centrifugal gravity concentrator**

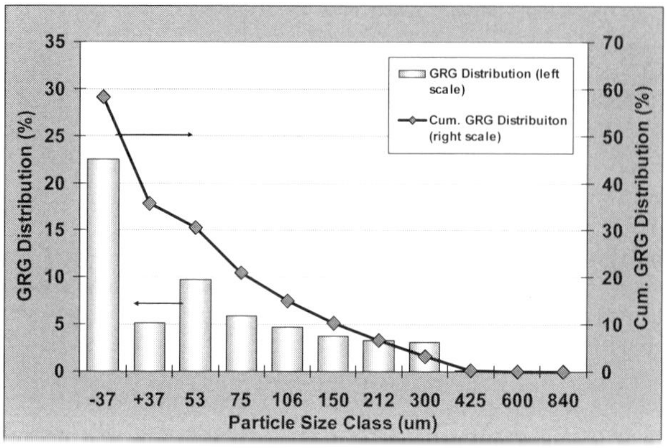


**Figure 2. Results from a GRG Test –Cumulative distribution as well as by individual particle size class**

**Cyclone Efficiency and Grinding Behaviour of Gold**

Previous studies of the behaviour of gold in grinding circuits have shown that cyclones are very effective at retaining gravity recoverable gold within the grinding circuit resulting in high circulating loads of gold (Banisi, Laplante and Marois, 1991). The retention of GRG within the grinding circuit allows for effective gravity recovery of gold from within the grinding circuit. A typical partition curve for both the ore and GRG, presented in Figure 3, illustrates the significant difference in the partition behaviour of the gold relative to the ore.

The cyclone partition curve for GRG can be generated from sampling the cyclone streams and measuring the GRG content of the cyclone overflow and underflow using lab scale BCC.

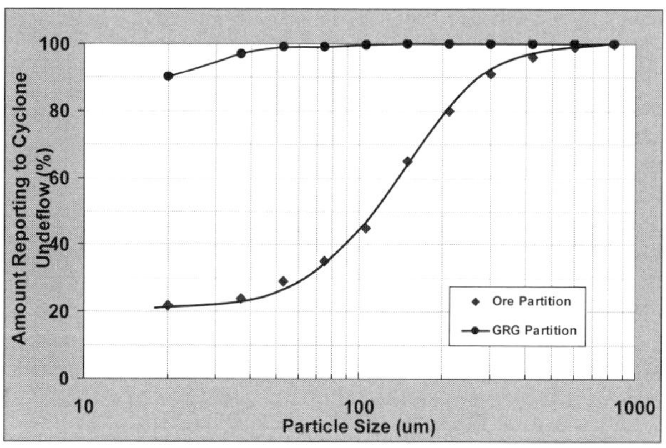


**Figure 3. Typical partition curves for ore and GRG**

Gold not only grinds six to twenty times slower than most minerals, it can even flatten and report to coarser size fractions. The long residence time of the gold within the grinding circuit can cause some of the GRG to become non-gravity recoverable due to smearing on to grinding media, mill liners and gangue minerals. This effect is minor as more than 98% of the GRG entering the ball mill remains gravity recoverable after grinding (LaPlante, 2000).

**Concentrator Efficiency and Concentrate Upgrading (Gold room efficiency)**

Batch centrifugal concentrator's provide high concentration ratios, >1000:1 typically, but operate with relatively low stage recoveries of 20-40%. They rely on the efficiency of the cyclones to retain GRG within the grinding circuit to be effective.

The concentrate produced by the primary BCC is often upgraded using a shaking table. Recently; intensive cyanide leaching of the concentrates has also become common. Specialized intensive leach units have been developed to treat concentrates generated by centrifugal gravity concentrators.

Modeling of gravity gold recovery includes the characteristic efficiency of the primary BCC and the concentrate processing method used. The data for the model inputs can be obtained from site surveys or from the manufacturers of the respective technologies.

**Modeling – incorporating tonnage treated by gravity**

A basic approach to the model incorporates the bulk GRG value or the un-weighted average of the result from the GRG test, cyclone efficiency and stage recoveries of the various centrifugal gravity concentration models. A more sophisticated approach incorporates the GRG results and the cyclone efficiency on a particle size class basis. A typical milling circuit incorporating gravity in the cyclone underflow presented in Figure 4.

A senstivity analysis from modeling provides a few important insights about the effect of various parameters on gold recovery. The single most important variable in effective recovery of gold is the efficiency of the cyclone to for retaining gold within the grinding circuit. Poor partition behaviour of the cyclone combined with upsets in the cyclone feed stream ~~can~~ is the most significant cause of reduced gravity gold recovery. This impact of cyclone inefficiency is illustrated in Figure 5 where a drop of 2% in the efficiency of the cyclone significantly reduces gravity gold recovery. It is important to note that the slope of the gravity recovery line is significantly steeper at the higher cyclone efficiency values. When minor fractions of the circulating load are treated, high recovery of gravity gold is only possible when cycloning is extremely efficient. The actual amount of the circulating load treated is dictated not only by metallurgical requirements but also economic and practical considerations discussed in the following sections.

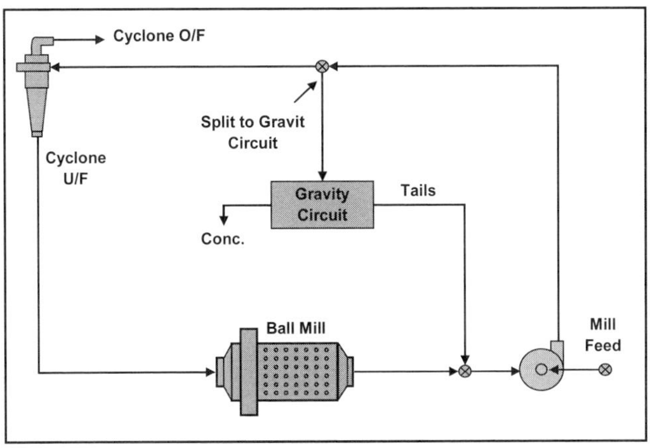


**Figure 4. Typical grinding circuit incorporating gravity concentration**

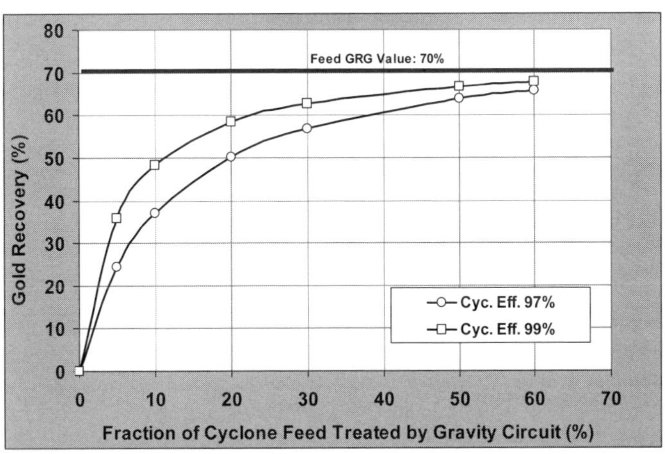


**Figure 5. Modeling result for gold recovery as a function of the fraction of the cyclone feed; assuming 1% loss of GRG in the ball mil, 70% feed GRG and 250% circulating load and BCC stage recovery of 40%. No upgrading of the primary gravity concentration is considered in this analysis.**

## ECONOMIC ANALYSIS

Gravity concentration has undergone a transformation in recent decades due to rapid advancements in technology. As technology has changed, so too has the definition of where gravity technology can be economically applied. The economic benefits of integrating BCC's into the grinding circuit ahead of flotation or cyanidation are sometimes difficult to measure. The difficulty has never arisen from the calculations themselves, which are relatively straightforward, but rather with the understanding of gold's behavior in mills and in obtaining accurate plant data. With the advent of modern circuit modeling techniques for gravity, the understanding of gold's behavior in milling circuits has improved significantly, however obtaining accurate circuit data is still problematic and can render the calculations meaningless in some cases.

In cases where a gravity circuit accounts for a very large change in overall gold recovery, the error in the measured quantities used in the calculations is negligible when compared to the benefit achieved and therefore of limited concern. Conversely, when the change in overall recovery attributable to gravity is small, extremely accurate circuit measurements are required in order to quantify a statistically significant economic benefit. Even in cases when the overall efficiency increase attributed to gravity is very small (a fraction of a percent), the addition of a gravity circuit can still have a surprisingly large benefit due to the low cost associated with the installation and operation of a gravity circuit. It is because of the potential for a high benefit given a very small efficiency change that the need for increasingly accurate benefit quantification has arisen.

In the case of a Greenfield project or any other project where both capital and operating costs need to be considered, the Rate of Return (ROR) or Net Present Value (NPV) methods can be used. In some cases where the capital costs have been fully depreciated, an even simpler calculation consisting of subtracting the costs from the revenue per time period can be used although in this case NPV is equally usable and perhaps a more sophisticated approach .

## ECONOMICS OF VARIOUS CIRCUIT TYPES

### Gravity Only Circuit

The simplest economic case is the "Gravity Only Circuit". Historically, this was the only method of gold extraction (along with mercury) until the introduction of chemical alternatives (cyanidation, flotation) in the last century. In this case, gravity technology is the only technology applied, and thus the economic calculation is a simple NPV calculation in the case of a Greenfield project or, in older operations where the capital costs have been fully depreciated, it is simply the revenue minus the costs for a given time period. This type of circuit was largely replaced by flotation and cyanidation recovery methods although environmental pressure against chemical systems has resulted in a resurgence of interest in these types of "environmentally friendly" circuits.

### Flotation Plus Gravity

In circuits that utilize flotation plus gravity, the economics of the gravity portion of the circuit are generally very strong. According to Dr. André Laplante (Laplante, 2005) the benefit for every 10% of recovery by gravity usually falls in the range of +0.5-5% of additional overall gold recovery. This rule of thumb is based on a large database of mills from around the world and indicates an extremely strong economic incentive for the use of gravity in flotation mills.

While the number provided by Laplante is an interesting start, one can calculate the gravity circuit benefit in far more detail. When calculating economics in flotation mills, one generally takes a flotation-only circuit as a base case, and then utilizes laboratory or pilot scale gravity test work to consider the effect of adding gravity recovery in various locations. Not all of the gold recovered to a gravity concentrate is gold that would have been lost to final tails. The gold recovered to a gravity concentrate will be a mixture of gold that would have reported to tails, and gold that would have reported to the various final concentrate products.

Gold contained in final tailings results in zero benefit and thus any gold shown to be recovered from that stream and into a gravity concentrate can be factored 100% as revenue gain. For the other product streams, any gold removed into a gravity concentrate will generally not be a 100% revenue gain. Most smelter contracts for flotation concentrate will pay a certain percentage for the gold contained therein, subject to various conditions and penalties. Each smelter contract will vary widely and have a significant effect on the overall economics of a gravity circuit. The smelter contract should be examined in detail even prior to initial test work or pilot work in order to determine higher value gravity recovery targets. Some products, such as zinc concentrates, may pay as little as zero for contained gold whereas gold in copper concentrates can pay well over 90%. When calculating the economic benefit in relocating gold from flotation concentrates to gravity concentrates, we must calculate the total value of gold recovered in the gravity concentrate and subtract the value that would have been paid for the gold by the smelter purchasing the flotation concentrate. This residual is the gravity benefit and must be calculated for each flotation product.

The final benefit consideration is associated with the speed with which payment is received for the gold product. Payment for gold in a flotation concentrate will take a number of weeks or even months depending on the distance to the smelter and the contract terms. Payment for gold as a gravity concentrate will be much faster, resulting in a gain that can be calculated using the appropriate capital discount rate for the project in question. Once the benefit for each stream has been calculated, including the speed with which the revenue will be received, one can again apply ROR or NPV methods to calculate overall benefit.

When choosing the placement of gravity equipment in a flotation mill there are a number of options. These options are not necessarily mutually exclusive and each potential circuit will have its own economic calculation. These circuits should be examined in the same order as presented here to ensure downstream effects are factored in.

- Primary Milling Circuit – In high grade mills this can be a good target, however it is much less likely to be economic in the large lower grade mills (<0.3 g/t). A gravity installation in the primary milling circuit will reduce the amount of gold in any downstream recovery circuit; therefore, this must be taken into account when performing downstream economic calculations.
- Flotation Concentrate Regrind Milling Circuit – If the primary milling circuit proves to be economic for a gravity circuit, the regrind circuit will almost always show an even stronger economic incentive, even when factoring in the gold removed by the primary circuit. This location will almost always have the highest return of any installation option. In fact, this location is generally among the most lucrative of any gravity circuits in any type of mill. If gravity is not viable in this location, it is very unlikely to be viable anywhere else in a flotation circuit, although there are exceptions depending on mineral associations and the deportment of gold in differential flotation circuits.
- Scalping gold from concentrate products – Placing gravity recovery in a single pass scalping application (not within a grinding circuit) can sometimes be very economic. Any gold removed previously by gravity in the primary or regrind milling circuits will affect the content and nature of gold in the final concentrate products; therefore care should be taken to reflect this in any laboratory or pilot testing. Generally, this type of application is only pursued after the possibilities in the milling circuits have been exhausted.

**Cyanidation Plus Gravity**

In circuits that utilize cyanidation plus gravity, the economics of the gravity portion of the circuit can often be very strong. According to Laplante (Laplante 2005) the benefit for every 10% of recovery by gravity usually falls in the range of +0.2-1.0% of additional overall gold recovery. Like the previous rule of thumb for flotation, these numbers are based on a large database of mills from around the world and although the percentage benefit average is slightly less than in flotation mills, grades are generally higher in cyanide operations (cyanide operations are generally pure gold mines whereas in flotation mills gold is often a by product) and therefore the cash incentive may be even greater. In cases where the ore has a preg-robbing component, the gravity benefit can be far higher than the rule of thumb.

There are additional incentives in a cyanide circuit that do not occur in a flotation circuit. A gravity circuit in a mill utilizing cyanide can see benefits from lower carbon costs, lower gold inventory, lower reagent costs and reduced carbon stripping frequency. In Greenfield projects there can be huge capital cost savings where the leach residence time can be reduced with the inclusion of a gravity circuit.

In cyanide circuits there are generally only two placement options for gravity recovery. The most common is the standard batch concentrator installation within the milling circuit. A second option, usually in addition to the batch concentrator in the milling circuit and generally only in refractory ores, is the installation of a continuous concentrator in the cyclone overflow. In either case, as with flotation, one uses the standard cyanide only circuit as the base case and then, utilizing laboratory or pilot plant data; one calculates the effect of adding the gravity recovery units. As with other circuits, one can then use NPV or ROR methods to calculate the final benefit.

## PRACTICAL CONSIDERATIONS

Once the decision to install a gravity section within a gold grinding circuit is taken, there are many practical issues that must be considered in the actual placement of the BCC within the grinding circuit.

**Location Within the Milling Circuit**

Given the elevated grade, most of the early BCC installations were placed in the cyclone underflow, however further research has shown that cyclone feed or mill discharge is often the best placement option. While the higher grade in cyclone underflow is an advantage, this is outweighed by the high density of the cyclone underflow. As with other gravity devices, higher feed densities will lower the unit efficiency of the device. In the case of BCC's the best efficiency point is less than 55% solids with the optimum being in the 45%-50% range. When considering the need for lower density feed and the constraints of the mill water balance, mill discharge or cyclone feed is usually the best economic option as far more feed can be treated without compromising the circuit water balance. The benefits associated with proper feed density increase with decreasing gold particle size.

**Other Placement Considerations**

Prior to the advent of BCC's, it was usual to treat mill discharge with a mineral jig for gravity recovery. There are other considerations that lead to treating cyclone feed:

***Plant Height***. Unless cyclone underflow is pumped (**not** recommended), overall plant height can be reduced by treating cyclone feed. One saving for reduced height is lower capital costs. Where the plant is inside a cladded building, this height saving can be large.

***Pumping Costs***. Cyclone feed pumps must work against lower physical and friction heads if the plant is shorter. These lower heads result in lower pumping costs, another advantage of reduced height.

***Grade Difference***. As shown in Figure 6, cyclone feed grade is only slightly lower than cyclone underflow in most grinding situations. However, this grade deficit is more than compensated for since efficiencies are higher when treating cyclone feed.

In evaluating grade differences between cyclone feed and underflow, it is important to remember that the CUF/COF ratios as shown in Figure 6 will shift left when a gravity section starts operating in a given circuit. This shift will continue until a new steady state is achieved that reduces the difference between CUF and cyclone feed grades, adding further weight to the cyclone feed argument.

***Efficiency***. VSMA guidelines dictate that both cyclone feed and underflow should be diluted for optimum screening efficiency. As produced, screening efficiency will be higher with cyclone feed. Optimum solids concentration for BCC's is ~45% by weight. Although BCC's can process high percentage solids slurries, the natural pulp density of screened cyclone feed is closer to optimum than that of cyclone underflow. The increase in efficiency gained by treating cyclone feed more than compensates for the deficit in grade.

| CIRC. LOAD (%) | CUF/COF RATIO | | | | | | | |
|---|---|---|---|---|---|---|---|---|
| | 1 | 2 | 3 | 4 | 5 | 10 | 15 | 20 |
| 100 | 100 | 75 | 67 | 63 | 60 | 55 | 53 | 53 |
| 150 | 100 | 80 | 73 | 70 | 68 | 64 | 63 | 62 |
| 200 | 100 | 83 | 78 | 75 | 73 | 70 | 69 | 68 |
| 250 | 100 | 86 | 81 | 79 | 77 | 74 | 73 | 73 |
| 300 | 100 | 88 | 83 | 81 | 80 | 78 | 77 | 76 |
| 350 | 100 | 89 | 85 | 83 | 82 | 80 | 79 | 79 |
| 400 | 100 | 90 | 87 | 85 | 84 | 82 | 81 | 81 |
| 450 | 100 | 91 | 88 | 86 | 85 | 84 | 83 | 83 |
| 500 | 100 | 92 | 89 | 88 | 87 | 85 | 84 | 84 |
| | CYCLONE FEED GRADE AS A PERCENTAGE OF CYCLONE UNDERFLOW GRADE | | | | | | | |

**Figure 6. Cyclone Feed Grade as a Percentage of Cyclone Underflow Grade for Various Circulating Loads**

**Screening and G-force**

World practice is converging on the use of slotted 2 mm apertures for preparation of BCC feed in gold grinding circuits rather than the coarser screens used in the past. Larger openings significantly increase BCC operating costs on account of increased wear rates while smaller apertures require significant increases in screen area and tend to blind. The reason for the modern tendency towards finer screening is twofold. Firstly, finer screening in front of a BCC has been shown to increase the unit efficiency. Secondly, the newest generation of BCC's maximize recovery by utilizing variable frequency drives to operate at g-forces of up to 200g. To operate at high g-force the maximum particle size should be minimized to reduce wear on the BCC. In most applications, the cost of screening down to 2mm will be outweighed by the increased benefit due to high g-force. If extremely coarse gold is present, screening coarser may be warranted, but BCC operating costs will be higher than for 2 mm. If substantial amounts of free gold are known to be less than 53 um (270 mesh) higher g-forces are certainly required and screening at 2 mm (or even finer in some applications) will show a very large economic benefit.

Screens should be offset from the centerline of the concentrator so as to allow overhead crane access to the BCC. Please refer to Figure 7.

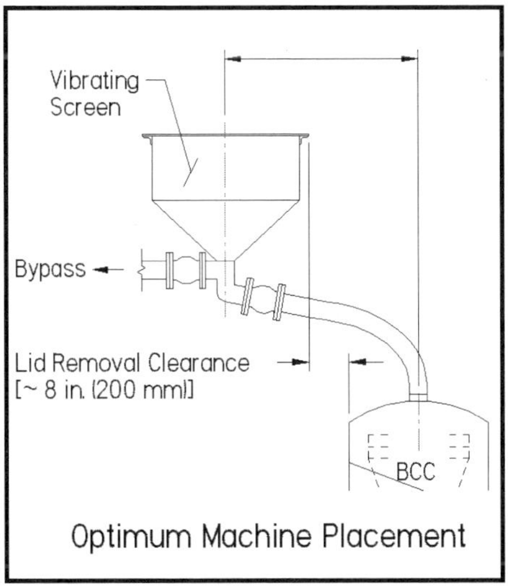


**Figure 7. BCC placement relative to sizing screen**

**Bypassing**

BCC's require a bypass that is utilized when concentrate is periodically rinsed from the concentrating bowl. Bypass frequency is typically every ~30 minutes in a grinding circuit application. Based on 8,000 operating hours per year, feed to the BCC will bypass ~16,000 times annually. Since the BCC is not concentrating while bypassing, it is important to minimize the time spent rinsing. Off line time required for rinsing depends to some extent on the design of the bypass. Here are some guidelines:

1. Keep the bypass as close as practical to the entrance of the concentrator, certainly downstream of the sizing screen. This minimizes delays required for clearing the feed system of incoming slurry.
2. Offset valves from the centerline of the concentrator and support them independently. This reduces down time when servicing concentrator internals.
3. Make certain that the bypass is easy to access by maintenance personnel. Consider how the various wear components such as sleeves and solenoids will be replaced.
4. Avoid creation of gold traps that will force coarse gold to bypass, as opposed to going through the concentrator. Please refer to Figure 8.
5. Cause the feed to bypass the BCC in the event of a power failure.
6. Use fast-acting valves (pinch or dart) with adequately sized compressed air supply lines. Use large solenoid valves to minimize fill and exhaust times.

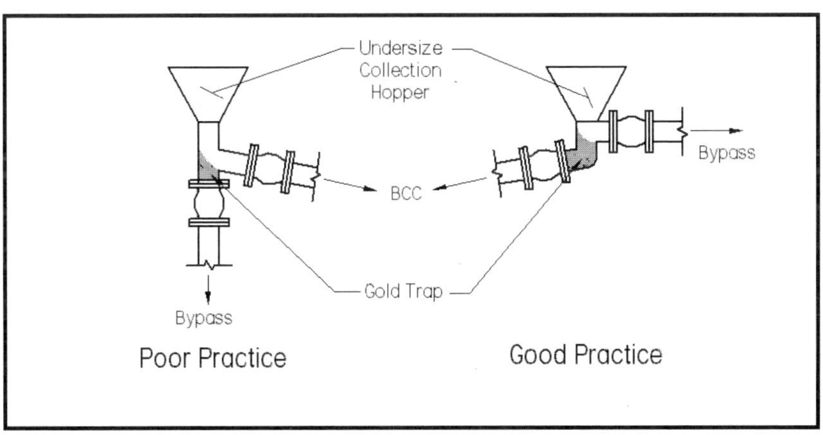


**Figure 8. Preferred bypass details**

**Rinsing**

The time taken to rinse concentrate out is an important consideration when designing BCC circuits since it is non-productive time. For example, if Concentrator 'A' uses a Variable Frequency Drive equipped with dynamic braking when rinsing, and Concentrator 'B' coasts to rest, the difference in rinsing time can be 2 minutes or more on each cycle. Using a delta of 2 minutes in a continuous duty application with 16,000 rinses annually translates to an advantage for the VFD-equipped machine of approximately 11 extra operating days or ~7% better concentration availability on account of reduced rinsing time alone. VFD's are recommended for minimizing rinsing time.

The quantity of water associated with each rinse needs to be considered carefully especially if the following process is intensive leaching. Provision needs to be made for receiving dilute slurry without losing fine gold. Rinse water quantity varies widely with machine size and manufacturer.

**Pumping**

BCC concentrate is notoriously difficult to pump on account of high S.G. solids, relatively large top size and lack of fines. Pumping BCC concentrates should be avoided wherever possible. It is preferable to use a gravity flow system. Pumping of cyclone underflows should also be avoided for the same reason.

**Fluidization Water Quality**

Fluidization water injection hole maintenance can be a significant source of downtime and maintenance labor in BCC circuits. These holes require periodic cleaning. The frequency of this operation is dependent on the design of the holes (Each manufacturer has it own approach.) and also the quality of the fluidizing water. 'High' quality fluidizing water has few, if any, suspended solids and little dissolved solids. It is difficult to provide a water quality specification since the equipment will operate for albeit short intervals when suspended solids are present. Bearing in mind that BCC's can operate outside the following limits, here are some quantitative guidelines for what constitutes 'good' fluidization water.

Total Suspended Solids (TSS) : <1,000 mg/l
Total Dissolved Solids (TDS) : <1,000mg/l
P100: ~ 20 Microns

Anti-scalants and Dispersants are available for higher concentrations. Chemical composition and dosage are highly dependent on variables such as pH and temperature gradient. Reagent vendors should be consulted for these situations."

**SUMMARY**

Recovering gold from grinding circuits using batch centrifugal gravity concentrators is now a well-established, powerful tool in the mineral processing engineer's toolbox. The general approach used to model and predict gravity gold recovery using BCC's from within the grinding circuit was presented. The approach utilizes results from a laboratory test for measuring the gravity-recoverable-gold content of an ore combined with the characteristic behaviour of gold with grinding circuits. While the modeling results may show that gold can be recovered from within grinding circuits, the economic case may require a full understanding of the influence of gravity gold on the overall gold recovery when downstream processes are considered. An increase in overall gold recovery may not be the only important factor as net impacts on smelter returns also need to be considered. The various factors influencing the economic justification for gravity gold recovery were discussed as well as the many practical considerations necessary for a successful installation.

**REFERENCES**

Banisi, S., Laplante, A.R. and Marois, J., 1991, The behaviour of gold in Hemlo Mines Ltd. Grinding Circuit, CIM Bulletin, 84(955), 72-78

Laplante, A.R., Woodcock, F. and Noaparast, M., 1995, Predicting gravity separation gold recoveries, Minerals and Metallurgical Processing, 74-79

Laplante, A.R. ., 2000, Testing requirements and insight for gravity gold circuit design, Randol Gold and Silver Forum, Vancouver

Laplante, A.R., 2005, Gravity Gold Recovery - Introductory presentation, AJ Parker Cooperative Research Centre for Hydrometallurgy Gravity Gold Workshop, Perth, Western Australia.

# DEVELOPMENT OF THE ORC COOS BAY MINERAL SANDS DEPOSIT IN OREGON, USA

Peter Dunn[1] and Todd Lessard[2]

[1] Outotec (USA), Inc., 6100 Philips Hwy, Jacksonville, FL, USA
[2] Oregon Resources Corp., 125 W. Central Ave. Suite 250, Coos Bay, OR, USA

## ABSTRACT

A unique mineral sands operation is being developed in southwest Oregon by Oregon Resources Corporation. Unique and variable mineralogy, and various mining and processing constraints required careful testing and process development efforts to ensure a robust separation process was designed. The heavy mineral content of the localized deposits varies significantly, as does the specific assemblage of heavy minerals.

The wet and dry separation processes were designed by Outotec based on testing efforts conducted on four separate and varied samples from different deposits. The resulting conceptual process flowsheet was also evaluated at pilot-scale, which resulted in the production of several tonnes of chromite foundry sand for customer evaluations. The final process development and circuit design efforts utilized all the testing data to design the plant process to reliably produce consistent recoveries and grades of chromite foundry sand and zircon sand.

The flowsheet utilizes traditional physical separation equipment, however certain applications and arrangements of the equipment are unique compared to other typical mineral sands operations. In this paper the testing, piloting and flowsheet design efforts will be described and explained. Conclusions will be drawn regarding the importance of careful testing and piloting in process development, and the steps needed to design flowsheets for feed quality variability.

## INTRODUCTION

### Background

Oregon Resources Corporation (ORC) is an industrial minerals mining company engaged in the development of the unique paleo-beach placer deposits of southwestern Oregon to recover mainly chromite, garnet and zircon sands. Engineering design has been guided by the varying geologoy and mineralogy of the deposits, as well as water limitations at the mining location. Processing studies were conducted on four distinct samples based on deposition and related mineralogy to represent the extremes likely to be encountered in the plant feed stream. The process design was then carried out with the critical constraint that the design allow for dynamic operation, with the ability to adjust successfully to handle feed variability and meet production needs. For various reasons, ORC plans to construct the ore processing facilities, both wet gravity and downstream dry processing operations, near Coos Bay, Oregon and will use municipal water as required (see Figure 1). Ore will be transported approximately twenty miles to the processing plant, with tailings returning to the active mining location.

**Figure 1. ORC proposed mining and processing locations**

### Geology and Mineralogy

Economic concentrations of placer heavy minerals have been evaluated and studied along the Oregon coastline for over a century, beginning with exploitation for gold. The beaches were small and irregular, routinely washed away by harsh winter storm events (Hornor 1918). In the 1920's, the beach deposits were followed upstream to their terrace origins and mined with little success due to high mining costs versus beach extraction. The most significant efforts to understand and develop these deposits were warranted, with support from the US government, by the demand for chromite in steel production during World War II. The first significant mining efforts began in 1943 by Humphrey's

Gold Corporation and Krome Corporation, which produced 25% $Cr_2O_3$ concentrate by wet gravity processing (Griggs 1945). This concentrate was then transferred to a downstream facility that further upgraded the material to approximately 40% $Cr_2O_3$. This processing operation was pioneering, in that Humphreys Gold Corporation developed and placed in operation the first "spiral concentrator" that is now broadly used in gravity concentration plants for sands, iron ore, coal and other minerals.

The heavy minerals of the placer deposits are sourced from the metamorphic and ultramafic rock of the Klamath mountains that hosted podiform chromite deposits (Twenhofel 1943). Several small operations attempted to exploit these deposits with little success due to the podiform variability (Libbey 1963). Several major rivers, including the Chetco, Rogue, Elk, Sixes and the Coquille, drain from the Klamath Mountains and, as the weathered sediments reach the Pacific Ocean, predominant longshore currents transport the material northward. Preferential deposition of the heavier minerals then occurs and the heavy storms, characteristic of the northwest Pacific coast, serve to further concentrate the heavy minerals through wind and wave action. Deposit preservation is suggested to be a result of transgressive sequences encroaching inland, followed by regression, sea standing and progradational beach formation (Petersen 1986).

There are several terraces that exist; ORC will be developing the reserves within the Seven Devils and Pioneer terraces (see Figure 2).

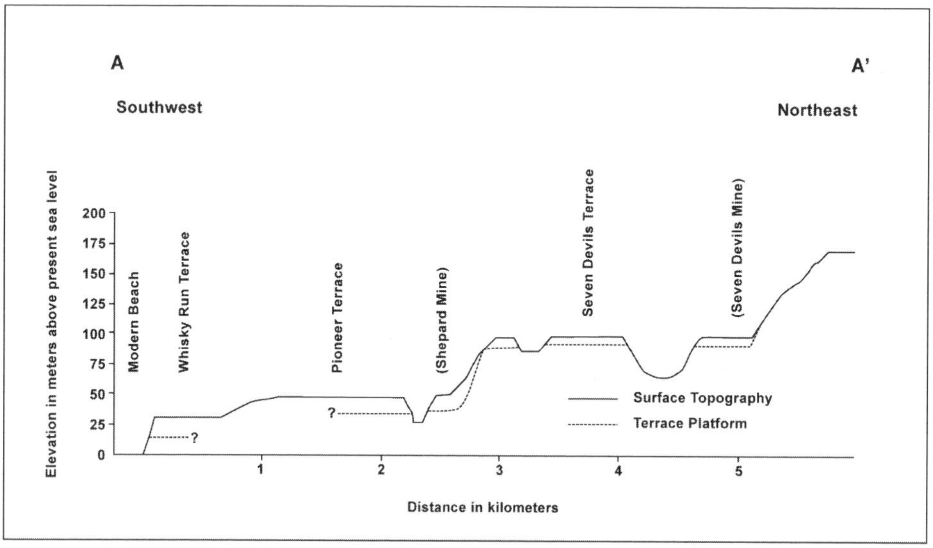


**Figure 2. Terrace geometry, cross section**

The Seven Devils terrace is at an elevation of approximately 80 meters above sea level, is five miles inland and is roughly parallel with the current beach. The Seven Devils orebodies (North and South) have mineralized depths up to 30 meters, averaging 15 meters. West of these orebodies on the Seven Devils terrace, ORC has further investigated a few shallower orebodies (Westbrook, West Bohemia, Section 33) that are at or near surface averaging 6 meters in depth. The younger Pioneer terrace contains the Shepard deposit, which is also being delineated by ORC.

The primary minerals of interest to the project are chromite, garnet and zircon with other recoverable minerals including ilmenite, staurolite and kyanite. The deposits on the Seven Devils terrace contain higher concentrations of chromite and zircon in the heavy mineral assemblage than does the deposit of the Pioneer terrace. ORC's reserves analysis indicates that chromite concentrations on the South Seven Devils terrace range from 33 to 43% in contrast to an average of 23% on the Pioneer terrace (Drew 2008). Zircon is similar, averaging 1.5 to 2.9% on the Seven Devils terrace while averaging 1.4% on the Pioneer terrace. Interestingly, the Pioneer terrace contains a greater percentage of garnet, averaging 12.6% versus the range of 4.6 to 9.8% found on the Seven Devils terrace (Drew 2008). While there is significant deposit-to-deposit and terrace-to-terrace variability in the concentration of these minerals, the sphericity, rounding and sizing is very consistent (Hoyt 2006).

## PROCESS DEVELOPMENT

### Sample Preparation and Characterization

Bulk samples from four different areas under consideration for future mining activity were selected based on a review of terrace deposition and mineralogy. The Shepard sample (SH) was from the lone delineated deposit area on the Pioneer terrace and represents the largest shift in mineralogy from the others. Two additional areas for samples on the Eastern side of the Seven Devils terrace were selected on the basis a north-south fault and associated mineralogical variation. They were prepared as "North" and "South" Seven Devils samples (N7D and S7D). Lastly, also from the Seven Devils terrace, the Westbrook sample (WB) was prepared representing the Western orebodies.

Each bulk sample was prepared by splitting drillhole interval samples using an in-ground average cutoff grade of 4% chromite. Some interbedded sections of drillholes were included in the sample preparation process that had less than 4% chromite to ensure the sample was realistic in representing as-mined material. Table 1 presents the heavy mineral (HM) content and associated mineralogy for the four samples selected for flowsheet development testing. Table 2 outlines the practical particle size distribution of the four feed test samples, after appropriate oversize and undersize fraction removal.

Considerable effort was made by ORC and Outotec to develop a robust process design. The goal of the flowsheet testing and design work was to yield a process plant capable of producing 70,000 tons/year of high quality chromite foundry sand, and subsequent secondary products (garnet and zircon), regardless of what deposit was being worked. Characterizing and testing representative samples for each deposit was critical for understanding the range of processing requirements necessary to reach design goals with all potential feed variability kept in mind. Simply taking all the drill hole samples and combining them into one large sample for testing would have resulted in some sort of design not suited to efficient treatment of the specific individual deposits. Table 1 demonstrates both the variability of heavy mineral concentration overall and within the specific mineral products, by terrace as well as deposit. Table 1 illustrates that S7D contains almost twice as much HM and chromite as WB. An inflexible design based on one deposit would have been inappropriate when mining the next deposit. For this reason, the solids handling systems and wet plant feed capacity has an operating range of 70 to 140 tph. Table 1 also illustrates that SH contains approximately three times as much HM as WB. A dry plant design based mining on an average HM value would not have the required capacity when processing ore from a deposit with a large amount of low value "light-heavy" minerals, so for this reason the dry plant was designed with an operating range of 20-40 tph. This design process included a plant wide bottleneck analysis. An example need for this analysis is found by reviewing garnet content in Table 1. If the solids handling capacity for the garnet circuit were sized for WB feed, the entire process would suffer from capacity restrictions resulting in production loss and wasted product. Review of Table 2 shows that variation in oversize and slimes content requires appropriate design capacity and flexibility in feeding capacity, and slimes handling systems to be viably operable across the deposits.

**Table 1. ORC bulk samples HM and mineralogy**

| As-Received/In-Ground | SH | WB | S7D | N7D | Avg. |
|---|---|---|---|---|---|
| **% HM** | **62.3** | **21.9** | **43.4** | **34.8** | **40.6** |
| % Epidote/Clinozoisite (petro) | 30.5 | 12.3 | 18.6 | 20.9 | 20.6 |
| **% Chromite (calc, AA)** | **14.5** | **10.3** | **18.9** | **11.9** | **13.9** |
| **% Garnet (petro)** | **10.1** | **1.0** | **6.0** | **1.7** | **4.7** |
| **% Zircon (calc, XRF)** | **1.2** | **0.6** | **1.8** | **1.2** | **1.2** |
| % Staurolite (petro) | 1.5 | 0.4 | 0.3 | 0.7 | 0.7 |
| % Ilmenite (petro) | 1.0 | 1.1 | 4.3 | 0.5 | 1.7 |
| % Leucoxene (petro) | 0.5 | 0.5 | 0.3 | 0.8 | 0.5 |
| % Rutile (petro) | 0.2 | 0.2 | 0.5 | 0.4 | 0.3 |
| % Magnetite (petro) | 0.0 | 0.1 | 0.4 | 0.1 | 0.2 |
| % Misc. Light Heavies (petro) | 3.6 | 0.4 | 1.0 | 0.9 | 1.5 |

**Table 2. ORC bulk samples particle size distribution**

| | SH | WB | S7D | N7D | Avg. |
|---|---|---|---|---|---|
| % Pit Oversize (+4#, + 4.75mm) | 1.3 | 4.0 | 3.1 | 2.6 | 3.1 |
| % Plant Oversize (4x18#, 4.75mm x 1.0mm) | 2.1 | 6.1 | 4.8 | 4.0 | 3.9 |
| % Slimes (-230 mesh, -63 μm) | 6.7 | 14.8 | 14.0 | 14.8 | 12.6 |
| **% → Deslime/Gravity Circuit** | **89.9** | **75.1** | **78.1** | **78.6** | **80.4** |
| **Screen analysis of deslime to Gravity Circuit** | **100.0** | **100.0** | **100.0** | **100.0** | **100.0** |
| +20 mesh (850 μm) | 0.0 | 0.1 | 0.0 | 0.1 | 0.1 |
| -20+30 mesh (600 μm) | 0.2 | 2.7 | 0.8 | 0.4 | 1.0 |
| -30+40 mesh (425 μm) | 0.7 | 8.4 | 1.9 | 1.1 | 3.0 |
| -40+50 mesh (300 μm) | 5.6 | 28.3 | 6.9 | 8.1 | 12.2 |
| -50+70 mesh (212 μm) | 26.6 | 29.4 | 29.1 | 41.2 | 31.6 |
| -70+100 mesh (150 μm) | 43.4 | 17.8 | 33.7 | 30.7 | 31.4 |
| -100+140 mesh (106 μm) | 18.9 | 8.1 | 20.5 | 13.0 | 15.1 |
| -140+200 mesh (75 μm) | 3.7 | 3.3 | 5.5 | 3.7 | 4.1 |
| -200+230 mesh (63 μm) | 1.0 | 1.8 | 1.7 | 1.9 | 1.6 |

**Testing and Flowsheet Development**

Pathfinding testwork for the entire flowsheet was conducted separately for each of the four deposit samples to obtain a clear understanding of the unique plant requirements for accepting feed from any deposit and still producing at the required capacity and grade.

The first area of focus was the spiral pathfinding test program. Many tests were conducted on a variety of different spirals to determine which spiral was most effective and at what specific operating conditions (feed rate and percent solids). Similar spiral testing was repeated over and over again to determine what sort of spiral circuit configuration would be required to meet the capacity and recovery objectives for all four deposits. It was noted that, while rougher spiral circuit performance did show significant differences in first-stage separation between deposits, the simple three-stage circuit shown in Figure 3 functioned effectively for all deposits in terms of valuable heavy mineral recovery (VHM, in this case chromite and zircon minerals). The spiral circuit employed is a straightforward low-recirculation circuit consisting of rougher, middlings cleaner, and tailings cleaner stages. The circuit allows concentrate from all three stages to go to final gravity concentrate, and tailings from both middling and tailings circuits to go directly to final tailings without recirculation, resulting in a streamlined, simple to manage gravity circuit. Table 3 shows the resulting valuable heavy mineral recoveries for the wet processing circuit.

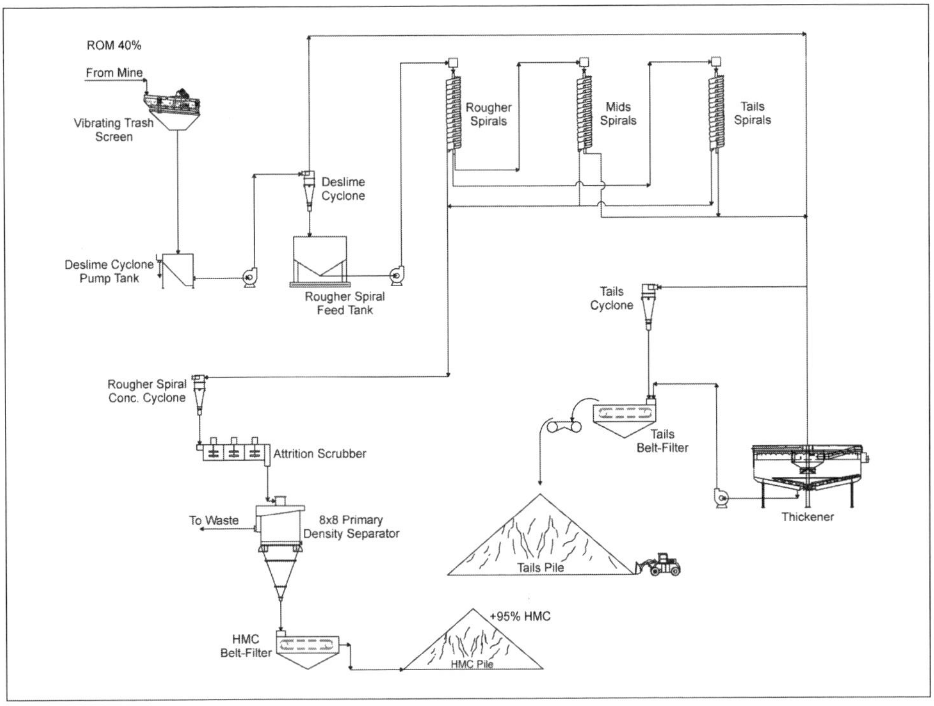


**Figure 3. Wet process flowsheet**

**Table 3. Summary of wet plant VHM recoveries**

| Deposit | $Cr_2O_3$ | $TiO_2$ | $ZrO_2$ |
|---|---|---|---|
| SH | 97.4 | 74.5 | 92.7 |
| WB | 100.7 | 87.3 | 99.3 |
| S7D | 99.2 | 84.7 | 98.8 |
| N7D | 100.7 | 91.6 | 99.8 |

The next area of study focused on preparing the heavy mineral concentrate (HMC) for efficient dry separations. A series of attrition scrubbing and final classification tests using the Floatex Density Separator were conducted to ensure that the grain surfaces were best prepared for the downstream high-tension electrostatic separation process, with a bonus of rejecting any additional unwanted fine silica or "light-heavy" undesirable minerals prior to feeding the dry plant. The testing revealed that high-density attrition scrubbing, without the use of any additive chemicals, was all that was required to ensure effective electrostatic response in the dry process steps. The rinsing step conducted by the Floatex Density Separator was effective in removing the liberated clays as well as rejecting unwanted fine silica to improve the quality of the HMC, before the dry process, with essentially zero VHM loss. These process steps are shown in Figure 3.

Pathfinding work for the dry process flowsheet was carried out in a detailed manner very similar to the pathfinding efforts carried out for the wet gravity circuit, with all four deposits still being treated individually. The first separations step in the dry processing flowsheet, following drying and screening to remove deleterious oversize particles, is high-tension electrostatic separation (HT). This circuit is the heart of the dry processing plant with the main purpose of generating a very clean conductor fraction, void of non-conductive minerals to feed the chromite concentration circuit. The HT circuit is essentially a three-stage circuit in a rougher, scavenger and scavenger-cleaner configuration. The chromite circuit is a series of magnetic separation steps using rare-earth drum (RED) magnetic separators to isolate a high-grade chromite foundry sand product as a non-magnetic product. As most non-conductor minerals would also report to the non-magnetic product in the chromite circuit, it is critical that the HT circuit conductor fraction is managed carefully to ensure final chromite quality. On the non-conductor side of the HT circuit, additional stages of magnetic separation, both REDs and rare-earth rolls (RER), are used to generate a garnet product and to prepare a final non-conductor, non-magnetic stream rich in zircon for additional downstream processing. The overall dry processing flowsheet is shown in Figure 4.

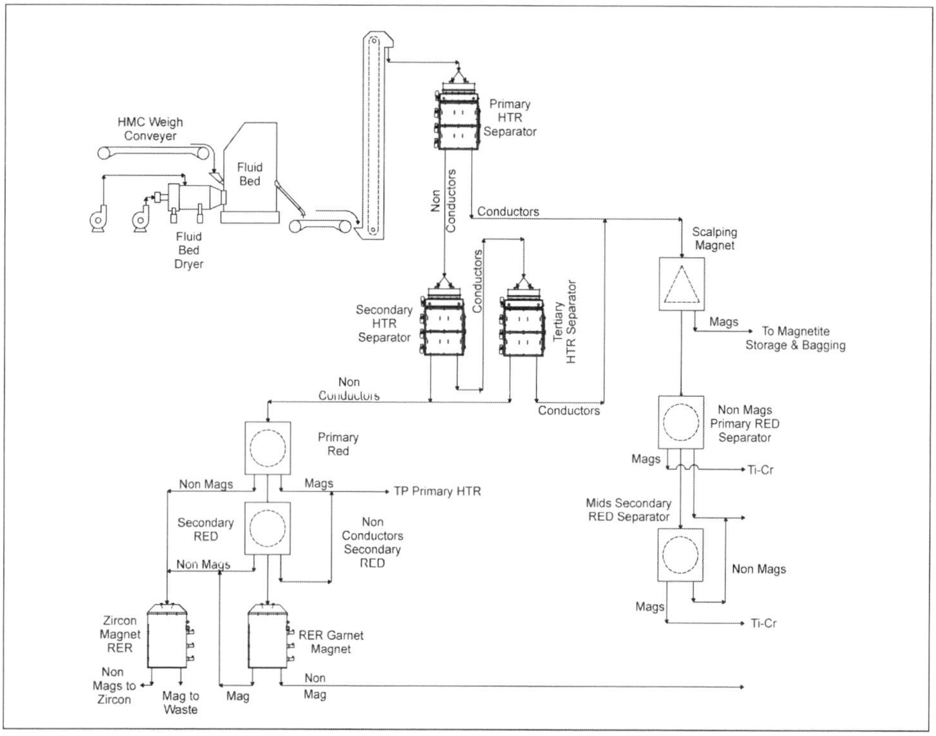


**Figure 4. Dry process flowsheet**

Once the basic flowsheet scenarios for each deposit sample were worked entirely through the testing program the flowsheets reflected in Figures 3 and 4 became firm. The next step was to undergo a modeling exercise where unit operation performances were modeled (conditional recovery models) and validated against the bulk test runs conducted for each deposit sample. The main purpose of the modeling exercise was to ensure that a single practical flowsheet could be designed to effectively accept feed at varying capacities from any of the four deposits, and, with only minor adjustments, still be able to produce the required quantity and quality of chromite foundry sand. The modeling work permitted study of recirculating loads and de-bottlenecking studies at various feed rates to ensure the number and sizing of the equipment was appropriate for the various "stretch" scenarios. Figure 5 shows an example of the model flowsheet and Table 4 contains example model output during the de-bottlenecking studies. Once the modeling exercises were complete, all relevant engineering process data was extracted and carried forward as the process design basis for detailed design efforts. Table 5 shows the final flowsheet predicted mineral recoveries based on the overall flowsheet design.

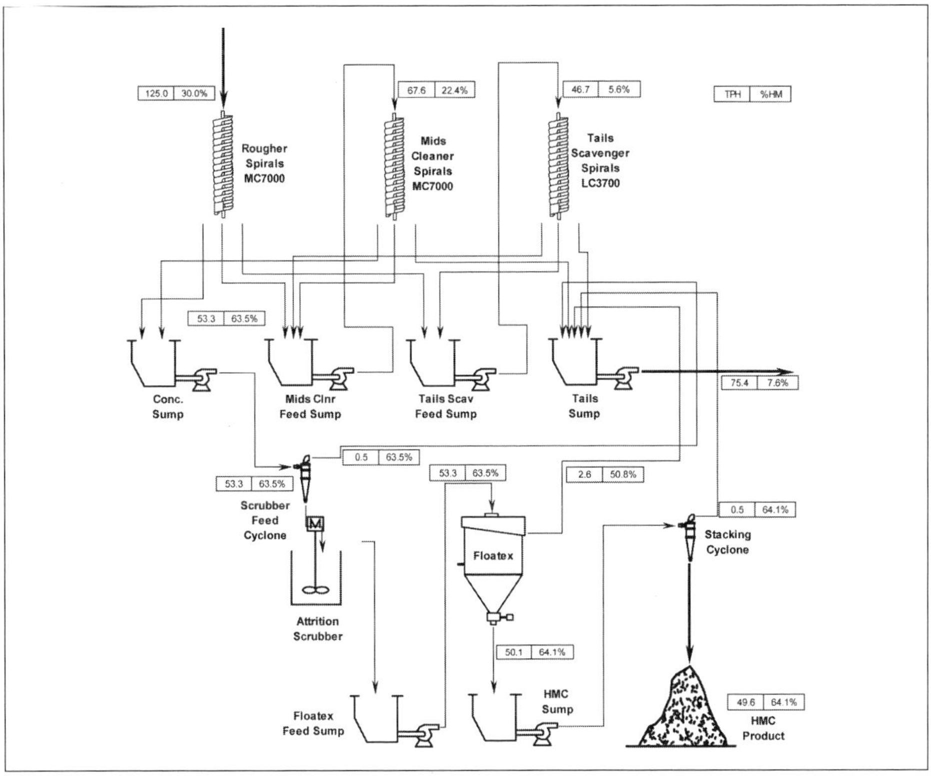


**Figure 5. Example wet plant model sheet**

**Table 4. Example model output sheet, de-bottlenecking**

| Stream | ROM Plant Feed at 125 tph | | | WB Case | SH Case |
|---|---|---|---|---|---|
| | Max (TPH Solids) | Min. (TPH Solids) | Avg. (TPH Solids) | 40 tph Product (TPH Solids) | 40 tph Product (TPH Solids) |
| **From Rougher Feed Sump** | **112.8** | **96.3** | **102.2** | **150.0** | **70.0** |
| Rougher Con | 58.7 | 22.1 | 38.9 | 33.8 | 36.4 |
| Rougher Mids | 34.8 | 14.0 | 27.8 | 52.5 | 14.0 |
| Rougher Tails | 40.4 | 19.6 | 29.1 | 63.8 | 19.6 |
| **Mids Cleaner Feed** | **54.2** | **28.8** | **46.8** | **90.2** | **28.8** |
| Mids Clnr Con | 13.9 | 4.6 | 9.0 | 9.0 | 8.6 |
| Mids Clnr Mids | 18.5 | 11.5 | 15.0 | 27.1 | 11.5 |
| Mids Clnr Tails | 33.6 | 8.6 | 22.8 | 54.1 | 8.6 |
| **Tails Scav Feed** | **44.9** | **21.8** | **33.0** | **106.2** | **21.8** |
| Tails Scav Con | 5.3 | 2.8 | 4.0 | 10.6 | 3.3 |
| Tails Scav Mids | 5.7 | 2.2 | 4.0 | 42.5 | 2.2 |
| Tails Scav Tails | 36.0 | 16.3 | 25.1 | 53.1 | 16.3 |
| **Conc Pump Discharge** | **72.6** | **26.7** | **47.9** | **42.8** | **45.0** |
| Scrubber Feed | 71.8 | 26.5 | 47.4 | 42.3 | 44.6 |
| Scrubber Feed Cyc OF | 0.7 | 0.3 | 0.5 | 0.4 | 0.5 |
| Srubber Product | 71.8 | 26.5 | 47.4 | 42.3 | 44.6 |
| Floatex Feed | 71.8 | 26.5 | 47.4 | 42.3 | 44.6 |
| **Floatex Underflow** | **68.3** | **25.1** | **45.0** | **40.2** | **42.4** |
| Floatex OF | 3.6 | 1.3 | 2.4 | 2.1 | 2.2 |
| **Product Cyclone Feed** | **68.3** | **25.1** | **45.0** | **40.2** | **42.4** |
| Product Cyclone UF | 67.6 | 24.9 | 44.6 | 39.8 | 41.9 |
| Product Cyclone OF | 0.7 | 0.3 | 0.5 | 0.4 | 0.4 |
| **Combined Wet Plant Tails** | **71.4** | **28.1** | **51.1** | **110.2** | **28.1** |

**Table 5. Predicted valuable mineral recoveries**

| 2.14 x % $Cr_2O_3$ = %Chromite Mineral | $Cr_2O_3$/Chromite Recovery to FS | 1.52 x %$ZrO_2$ = %Zircon Mineral | $ZrO_2$/Zircon Recovery |
|---|---|---|---|
| SH | 74.2% | SH | 67.3% |
| WB | 79.8% | WB | 80.6% |
| S7D | 88.1% | S7D | 71.5% |
| N7D | 83.5% | N7D | 74.0% |

**Piloting Confirmation**

For the purpose of generating several tons of chromite foundry sand for market testing, ORC contracted with Outotec to manage a bulk testing campaign involving 100 tons of heavy mineral concentrate. In late 2007, the bulk program commenced at the ORC location where a portable spiral plant, complete with oversize removal by trommel and cyclone desliming, was configured with the appropriate spiral circuit defined by earlier testwork. The pilot plant was run at approximately 5-6 tph with feed material from the S7D deposit and generated roughly 100 tons of HMC.

The HMC was transferred to Hazen's facility in Golden, CO and Outotec provided the necessary process equipment to enable the dry piloting campaign setup. After attrition scrubbing and Floatex Density separation, the material was dried by rotary dryer and fed to Outotec's HT and RED equipment to accomplish the required separations for producing clean finished product chromite foundry sand. In total, 34 tons of product were produced and loaded into drums for market evaluation shipments. Figures 6-8 show the gravity pilot plant, the chromite product being produced on a RED and finished product being loaded into drums.

The non-conductor fraction was also further processed. The residual non-conductor material was shipped to NCMRL in Asheville, NC where the required wet gravity and dry HT and RER separation steps were carried out to produce a clean zircon product. Table 6 presents the finished product quality of the chromite and zircon products from the pilot program. Throughout the piloting campaign, the flowsheet design was followed and validated to be effective for the treatment of the ORC deposit material.

Work on the project has continued; presented in Figure 9 is a 3-D rendition of the processing plant facility intended for construction at the ORC site in Coos Bay, OR.

**Figure 6. Gravity pilot plant**

Figure 7. Chromite product on RED

Figure 8. Chromite foundry sand product being loaded for shipment

**Table 6. Pilot trial finished product quality**

| Zircon Product | % DISTRIBUTION |
|---|---|
| $ZrO_2$ | 65.24 |
| $TiO_2$ | 0.70 |
| $Al_2O_3$ | 0.26 |
| $Fe_2O_3$ | 0.10 |
| **Chromite Product** | **% Distribution** |
| $Cr_2O_3$ | 46.7 |
| $TiO_2$ | 0.94 |
| $ZrO_2$ | 0.00 |
| $SiO_2$ | 0.58 |

**Figure 9. 3D model of ORC processing facility**

## CONCLUSION

The ORC flowsheet development testwork initiative included basic pathfinding testing, flowsheet development testing and flowsheet refining into a practical design, finally concluding with confirmation efforts at pilot scale. The process was also complicated by the design requirement of a flexible flowsheet capable of handling the required capacity (both head feed and internal flowsheet circuitry) to handle highly variable HM content and specific valuable mineral content (chromite, zircon and garnet) while maintaining finished product quality and throughput requirements.

In conclusion the following general guidelines are presented to summarize some key areas of proper flowsheet development efforts.

### Before Starting

Understanding the origin of the samples to be tested is important. Samples must adequately represent "design basis" feed material, and be of sufficient quantity to allow for pathfinding efforts as well as larger bulk confirmation testing. Any feed preparation work that has been performed prior to sample receipt needs to be clearly understood, and if necessary, accounted for in later process design efforts.

Adequate characterization of the feed sample should be completed properly before separation tests begin and accepted sampling techniques should be used.

Any project-specific guidelines should be defined at the onset – the specific scope of testing, which component recoveries are important, measurement methods to be used, and specific goals (recoveries, product grades, etc.) set. These guidelines are sometimes naturally set by the type of material being tested, but are often influenced and or dictated by specific project constraints.

### General Testing Practices

Good lab management protocols must be used to manage and log the flow of material through the testing process. Weights need to be carefully tracked, including small subsamples from each process step that are removed and held as "retains." The data logging must always allow for accurate flowsheet understanding on a "percentage of head feed" basis.

Care must be taken with making flowsheet assumptions based on prior experience and "typical market flowsheets" – always consider the specific sample and project specifics. Alternatives should be evaluated and tested with results guiding the process.

Test the assumptions around specific equipment capability for certain process steps: Do the testing. An example would be selection of rare-earth drum or rare-earth roll magnetic separators. There are cases where the drum seems the logical choice, but actual testing would reveal the roll separator performs better, and vice versa.

Consider also the very practical items, such as the order of unit operations (i.e. put more expensive process equipment steps as far back in the flowsheet as possible, dry once, minimize drying of material that will only be discarded later, etc).

**As you go along keep in mind**

Complete the smaller pathfinding tests, and analyze the results, before continuing with the larger bulk runs. If needed, a few small path finding efforts that extend a few process steps ahead (or sometimes the whole way through a flowsheet) may need to occur to answer difficult questions – two examples of this would be "game changing" flowsheet decisions such as use of wet magnetic separation in the wet gravity circuit, or which first unit operation is best in the dry plant, electrostatic or magnetic separation.

Once the basic flow and unit operations are defined in the flowsheet, results from scoping tests should determine the optimum operation conditions and settings for each unit operation (feed % solids, loading, roll speeds, magnetic configurations etc). Later, bulk sample runs should be focused on validating the flowsheet and suggesting smaller modifications for efficiency enhancement, not making massive changes to the design.

**Moving from pathfinding to runner larger bulk samples:**

Take samples during bulk processing runs to confirm and validate (and occasionally improve) the pathfinding test results. Occasionally, through careful attention during bulk processing efforts, ways to simplify and improve performance of the initial flowsheet design can be found and implemented.

The "scavenging" sections of flowsheets should be tested out first before moving towards to critical concentrating stages testing. This allows scavenged material to be combined and moved forward towards product generation, providing a more realistic feed consistency and most closely approximating the working flowsheet design with recirculation.

If time permits, various misplaced middlings streams that result from bulk testing programs can also be carefully reviewed, allowing these streams to be re-introduced into the bulk processing efforts at the proper location to more correctly approximate plant feed conditions.

**Wrapping Up**

Prepare the overall mass balance for testing programs using "percentage of head feed" as the main mass tracking metric to eliminate confusion with weights removed for retain samples, and of various stranded materials (middlings streams etc). Various modeling tools can then be used to further test and refine the flowsheet design, particularly in cases where adequate pathfinding data exists to generate conditional recovery models for critical unit operations in the flowsheet. The working model should be validated by the results of the bulk processing runs, and then used to size the unit operations, identify areas of concern with recirculating loads, and understand flowsheet robustness and specific areas of concern.

Once the flowsheet is set in terms of sized unit operations (equipment is sized, natural capacity constraints now exist), recirculating load analyses in conjunction with de-bottlenecking studies should be carried out to test the flexibility and performance of the flowsheet across the range of feed capacity and mineral content. This may result in some final modifications to ensure that the final flowsheet design is efficient and can handle expected variability with predictable results.

## REFERENCES

Drew, J.D., 2008. Oregon Resources Corporation JORC compliant ore reserve and min. resource statement. *http://www.imcl.com.au/productionandreserves.htm*

Griggs, A.B., 1945. Chromite bearing sands of the southern part of the coast of Oregon. U.S. Geological Survey Bulletin, 945(E): 113-150.

Hornor, R.R., 1918. Notes on the black sand deposits of southern Oregon and northern California. U.S. Bureau of Mines, Technical Paper, 196: 33pp.

Hoyt, D., 2006. Mining a domestic source of chromite. Modern Casting, v 96, n 2: 42-45.

Libbey, F.W., 1963. Lest We Forget. Ore Bin, v 25, n 6: 93-112.

Peterson, C.D., Komar, P.D., and Scheidegger, K.F., 1986. Distribution, geometry, and origin of heavy mineral placer deposits on Oregon beaches. Journal of Sedimentary Petrology, 56: 67-77.

Twenhofel, W.H., 1943. Origin of the black sands of the coast of southwestern Oregon. Oregon Department of Geology and Mineral Industries (DOGAMI) Bulletin, 24: 25pp.

# Process Development

# INNOVATIVE SURGE BIN DESIGN FOR MINERAL SANDS PROCESSING PLANT

John Lyons[1], Greg Hill[1], Craig Vadeikis[1], and Stephen Wiche[2]

[1]Downer EDI Mining – Mineral Technologies, 11 Elysium Road, Carrara, Queensland, Australia
[2]TUNRA Bulk Solids, TA Building, University Drive, Callaghan, New South Wales, Australia

## ABSTRACT

Mineral Technologies carried out the mineral processing test work and developed the flow sheets for a mineral sands project in the Murray Basin, Australia. Mineral Technologies then completed the detailed engineering design for the project.

The ore body required a processing plant that could operate efficiently over a wide range of conditions:

- Slimes up to 30%
- Feed Rate from 500 to 750tph
- Heavy Mineral in situ from 12 to 19% w/w

The mined ore was to be stockpiled and blended to achieve this operating range. The material was to be screened and scrubbed to liberate the HM from the clays. The feed was then pumped to the Feed Preparation area of the Concentrator. The concentrator utilised gravity separation and wet magnetic separation to produce a Heavy Mineral Concentrate. The separation can cater for some slimes but anything more than about 3% will affect mineral recovery. The feed preparation utilised a set of cyclones to remove the majority of the clays and presented the upgraded feed to the surge bin. The cyclone overflow was then treated in thickeners for disposal with the tailings.

The Surge Bin was designed to have mass flow characteristics allowing material to be stored for up to an hour with a consolidation in excess of 80%w/w. The material is withdrawn using slurry pumps to feed to the spiral separators. Various forms of dilution water are injected into the bin to assist in the density control of the slurry.

The bin has a storage capacity of 350 tonnes and delivers the slurry to the spirals at a constant density.

This bin design utilises the work of Jenike and the University of Newcastle, Australia to determine the critical dimensions of the Surge Bin to ensure that there is mass flow. This work has been applied to particulate dry solids in silos previously but not to slurries. However, in fact, the slurry acts more like a solid in this situation rather than a liquid.

## INTRODUCTION

In Australia, mineral sands mining began on the beaches of the east coast during the 1950's and 60's. These were generally dredging operations, consisting of a floating dredge pumping the sand slurry along a float line to the concentrator plant. At the concentrator plant the feed slurry would be screened in a trommel from which the underflow would fall directly into a surge bin and from where it is pumped to the separation equipment. It is characteristic of dredge mining operations that they result in substantial fluctuations in feed rate to the plant, including periods of interrupted feed. By making the surge bins relatively large it was expected that they would contain enough material to even out the variations in feed rate to the plant and maintain feed when dredge feed is interrupted. Early designs gave little consideration to the physical flow properties of the material in the bin, and how this would affect discharge rate. Initially the bins were designed with relatively low wall angles of around 45 degrees above the horizontal. This resulted in a number of material flow problems regularly occurring in surge bins, including material hang-ups, slumping and rat-holing. These issues caused the feed rate and slurry density to the separation equipment to vary considerably resulting in reduced concentrate grades and reduced mineral recoveries in the concentrator plants. Furthermore the separation equipment; Reichert cones and trays, and later spirals required slurry be presented at relatively high density for optimum performance, which was not always achievable with conventional bin designs.

Conventional slurry bins weren't designed for mass flow there would be funnel flow with a rat-hole in the middle of the bin allowing the incoming feed to pass straight through the bin. Interruptions in feed caused the consolidated sand hanging up on the bin sides to slough off sending high density slugs forward to the processing plant. If these were severe enough it would result in slurry line blockages in the plant. The plant would have to be shut down to clear the blockages resulting in excessive downtime. Generally the operators were very reluctant to carry excess sand in the bin for fear of material slumps and as a result the full benefit of having a surge bin was not being realised.

To overcome the identified material flow problems the surge bin designs were changed in a number ways over the years. The bin side walls were steepened to around 60 degrees and agitation water injection was added around the base of the pump suction to fluidise the slurry. Later a system was introduced to draw water off the top of the bin and into the pump suction. In principle the density of the slurry could be regulated by throttling the water flow through this bypass using valves. It was found that this system could operate reasonably well for a narrow range of surge bin levels, but that outside of this range the pressure differential between the slurry suction and the bypass suction made density control difficult. Furthermore, as technology in variable speed pump drives, flow meters and nucleonic density meters developed these were introduced along with control systems to improve plant stability.

Even with these improvements steady state operation of surge bins was not readily achieved using conventional surge bin designs. Having reliable discharge from the surge bin to the concentrator plant was identified as fundamental to improving the overall throughput capacity and mineral recovery rates.

## MASS FLOW BIN DESIGN

Conventional thinking tended to treat bins containing a water sand slurry mixture as a viscous liquid storage system. In fact storage of slurry in a surge bin is better described as the storage of a bulk solid under saturated conditions rather than the containment of a liquid. This change of thinking resulted in the investigation of the behaviour of saturated bulk solids as analogous to the well studied and documented behaviour of dry bulk solids in bins.

There are two types of flow patterns that can form when the solids are withdrawn from a bin or a hopper. (Refer Figure 1). If on withdrawal all particles commence and maintain motion, including those along the bin wall, the bin can be defined as a mass flow bin. If on the other hand a "rat hole" forms within the stored material itself the bin would be classified as funnel flow. Generally the wall angles of mass flow bins are much steeper than those for funnel flow bins.

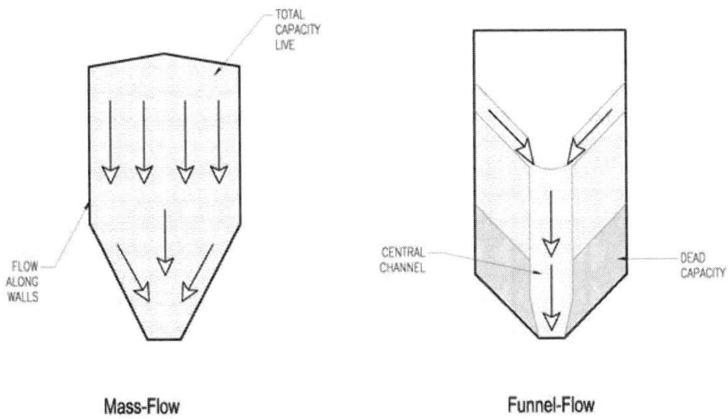


**Figure 1.**

The design parameters for bins storing dry solids were investigated by Jenike[i] and involved the evaluation and comparison of both the consolidation stresses and the stresses causing the stored material to fail.

A water media affects the behaviour of the solid particles during flow by the effects of buoyancy and viscosity of the liquid. As the sand settles and consolidates in the bin it causes water to be displaced from the particle interstices and to rise towards the bin overflow. Similarly as the solids move through the bin towards the outlet the particles will need to dilate from each other. (McLean[ii])

The primary objective in the design of wet solids storage containers is to ensure that the contents will gravity flow at a consistent rate without flow obstructions occurring. The material flow properties need to be determined to allow selection of suitable bin geometry that will ensure that mass flow pattern is promoted (Arnold[iii] and Roberts[iv]). These will depend on whether the bin is axi-symmetrical, such as a conical bin, or is symmetrical about a plane, such as a chisel or wedge shaped bin. (Refer Figure 2) Generally wall angles in axi-symmetric bins need to be steeper to achieve mass flow than for plane flow bins.

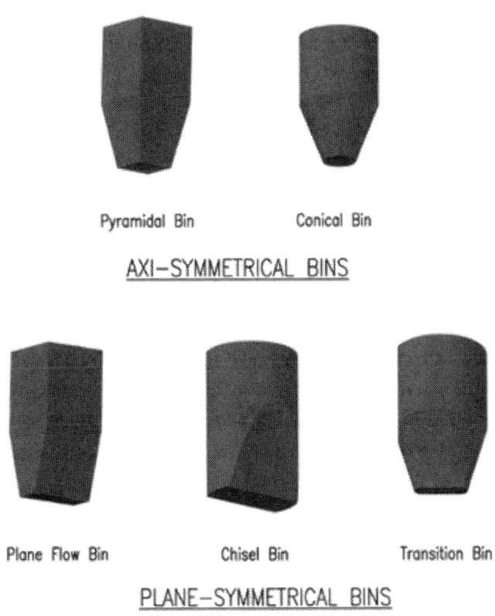


**Figure 2.**

## MINE OREBODY

The mine is located in the South West corner in the State of Victoria, Australia.

The ore body is a series of mineralised sand strand lines containing quantities of the Valuable Heavy Minerals (VHM) Rutile, Zircon, Ilmenite, Leucoxene and minor quantities Cassiterite in a host of non-economic materials of silica sand, rock and clays. Total Heavy Mineral (HM) is defined as the fraction of material greater than a specific gravity of 2.85gm/cm$^3$. The in-situ average grade of heavy mineral was 13.9% HM, the heavy mineral assemblage consists, on average, of 44% Ilmenite, 12% zircon and 8% rutile & leucoxene.

The ore processing operations at the Douglas mine site consist of two main sections, the Mining Unit plant and the Wet Concentrator Plant (WCP). (Refer Figure 3). The mining unit plant was designed to slurry the run of mine ore at a rate upto 750tph and remove oversize rock before being pumped overland to the WCP. The WCP consists of a number of gravity, magnetic and classification stages designed to separate and upgrade the zircon, rutile, leucoxene and cassiterite minerals into a concentrate for downstream processing at the Hamilton mineral separation plant.

The Surge-bin (Feed bin) is central to the operation and control of the entire beneficiation process. It receives slurried ore from the MUP and is required to perform three major operational functions:

1. Provide surge capacity between the two remotely located and independently operated mining and wet concentrator plants,
2. Provide, a stable feed rate and density control to the gravity, magnetic and classifications unit operations of the WCP,
3. Provide a secondary classification for the continuous removal of clays from the spiral concentrators feed.

The practical result of combining the above functionality is the stable operation of the key mineral separation equipment ie the Spiral separators and the wet high intensity magnetic separators (WHIMS), thereby promoting conditions for optimum mineral separating efficiency.

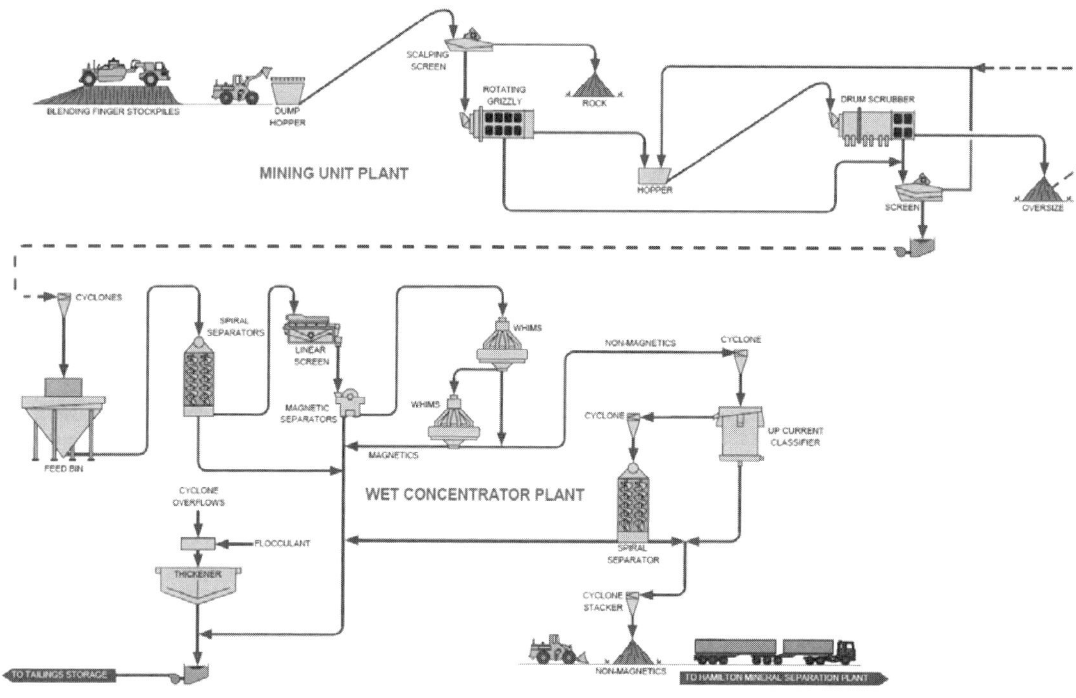


**Figure 3. Schematic Flow Diagram of the Douglas Mining and Wet Concentration Plants**

## SURGE BIN DESIGN

The surge bin had the following basic design requirements:

- It had to accept screened slurry nominally in the size range +53μm to -2500μm.
- The feed slurry was delivered as a cyclone underflow at 75% solids w/w.
- Residual -53μm material in the cyclone underflow had to report to the bin overflow.
- The bin had to have a live capacity of 350 tonnes of solids.
- The discharge pump had to deliver the solids to the plant at a consistent 35% w/w density.
- A nominal discharge rate of 350tph was required.

In the bin design the incoming feed from the cyclones is spread out over a 3 metre by 1.7 metre feed box then tumbles through a series baffles into the bin. There the wash water jets wash the slimes free of the sand and it is carried toward the bin overflow. The bin overflow water containing the slimes reports to the thickener. The sand is then left to fall through the bin and pass through the consolidating zone to a layer of maximum consolidation. In doing this water is squeezed out of the slurry rising towards the overflow joining with the wash water. Further down the bin the sand must dilate to allow it to flow towards the outlet. As the sand dilates fluid must be drawn into fill the expanding voids. This may be drawn from the water above or below if the sand is permeable enough. However dilation water is injected at this level to ensure that the sand movement is not obstructed.

The main parameters to be used for the surge bin design were determined from the establishment of material flow properties by TUNRA Bulk Solids Handling Research Associates. (Refer Figure 4 and Figure 5). The bulk density value is used to determine the volume required to be contained within the bin (Refer Figure 6 and Figure 7). The storage volume required was determined to be about 180m$^3$.

The final shape of the bin was determined by using a plane flow insert allowing the bin angle to be 63 degrees above the horizontal (refer Figure 5). An axi-symmetrical section was placed above this which required the wall angle to be 73 degrees above the horizontal. The slurry is withdrawn from the bottom of the bin by a pump suction. The consolidated sand is fluidised with water injected through a plenum. This assists in moving the sand towards the outlet and reduces the % solids to a more dilute mixture.

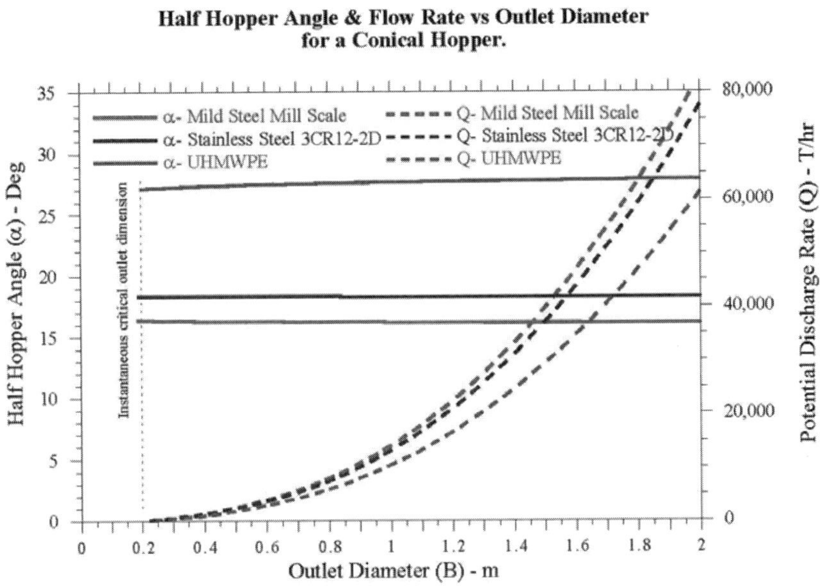


**Figure 4.**

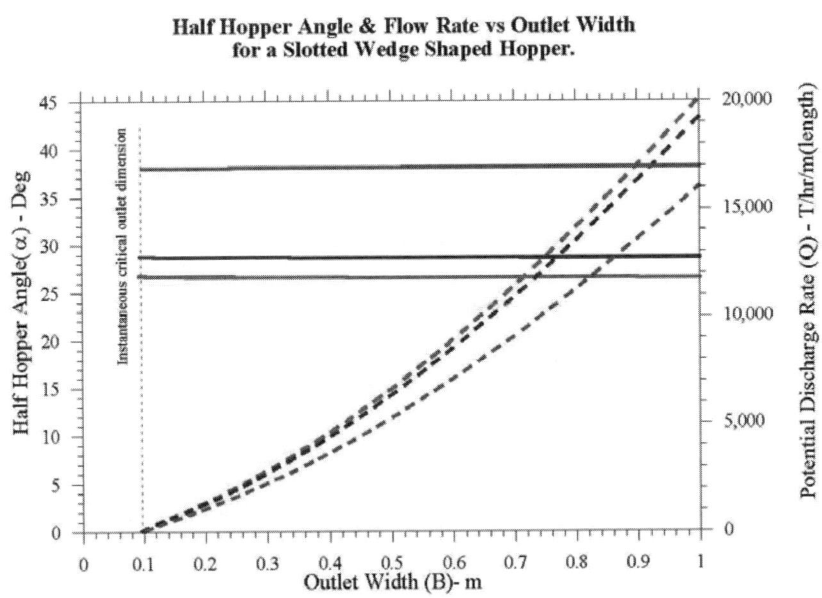


**Figure 5.**

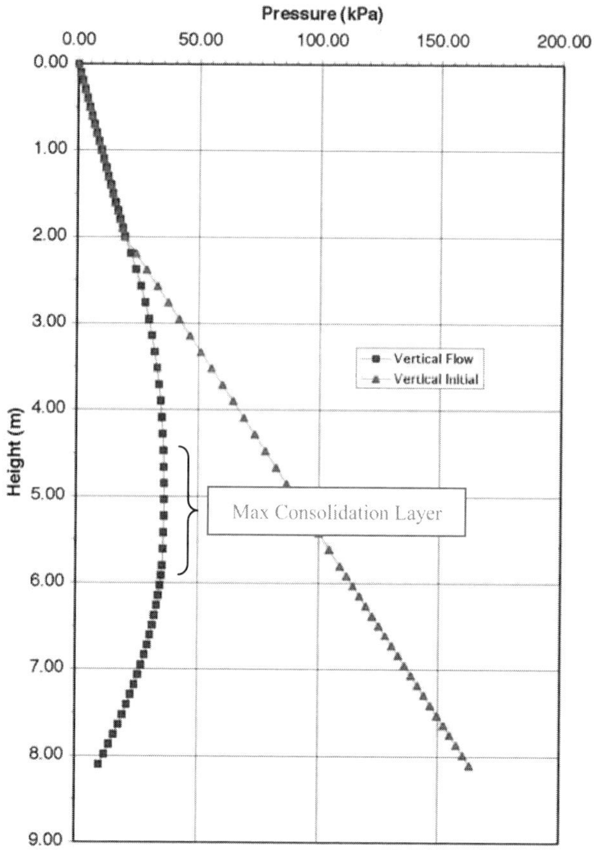


**Figure 6.**

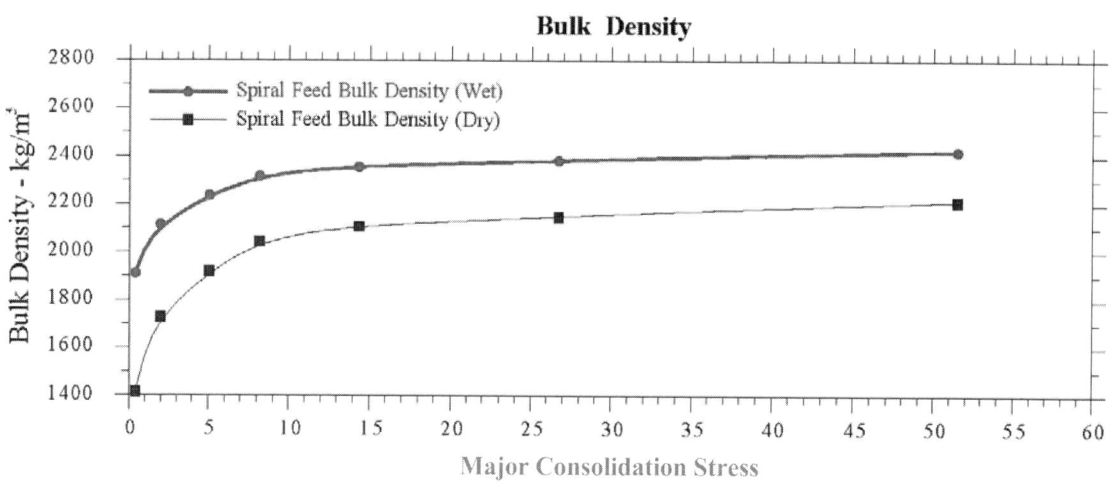


**Figure 7.**

The operation of the bin is controlled by various water injection functions. Successful bin operation relies on careful use of four dilution waters, which must be individually controlled and are drawn from the clean process water tank. The four points of water injection into the bin are:

**Overflow Control**

This water ensures the bin is full of water and maintains a constant overflow from the bin. The flow rate of water is controlled by measuring the bin water level. The bin is fed with high density slurry directly from the de-sliming cyclone underflow. The overflow water is injected directly under the feed point to assist in washing residual slimes to the bin overflow. As the slurry further consolidates within the bin, some of the water from the feed should also report to the overflow.

**Dilation Water**

This water is required to fill the voids formed as the material dilates further down in the bin, going from a higher to a lower level of consolidation. The quantity of water required for dilation is related to the throughput and the expected consolidation stresses lower in the bin. In low permeability bulk solids insufficient dilation water can cause the material to hang up in the bin.

**Fluidisation Water**

This water is injected at the base of the bin and is required to keep the bin active and the material moving towards the pump suction. This system supplies the majority of the dilution water required to achieve the final discharge slurry density. The flow requirement for fluidisation is kept relatively constant over the anticipated operating range of the plant.

**Density Control**

This water is fed directly in to the pump suction pipe work. This water is to be supplied by a variable speed pump controlled by flow and density meters in the pump discharge line. In this way the slurry density and feed rate to the spiral plant can be accurately controlled.

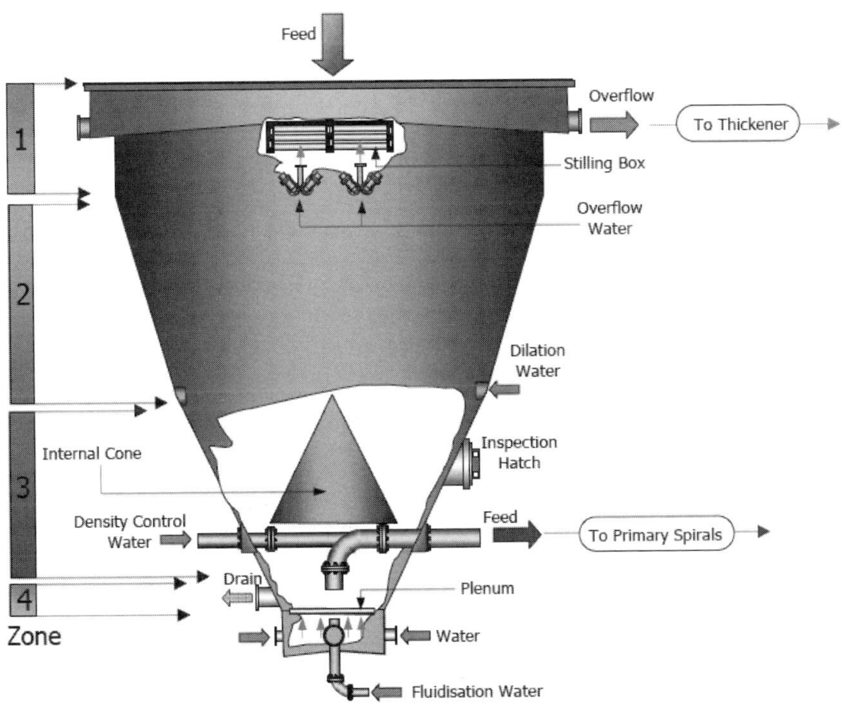


**Figure 8.**

## OPERATIONAL PERFORMANCE

The success of this unit is best illustrated by the spiral feed pump slurry density. (Refer Figure 9) The slurry density at the inlet to the pump suction obviously varies over time but the response of the density control water pump is more than adequate to maintain the density at the set point. If the bin was rat-holing and the solids slumping then a stable output would not be achieved. This consistency of feed allows the plant settings to be optimised to provide maximum recovery (MacHunter[v]). The slurry level in the bin falls over the first 2 hours. During this period the density control water reduces suggesting that the slurry density at the pump suction has reduced. Similarly from 3 hours to 6 hours the slurry level in the bin increases the density control water requirement also increases. However over the range of slurry levels the spiral feed slurry density is held to an SG of 1.3 +/- .001. At hour 7 the operator increased the feed rate set point from 330 tph to 350 tph which caused the feed density to rise appropriately with no change in overall flow rate.

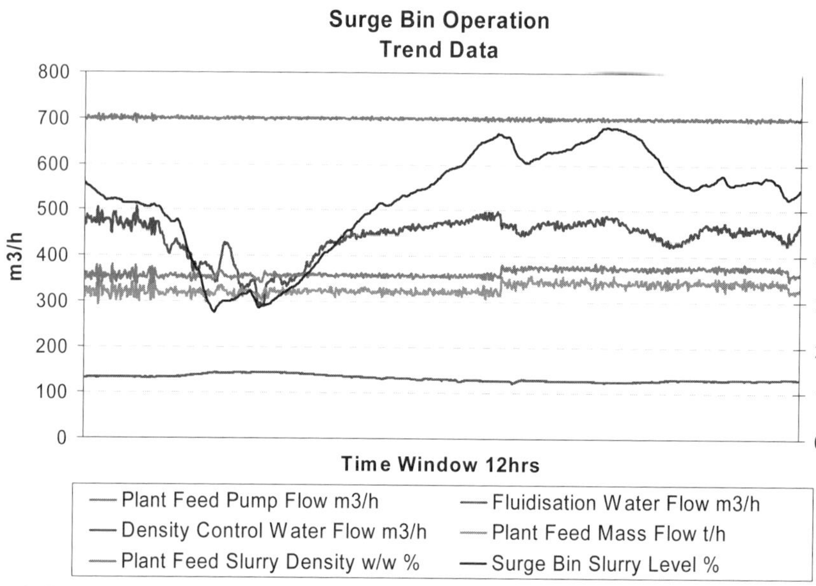


**Figure 9. Surge Bin Operation Over a 12 Hour Period**

## CONCLUSION

This paper has given a brief outline of the application of the bulk solids flow design procedure for providing reliable discharge for a surge bin feeding a mineral sands processing plant. In comparison to a number of alternative designs the mass flow surge bin design provides significantly improved feed stability and reliability across a range of bin levels. The bin performance is enhanced by the responsiveness of the control system.

## REFERENCES

[i] Jenike, A.W., Storage and flow of solids,. Bul. 123 Utah Engineering Experiment Station, University of Utah, Utah,1964

[ii] McLean, A.G., System Design Requirements for Wet Solids Handling., International Conference on Bulk Handling and Transportation; Symposium on freight pipelines, Wollongong, Australia 6-8 July 1992

[iii] Arnold P.C., McLean A.G. and Roberts A.W. 'Bulk Solids: Storage Flow and Handling', TUNRA, The University of Newcastle, 2nd Edition, 1982

[iv] Roberts A.W., 'Basic Principles of Bulk Solids Storage, Flow and Handling'. The Institute for Bulk Materials Handling Research, University of Newcastle, Australia, 1993

[v] MacHunter, R.M.G. and Pax, R.A. Studies of the dynamic characteristics of plant circuits for the development of control strategies. The 6th International Heavy Minerals Conference 'Back to Basics', The Southern African Institute of Mining and Metallurgy, 2007

# PROCESS DESIGN STUDIES FOR NORTHLAND RESOURCES' FENNOSCANDINAVIAN IOCG PROPERTIES

P. Marsden[1], E. Spiller[2], L. Hooey[3], B. Arvidson[4], E. Olsson[5], and G. Chiasson[6]

[1,5] Northland Resources Inc., Datavägen 14, 97754, Luleå, Sweden
[2] Spiller Consultants LLC, 3255 Yarrow Ct., Wheat Ridge, CO 80033 USA
[3]Swerea MEFOS AB, Box 812, SE-971 25, Luleå, Sweden
[4] Bo Arvidson Consulting LLC, 8159 S. Columbine Dr., Morrison, CO 80465, USA
[6] SGS Lakefield Research Limited, 185 Concession St, Lakefield, Ontario K0L 2H0, Canada

## ABSTRACT

Northland Resources Inc. and subsidiary companies Northland Exploration Sweden AB and Northland Exploration Finland Oy are currently developing four Iron-Oxide Copper Gold mineral resource properties in northern Sweden and Finland. The Tapuli, Stora Sahavaara, and Pellivuoma occurrences lie in the Pajala region of Sweden, and the Hannukainen occurrence is in the Kolari region of Finland about 30 km from Pajala. Significant metallurgical development work has already been conducted on all four of these mineral resources. Process studies have shown that high grade iron ore concentrates can be produced from all four mineral resources. In terms of metallurgical processing Tapuli is the least complex, with grinding and low-intensity magnetic separation (LIMS) required to produce a saleable iron ore concentrate. The Stora Sahavaara, Pellivuoma and Hannukainen mineral resources contain pyrite, pyrrhotite and chalcopyrite which require flotation stages to remove the sulfides. Hannukainen has the additional potential to produce a copper-gold bearing concentrate. Geometallurgical work on Hannukainen core samples has shown that strong predictive models for final product grade and recovery can be developed from typical chemical/mineralogical exploration data on drill hole increments.

## INTRODUCTION

Northland Resources Inc. and subsidiary companies are conducting advanced metallurgical testwork programs for the development of four properties in the Pajala and Kolari districts of Sweden and Finland. The Tapuli, Stora Sahavaara, and Pellivuoma resources lie in the Pajala region of Sweden, while the Hannukainen property is in the Kolari region of Finland, around 30 km from the Swedish mineral resource properties, see Figure 1. The four mineral resources have been classified as IOCG (Iron-oxide copper gold) type and lie within the Pajala Shear Zone. The dominant iron-bearing mineral in all four is magnetite; however there are some significant differences in the metallurgical processing characteristics. Tapuli ore contains mostly low sulphur ore and is amenable to upgrade by low intensity magnetic separation only. Stora Sahavaara, Pellivuoma, and Hannukainen contain enough multi-phase pyrrhotite to necessitate flotation for sulfur removal. Hannukainen also contains significant recoverable chalcopyrite and associated gold. Whilst Northland management is currently focused on development of the three Swedish resources (Tapuli, Stora Sahavaara, and Pellivuoma) for treatment in a common beneficiation plant at Kaunisvaara, development work continues on the Hannukainen project (with its by-product metal values) with a view to bringing the mine to production as soon as possible.

## RESOURCE

The NI43-101 compliant mineral resource estimates as of April 29, 2009 for each of the occurrences are given in Table 1 (News Release 2009). Significant additional drilling is ongoing in the region with the primary purpose to increase the total NI 43-101 compliant resources in the area.

## METALLURGICAL FLOWSHEET DEVELOPMENT

There has been significant flowsheet development work done on all four of the resources. The development of the Stora Sahavaara and Hannukainen mineral processing flowsheets was described by Lascelles et al (Lascelles 2008). More recently Hooey et al updated the metallurgical process development of these same properties (Hooey 2009). Since then laboratory and pilot scale testwork has been conducted on ore from Pellivuoma and geometallurgical orebody variability testing has been completed on ore drill core from Hannukainen.

### Tapuli Flowsheet Development

The objective of Tapuli ore processing was to produce a high-grade iron ore (magnetite) concentrate suitable for pelletizing. During the course of early study the possibility of producing a concentrate suitable for sintering was rejected because magnetite liberation size was determined to be too fine for acceptance by the steel industry for use in their sinter plants.

One of the challenges in the Tapuli mineralization is the presence of sulfur, although at low levels. Early exploration work determined that high sulfur zones are quite limited and generally tend to be located near the footwall. Therefore, by appropriate mine planning these high-sulfur pockets would not be forwarded to the main process plant during the early mining phases, thereby eliminiating the need for flotation for most of the ore. Higher sulfur ore will be investigated for potential beneficiation through a Stora Sahavaara and Pellivuoma process flowsheet circuit being considered in the central Kaunisvaara plant.

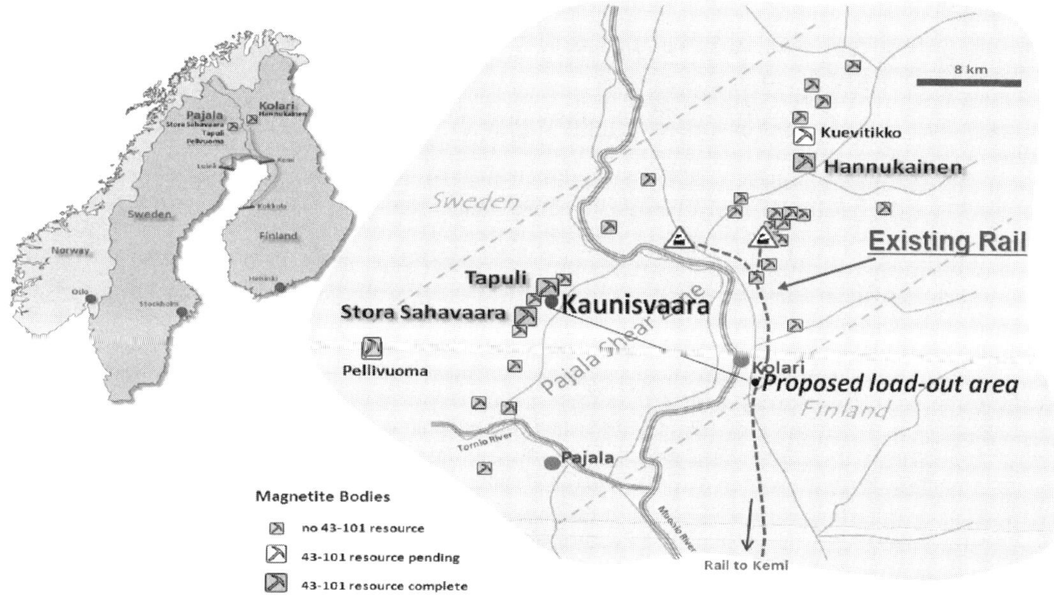


**Figure 1. Location of Northlands' mineral properties in the Pajala shear zone**

**Table 1. N43-101 Compliant resource estimates for Tapuli, Stora Sahavaara, Hannukainen, and Pellivuoma**

| **Tapuli 15% Cut off, Dec 2008** | **Mt** | **Fe%** | **Cu%** | **Au g/t** |
|---|---|---|---|---|
| Measured | 59.33 | 27.8 | -- | -- |
| Indicated | 34.85 | 24.1 | -- | -- |
| Total Measured and Indicated | 94.18 | 26.4 | -- | -- |
| Inferred | 9.74 | 23.7 | -- | -- |
| **Stora Sahavaara 25% cut off, June 2006** | **Mt** | **Fe%** | **Cu%** | **Au g/t** |
| Measured | 77.06 | 43.3 | 0.08 | -- |
| Indicated | 44.64 | 43.3 | 0.08 | -- |
| Total Measured and Indicated | 121.70 | 43.3 | 0.08 | -- |
| Inferred | 23.00 | 42.0 | 0.05 | -- |
| **Hannukainen 15% cut off, Sept 2007** | **Mt** | **Fe%** | **Cu%** | **Au g/t** |
| Measured | 51.14 | 35.6 | 0.25 | 0.12 |
| Indicated | 31.46 | 32.9 | 0.11 | 0.04 |
| Total Measured and Indicated | 82.60 | 34.6 | 0.20 | 0.09 |
| Inferred | 81.63 | 34.6 | 0.13 | 0.04 |
| **Pellivuoma 20% cut off, April 2009** | **Mt** | **Fe%** | **Cu%** | **Au g/t** |
| Measured | -- | -- | -- | -- |
| Indicated | 33.80 | 30.1 | -- | -- |
| Total Measured and Indicated | 33.80 | 30.1 | -- | -- |
| Inferred | 57.00 | 29.8 | -- | -- |

**Preliminary Bench Scale Testwork.** Preliminary bench scale testwork was performed on six drill cores from Tapuli by SGS Cornwall, UK. This work established that a high grade magnetite concentrate could be produced at a liberation grind size of 106–250 μm. Of particular interest was liberation of non-magnetic sulfur-bearing gangue which was accomplished at 106 μm. This work concluded that a magnetite concentrate could be produced with a sulfur content of

**Table 2. Summary of 50 MLA mineralogical assessments of magnetic and non-magnetic products after Davis tube separation of Tapuli drill core composites**

| | Feed Wt % | Non-magnetic Wt % | Magnetic Wt % |
|---|---|---|---|
| Serpentines | 14.4 | 23.1 | 0.8 |
| Amphiboles | 9.5 | 15.2 | 0.7 |
| Clinopyroxene | 6.3 | 9.9 | 0.7 |
| Micas | 11.7 | 18.5 | 1.1 |
| Silicates | 9.8 | 15.7 | 0.5 |
| Carbonates | 7.6 | 12.2 | 0.2 |
| Magnetite | 38.7 | 2.5 | 95.4 |
| Pyrrhotite | 0.30 | 0.3 | 0.3 |
| Chalcopyrite | 0.003 | 0.005 | 0.001 |
| **Total (Sum)** | **98.2** | **97.3** | **99.7** |

less than 0.05%. Dry cobbing results showed that the material was amenable to dry cobbing which was tested at <6.3 mm and <2.36 mm sizes. Ball mill grindability testing showed the ore to be in the soft-medium range.

Further bench-scale testwork was done by the Geological Survey of Finland (GTK) on reject pulps ($P_{85}$ 75 µm) from seven drill cores. One meter intervals were composited into about 5 meter composites and Davis tube separations made. These tests confirmed the necessity to selectively exclude the higher sulfur ore pockets if the ore is processed with only magnetic separation. Preliminary iron recovery correlations were also made. Quantitative mineralogical assessment of the concentrate and tailings products showed a mixture of gangue minerals including serpentine and micas, see Table 2.

From the information provided by the preliminary testing program a provisional flowsheet was envisioned using conventional crushing to 50 mm, HPGR to 2.8 mm, coarse dry (natural moisture) cobbing at <2.8 mm, wet ball mill grinding to nominal 120 µm followed by magnetic separation. Subsequent testing determined that finer grinding was required for adequate liberation (see subsequent discussion).

**Pilot Plant Testwork.** In the winter of 2007- 08 a drilling program provided 25 tonnes of drill core for further bench scale and pilot plant testwork. Drilling and logging of 35 holes was completed in August 2008, following which the drill core was sent to GTK where the core was composited according to mine planning estimates to provide a representative average sample of the mineral resource. The iron grade of the composite sample was extremely close to the NI43-101 compliant estimates: 27.1% $Fe^T$ in the composite sample compared to the first NI43-101 estimate of indicated resources of 27.7% $Fe^T$; and the most recent NI43-101 complaint estimate of measured and indicated resources at 26.4% $Fe^T$.

During pre-pilot plant testwork (which aimed to establish the preferred flowsheet) it was assumed that the moisture content of the ROM ore to the cobbing plant would be >3% at 11 mm and that this stage would be followed by wet ball milling. This process would use crushing and wet milling only and eliminates the original idea of using a HPGR, which, it was concluded, would be a risky practice when the efficiency of the rolls could potentially be compromised by feed moistures >2%.

The cobber concentrate was ground to finer sizes and subjected to Davis tube separations to determine if a wet cobbing stage was warranted. The results showed that the first grinding stage should be at <0.5 mm, giving a nominal $P_{80}$ 340 µm. Grinding and LIMS separation were repeated in order to achieve a high-grade concentrate, Fe >68%, as it was necessary to grind to about $P_{80}$ 60 µm. This more representative sample of Tapuli ore showed that it was necessary to grind the ore to a significantly finer level than the preliminary testwork had estimated.

The pilot plant was then set up as shown in Figure 2, using a rod mill combined with a ball mill in closed circuit with a 0.5 mm vibrating screen. This was used in lieu of a ball mill circuit in the first stage grinding to better simulate industrial performance. Wet cobbing was followed by the second grinding stage to $P_{80}$ 60 µm via closed circuit ball mill/hydrocyclone and three stage LIMS, see Figure 3. Total iron recovery was approximately 85%. The final product analyses are shown in Table 3.

**Summary Tapuli Processing Considerations.** Pilot plant study resulted in a quality concentrate product and the data can be used for prefeasibility study. The relatively simple process comprising comminution and magnetic separation should be easily commercialized. Autogenous milling may offer economic advantages and should be considered in future trade-off studies. However, alternative comminution systems and/or flowsheet configurations are not expected to change or improve the overall metallurgical performance (recovery and grade).

**Stora Sahavaara Flowsheet Development**

Stora Sahavaara ore has been studied to a considerable extent on drill core and a 2200 tonne mined bulk sample taken in 2007. The mined sample was reduced to 227 tonnes and provided to SGS Mineral Services for pilot testwork. Flowsheet development was primarily the work of SGS Mineral Services (SGS Lakefield Research), the results of which were recently published (Lascelles 2008). Process development work included bench and pilot scale studies. In general the ore required both magnetic separation and flotation, at a relatively low (acidic) pH, to produce an acceptable iron

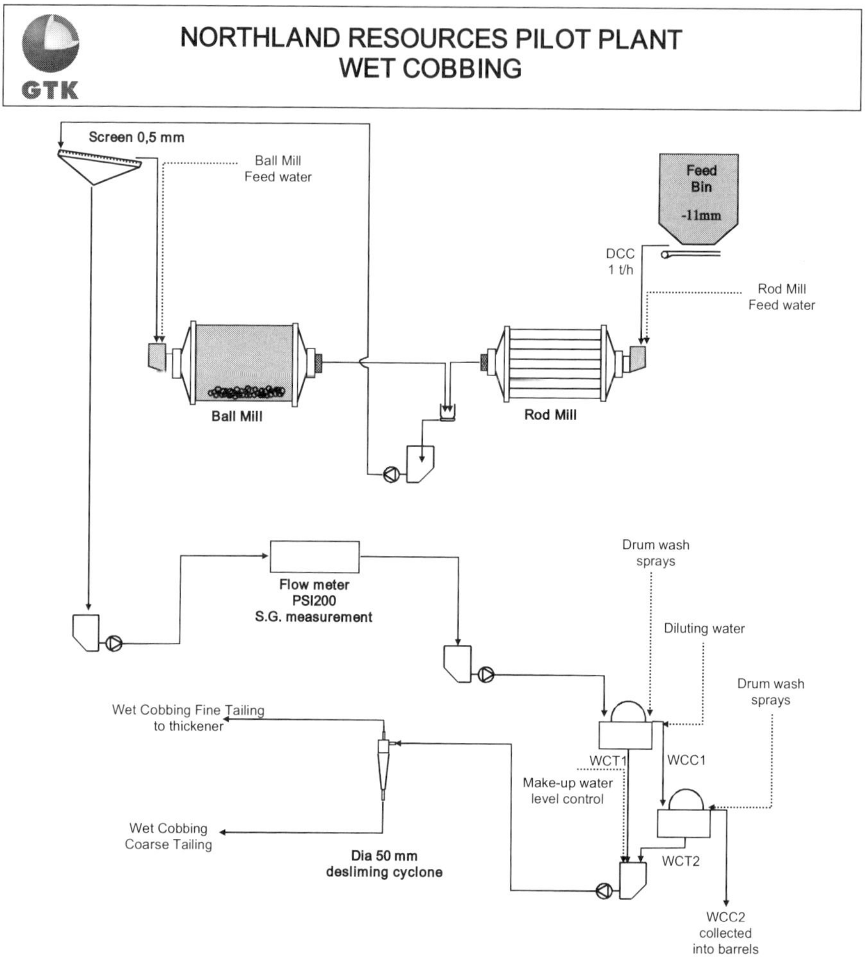


**Figure 2. Tapuli Pilot Plant Wet Cobbing Flow Diagram**

concentrate. Indications were that autogenous grinding resulted in a process with the lowest energy consumption and cleanest concentrate. The flowsheets investigated are shown in Figure 4.

Early study established that achieving a high quality Fe concentrate low in sulfur required fine liberation grinding to $P_{80}$ 90 μm and that flotation was needed to remove the magnetic pyrrhotite. Although the silica and alumina content of the concentrate produced from the bulk sample and pilot plant processing was low at <1.5% $SiO_2$ + $Al_2O_3$, the MgO content was found to be quite high at >4%. Further grinding to about $P_{80}$ 40 μm (approximately pellet feed size) and additional LIMS separation improved the quality of the concentrate by lowering the MgO, $SiO_2$ and $Al_2O_3$ content and increasing the Fe content.

The issue of MgO in the pilot plant concentrate prompted additional detail study of the orebody with respect to its occurrence. This detail study used Davis tube separation and mineralogical characterization of the resulting magnetic products. In total 233 Davis tube tests were conducted across the orebody. From the Davis tube products 264 magnetite grains were selected for EPMA-WDS (microprobe) analyses. Using this technique it was determined that the weighted average of MgO within the magnetite grains averaged 2.60%. However, within the Stora Sahavaara orebody there is the minor ore lens "Södra Sahavaara" and magnetite grains specific to this lens contained a weighted average of only 0.37% MgO. Thus the variability of MgO in magnetite is an issue that will have to be addressed by process flexibility in the production facility. Additional laboratory and pilot scale testwork is ongoing in conjunction with metallurgical process testing of Pellivuoma ore, to better establish grind liberation size and flotation schemes to address the MgO issue and its variability across the orebody.

**Pellivuoma Flowsheet Development**

The Pellivuoma orebody is very similar to Stora Sahavaara because it contains magnetic pyrrhotite that reports directly to a low intensity magnetic separation (LIMS) product. Fortunately, the pyrrhotite responds to flotation and ongoing testwork has produced a high quality magnetite concentrate.

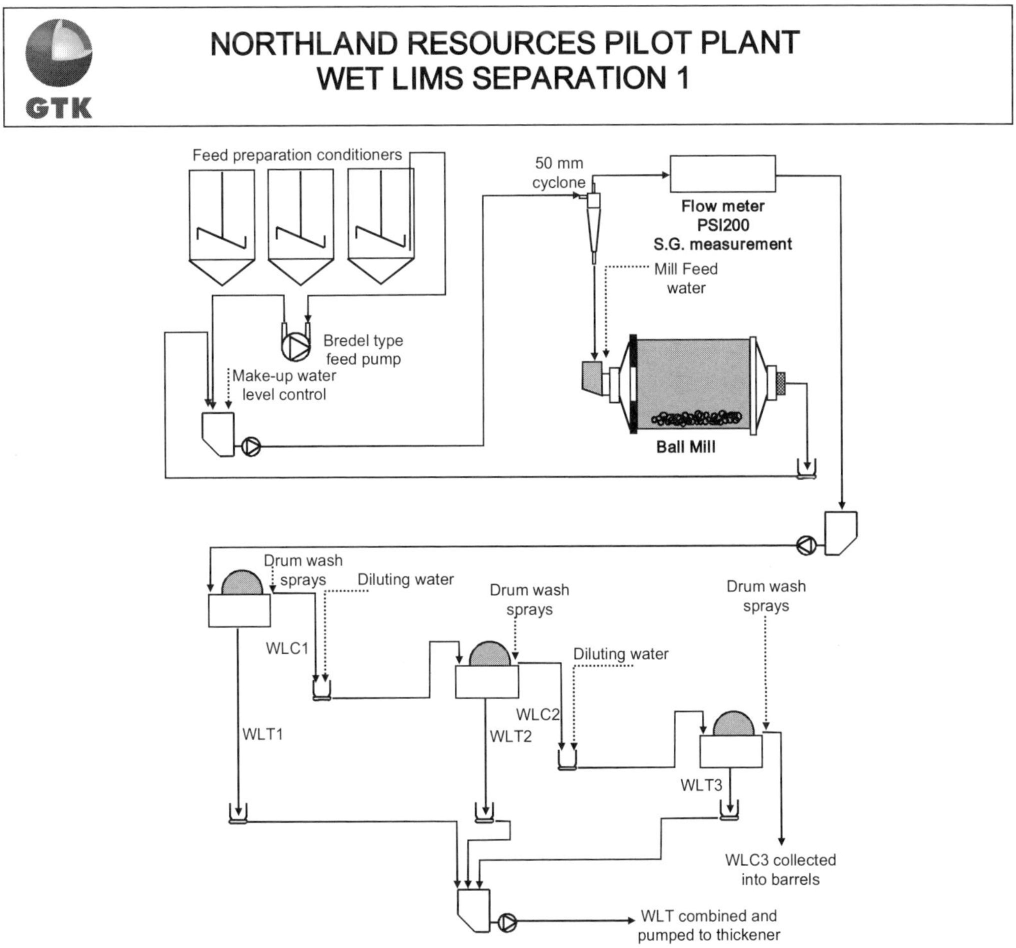


**Figure 3. Tapuli Pilot Plant LIMS flow diagram**

**Table 3. Tapuli Pilot Plant final concentrate analyses.**

| Analyses wt% | FeT | SiO2 | Al2O3 | CaO | MgO | S | Na | K | Blaine $cm^2/g$ |
|---|---|---|---|---|---|---|---|---|---|
| Tapuli Concentrate | 69.3 | 0.95 | 0.14 | 0.10 | 1.99 | 0.045 | 0.007 | 0.013 | 1700 |

Laboratory scale testwork on core sample material has been conducted by GTK following a simplified process flowsheet (see Figure 5) compared to earlier Stora Sahavaara process (see Figure 4). The concentrate produced from average Pellivuoma ore contained 68.4% $Fe^T$, 0.03% S, 0.90% $SiO_2$, 0.10% $Al_2O_3$, 0.03% CaO, 1.94% MgO, and 0.04% $TiO_2$. This product is very similar to concentrates produced from the Tapuli orebody using only comminution followed by LIMS (News Release, 2009).

**Hannukainen Flowsheet Development**

The Hannukainen resource is located in the Kolori area of northern Finland and about 30 km from the Pajala region of Sweden where the three previously described resources are located. Similar to the Pajala area resources the iron ore product is magnetite, and similar to both Stora Sahavaara and Pellivuoma ores, it contains magnetic pyrrhotite that requires flotation for its removal/rejection, hence sulfur reduction. However, and different from all three Pajala area resources, the Hannukainen resource contains both copper and gold in potentially economic quantities. Thus, the process for Hannukainen must account for these by-product metals's recovery. There have been two significant phases of process design studies on the Hannukainen resource, both conducted by SGS Mineral Services; initial (phase 1) laboratory scale studies reported as "The Beneficiation Characteristics of Samples from the Hannukainen Deposit" September 9, 2008 has

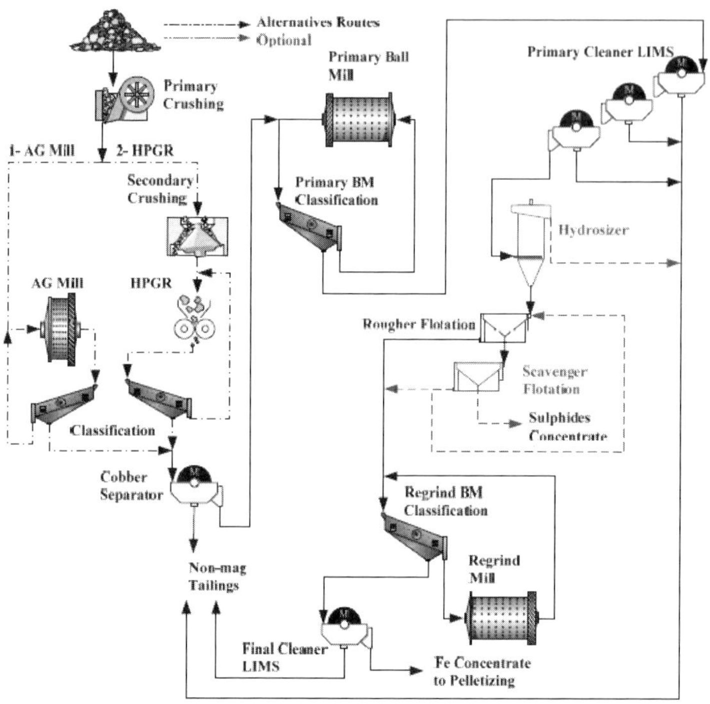


Figure 4. Stora Sahavaara Flowsheet Options (SGS Mineral Services)

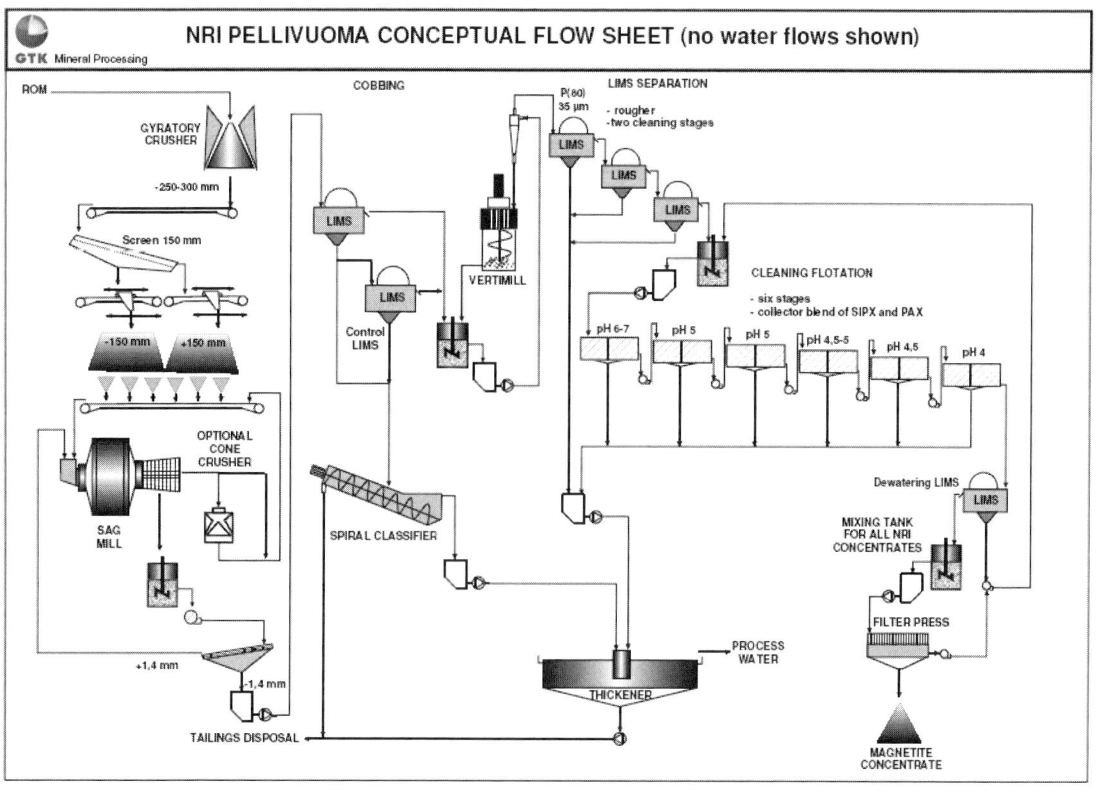


Figure 5. Pellivuoma Simplified Conceptual Flowsheet

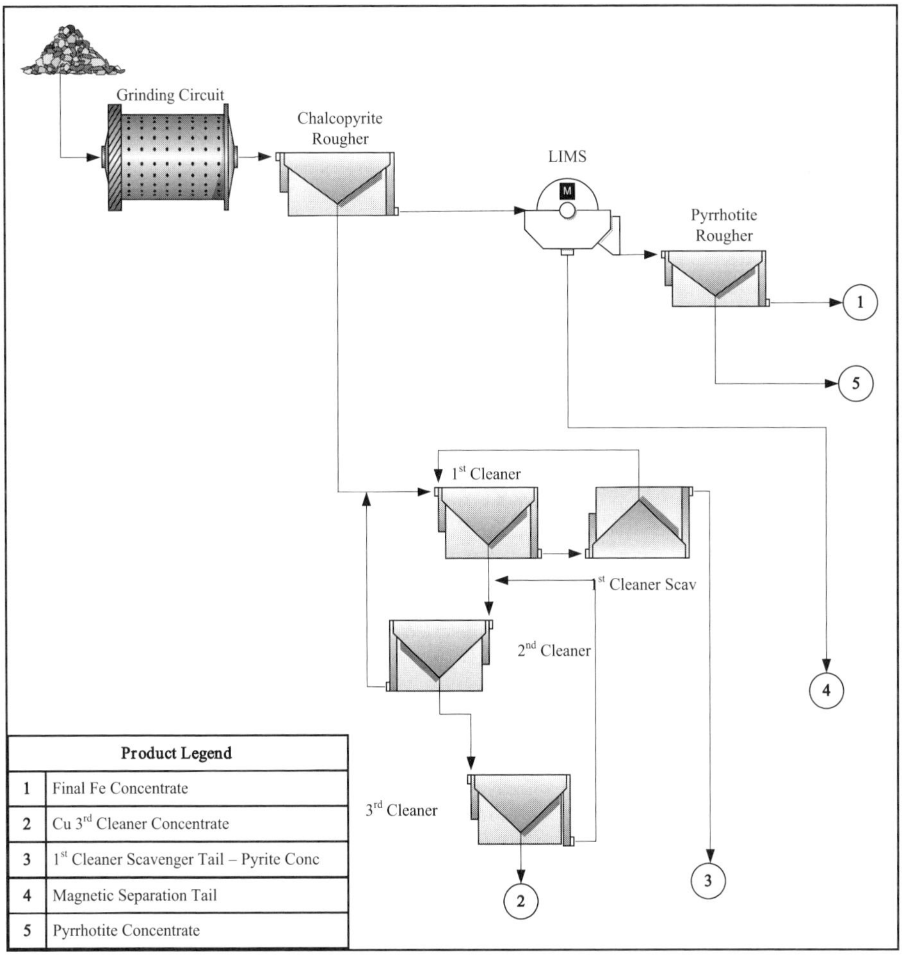


**Figure 6. Flowsheet 3 for Hannukainen ore, developed in phase 1 study at SGS Mineral Services**

been summarized by Lascelles et al. The second phase of study, also at laboratory scale, was reported as "The Baseline Characterization and Metallurgical Response of Hannukainen Cluster Samples" July 2, 2009. This paper will primarily present findings from this second phase of process design study.

**Beneficiation Characteristics (Phase 1).** The feed to phase 1 testing was composited increments from a twin 'typical' drill hole (HAN07009 A and B) in the Hannukainen resource. Early testwork showed that if significant copper and gold were to be recovered then the whole ore needed to be ground and beneficiated, i.e., coarse cobbing presented too much metal loss. In the first phase of study three flowsheets were examined in detail at bench scale. All three involved flotation of pyrite, chalcopyrite, and pyrrhotite, and magnetic separation of magnetite in slightly differing configurations. Flowsheet 3, presented here as Figure 6, emerged from the first phase of study (Lascelles 2008) (McKen 2008) as most efficient on the basis of iron, copper, and gold recovery.

Flowsheet 3 resulted in five product streams; magnetite ($Fe_3O_4$) concentrate, chalcopyrite ($CuFeS_2$) copper concentrate enriched in gold, pyrrhotite concentrate, pyrite concentrate, and a low sulfur tailing. Typical product stream analyses produced from Flowsheet 3 are presented in Table 4.

Magnetite losses from Flowsheet 3 were eventually identified as primarily attributed to entrained carryover during pyrrhotite flotation. Flowsheet 2 (not presented herein) differed from Flowsheet 3 in that pyrrhotite flotation was done ahead of LIMS separation which should reduce carryover and could improve magnetite beneficiation recovery efficiency. In addition, it is also possible that cleaning (scavenging magnetite) from the pyrrhotite froth product could also recover additional iron units. Further investigation of these flowsheet alternatives has been recommended for future consideration.

**Baseline Characterization of Cluster Samples (Geometallurgical—Phase 2).** The initial purpose of the geometallurgical approach to phase 2 testwork was to aid in the selection of an appropriate location for mining a bulk sample for pilot scale testwork. To this end the SGS Mineral Services Geometallurgical team analyzed the Northland drill core database and identified three groups of chemically comparable intervals (ore types) in the deposit; referred to as "Clusters 1, 2, and 3" (Turner-Saad 2008). The clusters were developed without benefit of mineralogical information on the drill core intervals. Once identified, the clusters were used to identify a target area with a contained existing drill hole, from which near surface ore could be selectively mined to produce sample materials of all three clusters for pilot testing and eventually to aid in modeling of the resource for mine planning. HAN07009 drill hole was selected and laboratory

**Table 4. Flowsheet 3 product stream analyses, Hannukainen ore**

| Element | Unit | Head | Fe Conc | Cu Conc | Po Conc | Py Conc | Low S tails |
|---|---|---|---|---|---|---|---|
| Cu | % | 0.26 | 0.002 | 26.2 | 0.15 | 0.33 | 0.012 |
| Fe | % | 35.2 | 70.6 | 31.8 | 63.3 | 39.7 | 8.53 |
| Au | g/t | 0.10 | 0.02 | 7.06 | 0.11 | 0.74 | 0.07 |
| S | % | 2.67 | 0.05 | 28.2 | 22.6 | 35.3 | 0.12 |
| SiO2 | % | 27.9 | 0.98 | 14.7 | 0.94 | --- | 49.3 |
| Al2O3 | % | 4.25 | 0.39 | 1.04 | 0.27 | --- | 6.83 |
| MgO | % | 4.42 | 0.35 | 0.60 | 0.24 | --- | 7.76 |
| CaO | % | 9.10 | 0.29 | 1.74 | 0.19 | --- | 16.1 |
| Na2O | % | 1.25 | <0.01 | 0.21 | 0.02 | --- | 2.05 |
| K2O | % | 0.41 | 0.01 | 0.26 | 0.02 | --- | 0.76 |
| TiO2 | % | 0.19 | 0.11 | 0.06 | 0.05 | --- | 0.25 |
| P2O5 | % | 0.12 | <0.01 | <0.02 | <0.01 | --- | 0.18 |
| MnO | % | 0.19 | 0.12 | 0.04 | 0.05 | --- | 0.27 |
| Cr2O3 | % | <0.01 | 0.02 | 0.003 | 0.05 | --- | 0.01 |
| V2O5 | % | 0.01 | 0.02 | 0.002 | <0.01 | --- | <0.01 |
| LOI | % | 0.49 | -2.97 | 14.0 | 8.04 | --- | 0.82 |
| Sum | % | 98.7 | 100.3 | N/A | 100.4 | --- | 99.0 |
| wt yield | % | 100 | 31.5 | 0.91 | 4.69 | 3.04 | 59.8 |

testwork was conducted on cluster composite material from drill hole interval samples; identified as Bulk Clusters 1, 2, and 3. Ultimately testwork was done on combined Bulk Clusters 1 and 3 (1/3), and on Bulk Cluster 2. It should be noted that although the bulk clusters from drill hole HAN07009 were all within the geometallurgical limits of the cluster definitions, their average chemical compositions significantly diverged from the average orebody cluster groupings. The orebody wide cluster groupings were termed "global" clusters (Chiasson 2009).

To better understand the relationship of the Bulk Clusters to Global Clusters additional laboratory testwork was conducted to compare mineralogical and metallurgical process characteristics of the *bulk clusters*, representing the potential bulk sample for pilot scale testwork, and the *global clusters*, representing three ore types throughout the orebody. Additional detail was gathered by individual characterization and beneficiation testing (by Davis tube) of the 103 drill core increment samples that comprised the three global clusters. Finally, Bulk Clusters 1/3 and 2 and Global Clusters 1, 2, and 3 were subjected to bench scale tests following Flowsheet 3 as previously described. The testwork included both full grindability and metallurgical response testwork.

The data showed the bulk cluster samples to be significantly different than the global cluster samples and the original HAN07009 drill hole sample composite material. Thus it was determined that the proposed target area for bulk mining/sampling was not appropriate. Positively, four of the five cluster samples achieved very high grade iron ore concentrates (> 69% $Fe^T$) and all contained less than 0.06% S; note Bulk Cluster 2 concentrate contained an acceptable 66.4% $Fe^T$. The chemistry and mineralogical differences between the cluster samples themselves and the drill hole HAN07172 ore composite, on which the initial process flowsheet was developed, has indicated a need to further optimize both the pyrrhotite flotation for entrained magnetite and chalcopyrite flotation for Cu and Au recovery.

The following Table 5 presents details of the metallurgical testing conducted on the cluster samples as well as the initial drill hole HAN07172 ore composite. In general, the data show significant response differences from the clusters while producing quality iron ore concentrates.

Of particular interest from the geometallurgical study in phase 2 was the relationships developed for each of the 103 interval drill core samples. The objective of this work was to improve the reliability of the resource model by predicting metallurgical process performance of various ore units from basic chemical data contained in the exploration data base. Relationships were developed between the routine exploration chemical/mineralogical analyses and magnetic (Davis tube) iron and weight recovery and other important concentrate attributes. Each of the 103 increment sample was subjected to Davis tube magnetic separation and QEMSCAN (automated mineralogy) analyses, and in addition (as describe above) each cluster was analyzed for its response to the developed Flowsheet 3. From the initial chemical analyses for each increment, the data were then used to predict the weight recovery, the recoverable Fe, as well as the S and $SiO_2$ grades in the magnetic iron concentrates. Although many relationships were examined, three are presented here as examples of what might be possible from this type of data analyses.

***Weight Recovery to Davis tube magnetic product.*** A clear relationship was established (overall $R^2$ of 0.97) between Satmagan magnetic iron in the drill hole increment and weight recovery to the Davis tube magnetic product as shown in Figure 7.

**Table 5. Metallurgical Testing Summary, Hannukainen Cluster Samples**

| Sample/ | Products | Weight | Assays, %, g/t | | | | | | | | % Distribution | | | | |
|---|---|---|---|---|---|---|---|---|---|---|---|---|---|---|---|
| Reagent Dosages | | % | $Fe_{Tot}$ | S | Cu | Au | Co | $SiO_2$ | $Al_2O_3$ | MgO | Fe | S | Cu | Au | Co |
| HAN07009 | Fe Concentrate | 31.5 | 70.6 | 0.05 | 0.002 | 0.02 | 0.006 | 0.98 | 0.39 | 0.35 | 67.8 | 0.63 | 0.24 | 4.33 | 7.62 |
| 1275g/t Ca(OH)2 | Cu Concentrate | 0.91 | 32.6 | 28.7 | 26.2 | 7.66 | 0.082 | 14.7 | 1.04 | 0.60 | 0.90 | 10.5 | 90.6 | 48.0 | 3.01 |
| 25g/t PEX, 17.5g/t 3418A | Py Concentrate | 3.04 | 39.7 | 35.3 | 0.33 | 0.74 | 0.37 | - | - | - | 3.68 | 43.3 | 3.82 | 15.4 | 45.4 |
| 180g/t PAX | Po Concentrate | 4.69 | 63.3 | 22.6 | 0.15 | 0.11 | 0.15 | - | - | - | 9.03 | 42.7 | 2.66 | 3.54 | 28.3 |
| 5g/t CMC | Low 'S' Tail | 59.8 | 10.2 | 0.12 | 0.012 | 0.07 | 0.007 | 49.3 | 6.83 | 7.76 | 18.6 | 2.89 | 2.72 | 28.7 | 15.7 |
| 33g/t MIBC | Head (calc) | 100.0 | 32.9 | 2.48 | 0.26 | 0.15 | 0.025 | - | - | - | 100.0 | 100.0 | 100.0 | 100.0 | 100.0 |
| F3/Bulk Clusters 1/3 | Fe Concentrate | 33.0 | 69.9 | 0.03 | <0.002 | 0.06 | 0.020 | 1.59 | 0.38 | 0.51 | 65.7 | 0.27 | 0.89 | 42.4 | 23.1 |
| 20g/t PEX, 15g/t 3418A | Cu Concentrate | 0.29 | 40.7 | 35.5 | 15.9 | 1.79 | 0.31 | 2.80 | 0.53 | 0.77 | 0.34 | 2.84 | 62.7 | 11.2 | 3.20 |
| 430g/t PAX | Py Concentrate | 1.42 | 55.8 | 35.3 | 1.14 | 0.46 | 0.18 | 3.34 | 0.60 | 0.84 | 2.25 | 13.7 | 21.9 | 14.0 | 8.81 |
| 10g/t CMC | Po Concentrate | 10.4 | 62.7 | 24.2 | 0.023 | 0.04 | 0.12 | 0.89 | 0.50 | 0.21 | 18.6 | 69.1 | 3.25 | 8.92 | 43.8 |
| 113g/t MIBC | Low 'S' Tail | 54.9 | 8.34 | 0.94 | 0.015 | <0.02 | <0.01 | 47.8 | 8.77 | 8.32 | 13.1 | 14.1 | 11.2 | 23.5 | 21.1 |
| 290g/t H2SO4 | Head (calc) | 100.0 | 35.1 | 3.65 | 0.074 | 0.05 | 0.029 | 26.9 | 5.00 | 4.78 | 100.0 | 100.0 | 100.0 | 100.0 | 100.0 |
| F4/Bulk Cluster 2 | Fe Concentrate | 19.8 | 66.4 | 0.06 | <0.002 | 0.04 | 0.020 | 4.47 | 0.85 | 1.12 | 39.7 | 0.17 | 0.23 | 15.5 | 10.1 |
| 30g/t PEX, 20g/t 3418A | Cu Concentrate | 0.84 | 42.0 | 36.9 | 14.5 | 1.83 | 0.28 | 2.51 | 0.69 | 0.43 | 1.06 | 4.42 | 72.1 | 30.0 | 6.05 |
| 950g/t PAX | Py Concentrate | 3.91 | 56.6 | 33.1 | 0.69 | 0.11 | 0.15 | 3.12 | 0.81 | 0.64 | 6.68 | 18.5 | 15.9 | 8.20 | 14.6 |
| 10g/t CMC | Po Concentrate | 19.6 | 61.8 | 25.5 | 0.043 | 0.05 | 0.11 | 1.07 | 0.51 | 0.22 | 36.6 | 71.3 | 4.99 | 19.2 | 55.0 |
| 113g/t MIBC | Low 'S' Tail | 55.9 | 9.43 | 0.70 | 0.021 | <0.02 | 0.010 | 45.4 | 8.78 | 9.80 | 15.9 | 5.61 | 6.81 | 27.1 | 14.3 |
| 175g/t H2SO4 | Head (calc) | 100.0 | 33.1 | 7.01 | 0.17 | - | 0.039 | 26.6 | 5.21 | 5.77 | 100.0 | 100.0 | 100.0 | 100.0 | 100.0 |
| F5/Global Cluster 1 | Fe Concentrate | 50.2 | 70.6 | 0.05 | <0.002 | <0.02 | 0.022 | 0.80 | 0.24 | 0.40 | 73.0 | 1.02 | 0.43 | 3.81 | 40.2 |
| 20g/t PEX, 15g/t 3418A | Cu Concentrate | 0.64 | 32.8 | 34.6 | 26.4 | 14.9 | 0.088 | 2.25 | 0.39 | 0.51 | 0.43 | 8.94 | 71.2 | 35.8 | 2.04 |
| 130g/t PAX | Py Concentrate | 1.66 | 38.8 | 42.7 | 1.83 | 3.19 | 0.36 | 2.59 | 0.43 | 0.63 | 1.32 | 28.7 | 12.8 | 20.1 | 21.7 |
| 10g/t CMC | Po Concentrate | 11.6 | 67.4 | 9.64 | 0.079 | 0.57 | 0.044 | 1.45 | 0.30 | 0.48 | 16.1 | 45.4 | 3.88 | 25.1 | 18.6 |
| 68g/t MIBC | Low 'S' Tail | 35.9 | 12.3 | 1.09 | 0.077 | 0.11 | 0.013 | 41.3 | 5.77 | 8.60 | 9.10 | 15.9 | 11.7 | 15.2 | 17.5 |
| 185g/t H2SO4 | Head (calc) | 100.0 | 48.6 | 2.46 | 0.24 | 0.26 | 0.027 | 15.5 | 2.24 | 3.36 | 100.0 | 100.0 | 100.0 | 100.0 | 100.0 |
| F6/Global Cluster 2 | Fe Concentrate | 26.5 | 70.6 | 0.02 | 0.021 | <0.02 | 0.017 | 1.42 | 0.36 | 0.38 | 61.6 | 0.22 | 2.91 | 4.69 | 21.0 |
| 20g/t PEX, 15g/t 3418A | Cu Concentrate | 0.62 | 35.3 | 37.1 | 19.9 | 6.26 | 0.16 | 2.95 | 0.58 | 0.48 | 0.72 | 9.36 | 64.7 | 34.3 | 4.60 |
| 400g/t PAX | Py Concentrate | 1.35 | 42.7 | 42.6 | 1.80 | 2.98 | 0.30 | 4.42 | 0.90 | 0.70 | 1.90 | 23.4 | 12.7 | 35.7 | 19.2 |
| 10g/t CMC | Po Concentrate | 6.87 | 64.3 | 19.1 | 0.110 | 0.11 | 0.077 | 1.34 | 0.28 | 0.27 | 14.6 | 53.5 | 3.96 | 6.70 | 24.7 |
| 65g/t MIBC | Low 'S' Tail | 64.7 | 9.92 | 0.51 | 0.046 | 0.03 | <0.01 | 50.7 | 7.77 | 7.67 | 21.2 | 13.5 | 15.7 | 18.7 | 30.4 |
| 295g/t H2SO4 | Head (calc) | 100.0 | 30.3 | 2.45 | 0.19 | 0.11 | 0.021 | - | - | - | 100.0 | 100.0 | 100.0 | 100.0 | 100.0 |
| F7/Global Cluster 3 | Fe Concentrate | 5.57 | 69.1 | 0.06 | <0.002 | <0.02 | 0.014 | 2.66 | 0.43 | 0.51 | 27.5 | 0.22 | 0.12 | 3.02 | 5.00 |
| 20g/t PEX, 15g/t 3418A | Cu Concentrate | 0.24 | 34.9 | 36.0 | 18.1 | 2.00 | 0.15 | 4.32 | 0.98 | 0.47 | 0.59 | 5.54 | 45.4 | 12.8 | 2.23 |
| 200g/t PAX | Py Concentrate | 1.32 | 39.7 | 40.8 | 1.90 | 0.79 | 0.25 | 7.06 | 1.63 | 0.77 | 3.72 | 35.0 | 26.6 | 28.2 | 20.9 |
| 10g/t CMC | Po Concentrate | 4.05 | 64.6 | 15.8 | 0.066 | 0.07 | 0.057 | 1.57 | 0.34 | 0.35 | 18.7 | 41.7 | 2.84 | 7.69 | 14.8 |
| 60g/t MIBC | Low 'S' Tail | 88.8 | 7.82 | 0.30 | 0.026 | <0.02 | <0.01 | 53.4 | 9.70 | 5.95 | 49.6 | 17.6 | 25.0 | 48.2 | 57.1 |
| 185g/t H2SO4 | Head (calc) | 100.0 | 14.0 | 1.53 | 0.094 | 0.04 | 0.016 | - | - | - | 100.0 | 100.0 | 100.0 | 100.0 | 100.0 |

***Whole Rock Analyses (WRA) to Davis tube magnetic product.*** An excellent relationship was established (overall $R^2$ of 0.99) between WRA FeT in the drill core increment and weight recovery to the Davis tube magnetic product. See Figure 8.

***Pyrrhotite to Sulfur in the Davis tube magnetic product.*** A principle objective of the modelling was to understand the effect of pyrrhotite content in ore on sulfur content in the product, i.e., the Davis tube magnetic product. Pyrrhotite content was determined by QEMSCAN analyses. See Figure 9 for the relationship plot. A very satisfactory $R^2$ of 0.98 was established for Global Cluster 1 increments, $R^2$ of 0.98 for Global Cluster 2, and $R^2$ of 0.97 for Global Cluster 3.

***Satmagan iron in Globle Cluster increments related to SiO2 in the magnetic product.*** A model was generated for comparing Statmagan magnetic iron the the global cluster increments to the SiO2 content in the Davis tube magnetic product. See Figure 10 for the relationship which was relatively good with an $R^2$ of 0.63.

***Summary of Model Relationships.*** The relationships for weight, recoverable Fe, and S content were very good, while the SiO2 model was slightly more scattered. The data lends itself to investigating many relationships, many more than are included here. Grind size, an important varible, was not considered in the $SiO_2$ model.

A global model which predicts the overall Fe recovery, i.e., after pyrrhotite flotation,was developed using the results of a few beneficiation tests on composites which showed that the iron losses in sulfides were directly proportional to the S grade in the LIMS magnetic product. The model was very good except for Global Cluster 2. Futher work is required on this important relationship before it can be applied to the reserve model.

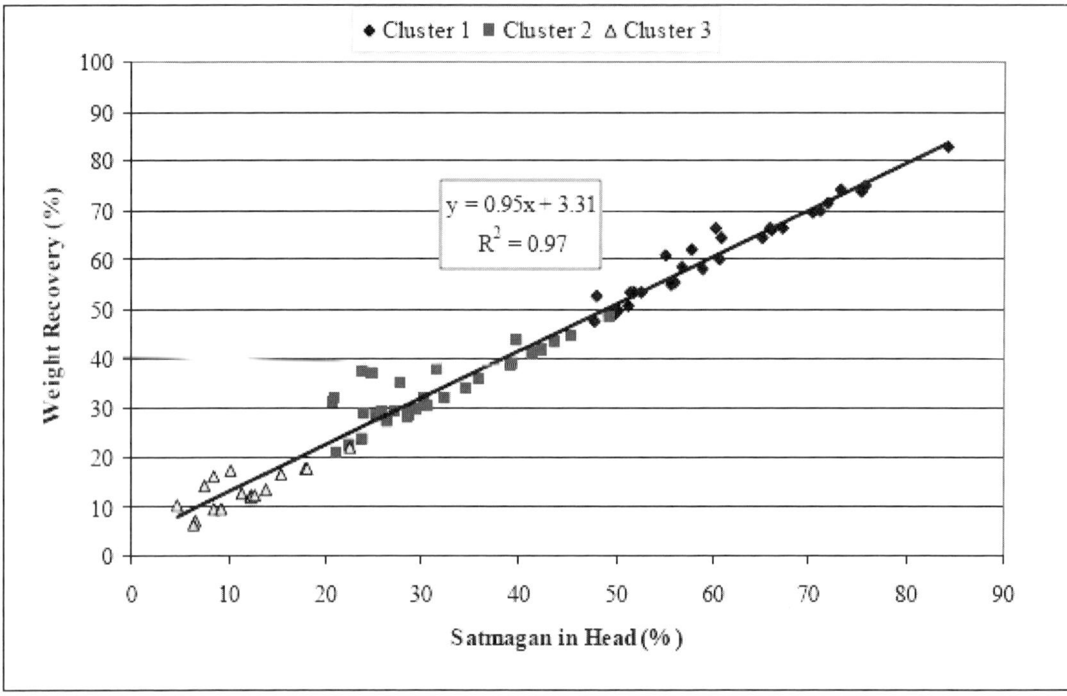


**Figure 7. Correlation between Satmagan iron content and weight recovery to Davis tube magnetic product**

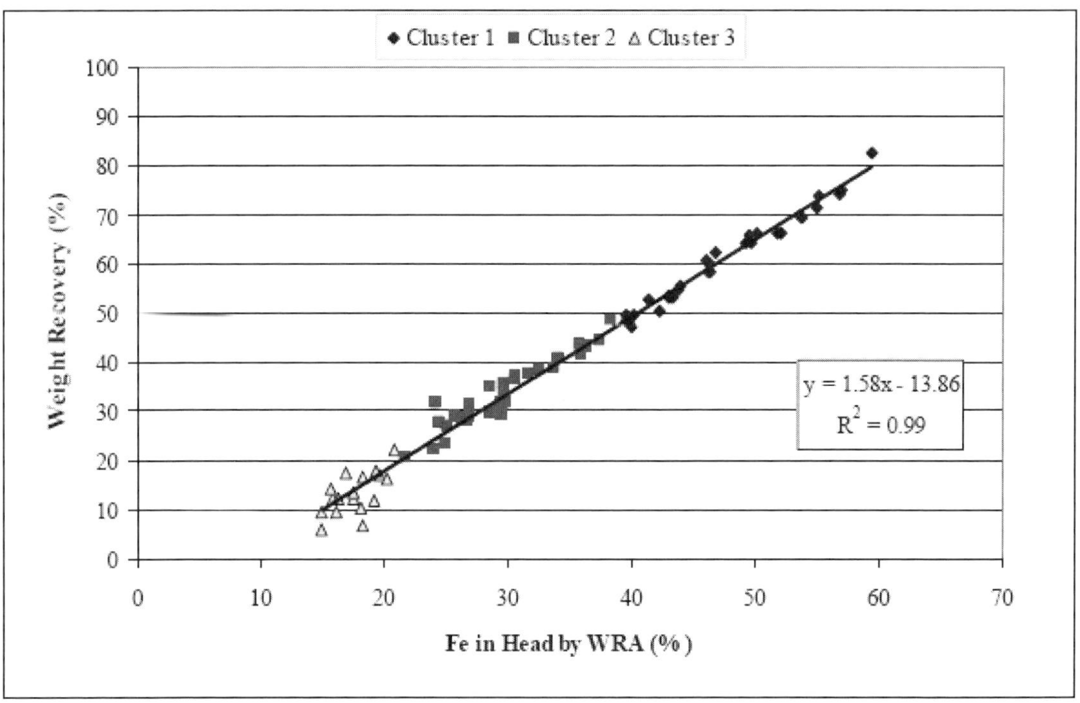


**Figure 8. Correlation between WRA $Fe^T$ and weight recovery to Davis tube magnetic product**

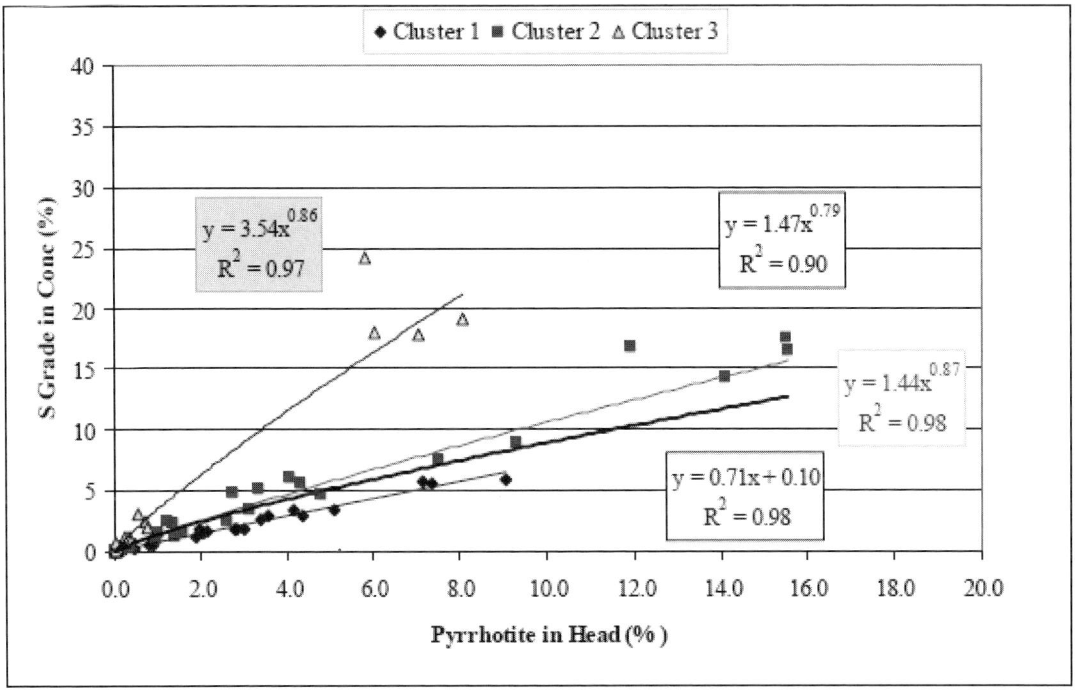


**Figure 9. Correlation between Pyrrhotite in the Global Cluster increments and sulfur grade in the Davis tube magnetic product**

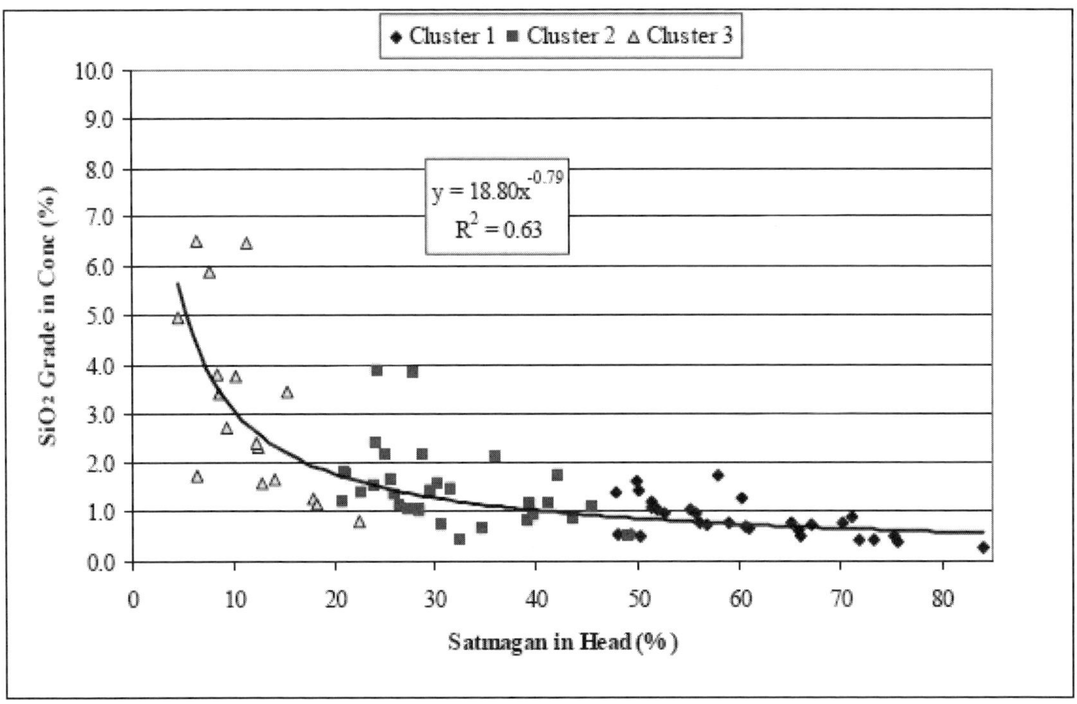


**Figure 10. Correlation between Satmagan magnetic iron content and SiO2 grade in Davis tube magnetic product**

## CONCLUSIONS

Process flowsheets have been developed for Tapuli, Stora Sahavaara, Pellivuoma and Hannukainen iron ores. All four ores require relatively fine grinding, to less than 100 μm to produce high grade magnetite concentrate. Additional studies are required to finalize comminution circuit alternatives. Laboratory and some pilot scale testing has produced high quality (low silica plus alumina) iron ore concentrates suitable for pelletizing from all four resource properties.

Stora Sahavaara, Pellivuoma and Hannukanien ores require flotation to remove pyrrhotite. While pyrrhotite flotation at low pH has shown to be successful, further testwork to reduce acid (reagent) consumption has been intitated. Hannukainen ore has the potential to produce a saleable copper gold concentrate.

A Preliminary Economic Assessment (PEA), commonly called a prefeasibility study, is currently progressing the development of the three Swedish resources in the Pajala region.

## ACKNOWLEDGEMENTS

The authors wish to thank Mr. Tapio Knuutinen, Mr. Markku Kuusisto, Mr. Markku Klemetti and Mr. Pekka Mörsky and their colleagues at the Geological Survey of Finland for their work on the Tapuli pilot plant; and Mr. André McKen, Ms. Dominique Lascelles, Mr. Guillaume Chiasson, Dr. Guillermo Turner-Saad and Dr. Cristian Dragusanu and their colleagues at SGS Mineral Services, Lakefield, for their work on the Hannukainen and Stora Sahavaara flowsheet development.

The assistance of Mr. Esko Pöyliö and Dr. Kyösti Heinänen is gratefully acknowledged.

## REFERENCES

Barry, J.P.; Sandefur, R.L., *Technical report, Stora Sahavaara Project, Northern Sweden*, Chlumsky, Armbrust & Meyer LLC, 2006.

Chiasson, G., McKen, A., *The Baseline Characterization and Metallurgical Response of Hannukainen Cluster Samples*, SGS Mineral Services Technical report, July 2, 2009

Hooey, L., Spiller, D.E., Arvidson, B.A., Marsden, P., Olsson, E., *Metallurgical Development of Northland Resources' IOCG Resources*, Conference In Minerals Engineering 2009, Luleå Sweden, February 2009

Lascelles, D., McKen, A., Chiasson, G., Marsden, P., Hooey, L., Dixon, A., Houseley, K., *The recovery of metals from the Stora Sahavaara and Hannukainen Iron Oxide Copper Gold (IOCG) Resources*, Procemin 2008, Oct. 22-24, 2008, Santiago, Chile, pp. 229-240

Mukhopadhyay, D.K., Steedman, J., Lindholm, T. *Mineral Resource Estimates for the Hannukainen Deposits, Kolari Area, Finland*, NI43-101 Technical Report, Micon International Co Limited, 2007.

McKen, A, Lascelles, D., *The Beneficiation Characteristics of Samples from the Hannukainen Deposit,* SGS Minera Services Technical report, September 9, 2008

News release, *Iron Concentrate from Tapuli and Pellivuoma Exceeds Expectation*, Northland Resources Inc., April 15, 2009

News release, *Pellivuoma Resource to Extend Life of Planned Kaunisvaara Mill Complex,* Northland Resources Inc., April 29, 2009

Turner-Saad, G, Dragusanu, C., *Northland Resources Incorporated; SGS Geometallurgy Framework, July 5, 2008*

# FRAC SAND PLANT DESIGN: A PROCESS SOLUTIONS PROVIDER PERSPECTIVE, FROM TESTWORK TO PROJECT COMPLETION

Peter Refsnider and Jim Sadowski

Outotec (USA) Inc., 6100 Philips Highway, Jacksonville, FL USA

## ABSTRACT

There is a diminishing number of premium, easily exploitable, frac sand deposits in the world. Exploiting many newly discovered deposits requires extensive and imaginative minerals processing design work to meet the needs of clients as they consider bringing their potential products to market.

After broad testing of the ore body, developing a basis of design document is a key step in defining a process as it relates to the quality of the deposit, the processing requirements, subsequent processing equipment and the expectations of the client. Conversely, the expectations of the client are typically grounded in the principles of project costs and scheduling.

This paper will address how a process solution provider develops the design document, and then moves to subsequent design processes through plant construction. How each of these phases impacts the others will also be evaluated. Some discussion on equipment selection for fracturing sand plants is also included.

## INTRODUCTION

Hydraulic fracturing sand plant design encompasses many facets of design and engineering. It is important that meticulousness be used in all areas to ensure a client's needs and expectations are satisfied and that an operator-friendly plant is constructed. The design phase is a rigorous exercise requiring tremendous amounts of input and time from both supplier and client. In the proper design of a plant, the basis of design document is the most important item required.

## BASIS OF DESIGN DOCUMENT

The basis of design document covers: sand deposit and geological characteristics, sand quality, geotechnical issues, climate, mine production capabilities, potential and existing infrastructure, sale of product, and process battery limits, just to name a few. While these items seem "obvious" and "easy," as the project is evolving through its initial exploratory stages, obtaining the necessary fundamental information and gaining agreement or consensus with the client is not always so. Typically, many man-hours can be wasted, with subsequent initial cost escalations to the project, through the many required meetings meant to solidify the vision, alleviate confusion and keep both parties current. Often, agreed to terms, items or basic practices for progress will change throughout these initial stages. What is often not realized is that perceived small and insignificant changes can affect the project schedule and the ultimate cost of the design, hence subsequent plant construction.

### Education Is Key

Success begins when project players on both sides of the table are knowledgeable, to a certain extent, of the skills and processes required to build the plant and fulfill its business needs. More than just an understanding of in-plant engineering and design areas, everyone must grasp the client's future markets and the specific products expected to be produced. The process solutions provider, as a collective of engineers and designers, must be sensitive to the client's ultimate desire of forming a profitable, long-term frac sand business. Conversely, they must lead the client to a true understanding of what exact services are going to be performed, especially if they do not possess the basic knowledge of processing plant design. Time spent early on educating a client of engineering challenges, and other realities of a process design project can equate to huge dividends as the project progresses. This critical step can potentially alleviate many hours of frustration downstream.

### Set the Stage Early

It is far better to agree on the fundamental project issues like product specs, storage, transportation and waste management, for example, during the initial development period. Detailed engineering must not be allowed to begin without clarity of even the most rudimentary changes. Once the detailed engineering phases of the project are begun, the smallest changes or misunderstandings between parties can lead to difficult and too often complicated and expensive fixes and change orders. A seemingly minor adjustment can influence multiple systems within a design, so for each small change requested during detailed engineering, a significant amount of evaluation is required. This evaluation period can quickly become a very timely and costly engineering exercise, and is only a precursor to actual reengineering efforts. Most important to remember is that someone will be required to pay for these excursions.

One example of a seemingly minor change might be the expected sales tonnage; another may be a different process feed rate. Both of these changes have a ripple effect on the overall design. A different plant feed rate could require adjustments to the mining operation, such as the size of the loader or mine haul truck required, or within the plant by requiring a change in the truck dump hopper, feed conveyor, scalping screen, etc. Changes to process equipment can further affect structure and foundation design, as well as overall plant layout. As the changes make their way through the

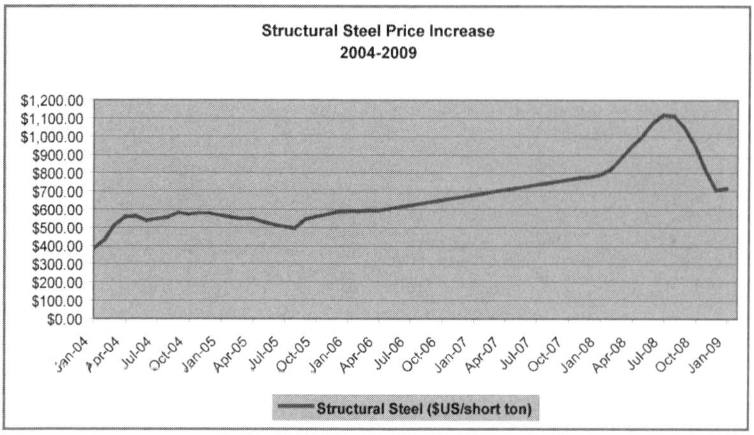


**Figure 1. Structural steel price increase since January 2004. (Pinkham, 2006 & MEPS (International) Ltd, 2009)**

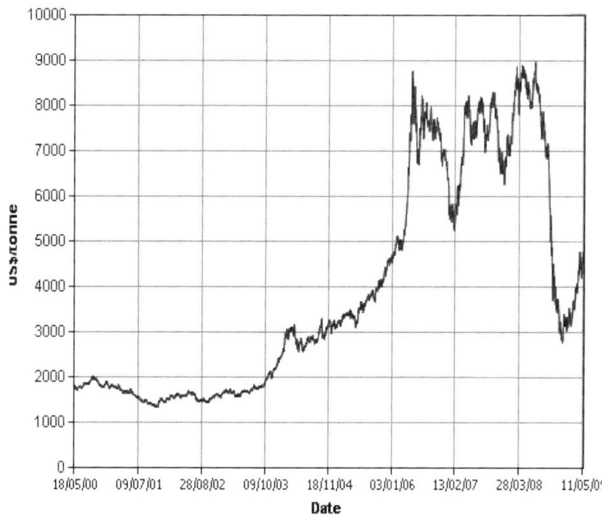


**Figure 2. Copper prices 2000-2009. (London Metal Exchange, 2009)**

process, the affects on the electrical system can easily force the modification of costly items like motor control centers, switchgear, substations and transformers, all expensive and long lead time items. Each piece is like a domino.

Ensuring the basis of design is correct from the very outset of conceptual design saves time, money and energy. It is imperative that all elements of design and engineering be checked to assure that no changes are required, and that all parties are in agreement before moving forward.

## TURBULENT TIME FOR COSTS

An unprecedented rise in commodity prices in recent years has affected the cost of all raw materials. Steel, concrete and copper wire comprise a substantial portion of total project costs.

Structural steel costs increased nearly 300% during 2008 when compared to 2004. (See Figure 1) If a plant design required 1,000 tons of structural steel, the increase could equate to an additional cost of $2,700,000. Add to this that most processing equipment and pumps are manufactured from steel or cast iron, and the cost increases even more dramatically. The increases in steel prices also affect the price of the concrete foundations because of the steel rebar required.

Copper prices have increased approximately four times since 2000. (See Figure 2) This increase in price increases the cost of not only wire and cable, but also motors, transformers, motor control centers and all switchgear. Depending on the size of the project, the increase in copper prices can increase the total cost of the project by as much as $2,000,000.

The costs for all fabricated equipment has also gone up substantially over the course of the last few years. Old 'rules of thumb' can be very difficult to apply without using significant cost escalating indexes. The recent increase in crude oil prices has increased costs for shipping of supplies and equipment like never before. The timing of deliveries to the job site can now be impacted if, at the time of transport, crude oil prices spike. Therefore, it is critical to schedule shipments so as to avoid these rises in cost, which may require delaying deliveries until the last possible moment to take advantage of a forecasted oil price decline.

Renegotiation of equipment supply purchases can be commonplace, especially if the project is delayed from one calendar year to the next. Equipment price increases are most likely to occur early in the year, so it is important to include the control of price increases when the original equipment purchases are being made.

Provisions for long lead-time equipment orders must be made while also considering the state of the global economy. Many pieces of equipment have individual components that are fabricated or cast in other countries, which may not be tied to the U.S. economy. For example, slurry pumps that are assembled in the U.S. might have major cast parts manufactured in Europe, China or South Africa. At the time of assembly, which is hardly ever immediately upon placing the pump order, parts sourced from other areas of the world may not be available. One trick is to develop a 'Plan B' with suppliers in case there is a shortfall of parts for the equipment. It is best to cover this issue prior to placing an order, rather than a week before the order is scheduled for shipment.

## SELECTION OF ENGINEERING FIRMS

It is most important to choose an engineering firm with experience in the design and layout of modern industrial minerals processing facilities. Many engineering companies can design the structural and civil elements that make up a process plant, but a well-designed frac sand plant is best designed by a team of professionals that include experienced process, project, civil, structural, mechanical, electrical and automation engineers. The opportunity to recruit operations personnel for the team should also be sought as they provide a vital, real-world view of the process, and so can aid the other engineering disciplines as well. It is an obvious benefit to search for a firm that has specific detailed experience in the design of frac sand plants. If a firm can be found that is willing to consider the project, request detailed drawings and calculations used to design various elements of past projects to confirm their experience.

When plant process and facilities are being designed, equipment being specified and purchased, structures being erected, commissioning being undertaken, and a plant eventually being operated, shortfalls in both the original and any ongoing designs become known. Unfortunately, many minor and major design flaws are not discovered until well into the process of design and construction. Often, they are found by the erecting contractor and not brought to the attention of the design firm until much later in the course of the project. This can multiply the effect of the issue, and transform a minor problem into a major headache. However, any engineering firm that has been involved in a complete project will benefit from the learning that coincides with making mistakes, and so continuously improves with each subsequent plant design. It is therefore important, when selecting an engineering company, to determine how many industrial sand projects they have actually been involved in, and to what extent. One or two projects are a good starting point, but four or five is better. Also, if the engineering company is familiar with the types of processing equipment that will be used in the plant, they also know better how to design the installation and ultimate layout of plant over all.

## CONSTRUCTION

The basis for construction is a combination of structural design and the project schedule. The general contractor (GC) will typically require assistance with these details from the engineering design firm on an as-needed basis throughout the construction phase. There is a tremendous amount of information, both engineering and otherwise, that must be transferred between all groups involved in the project, so orderly communication and transfer of knowledge is critical.

Two personnel functions should be maintained throughout all phases of the project, as they are critical to ensure a timely and successful completion of construction. An engineer knowledgeable in the civil or structural disciplines is required for design, erection and equipment installation. The engineers primary concern is to ensure that the process design and equipment layout mesh with the details of the structural design. The engineer's continuous involvement will prove invaluable throughout all project phases, including testing and final project commissioning. The other key function relating to construction is the on-site equipment expediter. The organization and proper receiving and reporting of equipment as it arrives on-site are critical. As equipment accumulates on a jobsite, this individual is charged with the organization and maintenance of the equipment laydown area thus enabling all equipment to be accounted for. This function is also of vital importance during the equipment installation period to liaise between the GC and the designers if modifications must be made, either to the equipment or to the structural design.

## SAND PLANT EQUIPMENT SELECTION—FORM VERSUS FUNCTION, A COST COMPARISON

Equipment selection is largely a case of pay now or pay later. Low priced capital equipment investments generally result in higher operating costs down the road. The same applies to low cost processing alternatives, which can also add to significant operating cost. Proper equipment selection results in an operation that utilizes the best available technology to achieve lowest cost production.

The following describes equipment types most often utilized in a frac sand plant processing operation and discusses their relative costs.

### Attrition Scrubbers

If clays are not liberated, the final frac sand product could fall outside required turbidity specifications. The general use of attrition scrubbers is to break down these clays, whether found in sand size particles or as coatings on the surface of sand grains, to allow for their rejection. (Attrition scrubbing cannot change or alter the grain shape. If the particles are not round, no economically justifiable amount of scrubbing will change that fact.)

Attrition scrubber's drives can either be gearbox driven, or via belts and sheaves. Gearbox driven units tend to be higher in capital costs, but offer long-term operation without any sacrifice in performance. Belt drives tend to be less expensive, but as the belts and sheaves wear, they need to be replaced, adding to maintenance cost and down time. Often, operators will add water to prevent plugging of the attrition cells rather than replacing the worn parts. The lower percent solids created by this added water prevents attrition scrubbers from operating as designed so product quality is ultimately sacrificed.

**Wet Preclassification**

For the most part, this process is conducted using up water classifiers like those produced by Outotec, McClanahan or CFS, to name a few. Other wet classifier tanks like those used primarily in the aggregate industries are not as efficient as those produced by the manufacturers listed above, and therefore result in the wasting of significant amounts of product or product misplacement.

Wet preclassification allows the operator to match the day-to-day market demand without the need to dry the sand prior to disposal of excess material. In some cases, there may be an excess amount of a particular size sand in the ore body when compared to what the market is demanding, so the majority of this size fraction can be removed and sent directly to waste prior to drying. The ability to dry only what is required for direct sale to the market saves considerable amounts of money in drying costs. The capital required to install these units is relatively high because of required civil and structural supports, as well as the pumps and stacking equipment to handle the various products. However, the cost savings in drying generally makes this an attractive investment.

**Thickeners**

Thickeners are only used in silica sand processing plants to combat limited space for water retention dams, some environmental issues or lack of water in the area. Thickeners allow for rapid return of water to the processing plant, and limit the need for new make-up water to volumes of approximately 100–200 GPM. The actual amount of required make-up water depends on a variety of factors including the types of minerals in the slimes, the amount of solid waste generated and the capacity of the operation.

In today's mining operations, high-rate thickeners are generally used since they offer a more compact footprint when compared to conventional thickeners, with generally 20 to 30% of the diameter of the thickener needed to handle the same water flows. The smaller footprint reduces the foundation work and total cost of installation. The high-rate thickeners also provide a cleaner overflow, higher percent solids underflow and reduced flocculent dosages compared to conventional thickeners.

**Dryers**

Dryers used in the frac sand market today are either rotary dryers, or fluid bed dryers. Fluid bed dryers are either static type or mechanical type. In recent years, both forms of fluid bed dryers have dominated the drying of sand with few rotary dryers being specified. Fluid bed dryers tend to be more expensive to purchase, but have reduced operating costs. Fluid bed systems also tend to use less thermal energy than rotary dryers, and the sand discharge temperature is lower. This typically results in lower maintenance costs in the immediate down stream material handling and processing equipment.

There is little difference in capital expenditure between the static type and mechanical type fluid bed dryer, but the mechanical type tends to have a smaller footprint and smaller fans. The oscillating motion performs some sand transport within the dryer; therefore the mechanical types have lower installed costs. However, the mechanical types tend to have higher operating costs because the units are in constant motion which translates into higher need for repair.

Mechanical fluid bed dryers are a better choice for larger particles, as found in aggregates, but not so much for silica sand. The mechanical conveyance is beneficial when conveying larger particles when compared to totally air transported systems. Since most frac sand types are less than 16 mesh (1 mm), there is little need for mechanical means to transport finer particles.

**Screens**

The most critical step in the production process of frac sand is final screening, because the market requires each product be greater than 90-wt% within a single size range. There are various screen manufacturers, and they fall into two general categories: the rotary type and the vibrating type. Most companies are specifying the sifter type rotary screen for tight, closely sized frac sand products, similar to those produced by Rotex, for example.

Rotex produces the highest capacity rotary type screener specifically made for processes such as those required to produce frac sand. The screen has up to three decks, allowing up to four products to be produced. The screener can also be configured to have the same size opening on all decks thereby tripling the capacity of the screen, if properly configured.

The high capacity of the Rotex screen allows for a more compact plant with a lower cost of installation when compared to lower capacity screens that typically have a larger footprint.

**Other Process Equipment**

At times, other processing is required to remove softer minerals to ensure products meet the API crush requirements. Magnetic separation can be used to remove iron-bearing minerals, which tend to be much softer than quartz. Magnetic separation can also be used to remove feldspar in some cases, when the feldspar contains high levels of iron. Testing is required to determine the amount of softer minerals that can be removed, and if there is any impact on the crush value.

Once the test work is performed, other separation technologies can be investigated and recommendations made. A rule of thumb for magnetic separator capital costs is US$10,000 per ton per hour processed. The type and amount of magnetic separation equipment can be determined at the conclusion of testing.

Feldspar removal can also be accomplished with flotation using an Outotec TankCell®, or in the dry process with triboelectric separation, such as the Outotec® T-Stat. Both methods require mineral treatment with hydrofluoric acid to be effective. Flotation generally costs about US$5000 per ton per hour processed as capital, plus ongoing operating costs of ~$1 per ton for reagents and HF. Capital costs for triboelectrostatic are generally US$10,000, with an additional ~$.25 per ton for HF during operation. There are new developments underway that may eliminate the need for hydrofluoric acid in the future, making this form of separation less costly. Until then, removal of the feldspar improves both the crush results, and improves the acid solubility of the sand products so there are instances where the added costs and extra steps for environmental security are out weighed by the processing benefits.

Finalizing the Project—dotting the i's and crossing the t's It is of vital importance that each of the different engineering processes utilized to complete a project are checked and verified before the project is considered truly complete. All teams, from the process design group through the construction and erection contractors, must perform final inspections and sign-off on project completion prior to handing the process over to the client for operation.

Process testing, once the civil and structural phases are complete and the system is ready for operation, is generally the ultimate measure of project completeness. All equipment must function as designed, and to the specifications laid out and agreed upon by the various engineering disciplines and the client at the beginning of the project. Control and instrumentation systems must also be checked to ensure that the process is controlled as expected so that final product specifications are met in a safe and efficient manner. This allows a process acceptance document to be signed by the client stating that they are satisfied with the process and the ability of the process to produce all products to necessary specifications. This final phase often includes an operational training period to ensure that the operators of the new process have the knowledge to competently operate all production related equipment and systems.

It is important to remember that other major issues can still arise, no matter how good the communication has been, so a documented final inspection plan is key. Some out-of-the-way connections can be missed, or that last bolt forgotten by the erecting contractor, and may only be discovered during a detailed inspection prior to the contractors leaving the site, and before the site is turned over to the client.

Attention to detail at all times, and in all of its forms, is what eventually makes for a successful project. Everything that goes into a project, from initial concepts, to final design and equipment selection and layout through process acceptance, must be checked and double-checked to ensure that nothing has been missed or misinterpreted. Problems that arise at the end of a project must be minimized, as these can become major costly issues once the project is considered complete.

## REFERENCES


London Metal Exchange Limited (n.d.). 2009 Copper price graph. Retrieved May 18, 2009, from https://secure.lme.com/Data/community/Dataprices_pricegraphs.aspx

MEPS (International) Ltd (n.d.). 2009 MEPS –World Carbon Steel Prices—With Individual Product Forecasts ($US/tonne). Retrieved May 18, 2009, from http://www.meps.co.uk/World%20Carbon%20Price.htm

Pinkham, M., 2006. Structurals Build Staying Power. Metals Center News. p 24. June.

# THE DESIGN AND BENEFITS OF CONTAINERISED MODULAR PLANT DESIGN

Andrew Foster

Downer EDI Mineral Technologies Pty LTD, 11 Elysium Road, Carrara, Queensland, Australia

## ABSTRACT

Modular mineral processing plants have been adopted in various applications for many years. Typically, a modular concept will be adopted in place of the traditional site constructed plant method when: the operating site is very remote, and / or, site access and space is limited, and / or, the project schedule requires plant modules to be built in parallel, and/ or, the plant has a short operating life at one site and will then be moved to another site

Containerised plants are modular plants that are further designed to fit into standard sea freight sized modules. This approach can yield great benefits where the plant will be moved frequently from site to site or the shipping cost to site is significant and the processing equipment can be made to fit inside standard container-sized blocks without constraining the operation and maintenance of the plant.

To create the building blocks, either standard sea containers can be suitably modified to accept all the equipment or a fit-for-purpose structure can be designed and built to the same size as standard sea containers complete with standard corner connection blocks.

When designing plants of this nature, effort is required to ensure best possible use of space and this often requires the engineering company to work with the relevant equipment suppliers to make some modifications to vendor equipment to allow the best layout to be achieved within the available space.

This paper presents in more detail the above concepts, as well as a case study of recently designed, constructed and now operating containerised Kelsey Centrifugal Jig plants.

## INTRODUCTION

Typically, a modular approach is followed where special circumstances make it more efficient from a project cost and schedule point of view. It is also important to look beyond the initial design, construction and commissioning parts of the project as there are also other factors regarding the nature of the operation and overall project lifecycle that can make a modular approach even more attractive.

Naturally, a modular plant can only be considered for plants (or parts of a plant) where the size of the process equipment involved allows assembly into appropriately sized modules to be conducted. Plants (or sections of a plant) comprising very large equipment, e.g., large capacity run of mine tips, large crushers, large mills, tanks, thickeners, large bins, etc., are clearly not good candidates for a modular plant approach. Some typical examples of project criteria that could make a modular approach attractive include:

- Very remote sites
- Short operating life
- Multiple mine sites each with relatively short life span
- Relatively small process equipment
- Site trial / pilot plants
- Sampling plants
- Fast track project schedules, requiring parallel off-site completion prior to assembly into plant on-site
- Phased increase in production capacity.

Some applications that have successfully adopted a modular approach include:

- Spiral concentrator plants
- Dense Media Separation (DMS) plants (both diamond and coal plants)
- Diamond final recovery plants
- Gold leaching plants
- Jigging plants
- Coal washing plants
- Cement and aggregate plants

The decision of modular or a traditional "all onsite construct plant" would typically be made during the pre-feasibility stage of a project. The evaluation process would include some initial conceptual designs and costing to determine if there are cost and schedule benefits to either option. A key step in this conceptual design stage would be to determine what size modules would need to be used to house all the processing equipment and ancillary systems, while still allowing sufficient operating and maintenance space in the plant. It follows that, in order to make operation and maintenance easier, more than less space would be asked for by future operators of the plant. However this drives up the cost to construct the plant due to additional steel and higher transport costs for shipping to site. Against this will be a trade off by the project team pushing for less space in order to reduce project costs and shorten the project schedule.

In special cases where the equipment sizes allow, consideration can be given to designing the modules to the size of standard shipping containers in order to reduce shipping costs as much as possible. This is particularly attractive for plants that require frequent moving from one location to another or transport into difficult to access areas. Designing a compact yet easy to operate / maintain plant around shipping container-sized modules requires "out of the box" innovative thinking during the design process (from flow sheet to plant layout).

This paper firstly reviews the basis for selecting a modular approach and then focuses on designing modular plants based on container sized modules. Finally a case study is presented on the design, construction and operation of containerised plants.

## THE MODULAR VERSUS TRADITIONAL PLANT DECISION

This section presents some key factors that should be considered at the start of a new process plant project which may result in a modular approach yielding some significant cost and schedule benefits to a project.

### Plant Equipment Size—Is a Modular Plant Really Feasible?

This is the first question that needs to be answered. Once the process flow sheet has been defined and basic equipment selection conducted, it is possible to determine if a modular plant approach might be feasible. The size of modules that the equipment has to fit into is generally determined by the constraints imposed by the method of transport to site as well as on-site lifting facilities (e.g., available cranes). The transport to site will invariably have road haulage as part of the journey and in some cases will also include sea freight and perhaps rail freight. Each mode of transport has its own shipping cost verses load size relationship as well as absolute size constraints determined by the relevant rules, codes and conventions. This paper focuses on both road and sea freight constraints (as these are the most likely modes of transport) and on module sizes that can be accommodated by conventional transport / shipping methods. Extra large modules, such as those involved in the offshore oil and gas industry that typically require specialist vehicles, vessels and crane barges, are excluded from the scope of this paper.

**Road freight** has width, height and length constraints. Although standard vehicle widths should be within 2.5m, in most western countries loads up to 3m wide can be transported without escorts. Wider loads, typically up to 5.5m, can be transported with suitable extra wide load escorts, while loads above 5.5m widths and up to 8m can also be road hauled, but at considerable cost due to the requirement for special transport carriers (multi-wheeled trailers) and large capacity cranes. In addition, special arrangements need to be made with state and local authorities which adds to the complexity and costs involved. From a height perspective, in most western countries a maximum height of 4.3m above ground is stipulated to clear bridges and overhead cables although the European Union limits total load heights to 4.0m. Based on a typical trailer height of 1.2m above the ground, this 4.3m total height gives a 3.1m maximum allowable module height. Load length is restricted to 14.6m in most western countries (semi trailer) with some restricting to 13.6m (Europe). It must be stressed that these maximum load dimensions are not specific for any one area—they are presented for indicative purposes. Maximum allowable load size will depend on which country the load is being road hauled and in most cases will also vary from state to state within that country.

**Sea freight** (like road freight) has dimension constraints linked to cost. The cheapest form of shipping is via a registered container. A registered container is a load carrying box or frame that is built to ISO Standards that has been approved and issued with a Container Safety Certificate for travel by road, rail or ship. This is discussed in detail in a later section. Loads larger than containers are shipped as "break bulk" where dimensions are relatively un-restricted however costs are orders of magnitude higher than shipping via registered container.

For general comparative purposes, escorted road freight (>3m and <5.5m wide loads) will cost double that of unescorted freight (<3m wide loads). Break-bulk sea freight will usually cost a factor of three (3) times as much as container freight for the same volume shipped.

It follows that a sensitivity analysis needs to be considered on transport costs versus overall project costs in conjunction with the size of module required to ensure good operability and maintainability in the design. The common perception is that modular plants are less easy to operate and maintain due to their inherent compact nature. This is not necessarily the case and if proper thought and planning is given during the layout stage of the design, a modular plant should not constrain good plant operation and maintenance. In fact, a modular design approach will invariably lead to a more compact design. This has a positive benefit for the project in that costs will be reduced during the design phase as the focus of the design team will be producing a compact and space efficient design. Many traditional plants have excessive / more than required space and so are not necessarily the most efficient designs. In order to use space efficiently in the design of a modular plant, it is critical that the design team adopt a different approach to conventional plant layouts. Some examples of innovation during the layout stage are presented later in this paper.

### Site Conditions and Site Works

The key consideration under this section is cost of site works. There will always be a site works component to every new plant project, however this can be significantly reduced (and in some special cases almost eliminated) by choosing a modular plant approach. The higher the cost to conduct work on site, the more attractive a modular approach will be. A trade off study needs to be conducted as part of the decision process. Some factors that drive site works costs include:

- Site location—the more remote the site, the higher the costs involved for many obvious reasons, e.g., transport and logistics costs, availability or choice of contractors and site labour hire firms, travel and accommodation costs, location of site with respect to prevailing construction costs in that area / country
- Available space onsite—in some cases, space is limited making off site construction of the plant in modules followed by a "just in time" approach to delivery and assembly onsite an attractive option
- Construction, assembly and commissioning costs—it is well known that the cost to perform site works will usually be significantly higher than conducting the same work in a workshop environment. This increase in

cost can range from a factor of two in the case of a relatively non remote site up to several times higher where very remote sites are involved.

- Skills level and quality of available site works contractors—bringing in highly skilled people to perform critical work on a project with a remote location will significantly drive up project costs
- Environmental considerations—it follows that risk to environment is higher onsite than in an established workshop where environmental controls are well established. Depending on the location of the site, this could yield advantages to projects where construction in an environmentally sensitive area is required. By reducing the amount of site works by pre-building and assembling the plant in modules in a workshop the environmental risk can be lowered

**Life of Plant and/or Mine Issues**

For cases where the life of plant / mine is relatively short or a mine comprises several deposits, a modular plant could yield the following attractive benefits:

- Reusing / selling entire plant
  - Where the required life of the plant is relatively short, a modular plant could be dismantled and sold or reused on another site in its entirety
  - Consider the obvious economic benefits of this approach versus the traditional plant where typically only process equipment (pumps, screens, crushers, etc.) are removed and sold via used equipment dealers
- Multiple operations
  - For projects where the mine plan involves multiple smaller operations, a relocatable modular plant could be an attractive option worthy of consideration
  - There are many examples of where this applies including coal, diamonds, stone and aggregate, etc.
- Reclamation and reprocessing of old tailings dumps
  - These operations typically consist of short duration operations (from several months to a few years) at each dump site before being moved onto the next dump, so they are a good example of where modular plants would have obvious advantages
- Pilot or site trial plants
  - These plants are often used for testing or demonstrating a process or items of equipment for a new project or client and are obvious candidates for modular design as they are only needed at each site for a short duration before relocation to other sites.
  - Where plants are to be trialled at multiple sites consideration should be given to containerization to reduce shipping costs (this is discussed in detail later in this paper)

**Other Factors**

Several other factors should be considered as they may yield benefits if a modular approach is adopted. They include:

- Project schedule advantages
  - Time savings can be made by making modules in parallel at different locations / constructors' facilities and then assembling onsite in a planned order
  - This is typical of offshore oil and gas projects where the entire plant or platform is constructed in modules at various sites in parallel and shipped to the final assembly port/site in the required sequence
- Direct cost benefits
  - Modules can be constructed, assembled and tested in countries where fabrication and general labour costs are the most attractive
  - In order for this benefit to be maximized, both construction costs and shipping costs need to be minimized
  - Sea and road freight will obviously be involved with this approach, so by designing the modules in container-sized units, the modules can be shipped as in gauge "break bulk" units, which attracts a lower sea freight cost than out of gauge freight, as well as making road freight very easy

## CONTAINERIZATION—BEYOND THE TRADITIONAL MODULAR PLANT

In special cases the benefits of plant modularization can be maximized by containerizing the plant modules. These cases would be where the shipping costs constitute a large portion of the overall costs over the life cycle of the plant. This approach seeks to minimize shipping costs and maximize ease of handling and site assembly, while maintaining good plant operability and maintainability.

**Applications and Project Criteria for Which Container Modules Provide Real Advantages—Unlocking the Potential**

The key advantages of containerised modules are:

- Minimizing shipping costs to the lowest possible—there is no cheaper form of road and sea freight than shipping containers
- Ease of handling onsite—faster and lower cost site works. It follows that handling container sized modules is easier than larger modules that would require larger vehicles and bigger cranes etc.
- A containerised plant will by its very nature be a compact plant with just enough space for operation and maintenance. This has advantages in ensuring that the design focuses on maximizing space utilisation and

**Figure 1. Kelsey Centrifugal Jig Pilot Plant**

**Figure 2. Spiral Concentrator Pilot Plants**

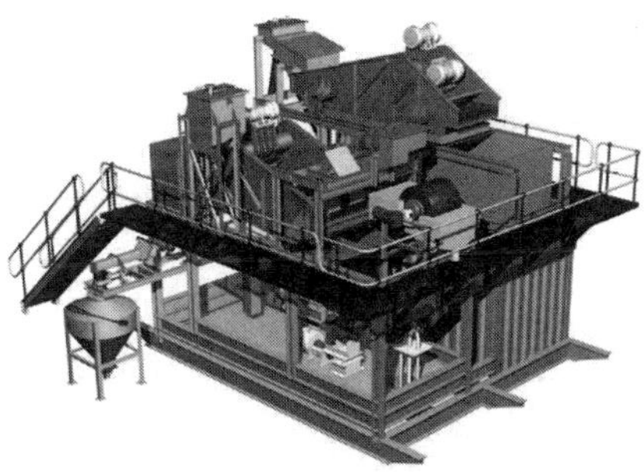

**Figure 3. Dense Media Separation Plants (courtesy ADP Projects)**

minimizing equipment both of which are key drivers in lowering project costs. It is unlikely that excessive space and associated unnecessary costs will results from a containerised design.

- Schedule benefits due to possible reductions in shipping time.

Against the background of these above advantages the following are examples of where containerised modular plants have previously been adopted:

**Site Trial or Pilot Plants**

- These are plants that are shipped to many locations for site trials and so minimizing shipping costs is important
- Furthermore, quick set up and commissioning time on-site is important to maximize utilization of capital equipment
- Maximum time can be spent on-site proving up equipment, rather than shipping and assembling the plant
- Some examples include:

**Figure 4. Charter vessel based marine diamond sampling containerised plant (Courtesy BDM)**

**Figure 5. Kelsey Centrifugal Jig Relocatable Plant for recovery of zircon from tailings dumps**

**Table 1. Standard Shipping Container Dimensions**

| Container Type | Length | Width | Height | Max Mass |
|---|---|---|---|---|
| 10 Foot Half GP | 3.0m | 2.438m | 2.591m | 12 tons |
| 10 Foot Half GP—high cube | 3.0m | 2.438m | 2.850m | 12 tons |
| 20 Foot GP | 6.058m | 2.438m | 2.591m | 24 tons |
| 40 Foot GP | 12.192m | 2.438m | 2.591m | 30 tons |
| 20 Foot High Cube | 6.058m | 2.438m | 2.850m | 24 tons |
| 40 Foot High Cube | 12.192m | 2.438m | 2.850m | 30 tons |

**Relocatable Short Life Cycle / Multi-Site Plants**

- As the name implies these applications would be where the plant is required to process material for a relatively short period of time before moving onto the next site
- Some examples include: tailings dump retreatment, sampling plants and toll treatment applications

**The Building Blocks—Standard Container Dimensions**

Table 1 shows the standard container dimensions in accordance with ISO Standards and the International Convention for Safe Containers which governs all sea freight container standards.

The dimensions and maximum allowable masses in Table 1 represent the plant module size options available to the containerised plant designer. In order to minimize freight costs it is necessary to either use certified containers or container structures that have been certified and thus carry a Safety Approval Plate or CSC (Container Safety certificate) in accordance with the International Convention for Safe Containers. If a container does not carry a valid Safety Approval

Plate or CSC then it will be shipped as "break bulk" at significantly higher costs. A separate section is dedicated to this subject later in this paper. Several options of module construction are available to the designer. These include:

- Utilizing standard containers and suitably modifying them
- Building the plant on top of a standard container flat rack or bolster platform
- Building new "container" structures to the dimensions above

The advantage of **modifying a standard container** is that the certification can be retained provided the modifications are approved by the original container designer / manufacturer. Designing modules using standard containers does have limitations as it is difficult to spread large concentrated loads typical of heavy equipment within the existing container structure. The plate thicknesses of standard containers is thin (typically 3-5mm) which makes structural engineering of additional point loads difficult. However, this said, if the load can be spread on a frame which is joined to the container over a large area then it can be done. In some instances where protection from the elements is required, using a fully clad container can offer a good solution. Due to structural constraints, modifications are restricted to cutting limited size access holes through the side walls, roof and deck. The added access openings cannot be placed close together or near the ends of the container structure. Typical examples where containers are used include: electrical switch rooms, control rooms, laboratories, x-ray sorting plants and other plants where smaller lightweight equipment is used—typically 1000kg or less per item.

Building a **plant module on top of a certified bolster or flat rack** is more flexible, as a purpose-designed structure can be built on top of the bolster or flat rack structure to house heavier equipment without jeopardising container certification. In order to stack modules on top of each other, a frame suitably designed to the correct size and dimensions can be attached to the flat rack / bolster. This frame would be used for stacking of modules onsite only—not shipping, otherwise the structure will require certification. With this method the original certification is maintained and a very substantial structure can be achieved that can house very heavy equipment. Unfortunately the structure of the flat rack / bolster cannot be changed without approval from the original designer and manufacturer and so this can cause some significant constraints for plants that require significant openings through the floor between modules for pipes and chute work.

Building an appropriately sized **purpose-built container module structure** equipped with twist lock corner connector blocks gives the most flexibility and allows the plant designer to fully optimize the plant design and layout without any constraints other than the physical size of the module.

The main drawback of using a purpose designed structure is that certification will need to be obtained to allow a Safety Approval Plate to be carried in accordance with the International Convention for Safe Containers.

**Container Certification—Minimizing Shipping Costs**

The International Conventions for Safe Containers was signed by members of the UN in 1972 to ensure the safety of all people involved with handling containers. This convention applies to all new and used containers used for international transportation. This convention defines all the design, construction, operation, security, inspection and control aspects of shipping containers.

To obtain certification for a new container module a Marine Classification Society needs to be engaged. Classification societies that certify new containers include: American Bureau of Shipping (ABS), Bureau Veritas (BV) and Germanischer Lloyd (GL).

The **Container Certification Procedure** encompasses:

- *A Review of all detailed design drawings* for the containerised module structure to verify that it complies with the requirements of the respective Classification Society, International CSC rules, and all other relevant codes and standards
- *A prototype structure is built and subjected to load tests* which are witnessed by the Classification Societies' surveyor. Tests under load include: lifting, stacking, concentrated loads, transverse racking, longitudinal restraint, end walls and side walls. This would typically be done at an established and approved container manufacturer –there are many of these in China and other South East Asian countries.
- *A survey of the container module manufacturer* is conducted as well as follow-up visits during manufacture. This survey is primarily focussed on manufacturing quality control procedures and specification with special reference to welding procedures and specifications.
- *Issue a certificate* attesting to the container modules compliance with the specified codes, regulations and standards as well as structural fitness for the specified service or use

The cost for the above process would typically be USD 20,000. Some containerised plant designers and builders have opted to have a standard module frame(s) fully certified and then use this certified frame as a basic building block during all further container plant designs, rather than having to certify every new design.

**Table 2. Typical Container Structural Member Sizes**

| Module Member | Conventional sections | Hollow sections |
|---|---|---|
| Columns | 152 x 152 x 30 UC | 150 x 150 x 6 SHS |
| Beams | 252 x 146 x 31 UB | 200 x 100 x 6 RHS |

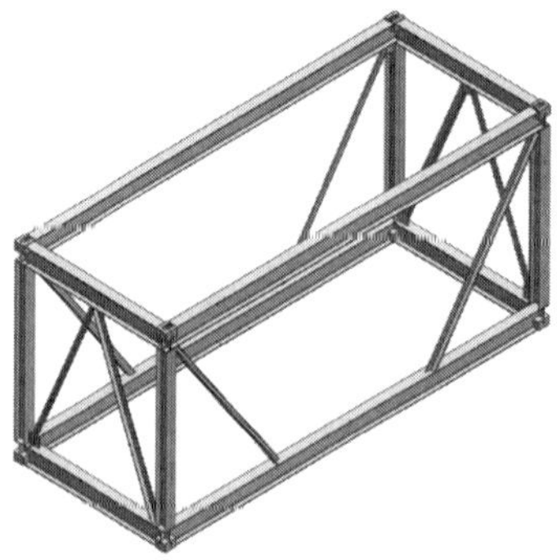

**Figure 6. Conventional structural members**

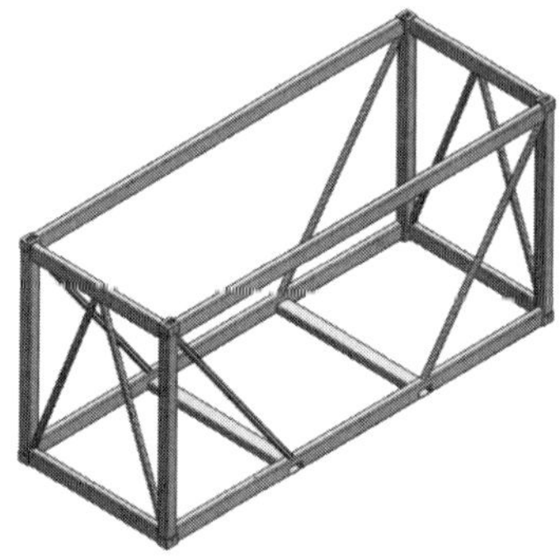

**Figure 7. Rectangular and square hollow sections**

**Design Considerations**

Some key design considerations are discussed in this section plus some innovation used to "create space" within compact modules.

**Structure**—the structure needs to be capable of withstanding the loads imposed during operation as well as shipping/transportation loads. The container module frame construction is generally based on either conventional structural steel or square / rectangular hollow section members. Table 2 shows typical members used and is illustrated in Figures 6 and 7.

Transport and handling loads are accommodated by designing for the required certification load tests. These load conditions are summarised below. The total mass of the container module plus rated load (gross mass) is defined as R, payload as P and g is gravitational acceleration:

- Static lift from all four corners when loaded to 2 x R
- Lifting from one end (two adjacent bottom corners) up to an angle of 45 degrees for the 20' container module and to an angle of 30 degrees for the 40' container module when loaded to 2 x R
- Container module bases capable of withstanding a longitudinal tensile and compressive load of 2 x R x g
- End walls / structure capable of withstanding a load of 0.4 x P x g
- Side walls / structure capable of withstanding a load of 0.6 x P x g

Structural steel mass will typically be around 2.5t per 20' module. One aspect that requires consideration is the connections between the container modules. Standard container twist lock connectors are designed for transport however they do not give a secure connection required for mineral process plants. It is necessary to design in bolted connection pads adjacent to the corner connectors for securing to the foundations (typically a concrete slab) and between the modules themselves.

**Process equipment** has the biggest impact on arriving at a space efficient design. In order to maximize space efficiency it is common that an iterative approach is applied between finalizing the flow sheet and equipment selection. To put things into perspective some suggested maximum equipment sizes are given below:

- Pumps—although cases of putting 20/20 dredge pumps in 20' container modules are known, generally pumps up to 10/8 size (and perhaps 12/10) will quite easily fit into container modules and still have sufficient space for maintenance and operations
- Vibrating Screens require careful consideration of the support structure and integrating this with the container module structure. Placing single and double deck screens onto the module with support fittings in the main structural members (with columns underneath) is a good solution. Typically single and double deck screens up to 1.8m wide by 4.2m long can be accommodated on container modules. Multi-deck screens such as the Derrick® Stack Sizer™ can be fitted within the module up to a maximum of 5 decks (more on this later)
- Kelsey Centrifugal Jigs—J1300 model fits within a container module
- Magnetic drum separators typically up to 1.8m wide will fit transversely within a container module

Other equipment such as pipe work, chutes, bins, sumps and electrical cabling systems can all be designed into the container modules, however consideration needs to be given to inter-module connections. Pipe work can be flanged at appropriate positions. Electrical cabling is either temporarily installed for workshop commissioning (and then removed and permanently installed onsite), or plug and socket arrangements are used for pilot / site trial plants. Several innovative design features are highlighted in the Case Study presented in the next section.

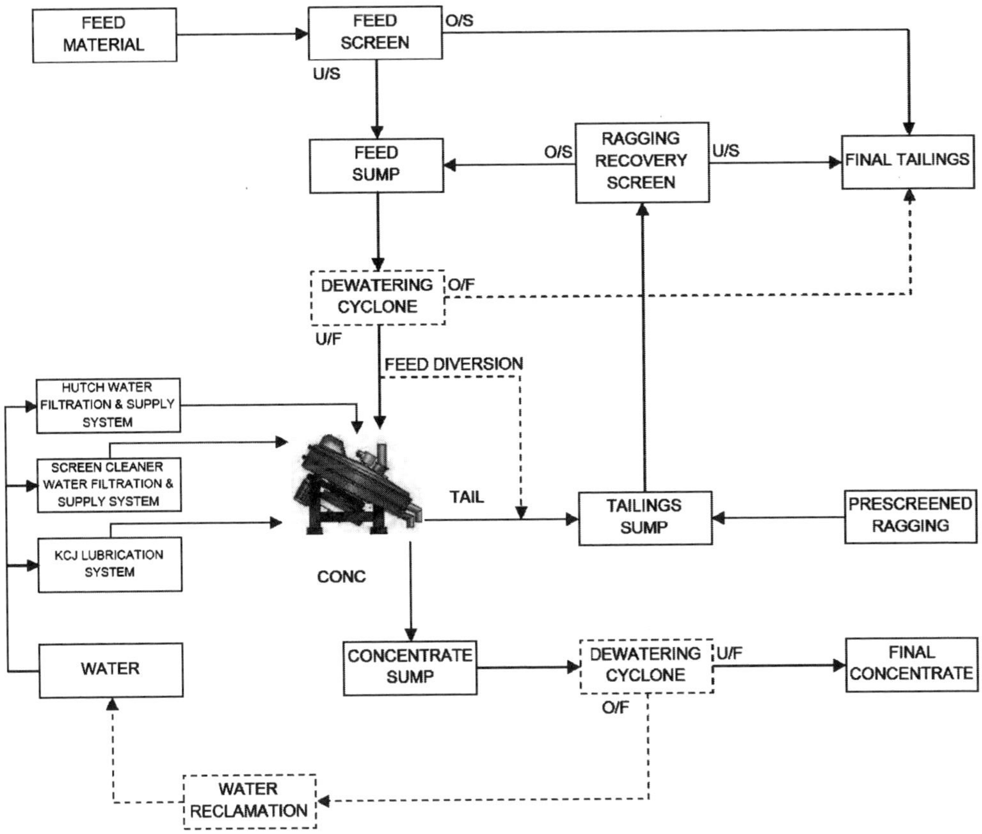


**Figure 8. Typical Kelsey Jig plant flow sheet**

## CASE STUDIES—CONTAINER MODULE KELSEY JIG PLANTS

Downer EDI—Mineral Technologies has designed and built a number of Kelsey Centrifugal Jig (KCJ) plants based on container-sized modules. KCJs are typically used in applications where fine heavy minerals (down to approximately 6 micron) are required to be separated from gangue material and/or where the specific gravity differential between heavy and gangue minerals is relatively low, such that conventional gravity separation techniques are unable to achieve target separations. Typical feed sizes to a KCJ will vary from -500 micron to -150 micron with -10 micron slimes generally removed prior to processing. Typical applications for a KCJ include processing of mineral sands, gold sulphides, tungsten, tin, tantalum, iron ore, chromite, etc. A typical KCJ circuit is shown in Figure 8.

Two such KCJ plants based on container-sized modules are discussed in these case studies. One plant was based on the smaller Model J1300 KCJ unit (built and used as a permanently installed plant for a customer in Australia) and the other based on the larger Model J1800 KCJ unit (designed as a relocatable site trial plant). These two plants had to be designed so that they could be effectively used as pilot or demonstration plants. Key design criteria included:

- Minimize shipping costs
- Minimize set up and commissioning time
- Maximise ability to transport into very remote locations and set up and commission without the need for any local engineering support or infrastructure

For these reasons a container module plant was selected. The two plants are shown in Figures 9 and 10.

### Model J1300 KCJ Modular Plant Design

The design premise was to fit this plant into two 20' container modules to keep shipping costs to a minimum and use a "plug and play" philosophy where possible. The major equipment included:

- Model J1300 KCJ unit (6 tons)
- 3-deck Derrick® Stack Sizer™ screen (1.5tons)
- Jig feed pump (Warman 3/2) and sump (0.8t)
- Jig tails and concentrate sumps and pumps (0.4t)
- Final tailings sump and pump (Warman 4/3) (1.0t)

Plant design throughput was 16 t/h solids.

**Figure 9. Model J1300 KCJ Pilot Plant**

**Figure 10. Model J1800 KCJ Pilot Plant**

**Figure 11. Standard side open container—LH side and end fitted with doors**

A container module was purpose designed for the lower module as this housed the KCJ unit which imposes significant dynamic loads (up to 10 kN at 200rpm 900mm above footings) onto its support structure. A modified side opening standard freight container was used for the top module—see Figure 11. This container was utilized because it already had certification in place, while offering a cover over the screen, a substantial base structure and side opening doors. Modifications to the floor to allow the screen chutework to be fitted were conducted in conjunction with the container designer Royal Wolf to retain certification. The intention was to certify the bottom purpose-designed container module structure, however this did not materialize as this design was used to fulfil an urgent client order and so certification was never completed (although structural design had been conducted to cater for both the operational loads and the certification load tests envisaged).

Due to the dynamic load constraints the jig was located in this lower module to keep the design as efficient as possible from a structural point of view. In order for the Jig to be positioned low down the concentrate and tailings sumps and pumps that the KCJ unit discharges into had to be as compact as possible. Jet pump hoppers (jet pump combined with sump bottom) were used as they can operate either flooded or with air in the suction thereby requiring no sump residence capacity, and because they are more compact than centrifugal pumps, allowing the sump bottom to be lowered to the floor—see Figures 12 and 13.

The Derrick® Stack Sizer™ screen was selected due to its superior performance on fine minerals combined with the layout of the Stack Sizer™ unit, where the multiple screen decks are stacked on top of each other, thereby saving significant space. An added advantage for KCJ modular plants is that selected decks can be used to perform both the feed and ragging recovery duties within one unit.

The electrical control room was designed to fit within a standard "mini-cube" insulated and air-conditioned container located nearby the main plant.

As part of ensuring the KCJ unit could be accessed during major maintenance activities, it was mounted on two slides. The sliding elements were made from four Vesconite™ pads attached to the KCJ frame running on stainless steel rails (6mm 316 stainless) attached to the tops of the beams. This is shown in Figure 14.

Also visible in Figure 14 are the cantilevered walkways attached to the base of the top container module. Note that no knee bracing is used, as each profiled support bracket which forms part of the walkway is bolted onto the top module prior to being lifted on top of the KCJ module, thereby making site assembly safe as no working at heights is involved. This plant was fully assembled and commissioned on water prior to being shipped to the customer's site.

Figure 12. Kelsey jig concentrate and tails jet pump hoppers

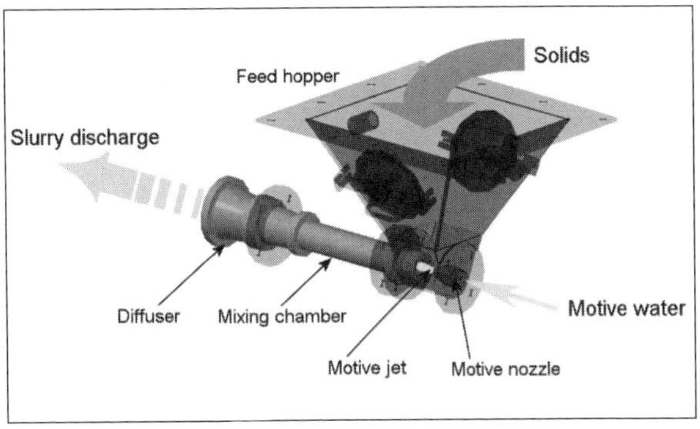


Figure 13. Jet Pump basics

Figure 14. KCJ unit maintenance slides—jig in maintenance position

**Model J1800 KCJ Modular Plant Design**

The Model J1800 KCJ modular plant has significantly higher design capacities (up to 50 t/h solids) than the smaller Model J1300 KCJ plant and so requires significantly larger supporting equipment. The design premise for this plant was to fit the plant into four 20' container modules again to keep shipping costs to a minimum and use a "plug and play" philosophy where possible. Major equipment included:

- Model J1800 KCJ unit (15 tons)
- 4-deck Derrick® Stack Sizer™ screen (2.1tons)
- Plant feed bin—jet pump delivery
- Jig feed pump (Warman 4/3) and sump (1.2t)
- Jig tails and concentrate sumps and pumps (Warman 4/3 and 2/1.5) (1.5t)
- Final tailings sump and pump (Warman 6/4) (1.5t)
- Water tank
- Floor spillage sump and pump
- Compressor
- MCC and PLC room.

The plant modules were a combination of modified 20' shipping containers and purpose built container frames. The purpose built container modules were again designed to comply with the certification test loads that the structure would be subjected to. Due to the schedule of a key site trial being moved forward, time did not allow certification to be finalized prior to shipping to the first test site.

**Design innovations** used with this plant to create a container module plant that has sufficient operating and maintenance space included:

- Elimination of bracing by designing sumps and bins as part of the container module structure (plate work within the structural members and suspending sumps from the top beam of the container module), thereby making equipment much easier to access (see Figure 15)
- Significantly reducing site civil works by designing an integral floor spillage sump into the floor of the bottom container module, such that only concrete support plinths are required under each corner
- As per the Model J1300 KCJ modular plant, a Derrick® Stack Sizer™ screen was selected (in this case with 4 decks to handle the higher throughput rates), but in order to fit this larger unit into a container module it was necessary to redesign the discharge chute work (see Figure 16). The redesign of the chute work was conducted by Downer EDI—Mineral Technologies with input from Derrick Corporation and, if the need arises in the future, a 5-deck Stack Sizer™ screen could be fitted into a container module with the redesigned chute work

**Figure 15. Suspended sumps integrated into container module structure eliminating the need for structural bracing**

**Figure 16. 4 deck Stack Sizer™**

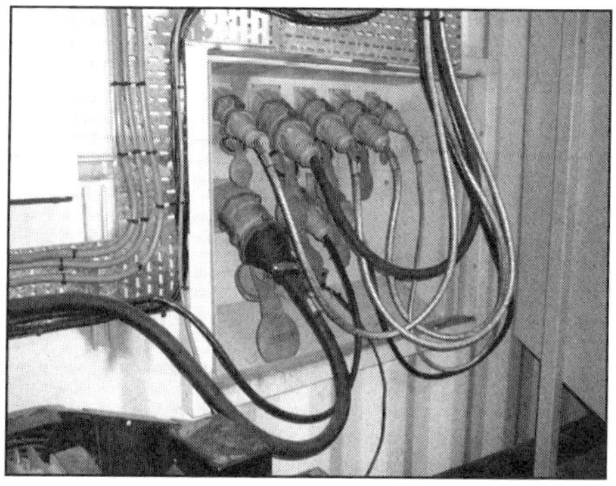

**Figure 17. Electrical connections between modules**

- In order to eliminate the need for electrical staff to travel to site for commissioning, all electrical cabling was connected via de-couplers into purpose built enclosures (see Figure 17)

An MCC / PLC control room was designed into a container module structure and fitted with fire resistance "Rockwool" thermal insulation and an air-conditioning system. The plant was fully assembled and workshop tested on water prior to being shipped to the first test / demonstration site. Site assembly takes approximately 3 days with commissioning taking a further 3-5 days depending on how ready the site is in terms of services and product & tailings discharge infrastructure.

## SUMMARY

In special circumstances, using a modular plant can offer both cost and schedule advantages to a project. In cases where the plant will be moved regularly from one site to another, using container modules can offer further advantages to the project by minimizing transport costs and site set up time. Container certification is a hurdle that has to be overcome if purposes built container structures are to be designed and built. By incorporating innovative design features into the plant, sufficient operating and maintenance space can be achieved. A container module plant offers significant advantages to a site trial / demonstration plant due to reduced shipping costs and ease of site assembly.

## ACKNOWLEDGEMENTS

American Bureau of Shipping—Container Certification
Germanischer Lloyd—Container Certification
Royal Wolf Pty Ltd
Derrick Corporation (NOTE: Derrick® is a registered trademark of Derrick Corporation)
Downer EDI Engineering

## REFERENCES

Australian Vehicle Standards Rules 1999
National Transport Commission (Australia)—Road Transport Reform (Oversize and Over mass)
International Convention for Safe Containers, Geneva 1972—Annexure I+II
US Federal Motor Carrier Safety Administration (Part 658), Size and Weight, Route Designations—Length, Width and Weight Limitations.
European Union Council Directive 96/53/EC (July 1996) Annex I

# LINERS AND COATINGS FOR MINERAL PROCESSING EQUIPMENT

Michael G. Nelson[1] and Rowan W. Truss[2]

[1]University of Utah Mining Engineering Department, 135 South 1460 East, Room 313, Salt Lake City, Utah, U.S.A.
[2]University of Queensland School of Engineering, Brisbane, Queensland 4072, Australia

## ABSTRACT

Liners and coatings are used to protect equipment from corrosion and abrasion. The appropriate use of these materials can greatly extend equipment lifetime, thus contributing markedly to the overall economics of any installation.

The design and specification of liners and coatings is covered in detail in the literature (Andrew 1997; Schnarr et al. 2002). In this paper, we review the materials most commonly in use and discuss typical applications for each. We then present two case studies in which we were directly involved, and suggest the lessons to be learned from each.

## CORROSION AND ABRASION

Corrosion is the deterioration of a substance (usually a metal) or its properties because of reaction with its environment. Corrosion can occur in liquid and gaseous environments, and even between two metals in direct contact.

Corrosion is *caused* by the presence or interaction of chemical species which may be unavoidable in a given process.

Abrasion is the removal of material from a surface, by the movement of solid particles across that surface. Abrasion can be beneficially used to shape or condition a surface for subsequent use. However, in processing equipment it is almost always undesirable, and can lead to accelerated corrosion, contamination of process flow streams, and outright equipment failure.

Both corrosion and abrasion often result from materials or conditions that are unavoidable in a given operation. However, both processes are often *exacerbated* by conditions that can be avoided, including

- poor design of infrastructure and equipment,
- incorrect or inadequate maintenance procedures,
- incorrect application or maintenance of protective coatings, and
- poor operating procedures.

Often the above conditions result from a lack of awareness on the part of management of the importance of corrosion and abrasion protection, which results in inadequate budgets for installation, inspection, and maintenance of systems designed to prevent these problems.

## PROTECTIVE COATINGS

Corrosion and abrasion may sometimes be avoided by constructing equipment of special materials. For example, a tank for holding hydrochloric acid may be fabricated of a high-alloy stainless steel or fiberglass. However, special materials are often expensive, and their fabrication can require special skills that may not be readily available. In such cases, it is often less expensive to use a liner or coating. For example, a steel slurry pipeline may be lined with natural basalt, elastomer, or fabricated ceramic material, either of which is highly resistant to abrasion. Similarly, the concrete floor around a concentrated sulfuric acid tank may be protected by a covering of acid resistant brick or tile, underlain by an impermeable membrane. Paints and elastomers are the two types of coatings typically used for mineral processing equipment.

## PAINT COATINGS

Paint coatings are used on steel substrates. They range from alkyd paints, which are primarily cosmetic, to paints that are highly corrosion resistant, based on epoxy, polyurethane, vinyl, or phenolic materials. Andrew (1997) provides recommendations for paint selection, based on general descriptions of the operating environment, as summarized in Table 1.

Successful application of paint coatings depends on correct surface preparation. Surface preparation should be rigorously defined in contract documents, and will normally refer to standards from the ISO, NACE International, or SSPC. Only experienced and proven contractors should be used. The contractor's capabilities for preparing schedules, managing staff, and providing required equipment on time should be verified.

Most protective paint coatings used in plant environments are not designed to last the full life of the installation. Thus periodic inspection and maintenance are required. Surface preparation procedures as rigorous as those used in the original application are often not possible in the plant environment; for example, abrasive blasting or high-pressure water washing may not be permitted. In such cases, special treatment with so-called *rust converters* or *rust transformers* may be considered. These treatments convert rust on the steel surface to inert substances, but they usually have no effect on other forms of surface contamination. Thus they should be used only after their effectiveness has been verified by in-plant tests.

Paint coatings are usually not highly resistant to abrasion, and are therefore not used in applications such as slurry lines and flotation machines.

**Table 1. Recommended Paint Coatings for Plant Environments**

| Environment | Examples | Equipment | Recommended Paint System |
|---|---|---|---|
| Dry, moderately corrosive | Wet ore handling, milling, and crushing; no influence from adjacent operations | Structural steel, piping, tanks, vessels, machines, conveyors<br>Cable racking, flooring, walkways | Inorganic zinc primer and epoxy intermediate and final coats (150 μm min. DFT1)<br>Epoxy or epoxy tar (200 μm min. DFT) |
| Wet, very corrosive | Wet ore handling, milling, and crushing; platinum and gold concentrators; coal wash plants; flotation plants; plants with nearby smelters | Structural steel, piping, tanks, vessels, machines, conveyors<br>Cable racking, flooring, walkways | Vinyl primer, vinyl enamel intermediate and final coats (150 μm min. DFT)<br>Epoxy or epoxy tar (200 μm min. DFT) |
| Wet, extremely corrosive | Sulfuric acid plants; acid handling areas; base and precious metals refineries; chemical storage areas | Structural steel, piping, tanks, vessels, machines, conveyors | 1. Vinyl primer, vinyl high-build intermediate, and vinyl enamel final coat (250 μm min. DFT)<br>2. Epoxy primer, epoxy intermediate and final coats (250 μm min. DFT)<br>3. Phenolic primer, phenolic intermediate and final coats (250-300 μm min. DFT) |
| | | Cable racking, flooring, walkways | 1. Phenolic primer, phenolic intermediate and final coats (250-300 μm min. DFT)<br>2. High-build, filled epoxy (250-300 μm min. DFT) |

[1]DFT is dry-film thickness

## ELASTOMERS

An elastomer is a polymeric material, such as a synthetic rubber, which at room temperature can be stretched under low stress to at least twice its original length and, upon immediate release of the stress, will return with force to its approximate original length (Parker, 1989). Elastomers include natural rubber compounds, and are important coatings for mineral processing because they provide both corrosion and abrasion resistance.

ASTM D2000 classifies elastomers for automotive applications. It uses two criteria: *type*, or resistance to heat aging, and *class*, or resistance to swelling in oil. These criteria are also useful for mineral processing applications, and Table 2 lists the types of elastomers commonly used in mineral processing.

Except for polyurethanes, elastomers are usually installed by hand laying, where uncured sheets of the elastomer are laid onto the steel surface, and then bonded to the metal during vulcanization. Several layers can be applied, and thicknesses up to 25 mm are common. Hand laying is very flexible, and designs can be easily modified because no molds are used. However, hand laying requires skilled labor and can be expensive.

Autoclave curing is preferred for hand laid applications, using steam pressure to provide temperatures of 140 to 150° C. If the parts are too large for an autoclave, an exhaust steam cure may be used. This requires a longer processing time, and thus there is a greater tendency for blisters to form, requiring post-cure repairs to the lining. When steam is not available, as is the case in some on-site lining jobs, a chemical cure may be used, but this is expensive and more difficult to apply.

Urethane elastomers are almost always applied by casting or spraying. In all cases—hand laying, casting, and spraying—the metal surface must be carefully cleaned and prepared, and the surface condition must be properly maintained during the application process.

Because of urethane's susceptibility to hydrolysis, urethane coatings are often installed in association with a corrosion-inhibiting primer coating. It is thought that the urethane will resist abrasive damage and provide some isolation of the metal substrate from the process slurry, and the primer will prevent corrosion by any process water that does penetrate the urethane.

In all cases, successful installation of elastomer liners requires careful and correct surface preparation. Surfaces must be clean, dry, and free from significant defects. In some cases, good bonding to the steel, or between layers of the elastomer coating, requires application in a fixed range of temperature and humidity. For sprayed-on linings, surface irregularities must be significantly smaller than the specified dry-film thickness of the primer coating.

**Table 2. Elastomers for Mineral Processing Applications**

| Elastomer | ASTM Designation | Processing Characteristics | Performance | Typical Applications |
|---|---|---|---|---|
| Natural Rubber | NR | Very good | Often preferred; additives required above 70° C | Sheeting; pump, tank, and pipe liners |
| Styrene Butadiene Rubber | SBR | Good; lacks tack and green strength | Resistant to sliding abrasion and impact; suitable up to 110° C | Belts, truck box and mill liners |
| Butadiene Rubber | BR | Difficult; usually blended with NR or SBR | Added to improve abrasion resistance, heat aging, resilience, fatigue resistance, low temperature flexibility, and crack resistance | Belts, hose, flanges, mill liners |
| Butyl Rubber | IIR | Fair; incompatible with NR, SBR, and BR | Good resistance to oxidation and ozone degradation; low gas permeability; resistant to acids and bases | Pump, tank, and pipe liners |
| Chlorobutyl and Bromobutyl Rubber | CIIR; BIIR | Similar to IIR | Better that IIR: low gas permeability, weather and ozone resistance, chemical and heat resistance; better adhesion to other rubbers | Belts, hose, tank liners |
| Ethylene and Propylene Rubber | EPM; EPDM | Good, but lacks building tack | Excellent ozone and oxygen resistance, good resistance to polar materials, fair resistance to chlorinated solvents, poor resistance to nonpolar solvents; abrasion resistance much less than NR, better than BR or SBR | Hot material belts, hose, pump liners |
| Polychloroprene (Neoprene) | CR | Fairly good; fast curing; good tack | Good combustion resistance (meets MSHA requirements); moderately good oil resistance | Belts, hose, pump liners |
| Polyurethane | AU; EU | Good, but application is by casting or spraying | High tensile and tear strength; high resilience; excellent resistance to abrasion and to nonpolar oils, fuel, and ozone; hardness and modulus much higher than other polymers; may be subject to hydrolysis at high temperatures | Mechanisms in mixing and flotation tanks; screens; wear liners in launders and chutes |

## CASE STUDIES

### Case A

Case A involved a series of flotation cells in a copper/gold mine. The flotation plant consisted of two banks of rougher/scavenger and retreatment cells based on U-shaped tanks of capacities 8.5, 17, and 28.3 m$^3$. The flotation plant was enclosed in a corrugated iron shed and the rougher/scavenger cells were located on either side of the plant adjacent to the uninsulated walls. These cells were linked together so that the end of each cell shared a common wall with its neighboring cell. The retreat lines were in the center of the processing shed. The mine site was inland rural Australia with warm to hot summers, moderate to low rainfall, and temperatures around freezing in winter.

**Coating system.** An elastomeric polyurethane (PU) was selected due to its resistance to fine particle erosion and abrasion. The specifications for the tanks required that the steel be grit-blasted to remove heavy rust, water washed to remove salts and other impurities, and then subjected to a further grit blasting to give a surface profile of 50 -100 µm. The surface was finally washed with solvents to remove grease and oil and other loose materials from the surface. The steel was primed to a dry film thickness of 50 µm with an epoxy-based primer. The final coating was 3 mm of sprayed polyurethane. This was a polyether/TDI based PU elastomer with a nominal Shore hardness of A80.

**Description of failures.** During the first year of operation, blisters were observed in the liners of some of the cells. Several attempts at repair and recoating of the tanks were made using similar or slightly different primer systems. Observations were made by numerous consultants over several years on both the original and repaired cells. Common observations are listed below:

- The cells showed extensive blistering on the bottom, front and back walls of the cell. These blisters ranged in size from < 1 cm to 10 cm in diameter. A quick count of blisters revealed from a few to as many as 10 blisters per 100 $cm^2$ of surface, depending on the size of the blisters. Blistering appeared first on the sides closest to the outside wall of the shed for the scavenger/rougher lines. In the retreat lines the blistering appeared first on the sides least protected by neighboring cells.

- The end walls were free from blisters with only a few blisters in the front and back corners near the outer walls.
- Blisters were found in the original lined cells and in cells that had repaired or recoated.
- Lining material away from the blisters was well adhered to the metal surface.
- The polyurethane stripped from blistered liners retained high flexibility and good toughness and strength, and showed no measurable change in hardness. There was little or no sign of reduced thickness due to wear. The lining material contained some close-cell porosity which is typical of sprayed PU materials, but no fine porosity that may have been associated with gas evolution on the reaction of isocyanates and moisture.
- Samples of liner stripped from the cells occasionally showed bits of the primer attached to the polyurethane. When these and a fresh sample of Polyurethane stripped from a coupon were examined by SEM and EDS, particles rich in aluminum powder and others rich in zinc and phosphorous (probably a zinc phosphate) were found. Particles rich in silica were found in the resin and were most likely silica used as a thixotropic agent.
- Liquor collected from the blisters was found to contain high concentrations of zinc. The liquid was otherwise clear and to be acidic (pH of 5.5).
- The metal below the liners showed regions of pitting corrosion. In some cases, corrosion products came off the surface with the liner.

**Response to failures.** As mentioned above, many consultants visited the site. Some of the liners were patched in spots. Eventually, the blistered areas seemed to stabilize. In one cell, the polyurethane was separated from the substrate over most of the area of the cell's interior. However, the liners in that cell were left in place and the flotation cells continued to function.

### Case B

Case B involved flotation cells in a zinc mine. The flotation cells were cylindrical, and were installed in a covered structure with one wall open to the outside. The mine site was in northern Australia, with moderate to very hot temperatures year round and moderate to low rainfall. Temperatures at night were significantly lower than those in the day.

**Coating system.** The cells originally were grit blasted with garnet and blown down, and selected areas cleaned with solvent. The steel was primed with a wash primer based on potassium zinc chromate. The primary wear coating was again 3 mm of a sprayed polyurethane elastomer of nominal shore hardness of 80 A.

**Description of failures.** These cells were factory coated, moved to site and filled with water. However, before commission of the plant, adhesion tests were conducted on the liner of one of the cells and it appeared that the polyurethane was not properly bonded to the substrate. On the basis of this test, the original polyurethane liners were not put into service.

SEM/EDS inspection of the PU liner after stripping from the tank wall showed embedded particles of garnet at a density of up to 50 particles per $mm^2$. Besides the garnet, zinc phosphate and chromium were present, from the corrosion inhibitor in the primer.

**Response to failures.** The owner decided to remove all of the PU liners. On stripping the polyurethane liner from the steel tanks, areas of poor adhesion of the liner to the steel were reported while other areas adhered to the steel tank so well that it had to be removed by hand grinding and grit blasting. When the liners were removed, the onset of corrosion of the steel was observed in some areas.

The cells were subsequently put into service with an alternative epoxy coating. The slurry in this mine was highly abrasive and after about three years of service, the epoxy lining had worn sufficiently to need replacement. Several sections of the epoxy coating had also been reported as suffering from blistering. In some areas of very high wear, such as the feed boxes and discharge areas, the epoxy was replaced with a natural rubber liner. The rest of the floor was replaced with a quick-setting polyurethane.

### Discussion of Case Studies

**Responsibility.** In both cases, several parties were responsible for the specification and application of the cell linings. The supplier of the flotation cells specified a polyurethane liner with a corrosion-inhibiting primer. Another company supplied the polyurethane and the primer. This second company provided guidelines for surface preparation and material application, and suggested that detailed, site-specific specifications be prepared. A third company fabricated the tanks, and grit-blasted in preparation for coating, and a fourth company applied the coatings. This complex chain of responsibility made it difficult to maintain quality control in the blasting, cleaning, and coating of the tanks. In both cases, detailed, site-specific specifications were not prepared by any of the companies involved. When problems occurred with the linings, a complicated set of legal proceedings resulted, in which each company tried to blame the other.

Both of these cases were grit blasted to an Australian standard that specifies a surface roughness of 75 μm or less. The surface was to be coated with a primer in Case A to a dry thickness of about 50 μm, while the wash primer in Case B was to be significantly thinner. Depending on the viscosity and wetting characteristics of the primer, this presents the possibility that high points of the surface profile might protrude through the primer or be thinly coated allowing localized corrosion. Further, the presence of garnet on samples of stripped polyurethane indicated that, in both cases, clean steel shot had not been used for blasting, as suggested by the material supplier. In Case B, the presence of embedded garnet in the primer constituted a condition that could cause the primer not to adhere to the steel.

**Chemistry.** In general, all polymers absorb some water when immersed or exposed to high humidity. Where there is a temperature gradient, between a warm process fluid or slurry and a cold substrate, the "cold wall effect" can accelerate the rate of water permeation (Kosek et al. 1995; DeVries 1970).

The cold-wall effect is a reasonable explanation for the blistering observed in Case A. The mechanism is consistent with the observed patterns of blistering where the fronts, backs, and bottoms of the cells were blistered and the end walls were not. The end walls of the cells had fluid on both sides, ensuring that a temperature gradient will not exist. Moreover, the rotors and stators in the flotation cell, which were also coated with PU and which were completely submerged in the fluid, did not show any blistering.

Once the water has condensed, an osmotic blister can develop relatively easily. The blister water was high in zinc, and zinc phosphate is slightly soluble in water. If the zinc phosphate in the primer partially dissolves, a concentration gradient is set up between the water at the primer surface and the fluid in the flotation cell. This drives water through the liner, which acts as a semi-permeable membrane, to the primer surface to counteract the concentration gradient, and a blister is formed. The pressures generated by osmotic cells can be substantial and are quite sufficient to break most interfaces. The interface that breaks will be the weakest of the interfaces in the liner and slight variations in the strength of the various interfaces in the coating will explain the variation in position of the blisters.

Note that zinc phosphates form during the phosphating processes commonly used in coating of steels in many applications. The literature suggests that such coatings improve the adherence of the coating to the steel (www.phosphating.net 2009). If such a mechanism is also present in the primers used here, partial dissolution of the zinc phosphate may reduce the interface adhesion.

Once a blister has formed, corrosion of the steel will eventually follow. This is likely to be pitting-type corrosion, localized in small areas. These areas can corrode quickly producing pits that can penetrate completely through the steel plate. As soon as corrosion occurs, the corrosion products may also develop a concentration gradient which could also force water into the blister via the osmotic mechanism.

A similar cold wall mechanism is also possible for Case B. The thermal mass of large tanks of water is sufficient that with daily temperature variation, a temperature gradient can exist between the exterior and the interior of the tank liner. However, there were not enough data available to confirm this possibility.

## CONCLUSIONS

There are many choices for the lining and coating of mineral processing equipment. Selection for a given application should be based on the combined experience of suppliers, manufacturers, and end users, and should take into account site-specific conditions.

Several companies may be involved in the selection, specification, and installation of a lining system. In all cases, one company should take the overall responsibility for quality control throughout the process. Specifications for materials, surface preparation, installation practices, and installation conditions should be prepared and agreed upon well before installation of the liners. The responsible party should keep careful records of all procedures and conditions during surface preparation and liner installation so that the causes of any later problems can be clearly identified and more easily corrected.

## REFERENCES


Andrew, R.H.C. 1997. *Practical Guidelines for Corrosion Protection in the Mining and Metallurgical Industry*, Houston, TX: NACE International.

DeVries, T. 1970. Blistering in Water-immersed Coatings under the Influence of a Temperature Gradient. *Anti-Corrosion Methods and Materials*, 17(11–19).

Kosek, J.R., J N. DuPont, and A.R. Marder. 1995. Effect of Porosity on Resistance of Epoxy Coatings to Cold-Wall Blistering, *Corrosion*, 51:11(861–871).

Parker, S.P., Ed. 1989. *McGraw-Hill Dictionary of Scientific and Technical Terms*, 6th ed. New York: McGraw Hill.

Schnarr, P., L E. Schaeffer, and H J. Weinend. 2002. Elastomers in the Mineral Processing Industry. In *Mineral Processing Plant Design, Practice, and Control*. Edited by A L. Mular, D.N. Halbe, and D.J. Barratt. Littleton, CO:SME, pp. 1932–1952.

www.phosphating.net, accessed May 2009.

# DESIGNING FOR A REDUCED CARBON FOOTPRINT AT GREENFIELD AND OPERATING COMMINUTION PLANTS

Zeljka Pokrajcic[1], Rob Morrison[1], and Bill Johnson[2]

[1] Julius Kruttschnitt Mineral Research Centre, 40 Isles Road, the University of Queensland, Australia
[2] Mineralurgy Pty Ltd, 2/42 Morrow Street, Brisbane, Queensland, Australia

## ABSTRACT

New comminution circuits design strategies are presented. The aim of the new strategies is to improve the overall efficiency of the comminution process and reduce the total energy consumption of a comminution circuit. The strategies described can be applied during the design of a greenfield comminution circuit and be applied to improve the performance and energy efficiency of an operating comminution circuit.

The energy efficient strategies can be summarized by:

1. Use of new technology to evaluate the potential for rejection of liberated gangue at coarse particle sizes.
2. A modified comminution strategy where more efficient grinding and crushing equipment is used to redistribute comminution energy and to minimize energy consumption as equipment power and embedded energy in media and wear materials. This strategy can provide the beneficial additional effect of a "tighter" size distribution more suitable for the following separation process.
3. A quantitative approach to the selection of target product size through thorough analysis of quantitative liberation data.

These factors can be combined to improve the performance of a comminution circuit by tailoring the design to suit better the properties of the orebody. This can significantly decrease the overall energy consumption and the resulting carbon footprint of the comminution circuit.

## INTRODUCTION

A review of the current comminution design practices reveals a recent trend that has seen the design and installation of low capital and uncomplicated crushing and grinding circuits. These normally include one or two stages of crushing, one or two large tumbling mills and a classification stage.

Grinding for the most part is carried out in tumbling mills, which are inefficient in their use of energy and the promotion of mineral liberation. It is understood that 85% of the energy used in comminution is dissipated as heat, 12% is attributed to mechanical losses and only 1% of the total energy input is used in size reduction of feed material (Alvarado et al, 1998). The indiscriminate nature of the breakage process in tumbling mills leads to high energy consumption and the generation of critical size material and ultra fine slimes, both of which are difficult to process.

Another aspect of comminution which contributes significantly to the energy intensity of the process is the consumption of media and mill liners. The quantity of energy necessary for the fabrication of the media and/or liners from extraction, to transport, to manufacture and assembly is considered part of the total energy utilization of the comminution process. This is also quite an energy intensive aspect of the process and opportunities to minimize steel consumption in comminution circuits should be realized.

The chosen target product size, or grind size, can also significantly contribute to the energy intensity of a comminution circuit. As the target product size becomes finer the energy required to achieve the product size increases. As the particles become smaller their strength and therefore their resistance to breakage increases. For small particles as internal flaws are depleted, the observed strength approaches the high intrinsic strength of the solid. The target product size and the throughput rate of a proposed comminution circuit dictate the size of the tumbling mill(s).

The need to grind to finer and more energy intensive target product sizes is increasing as a result of the decline in high grade ore reserves. As higher grade ore reserves are gradually becoming more difficult to locate and excavate, the need to mine and process ever lower grade, hard, finely disseminated ores at much higher throughput rates and finer product sizes is increasing. These lower grade orebodies need to be processed economically and with minimal impact to the environment to meet the increasing demand for minerals and the requirements of an environmentally conscious industry.

All these factors and the increasing likelihood of higher energy costs, as well as the commitment to reduce greenhouse gas (GHG) emissions requires that power consumption of comminution circuits be minimized.

This is the prospect facing all comminution circuit design and operating engineers. Hence, there is a driving need to implement more efficient size reduction strategies and innovative comminution technologies that not only increase the efficiency of the size reduction process but also act to minimize the carbon footprint of both greenfield and operating comminution circuits.

This paper aims to show how the following concepts, presented as strategies, can be used in the design of new comminution circuits or in the modification of operating comminution circuits for improved energy efficiency, where the focus is a better understanding of the inherent ore characteristics and the incorporation of these characteristics in the design for reduced energy consumption and carbon footprint.

- Pre-concentration for gangue rejection at coarse particle sizes. A better understanding of ore texture and ore characteristics to enable the selection of a separation method for the removal of liberated gangue from the feed before energy intensive size reduction commences.
- A modified grinding strategy for improved eco-efficiency[1]. The use of more efficient size reduction equipment and the reduced requirement for grinding consumables is illustrated using a simulation and modeling software package.
- A progressive liberation strategy for the selection of target product size(s). A method for the interpretation and analysis of mineral liberation data for the selection of target product size(s) and its effect on the configuration of the downstream size reduction and separation circuits.

## DESCRIPTION OF THE ENERGY EFFICINET STRATEGIES

### Pre-concentration for Gangue Rejection at Coarse Particle Sizes

The aim of any pre-concentration method, regardless of scale, is to improve the economics of mining, comminution and concentration processes by reducing ore dilution at each stage of the process. The total cost and energy consumed per unit of concentrate can be reduced significantly by avoiding the mining, comminution and concentration of gangue and material below metal cut-off grade. Gangue is usually high in silicates and typically harder and more competent than the valuable minerals, adding to energy and processing costs.

Gangue usually occurs in the orebody as large mineral groupings or "clumps" containing little or no valuable minerals, allowing gangue to be readily liberated at coarser sizes. Gangue can be separated from regions of high mineralisation using selective mining techniques, ore sorting devices, screens, density separation such as heavy media circuits or drum separators, magnetic separators, etc. Such applications are dependant on the properties of the orebody including the mineralogy, mineral grain size and shape and association between minerals including gangue, collectively termed ore texture (Vink, 1997).

Pre-concentration techniques can be applied in a comminution circuit through a better understanding of ore texture variations at different scales. It is a valuable consideration for the design of greenfield comminution circuits. Differences in ore texture at large scale, termed macro-texture, imply differences in the mineralisation pattern can lead to the use of selective mining techniques for zones of high mineralization. Differences in ore texture at a medium scale, termed meso-texture, like the macro-texture observations are visible by the naked eye but at a smaller scale and are not limited to visual differences, rather can include differences in ore properties such as density and magnetic susceptibility.

The key element in determining whether an opportunity for pre-concentration at coarse sizes (macro and/or meso scale) exists is the method or strategy used to assess ore characteristics and mineralization properties.

Differences in ore properties are the basis for pre-concentration techniques. Inherent differences in ore properties can be explored using conventional techniques such as visual analysis, wet and dry density investigations, and specialist core logging systems where the ability to recognize changes in the frequency of occurrence of meso-textures can be very useful in the modeling of the orebody and in the prediction of processing performance. However, presented in this paper is a novel non-destructive x-ray method, called micro Cone Beam Tomography (CBT). CBT is a non-destructive evaluation technique for producing two-dimensional and three-dimensional images of a particle, providing information on mineral dispersion, dimensions, shape, and internal defects within a particle.

The imaging system of CBT provides a three-dimensional image of the particle suitable for understanding the properties of particles produced from an orebody, particularly those related to density variations. CBT analysis reveals areas of high and low density mineral within the particle, low density regions being typically gangue related and high density regions being mainly valuable minerals. Hence, in the early development of a comminution flowsheet for an orebody, the CBT method can be used to study coarse particles produced from the ore to allow recognition of opportunities for gangue rejection.

The benefits of CBT analysis are that it can be used to measure large numbers of individual coarse particles quickly for ore characterisation. It can provide information on mineral composition and spatial mineral distribution within the particle. This non-destructive analysis method is being developed to provide information quickly, relatively cheaply and potentially at a larger scale (CBT devices now can analyze batches of drill core) on the valuable mineral location and more importantly the gangue location and its properties and associations within an orebody.

Pre-concentration techniques when employed early in the comminution circuit can have positive implications for resource recovery and utilization by allowing lower-grade ores to be upgraded. This increases the net output of the valuable mineral per tonne of ore processed. For example, if a copper operation with a capacity of 100,000 tpd and a head grade of 0.5% Cu, increased its mining capacity to 110,000 tpd at a lower head grade, say 0.48 %Cu, and installed an ore sorting facility to reject 10,000tpd of liberated gangue at or below 0.1% Cu, the copper grade following the separation step would be 0.52 %Cu at a feed rate of 100,000 tpd. This extra 0.02 %Cu in the feed at an 85 %Cu recovery and 36% Cu concentrate grade means an increase in revenue of over $100,000/day (with upper prices of copper at $AUD6000/t). Additional costs of mining the extra 10,000 tpd of ore at $AUD1.3/t and initial outlay for the separation system and its operating costs only slightly detract from this increased revenue estimate.

The second strategy used in the design of reduced carbon footprint comminution circuits is the modified grinding strategy, where the simulation and modeling software package—JKSimMet, is used to predict circuit outcomes of an eco-efficient comminution circuit.

[1] See Figure 1 for definition.

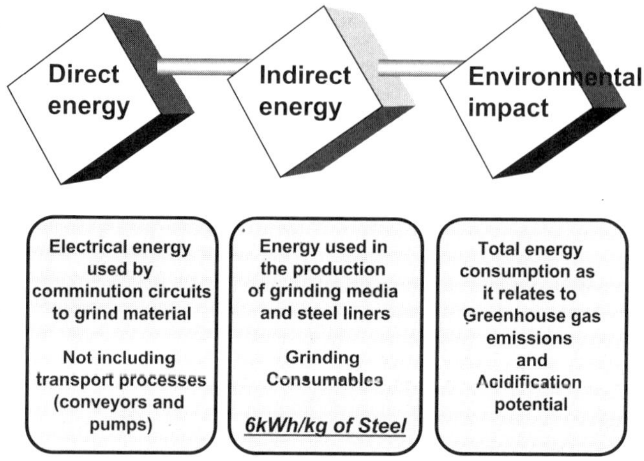


**Figure 1. Eco-Efficiency in Comminution Circuits**

**A Modified Grinding Strategy**

Eco-efficiency is defined simply as "doing more with less." It incorporates the environmental and economic aspects of sustainability. Figure 1 shows how eco-efficiency is applied to comminution practice. It defines the reduction in both the amount of energy consumed in size reduction (direct energy) and as well as the amount of energy consumed in the manufacture of comminution consumables such as steel grinding media and mill liners (indirect energy). These act together to reduce the impact on the environment and reduce the acidification potential of the comminution circuit.

The aim of the strategy is to:

- Reduce the direct energy usage. That is, decrease the grinding power used in the size reduction process.
- Reduce the indirect energy usage. That is, reduce the amount of grinding mill consumables including steel media and mill liners.
- Generate a product size distribution that is narrow in shape, containing less material in both the finer and coarser size fractions for improved liberation and ease of downstream processing.

The strategy is best applied to the design of greenfield comminution circuits or when an increase in capacity is required or when a change is ore hardness is expected for an existing operating circuit. The strategy involves a better utilization of comminution energy. This can be achieved by purposely focusing the comminution energy at problematic size fractions such as the critical size material in the primary mill and the very fine fraction in the later stages of the comminution circuit. Through the use of more energy efficient comminution devices such as the HPGR (high pressure grinding roll), tower mill and IsaMill to treat specific size fractions where they are proven to be inherently more efficient, it is possible to reduce energy requirements of the primary and secondary grinding circuits. This does not take into account the additional energy requirements of auxiliary equipment such as pumps and conveyors. The strategy also exploits opportunities to decrease the mass of solid and grinding media (total load) in grinding mills and, in some cases, employs autogenous grinding techniques to decrease indirect energy consumption.

Typically, the new product size distribution generated from this modified grinding strategy contains less material in the coarse fraction, to aid overall mineral liberation, and less material in the fine fraction, to minimize overgrinding and the generation of slimes. A product size distribution of this type is expected to produce superior flotation performance (Crosbie et al, 2005).

This paper presents simulation models in the JKSimMet package to assess the effect the modified grinding strategy for improved eco-efficiency has on two typical comminution circuits. The mathematical models and ore testing techniques have been described (Napier-Munn et al, 1999). The data analysis and simulations were carried out using the JKSimMet software package (Napier-Munn, and Lynch, 1992; Morrison and Richardson, 2002).

Although the JKSimMet package is very useful in predicting and simulating comminution circuit performance given certain adjustments and changes to the circuit, the product sizings and power predictions are only as reliable and as accurate as the site data used to generate the grinding models. Some of the models presented in this paper have been adjusted and some data have been imported from other sites. Therefore, observations and discussions made should only be viewed as examples of the potential benefits given such circuit modifications, rather than actual results expected if the changes were implemented.

The third strategy has a significant impact on energy consumption of the comminution circuit as it is used in the selection of a target product size for a circuit. It can indirectly influence the design of the initial stages of the separation circuit.

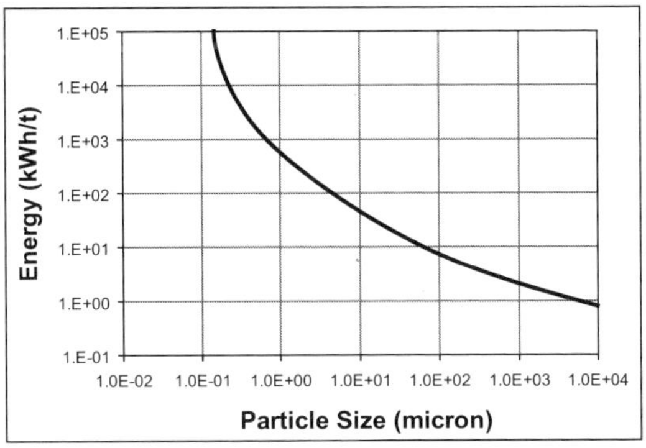


**Figure 2. Relationship between energy input and product size in comminution (after Hukki, 1961)**

### A Progressive Liberation Strategy for the Selection of Target Product Size(s)

The target product size chosen for a comminution circuit has a significant impact on the energy requirement of that circuit. The finer the product size, the greater the energy required to achieve that product size. In fact, the energy requirement increases significantly as the target particle size drops below 100μm, see Figure 2. As the particles become smaller their effective strength and therefore their resistance to breakage increases. For small particles, as internal flaws are depleted, the observed strength approaches the high intrinsic strength of the solid.

A major determinant in the selection of a target product size is mineral liberation. A common premise applied to the selection of a product size is that the mineral of interest must be at least 80% liberated at the target product size for efficient downstream separation.

A mineral is termed fully liberated when that mineral is the only mineral in a particle. It implies the particles must be smaller than, or the same size, as the mineral grains of the unbroken ore, meaning that mineral liberation increases as particle size decreases. Although fully liberated minerals may be ideal for separation processes, in many instances it is not economic given the high energy required to achieve the fine grind size.

An alternative approach for the selection of a target product size is presented in this paper. It is termed the progressive liberation strategy and can be applied to multi-mineral ores. It involves acknowledging the liberation of one mineral or a group of minerals before the others. Particularly, the following concepts are applied:

- minerals are grouped to form larger entities and therefore deemed liberated at coarser sizes
- fully liberated particles are recoverable in a flotation process
- particles containing greater than 15% by sectional area of the mineral of interest are recoverable in a flotation process using the appropriate flotation conditions and flotation reagents

The aim of the progressive liberation strategy is to determine whether a mineral, or group of minerals, is sufficiently liberated at a coarser size, or more importantly, exists in a form that can be separated from the bulk at coarser particle sizes. The mineral of interest can either be a gangue or a valuable mineral, it may have a coarser grain size or it may be sparsely disseminated within another mineral, however, an opportunity may exist to recover and remove it based on its liberation characteristics prior to the next stage of processing.

In order to implement the progressive liberation strategy a better understanding of the composition of particles at different product sizes is required.

Progressive liberation methods can be applied in two ways:

- if gangue or valuable particles are liberated, then separate the liberated particles from the bulk before further size reduction
- if gangue or valuable minerals are recoverable (in composite particles), then separate the recoverable particles from the bulk before further size reduction

Together the above two points act not only to increase the target product size but they also enable the separation and rejection of material (typically liberated gangue) from the circuit. The rejection of gangue from the feed stream can limit the amount of over-grinding but more importantly it will reduce the mass of material proceeding to the next stage of size reduction and/or separation. This allows for more efficient separation and significantly lower energy requirements for downstream size reduction.

## ILLUSTRATION OF THE STRATEGIES FOR DESIGN OF AN REDUCED CARBON FOOT PRINT COMMINUTION CIRCUIT

The previously described strategies can be used in combination for the design of new comminution circuits or can be used singularly to improve the energy efficiency of an operating comminution circuit. A methodology is developed where these energy efficient design strategies are linked to provide a frame work for new comminution circuit design (Pokrajcic, 2009). However, for the purposes of this paper and in an attempt to minimize complexity, each strategy is illustrated in separate case studies. This also highlights design features and ore properties where the strategies can be most readily applied to modify an existing circuit for a reduced carbon footprint.

**Table 1. Meso-texture group identified by Cone Beam Tomographic (CBT) analysis of the orebody**

| Meso-texture Types Observed Using CBT | | Description |
|---|---|---|
| (A) | Gangue | The majority of the particle volume consists of gangue silicates |
| (B) | Silicates with trace chalcopyrite and gold/silver | A small amount of precious metals (Ag, Au) associated with chalcopyrite appear near the surface of a mainly silicate particle |
| (C) | Pyrite with trace chalcopyrite | The particles consist of mainly of pyrite with some regions of chalcopyrite evident |

**Pre-concentration at Coarse Particle Size**

The aim of this strategy is two-fold as determined by scale:

- identify the large regions of gangue material in the ore deposit (macro-scale) and avoid these during the mining and excavation process
- exploit inherent ore properties at the meso-scale within the deposit for pre-concentration via liberated gangue rejection

Opportunities for gangue rejection at the macro-scale and meso-scale can best be realized through a better understanding of:

- the variation in macro-texture and meso-texture of the ore
- differences in ore properties such as density, magnetism, etc. between mineral rich and poor macro-textures and meso-textures
- methods available for ore characterisation
- options available for valuable and gangue mineral separation, such as mining method and ore sorting.

A simple yet effective means of non-destructive ore characterisation is the use of wet-dry density investigations to reveal the extent of particle density variation within a sample. This can be followed by CBT analysis to shed light on the meso-texture properties, including mineral composition and spatial distribution of minerals within the sample. Once the sample is characterized using this method, a radio frequency identification (RFID) tag tracker system (La Rosa, D. 2008), could be used to assess the efficiency of the potential separation process, either ore sorting or other separation options pertaining to the orebody properties.

**The Application of Selective Mining Techniques.** Based on the description of the macro-texture and the geological block model developed showing the three dimensional data inventory of the orebody, it is possible to assess whether selective mining techniques can be employed to minimize the amount of gangue excavation during mining. To return the greatest profit, the chosen method must best match the unique characteristics of the deposit, as well as limits imposed by safety technology and economics (Camus, 2002).

If the orebody is a low grade massive sulphide deposit where pyrite and chalcopyrite appear in closely spaced veins, often in multiple orientations and directions and are disseminated through the quartz host rock, the application of selective mining practices may not be obvious and bulk mining techniques may be the best option for this type of orebody.

However, if the orebody is a stratiform, lens-shaped or tabular ore deposit consisting of at least 60% sulphide minerals, such as high grade pyrite, sphalerite, galena and chalcopyrite and the host rock is a volcanic rock., selective mining techniques can be used to excavate sulphide bands that are greater than 1m in width at some penalty in operating costs because of the use of smaller scale excavation equipment. The sulphide bands should be easily recognizable within the vitreous volcanic rock.

The description of each selective mining process, and details of the macro-texture properties that suit such techniques, are the beyond the scope of this paper. It is sufficient to emphasize an opportunity to reject gangue at this point of the process through an examination of ore texture variations revealed in the geological block model and observed at the mine face.

**Gangue Rejection Technique in the Comminution Circuit.** Analysis of the CBT results may show that the particles in the sample could be divided further into meso-textural categories based on CBT observations of the internal properties of the particle, as described in Table 1.

The information in Table 1 suggests that the entire orebody can be grouped according to these meso-textures. This signifies the potential for sorting the ore on this basis. Density difference investigations were ruled out as the density difference between the gangue and sulphide minerals is not pronounced and it will prove difficult to separate on this basis.

A sequence of ore sorting techniques can be employed to generate three distinct ore streams based on the CBT investigations. The financial viability of processing the orebody via three different processing routes to produce three products would need to be investigated and validated.

Ore sorting is a niche mineral processing technology that requires optimization for each application. Existing ore-sorting systems are largely based on sensors that differentiate according to colour or intensity of reflected light. However, research into sensing technologies such as lasers, microwaves, x-rays and neutron sensors, as well as new approaches to signal processing is continuing. The most common and proven ore characteristics that can be used by an ore sorter to separate ore into distinct classes are:

- color
- magnetic susceptibility

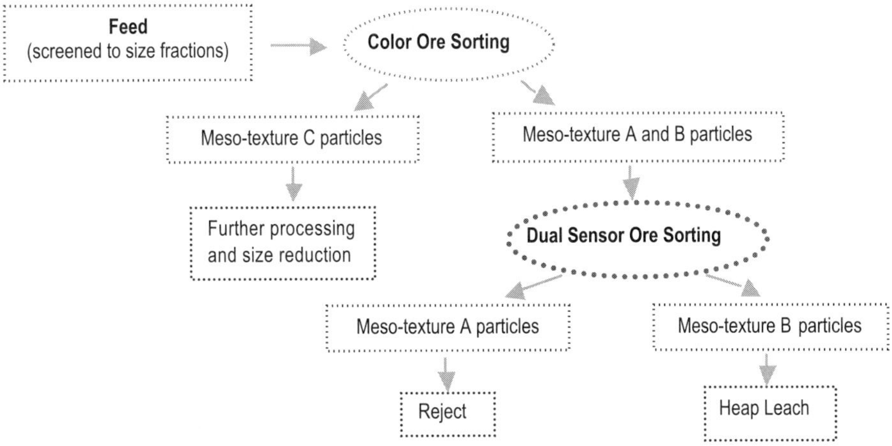


**Figure 3. Summary of the separation steps suggested for the orebody**

- conductivity
- x-ray transmission.

It is important to note that consumption of compressed air is a major feature of most ore sorting operations and indirect energy (see Figure 1 for definition) is consumed in the supply and use of compressed air.

The first stage separation for this orebody could be based on the color difference of the mainly pyrite particles (meso-texture C) verses the mainly silicate particles (meso-texture A and B). There are numerous commercial ore sorting devices available which separate based on color. These can treat a feed material up to 250mm top size at a rate of 300tph.

The separated (mainly pyrite) particles of meso-texture C can proceed to further size reduction and processing to recover valuable minerals. The remainder of the mainly silicate particles (meso-texture A and B) can proceed to further sorting to be separated based on grade. Particles of meso-texture B contain some gold, silver and chalcopyrite near the surface of the particle. These particles can be separated from the mainly silicate particles (meso-texture A) by the use of dual sensor sorting device. The dual sensor may consist of a metal detector, detecting differences in conductivity, followed by a LIBS (Laser Induced Breakdown Spectroscopy) system which can analyze for differences in metal grade.

LIBS is a simultaneous multi-elemental analysis technique which uses a pulsed laser to initiate a plasma spark at the surface of the sample to be analyzed. Light radiated from atoms and ions in the spark are collected and recorded using an optical multi-channel analyzer (OMA) for subsequent spectral analysis to determine which elements are present and their relative concentrations (Pollard and Death, 2007).

The separated meso-texture B particles which possess low grade gold, silver and chalcopyrite may be suitable for processing with meso-texture C or by another separation method such as heap leaching. Financial analysis is also required to validate the feasibility of all the options including a heap leach application.

At this point, the particles of meso-texture A, mostly gangue silicates, can be removed from further processing. Figure 3 provides a summary of the separation steps proposed for the coarse size feed material from the orebody.

**A Modified Grinding Strategy**

The modified grinding strategy can be applied to an operating SABC comminution circuit, featuring a SAG mill with recycle crushing followed by a closed circuit ball mill. The modifications to improve the eco-efficiency of the circuit include:

- SAG mill conversion to an AG mill and the ball mill conversion to a *semi*-pebble mill to decrease liner and media consumption. This enabled existing mill sizes to be reduced in the simulation.
- Insertion of a HPGR for partial pre-crushing to minimize the generation of critical size material in the AG mill. The aim is to reduce a portion of the feed material below critical size and generate more fines to decrease the load in the AG mill. Not all of the feed material is pre-crushed (only 50% of the feed) to allow for sufficient large media for abrasion breakage in the AG mill and to maintain effective operation of the AG mill.
- A *semi*-pebble mill is introduced to minimize the steel ball load and the power draw in the secondary mill, where the AG mill critical size material together with a small amount of steel balls are used as media.

This new circuit is denoted ABC-HPGR-PM. Both circuits are shown in Figure 4.

The new eco-efficient comminution circuit is now an ABC circuit with a HPGR pre-crush stage and a pebble mill for secondary grinding. The required AG mill length is reduced as the HPGR generates more feed material which is already at the product size and therefore reduces the power draw of the AG mill. The bimodal feed size distribution also improves AG mill energy efficiency. Converting the ball mill to a pebble mill also reduced its power draw as the media density in the mill is lower than that of a conventional ball mill, 40% lower in this instance, 4.5kg/l compared to 7.8kg/l. This decreases the power draw of the mill and significantly reduces the requirement for grinding consumables.

These factors allow increased throughput for the eco-efficient autogenous circuit at a coarser feed size, 740tph and $F_{80}$ of 97mm compared to 689tph and $F_{80}$ of 64mm for the original SABC circuit.

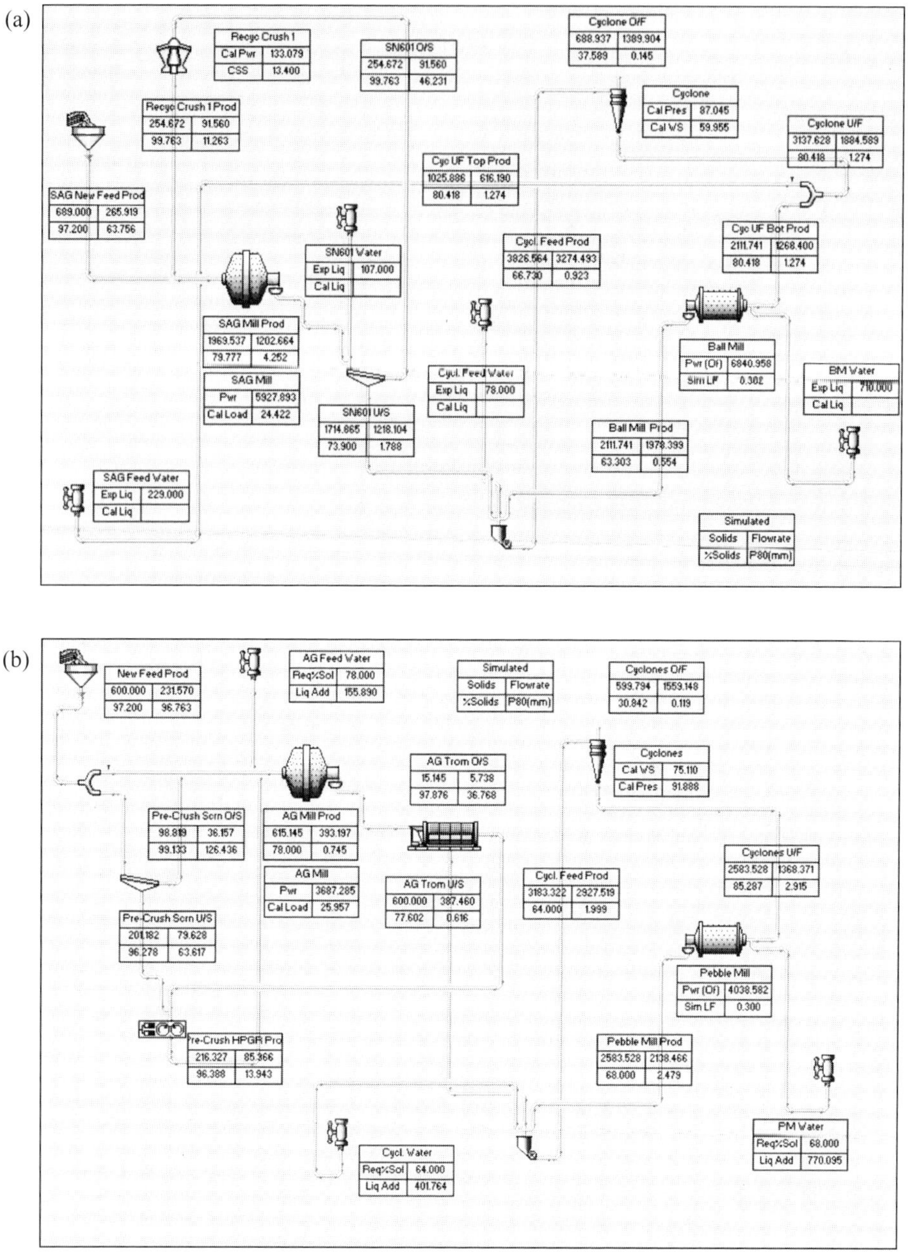


**Figure 4. Configuration and performance of the (a) SABC circuit and the modified for eco-efficiency (b) ABC-HPGR-PM circuit**

Direct energy comparisons, energy used in size reduction, show that the ABC-HPGR-PM circuit has an operating work index of 15.9kWh/t compared to 24.3kWh/t for the base case SABC circuit. This is an energy saving greater than 8kWh/t for the whole circuit, see Table 2. Based on natural gas as the source of power at a rate of A\$0.15/kWh, the energy saving is greater than A\$1.20 per tonne of ore processed when using the ABC-HPGR-PM circuit.

SAG/AG mill and ball/pebble mill power draws are calculated using a method similar to that in JKSimMet. HPGR and tower mill power draws are taken from experimental work carried out at JKMRC and CSIRO Minerals. Pebble mill power draw is calculated based on density ratios of the steel media and pebbles, the length of the mill in the power model has been adjusted accordingly. The media density required to achieve sufficient grinding in the pebble mill is assumed to be 4.5kg/l. The ore density is 2.8kg/l and the density of steel media is 7.8kg/l. Therefore some steels balls are required in the pebble mill for efficient grinding.

Comparison of indirect energy consumption for the two circuits showed that again the ABC-HPGR-PM out performed the original SABC circuit with a 79% saving in indirect energy based on media and liner wear alone. The

**Table 2. Direct energy comparisons for the SABC and ABC-HPGR-PM circuits**

| | | SABC Circuit | ABC-HPGR-PM Circuit |
|---|---|---|---|
| Throughput | tph | 689 | 740 |
| Feed Size F80 | mm | 64 | 97 |
| Product Size P80 | mm | 0.145 | 0.119 |
| Total Power | kW | 13,262 | 9,710 |
| Operating Work Index | kWh/t | 24.3 | 15.9 |

**Table 3. Indirect energy comparisons for the SABC and ABC-HPGR-PM circuits**

| Indirect Energy | | SABC | ABC-HPGR-PM | Difference |
|---|---|---|---|---|
| **Throughput** | tph | 689 | 740 | |
| **Power to Gen. Media** | kWh/kg steel | 6 | 6 | |
| **SAG/AG** | | | | |
| Ball Load | % | 9.5 | 0.0 | |
| Diameter | m | 9.6 | 9.6 | |
| Length | m | 4.3 | 4.0 | |
| Steel Media Consumption | kg/t | 0.23 | 0.00 | |
| Steel Liner Consumption | kg/t | 0.12 | 0.10 | |
| Indirect Power | kW | 1465 | 438 | -70% |
| **Ball Mill/Pebble Mill** | | | | |
| Media Load | % | 30 | 30 | |
| Diameter | m | 6.6 | 6.6 | |
| Length | m | 8.5 | 8.5 | |
| Steel Media Consumption | kg/t | 0.38 | 0.00 | |
| Steel Liner Consumption | kg/t | 0.12 | 0.07 | |
| Indirect Power | kW | **2070** | **315** | **-85%** |
| **Total Indirect Power** | **kW** | **3535** | **753** | **-79%** |
| **Cost per Year** | A$AUDt steel | **$6,967,043** | **$1,484,783** | **$5,482,260** |

SABC circuit consumed 3,535kW while the ABC-HPGR-PM circuit consumed 753kW, see Table 3. Based on energy required to manufacture the grinding consumables at 6kWh/kg, more than A$5M per year can be saved on the purchase of grinding consumables.

Indirect energy figures are derived from ball mill and SAG mill steel media consumption figures provided by the site. The mill liner wear for the SAG mill and ball mill is based on the media consumption of the two mills and the decrease in power draw.

Based on mass balance predictions for flotation circuit performance, the ABC-HPGR-PM circuit produced a finer and steeper product size distribution which promoted better flotation performance due to the reduced amount of material in more difficult to float very fine and very coarse size fractions of the distribution. This provided about a 2% increase in gold recovery.

The use of the HPGR and the conversion of the circuit to mainly autogenous grinding reduced the total energy consumption (direct and indirect) of the circuit by limiting the re-circulating load and the generation of critical size material. This in turn decreases the mill power draw, media load and at times the actual size of the mill required. These modifications also assist in the generation a steeper product size distribution for improved downstream processing.

**Selection of a Target Product Size**

The concept of progressive liberation for the selection of a target product size can be illustrated using a multi-component orebody such as a lead-zinc deposit, where galena and sphalerite are the valuable minerals in association with iron sulphide gangue (pyrite and pyrrhotite) and numerous other non-sulphide gangue minerals.

The sample from the lead-zinc orebody underwent staged grinding to generate samples of decreasing particle size, product sizes ranging from $P_{80}$ of 1.50mm to a $P_{80}$ of 0.015mm. The sub-samples were then screened to determine a product size distribution and a sample from each screen fraction analyzed using MLA (Mineral Liberation Analysis)

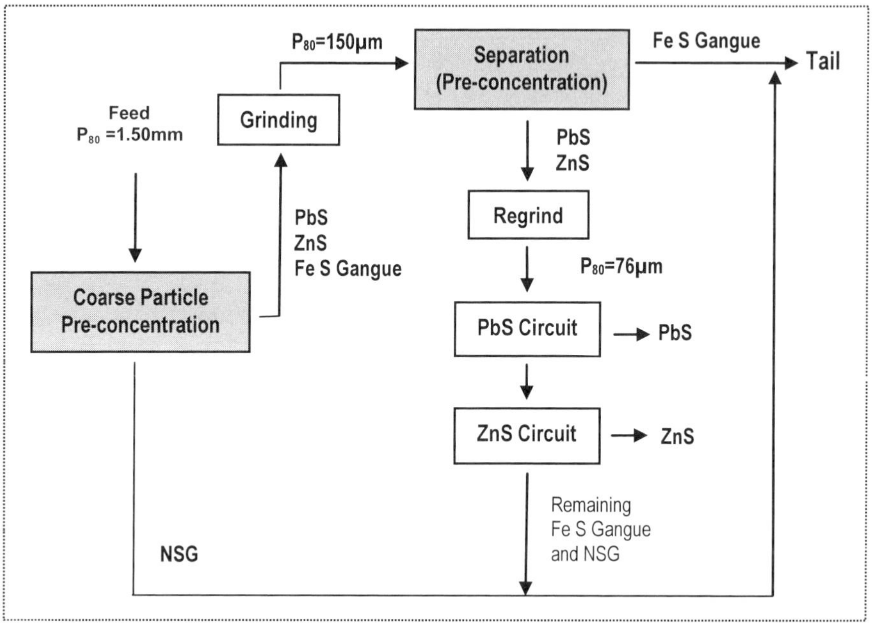


**Figure 5. Conceptual design of separation circuit post primary grinding**

techniques to determine the liberation of individual and grouped mineral components. All liberation values are based on two-dimensional measurement of polished sections.

Analysis of the MLA data for the various product sizes to include liberated and recoverable particles, showed that:

1. 70-80% of galena and sphalerite is liberated at a $P_{80}$ of 0.076mm. That is, at 0.076mm, 70-80% of the galena and sphalerite exists as mono-mineral particles.
2. 70-80% of galena and sphalerite is recoverable at a $P_{80}$ of 0.150mm. This means 70-80% of the galena or sphalerite at this size exist as either fully liberated particles or make up at least 15% by area of a composite particle, which can be recovered in a flotation process.
3. When galena and sphalerite are combined and analyzed as one phase, 85% of this combined phase is recoverable at a $P_{80}$ of 0.150mm.
4. >95% of the sulphides (combined galena, sphalerite, pyrite and pyrrhotite) are recoverable at a $P_{80}$ of 1.50mm.
5. 45% of non-sulphide gangue is liberated at a $P_{80}$ of 1.50mm.

These points can be used in the conceptual design a comminution circuit where the product size is finer than 1.5mm. The aim of the design is to exploit the recoverability of valuable minerals and the rejection of liberated gangue in a separation step at the coarsest product size. The separation step could be a flotation stage or a relative density separation stage. Ore properties will determine the most suitable method for separation. For a sulphide flotation process on this coarse feed, improvements in technology for the recovery of liberated and composite particles would be required to implement the process.

The chosen circuit may resemble that shown in Figure 5.

The circuit initially features a separation step to treat feed material with a $P_{80}$ of 1.50mm. The aim is to remove the non-sulphide gangue to tail by recovering greater than 95% of the sulphides (galena, sphalerite and iron sulphide gangue), see point 4. This can be achieved using coarse particle flotation or relative density separation techniques.

Standard practice in some applications has been the successful flotation of coarse particles up to 2mm in size. The phosphate and potash industries are examples of this. The biggest issues in coarse particle flotation are the adhesion of the coarse particle to the bubble and the transportation of the bubble-coarse particle though the pulp phase and the froth phase. For the mentioned non sulphide minerals, this had led to the use of multiple banks of small mechanical cells that limit the transport distance and promote high weight recovery of concentrate and the use of column cells that have low turbulence in the pulp phase. Separate reagent conditioning for coarse particles has also improved their flotation performance. Recent research has shown that coarse particle flotation is more effective when the particle bubble contact angle is high and sulphide particles as coarse as 2mm in diameter can be floated under laboratory conditions when the critical angle exceeds 50° (Grano, 2006). New flotation devices specifically designed for coarse particle flotation have been developed, these include the Hydrofloat cell which is a combination of teeter bed and flotation technologies (Kohmuench et al,1991).

The second option for coarse particle pre-concentration is the use of density separation. This is applied when a marked density difference exists between the valuable mineral and the gangue minerals. Typically for lead-zinc ores relative density separation is feasible, which include the use of jig or spiral concentrator. Absolute density separation such as heavy media may be an option when a better separation is required and sufficient capital is warranted.

Regardless of which mode of separation is employed, throughput following each stage of separation will be greatly reduced as each stage of gangue removal decreases the mass of material proceeding to the next stage of processing.

Progressive liberation is not completely foreign to the industry as it is often used for platinum and the George Fisher (lead zinc) circuit at Mt Isa (Young et al, 2000) provides another example. However, this strategy has great potential for reducing energy and improving resource utilization. It deserves to be more widely considered.

As shown in Figure 5, the sulphides then pass to a size reduction step to achieve a target product size $P_{80}$ of 0.150mm. At this size, the combined galena and sphalerite particles can be separated from the iron sulphide gangue (which reports to the tail) using standard flotation techniques, (see point 3). The combined galena and sphalerite particles then proceed to a third stage of size reduction (regrind) to achieve a target product size $P_{80}$ of 0.076mm, (see point 1), where the valuable minerals can be separated from each other and from the remaining gangue minerals they may be in composites with to produce individual concentrates. The details of the galena and sphalerite flotation circuits are beyond the scope of this paper—suffice it to say a number of circuit options with or without regrind applications are available.

For this example, it is assumed that the separation efficiency of each separation step (flotation or relative density) is 100%, that is, all recoverable minerals of interest will be separated from the gangue and only those particles containing less than 15% by of the area mineral of interest will report to the tail for rejection from the circuit together with the gangue. In fact, approximately 15%[2] of the combined galena and sphalerite is not recoverable at a $P_{80}$ of 150µm and will be lost to tail following the second separation process, even if the separation process is 100% efficient. This means the maximum possible recovery for galena and sphalerite in this second stage of separation, assuming the separation step is 100% efficient, is 85%. The economic viability of such a loss has to assessed during flow sheet testing at laboratory scale..

In essence, the progressive liberation strategy involves simultaneous recovery of valuable minerals (firstly as grouped sulphides, then grouped valuables and then ultimately as separate entities) to the next stage of size reduction and separation and rejection of liberated gangue minerals. It promotes the selection of more than one target product size where only the valuable minerals that need to be liberated proceed to further stages of size reduction. By the removal of gangue material when the percentage of recoverable mineral exceeds approximately 95%, which is achieved by separating the recoverable minerals of interest, it is possible to significantly decrease the amount of material continuing to downstream size reduction and flotation processes. This can have a positive impact on the operating and capital costs as well as overall power requirement for size reduction.

## IMPLICATIONS OF THE DESIGN CHANGES

Implementation of these strategies can generate comminution circuits with a reduced carbon footprint and will impact on several aspects of the operation and performance of the comminution circuit as well as the project economics.

### Circuit Performance

Selective mining, ore sorting and separation techniques together with the progressive liberation strategy for the selection of target product size(s), can collectively be grouped as pre-concentration techniques as each of these concepts involves the identification and removal of gangue material from the comminution circuit. The common implication of any pre-concentration technique is the reduction in the mass of material proceeding to the next stage of processing; this has the following processing effects:

- reduction in energy requirements for downstream size reduction
- reduction in the size of downstream comminution equipment
- reduction in the size of downstream separation equipment
- increase in the grade of material proceeding to the next stage of processing
- more efficient operation of downstream equipment

The modified grinding strategy can also have benefits for downstream separation, such as a narrower product size distribution which is favorable to most separation methods including flotation. The modified grinding strategy has the greatest impact on energy consumption, both in terms of energy used in size reduction and energy used in the manufacture of grinding consumables. The strategy promotes a re-distribution of the grinding energy around the comminution circuit through the use of more efficient equipment and autogenous grinding techniques.

### Project Economics

In the same way as mineral resources, energy resources are being depleted and as a result energy costs are likely to increase—with or without the imposition of carbon taxes in various forms. These factors coupled with the increased capital required for the multi-stage comminution circuits that result from the implementation of the described energy efficient strategies, means that in the future, comminution, is likely to become much more costly.

Although the flow sheets to reduce carbon footprint comminution circuits will probably require greater capital and may be more complex to operate, such circuits will reduce total energy consumption and can increase resource recovery and utilization by allowing lower grade ores to be progressively upgraded within the comminution process. This increases the net output of the valuable mineral per tonne of ore processed and should contribute to a more profitable operation.

The Australian government has committed to an emission trading scheme called the Carbon Pollution Reduction Scheme (CPRS), which will be implemented by 2010. It requires industries which generate carbon dioxide in large quantities to purchase pollution permits for each tonne of carbon they contribute to the atmosphere. The aim of the

[2] Obtained by subtracting the recoverable combined galena and sphalerite (85%—from point 3) from 100% to determine the non-recoverable component

CPRS scheme is to achieve 60% reduction in emissions by 2050. This means the minerals industry will be subject to the mandatory purchase of energy permits, greenhouse gas (GHG) emissions reporting and auditing and potentially a reduction in output if targets are exceeded. Comminution is one of the major sources of energy use and GHG emissions in the mineral processing industry. Although it is not essential that energy efficient techniques be implemented under the new CPRS laws, it is important that the minerals industry be proactive in maintaining a reputation of reducing its carbon emissions or the potential repercussions could be costly in term of energy costs and carbon emission permits.

## CONCLUSIONS

Comminution circuits can be designed for or modified to achieve a reduced carbon footprint. A holistic approach where the comminution circuit features and equipment are selected based on the inherent liberation properties of the orebody can achieve further reductions in energy consumption.

The design process of comminution circuits with a reduced carbon footprint, can be summarized as follows:

- Use more energy efficient size reduction equipment and apply autogenous grinding techniques where the ore properties are suitable
- Apply progressive ore upgrading strategies to remove liberated waste from the comminution circuit and reduce the required throughput for grinding
- Select the coarsest target product size(s) based on improved knowledge of the liberation of the valuable and gangue minerals

When applied as outlined, to either greenfield circuits or to modifying operating comminution circuits, these strategies should decrease the energy consumption of a circuit, improve circuit operation and performance as the design of the circuit will be more tailored to the ore properties. Although the capital cost of such circuits is likely to increase to some degree, the operating costs should be reduced in the medium and possibly even the short term. In the longer term, the impact of the climate change legislation and associated carbon reduction schemes are likely to provide an overwhelming incentive for the mineral industry to move in this direction.

## REFERENCES


Alvarado, S., Alguerno, J., Auracher, H., and Casali, A., 1998. Energy-exergy optimization of comminution, in *Energy Oxford*, Vol 23, No 2: pp153-158.

Camus, J.P., 2002. *Management of Mineral Resources—Creating Value in the Mining Business*, Colorado, USA.

Crosbie, R., Robertson, C., Smit, I., and Ser, V., 2005. The benefits of inter-particle comminution on flotation, in *Conference Proceedings Centenary of Flotation Symposium*, Brisbane, Australia, pp823-828.

Grano, S., 2006. *Ian Wark Research Institute–Annual Report 2006,* University of South Australia, Adelaide, Australia.

Hukki, R.T., 1961. Proposal for a Solomonic settlement between the theories of von Rittinger, Kick and Bond., in *Trans SME/AIME*, Vol, No 220: pp403-408.

Kohmuench, J.N., Luttrell, G.H., and Mankosa, M.J., 1991. Testing of the Hydrofloat cell for recovery of coarse phosphate, in *Beneficiation of Phosphates Advances in Research and Practice*, Zhang, P., El-Shall, P., and Weigel, R., eds., Society for Mining, Metallurgy, and Exploration, Inc.

La Rosa, D., 2008. *Smart Ore Tracking,* Mining Magazine, 110-113.

Morrison , R.D and Richardson, J.M., 2002. JKSimMet a simulator for analysis, optimisation and design of comminution circuits. *Mineral processing plant design, practice and control, proceedings* Society for Mining, Metallurgy and Exploration, Vancouver, Volume 1 :pp 442-460

Napier-Munn, T.J., and Lynch, A.J., 1992. The modelling and computer simulation of mineral treatment processes-current status and future trends, in *Minerals Engineering*, Vol 5, No 2: pp143-167.

Napier-Munn, T.J., Morrell, S., Morrison, R.D., and Kojovic, T., 1999. *Mineral Comminution Circuits—Their Operation and Optimisation*, JKMRC Monograph Series in Mining and Mineral Processing 2, JKMRC, Brisbane.

Pokrajcic, Z., 2009. *A Methodology for the Design of Energy Efficient Comminution Circuits—PhD Thesis,* PhD Thesis, University of Queensland , Unpublished.

Pollard, L.J., and Death, D.L., 2007. *LIBS Measurements of Selected Samples—CSRP Report*, (DMR-3242), CSIRO Minerals, (Unpublished).

Vink, L., 1997. *Textures of the Hilton North deposit, Queensland, Australia and their relationships to liberation—PhD Thesis,* PhD Thesis, The University of Queensland, Brisbane.

Young, M.F., Pease, J.D. and Fisher K.S., 2000. The George Fisher project to increase recovery in the Mt Isa lead/zinc concentrator. 7th Mill Operators Conference, AusIMM, Kalgoolie, WA, pp 157-166.

# Index